A BRIEF GUIDE TO GETTING THE MOST FROM THIS BOOK

1 Read the Book

FEATURE	DESCRIPTION	BENEFIT
Applications Using Real-World Data	From the chapter and section openers through the examples and exercises, interesting applications from nearly every discipline, supported by up-to-date real-world data, are included in every section.	Ever wondered how you'll use algebra? This feature will show you how algebra can solve real problems.
Detailed Worked-Out Examples	Examples are clearly written and provide step-by-step solutions. No steps are omitted, and key steps are thoroughly explained to the right of the mathematics.	The blue annotations will help you to understand the solutions by providing the reason why the algebraic steps are true.
Explanatory Voice Balloons	Voice balloons help to demystify algebra. They translate algebraic language into plain English, clarify problem-solving procedures, and present alternative ways of understanding.	Does math ever look foreign to you? This feature translates math into everyday English.
Study Tips	Study Tip boxes offer suggestions for problem solving, point out common errors to avoid, and provide informal hints and suggestions.	By seeing common mistakes, you'll be able to avoid them.

2 Work the Problems

FEATURE	DESCRIPTION	BENEFIT
Check Point Examples	Each example is followed by a similar problem, called a Check Point, that offers you the opportunity to work a similar exercise. Answers to all Check Points are provided in the answer section.	You learn best by doing. You'll solidify your understanding of worked examples if you try a similar problem right away to be sure you understand what you've just read.
Extensive and Varied Exercise Sets	An abundant collection of exercises is included in an exercise set at the end of each section. Exercises are organized within several categories. Practice Exercises follow the same order as the section's worked examples. Practice PLUS exercises contain more challenging problems that often require you to combine several skills or concepts.	The parallel order of the Practice Exercises lets you refer to the worked examples and use them as models for solving these problems. Practice PLUS provides you with ample opportunity to dig in and develop your problem-solving skills.

3 Review for Quizzes and Tests

FEATURE	DESCRIPTION	BENEFIT
Mid-Chapter Check Points	Near the midway point in the chapter, an integrated set of review exercises allows you to review the skills and concepts you learned separately over several sections.	Combining exercises from the first half of the chapter gives you a comprehensive review before you continue on.
Chapter Review Chart	Each chapter contains a review chart that summarizes the definitions and concepts in every section of the chapter, complete with examples.	Review this chart and you'll know the most important material in the chapter!
Chapter Tests	Each chapter contains a practice test with problems that cover the important concepts in the chapter. Take the test, check your answers, and then watch the Chapter Test Prep Video CD to see worked-out solutions for any exercises you miss.	You can use the chapter test to determine whether you have mastered the material covered in the chapter.
Chapter Test Prep Video CD	This video CD found at the front of your text contains worked-out solutions to every exercise in each chapter test.	The video lets you review any exercises you miss on the chapter test.

Licensing Agreement
Robert Blitzer
Chapter Test Prep Video CD to accompany Introductory Algebra for College Students, 5e
ISBN: 0-13-602926-4/978-0-13-602926-4
© 2009 Pearson Education, Inc.
Pearson Prentice Hall
Pearson Education, Inc.
Upper Saddle River, NJ 07458
Pearson Prentice Hall™ is a trademark of Pearson Education, Inc.

YOU SHOULD CAREFULLY READ THE TERMS AND CONDITIONS BEFORE USING THE CD-ROM PACKAGE. USING THIS CD-ROM PACKAGE INDICATES YOUR ACCEPTANCE OF THESE TERMS AND CONDITIONS.

Pearson Education, Inc. provides this program and licenses its use. You assume responsibility for the selection of the program to achieve your intended results, and for the installation, use, and results obtained from the program. This license extends only to use of the program in the United States or countries in which the program is marketed by authorized distributors.

LICENSE GRANT

You hereby accept a nonexclusive, nontransferable, permanent license to install and use the program ON A SINGLE COMPUTER at any given time. You may copy the program solely for backup or archival purposes in support of your use of the program on the single computer. You may not modify, translate, disassemble, decompile, or reverse engineer the program, in whole or in part.

TERM

The License is effective until terminated. Pearson Education, Inc. reserves the right to terminate this License automatically if any provision of the License is violated. You may terminate the License at any time. To terminate this License, you must return the program, including documentation, along with a written warranty stating that all copies in your possession have been returned or destroyed.

LIMITED WARRANTY

THE PROGRAM IS PROVIDED "AS IS" WITHOUT WARRANTY OF ANY KIND, EITHER EXPRESSED OR IMPLIED, INCLUDING, BUT NOT LIMITED TO, THE IMPLIED WARRANTIES OF MERCHANTABILITY AND FITNESS FOR A PARTICULAR PURPOSE. THE ENTIRE RISK AS TO THE QUALITY AND PERFORMANCE OF THE PROGRAM IS WITH YOU. SHOULD THE PROGRAM PROVE DEFECTIVE, YOU (AND NOT PEARSON EDUCATION, INC. OR ANY AUTHORIZED DEALER) ASSUME THE ENTIRE COST OF ALL NECESSARY SERVICING, REPAIR, OR CORRECTION. NO ORAL OR WRITTEN INFORMATION OR ADVICE GIVEN BY PEARSON EDUCATION, INC., ITS DEALERS, DISTRIBUTORS, OR AGENTS SHALL CREATE A WARRANTY OR INCREASE THE SCOPE OF THIS WARRANTY. SOME STATES DO NOT ALLOW THE EXCLUSION OF IMPLIED WARRANTIES, SO THE ABOVE EXCLUSION MAY NOT APPLY TO YOU. THIS WARRANTY GIVES YOU SPECIFIC LEGAL RIGHTS AND YOU MAY ALSO HAVE OTHER LEGAL RIGHTS THAT VARY FROM STATE TO STATE.

Pearson Education, Inc. does not warrant that the functions contained in the program will meet your requirements or that the operation of the program will be uninterrupted or error-free. However, Pearson Education, Inc. warrants the CD-ROM(s) on which the program is furnished to be free from defects in material and workmanship under normal use for a period of ninety (90) days from the date of delivery to you as evidenced by a copy of your receipt. The program should not be relied on as the sole basis to solve a problem whose incorrect solution could result in injury to person or property. If the program is employed in such a manner, it is at the user's own risk and Pearson Education, Inc. explicitly disclaims all liability for such misuse.

LIMITATION OF REMEDIES

Pearson Education, Inc.'s entire liability and your exclusive remedy shall be:
1. the replacement of any CD-ROM not meeting Pearson Education, Inc.'s "LIMITED WARRANTY" and that is returned to Pearson Education, or
2. if Pearson Education is unable to deliver a replacement CD-ROM that is free of defects in materials or workmanship, you may terminate this agreement by returning the program.

IN NO EVENT WILL PEARSON EDUCATION, INC. BE LIABLE TO YOU FOR ANY DAMAGES, INCLUDING ANY LOST PROFITS, LOST SAVINGS, OR OTHER INCIDENTAL OR CONSEQUENTIAL DAMAGES ARISING OUT OF THE USE OR INABILITY TO USE SUCH PROGRAM EVEN IF PEARSON EDUCATION, INC. OR AN AUTHORIZED DISTRIBUTOR HAS BEEN ADVISED OF THE POSSIBILITY OF SUCH DAMAGES, OR FOR ANY CLAIM BY ANY OTHER PARTY.

SOME STATES DO NOT ALLOW FOR THE LIMITATION OR EXCLUSION OF LIABILITY FOR INCIDENTAL OR CONSEQUENTIAL DAMAGES, SO THE ABOVE LIMITATION OR EXCLUSION MAY NOT APPLY TO YOU.

GENERAL

You may not sublicense, assign, or transfer the license of the program. Any attempt to sublicense, assign or transfer any of the rights, duties, or obligations hereunder is void.

This Agreement will be governed by the laws of the State of New York. Should you have any questions concerning this Agreement, you may contact Pearson Education, Inc. by writing to:

ESM Media Development
Higher Education Division
Pearson Education, Inc.
1 Lake Street
Upper Saddle River, NJ 07458

Should you have any questions concerning technical support, you may write to:

New Media Production
Higher Education Division
Pearson Education, Inc.
1 Lake Street, Upper Saddle River, NJ 07458

YOU ACKNOWLEDGE THAT YOU HAVE READ THIS AGREEMENT, UNDERSTAND IT, AND AGREE TO BE BOUND BY ITS TERMS AND CONDITIONS. YOU FURTHER AGREE THAT IT IS THE COMPLETE AND EXCLUSIVE STATEMENT OF THE AGREEMENT BETWEEN US THAT SUPERSEDES ANY PROPOSAL OR PRIOR AGREEMENT, ORAL OR WRITTEN, AND ANY OTHER COMMUNICATIONS BETWEEN US RELATING TO THE SUBJECT MATTER OF THIS AGREEMENT.

System Requirements

Macintosh:
Power PC G3 233 MHz or better
Mac OS 10.x
64 MB RAM
19 MB of additional hard drive space (if QuickTime installation is needed)
800 x 600 resolution
8x CD-ROM drive
QuickTime 7.x
Internet browser

Windows:

Pentium II 300 MHz processor-based computer
Windows 2000, XP, Vista Ultimate
64 MB RAM (128 MB RAM required for Windows XP)
7.2 MB additional hard drive space (if QuickTime installation is needed)
800 x 600 resolution
8x CD-ROM drive
QuickTime 7.x
Sound Card
Internet browser

Support Information

If you are having problems with this software, you can get support by filling out the web form located at: http://247pearsoned.custhelp.com/

Our technical staff will need to know certain things about your system in order to help us solve your problems more quickly and efficiently. If possible, please be at your computer when you call for support. You should have the following information ready:

- ISBN
- CD-ROM ISBN
- corresponding product and title
- computer make and model
- Operating System (Windows or Macintosh) and Version
- RAM available
- hard disk space available
- Sound card? Yes or No
- printer make and model
- network connection
- detailed description of the problem, including the exact wording of any error messages.

NOTE: Pearson does not support and/or assist with the following:
- third-party software (i.e. Microsoft including Microsoft Office suite, Apple, Borland, etc.)
- homework assistance
- Textbooks and CD-ROMs purchased used are not supported and are non-replaceable. To purchase a new CD-ROM, contact Pearson Individual Order Copies at 1-800-282-0693.

Introductory Algebra
for College Students

Introductory Algebra
for College Students

Fifth Edition

Robert Blitzer
Miami Dade College

PEARSON

Prentice Hall

Upper Saddle River, New Jersey 07458

Library of Congress Cataloging-in-Publication Data

Blitzer, Robert.
　Introductory algebra for college students / Robert Blitzer.-5th ed.
　　p. cm.
　Includes index.
　ISBN 0-13-235679-1
　　1. Algebra—Textbooks. I. Title.

QA152.3.B648 2008
512.9—dc22　　　　　　　　　　　　　　2007046108

President: *Greg Tobin*
Editor in Chief: *Paul Murphy*
Vice President and Editorial Director: *Christine Hoag*
Editorial Project Manager: *Dawn Nuttall*
Assistant Editor: *Christine Whitlock*
Editorial Assistant: *Georgina Brown*
Production Project Manager: *Barbara Mack*
Associate Managing Editor: *Bayani Mendoza de Leon*
Senior Managing Editor: *Linda Mihatov Behrens*
Operations Specialist: *Ilene Kahn*
Senior Operations Supervisor: *Diane Peirano*
Vice President and Executive Director of Development: *Carol Trueheart*
Media Producer: *Audra J. Walsh*
Media Project Manager: *John Cassar*
Software Development: *Michael Duguay and Mary Durnwald*
Vice President and Director of Marketing: *Amy Cronin*
Executive Marketing Manager: *Kate Valentine*
Senior Marketing Manager: *Michella Renda*
Marketing Manager: *Marlana Voerster*
Marketing Assistants: *Jill Kapinus and Nathaniel Koven*
Senior Art Director: *Juan R. López*
Interior Designer: *Juan R. López*
Cover Art Director: *Maureen Eide*
Cover Designer: *John Christiana*
AV Project Manager: *Thomas Benfatti*
Art Manuscript Coordinator: *Dan Missildine*
Cover Photo: *Shutterstock Images*
Manager, Cover Visual Research and Permissions: *Karen Sanatar*
Director, Image Resource Center: *Melinda Patelli*
Manager, Rights and Permissions: *Zina Arabia*
Manager, Visual Research: *Beth Brenzel*
Image Permission Coordinator: *Debbie Hewitson*
Photo Researcher: *Rachel Lucas*
Art Studios: *Scientific Illustrators/Laserwords*
Compositor: *Prepare, Inc.*

© 2009, 2006, 2002, 1998, 1995 Pearson Education, Inc.
Pearson Prentice Hall
Pearson Education, Inc.
Upper Saddle River, New Jersey 07458

Printed in the United States of America
10 9 8 7 6 5 4 3 2 1

ISBN-13: 978-0-13-235679-4
ISBN-10:　　0-13-235679-1

Pearson Education Ltd., *London*
Pearson Education Australia PTY, Limited, *Sydney*
Pearson Education Singapore, Pte. Ltd
Pearson Education North Asia Ltd, *Hong Kong*
Pearson Education Canada, Ltd., *Toronto*
Pearson Educación de Mexico, S.A. de C.V.
Pearson Education—Japan, *Tokyo*
Pearson Education Malaysia, Pte. Ltd

CONTENTS

PREFACE

Introductory Algebra for College Students, Fifth Edition, provides comprehensive, in-depth coverage of the topics required in a one-term course in beginning or introductory algebra. The book is written for college students who have no previous experience in algebra and for those who need a review of basic algebra concepts. I wrote the book to help diverse students, with different backgrounds and career plans, to succeed in beginning algebra. *Introductory Algebra for College Students,* Fifth Edition, has two primary goals:

1. To help students acquire a solid foundation in the basic skills of algebra.
2. To show students how algebra can model and solve authentic real-world problems.

One major obstacle in the way of achieving these goals is the fact that very few students actually read their textbook. This has been a regular source of frustration for me and for my colleagues in the classroom. Anecdotal evidence gathered over years highlights two basic reasons why students do not take advantage of their textbook:

- "I'll never use this information."
- "I can't follow the explanations."

I've written every page of the Fifth Edition with the intent of eliminating these two objections. The ideas and tools I've used to do so are described in the features that follow. These features and their benefits are highlighted for the student in "A Brief Guide to Getting the Most from This Book," which appears inside the front cover.

What's New in the Fifth Edition?

- **New Applications and Real World Data.** I'm on a constant search for real-world data that can be used to illustrate unique algebraic applications. I researched hundreds of books, magazines, newspapers, almanacs, and online sites to prepare the Fifth Edition. With 214 worked-out examples and application exercises based on new data sets, the Fifth Edition contains a greater array of applications than any previous revision of this book.

- **"Make Sense?" Classroom Discussion Exercises.** Each exercise set contains four Critical Thinking exercises intended for classroom discussion in order to engage participation in the learning process. These items test conceptual understanding by asking students to determine whether statements are sensible, and to explain why or why not. Although sample answers are provided, students have skills and perspectives that frequently differ from those of math teachers, so answers and explanations may vary. The important part of this new feature is to let you ask students what they think about selected statements, determine whether they understand the concepts, and give them feedback to clarify any misunderstandings.

- **New Directions for the True/False Critical Thinking Exercises.** The Fifth Edition asks students to determine whether each statement in an itemized list is true or false. If the statement is false, students are then asked to make the necessary change or changes to produce a true statement.

- **Preview Exercises.** Each exercise set concludes with three problems to help students prepare for the next section. Some of these problems review previously covered material that students will need to be successful in the forthcoming section. Other problems are designed to get students thinking about concepts they will soon encounter.

- **More Detailed Directions When Comparing Mathematical Models with Actual Data.** The Fifth Edition asks students if values obtained from mathematical models underestimate or overestimate data displayed by graphs, and, if so, by how much.

- **Increased Study Tip Boxes.** The book's Study Tip boxes offer suggestions for problem solving, point out common errors to avoid, and provide informal hints and suggestions. These invaluable hints appear in greater abundance in the Fifth Edition. Within the book's first few chapters, many of the new study tips involve general study skills and confidence building, encouraging students with negative past experiences to open their minds and efforts to math. Some of these tips are based on suggestions given by Cheryl Ooten in *Managing the Mean Math Blues* (Prentice Hall, 2003) and Cal Newport in *How to Win at College* (Broadway Books, 2005).

- **New Chapter-Opening and Section-Opening Scenarios.** Every chapter and every section open with a scenario based on an application, the majority of which are unique to the Fifth Edition. These scenarios are revisited in the course of the chapter or section in one of the book's new examples, exercises, or discussions. The often-humorous tone of these openers is intended to help fearful and reluctant students overcome their negative perceptions about math.

- **922 New Examples and Exercises.** The Fifth Edition contains 23 detailed worked-out examples involving new data, 191 new application exercises, 240 "make sense" discussion exercises, 177 preview exercises, and 291 new exercises that appear in the various other categories of the exercise sets.

- **New Blitzer Bonuses.** The Fifth Edition contains a variety of new, but optional, enrichment essays. Examples include "Deceptions in Visual Displays of Data," which appears in Section 3.3 (Problem Solving in Geometry) and is an outgrowth of the section's discussion on area and volume, and "Art, Nature, and Quadratic Equations," which appears in Section 10.3 (The Quadratic Formula).

What Content and Organizational Changes Have Been Made to the Fifth Edition?

- **Section 1.1**(Introduction to Algebra: Variables and Mathematical Models) is a new section that sets the stage for the entire book. The implications of using letters to represent numbers allows students to understand how algebra is used to model data, with connections to graphing through data displayed by bar and line graphs. This early introduction to algebraic expressions, equations, and mathematical models changes the chapter's focus from real numbers, a topic for a pre-algebra class, to a broader perspective reflected by the chapter's new title: **Variables, Real Numbers, and Mathematical Models.** Although the rectangular coordinate system is no longer introduced in Chapter 1, the models that appear throughout the book's early chapters are all connected to bar and line graphs. This leads to a fluid transition into rectangular coordinates, introduced in Chapter 4 when students begin graphing linear equations.

- **Section 1.2** (Fractions in Algebra) still reviews fractions and their operations, but now discusses problems involving fractions that appear in algebraic expressions, equations, and mathematical models.

- **Section 2.6** (Solving Linear Inequalities) introduces interval notation, using both interval notation and set-builder notation to express solution sets of inequalities.

- **Section 4.1** (Graphing Linear Equations in Two Variables) introduces the rectangular coordinate system and uses point plotting to graph linear equations. The section focuses exclusively on linear equations, postponing discussions on graphing of nonlinear equations until Chapter 6 (Exponents and Polynomials).

- **Section 4.3** (Slope) contains a new discussion of slope and perpendicular lines.

- **Section 5.2** (Solving Systems of Linear Equations by the Substitution Method) presents supply and demand models in forms students will encounter if they take an economics course.

- **Section 6.1** (Adding and Subtracting Polynomials) includes a brief discussion on graphing equations defined by polynomials of degree 2, using point plotting to graph $y = x^2 - 4$.
- **Section 6.6** (Dividing Polynomials by Binomials) illustrates the steps involved in long division of whole numbers before considering the similar process for dividing a polynomial by a binomial.
- **Appendix** (Mean, Median, and Mode) has been added to familiarize students with common statistical measures of central tendency.

What Familiar Features Have Been Retained in the Fifth Edition?

- **Detailed Worked-Out Examples.** Each worked example is titled, making clear the purpose of the example. Examples are clearly written and provide students with detailed step-by-step solutions. No steps are omitted and key steps are thoroughly explained to the right of the mathematics.
- **Explanatory Voice Balloons.** Voice balloons are used in a variety of ways to demystify mathematics. They translate algebraic ideas into everyday English, help clarify problem-solving procedures, present alternative ways of understanding concepts, and connect problem solving to concepts students have already learned.
- **Check Point Examples.** Each example is followed by a similar matched problem, called a Check Point, offering students the opportunity to test their understanding of the example by working a similar exercise. The answers to the Check Points are provided in the answer section.
- **Extensive and Varied Exercise Sets.** An abundant collection of exercises is included in an exercise set at the end of each section. Exercises are organized within eight category types: Practice Exercises, Practice Plus Exercises, Application Exercises, Writing in Mathematics, Critical Thinking Exercises, Technology Exercises, Review Exercises, and Preview Exercises. This format makes it easy to create well-rounded homework assignments. The order of the practice exercises is exactly the same as the order of the section's worked examples. This parallel order enables students to refer to the titled examples and their detailed explanations to achieve success working the practice exercises.
- **Practice Plus Problems.** This category of exercises contains more challenging practice problems that often require students to combine several skills or concepts. With an average of ten practice plus problems per exercise set, instructors are provided with the option of creating assignments that take practice exercises to a more challenging level.
- **Mid-Chapter Check Points.** At approximately the midway point in each chapter, an integrated set of review exercises allows students to review and assimilate the skills and concepts they learned separately over several sections.
- **Early Graphing.** Chapter 1 connects formulas and mathematical models to data displayed by bar and line graphs. The rectangular coordinate system is introduced in Chapter 4. Graphs appear in nearly every section and exercise set. Examples and exercises use graphs to explore relationships between data and to provide ways of visualizing a problem's solution.
- **Geometric Problem Solving.** Chapter 3 on problem solving contains a section that teaches geometric concepts that are important to a student's understanding of algebra. There is frequent emphasis on problem solving in geometric situations, as well as on geometric models that allow students to visualize algebraic formulas.
- **Section Objectives.** Learning objectives are clearly stated at the beginning of each section. These objectives help students recognize and focus on the section's most important ideas. The objectives are restated in the margin at their point of use.

- **Thorough, Yet Optional Technology.** Although the use of graphing utilities is optional, they are utilized in Using Technology boxes to enable students to visualize and gain numerical insight into algebraic concepts. The use of graphing utilities is also reinforced in the technology exercises appearing in the exercise sets for those who want this option. With the book's early introduction to graphing, students can look at the calculator screens in the Using Technology boxes and gain an increased understanding of an example's solution even if they are not using a graphing utility in the course.

- **Chapter Review Grids.** Each chapter contains a review chart that summarizes the definitions and concepts in every section of the chapter. Examples that illustrate these key concepts are also included in the chart.

- **End-of-Chapter Materials.** A comprehensive collection of review exercises for each of the chapter's sections follows the review grid. This is followed by a chapter test that enables students to test their understanding of the material covered in the chapter. Beginning with Chapter 2, each chapter concludes with a comprehensive collection of mixed cumulative review exercises.

- **Chapter Test Prep Video CD.** Packaged at the front of the text, this video CD provides students with step-by-step solutions for each of the exercises in the book's chapter tests.

- **Blitzer Bonuses.** These enrichment essays provide historical, interdisciplinary, and otherwise interesting connections to the algebra under study, showing students that math is an interesting and dynamic discipline.

- **Discovery.** Discover for Yourself boxes, found throughout the text, encourage students to further explore algebraic concepts. These explorations are optional and their omission does not interfere with the continuity of the topic under consideration.

- **Chapter Projects.** At the end of each chapter is a collaborative activity that gives students the opportunity to work cooperatively as they think and talk about mathematics. Additional group projects can be found in the *Instructor's Resource Manual*. Many of these exercises should result in interesting group discussions.

I hope that my passion for teaching, as well as my respect for the diversity of students I have taught and learned from over the years, is apparent throughout this new edition. By connecting algebra to the whole spectrum of learning, it is my intent to show students that their world is profoundly mathematical, and indeed, π is in the sky.

Robert Blitzer

Resources for the Fifth Edition

FOR STUDENTS

Student Solutions Manual Fully worked solutions to the odd-numbered section exercises plus all Check Points, Review/Preview Exercises, Mid-Chapter Check Points, Chapter Reviews, Chapter Tests, and Cumulative Reviews.

Worksheets Provide a ready-to-use lesson and exercise set for every section of the text with ample student work space.

CD Lecture Series A comprehensive set of CD-ROMS, tied to the textbook, containing short video clips of an instructor working key text examples/exercises. (Also available separately on DVD.)

MathXL® Tutorials on CD This interactive tutorial CD-ROM provides algorithmically generated practice exercises that are correlated at the objective level to the exercises in the textbook. Every practice exercise is accompanied by an example and a guided solution designed to involve students in the solution process.

Chapter Test Prep Video CD Provides step-by-step video solutions to each problem in each Chapter Test in the textbook. Packaged with a new text, inside the front cover.

FOR INSTRUCTORS

Instructor Resource Distribution Most instructor resources can be downloaded from the Web site, *www.prenhall.com.* Select "Browse our catalog," then click on "mathematics," select your course and choose your text. Under "resources," on the left side, select "instructor" and choose the supplement you need to download. You will be required to run through a one time registration before you can complete this process.

Math Adjunct Support Center The Pearson Math Adjunct Support Center is staffed by qualified mathematics instructors with over 50 years of combined experience at both the community college and university level. Assistance is provided for faculty in the following areas:

- Suggested syllabus consultation
- Tips on using materials packed with your book
- Book-specific content assistance
- Teaching suggestions including advice on classroom strategies

Instructor's Solutions Manual Fully worked solutions to every exercise in the text.

Instructor's Resource Manual with Tests Includes a Mini-Lecture, Skill Builder, and Additional Exercises for every section of the text; two short group Activities per chapter, several chapter test forms, both free-response and multiple-choice, as well as cumulative tests and final exams. Answers to all items also included.

TestGen® Easily create tests from section objectives. Questions are algorithmically generated allowing for unlimited versions. Edit problems or create your own. There's a chapter test file for each Chapter Test in the text.

Annotated Instructor's Edition Answers to exercises are printed on the same text page with graphing answers in a special Graphing Answer Section in the back of the text.

FOR BOTH

MathXL®

MathXL® is a powerful online homework, tutorial, and assessment system that accompanies this textbook. Instructors can create, edit, and assign online homework and tests using algorithmically generated exercises correlated at the objective level to the textbook. Student work is tracked in an online gradebook. Students can take chapter tests and receive personalized study plans based on their results. The study plan diagnoses weaknesses and links students directly to tutorial exercises for objectives they need to study. Students can also access video clips directly from selected exercises. MathXL is available to qualified adopters. For more information, visit our website at *www.mathxl.com* or contact your Prentice Hall sales representative.

MyMathLab®

MyMathLab® is a series of text-specific, customizable online courses for your textbooks. Powered by CourseCompass™ (Pearson Education's online teaching and learning environment) and MathXL® (our online homework, tutorial, and assessment system), MyMathLab gives you the tools you need to deliver all or a portion of your course online, whether students are in a lab setting or working from home. MyMathLab provides a rich and flexible set of course materials, featuring free-response exercises that are algorithmically generated for unlimited practice. Students can also use online tools, such as video lectures and a multimedia textbook, to improve their performance. Instructors can use MyMathLab's homework and test managers to select and assign online exercises correlated to the textbook, and can import TestGen tests for added flexibility. The online gradebook—designed specifically for mathematics—automatically tracks students' homework and test results and gives the instructor control over how to calculate final grades. MyMathLab also includes access to **Pearson's Tutor Center**, which provides students with tutoring via toll-free phone, fax, email, and interactive Web sessions. MyMathLab is available to qualified adopters. For more information, visit our website at *www.mymathlab.com* or contact your Pearson sales representative.

Acknowledgments

An enormous benefit of authoring a successful series is the broad-based feedback I receive from the students, dedicated users, and reviewers. Every change to this edition is the result of their thoughtful comments and suggestions. I would like to express my appreciation to all the reviewers, whose collective insights form the backbone of this revision. In particular, I would like to thank the following people for reviewing *Introductory Algebra for College Students.*

Gwen P. Aldridge	*Northwest Mississippi Community College*
Howard Anderson	*Skagit Valley College*
John Anderson	*Illinois Valley Community College*
Michael H. Andreoli	*Miami Dade College–North Campus*
Jana Barnard	*Angelo State University*
Gale Brewer	*Amarillo College*
Carmen Buhler	*Minneapolis Community & Technical College*
Warren J. Burch	*Brevard Community College*
Alice Burstein	*Middlesex Community College*
Edie Carter	*Amarillo College*
Sandra Pryor Clarkson	*Hunter College*
Sally Copeland	*Johnson County Community College*
Carol Curtis	*Fresno City College*
Robert A. Davies	*Cuyahoga Community College*
Ben Divers, Jr.	*Ferrum College*
Irene Doo	*Austin Community College*
Charles C. Edgar	*Onondaga Community College*
Karen Edwards	*Diablo Valley College*
Susan Forman	*Bronx Community College*
Wendy Fresh	*Portland Community College*
Gary Glaze	*Eastern Washington University*
Jay Graening	*University of Arkansas*
Robert B. Hafer	*Brevard Community College*
Mary Lou Hammond	*Spokane Community College*
Andrea Hendricks	*Georgia Perimeter College*
Donald Herrick	*Northern Illinois University*
Beth Hooper	*Golden West College*
Sandee House	*Georgia Perimeter College*
Tracy Hoy	*College of Lake County*
Laura Hoye	*Trident Community College*
Margaret Huddleston	*Schreiner University*
Marcella Jones	*Minneapolis Community and Technical College*
Shelbra B. Jones	*Wake Technical Community College*
Sharon Keenee	*Georgia Perimeter College*
Gary Kersting	*North Central Michigan College*
Kandace Kling	*Portland Community College*
Gray Knippenberg	*Lansing Community College*
Mary Kochler	*Cuyahoga Community College*
Scot Leavitt	*Portland Community College*
Robert Leibman	*Austin Community College*
Jennifer Lempke	*North Central Michigan College*

Ann M. Loving	*J. Sargent Reynolds Community College*
Kent MacDougall	*Temple College*
Hank Martel	*Broward Community College*
Kim Martin	*Southeastern Illinois College*
John Robert Martin	*Tarrant County Jounior College*
Irwin Metviner	*State University of New York at Old Westbury*
Jean P. Millen	*Georgia Perimeter College*
Lawrence Morales	*Seattle Central Community College*
Lois Jean Nieme	*Minneapolis Community and Technical College*
Allen R. Newhart	*Parkersburg Community College*
Peg Pankowski	*Community College of Allegheny County–South Campus*
Robert Patenaude	*College of the Canyons*
Jill Rafael	*Sierra College*
James Razavi	*Sierra College*
Christopher Reisch	*The State University of New York at Buffalo*
Nancy Ressler	*Oakton Community College*
Katalin Rozsa	*Mesa Community College*
Haazim Sabree	*Georgia Perimeter College*
Shannon Schumann	*University of Phoenix*
Gayle Smith	*Lane Community College*
Dick Spangler	*Tacoma Community College*
Janette Summers	*University of Arkansas*
Robert Thornton	*Loyola University*
Lucy C. Thrower	*Francis Marion College*
Richard Townsend	*North Carolina Central University*
Andrew Walker	*North Seattle Community College*
Kathryn Wetzel	*Amarillo College*

Additional acknowledgments are extended to Dan Miller and Kelly Barber, for preparing the solutions manuals, Brad Davis, for preparing the answer section and serving as accuracy checker, the Preparè, Inc. formatting team, for the book's brilliant paging, Aaron Darnall at Scientific Illustrators, for superbly illustrating the book, Rachel Lucas, photo researcher, for obtaining the book's new photographs, and Barbara Mack, whose talents as production editor kept every aspect of this complex project moving through its many stages.

I would like to thank my editors at Prentice Hall, Paul Murphy and Chris Hoag, and Project Manager, Dawn Nuttall, who guided and coordinated the book from manuscript through production. Thanks to Maureen Eide and John Christiana for the beautiful cover, Juan López for the wonderful interior design, Kate Valentine and Patrice Jones for their innovative marketing efforts, and to the entire Pearson Education sales force for their confidence and enthusiasm about the book.

Robert Blitzer

TO THE STUDENT

I have written this book so that you can learn about the power of algebra and how it relates directly to your life outside the classroom. All concepts are carefully explained, important definitions and procedures are set off in boxes, and worked-out examples that present solutions in a step-by-step manner appear in every section. Each example is followed by a similar matched problem, called a Check Point, for you to try so that you can actively participate in the learning process as you read the book. (Answers to all Check Points appear in the back of the book.) Study Tips offer hints and suggestions and often point out common errors to avoid. A great deal of attention has been given to applying algebra to your life to make your learning experience both interesting and relevant.

As you begin your studies, I would like to offer some specific suggestions for using this book and for being successful in this course:

- **Attend all lectures.** No book is intended to be a substitute for valuable insights and interactions that occur in the classroom. In addition to arriving for lecture on time and being prepared, you will find it useful to read the section before it is covered in lecture. This will give you a clear idea of the new material that will be discussed.

- **Read the book.** Read each section with pen (or pencil) in hand. Move through the worked-out examples with great care. These examples provide a model for doing exercises in the exercise sets. As you proceed through the reading, do not give up if you do not understand every single word. Things will become clearer as you read on and see how various procedures are applied to specific worked-out examples.

- **Work problems every day and check your answers.** The way to learn mathematics is by doing mathematics, which means working the Check Points and assigned exercises in the exercise sets. The more exercises you work, the better you will understand the material.

- **Review for quizzes and tests.** After completing a chapter, study the chapter review chart, work the exercises in the Chapter Review, and work the exercises in the Chapter Test. Answers to all these exercises are given in the back of the book.

  The methods that I've used to help you read the book, work the problems, and review for tests are described in "A Brief Guide to Getting the Most from This Book" that appears inside the front cover. Spend a few minutes reviewing the guide to familiarize yourself with the book's features and their benefits.

- **Use the resources available with this book.** Additional resources to aid your study are described on page xiii. These resources include a Solutions Manual; a Chapter Test Prep Video CD; MyMathLab®, an online version of the book with links to multimedia resources; MathXL®, an online homework, tutorial, and assessment system of the text; and tutorial support at the Pearson Tutor Services.

I wrote this book in Point Reyes National Seashore, 40 miles north of San Francisco. The park consists of 75,000 acres with miles of pristine surf-washed beaches, forested ridges, and bays bordered by white cliffs. It was my hope to convey the beauty and excitement of mathematics using nature's unspoiled beauty as a source of inspiration and creativity. Enjoy the pages that follow as you empower yourself with the algebra needed to succeed in college, your career, and in your life.

Regards,

Bob

Robert Blitzer

ABOUT THE AUTHOR

Bob and his horse Jerid

Bob Blitzer is a native of Manhattan and received a Bachelor of Arts degree with dual majors in mathematics and psychology (minor: English literature) from the City College of New York. His unusual combination of academic interests led him toward a Master of Arts in mathematics from the University of Miami and a doctorate in behavioral sciences from Nova University. Bob's love for teaching mathematics was nourished for nearly 30 years at Miami Dade College, where he received numerous teaching awards, including Innovator of the Year from the League for Innovations in the Community College and an endowed chair based on excellence in the classroom. In addition to *Introductory Algebra for College Students,* Bob has written textbooks covering intermediate algebra, college algebra, algebra and trigonometry, precalculus, and liberal arts mathematics, all published by Prentice Hall. When not secluded in his Northern California writer's cabin, Bob can be found hiking the beaches and trails of Point Reyes National Seashore, and tending to the chores required by his beloved entourage of horses, chickens, and irritable roosters.

Variables, Real Numbers, and Mathematical Models

What can algebra possibly have to tell me about
- the decline in the percentage of African-American Major League Baseball players?
- whether I'm overdoing my workouts or not pushing myself hard enough?
- the widening imbalance between numbers of women and men on college campuses?
- a relationship between age and number of car accidents?
- how rapidly cyberspace is reshaping our lives?
- how many people watched *The Sopranos,* cable's violent and satiric profile of an angst-ridden criminal family man and his dysfunctional domestic clan?

In this chapter, you will learn how the special language of algebra describes your world.

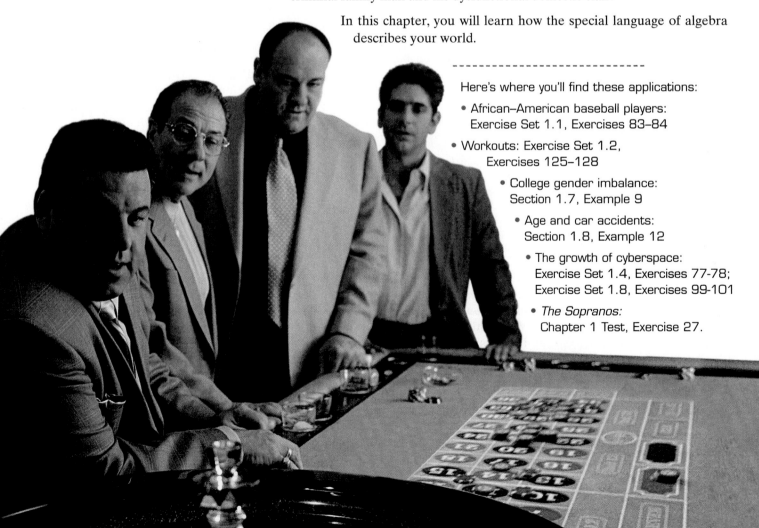

Here's where you'll find these applications:

- African–American baseball players: Exercise Set 1.1, Exercises 83–84

- Workouts: Exercise Set 1.2, Exercises 125–128

- College gender imbalance: Section 1.7, Example 9

- Age and car accidents: Section 1.8, Example 12

- The growth of cyberspace: Exercise Set 1.4, Exercises 77-78; Exercise Set 1.8, Exercises 99-101

- *The Sopranos:* Chapter 1 Test, Exercise 27.

SECTION

1.1

Introduction to Algebra: Variables and Mathematical Models

Objectives

1 Evaluate algebraic expressions.

2 Translate English phrases into algebraic expressions.

3 Determine whether a number is a solution of an equation.

4 Translate English sentences into algebraic equations.

5 Evaluate formulas.

Feeling attractive with a suntan that gives you a "healthy glow"? Think again. Direct sunlight is known to promote skin cancer. Although sunscreens protect you from burning, dermatologists are concerned with the long-term damage that results from the sun even without sunburn.

Algebraic Expressions

Let's see what this man's "healthy glow" has to do with algebra. The biggest difference between arithmetic and algebra is the use of *variables* in algebra. A **variable** is a letter that represents a variety of different numbers. For example, we can let x represent the number of minutes that a person can stay in the sun without burning with no sunscreen. With a number 6 sunscreen, exposure time without burning is six times as long, or 6 times x. This can be written $6 \cdot x$, but it is usually expressed as $6x$. Placing a number and a letter next to one another indicates multiplication.

Notice that $6x$ combines the number 6 and the variable x using the operation of multiplication. A combination of variables and numbers using the operations of addition, subtraction, multiplication, or division, as well as powers or roots, is called an **algebraic expression**. Here are some examples of algebraic expressions:

$$x + 6 \qquad x - 6 \qquad 6x \qquad \frac{x}{6} \qquad 3x + 5 \qquad \sqrt{x} + 7.$$

| The variable x increased by 6 | The variable x decreased by 6 | 6 times the variable x | The variable x divided by 6 | 5 more than 3 times the variable x | 7 more than the square root of the variable x |

1 Evaluate algebraic expressions.

Evaluating Algebraic Expressions

We can replace a variable that appears in an algebraic expression by a number. We are **substituting** the number for the variable. The process is called **evaluating the expression**. For example, we can evaluate $6x$ (from the sunscreen example) for $x = 15$. We substitute 15 for x. We obtain $6 \cdot 15$, or 90. This means that if you can stay in the sun for 15 minutes without burning when you don't put on any lotion, then with a number 6 lotion, you can sunbathe for 90 minutes without burning.

Many algebraic expressions involve more than one operation. The order in which we add, subtract, multiply, and divide is important. In Section 1.8, we will discuss the rules for the order in which operations should be done. For now, follow this order:

A First Look at Order of Operations

1. Perform all operations within grouping symbols, such as parentheses.
2. Do all multiplications in the order in which they occur from left to right.
3. Do all additions and subtractions in the order in which they occcur from left to right.

Study Tip

Study the step-by-step solutions in the worked examples. Reading the solutions with great care will prepare you for success with the exercises in the Exercise Sets.

EXAMPLE 1   Evaluating Expressions

Evaluate each algebraic expression for $x = 5$:

a. $3 + 4x$ **b.** $4(x + 3)$.

Solution

a. We begin by substituting 5 for x in $3 + 4x$. Then we follow the order of operations: Multiply first, and then add.

$$3 + 4x$$

Replace x with 5.

$$= 3 + 4 \cdot 5$$
$$= 3 + 20 \qquad \text{Perform the multiplication: } 4 \cdot 5 = 20.$$
$$= 23 \qquad \text{Perform the addition.}$$

b. We begin by substituting 5 for x in $4(x + 3)$. Then we follow the order of operations: Perform the addition in parentheses first, and then multiply.

$$4(x + 3)$$

Replace x with 5.

$$= 4(5 + 3)$$
$$= 4(8) \qquad \text{Perform the addition inside}$$
$$\qquad\qquad \text{the parentheses: } 5 + 3 = 8.$$
$$\qquad\qquad 4(8) \text{ can also be written as } 4 \cdot 8.$$
$$= 32 \qquad \text{Multiply.}$$

Study Tip

You learn best by doing. To be sure you understand the worked examples, try each Check Point. Check your answer in the answer section before continuing your reading. Expect to read this book with pencil and paper handy to work the Check Points.

✓ **CHECK POINT 1**   Evaluate each expression for $x = 10$:

a. $6 + 2x$ **b.** $2(x + 6)$.

Example 2 illustrates that algebraic expressions can contain more than one variable.

EXAMPLE 2   Evaluating Expressions

Evaluate each algebraic expression for $x = 6$ and $y = 4$:

a. $5x - 3y$ **b.** $\dfrac{3x + 5y + 2}{2x - y}$.

Solution

a.
$$5x - 3y \qquad \text{This is the given algebraic expression.}$$

Replace x with 6.   Replace y with 4.

$$= 5 \cdot 6 - 3 \cdot 4 \qquad \text{We are evaluating the expression for}$$
$$\qquad\qquad\qquad x = 6 \text{ and } y = 4.$$
$$= 30 - 12 \qquad \text{Multiply: } 5 \cdot 6 = 30 \text{ and } 3 \cdot 4 = 12.$$
$$= 18 \qquad \text{Subtract.}$$

b.
$$\frac{3x + 5y + 2}{2x - y}$$

This is the given algebraic expression.

Replace x with 6.   Replace y with 4.

$$= \frac{3 \cdot 6 + 5 \cdot 4 + 2}{2 \cdot 6 - 4}$$

We are evaluating the expression for $x = 6$ and $y = 4$.

$$= \frac{18 + 20 + 2}{12 - 4}$$

Multiply: $3 \cdot 6 = 18, 5 \cdot 4 = 20$, and $2 \cdot 6 = 12$.

$$= \frac{40}{8}$$

Add in the numerator.
Subtract in the denominator.

$$= 5$$

Simplify by dividing 40 by 8.   ■

☑ **CHECK POINT 2**   Evaluate each algebraic expression for $x = 3$ and $y = 8$:

   a. $7x + 2y$ **b.** $\dfrac{6x - y}{2y - x - 8}$.

---

2   Translate English phrases into algebraic expressions.

Translating to Algebraic Expressions

Problem solving in algebra often requires the ability to translate word phrases into algebraic expressions. **Table 1.1** lists some key words associated with the operations of addition, subtraction, multiplication, and division.

Table 1.1	Key Words for Addition, Subtraction, Multiplication, and Division			
Operation	**Addition (+)**	**Subtraction (−)**	**Multiplication (·)**	**Division (÷)**
Key Words	plus	minus	times	divide
	sum	difference	product	quotient
	more than	less than	twice	ratio
	increased by	decreased by	multiplied by	divided by

EXAMPLE 3   **Translating English Phrases into Algebraic Expressions**

Write each English phrase as an algebraic expression. Let the variable x represent the number.

   a. the sum of a number and 7

   b. ten less than a number

   c. twice a number, decreased by 6

   d. the product of 8 and a number

   e. three more than the quotient of a number and 11

Solution

a.
   The sum of   a number and 7

$$x + 7$$

The algebraic expression for "the sum of a number and 7" is $x + 7$.

b.

Ten less than	a number

$$x - 10$$

The algebraic expression for "ten less than a number" is $x - 10$.

c.

Twice a number,	decreased by 6

$$2x - 6$$

The algebraic expression for "twice a number, decreased by 6" is $2x - 6$.

d.

The product of	8 and a number

$$8 \cdot x$$

The algebraic expression for "the product of 8 and a number" is $8 \cdot x$, or $8x$.

e.

Three more than	the quotient of a number and 11

$$\frac{x}{11} + 3$$

The algebraic expression for "three more than the quotient of a number and 11" is $\dfrac{x}{11} + 3$.

Study Tip

Pay close attention to order when translating phrases involving subtraction.

Phrase	Translation
A number decreased by 7	$x - 7$
A number subtracted from 7	$7 - x$
Seven less than a number	$x - 7$
Seven less a number	$7 - x$

Think carefully about what is expressed in English before you translate into the language of algebra.

✓ **CHECK POINT 3**   Write each English phrase as an algebraic expression. Let the variable x represent the number.

a. the product of 6 and a number

b. a number added to 4

c. three times a number, increased by 5

d. twice a number subtracted from 12

e. the quotient of 15 and a number

3 Determine whether a number is a solution of an equation.

Equations

An **equation** is a statement that two algebraic expressions are equal. **An equation always contains the equality symbol =.** Here are some examples of equations:

$$6x + 16 = 46 \qquad 4y + 2 = 2y + 6 \qquad 2(z + 1) = 5(z - 2).$$

Solutions of an equation are values of the variable that make the equation a true statement. To determine whether a number is a solution, substitute that number for the variable and evaluate each side of the equation. If the values on both sides of the equation are the same, the number is a solution.

> **EXAMPLE 4** Determining Whether Numbers Are Solutions of Equations

Determine whether the given number is a solution of the equation.

a. $6x + 16 = 46; 5$ b. $2(z + 1) = 5(z - 2); 7$

Solution

a.
$$6x + 16 = 46 \quad \text{This is the given equation.}$$

Is 5 a solution?

$$6 \cdot 5 + 16 \stackrel{?}{=} 46 \quad \text{To determine whether 5 is a solution, substitute 5 for x. The question mark over the equal sign indicates that we do not yet know if the statement is true.}$$

$$30 + 16 \stackrel{?}{=} 46 \quad \text{Multiply: } 6 \cdot 5 = 30.$$

This statement is true. $\quad 46 = 46 \quad \text{Add: } 30 + 16 = 46.$

Because the values on both sides of the equation are the same, the number 5 is a solution of the equation.

b.
$$2(z + 1) = 5(z - 2) \quad \text{This is the given equation.}$$

Is 7 a solution?

$$2(7 + 1) \stackrel{?}{=} 5(7 - 2) \quad \text{To determine whether 7 is a solution, substitute 7 for z.}$$

$$2(8) \stackrel{?}{=} 5(5) \quad \text{Perform operations in parentheses: } 7 + 1 = 8 \text{ and } 7 - 2 = 5.$$

This statement is false. $\quad 16 = 25 \quad \text{Multiply: } 2 \cdot 8 = 16 \text{ and } 5 \cdot 5 = 25.$

Because the values on both sides of the equation are not the same, the number 7 is not a solution of the equation. ▬

☑ **CHECK POINT 4** Determine whether the given number is a solution of the equation.

a. $9x - 3 = 42; 6$ b. $2(y + 3) = 5y - 3; 3$

---

4 Translate English sentences into algebraic equations.

Translating to Equations

Earlier in the section, we translated English phrases into algebraic expressions. Now we will translate English sentences into equations. You'll find that there are a number of different words and phrases for an equation's equality symbol.

| equals
gives
yields | $=$ | is the same as
is/was/will be |

| EXAMPLE 5 | **Translating English Sentences into Algebraic Equations** |

Write each sentence as an equation. Let the variable x represent the number.

   a. The product of 6 and a number is 30.

   b. Seven less than 3 times a number gives 17.

Study Tip

Do not be concerned if you don't know how to find the solutions of the equations in Example 5. We'll discuss how to solve these equations in Chapter 2.

Solution

a.

$$6x = 30$$

The equation for "the product of 6 and a number is 30" is $6x = 30$.

b.

| Seven less than | 3 times a number | gives | 17. |

$$3x - 7 = 17$$

The equation for "seven less than 3 times a number gives 17" is $3x - 7 = 17$. ∎

☑ **CHECK POINT 5** Write each sentence as an equation. Let the variable x represent the number.

   a. The quotient of a number and 6 is 5.

   b. Seven decreased by twice a number yields 1.

Study Tip

Commas make a difference. In English, sentences and phrases can take on different meanings depending on the way words are grouped with commas. Some examples:

These are meant to be amusing.

- What's the latest dope?
  What's the latest, dope?

- Population of Amsterdam broken down by age and sex
  Population of Amsterdam, broken down by age and sex

Algebraically, this is the important item.

- The product of 6 and a number increased by 5 is 30: $6(x + 5) = 30$.
  The product of 6 and a number, increased by 5, is 30: $6x + 5 = 30$.

5 Evaluate formulas.

Formulas and Mathematical Models

One aim of algebra is to provide a compact, symbolic description of the world. These descriptions involve the use of *formulas*. A **formula** is an equation that expresses a relationship between two or more variables. For example, one variety of crickets chirps faster as the temperature rises. You can calculate the temperature by counting the number of times a cricket chirps per minute and applying the following formula:

$$T = 0.3n + 40.$$

In the formula, T is the temperature, in degrees Fahrenheit, and n is the number of cricket chirps per minute. We can use this formula to determine the temperature if you are sitting on your porch and count 80 chirps per minute. Here is how to do so:

$T = 0.3n + 40$	This is the given formula.
$T = 0.3(80) + 40$	Substitute 80 for n.
$T = 24 + 40$	Multiply: $0.3(80) = 24$.
$T = 64.$	Add.

When there are 80 cricket chirps per minute, the temperature is 64 degrees.

The process of finding formulas to describe real-world phenomena is called **mathematical modeling**. Such formulas, together with the meaning assigned to the variables, are called **mathematical models**. We often say that these formulas model, or describe, the relationship among the variables.

In creating mathematical models, we strive for both accuracy and simplicity. For example, the formula $T = 0.3n + 40$ is relatively simple to use. However, you should not get upset if you count 80 cricket chirps and the actual temperature is 62 degrees, rather than 64 degrees, as predicted by the formula. Many mathematical models give an approximate, rather than an exact, description of the relationship between variables.

Sometimes a mathematical model gives an estimate that is not a good approximation or is extended to include values of the variable that do not make sense. In these cases, we say that **model breakdown** has occurred. Here is an example:

Use the mathematical model $T = 0.3n + 40$ with $n = 1200$ (1200 cricket chirps per minute).

$$T = 0.3(1200) + 40 = 360 + 40 = 400$$

At 400° F, forget about 1200 chirps per minute! At this temperature, the cricket would "cook" and, alas, all chirping would cease.

EXAMPLE 6   Age at Marriage and the Probability of Divorce

Divorce rates are considerably higher for couples who marry in their teens. The line graphs in **Figure 1.1** show the percentages of marriages ending in divorce based on the wife's age at marriage.

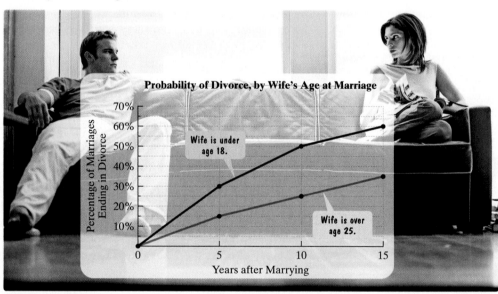

FIGURE 1.1

Source: B. E. Pruitt et al., *Human Sexuality*, Prentice Hall, 2007

Here are two mathematical models that approximate the data displayed by the line graphs:

Wife is under 18 at time of marriage.   Wife is over 25 at time of marriage.

$$d = 4n + 5 \qquad\qquad d = 2.3n + 1.5.$$

In each model, the variable n is the number of years after marriage and the variable d is the percentage of marriages ending in divorce.

  a. Use the appropriate formula to determine the percentage of marriages ending in divorce after 10 years when the wife is over 25 at the time of marriage.

  b. Use the appropriate line graph in **Figure 1.1** to determine the percentage of marriages ending in divorce after 10 years when the wife is over 25 at the time of marriage.

  c. Does the value given by the mathematical model underestimate or overestimate the actual percentage of marriages ending in divorce after 10 years as shown by the graph? By how much?

Solution

  a. Because the wife is over 25 at the time of marriage, we use the formula on the right, $d = 2.3n + 1.5$. To find the percentage of marriages ending in divorce after 10 years, we substitute 10 for n and evaluate the formula.

$$d = 2.3n + 1.5 \qquad \text{This is one of the two given}$$
$$\text{mathematical models.}$$

$$d = 2.3(10) + 1.5 \qquad \text{Replace } n \text{ with 10.}$$
$$d = 23 + 1.5 \qquad \text{Multiply: } 2.3(10) = 23.$$
$$d = 24.5 \qquad \text{Add.}$$

  The model indicates that 24.5% of marriages end in divorce after 10 years when the wife is over 25 at the time of marriage.

  b. Now let's use the line graph that shows the percentage of marriages ending in divorce when the wife is over 25 at the time of marriage. The graph is shown again in **Figure 1.2**. To find the percentage of marriages ending in divorce after 10 years:

  • Locate 10 on the horizontal axis and locate the point above 10.

  • Read across to the corresponding percent on the vertical axis.

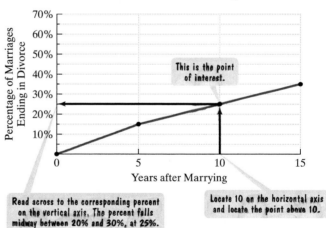

FIGURE 1.2

  The actual data displayed by the graph indicate that 25% of these marriages end in divorce after 10 years.

  c. Here's a summary of what we found in parts (a) and (b).

Percentage of Marriages Ending in Divorce after 10 years (wife over 25 at Marriage)
Mathematical model: 24.5%
Actual data displayed by graph: 25.0%

  The value obtained by evaluating the mathematical model, 24.5%, is close to, but slightly less than, the actual percentage of divorces, 25.0%. The difference between these percents is 25.0% − 24.5%, or 0.5%. The value given by the mathematical model underestimates the actual percent by only 0.5%, providing a fairly accurate description of the data.

✓ **CHECK POINT 6**

   a. Use the appropriate formula from Example 6 to determine the percentage of marriages ending in divorce after 15 years when the wife is under 18 at the time of marriage.

   b. Use the appropriate line graph in **Figure 1.1** to determine the percentage of marriages ending in divorce after 15 years when the wife is under 18 at the time of marriage.

   c. Does the value given by the mathematical model underestimate or overestimate the actual percentage of marriages ending in divorce after 15 years as shown by the graph? By how much?

1.1 EXERCISE SET   *MyMathLab*

PRACTICE   WATCH   DOWNLOAD   READ   REVIEW

Study Tip

Do homework soon. It's okay to take a short break after class, but start reviewing and working the assigned homework in the Exercise Sets as soon as possible.

Practice Exercises

In Exercises 1–14, evaluate each expression for x = 4.

1. $x + 8$

2. $x + 10$

3. $12 - x$

4. $16 - x$

5. $5x$

6. $6x$

7. $\dfrac{28}{x}$

8. $\dfrac{36}{x}$

9. $5 + 3x$

10. $3 + 5x$

11. $2(x + 5)$

12. $5(x + 3)$

13. $\dfrac{12x - 8}{2x}$

14. $\dfrac{5x + 52}{3x}$

In Exercises 15–24, evaluate each expression for x = 7 and y = 5.

15. $2x + y$

16. $3x + y$

17. $2(x + y)$

18. $3(x + y)$

19. $4x - 3y$

20. $5x - 4y$

21. $\dfrac{21}{x} + \dfrac{35}{y}$

22. $\dfrac{50}{y} - \dfrac{14}{x}$

23. $\dfrac{2x - y + 6}{2y - x}$

24. $\dfrac{2y - x + 24}{2x - y}$

In Exercises 25–42, write each English phrase as an algebraic expression. Let the variable x represent the number.

25. four more than a number

26. six more than a number

27. four less than a number

28. six less than a number

29. the sum of a number and 4

30. the sum of a number and 6

31. nine subtracted from a number

32. three subtracted from a number

33. nine decreased by a number

34. three decreased by a number

35. three times a number, decreased by 5

36. five times a number, decreased by 3

37. one less than the product of 12 and a number

38. three less than the product of 13 and a number

39. the sum of 10 divided by a number and that number divided by 10

40. the sum of 20 divided by a number and that number divided by 20

41. six more than the quotient of a number and 30

42. four more than the quotient of 30 and a number

In Exercises 43–58, determine whether the given number is a solution of the equation.

43. $x + 14 = 20; 6$

44. $x + 17 = 22; 5$

45. $30 - y = 10; 20$

46. $50 - y = 20; 30$

47. $4z = 20; 10$

48. $5z = 30; 8$

49. $\dfrac{r}{6} = 8; 48$

50. $\dfrac{r}{9} = 7; 63$

51. $4m + 3 = 23; 6$

52. $3m + 4 = 19; 6$

53. $5a - 4 = 2a + 5; 3$

54. $5a - 3 = 2a + 6; 3$

55. $6(p - 4) = 3p; 8$

56. $4(p + 3) = 6p; 6$

57. $2(w + 1) = 3(w - 1); 7$

58. $3(w + 2) = 4(w - 3); 10$

In Exercises 59–74, write each sentence as an equation. Let the variable x represent the number.

59. Four times a number is 28.

60. Five times a number is 35.

61. The quotient of 14 and a number is $\frac{1}{2}$.

62. The quotient of a number and 8 is $\frac{1}{4}$.

63. The difference between 20 and a number is 5.

64. The difference between 40 and a number is 10.

65. The sum of twice a number and 6 is 16.

66. The sum of twice a number and 9 is 29.

67. Five less than 3 times a number gives 7.

68. Three less than 4 times a number gives 29.

69. The product of 4 and a number, increased by 5, is 33.

70. The product of 6 and a number, increased by 3, is 33.

71. The product of 4 and a number increased by 5 is 33.

72. The product of 6 and a number increased by 3 is 33.

73. Five times a number is equal to 24 decreased by the number.

74. Four times a number is equal to 25 decreased by the number.

Practice PLUS

75. Evaluate $\dfrac{x - y}{4}$ when x is 2 more than 7 times y and $y = 5$.

76. Evaluate $\dfrac{x - y}{3}$ when x is 2 more than 5 times y and $y = 4$.

77. Evaluate $4x + 3(y + 5)$ when x is 1 less than the quotient of y and 4 and $y = 12$.

78. Evaluate $3x + 4(y + 6)$ when x is 1 less than the quotient of y and 3, and $y = 15$.

79. a. Evaluate $2(x + 3y)$ for $x = 4$ and $y = 1$.
    b. Is the number you obtained in part (a) a solution of $5z - 30 = 40$?

80. a. Evaluate $3(2x + y)$ for $x = 1$ and $y = 5$.
    b. Is the number you obtained in part (a) a solution of $4z - 30 = 54$?

81. a. Evaluate $6x - 2y$ for $x = 3$ and $y = 6$.
    b. Is the number you obtained in part (a) a solution of $7w = 45 - 2w$?

82. a. Evaluate $5x - 14y$ for $x = 3$ and $y = \dfrac{1}{2}$.
    b. Is the number you obtained in part (a) a solution of $4w = 54 - 5w$?

Application Exercises

In 2006, 60 years after Jackie Robinson broke the sports color barrier, only 8% of Major League Baseball players were African American. The bar graph shows the decline in the percentage of African-American Major League Baseball players from 1997 through 2006.

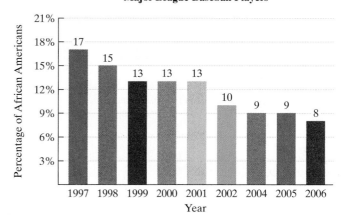

Percentage of African-American Major League Baseball Players

Source: University of Central Florida's Institute for Diversity and Ethics in Sports (Data for 2003 unavailable.)

Here is a mathematical model that approximates the data displayed by the bar graph:

$$p = 16 - n.$$

Number of years after 1997

Percentage of African-American Major League Baseball players

Use this formula to solve Exercises 83–84.

83. a. Use the formula to find the percentage of African-American players 5 years after 1997, or in 2002. Does the mathematical model underestimate or overestimate the actual percent shown by the bar graph for 2002? By how much?
    b. Use the formula to find the percentage of African-American players in 2004. How well does the model describe the percent in the bar graph for this year?

84. a. Use the formula to find the percentage of African-American players 8 years after 1997, or in 2005. Does the mathematical model underestimate or overestimate the actual percent shown by the bar graph for 2005? By how much?
    b. Use the formula to find the percentage of African-American players in 2000. How well does the model describe the percent in the bar graph for this year?

Although lack of sleep affects the way you feel and perform, fewer Americans are getting enough sleep. The graph compares hours of sleep in 1998 and 2005. For each year, it shows the number of hours of sleep per night on weekdays and the percentage of U.S. adults sleeping for this number of hours.

Number of Hours of Sleep per Night on Weekdays for U.S. Adults

Source: National Sleep Foundation poll of 1506 adults (percents do not add up to 100% due to rounding.)

Here are two mathematical models that approximate the data displayed by the graph:

Group number – 1, 2, 3, or 4 – for hours of sleep per night on weekdays

1998: $p = 7.4h + 6$ 2005: $p = 3.7h + 15$

Percentage of U.S. adults in the group

Use these formulas to solve Exercises 85–86.

85. a. Use the formula for 1998 to find the percentage of U.S. adults in group 4—that is, the percentage who got 8 or more hours of sleep per night on weekdays in 1998. Does the mathematical model underestimate or overestimate the percent shown for 1998 for this group? By how much?

b. Use the formula for 2005 to find the percentage of U.S. adults who slept 8 or more hours on weekdays in 2005. Does this overestimate or underestimate the percent shown for 2005 for this group? By how much?

86. a. Use the formula for 1998 to find the percentage of U.S. adults in group 3—that is, the percentage who got between 7 and 7.9 hours of sleep per night on weekdays in 1998. Does the mathematical model underestimate or overestimate the percent shown for 1998 for this group? By how much?

b. Use the formula for 2005 to find the percentage of U.S. adults who slept between 7 and 7.9 hours per night on weekdays in 2005. Does this overestimate or under-estimate the percent shown for 2005 for this group? By how much?

A bowler's handicap, H, is often found using the following formula:

$$H = 0.8(200 - A).$$

Bowler's handicap   Bowler's average score

A bowler's final score for a game is the score for that game increased by the handicap.

Use this information to solve Exercises 87–88.

87. a. If your average bowling score is 145, what is your handicap?

b. What would your final score be if you bowled 120 in a game?

88. a. If your average bowling score is 165, what is your handicap?

b. What would your final score be if you bowled 140 in a game?

Writing in Mathematics

Writing about mathematics will help you to learn mathematics. For all writing exercises in this book, use complete sentences to respond to the questions. Some writing exercises can be answered in a sentence; others require a paragraph or two. You can decide how much you need to write as long as your writing clearly and directly answers the question in the exercise. Standard references such as a dictionary and a thesaurus should be helpful.

89. What is a variable?

90. What is an algebraic expression?

91. Explain how to evaluate $2 + 5x$ for $x = 3$.

92. If x represents the number, explain the difference between translating the following phrases:

a number decreased by 5

a number subtracted from 5.

93. What is an equation?

94. How do you tell the difference between an algebraic expression and an equation?

95. How do you determine whether a given number is a solution of an equation?

96. What is a mathematical model?

97. The bar graph for Exercises 83–84 shows a decline in the percentage of African-American Major League Baseball players from 1997 through 2006. What explanations can you offer for this trend?

98. In Exercises 87–88, we used the formula $H = 0.8(200 - A)$ to find a bowler's handicap, H, where the variable A represents the bowler's average score. Describe what happens to the handicap when the average score is 200.

Critical Thinking Exercises

Make Sense? *In Exercises 99–102, determine whether each statement "makes sense" or "does not make sense" and explain your reasoning.*

99. As I read this book, I write questions in the margins that I might ask in class.

100. I'm solving a problem that requires me to determine if 5 is a solution of $4x + 7$.

101. The model $p = 16 - n$ describes the percentage of African-American Major League Baseball players n years after 1997, so I can use it to estimate the percentage of African-American players in 1997.

102. Because there are four quarters in a dollar, I can use the formula $q = 4d$ to determine the number of quarters, q, in d dollars.

In Exercises 103–106, determine whether each statement is true or false. If the statement is false, make the necessary change(s) to produce a true statement.

103. The algebraic expression for "3 less than a number" is the same as the algebraic expression for "a number decreased by 3."

104. Some algebraic expressions contain the equality symbol, =.

105. The algebraic expressions $3 + 2x$ and $(3 + 2)x$ do not mean the same thing.

106. The algebraic expression for "the quotient of a number and 6" is the same as the algebraic expression for "the quotient of 6 and a number."

In Exercises 107–108, define variables and write a formula that describes the relationship in each table.

107.

Number of Hours Worked	Salary
3	$60
4	$80
5	$100
6	$120

108.

Number of Workers	Number of Televisions Built
3	30
4	40
5	50
6	60

Technology Exercise

109. In 2006, a month into their new jobs, *Entertainment Weekly* magazine wanted to know which of three TV ladies was the "best buy." The data:

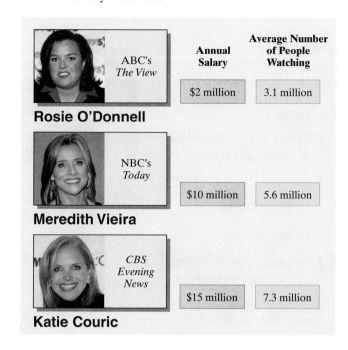

Source: *Entertainment Weekly*, November 3, 2006

Here's the formula used by *Entertainment Weekly*:

Price per viewer
$$p = \frac{s}{w}$$
Annual salary, in millions
Number of people watching, in millions

In the words of the magazine, "the lower the price per viewer, the better the bargain."

a. Use a calculator to find the price per viewer, correct to the nearest cent, for each TV lady.

b. Using the magazine's criterion, which of the three women was the best buy?

Preview Exercises

Exercises 110–112 will help you prepare for the material covered in the next section. In each exercise, use the given formula to perform the indicated operation with the two fractions.

110. $\dfrac{a}{b} \cdot \dfrac{c}{d} = \dfrac{a \cdot c}{b \cdot d}; \quad \dfrac{3}{7} \cdot \dfrac{2}{5}$

111. $\dfrac{a}{b} \div \dfrac{c}{d} = \dfrac{a}{b} \cdot \dfrac{d}{c} = \dfrac{a \cdot d}{b \cdot c}; \quad \dfrac{2}{3} \div \dfrac{7}{5}$

112. $\dfrac{a}{b} - \dfrac{c}{b} = \dfrac{a - c}{b}; \quad \dfrac{9}{17} - \dfrac{5}{17}$

SECTION

1.2

Objectives

1 Convert between mixed numbers and improper fractions.

2 Write the prime factorization of a composite number.

3 Reduce or simplify fractions.

4 Multiply fractions.

5 Divide fractions.

6 Add and subtract fractions with identical denominators.

7 Add and subtract fractions with unlike denominators.

8 Solve problems involving fractions in algebra.

Fractions in Algebra

 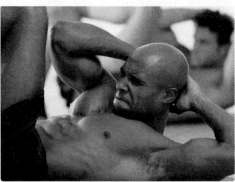

Had a good workout lately? If so, could you tell from your heart rate if you were overdoing it or not pushing yourself hard enough?

Couch-Potato Exercise	**Working It**

$$H = \frac{2}{5}(220 - a)$$ $$H = \frac{9}{10}(220 - a)$$

Heart rate, in beats per minute   Age   Heart rate, in beats per minute   Age

The fractions $\frac{2}{5}$ and $\frac{9}{10}$ provide the difference between these formulas. Recall that in a fraction, the number that is written above the fraction bar is called the **numerator**. The number below the fraction bar is called the **denominator**.

Numerator $\frac{2}{5}$ Fraction bar $\frac{9}{10}$ Numerator

Denominator   Denominator

The numerators and denominators of these fractions, 2, 5, 9, and 10, are examples of *natural numbers*. The **natural numbers** are the numbers that we use for counting.

Natural numbers   1, 2, 3, 4, 5, . . .

The three dots after the 5 indicate that the list continues in the same manner without ending.

Fractions appear throughout algebra. The first part of this section provides a review of the arithmetic of fractions. Later in the section, we focus on fractions in algebra.

1 Convert between mixed numbers and improper fractions.

Mixed Numbers and Improper Fractions

A **mixed number** consists of the sum of a natural number and a fraction, expressed without the use of an addition sign. Here is an example of a mixed number:

$$3\frac{4}{5}.$$ The natural number is 3 and the fraction is $\frac{4}{5}$. $3\frac{4}{5}$ means $3 + \frac{4}{5}$.

An **improper fraction** is a fraction whose numerator is greater than its denominator. An example of an improper fraction is $\frac{19}{5}$.

The mixed number $3\frac{4}{5}$ can be converted to the improper fraction $\frac{19}{5}$ using the following procedure:

> **Converting a Mixed Number to an Improper Fraction**
> 1. Multiply the denominator of the fraction by the natural number and add the numerator to this product.
> 2. Place the result from step 1 over the denominator of the original mixed number.

EXAMPLE 1 Converting from Mixed Number to Improper Fraction

Convert $3\frac{4}{5}$ to an improper fraction.

Solution

$$3\frac{4}{5} = \frac{5 \cdot 3 + 4}{5} \quad \text{Multiply the denominator by the natural number and add the numerator.}$$

$$= \frac{15 + 4}{5} = \frac{19}{5} \quad \text{Place the result over the mixed number's denominator.}$$

☑ **CHECK POINT 1** Convert $2\frac{5}{8}$ to an improper fraction.

An improper fraction can be converted to a mixed number using the following procedure:

> **Converting an Improper Fraction to a Mixed Number**
> 1. Divide the denominator into the numerator. Record the quotient (the result of the division) and the remainder.
> 2. Write the mixed number using the following form:
>
>
>
> $$\text{quotient } \frac{\text{remainder}}{\text{original denominator}}.$$
>
> natural number part   fraction part

EXAMPLE 2 Converting from an Improper Fraction to a Mixed Number

Convert $\frac{42}{5}$ to a mixed number.

Solution We use two steps to convert $\frac{42}{5}$ to a mixed number.

Step 1. Divide the denominator into the numerator.

$$\begin{array}{r} 8 \leftarrow \text{quotient} \\ 5\overline{)42} \\ \underline{40} \\ 2 \leftarrow \text{remainder} \end{array}$$

Step 2. Write the mixed number using quotient $\dfrac{\text{remainder}}{\text{original denominator}}$. Thus,

$$\frac{42}{5} = 8\frac{2}{5}.$$

with labels: quotient, remainder, original denominator

☑ **CHECK POINT 2** Convert $\frac{5}{3}$ to a mixed number.

2 Write the prime factorization of a composite number.

Factors and Prime Factorizations

Fractions can be simplified by first factoring the natural numbers that make up the numerator and the denominator. To **factor** a natural number means to write it as two or more natural numbers being multiplied. For example, 21 can be factored as $7 \cdot 3$. In the statement $7 \cdot 3 = 21$, 7 and 3 are called the **factors** and 21 is the **product**.

7 is a factor of 21. $7 \cdot 3 = 21$ The product of 7 and 3 is 21.

3 is a factor of 21.

Are 7 and 3 the only factors of 21? The answer is no because 21 can also be factored as $1 \cdot 21$. Thus, 1 and 21 are also factors of 21. The factors of 21 are 1, 3, 7, and 21.

Unlike the number 21, some natural numbers have only two factors: the number itself and 1. For example, the number 7 has only two factors: 7 (the number itself) and 1. The only way to factor 7 is $1 \cdot 7$ or, equivalently, $7 \cdot 1$. For this reason, 7 is called a *prime number*.

Prime Numbers

A **prime number** is a natural number greater than 1 that has only itself and 1 as factors.

The first ten prime numbers are

2, 3, 5, 7, 11, 13, 17, 19, 23, and 29.

Can you see why the natural number 15 is not in this list? In addition to having 15 and 1 as factors $(15 = 1 \cdot 15)$, it also has factors of 3 and $5 (15 = 3 \cdot 5)$. The number 15 is an example of a *composite number*.

Composite Numbers

A **composite number** is a natural number greater than 1 that is not a prime number.

Every composite number can be expressed as the product of prime numbers. For example, the composite number 45 can be expressed as

$$45 = 3 \cdot 3 \cdot 5.$$

This product contains only prime numbers: 3 and 5.

Expressing a composite number as the product of prime numbers is called the **prime factorization** of that composite number. The prime factorization of 45 is $3 \cdot 3 \cdot 5$. The order in which we write these factors does not matter. This means that

$$45 = 3 \cdot 3 \cdot 5 \quad \text{or} \quad 45 = 5 \cdot 3 \cdot 3 \quad \text{or} \quad 45 = 3 \cdot 5 \cdot 3.$$

To find the prime factorization of a composite number, begin by selecting any two numbers, excluding 1 and the number itself, whose product is the number to be factored. If one or both of the factors are not prime numbers, continue by factoring each composite number. Stop when all numbers in the factorization are prime.

EXAMPLE 3 Prime Factorization of a Composite Number

Find the prime factorization of 100.

Solution Begin by selecting any two numbers, excluding 1 and 100, whose product is 100. Here is one possibility:

$$100 = 4 \cdot 25.$$

Because the factors 4 and 25 are not prime, we factor each of these composite numbers.

$$100 = 4 \cdot 25 \qquad \text{This is our first factorization.}$$
$$= 2 \cdot 2 \cdot 5 \cdot 5 \qquad \text{Factor 4 and 25.}$$

Notice that 2 and 5 are both prime. The prime factorization of 100 is $2 \cdot 2 \cdot 5 \cdot 5$. ∎

✓ **CHECK POINT 3** Find the prime factorization of 36.

---

3 Reduce or simplify fractions.

FIGURE 1.3

Reducing Fractions

Two fractions are **equivalent** if they represent the same value. Writing a fraction as an equivalent fraction with a smaller denominator is called **reducing a fraction**. A fraction is **reduced to its lowest terms** when the numerator and denominator have no common factors other than 1.

Look at the rectangle in **Figure 1.3**. Can you see that it is divided into 6 equal parts? Of these 6 parts, 4 of the parts are red. Thus, $\frac{4}{6}$ of the rectangle is red.

The rectangle in **Figure 1.3** is also divided into 3 equal stacks and 2 of the stacks are red. Thus, $\frac{2}{3}$ of the rectangle is red. Because both $\frac{4}{6}$ and $\frac{2}{3}$ of the rectangle are red, we can conclude that $\frac{4}{6}$ and $\frac{2}{3}$ are equivalent fractions.

How can we show that $\frac{4}{6} = \frac{2}{3}$ without using **Figure 1.3**? Prime factorizations of 4 and 6 play an important role in the process. So does the **Fundamental Principle of Fractions**.

Fundamental Principle of Fractions

In words: The value of a fraction does not change if both the numerator and the denominator are divided (or multiplied) by the same nonzero number.

In algebraic language: If $\frac{a}{b}$ is a fraction and c is a nonzero number, then

$$\frac{a \cdot c}{b \cdot c} = \frac{a}{b}.$$

We use prime factorizations and the Fundamental Principle to reduce $\frac{4}{6}$ to its lowest terms as follows:

$$\frac{4}{6} = \frac{2 \cdot 2}{3 \cdot 2} = \frac{2}{3}$$

| Write prime factorizations of 4 and 6. | Divide the numerator and the denominator by the common prime factor, 2. |

Here is a procedure for writing a fraction in lowest terms:

Reducing a Fraction to its Lowest Terms

1. Write the prime factorizations of the numerator and the denominator.
2. Divide the numerator and the denominator by the greatest common factor, the product of all factors common to both.

Division lines can be used to show dividing out common factors from a fraction's numerator and denominator:

$$\frac{4}{6} = \frac{2 \cdot \cancel{2}}{3 \cdot \cancel{2}} = \frac{2}{3}.$$

Study Tip

When reducing a fraction to its lowest terms, only *factors* that are common to the numerator and the denominator can be divided out. **If you have not factored** and expressed the numerator and denominator in terms of multiplication, **do not divide out**.

Correct:

$$\frac{2 \cdot \cancel{2}}{3 \cdot \cancel{2}} = \frac{2}{3}$$

Incorrect:

$$\frac{2 + \cancel{2}}{3 + \cancel{2}} = \frac{2}{3}$$

Note that $\frac{2+2}{3+2} = \frac{4}{5}$, not $\frac{2}{3}$.

EXAMPLE 4   Reducing Fractions

Reduce each fraction to its lowest terms:

a. $\dfrac{6}{14}$ b. $\dfrac{15}{75}$ c. $\dfrac{25}{11}$ d. $\dfrac{11}{33}$.

Solution   For each fraction, begin with the prime factorization of the numerator and the denominator.

a. $\dfrac{6}{14} = \dfrac{3 \cdot 2}{7 \cdot 2} = \dfrac{3}{7}$ 2 is the greatest common factor of 6 and 14. Divide the numerator and the denominator by 2.

Including 1 as a factor is helpful when all other factors can be divided out.

b. $\dfrac{15}{75} = \dfrac{3 \cdot 5}{3 \cdot 25} = \dfrac{1 \cdot \cancel{3} \cdot \cancel{5}}{\cancel{3} \cdot \cancel{5} \cdot 5} = \dfrac{1}{5}$ 3 · 5, or 15, is the greatest common factor of 15 and 75. Divide the numerator and the denominator by 3 · 5.

c. $\dfrac{25}{11} = \dfrac{5 \cdot 5}{1 \cdot 11}$

Because 11 and 25 share no common factor (other than 1), $\frac{11}{25}$ is already reduced to its lowest terms.

d. $\dfrac{11}{33} = \dfrac{1 \cdot \cancel{11}}{3 \cdot \cancel{11}} = \dfrac{1}{3}$ 11 is the greatest common factor of 11 and 33. Divide the numerator and denominator by 11.

When reducing fractions, it may not always be necessary to write prime factorizations. In some cases, you can use inspection to find the greatest common factor of the numerator and the denominator. For example, when reducing $\frac{15}{75}$, you can use 15 rather than $3 \cdot 5$:

$$\frac{15}{75} = \frac{1 \cdot \cancel{15}}{5 \cdot \cancel{15}} = \frac{1}{5}.$$

☑ **CHECK POINT 4** Reduce each fraction to its lowest terms:

a. $\dfrac{10}{15}$ **b.** $\dfrac{42}{24}$ **c.** $\dfrac{13}{15}$ **d.** $\dfrac{9}{45}$.

4 Multiply fractions.

Multiplying Fractions

The result of multiplying two fractions is called their **product**.

> ### Multiplying Fractions
>
> **In words:** The product of two or more fractions is the product of their numerators divided by the product of their denominators.
>
> **In algebraic language:** If $\frac{a}{b}$ and $\frac{c}{d}$ are fractions, then
>
> $$\frac{a}{b} \cdot \frac{c}{d} = \frac{a \cdot c}{b \cdot d}.$$

Here is an example that illustrates the rule in the previous box:

$$\frac{3}{8} \cdot \frac{5}{11} = \frac{3 \cdot 5}{8 \cdot 11} = \frac{15}{88}.$$

The product of $\frac{3}{8}$ and $\frac{5}{11}$ is $\frac{15}{88}$.

Multiply numerators and multiply denominators.

EXAMPLE 5 **Multiplying Fractions**

Multiply. If possible, reduce the product to its lowest terms:

a. $\dfrac{3}{7} \cdot \dfrac{2}{5}$ **b.** $5 \cdot \dfrac{7}{12}$ **c.** $\dfrac{2}{3} \cdot \dfrac{9}{4}$ **d.** $\left(3\dfrac{2}{3}\right)\left(1\dfrac{1}{4}\right)$.

Solution

a. $\dfrac{3}{7} \cdot \dfrac{2}{5} = \dfrac{3 \cdot 2}{7 \cdot 5} = \dfrac{6}{35}$ Multiply numerators and multiply denominators.

b. $5 \cdot \dfrac{7}{12} = \dfrac{5}{1} \cdot \dfrac{7}{12} = \dfrac{5 \cdot 7}{1 \cdot 12} = \dfrac{35}{12}$ or $2\dfrac{11}{12}$ Write 5 as $\frac{5}{1}$. Then multiply numerators and multiply denominators.

c. $\dfrac{2}{3} \cdot \dfrac{9}{4} = \dfrac{2 \cdot 9}{3 \cdot 4} = \dfrac{18}{12} = \dfrac{3 \cdot \cancel{6}}{2 \cdot \cancel{6}} = \dfrac{3}{2}$ or $1\dfrac{1}{2}$

Simplify $\frac{18}{12}$; 6 is the greatest common factor of 18 and 12.

d. $\left(3\dfrac{2}{3}\right)\left(1\dfrac{1}{4}\right) = \dfrac{11}{3} \cdot \dfrac{5}{4} = \dfrac{11 \cdot 5}{3 \cdot 4} = \dfrac{55}{12}$ or $4\dfrac{7}{12}$

☑ **CHECK POINT 5** Multiply. If possible, reduce the product to its lowest terms:

a. $\dfrac{4}{11} \cdot \dfrac{2}{3}$

b. $6 \cdot \dfrac{3}{5}$

c. $\dfrac{3}{7} \cdot \dfrac{2}{3}$

d. $\left(3\dfrac{2}{5}\right)\left(1\dfrac{1}{2}\right).$

Study Tip

You can divide numerators and denominators by common factors before performing multiplication. Then multiply the remaining factors in the numerators and multiply the remaining factors in the denominators. For example,

$$\frac{7}{15} \cdot \frac{20}{21} = \frac{\cancel{7} \cdot 1}{5 \cdot 3} \cdot \frac{\cancel{5} \cdot 4}{\cancel{7} \cdot 3} = \frac{1 \cdot 4}{3 \cdot 3} = \frac{4}{9}.$$

| 7 is the greatest common factor of 7 and 21. | 5 is the greatest common factor of 15 and 20. |

The divisions involving the common factors, 7 and 5, are often shown as follows:

$$\frac{7}{15} \cdot \frac{20}{21} = \frac{\overset{1}{\cancel{7}}}{\underset{3}{15}} \cdot \frac{\overset{4}{\cancel{20}}}{\underset{3}{21}} = \frac{1 \cdot 4}{3 \cdot 3} = \frac{4}{9}.$$

| Divide by 7. | Divide by 5. |

5 Divide fractions.

Dividing Fractions

The result of dividing two fractions is called their **quotient**. A geometric figure is useful for developing a process for determining the quotient of two fractions.

Consider the division

$$\frac{4}{5} \div \frac{1}{10}.$$

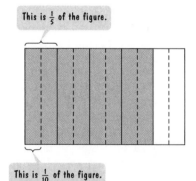

This is $\frac{1}{5}$ of the figure.

This is $\frac{1}{10}$ of the figure.

FIGURE 1.4

We want to know how many $\frac{1}{10}$'s are in $\frac{4}{5}$. We can use **Figure 1.4** to find this quotient. The rectangle is divided into fifths. The dashed lines further divide the rectangle into tenths.

Figure 1.4 shows that $\frac{4}{5}$ of the rectangle is red. How many $\frac{1}{10}$'s of the rectangle does this include? Can you see that this includes eight of the $\frac{1}{10}$ pieces? Thus, there are eight $\frac{1}{10}$'s in $\frac{4}{5}$:

$$\frac{4}{5} \div \frac{1}{10} = 8.$$

We can obtain the quotient 8 in the following way:

$$\frac{4}{5} \div \frac{1}{10} = \frac{4}{5} \cdot \frac{10}{1} = \frac{4 \cdot 10}{5 \cdot 1} = \frac{40}{5} = 8.$$

| Change the division to multiplication. | Invert the divisor, $\frac{1}{10}$. |

By inverting the divisor, $\frac{1}{10}$, and obtaining $\frac{10}{1}$, we are writing the divisor's *reciprocal*. Two fractions are **reciprocals** of each other if their product is 1. Thus, $\frac{1}{10}$ and $\frac{10}{1}$ are reciprocals because $\frac{1}{10} \cdot \frac{10}{1} = 1$.

Generalizing from the result shown on the previous page and using the word *reciprocal*, we obtain the following rule for dividing fractions:

Dividing Fractions

In words: The quotient of two fractions is the first fraction multiplied by the reciprocal of the second fraction.

In algebraic language: If $\frac{a}{b}$ and $\frac{c}{d}$ are fractions and $\frac{c}{d}$ is not 0, then

$$\frac{a}{b} \div \frac{c}{d} = \frac{a}{b} \cdot \frac{d}{c}.$$

> Change division to multiplication.   Invert $\frac{c}{d}$ and write its reciprocal.

EXAMPLE 6   Dividing Fractions

Divide:

a. $\frac{2}{3} \div \frac{7}{15}$ **b.** $\frac{3}{4} \div 5$ **c.** $4\frac{3}{4} \div 1\frac{1}{2}.$

Solution

a. $\dfrac{2}{3} \div \dfrac{7}{15} = \dfrac{2}{3} \cdot \dfrac{15}{7} = \dfrac{2 \cdot 15}{3 \cdot 7} = \dfrac{30}{21} = \dfrac{10 \cdot 3}{7 \cdot 3} = \dfrac{10}{7}$ or $1\dfrac{3}{7}$

> Change division to multiplication.   Invert $\frac{7}{15}$ and write its reciprocal.   Simplify: 3 is the greatest common factor of 30 and 21.

b. $\dfrac{3}{4} \div 5 = \dfrac{3}{4} \div \dfrac{5}{1} = \dfrac{3}{4} \cdot \dfrac{1}{5} = \dfrac{3 \cdot 1}{4 \cdot 5} = \dfrac{3}{20}$

> Change division to multiplication.   Invert $\frac{5}{1}$ and write its reciprocal.

c. $4\dfrac{3}{4} \div 1\dfrac{1}{2} = \dfrac{19}{4} \div \dfrac{3}{2} = \dfrac{19}{4} \cdot \dfrac{2}{3} = \dfrac{19 \cdot 2}{4 \cdot 3} = \dfrac{38}{12} = \dfrac{19 \cdot 2}{6 \cdot 2} = \dfrac{19}{6}$ or $3\dfrac{1}{6}$

✓ CHECK POINT 6   Divide:

a. $\dfrac{5}{4} \div \dfrac{3}{8}$ **b.** $\dfrac{2}{3} \div 3$ **c.** $3\dfrac{3}{8} \div 2\dfrac{1}{4}.$

---

6 Add and subtract fractions with identical denominators.

Adding and Subtracting Fractions with Identical Denominators

The result of adding two fractions is called their **sum**. The result of subtracting two fractions is called their **difference**. A geometric figure is useful for developing a process for determining the sum or difference of two fractions with identical denominators.
    Consider the addition

$$\frac{3}{7} + \frac{2}{7}.$$

We can use **Figure 1.5** to find this sum. The rectangle is divided into sevenths. On the left, $\frac{3}{7}$ of the rectangle is red. On the right, $\frac{2}{7}$ of the rectangle is red. Including both the left and the right, a total of $\frac{5}{7}$ of the rectangle is red. Thus,

$$\frac{3}{7} + \frac{2}{7} = \frac{5}{7}.$$

FIGURE 1.5

We can obtain the sum $\frac{5}{7}$ in the following way:

$$\frac{3}{7} + \frac{2}{7} = \frac{3 + 2}{7} = \frac{5}{7}.$$

Add numerators and put this result over the common denominator.

Generalizing from this result gives us the following rule:

Adding and Subtracting Fractions with Identical Denominators

In words: The sum or difference of two fractions with identical denominators is the sum or difference of their numerators over the common denominator.

In algebraic language: If $\frac{a}{b}$ and $\frac{c}{b}$ are fractions, then

$$\frac{a}{b} + \frac{c}{b} = \frac{a + c}{b} \quad \text{and} \quad \frac{a}{b} - \frac{c}{b} = \frac{a - c}{b}.$$

EXAMPLE 7 Adding and Subtracting Fractions with Identical Denominators

Perform the indicated operations:

a. $\dfrac{3}{11} + \dfrac{4}{11}$ **b.** $\dfrac{11}{12} - \dfrac{5}{12}$ **c.** $5\dfrac{1}{4} - 2\dfrac{3}{4}.$

Solution

a. $\dfrac{3}{11} + \dfrac{4}{11} = \dfrac{3 + 4}{11} = \dfrac{7}{11}$

b. $\dfrac{11}{12} - \dfrac{5}{12} = \dfrac{11 - 5}{12} = \dfrac{6}{12} = \dfrac{1 \cdot 6}{2 \cdot 6} = \dfrac{1}{2}$

c. $5\dfrac{1}{4} - 2\dfrac{3}{4} = \dfrac{21}{4} - \dfrac{11}{4} = \dfrac{21 - 11}{4} = \dfrac{10}{4} = \dfrac{2 \cdot 5}{2 \cdot 2} = \dfrac{5}{2}$ or $2\dfrac{1}{2}$

☑ **CHECK POINT 7** Perform the indicated operations:

a. $\dfrac{2}{11} + \dfrac{3}{11}$ **b.** $\dfrac{5}{6} - \dfrac{1}{6}$ **c.** $3\dfrac{3}{8} - 1\dfrac{1}{8}.$

---

7 Add and subtract fractions with unlike denominators.

Adding and Subtracting Fractions with Unlike Denominators

How do we add or subtract fractions with different denominators? We must first rewrite them as equivalent fractions with the same denominator. We do this by using the Fundamental Principle of Fractions: The value of a fraction does not change if the numerator and the denominator are multiplied by the same nonzero number. Thus, if $\frac{a}{b}$ is a fraction and c is a nonzero number, then

$$\frac{a}{b} = \frac{a \cdot c}{b \cdot c}.$$

EXAMPLE 8   Writing an Equivalent Fraction

Write $\frac{3}{4}$ as an equivalent fraction with a denominator of 16.

Solution   To obtain a denominator of 16, we must multiply the denominator of the given fraction, $\frac{3}{4}$, by 4. So that we do not change the value of the fraction, we also multiply the numerator by 4.

$$\frac{3}{4} = \frac{3 \cdot 4}{4 \cdot 4} = \frac{12}{16}$$

☑ **CHECK POINT 8**   Write $\frac{2}{3}$ as an equivalent fraction with a denominator of 21.

Equivalent fractions can be used to add fractions with different denominators, such as $\frac{1}{2}$ and $\frac{1}{3}$. **Figure 1.6** indicates that the sum of half the whole figure and one-third of the whole figure results in 5 parts out of 6, or $\frac{5}{6}$, of the figure. Thus,

$$\frac{1}{2} + \frac{1}{3} = \frac{5}{6}.$$

We can obtain the sum $\frac{5}{6}$ if we rewrite each fraction as an equivalent fraction with a denominator of 6.

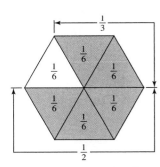

FIGURE 1.6 $\frac{1}{2} + \frac{1}{3} = \frac{5}{6}$

$$\frac{1}{2} + \frac{1}{3} = \frac{1 \cdot 3}{2 \cdot 3} + \frac{1 \cdot 2}{3 \cdot 2}$$ Rewrite each fraction as an equivalent fraction with a denominator of 6.

$$= \frac{3}{6} + \frac{2}{6}$$ Perform the multiplications. We now have a common denominator.

$$= \frac{3 + 2}{6}$$ Add the numerators and place this sum over the common denominator.

$$= \frac{5}{6}$$ Perform the addition.

When adding $\frac{1}{2}$ and $\frac{1}{3}$, there are many common denominators that we can use, such as 6, 12, 18, and so on. The given denominators, 2 and 3, divide into all of these numbers. However, the denominator 6 is the smallest number that 2 and 3 divide into. For this reason, 6 is called the *least common denominator*, abbreviated LCD.

> **Adding and Subtracting Fractions with Unlike Denominators**
> 1. Rewrite the fractions as equivalent fractions with the least common denominator.
> 2. Add or subtract the numerators, putting this result over the common denominator.

EXAMPLE 9   Adding and Subtracting Fractions
with Unlike Denominators

Perform the indicated operation:

a. $\dfrac{1}{5} + \dfrac{3}{4}$ b. $\dfrac{3}{4} - \dfrac{1}{6}$ c. $2\dfrac{7}{15} - 1\dfrac{4}{5}.$

Solution

a. Just by looking, can you tell that the smallest number divisible by both 5 and 4 is 20? Thus, the least common denominator for the denominators 5 and 4 is 20. We rewrite both fractions as equivalent fractions with the least common denominator, 20.

Discover for Yourself

Try Example 9(a), $\frac{1}{5} + \frac{3}{4}$, using a common denominator of 40. Because both 5 and 4 divide into 40, 40 is a common denominator, although not the *least* common denominator. Describe what happens. What is the advantage of using the least common denominator?

$$\frac{1}{5} + \frac{3}{4} = \frac{1 \cdot 4}{5 \cdot 4} + \frac{3 \cdot 5}{4 \cdot 5}$$

To obtain denominators of 20, multiply the numerator and denominator of the first fraction by 4 and the second fraction by 5.

$$= \frac{4}{20} + \frac{15}{20}$$

Perform the multiplications.

$$= \frac{4 + 15}{20}$$

Add the numerators and put this sum over the least common denominator.

$$= \frac{19}{20}$$

Perform the addition.

b. By looking at $\frac{3}{4} - \frac{1}{6}$, can you tell that the smallest number divisible by both 4 and 6 is 12? Thus, the least common denominator for the denominators 4 and 6 is 12. We rewrite both fractions as equivalent fractions with the least common denominator, 12.

$$\frac{3}{4} - \frac{1}{6} = \frac{3 \cdot 3}{4 \cdot 3} - \frac{1 \cdot 2}{6 \cdot 2}$$

To obtain denominators of 12, multiply the numerator and denominator of the first fraction by 3 and the second fraction by 2.

$$= \frac{9}{12} - \frac{2}{12}$$

Perform the multiplications.

$$= \frac{9 - 2}{12}$$

Subtract the numerators and put this difference over the least common denominator.

$$= \frac{7}{12}$$

Perform the subtraction.

c. $2\frac{7}{15} - 1\frac{4}{5} = \frac{37}{15} - \frac{9}{5}$ Convert each mixed number to an improper fraction.

The smallest number divisible by both 15 and 5 is 15. Thus, the least common denominator for the denominators 15 and 5 is 15. Because the first fraction already has a denominator of 15, we only have to rewrite the second fraction.

$$= \frac{37}{15} - \frac{9 \cdot 3}{5 \cdot 3}$$

To obtain denominators of 15, multiply the numerator and denominator of the second fraction by 3.

$$= \frac{37}{15} - \frac{27}{15}$$

Perform the multiplications.

$$= \frac{37 - 27}{15}$$

Subtract the numerators and put this difference over the common denominator.

$$= \frac{10}{15}$$

Perform the subtraction.

$$= \frac{2 \cdot 5}{3 \cdot 5} = \frac{2}{3}$$

Reduce to lowest terms.

✓ **CHECK POINT 9**   Perform the indicated operation:

a. $\frac{1}{2} + \frac{3}{5}$ **b.** $\frac{4}{3} - \frac{3}{4}$ **c.** $3\frac{1}{6} - 1\frac{11}{12}$.

EXAMPLE 10   Using Prime Factorizations to Find the LCD

Perform the indicated operation: $\frac{1}{15} + \frac{7}{24}$.

Solution   We need to first find the least common denominator. Using inspection, it is difficult to determine the smallest number divisible by both 15 and 24. We will use their prime factorizations to find the least common denominator:

$$15 = 5 \cdot 3 \quad \text{and} \quad 24 = 8 \cdot 3 = 2 \cdot 2 \cdot 2 \cdot 3.$$

The different prime factors are 5, 3, and 2. The least common denominator is obtained by using the greatest number of times each factor appears in any prime factorization. Because 5 and 3 appear as prime factors and 2 is a factor of 24 three times, the least common denominator is

$$5 \cdot 3 \cdot 2 \cdot 2 \cdot 2 = 5 \cdot 3 \cdot 8 = 120.$$

Now we can rewrite both fractions as equivalent fractions with the least common denominator, 120.

$$\frac{1}{15} + \frac{7}{24} = \frac{1 \cdot 8}{15 \cdot 8} + \frac{7 \cdot 5}{24 \cdot 5}$$

To obtain denominators of 120, multiply the numerator and denominator of the first fraction by 8 and the second fraction by 5.

$$= \frac{8}{120} + \frac{35}{120}$$

Perform the multiplications.

$$= \frac{8 + 35}{120}$$

Add the numerators and put this sum over the least common denominator.

$$= \frac{43}{120}$$

Perform the addition.

☑ **CHECK POINT 10**   Perform the indicated operation: $\dfrac{3}{10} + \dfrac{7}{12}$.

---

8 Solve problems involving fractions in algebra.

Fractions in Algebra

Fractions appear throughout algebra. Operations with fractions can be used to determine whether a particular fraction is a solution of an equation.

EXAMPLE 11   **Determining Whether Fractions Are Solutions of Equations**

Determine whether the given number is a solution of the equation.

a. $x + \frac{1}{4}x = 7; 6\frac{2}{5}$ **b.** $\frac{1}{7} - w = \frac{1}{2}w; \frac{2}{21}$

Solution

a. To determine whether $6\frac{2}{5}$ is a solution of $x + \frac{1}{4}x = 7$, we begin by converting $6\frac{2}{5}$ from a mixed number to an improper fraction.

$$6\frac{2}{5} = \frac{5 \cdot 6 + 2}{5} = \frac{30 + 2}{5} = \frac{32}{5}$$

Now we substitute $\frac{32}{5}$ for x.

$$x + \frac{1}{4}x = 7 \qquad \text{This is the given equation.}$$

Is $\frac{32}{5}$ a solution?

$$\frac{32}{5} + \frac{1}{4} \cdot \frac{32}{5} \overset{?}{=} 7 \qquad \text{Substitute } \frac{32}{5} \text{ for } x.$$

$$\frac{32}{5} + \frac{8}{5} \overset{?}{=} 7 \qquad \text{Multiply: } \frac{1}{\underset{1}{\cancel{4}}} \cdot \frac{\overset{8}{\cancel{32}}}{5} = \frac{1 \cdot 8}{5} = \frac{8}{5}.$$

$$\frac{40}{5} \overset{?}{=} 7 \qquad \text{Add: } \frac{32}{5} + \frac{8}{5} = \frac{32 + 8}{5} = \frac{40}{5}.$$

This statement is false. $8 = 7$ Simplify: $\frac{40}{5} = 8.$

Because the values on both sides of the equation are not the same, the fraction $\frac{32}{5}$, or equivalently $6\frac{2}{5}$, is not a solution of the equation.

b.

$$\frac{1}{7} - w = \frac{1}{2}w \qquad \text{This is the given equation.}$$

Is $\frac{2}{21}$ a solution?

$$\frac{1}{7} - \frac{2}{21} \overset{?}{=} \frac{1}{2} \cdot \frac{2}{21} \qquad \text{Substitute } \frac{2}{21} \text{ for } w.$$

$$\frac{1}{7} - \frac{2}{21} \overset{?}{=} \frac{1}{21} \qquad \text{Multiply: } \frac{1}{\cancel{2}} \cdot \frac{\overset{1}{\cancel{2}}}{21} = \frac{1 \cdot 1}{1 \cdot 21} = \frac{1}{21}.$$

$$\frac{1 \cdot 3}{7 \cdot 3} - \frac{2}{21} \overset{?}{=} \frac{1}{21} \qquad \begin{array}{l}\text{Turn to the subtraction. The LCD is 21, so multiply}\\\text{the numerator and denominator of the first}\\\text{fraction by 3.}\end{array}$$

$$\frac{3}{21} - \frac{2}{21} \overset{?}{=} \frac{1}{21} \qquad \text{Perform the multiplications.}$$

This statement is true.
$$\frac{1}{21} = \frac{1}{21} \qquad \text{Subtract: } \frac{3}{21} - \frac{2}{21} = \frac{3-2}{21} = \frac{1}{21}.$$

Because the values on both sides of the equation are the same, the fraction $\frac{2}{21}$ is a solution of the equation. ■

✓ **CHECK POINT 11**   Determine whether the given number is a solution of the equation.

a. $x - \frac{2}{9}x = 1; 1\frac{2}{7}$

b. $\frac{1}{5} - w = \frac{1}{3}w; \frac{3}{20}$

In Section 1.1, we translated phrases into algebraic expressions and sentences into equations. When these phrases and sentences involve fractions, the word *of* frequently appears. **When used with fractions, the word *of* represents multiplication.** For example, the phrase "$\frac{2}{5}$ of a number" can be represented by the algebraic expression $\frac{2}{5} \cdot x$, or $\frac{2}{5}x$.

EXAMPLE 12   Algebraic Representations of Phrases and Sentences with Fractions

Translate from English to an algebraic expression or equation, whichever is appropriate. Let the variable x represent the number.

a. $\frac{1}{3}$ of a number increased by 5 is half of that number.

b. $\frac{1}{4}$ of a number, decreased by 7

Solution   Part (a) is a sentence, so we will translate into an equation. Part (b) is a phrase, so we will translate into an algebraic expression.

a.

$\frac{1}{3}$	of	a number increased by 5	is	half	of	that number.
$\frac{1}{3}$	$\cdot$	$(x + 5)$	$=$	$\frac{1}{2}$	$\cdot$	x

The equation for "$\frac{1}{3}$ of a number increased by 5 is half of that number" is $\frac{1}{3}(x + 5) = \frac{1}{2}x$.

b.

$$\frac{1}{4} \quad \cdot \quad x \quad - \quad 7$$

The algebraic expression for "$\frac{1}{4}$ of a number, decreased by 7" is $\frac{1}{4}x - 7$.

✓ **CHECK POINT 12** Translate from English to an algebraic expression or equation, whichever is appropriate. Let the variable x represent the number.

a. $\frac{2}{3}$ of a number decreased by 6

b. $\frac{3}{4}$ of a number, decreased by 2, is $\frac{1}{5}$ of that number.

Many formulas and mathematical models contain fractions. For example, consider temperatures on the Celsius scale and on the Fahrenheit scale, as shown in **Figure 1.7**. The formula $C = \frac{5}{9}(F - 32)$ expresses the relationship between Fahrenheit temperature, F, and Celsius temperature, C.

FIGURE 1.7 The Celsius scale is on the left and the Fahrenheit scale is on the right.

The Formula	**What the Formula Tells Us**
$C = \dfrac{5}{9}(F - 32)$	If 32 is subtracted from the Fahrenheit temperature, $F - 32$, and this difference is multiplied by $\frac{5}{9}$, the resulting product, $\frac{5}{9}(F - 32)$, gives the Celsius temperature.

EXAMPLE 13 Evaluating a Formula Containing a Fraction

The temperature on a warm summer day is 86°F. Use the formula $C = \frac{5}{9}(F - 32)$ to find the equivalent temperature on the Celsius scale.

Solution Because the temperature is 86°F, we substitute 86 for F in the given formula. Then we determine the value of C.

$$C = \frac{5}{9}(F - 32) \qquad \text{This is the given formula.}$$

$$C = \frac{5}{9}(86 - 32) \qquad \text{Replace F with 86.}$$

$$C = \frac{5}{9}(54) \qquad \text{Work inside parentheses first: } 86 - 32 = 54.$$

$$C = 30 \qquad \text{Multiply: } \frac{5}{9}(54) = \frac{5}{\cancel{9}} \cdot \frac{\cancel{54}^{6}}{1} = \frac{5 \cdot 6}{1 \cdot 1} = \frac{30}{1} = 30.$$

Thus, 86°F is equivalent to 30°C.

✓ **CHECK POINT 13** The temperature on a warm spring day is 77°F. Use the formula $C = \frac{5}{9}(F - 32)$ to find the equivalent temperature on the Celsius scale.

1.2 EXERCISE SET **MyMathLab** Math XL
PRACTICE   WATCH   DOWNLOAD   READ   REVIEW

Study Tip

Don't panic at the length of the Exercise Sets. You are not expected to work all, or even most, of the problems. Your instructor will provide guidance on which exercises to work by assigning those problems that are consistent with the goals and objectives of your course.

Practice Exercises

In Exercises 1–6, convert each mixed number to an improper fraction.

1. $2\frac{3}{8}$ **2.** $2\frac{7}{9}$ **3.** $7\frac{3}{5}$

4. $6\frac{2}{5}$ **5.** $8\frac{7}{16}$ **6.** $9\frac{5}{16}$

In Exercises 7–12, convert each improper fraction to a mixed number.

7. $\frac{23}{5}$ **8.** $\frac{47}{8}$ **9.** $\frac{76}{9}$

10. $\frac{59}{9}$ **11.** $\frac{711}{20}$ **12.** $\frac{788}{25}$

In Exercises 13–28, identify each natural number as prime or composite. If the number is composite, find its prime factorization.

13. 22   **14.** 15   **15.** 20

16. 75   **17.** 37   **18.** 23

19. 36   **20.** 100   **21.** 140

22. 110   **23.** 79   **24.** 83

25. 81   **26.** 64

27. 240   **28.** 360

In Exercises 29–40, simplify each fraction by reducing it to its lowest terms.

29. $\frac{10}{16}$ **30.** $\frac{8}{14}$ **31.** $\frac{15}{18}$ **32.** $\frac{18}{45}$

33. $\frac{35}{50}$ **34.** $\frac{45}{50}$ **35.** $\frac{32}{80}$ **36.** $\frac{75}{80}$

37. $\frac{44}{50}$ **38.** $\frac{38}{50}$ **39.** $\frac{120}{86}$ **40.** $\frac{116}{86}$

In Exercises 41–90, perform the indicated operation. Where possible, reduce the answer to its lowest terms.

41. $\frac{2}{5} \cdot \frac{1}{3}$ **42.** $\frac{3}{7} \cdot \frac{1}{4}$ **43.** $\frac{3}{8} \cdot \frac{7}{11}$

44. $\frac{5}{8} \cdot \frac{3}{11}$ **45.** $9 \cdot \frac{4}{7}$ **46.** $8 \cdot \frac{3}{7}$

47. $\frac{1}{10} \cdot \frac{5}{6}$ **48.** $\frac{1}{8} \cdot \frac{2}{3}$ **49.** $\frac{5}{4} \cdot \frac{6}{7}$

50. $\frac{7}{4} \cdot \frac{6}{11}$ **51.** $\left(3\frac{3}{4}\right)\left(1\frac{3}{5}\right)$

52. $\left(2\frac{4}{5}\right)\left(1\frac{1}{4}\right)$ **53.** $\frac{5}{4} \div \frac{4}{3}$

54. $\frac{7}{8} \div \frac{2}{3}$ **55.** $\frac{18}{5} \div 2$ **56.** $\frac{12}{7} \div 3$

57. $2 \div \frac{18}{5}$ **58.** $3 \div \frac{12}{7}$ **59.** $\frac{3}{4} \div \frac{1}{4}$

60. $\frac{3}{7} \div \frac{1}{7}$ **61.** $\frac{7}{6} \div \frac{5}{3}$ **62.** $\frac{7}{4} \div \frac{3}{8}$

63. $\frac{1}{14} \div \frac{1}{7}$ **64.** $\frac{1}{8} \div \frac{1}{4}$ **65.** $6\frac{3}{5} \div 1\frac{1}{10}$

66. $1\frac{3}{4} \div 2\frac{5}{8}$ **67.** $\frac{2}{11} + \frac{4}{11}$ **68.** $\frac{5}{13} + \frac{2}{13}$

69. $\frac{7}{12} + \frac{1}{12}$ **70.** $\frac{5}{16} + \frac{1}{16}$ **71.** $\frac{5}{8} + \frac{5}{8}$

72. $\frac{3}{8} + \frac{3}{8}$ **73.** $\frac{7}{12} - \frac{5}{12}$ **74.** $\frac{13}{18} - \frac{5}{18}$

75. $\frac{16}{7} - \frac{2}{7}$ **76.** $\frac{17}{5} - \frac{2}{5}$ **77.** $\frac{1}{2} + \frac{1}{5}$

78. $\frac{1}{3} + \frac{1}{5}$ **79.** $\frac{3}{4} + \frac{3}{20}$ **80.** $\frac{2}{5} + \frac{2}{15}$

81. $\frac{3}{8} + \frac{5}{12}$ **82.** $\frac{3}{10} + \frac{2}{15}$ **83.** $\frac{11}{18} - \frac{2}{9}$

84. $\frac{17}{18} - \frac{4}{9}$ **85.** $\frac{4}{3} - \frac{3}{4}$ **86.** $\frac{3}{2} - \frac{2}{3}$

87. $\frac{7}{10} - \frac{3}{16}$ **88.** $\frac{7}{30} - \frac{5}{24}$

89. $3\frac{3}{4} - 2\frac{1}{3}$ **90.** $3\frac{2}{3} - 2\frac{1}{2}$

In Exercises 91–102, determine whether the given number is a solution of the equation.

91. $\frac{7}{2}x = 28; 8$ **92.** $\frac{5}{3}x = 30; 18$

93. $w - \frac{2}{3} = \frac{3}{4}; 1\frac{5}{12}$

94. $w - \frac{3}{4} = \frac{7}{4}; 2\frac{1}{2}$

95. $20 - \frac{1}{3}z = \frac{1}{2}z; 12$

96. $12 - \frac{1}{4}z = \frac{1}{2}z; 20$

97. $\frac{2}{9}y + \frac{1}{3}y = \frac{3}{7}; \frac{27}{35}$

98. $\frac{2}{3}y + \frac{5}{6}y = 2; 1\frac{1}{3}$

99. $\frac{1}{3}(x - 2) = \frac{1}{5}(x + 4) + 3; 26$

100. $\frac{1}{2}(x - 2) + 3 = \frac{3}{8}(3x - 4); 4$

101. $(y \div 6) + \frac{2}{3} = (y \div 2) - \frac{7}{9}; 4\frac{1}{3}$

102. $(y \div 6) + \frac{1}{3} = (y \div 2) - \frac{5}{9}; 2\frac{2}{3}$

In Exercises 103–114, translate from English to an algebraic expression or equation, whichever is appropriate. Let the variable x represent the number.

103. $\frac{1}{5}$ of a number

104. $\frac{1}{6}$ of a number

105. A number decreased by $\frac{1}{4}$ of itself

106. A number decreased by $\frac{1}{3}$ of itself

107. A number decreased by $\frac{1}{4}$ is half of that number.

108. A number decreased by $\frac{1}{3}$ is half of that number.

109. The sum of $\frac{1}{7}$ of a number and $\frac{1}{8}$ of that number gives 12.

110. The sum of $\frac{1}{9}$ of a number and $\frac{1}{10}$ of that number gives 15.

111. The product of $\frac{2}{3}$ and a number increased by 6

112. The product of $\frac{3}{4}$ and a number increased by 9

113. The product of $\frac{2}{3}$ and a number, increased by 6, is 3 less than the number.

114. The product of $\frac{3}{4}$ and a number, increased by 9, is 2 less than the number.

Practice PLUS

In Exercises 115–118, perform the indicated operation. Write the answer as an algebraic expression.

115. $\frac{3}{4} \cdot \frac{a}{5}$ **116.** $\frac{2}{3} \div \frac{a}{7}$

117. $\frac{11}{x} + \frac{9}{x}$ **118.** $\frac{10}{y} - \frac{6}{y}$

In Exercises 119–120, perform the indicated operations. Begin by performing operations in parentheses.

119. $\left(\frac{1}{2} - \frac{1}{3}\right) \div \frac{5}{8}$ **120.** $\left(\frac{1}{2} + \frac{1}{4}\right) \div \left(\frac{1}{2} + \frac{1}{3}\right)$

In Exercises 121–122, determine whether the given number is a solution of the equation.

121. $\frac{1}{5}(x + 2) = \frac{1}{2}\left(x - \frac{1}{5}\right); \frac{5}{8}$

122. $12 - 3(x - 2) = 4x - (x + 3); 3\frac{1}{2}$

Application Exercises

The formula

$$C = \frac{5}{9}(F - 32)$$

expresses the relationship between Fahrenheit temperature, F, and Celsius temperature, C. In Exercises 123–124, use the formula to convert the given Fahrenheit temperature to its equivalent temperature on the Celsius scale.

123. 68°F   **124.** 41°F

The maximum heart rate, in beats per minute, that you should achieve during exercise is 220 minus your age:

$$220 - a.$$

> This algebraic expression gives maximum heart rate in terms of age, a.

The bar graph at the top of the next column shows the target heart rate ranges for four types of exercise goals. The lower and

upper limits of these ranges are fractions of the maximum heart rate, 220 – a. Exercises 125–128 are based on the information in the graph.

Target Heart Rate Ranges for Exercise Goals

125. If your exercise goal is to improve cardiovascular conditioning, the graph shows the following range for target heart rate, H, in beats per minute:

Lower limit of range $H = \dfrac{7}{10}(220 - a)$

Upper limit of range $H = \dfrac{4}{5}(220 - a).$

 a. What is the lower limit of the heart range, in beats per minute, for a 20-year-old with this exercise goal?

 b. What is the upper limit of the heart range, in beats per minute, for a 20-year-old with this exercise goal?

126. If your exercise goal is to improve overall health, the graph shows the following range for target heart rate, H, in beats per minute:

Lower limit of range $H = \dfrac{1}{2}(220 - a)$

Upper limit of range $H = \dfrac{3}{5}(220 - a).$

 a. What is the lower limit of the heart range, in beats per minute, for a 30-year-old with this exercise goal?

 b. What is the upper limit of the heart range, in beats per minute, for a 30-year-old with this exercise goal?

127. a. Write a formula that models the heart rate, H, in beats per minute, for a person who is a years old and would like to achieve $\frac{9}{10}$ of maximum heart rate during exercise.

 b. Use your formula from part (a) to find the heart rate during exercise for a 40-year-old with this goal.

128. a. Write a formula that models the heart rate, H, in beats per minute, for a person who is a years old and would like to achieve $\frac{7}{8}$ of maximum heart rate during exercise.

 b. Use your formula from part (a) to find the heart rate during exercise for a 20-year-old with this goal.

The graph shows the number of cigarettes smoked, in billions, in the United States for selected years from 1980 through 2005.

Butt Out: Number of Cigarettes Smoked, in Billions, in the United States

Source: Orzechowski and Walker

Here is a mathematical model that approximates the data displayed in the bar graph:

$$c = 615 - \frac{47}{5}n.$$

Number of cigarettes smoked, in billions   Number of years after 1980

Use this formula to solve Exercises 129–130.

129. Use the formula to find the number of cigarettes smoked, in billions, in 1995. Does the mathematical model underestimate or overestimate the actual number shown in the bar graph for 1995? By how much?

130. Use the formula to find the number of cigarettes smoked, in billions, in 2000. How well does the mathematical model describe the actual number shown in the bar graph for 2000?

Writing in Mathematics

131. Explain how to convert a mixed number to an improper fraction and give an example.

132. Explain how to convert an improper fraction to a mixed number and give an example.

133. Describe the difference between a prime number and a composite number.

134. What is meant by the prime factorization of a composite number?

135. What is the Fundamental Principle of Fractions?

136. Explain how to reduce a fraction to its lowest terms. Give an example with your explanation.

137. Explain how to multiply fractions and give an example.

138. Explain how to divide fractions and give an example.

139. Describe how to add or subtract fractions with identical denominators. Provide an example with your description.

140. Explain how to add fractions with different denominators. Use $\frac{5}{6} + \frac{1}{2}$ as an example.

Critical Thinking Exercises

Make Sense? *In Exercises 141–144, determine whether each statement "makes sense" or "does not make sense" and explain your reasoning.*

141. I find it easier to multiply $\frac{1}{5}$ and $\frac{3}{4}$ than to add them.

142. Fractions frustrated me in arithmetic, so I'm glad I won't have to use them in algebra.

143. I need to be able to perform operations with fractions to determine whether $\frac{3}{2}$ is a solution of $8x = 12\left(x - \frac{1}{2}\right)$.

144. I saved money by buying a computer for $\frac{3}{2}$ of its original price.

In Exercises 145–148, determine whether each statement is true or false. If the statement is false, make the necessary change(s) to produce a true statement.

145. $\frac{1}{2} + \frac{1}{5} = \frac{2}{7}$

146. $\frac{1}{2} \div 4 = 2$

147. Every fraction has infinitely many equivalent fractions.

148. $\dfrac{3+7}{30} = \dfrac{\overset{1}{\cancel{3}}+7}{\underset{10}{\cancel{30}}} = \dfrac{8}{10} = \dfrac{4}{5}$

149. Shown below is a short excerpt from "The Star-Spangled Banner." The time is $\frac{3}{4}$, which means that each measure must contain notes that add up to $\frac{3}{4}$. The values of the different notes tell musicians how long to hold each note.

Use vertical lines to divide this line of "The Star-Spangled Banner" into measures.

Preview Exercises

Exercises 150–152 will help you prepare for the material covered in the next section. Consider the following "infinite ruler" that shows numbers that lie to the left and to the right of zero.

150. What number is represented by point (a)?

151. What number is represented by point (b)? Express the number as an improper fraction.

152. What number is represented by point (c)?

SECTION **1.3**

The Real Numbers

Objectives

1. Define the sets that make up the real numbers.
2. Graph numbers on a number line.
3. Express rational numbers as decimals.
4. Classify numbers as belonging to one or more sets of the real numbers.
5. Understand and use inequality symbols.
6. Find the absolute value of a real number.

The U.N. building is designed with three golden rectangles.

The United Nations Building in New York was designed to represent its mission of promoting world harmony. Viewed from the front, the building looks like three rectangles stacked upon each other. In each rectangle, the ratio of the width to height is $\sqrt{5} + 1$ to 2, approximately 1.618 to 1. The ancient Greeks believed that such a rectangle, called a **golden rectangle**, was the most visually pleasing of all rectangles.

The ratio 1.618 to 1 is approximate because $\sqrt{5}$ is an irrational number, a special kind of real number. Irrational? Real? Let's make sense of all this by describing the kinds of numbers you will encounter in this course.

1. Define the sets that make up the real numbers.

Natural Numbers and Whole Numbers

Before we describe the set of real numbers, let's be sure you are familiar with some basic ideas about sets. A **set** is a collection of objects whose contents can be clearly determined. The objects in a set are called the **elements** of the set. For example, the set of numbers used for counting can be represented by

$$\{1, 2, 3, 4, 5, \ldots\}.$$

The braces, { }, indicate that we are representing a set. This form of representing a set uses commas to separate the elements of the set. Remember that the three dots after the 5 indicate that there is no final element and that the listing goes on forever.

We have seen that the set of numbers used for counting is called the set of **natural numbers**. When we combine the number 0 with the natural numbers, we obtain the set of **whole numbers**.

Natural Numbers and Whole Numbers

The set of **natural numbers** is $\{1, 2, 3, 4, 5, \ldots\}$.
The set of **whole numbers** is $\{0, 1, 2, 3, 4, 5, \ldots\}$.

Integers and the Number Line

The whole numbers do not allow us to describe certain everyday situations. For example, if the balance in your checking account is $30 and you write a check for $35, your checking account is overdrawn by $5. We can write this as −5, read *negative* 5. The set consisting of the natural numbers, 0, and the negatives of the natural numbers is called the set of **integers**.

Integers

The set of **integers** is

$$\{\underbrace{\ldots, -4, -3, -2, -1,}_{\text{Negative integers}} \; 0, \; \underbrace{1, 2, 3, 4, \ldots}_{\text{Positive integers}}\}.$$

Notice that the term **positive integers** is another name for the natural numbers. The positive integers can be written in two ways:

1. Use a "+" sign. For example, +4 is "positive four."
2. Do not write any sign. For example, 4 is assumed to be "positive four."

EXAMPLE 1   Practical Examples of Negative Integers

Write a negative integer that describes each of the following situations:

a. A debt of $10

b. The shore surrounding the Dead Sea is 1312 feet below sea level.

Solution

a. A debt of $10 can be expressed by the negative integer −10 (negative ten).

b. The shore surrounding the Dead Sea is 1312 feet below sea level, expressed as −1312.   ■

☑ **CHECK POINT 1**   Write a negative integer that describes each of the following situations:

a. A debt of $500

b. Death Valley, the lowest point in North America, is 282 feet below sea level.

---

2 Graph numbers on a number line.

The **number line** is a graph we use to visualize the set of integers, as well as other sets of numbers. The number line is shown in **Figure 1.8**.

FIGURE 1.8   The number line

The number line extends indefinitely in both directions. Zero separates the positive numbers from the negative numbers on the number line. The positive integers are located to the right of 0, and the negative integers are located to the left of 0. Zero is neither positive nor negative. For every positive integer on a number line, there is a corresponding negative integer on the opposite side of 0.

Integers are graphed on a number line by placing a dot at the correct location for each number.

EXAMPLE 2   Graphing Integers on a Number Line

Graph:

a. −3   b. 4   c. 0.

Solution   Place a dot at the correct location for each integer.

(a)   (c)   (b)
−5 −4 −3 −2 −1   0   1   2   3   4   5   ■

✓ **CHECK POINT 2** Graph: **a.** −4 **b.** 0 **c.** 3.

Rational Numbers

If two integers are added, subtracted, or multiplied, the result is always another integer. This, however, is not always the case with division. For example, 10 divided by 5 is the integer 2. By contrast, 5 divided by 10 is $\frac{1}{2}$, and $\frac{1}{2}$ is not an integer. To permit divisions such as $\frac{5}{10}$, we enlarge the set of integers, calling the new collection the *rational numbers*. The set of **rational numbers** consists of all the numbers that can be expressed as a quotient of two integers, with the denominator not 0.

> ### The Rational Numbers
>
> The set of **rational numbers** is the set of all numbers that can be expressed in the form $\frac{a}{b}$, where a and b are integers and b is not equal to 0, written $b \neq 0$. The integer a is called the **numerator** and the integer b is called the **denominator**.

Here are two examples of rational numbers:

$$\bullet \ \frac{1}{2} \quad \boxed{a = 1} \ \boxed{b = 2} \qquad \bullet \ \frac{-3}{4} \quad \boxed{a = -3} \ \boxed{b = 4}$$

Is the integer 5 another example of a rational number? Yes. The integer 5 can be written with a denominator of 1.

$$5 = \frac{5}{1} \quad \boxed{a = 5} \ \boxed{b = 1}$$

All integers are also rational numbers because they can be written with a denominator of 1.
 How can we express a negative mixed number, such as $-2\frac{3}{4}$, in the form $\frac{a}{b}$? Copy the negative sign and then follow the procedure discussed in the previous section:

$$-2\frac{3}{4} = -\frac{4 \cdot 2 + 3}{4} = -\frac{8 + 3}{4} = -\frac{11}{4} = \frac{-11}{4}. \quad \boxed{a = -11} \ \boxed{b = 4}$$

Copy the negative sign from step to step and convert $2\frac{3}{4}$ to an improper fraction.

Rational numbers are graphed on a number line by placing a dot at the correct location for each number.

EXAMPLE 3 Graphing Rational Numbers on a Number Line

Graph: **a.** $\frac{7}{2}$ **b.** −4.6.

Solution Place a dot at the correct location for each rational number.

a. Because $\frac{7}{2} = 3\frac{1}{2}$, its graph is midway between 3 and 4.

b. Because $-4.6 = -4\frac{6}{10}$, its graph is $\frac{6}{10}$, or $\frac{3}{5}$, of a unit to the left of −4.

✓ **CHECK POINT 3** Graph: **a.** $\frac{9}{2}$ **b.** −1.2.

3 Express rational numbers as decimals.

Every rational number can be expressed as a fraction and as a decimal. To express the fraction $\frac{a}{b}$ as a decimal, divide the denominator, b, into the numerator, a.

EXAMPLE 4 Expressing Rational Numbers as Decimals

Express each rational number as a decimal:

a. $\frac{5}{8}$　　　　　　　b. $\frac{7}{11}$.

Solution In each case, divide the denominator into the numerator.

a.
$$\begin{array}{r} 0.625 \\ 8\overline{)5.000} \\ \underline{4\,8} \\ 20 \\ \underline{16} \\ 40 \\ \underline{40} \\ 0 \end{array}$$

b.
$$\begin{array}{r} 0.6363\ldots \\ 11\overline{)7.0000\ldots} \\ \underline{6\,6} \\ 40 \\ \underline{33} \\ 70 \\ \underline{66} \\ 40 \\ \underline{33} \\ 70 \\ \vdots \end{array}$$

Using Technology

Calculators and Decimals

Given a rational number, $\frac{a}{b}$, you can express it as a decimal by entering

$$a \boxed{\div} b$$

into a calculator. In the case of a repeating decimal, the calculator rounds off in the final decimal place displayed.

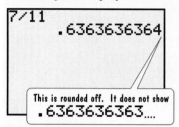

This is rounded off. It does not show .6363636363....

In Example 4, the decimal for $\frac{5}{8}$, namely 0.625, stops and is called a **terminating decimal**. Other examples of terminating decimals are

$$\frac{1}{4} = 0.25, \quad \frac{2}{5} = 0.4, \quad \text{and} \quad \frac{7}{8} = 0.875.$$

By contrast, the division process for $\frac{7}{11}$ results in 0.6363..., with the digits 63 repeating over and over indefinitely. To indicate this, write a bar over the digits that repeat. Thus,

$$\frac{7}{11} = 0.\overline{63}.$$

The decimal for $\frac{7}{11}$, $0.\overline{63}$, is called a **repeating decimal**. Other examples of repeating decimals are

$$\frac{1}{3} = 0.333\ldots = 0.\overline{3} \quad \text{and} \quad \frac{2}{3} = 0.666\ldots = 0.\overline{6}.$$

Rational Numbers and Decimals

Any rational number can be expressed as a decimal. The resulting decimal will either terminate (stop), or it will have a digit that repeats or a block of digits that repeat.

✓ **CHECK POINT 4** Express each rational number as a decimal:

a. $\frac{3}{8}$　　　　　　b. $\frac{5}{11}$.

Irrational Numbers

Can you think of a number that, when written in decimal form, neither terminates nor repeats? An example of such a number is $\sqrt{2}$ (read: "the square root of 2"). The number $\sqrt{2}$ is a number that can be multiplied by itself to obtain 2. No terminating or repeating decimal can be multiplied by itself to get 2. However, some approximations come close to 2.

- 1.4 is an approximation of $\sqrt{2}$:

$$1.4 \times 1.4 = 1.96.$$

- 1.41 is an approximation of $\sqrt{2}$:

$$1.41 \times 1.41 = 1.9881.$$

- 1.4142 is an approximation of $\sqrt{2}$:

$$1.4142 \times 1.4142 = 1.99996164.$$

Can you see how each approximation in the list is getting better? This is because the products are getting closer and closer to 2.

The number $\sqrt{2}$, whose decimal representation does not come to an end and does not have a block of repeating digits, is an example of an **irrational number**.

The Irrational Numbers

Any number that can be represented on the number line that is not a rational number is called an **irrational number**. Thus, the set of irrational numbers is the set of numbers whose decimal representations are neither terminating nor repeating.

Perhaps the best known of all the irrational numbers is π (pi). This irrational number represents the distance around a circle (its circumference) divided by the diameter of the circle. In the *Star Trek* episode "Wolf in the Fold," Spock foils an evil computer by telling it to "compute the last digit in the value of π." Because π is an irrational number, there is no last digit in its decimal representation:

$$\pi = 3.14159265358979323846264338327950 \ldots.$$

Because irrational numbers cannot be represented by decimals that come to an end, mathematicians use symbols such as $\sqrt{2}$, $\sqrt{3}$, and π to represent these numbers. However, **not all square roots are irrational**. For example, $\sqrt{25} = 5$ because 5 multiplied by itself is 25. Thus, $\sqrt{25}$ is a natural number, a whole number, an integer, and a rational number $\left(\sqrt{25} = \frac{5}{1} \right)$.

Using Technology

You can obtain decimal approximations for irrational numbers using a calculator. For example, to approximate $\sqrt{2}$, use the following keystrokes:

Scientific Calculator   **Graphing Calculator**

2 $\boxed{\sqrt{}}$ or 2 $\boxed{\begin{array}{c}\text{2ND}\\\text{INV}\end{array}}$ $\boxed{x^2}$ $\boxed{\sqrt{}}$ 2 $\boxed{\text{ENTER}}$ or $\boxed{\begin{array}{c}\text{2ND}\\\text{INV}\end{array}}$ $\boxed{x^2}$ 2 $\boxed{\text{ENTER}}$

Some graphing calculators show an open parenthesis after displaying $\sqrt{}$. In this case, enter a closed parenthesis, $\boxed{)}$, after 2.

The display may read 1.41421356237, although your calculator may show more or fewer digits. Between which two integers would you graph $\sqrt{2}$ on a number line?

4 Classify numbers as belonging to one or more sets of the real numbers.

The Set of Real Numbers

All numbers that can be represented by points on the number line are called **real numbers**. Thus, the set of real numbers is formed by combining the rational numbers and the irrational numbers. Every real number is either rational or irrational.

The sets that make up the real numbers are summarized in **Table 1.2**. Notice the use of the symbol $\approx$ in the examples of irrational numbers. The symbol $\approx$ means "is approximately equal to."

Real numbers

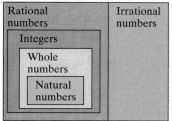

This diagram shows that every real number is rational or irrational.

Table 1.2 **The Sets that Make Up the Real Numbers**

Name	Description	Examples
Natural numbers	$\{1, 2, 3, 4, 5, \dots\}$ These numbers are used for counting.	$2, 3, 5, 17$
Whole numbers	$\{0, 1, 2, 3, 4, 5, \dots\}$ The set of whole numbers is formed by adding 0 to the set of natural numbers.	$0, 2, 3, 5, 17$
Integers	$\{\dots, -5, -4, -3, -2, -1, 0, 1, 2, 3, 4, 5, \dots\}$ The set of integers is formed by adding negatives of the natural numbers to the set of whole numbers.	$-17, -5, -3, -2, 0,$ $2, 3, 5, 17$
Rational numbers	The set of rational numbers is the set of all numbers that can be expressed in the form $\frac{a}{b}$, where a and b are integers and b is not equal to 0, written $b \neq 0$. Rational numbers can be expressed as terminating or repeating decimals.	$-17 = \frac{-17}{1}, -5 = \frac{-5}{1}, -3, -2,$ $0, 2, 3, 5, 17,$ $\frac{2}{5} = 0.4,$ $\frac{-2}{3} = -0.6666\cdots = -0.\overline{6}$
Irrational numbers	The set of irrational numbers is the set of all numbers whose decimal representations are neither terminating nor repeating. Irrational numbers cannot be expressed as a quotient of integers.	$\sqrt{2} \approx 1.414214$ $-\sqrt{3} \approx -1.73205$ $\pi \approx 3.142$ $-\frac{\pi}{2} \approx -1.571$

EXAMPLE 5   Classifying Real Numbers

Consider the following set of numbers:

$$\left\{ -7, -\frac{3}{4}, 0, 0.\overline{6}, \sqrt{5}, \pi, 7.3, \sqrt{81} \right\}.$$

List the numbers in the set that are

   a. natural numbers.   **b.** whole numbers.   **c.** integers.
   d. rational numbers.   **e.** irrational numbers.   **f.** real numbers.

Solution

   a. Natural numbers: The natural numbers are the numbers used for counting. The only natural number in the set is $\sqrt{81}$ because $\sqrt{81} = 9$. (9 multiplied by itself is 81.)

   b. Whole numbers: The whole numbers consist of the natural numbers and 0. The elements of the set that are whole numbers are 0 and $\sqrt{81}$.

   c. Integers: The integers consist of the natural numbers, 0, and the negatives of the natural numbers. The elements of the set that are integers are $\sqrt{81}$, 0, and -7.

   d. Rational numbers: All numbers in the set that can be expressed as the quotient of integers are rational numbers. These include $-7 \left(-7 = \frac{-7}{1} \right), -\frac{3}{4}, 0 \left(0 = \frac{0}{1} \right)$, and $\sqrt{81} \left(\sqrt{81} = \frac{9}{1} \right)$. Furthermore, all numbers in the set that are terminating or repeating decimals are also rational numbers. These include $0.\overline{6}$ and 7.3.

e. Irrational numbers: The irrational numbers in the set are $\sqrt{5}$ $\left(\sqrt{5} \approx 2.236\right)$ and π $(\pi \approx 3.14)$. Both $\sqrt{5}$ and π are only approximately equal to 2.236 and 3.14, respectively. In decimal form, $\sqrt{5}$ and π neither terminate nor have blocks of repeating digits.

f. Real numbers: All the numbers in the given set are real numbers. ▬

☑ **CHECK POINT 5**   Consider the following set of numbers:

$$\left\{-9, -1.3, 0, 0.\overline{3}, \frac{\pi}{2}, \sqrt{9}, \sqrt{10}\right\}.$$

List the numbers in the set that are

a. natural numbers.   b. whole numbers.

c. integers.   d. rational numbers.

e. irrational numbers.   f. real numbers.

---

5 Understand and use inequality symbols.

FIGURE 1.9

Ordering the Real Numbers

On the real number line, the real numbers increase from left to right. The lesser of two real numbers is the one farther to the left on a number line. The greater of two real numbers is the one farther to the right on a number line.

Look at the number line in **Figure 1.9**. The integers 2 and 5 are graphed. Observe that 2 is to the left of 5 on the number line. This means that 2 is less than 5.

$2 < 5$:   2 is less than 5 because 2 is to the *left* of 5 on the number line.

In **Figure 1.9**, we can also observe that 5 is to the right of 2 on the number line. This means that 5 is greater than 2.

$5 > 2$:   5 is greater than 2 because 5 is to the *right* of 2 on the number line.

The symbols $<$ and $>$ are called **inequality symbols**. These symbols always point to the lesser of the two real numbers when the inequality is true.

2 is less than 5.
$2 < 5$ The symbol points to 2, the lesser number.
$5 > 2$ The symbol points to 2, the lesser number.
5 is greater than 2.

EXAMPLE 6   **Using Inequality Symbols**

Insert either $<$ or $>$ in the shaded area between each pair of numbers to make a true statement:

a. 3 ▨ 17   b. -4.5 ▨ 1.2   c. -5 ▨ -83 d. $\frac{4}{5}$ ▨ $\frac{2}{3}$.

Solution   In each case, mentally compare the graph of the first number to the graph of the second number. If the first number is to the left of the second number, insert the symbol $<$ for "is less than." If the first number is to the right of the second number, insert the symbol $>$ for "is greater than."

a. Compare the graphs of 3 and 17 on the number line. Because 3 is to the left of 17, this means that 3 is less than 17: $3 < 17$.

b. Compare the graphs of -4.5 and 1.2. Because -4.5 is to the left of 1.2, this means that -4.5 is less than 1.2: $-4.5 < 1.2$.

c. Compare the graphs of -5 and -83. Because -5 is to the right of -83, this means that -5 is greater than -83: $-5 > -83$.

d. Compare the graphs of $\frac{4}{5}$ and $\frac{2}{3}$. To do so, convert to decimal notation or use a common denominator. Using decimal notation, $\frac{4}{5} = 0.8$ and $\frac{2}{3} = 0.\overline{6}$. Because 0.8 is to the right of $0.\overline{6}$, this means that $\frac{4}{5}$ is greater than $\frac{2}{3}$: $\frac{4}{5} > \frac{2}{3}$.   ▄

☑ **CHECK POINT 6**   Insert either $<$ or $>$ in the shaded area between each pair of numbers to make a true statement:

a. 14 ▨ 5   **b.** -5.4 ▨ 2.3   **c.** -19 ▨ -6 **d.** $\dfrac{1}{4}$ ▨ $\dfrac{1}{2}$.

The symbols $<$ and $>$ may be combined with an equal sign, as shown in the table.

	Symbols	Meaning	Examples	Explanation
This inequality is true if either the $<$ part or the $=$ part is true.	$a \le b$	a is less than or equal to b.	$3 \le 7$ $7 \le 7$	Because $3 < 7$ Because $7 = 7$
This inequality is true if either the $>$ part or the $=$ part is true.	$b \ge a$	b is greater than or equal to a.	$7 \ge 3$ $-5 \ge -5$	Because $7 > 3$ Because $-5 = -5$

When using the symbol $\le$ (is less than or equal to), the inequality is a true statement if either the $<$ part or the $=$ part is true. When using the symbol $\ge$ (is greater than or equal to), the inequality is a true statement if either the $>$ part or the $=$ part is true.

> **EXAMPLE 7**   Using Inequality Symbols

Determine whether each inequality is true or false:

a. $-7 \le 4$ **b.** $-7 \le -7$ **c.** $-9 \ge 6$.

Solution

a. $-7 \le 4$ is true because $-7 < 4$ is true.

b. $-7 \le -7$ is true because $-7 = -7$ is true.

c. $-9 \ge 6$ is false because neither $-9 > 6$ nor $-9 = 6$ is true.   ▄

☑ **CHECK POINT 7**   Determine whether each inequality is true or false:

a. $-2 \le 3$ **b.** $-2 \ge -2$ **c.** $-4 \ge 1$.

---

6 Find the absolute value of a real number.

Absolute Value

Absolute value describes distance from 0 on a number line. If a represents a real number, the symbol $|a|$ represents its absolute value, read "the absolute value of a." For example,

$$|-5| = 5.$$

> The absolute value of -5 is 5 because -5 is 5 units from 0 on a number line.

Absolute Value

The **absolute value** of a real number a, denoted by $|a|$, is the distance from 0 to a on a number line. Because absolute value describes a distance, it is never negative.

EXAMPLE 8 Finding Absolute Value

Find the absolute value:

a. $|-3|$ b. $|5|$ c. $|0|$.

Solution The solution is illustrated in **Figure 1.10**.

a. $|-3| = 3$ The absolute value of −3 is 3 because −3 is 3 units from 0.

b. $|5| = 5$ 5 is 5 units from 0.

c. $|0| = 0$ 0 is 0 units from itself.

FIGURE 1.10 Absolute value describes distance from 0 on a number line.

Example 8 illustrates that the absolute value of a positive real number or 0 is the number itself. The absolute value of a negative real number, such as −3, is the number without the negative sign. Zero is the only real number whose absolute value is 0: $|0| = 0$. **The absolute value of any real number other than 0 is always positive.**

☑ **CHECK POINT 8** Find the absolute value:

a. $|-4|$ b. $|6|$ c. $|-\sqrt{2}|$.

1.3 EXERCISE SET **MyMathLab** Math XL PRACTICE   WATCH   DOWNLOAD   READ   REVIEW

Practice Exercises

In Exercises 1–8, write a positive or negative integer that describes each situation.

1. Meteorology: 20° below zero
2. Navigation: 65 feet above sea level
3. Health: A gain of 8 pounds
4. Economics: A loss of $12,500.00
5. Banking: A withdrawal of $3000.00
6. Physics: An automobile slowing down at a rate of 3 meters per second each second
7. Economics: A budget deficit of 4 billion dollars
8. Football: A 14-yard loss

In Exercises 9–20, start by drawing a number line that shows integers from −5 to 5. Then graph each real number on your number line.

9. 2   10. 5   11. −5   12. −2   13. $3\frac{1}{2}$ 14. $2\frac{1}{4}$

15. $\frac{11}{3}$ 16. $\frac{7}{3}$ 17. −1.8   18. −3.4   19. $-\frac{16}{5}$ 20. $-\frac{11}{5}$

In Exercises 21–32, express each rational number as a decimal.

21. $\frac{3}{4}$ 22. $\frac{3}{5}$ 23. $\frac{7}{20}$

24. $\frac{3}{20}$ 25. $\frac{7}{8}$ 26. $\frac{5}{16}$

27. $\frac{9}{11}$ 28. $\frac{3}{11}$ 29. $-\frac{1}{2}$

30. $-\frac{1}{4}$ 31. $-\frac{5}{6}$ 32. $-\frac{7}{6}$

In Exercises 33–36, list all numbers from the given set that are:
a. natural numbers, b. whole numbers, c. integers, d. rational numbers, e. irrational numbers, f. real numbers.

33. $\left\{-9, -\frac{4}{5}, 0, 0.25, \sqrt{3}, 9.2, \sqrt{100}\right\}$

34. $\left\{-7, -0.\overline{6}, 0, \sqrt{49}, \sqrt{50}\right\}$

35. $\left\{-11, -\frac{5}{6}, 0, 0.75, \sqrt{5}, \pi, \sqrt{64}\right\}$

36. $\left\{-5, -0.\overline{3}, 0, \sqrt{2}, \sqrt{4}\right\}$

37. Give an example of a whole number that is not a natural number.
38. Give an example of an integer that is not a whole number.
39. Give an example of a rational number that is not an integer.
40. Give an example of a rational number that is not a natural number.
41. Give an example of a number that is an integer, a whole number, and a natural number.
42. Give an example of a number that is a rational number, an integer, and a real number.

43. Give an example of a number that is an irrational number and a real number.

44. Give an example of a number that is a real number, but not an irrational number.

In Exercises 45–62, insert either $<$ or $>$ in the shaded area between each pair of numbers to make a true statement.

45. $\dfrac{1}{2}$ ▨ 2 **46.** 4 ▨ -3

47. 3 ▨ $-\dfrac{5}{2}$ **48.** 3 ▨ $\dfrac{3}{2}$

49. -4 ▨ -6 **50.** $-\dfrac{5}{2}$ ▨ $-\dfrac{5}{3}$

51. -2.5 ▨ 1.5 **52.** -1.25 ▨ -0.5

53. $-\dfrac{3}{4}$ ▨ $-\dfrac{5}{4}$ **54.** 0 ▨ $-\dfrac{1}{2}$

55. -4.5 ▨ 3 **56.** -5.5 ▨ 2.5

57. $\sqrt{2}$ ▨ 1.5 **58.** $\sqrt{3}$ ▨ 2

59. $0.\overline{3}$ ▨ 0.3 **60.** 0.6 ▨ $0.\overline{6}$

61. $-\pi$ ▨ -3.5 **62.** $-\dfrac{\pi}{2}$ ▨ -2.3

In Exercises 63–70, determine whether each inequality is true or false.

63. $-5 \geq -13$ **64.** $-5 \leq -8$

65. $-9 \geq -9$ **66.** $-14 \leq -14$

67. $0 \geq -6$ **68.** $0 \geq -13$

69. $-17 \geq 6$ **70.** $-14 \geq 8$

In Exercises 71–78, find each absolute value.

71. $|6|$ **72.** $|3|$ **73.** $|-7|$

74. $|-9|$ **75.** $\left|\dfrac{5}{6}\right|$ **76.** $\left|\dfrac{4}{5}\right|$

77. $|-\sqrt{11}|$ **78.** $|-\sqrt{29}|$

Practice PLUS

In Exercises 79–86, insert either $<, >,$ or $=$ in the shaded area to make a true statement.

79. $|-6|$ ▨ $|-3|$ **80.** $|-20|$ ▨ $|-50|$

81. $\left|\dfrac{3}{5}\right|$ ▨ $|-0.6|$ **82.** $\left|\dfrac{5}{2}\right|$ ▨ $|-2.5|$

83. $\dfrac{30}{40} - \dfrac{3}{4}$ ▨ $\dfrac{14}{15} \cdot \dfrac{15}{14}$ **84.** $\dfrac{17}{18} \cdot \dfrac{18}{17}$ ▨ $\dfrac{50}{60} - \dfrac{5}{6}$

85. $\dfrac{8}{13} \div \dfrac{8}{13}$ ▨ $|-1|$ **86.** $|-2|$ ▨ $\dfrac{4}{17} \div \dfrac{4}{17}$

Application Exercises

In Exercises 87–94, determine whether natural numbers, whole numbers, integers, rational numbers, or all real numbers are appropriate for each situation.

87. Shoe sizes of students on campus

88. Heights of students on campus

89. Temperatures in weather reports

90. Class sizes of algebra courses

91. Values of d given by the formula $d = \sqrt{1.5h}$, where d is the distance, in miles, that you can see to the horizon from a height of h feet

92. Values of C given by the formula $C = 2\pi r$, where C is the circumference of a circle with radius r

93. The number of pets a person has

94. The number of siblings a person has

95. The table shows the record low temperatures for five U.S. states.

State	Record Low (°F)	Date
Florida	−2	Feb. 13, 1899
Georgia	−17	Jan. 27, 1940
Hawaii	12	May 17, 1979
Louisiana	−16	Feb. 13, 1899
Rhode Island	−25	Feb. 5, 1996

Source: National Climatic Data Center

a. Graph the five record low temperatures on a number line.

b. Write the names of the states in order from the coldest record low to the warmest record low.

96. The table shows the record low temperatures for five U.S. states.

State	Record Low (°F)	Date
Virginia	−30	Jan. 22, 1985
Washington	−48	Dec. 30, 1968
West Virginia	−37	Dec. 30, 1917
Wisconsin	−55	Feb. 4, 1996
Wyoming	−66	Feb. 9, 1933

Source: National Climatic Data Center

a. Graph the five record low temperatures on a number line.

b. Write the names of the states in order from the coldest record low to the warmest record low.

Writing in Mathematics

97. What is a set?

98. What are the natural numbers?

99. What are the whole numbers?

100. What are the integers?

101. How does the set of integers differ from the set of whole numbers?

102. Describe how to graph a number on the number line.

103. What is a rational number?

104. Explain how to express $\frac{3}{8}$ as a decimal.

105. Describe the difference between a rational number and an irrational number.

106. If you are given two different real numbers, explain how to determine which one is the lesser.

107. Describe what is meant by the absolute value of a number. Give an example with your explanation.

Critical Thinking Exercises

Make Sense? *In Exercises 108–111, determine whether each statement "makes sense" or "does not make sense" and explain your reasoning.*

108. The humor in this joke is based on the fact that the football will never be hiked.

FOXTROT © 2000 Bill Amend. Reprinted with permission of UNIVERSAL PRESS SYNDICATE. All rights reserved.

109. *Titanic* came to rest 12,500 feet below sea level and *Bismarck* came to rest 15,617 feet below sea level, so *Bismarck's* resting place is higher than *Titanic's*.

110. I expressed a rational number as a decimal and the decimal neither terminated nor repeated.

111. I evaluated the formula $d = \sqrt{1.5h}$ for a value of h that resulted in a rational number for d.

In Exercises 112–117, determine whether each statement is true or false. If the statement is false, make the necessary change(s) to produce a true statement.

112. Every rational number is an integer.

113. Some whole numbers are not integers.

114. Some rational numbers are not positive.

115. Irrational numbers cannot be negative

116. Some real numbers are not rational numbers.

117. Some integers are not rational numbers.

In Exercises 118–119, write each phrase as an algebraic expression.

118. a loss of $\frac{1}{3}$ of an investment of d dollars

119. a loss of half of an investment of d dollars

Technology Exercises

In Exercises 120–123, use a calculator to find a decimal approximation for each irrational number, correct to three decimal places. Between which two integers should you graph each of these numbers on the number line?

120. $\sqrt{3}$

121. $-\sqrt{12}$

122. $1 - \sqrt{2}$

123. $2 - \sqrt{5}$

Preview Exercises

Exercises 124–126 will help you prepare for the material covered in the next section. In each exercise, evaluate both expressions for $x = 4$. What do you observe?

124. $3(x + 5); 3x + 15$

125. $3x + 5x; 8x$

126. $9x - 2x; 7x$

Basic Rules of Algebra

Objectives

1. Understand and use the vocabulary of algebraic expressions.
2. Use commutative properties.
3. Use associative properties.
4. Use distributive properties.
5. Combine like terms.
6. Simplify algebraic expressions.

Starting as a link among U.S. research scientists, more than one billion people worldwide now use the Internet. Some random Internet factoids:

- Number of searches performed per day on Google: 200 million
- Peak time for sex-related searches: 11 P.M.
- Fraction of people who use the word "password" as their password: $\frac{1}{8}$

- Vanity searchers: Fraction of people who typed their own name into a search engine: $\frac{1}{4}$
- Speed at which an email travels from San Francisco to New York: 30,000 miles per second

(*Source*: Paul Grobman, *Vital Statistics*, Plume, 2005)

In this section, we move from these quirky tidbits to mathematical models that describe the remarkable growth of the Internet among U.S. men and women. To use these models efficiently (you'll work with them in the Exercise Set), they should be simplified using basic rules of algebra. Before turning to these rules, we open the section with a closer look at algebraic expressions.

1 Understand and use the vocabulary of algebraic expressions.

The Vocabulary of Algebraic Expressions

We have seen that an algebraic expression combines numbers and variables. Here is an example of an algebraic expression:

$$7x + 3.$$

The **terms** of an algebraic expression are those parts that are separated by addition. For example, the algebraic expression $7x + 3$ contains two terms, namely $7x$ and 3. Notice that a term is a number, a variable, or a number multiplied by one or more variables.

The numerical part of a term is called its **coefficient**. In the term $7x$, the 7 is the coefficient. If a term containing one or more variables is written without a coefficient, the coefficient is understood to be 1. Thus, x means $1x$ and ab means $1ab$.

A term that consists of just a number is called a **constant term**. The constant term of $7x + 3$ is 3.

The parts of each term that are multiplied are called the **factors of the term**. The factors of the term $7x$ are 7 and x.

Like terms are terms that have exactly the same variable factors. Here are two examples of like terms:

$7x$ and $3x$ These terms have the same variable factor, x.

$4y$ and $9y$.   These terms have the same variable factor, y.

By contrast, here are some examples of terms that are not like terms. These terms do not have the same variable factor.

$7x$ and 3   The variable factor of the first term is x.
The second term has no variable factor.

$7x$ and $3y$ The variable factor of the first term is x.
The variable factor of the second term is y.

Constant terms are like terms. Thus, the constant terms 7 and -12 are like terms.

EXAMPLE 1   Using the Vocabulary of Algebraic Expressions

Use the algebraic expression

$$4x + 7 + 5x$$

to answer the following questions:
  a. How many terms are there in the algebraic expression?
  b. What is the coefficient of the first term?
  c. What is the constant term?
  d. What are the like terms in the algebraic expression?

Solution
  a. Because terms are separated by addition, the algebraic expression $4x + 7 + 5x$ contains three terms.

$$4x + 7 + 5x$$

First term   Second term   Third term

b. The coefficient of the first term, $4x$, is 4.

c. The constant term in $4x + 7 + 5x$ is 7.

d. The like terms in $4x + 7 + 5x$ are $4x$ and $5x$. These terms have the same variable factor, x.

☑ **CHECK POINT 1**   Use the algebraic expression $6x + 2x + 11$ to answer each of the four questions in Example 1.

Equivalent Algebraic Expressions

In Example 1, we considered the algebraic expression

$$4x + 7 + 5x.$$

Let's compare this expression with a second algebraic expression

$$9x + 7.$$

Evaluate each expression for some choice of x. We will select $x = 2$.

$4x + 7 + 5x$	$9x + 7$
Replace x with 2.	Replace x with 2.
$= 4 \cdot 2 + 7 + 5 \cdot 2$	$= 9 \cdot 2 + 7$
$= 8 + 7 + 10$	$= 18 + 7$
$= 25$	$= 25$

Both algebraic expressions have the same value when $x = 2$. Regardless of what number you select for x, the algebraic expressions $4x + 7 + 5x$ and $9x + 7$ will have the same value. These expressions are called *equivalent algebraic expressions*. Two algebraic expressions that have the same value for all replacements are called **equivalent algebraic expressions**. Because $4x + 7 + 5x$ and $9x + 7$ are equivalent algebraic expressions, we write

$$4x + 7 + 5x = 9x + 7.$$

Properties of Real Numbers and Algebraic Expressions

We now turn to basic properties, or rules, that you know from past experiences in working with whole numbers and fractions. These properties will be extended to include all real numbers and algebraic expressions. We will give each property a name so that we can refer to it throughout the study of algebra.

2 Use commutative properties.

The Commutative Properties

The addition or multiplication of two real numbers can be done in any order. For example, $3 + 5 = 5 + 3$ and $3 \cdot 5 = 5 \cdot 3$. Changing the order does not change the answer of a sum or a product. These facts are called **commutative properties**.

Study Tip

The commutative property does not hold for subtraction or division.

$$6 - 1 \neq 1 - 6$$
$$8 \div 4 \neq 4 \div 8$$

The Commutative Properties

Let a and b represent real numbers, variables, or algebraic expressions.

Commutative Property of Addition

$$a + b = b + a$$

Changing order when adding does not affect the sum.

Commutative Property of Multiplication

$$ab = ba$$

Changing order when multiplying does not affect the product.

EXAMPLE 2 Using the Commutative Properties

Use the commutative properties to write an algebraic expression equivalent to each of the following:

    a. $y + 6$ **b.** $5x$.

Solution

    a. By the commutative property of addition, an algebraic expression equivalent to $y + 6$ is $6 + y$. Thus,
$$y + 6 = 6 + y.$$

    b. By the commutative property of multiplication, an algebraic expression equivalent to $5x$ is $x5$. Thus,
$$5x = x5.$$

    ☑ **CHECK POINT 2** Use the commutative properties to write an algebraic expression equivalent to each of the following:

    a. $x + 14$ **b.** $7y$.

EXAMPLE 3 Using the Commutative Properties

Write an algebraic expression equivalent to $13x + 8$ using

    a. the commutative property of addition.

    b. the commutative property of multiplication.

Solution

    a. By the commutative property of addition, we change the order of the terms being added. This means that an algebraic expression equivalent to $13x + 8$ is $8 + 13x$:
$$13x + 8 = 8 + 13x.$$

    b. By the commutative property of multiplication, we change the order of the factors being multiplied. This means that an algebraic expression equivalent to $13x + 8$ is $x13 + 8$:
$$13x + 8 = x13 + 8.$$

    ☑ **CHECK POINT 3** Write an algebraic expression equivalent to $5x + 17$ using

    a. the commutative property of addition.

    b. the commutative property of multiplication.

Blitzer Bonus

Commutative Words and Sentences

The commutative property states that a change in order produces no change in the answer. The words and sentences listed here suggest a characteristic of the commutative property; they read the same from left to right and from right to left!

- dad
- repaper
- never odd or even

- Go deliver a dare, vile dog!
- May a moody baby doom a yam?
- Madam, in Eden I'm Adam.

- Ma is a nun, as I am.
- A man, a plan, a canal: Panama
- Are we not drawn onward, we few, drawn onward to new era?

③ Use associative properties.

The Associative Properties

Parentheses indicate groupings. As we have seen, we perform operations within the parentheses first. For example,

$$(2 + 5) + 10 = 7 + 10 = 17$$

and

$$2 + (5 + 10) = 2 + 15 = 17.$$

In general, the way in which three numbers are grouped does not change their sum. It also does not change their product. These facts are called the **associative properties**.

Study Tip

The associative property does not hold for subtraction or division.

$$(6 - 3) - 1 \neq 6 - (3 - 1)$$
$$(8 \div 4) \div 2 \neq 8 \div (4 \div 2)$$

The Associative Properties

Let a, b, and c represent real numbers, variables, or algebraic expressions.

Associative Property of Addition

$$(a + b) + c = a + (b + c)$$

Changing grouping when adding does not affect the sum.

Associative Property of Multiplication

$$(ab)c = a(bc)$$

Changing grouping when multiplying does not affect the product.

The associative properties can be used to simplify some algebraic expressions by removing the parentheses.

EXAMPLE 4   Simplifying Using the Associative Properties

Simplify:

   a. $3 + (8 + x)$ **b.** $8(4x)$.

Solution

  a. $3 + (8 + x)$ This is the given algebraic expression.

     $= (3 + 8) + x$ Use the associative property of addition to group the first two numbers.

     $= 11 + x$ Add within parentheses.

Using the commutative property of addition, this simplified algebraic expression can also be written as $x + 11$.

  b. $8(4x)$ This is the given algebraic expression.

     $= (8 \cdot 4)x$ Use the associative property of multiplication to group the first two numbers.

     $= 32x$ Multiply within parentheses.

We can use the commutative property of multiplication and write this simplified algebraic expression as $x32$ or $x \cdot 32$. However, it is customary to express a term with its coefficient on the left. Thus, we use $32x$ as the simplified form of the algebraic expression. ◾

✓ **CHECK POINT 4**   Simplify:

  a. $8 + (12 + x)$ **b.** $6(5x)$.

The next example involves the use of both basic properties to simplify an algebraic expression.

EXAMPLE 5   **Using the Commutative and Associative Properties**

Simplify: $7 + (x + 2)$.

Solution

$$7 + (x + 2)$$ This is the given algebraic expression.

$$= 7 + (2 + x)$$ Use the commutative property to change the order of the addition.

$$= (7 + 2) + x$$ Use the associative property to group the first two numbers.

$$= 9 + x$$ Add within parentheses.

Using the commutative property of addition, an equivalent algebraic expression is $x + 9$. ∎

☑ **CHECK POINT 5**   Simplify: $8 + (x + 4)$.

Study Tip

Commutative: changes *order*
Associative: changes *grouping*

4 Use distributive properties.

The Distributive Properties

The **distributive property** involves both multiplication and addition. The property shows how to multiply the sum of two numbers by a third number. Consider, for example, $4(7 + 3)$, which can be calculated in two ways. One way is to perform the addition within the grouping symbols and then multiply:

$$4(7 + 3) = 4(10) = 40.$$

The other way is to *distribute* the multiplication by 4 over the addition by first multiplying each number within the parentheses by 4 and then adding:

$$4 \cdot 7 + 4 \cdot 3 = 28 + 12 = 40.$$

The result in both cases is 40. Thus,

$$4(\overset{\frown}{7 + 3}) = 4 \cdot 7 + 4 \cdot 3.$$ Multiplication distributes over addition.

The distributive property allows us to rewrite the product of a number and a sum as the sum of two products.

> **The Distributive Property**
>
> Let a, b, and c represent real numbers, variables, or algebraic expressions.
>
> $$a(\overset{\frown}{b + c}) = ab + ac$$
>
> Multiplication distributes over addition.

Study Tip

Do not confuse the distributive property with the associative property of multiplication.

Distributive:

$$4(5 + x) = 4 \cdot 5 + 4x$$
$$= 20 + 4x$$

Associative:

$$4(5 \cdot x) = (4 \cdot 5)x$$
$$= 20x$$

EXAMPLE 6   **Using the Distributive Property**

Multiply: $6(x + 4)$.

Solution   Multiply *each term* inside the parentheses, x and 4, by the multiplier outside, 6.

$$6(x + 4) = 6x + 6 \cdot 4$$ Use the distributive property to remove parentheses.

$$= 6x + 24$$ Multiply: $6 \cdot 4 = 24$. ∎

☑ **CHECK POINT 6**   Multiply: $5(x + 3)$.

EXAMPLE 7   **Using the Distributive Property**

Multiply: $5(3y + 7)$.

Solution   Multiply *each term* inside the parentheses, $3y$ and 7, by the multiplier outside, 5.

$$5(3y + 7) = 5 \cdot 3y + 5 \cdot 7$$ Use the distributive property to remove parentheses.

$$= 15y + 35$$ Multiply. Use the associative property of multiplication to find $5 \cdot 3y$: $5 \cdot 3y = (5 \cdot 3)y = 15y$.   ■

☑ **CHECK POINT 7**   Multiply: $6(4y + 7)$.

Study Tip

When using the distributive property to remove parentheses, be sure to multiply *each term* inside the parentheses by the multiplier outside.

Incorrect!

$$5(3y + 7) = 5 \cdot 3y + 7$$ 7 must also be multiplied by 5.

$$= 15y + 7$$

Table 1.3 shows a number of other forms of the distributive property.

Table 1.3   **Other Forms of the Distributive Property**

Property	Meaning	Example
$a(b - c) = ab - ac$	Multiplication distributes over subtraction.	$5(4x - 3) = 5 \cdot 4x - 5 \cdot 3$ $= 20x - 15$
$a(b + c + d) = ab + ac + ad$	Multiplication distributes over three or more terms in parentheses.	$4(x + 10 + 3y)$ $= 4x + 4 \cdot 10 + 4 \cdot 3y$ $= 4x + 40 + 12y$
$(b + c)a = ba + ca$	Multiplication on the right distributes over addition (or subtraction).	$(x + 7)9 = x \cdot 9 + 7 \cdot 9$ $= 9x + 63$

---

5 Combine like terms.

Combining Like Terms

The distributive property

$$a(b + c) = ab + ac$$

lets us add and subtract like terms. To do this, we will usually apply the property in the form

$$ax + bx = (a + b)x$$

and then combine a and b. For example,

$$3x + 7x = (3 + 7)x = 10x.$$

This process is called **combining like terms**.

EXAMPLE 8 Combining Like Terms

Combine like terms:

   a. $4x + 15x$ **b.** $7a - 2a$.

Solution

   a. $4x + 15x$ These are like terms because 4x and 15x have identical variable factors.

     $= (4 + 15)x$ Apply the distributive property.

     $= 19x$ Add within parentheses.

   b. $7a - 2a$ These are like terms because 7a and 2a have identical variable factors.

     $= (7 - 2)a$ Apply the distributive property.

     $= 5a$ Subtract within parentheses. ∎

✓ **CHECK POINT 8** Combine like terms:

   a. $7x + 3x$ **b.** $9a - 4a$.

When combining like terms, you may find yourself leaving out the details of the distributive property. For example, you may simply write

$$7x + 3x = 10x.$$

It might be useful to think along these lines: Seven things plus three of the (same) things give ten of those things. To add like terms, add the coefficients and copy the common variable.

Combining Like Terms Mentally

  1. Add or subtract the coefficients of the terms.

  2. Use the result of step 1 as the coefficient of the term's variable factor.

When an expression contains three or more terms, use the commutative and associative properties to group like terms. Then combine the like terms.

EXAMPLE 9 Grouping and Combining Like Terms

Simplify:

   a. $7x + 5 + 3x + 8$ **b.** $4x + 7y + 2x + 3y$.

Solution

   a. $7x + 5 + 3x + 8$

     $= (7x + 3x) + (5 + 8)$ Rearrange terms and group the like terms using the commutative and associative properties. This step is often done mentally.

     $= 10x + 13$ Combine like terms: 7x + 3x = 10x. Combine constant terms: 5 + 8 = 13.

   b. $4x + 7y + 2x + 3y$

     $= (4x + 2x) + (7y + 3y)$ Group like terms.

     $= 6x + 10y$ Combine like terms by adding coefficients and keeping the variable factor. ∎

✓ **CHECK POINT 9** Simplify:

   a. $8x + 7 + 10x + 3$ **b.** $9x + 6y + 5x + 2y$.

Study Tip

Combining like terms should remind you of adding and subtracting numbers of like objects.

$$7 \text{ apples} + 3 \text{ apples} = 10 \text{ apples} \quad \boxed{7a + 3a = 10a}$$

$$9 \text{ feet} - 5 \text{ feet} = 4 \text{ feet} \quad \boxed{9f - 5f = 4f}$$

$$6 \text{ apples} + 10 \text{ feet} = ? \quad \boxed{\begin{array}{l} 6a \text{ and } 10f \text{ are not like} \\ \text{terms and cannot be added.} \end{array}}$$

6 Simplify algebraic expressions.

Simplifying Algebraic Expressions

An algebraic expression is **simplified** when parentheses have been removed and like terms have been combined.

> ### Simplifying Algebraic Expressions
>
> 1. Use the distributive property to remove parentheses.
> 2. Rearrange terms and group like terms using the commutative and associative properties. This step may be done mentally.
> 3. Combine like terms by combining the coefficients of the terms and keeping the same variable factor.

EXAMPLE 10   Simplifying an Algebraic Expression

Simplify: $5(3x + 7) + 6x$.

Solution

$$5(3x + 7) + 6x$$
$$= 5 \cdot 3x + 5 \cdot 7 + 6x \quad \text{Use the distributive property to remove the parentheses.}$$
$$= 15x + 35 + 6x \quad \text{Multiply.}$$
$$= (15x + 6x) + 35 \quad \text{Group like terms.}$$
$$= 21x + 35 \quad \text{Combine like terms.} \quad \blacksquare$$

☑ **CHECK POINT 10**   Simplify: $7(2x + 3) + 11x$.

EXAMPLE 11   Simplifying an Algebraic Expression

Simplify: $6(2x + 4y) + 10(4x + 3y)$.

Solution

$$6(2x + 4y) + 10(4x + 3y)$$
$$= 6 \cdot 2x + 6 \cdot 4y + 10 \cdot 4x + 10 \cdot 3y \quad \text{Use the distributive property to remove the parentheses.}$$
$$= 12x + 24y + 40x + 30y \quad \text{Multiply.}$$
$$= (12x + 40x) + (24y + 30y) \quad \text{Group like terms.}$$
$$= 52x + 54y \quad \text{Combine like terms.} \quad \blacksquare$$

☑ **CHECK POINT 11**   Simplify: $7(4x + 3y) + 2(5x + y)$.

Study Tip

Use the distributive property to remove parentheses when the terms inside parentheses are not like terms:

$$4(3x + 5y) = 4 \cdot 3x + 4 \cdot 5y = 12x + 20y.$$

3x and 5y are not like terms.

It is not necessary to use the distributive property to remove parentheses when the terms inside parentheses are like terms:

$$4(3x + 5x) = 4(8x) = 32x.$$

3x and 5x are like terms and can be combined: 3x + 5x = 8x.

We mentally applied the associative property: 4(8x) = (4 · 8)x = 32x.

1.4 EXERCISE SET

MyMathLab
PRACTICE WATCH DOWNLOAD READ REVIEW

Practice Exercises

In Exercises 1–6, an algebraic expression is given. Use each expression to answer the following questions.

  a. *How many terms are there in the algebraic expression?*

  b. *What is the numerical coefficient of the first term?*

  c. *What is the constant term?*

  d. *Does the algebraic expression contain like terms? If so, what are the like terms?*

  1. $3x + 5$

  2. $9x + 4$

  3. $x + 2 + 5x$

  4. $x + 6 + 7x$

  5. $4y + 1 + 3x$

  6. $8y + 1 + 10x$

In Exercises 7–14, use the commutative property of addition to write an equivalent algebraic expression.

  7. $y + 4$ **8.** $x + 7$

  9. $5 + 3x$ **10.** $4 + 9x$

  11. $4x + 5y$ **12.** $10x + 9y$

  13. $5(x + 3)$ **14.** $6(x + 4)$

In Exercises 15–22, use the commutative property of multiplication to write an equivalent algebraic expression.

  15. $9x$ **16.** $8x$

  17. $x + y6$ **18.** $x + y7$

  19. $7x + 23$ **20.** $13x + 11$

  21. $5(x + 3)$ **22.** $6(x + 4)$

In Exercises 23–26, use an associative property to rewrite each algebraic expression. Once the grouping has been changed, simplify the resulting algebraic expression.

  23. $7 + (5 + x)$

  24. $9 + (3 + x)$

  25. $7(4x)$

  26. $8(5x)$

In Exercises 27–46, use a form of the distributive property to rewrite each algebraic expression without parentheses.

  27. $3(x + 5)$ **28.** $4(x + 6)$

  29. $8(2x + 3)$ **30.** $9(2x + 5)$

  31. $\frac{1}{3}(12 + 6r)$ **32.** $\frac{1}{4}(12 + 8r)$

  33. $5(x + y)$ **34.** $7(x + y)$

  35. $3(x - 2)$ **36.** $4(x - 5)$

  37. $2(4x - 5)$ **38.** $6(3x - 2)$

  39. $\frac{1}{2}(5x - 12)$ **40.** $\frac{1}{3}(7x - 21)$

  41. $(2x + 7)4$ **42.** $(5x + 3)6$

  43. $6(x + 3 + 2y)$

  44. $7(2x + 4 + y)$

  45. $5(3x - 2 + 4y)$

  46. $4(5x - 3 + 7y)$

In Exercises 47–64, simplify each algebraic expression.

  47. $7x + 10x$ **48.** $5x + 13x$

  49. $11a - 3a$ **50.** $14b - 5b$

  51. $3 + (x + 11)$ **52.** $7 + (x + 10)$

  53. $5y + 3 + 6y$ **54.** $8y + 7 + 10y$

  55. $2x + 5 + 7x - 4$ **56.** $7x + 8 + 2x - 3$

  57. $11a + 12 + 3a + 2$

  58. $13a + 15 + 2a + 11$

  59. $5(3x + 2) - 4$ **60.** $2(5x + 4) - 3$

  61. $12 + 5(3x - 2)$ **62.** $14 + 2(5x - 1)$

  63. $7(3a + 2b) + 5(4a + 2b)$

  64. $11(6a + 3b) + 4(12a + 5b)$

Practice PLUS

In Exercises 65–66, name the property used to go from step to step each time that "(why?)" occurs.

65. $7 + 2(x + 9)$

$= 7 + (2x + 18)$ (why?)

$= 7 + (18 + 2x)$ (why?)

$= (7 + 18) + 2x$ (why?)

$= 25 + 2x$

$= 2x + 25$ (why?)

66. $5(x + 4) + 3x$

$= (5x + 20) + 3x$ (why?)

$= (20 + 5x) + 3x$ (why?)

$= 20 + (5x + 3x)$ (why?)

$= 20 + (5 + 3)x$ (why?)

$= 20 + 8x$

$= 8x + 20$ (why?)

In Exercises 67–76, write each English phrase as an algebraic expression. Then simplify the expression. Let x represent the number.

67. the sum of 7 times a number and twice the number

68. the sum of 8 times a number and twice the number

69. the product of 3 and a number, which is then subtracted from the product of 12 and a number

70. the product of 5 and a number, which is then subtracted from the product of 11 and a number

71. six times the product of 4 and a number

72. nine times the product of 3 and a number

73. six times the sum of 4 and a number

74. nine times the sum of 3 and a number

75. eight increased by the product of 5 and one less than a number

76. nine increased by the product of 3 and 2 less than a number

Application Exercises

The graph shows the percentage of U.S. men and women who used the Internet for four selected years. Exercises 77–78 involve mathematical models for the data.

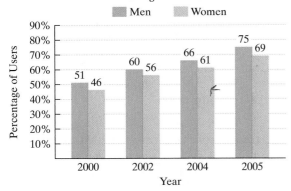

Percentage of United States Adults Using the Internet

Source: Pew Internet and American Life Project

77. The percentage of U.S. men, M, who used the Internet n years after 2000 can be modeled by the formula

$$M = 2(2n + 25) + 0.5(n + 2).$$

  a. Simplify the formula.

  b. Use the simplified form of the mathematical model to find the percentage of U.S. men who used the Internet in 2005. Does the model underestimate or overestimate the actual percent shown by the bar graph for 2005? By how much?

78. The percentage of U.S. women, W, who used the Internet n years after 2000 can be modeled by the formula

$$W = 2(2n + 20) + 0.3(n + 20).$$

  a. Simplify the formula.

  b. Use the simplified form of the mathematical model to find the percentage of U.S. women who used the Internet in 2005. Does the model underestimate or overestimate the actual percent shown by the bar graph for 2005? By how much?

Writing in Mathematics

79. What is a term? Provide an example with your description.

80. What are like terms? Provide an example with your description.

81. What are equivalent algebraic expressions?

82. State a commutative property and give an example.

83. State an associative property and give an example.

84. State a distributive property and give an example.

85. Explain how to add like terms. Give an example.

86. What does it mean to simplify an algebraic expression?

87. An algebra student incorrectly used the distributive property and wrote $3(5x + 7) = 15x + 7$. If you were that student's teacher, what would you say to help the student avoid this kind of error?

88. You can transpose the letters in the word "conversation" to form the phrase "voices rant on." From "total abstainers" we can form "sit not at ale bars." What two algebraic properties do each of these transpositions (called *anagrams*) remind you of? Explain your answer.

Critical Thinking Exercises

Make Sense? *In Exercises 89–92, determine whether each statement "makes sense" or "does not make sense" and explain your reasoning.*

89. I applied the commutative property and rewrote $x - 4$ as $4 - x$.

90. Just as the commutative properties change groupings, the associative properties change order.

91. I did not use the distributive property to simplify $3(2x + 5x)$.

92. The commutative, associative, and distributive properties remind me of the rules of a game.

In Exercises 93–96, determine whether each statement is true or false. If the statement is false, make the necessary change(s) to produce a true statement.

93. $(24 \div 6) \div 2 = 24 \div (6 \div 2)$

94. $2x + 5 = 5x + 2$

95. $a + (bc) = (a + b)(a + c)$; In words, addition can be distributed over multiplication.

96. Like terms contain the same coefficients.

Preview Exercises

Exercises 97–99 will help you prepare for the material covered in the next section. In each exercise, write an integer that is the result of the given situation.

97. You earn $150, but then you misplace $90.

98. You lose $30 and then you misplace $10.

99. The temperature is 30 degrees, and then it drops by 35 degrees.

MID-CHAPTER CHECK POINT   Section 1.1–Section 1.4

✓ **What You Know:** Algebra uses variables, or letters that represent a variety of different numbers. These variables appear in algebraic expressions, equations, and formulas. Mathematical models use variables to describe real-world phenomena. We reviewed operations with fractions and saw how fractions are used in algebra. We defined the real numbers and represented them as points on a number line. Finally, we introduced some basic rules of algebra and used the commutative, associative, and distributive properties to simplify algebraic expressions.

1. Evaluate for $x = 6$: $2 + 10x$.

2. Evaluate for $x = \frac{3}{5}$: $10x - 4$.

3. Evaluate for $x = 3$ and $y = 10$:

$$\frac{xy}{2} + 4(y - x).$$

In Exercises 4–5, translate from English to an algebraic expression or equation, whichever is appropriate. Let the variable x represent the number.

4. Two less than $\frac{1}{4}$ of a number

5. Five more than the quotient of a number and 6 gives 19.

In Exercises 6–7, determine whether the given number is a solution of the equation.

6. $3(x + 2) = 4x - 1$; 6

7. $8y = 12\left(y - \frac{1}{2}\right)$; $\frac{3}{4}$

8. Because the population of the United States continues to grow, criminologists use the crime rate per 100,000 population to compare crime statistics

from different years. The bar graph shows the number of violent crimes in the United States per 100,000 inhabitants for four selected years.

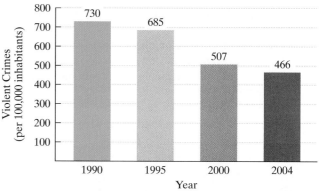

Violent Crime in the United States per 100,000 Inhabitants

Source: Federal Bureau of Investigation

Here is a mathematical model that approximates the data displayed by the bar graph:

$$V = 747 - 21n.$$

Number of years after 1990

Number of violent crimes per 100,000 inhabitants

a. Use the formula to find the number of violent crimes per 100,000 inhabitants in 2000. Does the mathematical model underestimate or overestimate the actual number shown by the bar graph for 2000? By how much?

b. If trends from 1990 through 2004 continue, use the formula to project the number of violent crimes per 100,000 inhabitants in 2010.

In Exercises 9–15, perform the indicated operation. Where possible, reduce answers to lowest terms.

9. $\dfrac{7}{10} - \dfrac{8}{15}$

10. $\dfrac{2}{3} \cdot \dfrac{3}{4}$

11. $\dfrac{5}{22} + \dfrac{5}{33}$

12. $\dfrac{3}{5} \div \dfrac{9}{10}$

13. $\dfrac{23}{105} - \dfrac{2}{105}$

14. $2\dfrac{7}{9} \div 3$

15. $5\dfrac{2}{9} - 3\dfrac{1}{6}$

16. The formula $C = \frac{5}{9}(F - 32)$ expresses the relationship between Fahrenheit temperature, F, and Celsius temperature, C. If the temperature is 50°F, use the formula to find the equivalent temperature on the Celsius scale.

17. Insert either $<$ or $>$ in the shaded area to make a true statement:

$$-8000 \quad \blacksquare \quad -8\dfrac{1}{4}.$$

18. Express $\frac{1}{11}$ as a decimal.

19. Find the absolute value: $|-19.3|$.

20. List all the rational numbers in this set:

$$\left\{ -11, -\dfrac{3}{7}, 0, 0.45, \sqrt{23}, \sqrt{25} \right\}.$$

In Exercises 21–23, rewrite $5(x + 3)$ as an equivalent expression using the given property.

21. the commutative property of multiplication

22. the commutative property of addition

23. the distributive property

In Exercises 24–25, simplify each algebraic expression.

24. $7(9x + 3) + \dfrac{1}{3}(6x)$

25. $2(3x + 5y) + 4(x + 6y)$

SECTION **1.5**

Addition of Real Numbers

Objectives

1 Add numbers with a number line.

2 Find sums using identity and inverse properties.

3 Add numbers without a number line.

4 Use addition rules to simplify algebraic expressions.

5 Solve applied problems using a series of additions.

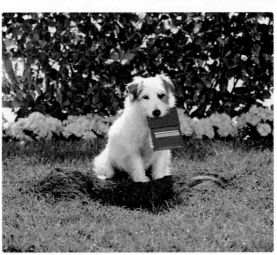

It has not been a good day! First, you lost your wallet with $30 in it. Then, to get through the day, you borrowed $10, which you somehow misplaced. Your loss of $30 followed by a loss of $10 is an overall loss of $40. This can be written

$$-30 + (-10) = -40.$$

The result of adding two or more numbers is called the **sum** of the numbers. The sum of -30 and -10 is -40. You can think of gains and losses of money to find sums. For example, to find $17 + (-13)$, think of a gain of $17 followed by a loss of $13. There is an overall gain of $4. Thus, $17 + (-13) = 4$. In the same way, to find $-17 + 13$, think of a loss of $17 followed by a gain of $13. There is an overall loss of $4, so $-17 + 13 = -4$.

1 Add numbers with a number line.

Adding with a Number Line

We use the number line to help picture the addition of real numbers. On the next page is the procedure for finding $a + b$, the sum of a and b, using the number line.

Using the Number Line to Find a Sum

Let a and b represent real numbers. To find $a + b$ using a number line,

1. Start at a.
2. a. If b is **positive**, move b units to the **right**.
   b. If b is **negative**, move b units to the **left**.
   c. If b is **0, stay** at a.
3. The number where we finish on the number line represents the sum of a and b.

This procedure is illustrated in Examples 1 and 2. Think of moving to the right as a gain and moving to the left as a loss.

EXAMPLE 1   Adding Real Numbers Using a Number Line

Find the sum using a number line:

$$3 + (-5).$$

Solution   We find $3 + (-5)$ using the number line in **Figure 1.11**.

Step 1. We consider 3 to be the first number, represented by a in the preceding box. We start at a, or 3.

Step 2. We consider -5 to be the second number, represented by b. Because this number is negative, we move 5 units to the left.

Step 3. We finish at -2 on the number line. The number where we finish represents the sum of 3 and -5. Thus,

$$3 + (-5) = -2.$$

FIGURE 1.11 $3 + (-5) = -2$

Observe that if there is a gain of \$3 followed by a loss of \$5, there is an overall loss of \$2.

☑ **CHECK POINT 1**   Find the sum using a number line:

$$4 + (-7).$$

EXAMPLE 2   Adding Real Numbers Using a Number Line

Find each sum using a number line:

   a. $-3 + (-4)$ **b.** $-6 + 2$.

Solution

   a. To find $-3 + (-4)$, start at -3. Move 4 units to the left. We finish at -7. Thus,

$$-3 + (-4) = -7.$$

Observe that if there is a loss of \$3 followed by a loss of \$4, there is an overall loss of \$7.

b. To find $-6 + 2$, start at -6. Move 2 units to the right because 2 is positive. We finish at -4. Thus,

$$-6 + 2 = -4.$$

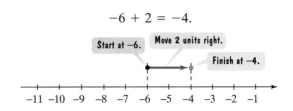

Observe that if there is a loss of $6 followed by a gain of $2, there is an overall loss of $4. ◼

✓ **CHECK POINT 2** Find each sum using a number line:

a. $-1 + (-3)$ **b.** $-5 + 3$.

---

2 Find sums using identity and inverse properties.

The Number Line and Properties of Addition

The number line can be used to picture some useful properties of addition. For example, let's see what happens if we add two numbers with different signs but the same absolute value. Two such numbers are 3 and -3. To find $3 + (-3)$ on a number line, we start at 3 and move 3 units to the left. We finish at 0. Thus,

$$3 + (-3) = 0.$$

Numbers that differ only in sign, such as 3 and -3, are called *additive inverses*. **Additive inverses**, also called **opposites**, are pairs of real numbers that are the same number of units from zero on the number line, but are on opposite sides of zero. Thus, -3 is the additive inverse, or opposite, of 3, and 5 is the additive inverse, or opposite, of -5. The additive inverse of 0 is 0. Other additive inverses come in pairs.

In general, the sum of any real number, denoted by a, and its additive inverse, denoted by $-a$, is zero:

$$a + (-a) = 0.$$

This property is called the **inverse property of addition**.

Table 1.4 shows the identity and inverse properties of addition.

Table 1.4	Identity and Inverse Properties of Addition	
Let *a* be a real number, a variable, or an algebraic expression.		
Property	**Meaning**	**Examples**
Identity Property of Addition	Zero can be deleted from a sum. $a + 0 = a$ $0 + a = a$	• $4 + 0 = 4$ • $-3x + 0 = -3x$ • $0 + (5a + b) = 5a + b$
Inverse Property of Addition	The sum of a real number and its additive inverse, or opposite, gives 0, the additive identity. $a + (-a) = 0$ $(-a) + a = 0$	• $6 + (-6) = 0$ • $3x + (-3x) = 0$ • $[-(2y + 1)] + (2y + 1) = 0$

3 Add numbers without a number line.

Adding without a Number Line

Now that we can picture the addition of real numbers, we look at two rules for using absolute value to add signed numbers.

> **Adding Two Numbers with the Same Sign**
>
> **1.** Add the absolute values.
>
> **2.** Use the common sign as the sign of the sum.

EXAMPLE 3 **Adding Real Numbers**

Add without using a number line:

a. $-11 + (-15)$ **b.** $-0.2 + (-0.8)$ **c.** $-\dfrac{3}{4} + \left(-\dfrac{1}{2}\right).$

Solution In each part of this example, we are adding numbers with the same sign.

a. $-11 + (-15) = -26$ Add absolute values: $11 + 15 = 26$.

Use the common sign.

b. $-0.2 + (-0.8) = -1$ Add absolute values: $0.2 + 0.8 = 1.0$ or 1.

Use the common sign.

c. $-\dfrac{3}{4} + \left(-\dfrac{1}{2}\right) = -\dfrac{5}{4}$ Add absolute values: $\dfrac{3}{4} + \dfrac{1}{2} = \dfrac{3}{4} + \dfrac{2}{4} = \dfrac{5}{4}$.

Use the common sign.

✓ **CHECK POINT 3** Add without using a number line:

a. $-10 + (-25)$ **b.** $-0.3 + (-1.2)$ **c.** $-\dfrac{2}{3} + \left(-\dfrac{1}{6}\right).$

We also use absolute value to add two real numbers with different signs.

> **Adding Two Numbers with Different Signs**
>
> **1.** Subtract the smaller absolute value from the greater absolute value.
>
> **2.** Use the sign of the number with the greater absolute value as the sign of the sum.

Using Technology

You can use a calculator to add signed numbers. Here are the keystrokes for finding $-11 + (-15)$:

Scientific Calculator

11 $\boxed{+/-}$ $\boxed{+}$ 15 $\boxed{+/-}$ $\boxed{=}$

Graphing Calculator

$\boxed{(-)}$ 11 $\boxed{+}$ $\boxed{(-)}$ 15 $\boxed{\text{ENTER}}$

Here are the keystrokes for finding $-13 + 4$:

Scientific Calculator

13 $\boxed{+/-}$ $\boxed{+}$ 4 $\boxed{=}$

Graphing Calculator

$\boxed{(-)}$ 13 $\boxed{+}$ 4 $\boxed{\text{ENTER}}$

EXAMPLE 4   Adding Real Numbers

Add without using a number line:

a. $-13 + 4$ **b.** $-0.2 + 0.8$ **c.** $-\dfrac{3}{4} + \dfrac{1}{2}$.

Solution   In each part of this example, we are adding numbers with different signs.

a. $-13 + 4 = -9$

Subtract absolute values: $13 - 4 = 9$.

Use the sign of the number with the greater absolute value.

b. $-0.2 + 0.8 = 0.6$

Subtract absolute values: $0.8 - 0.2 = 0.6$.

Use the sign of the number with the greater absolute value. The sign is assumed to be positive.

c. $-\dfrac{3}{4} + \dfrac{1}{2} = -\dfrac{1}{4}$

Subtract absolute values: $\dfrac{3}{4} - \dfrac{1}{2} = \dfrac{3}{4} - \dfrac{2}{4} = \dfrac{1}{4}$.

Use the sign of the number with the greater absolute value.

Study Tip

The sum of two numbers with different signs may be positive or negative. Keep in mind that the sign of the sum is the sign of the number with the greater absolute value.

✓ **CHECK POINT 4**   Add without using a number line:

a. $-15 + 2$ **b.** $-0.4 + 1.6$ **c.** $-\dfrac{2}{3} + \dfrac{1}{6}$.

---

4   Use addition rules to simplify algebraic expressions.

Algebraic Expressions

The rules for adding real numbers can be used to simplify certain algebraic expressions.

EXAMPLE 5   Simplifying Algebraic Expressions

Simplify:

a. $-11x + 7x$ **b.** $7y + (-12z) + (-9y) + 15z$ **c.** $3(7x + 8) + (-25x)$.

Solution

a. $-11x + 7x$ The given algebraic expression has two like terms: $-11x$ and $7x$ have identical variable factors.

$= (-11 + 7)x$ Apply the distributive property.

$= -4x$ Add within parentheses: $-11 + 7 = -4$.

b. $7y + (-12z) + (-9y) + 15z$ The colors indicate that there are two pairs of like terms.

$= 7y + (-9y) + (-12z) + 15z$ Arrange like terms so that they are next to one another.

$= [7 + (-9)]y + [(-12) + 15]z$ Apply the distributive property.

$= -2y + 3z$ Add within the grouping symbols: $7 + (-9) = -2$ and $-12 + 15 = 3$.

c. $3(7x + 8) + (-25x)$

$= 3 \cdot 7x + 3 \cdot 8 + (-25x)$ Use the distributive property to remove the parentheses.

$= 21x + 24 + (-25x)$ Multiply.

$= 21x + (-25x) + 24$ Arrange like terms so that they are next to one another.

$= [21 + (-25)]x + 24$ Apply the distributive property.

$= -4x + 24$ Add within the grouping symbols: $21 + (-25) = -4$.

☑ **CHECK POINT 5** Simplify:

a. $-20x + 3x$

b. $3y + (-10z) + (-10y) + 16z$

c. $5(2x + 3) + (-30x)$

---

5 Solve applied problems using a series of additions.

Applications

Positive and negative numbers are used in everyday life to represent such things as gains and losses in the stock market, rising and falling temperatures, deposits and withdrawals on bank statements, and ascending and descending motion. Positive and negative numbers are used to solve applied problems involving a series of additions.

One way to add a series of positive and negative numbers is to use the commutative and associative properties.

- Add all the positive numbers.
- Add all the negative numbers.
- Add the sums obtained in the first two steps.

The next example illustrates this idea.

EXAMPLE 6 An Application of Adding Signed Numbers

A glider was towed 1000 meters into the air and then let go. It descended 70 meters into a thermal (rising bubble of warm air), which took it up 2100 meters. At this point it dropped 230 meters into a second thermal. Then it rose 1200 meters. What was its altitude at that point?

Solution   We use the problem's conditions to write a sum. The altitude of the glider is expressed by the following sum:

Towed to 1000 meters	then	Descended 70 meters	then	Taken up 2100 meters	then	Dropped 230 meters	then	Rose 1200 meters
1000	+	(-70)	+	2100	+	(-230)	+	1200.

$1000 + (-70) + 2100 + (-230) + 1200$ This is the sum arising from the problem's conditions.

$= (1000 + 2100 + 1200) + [(-70) + (-230)]$ Use the commutative and associative properties to group the positive and negative numbers.

$= 4300 + (-300)$ Add the positive numbers. Add the negative numbers.

$= 4000$ Add the results.

The altitude of the glider is 4000 meters.   ∎

Discover for Yourself

Try working Example 6 by adding from left to right. You should still obtain 4000 for the sum. Which method do you find easier?

☑ **CHECK POINT 6**   The water level of a reservoir is measured over a five-month period. During this time, the level rose 2 feet, then fell 4 feet, then rose 1 foot, then fell 5 feet, and then rose 3 feet. What was the change in the water level at the end of the five months?

1.5 EXERCISE SET   *MyMathLab*

PRACTICE · WATCH · DOWNLOAD · READ · REVIEW

Practice Exercises

In Exercises 1–8, find each sum using a number line.

1. $7 + (-3)$ **2.** $7 + (-2)$ **3.** $-2 + (-5)$

4. $-1 + (-5)$ **5.** $-6 + 2$ **6.** $-8 + 3$

7. $3 + (-3)$ **8.** $5 + (-5)$

In Exercises 9–46, find each sum without the use of a number line.

9. $-7 + 0$ **10.** $-5 + 0$

11. $30 + (-30)$ **12.** $15 + (-15)$

13. $-30 + (-30)$ **14.** $-15 + (-15)$

15. $-8 + (-10)$ **16.** $-4 + (-6)$

17. $-0.4 + (-0.9)$ **18.** $-1.5 + (-5.3)$

19. $-\dfrac{7}{10} + \left(-\dfrac{3}{10}\right)$ **20.** $-\dfrac{7}{8} + \left(-\dfrac{1}{8}\right)$

21. $-9 + 4$ **22.** $-7 + 3$

23. $12 + (-8)$ **24.** $13 + (-5)$

25. $6 + (-9)$ **26.** $3 + (-11)$

27. $-3.6 + 2.1$ **28.** $-6.3 + 5.2$

29. $-3.6 + (-2.1)$ **30.** $-6.3 + (-5.2)$

31. $\dfrac{9}{10} + \left(-\dfrac{3}{5}\right)$ **32.** $\dfrac{7}{10} + \left(-\dfrac{2}{5}\right)$

33. $-\dfrac{5}{8} + \dfrac{3}{4}$ **34.** $-\dfrac{5}{6} + \dfrac{1}{3}$

35. $-\dfrac{3}{7} + \left(-\dfrac{4}{5}\right)$ **36.** $-\dfrac{3}{8} + \left(-\dfrac{2}{3}\right)$

37. $4 + (-7) + (-5)$ **38.** $10 + (-3) + (-8)$

39. $85 + (-15) + (-20) + 12$

40. $60 + (-50) + (-30) + 25$

41. $17 + (-4) + 2 + 3 + (-10)$

42. $19 + (-5) + 1 + 8 + (-13)$

43. $-45 + \left(-\dfrac{3}{7}\right) + 25 + \left(-\dfrac{4}{7}\right)$

44. $-50 + \left(-\dfrac{7}{9}\right) + 35 + \left(-\dfrac{11}{9}\right)$

45. $3.5 + (-45) + (-8.4) + 72$

46. $6.4 + (-35) + (-2.6) + 14$

In Exercises 47–60, simplify each algebraic expression.

47. $-10x + 2x$ **48.** $-19x + 10x$

49. $25y + (-12y)$ **50.** $26y + (-14y)$

51. $-8a + (-15a)$ **52.** $-9a + (-13a)$

53. $4y + (-13z) + (-10y) + 17z$

54. $5y + (-11z) + (-15y) + 20z$

55. $-7b + 10 + (-b) + (-6)$

56. $-10b + 13 + (-b) + (-4)$

57. $7x + (-5y) + (-9x) + 19y$

58. $13x + (-9y) + (-17x) + 20y$

59. $8(4y + 3) + (-35y)$

60. $7(3y + 5) + (-25y)$

Practice PLUS

In Exercises 61–64, find each sum.

61. $|-3 + (-5)| + |2 + (-6)|$

62. $|4 + (-11)| + |-3 + (-4)|$

63. $-20 + \left[-|15 + (-25)|\right]$

64. $-25 + \left[-|18 + (-26)|\right]$

In Exercises 65–66, insert either $<$, $>$, or $=$ in the shaded area to make a true statement.

65. $6 + [2 + (-13)]$ ▢ $-3 + [4 + (-8)]$

66. $[(-8) + (-6)] + 10$ ▢ $-8 + [9 + (-2)]$

In Exercises 67–70, write each English phrase as an algebraic expression. Then simplify the expression. Let x represent the number.

67. The product of -6 and a number, which is then increased by the product of -13 and a number

68. The product of -9 and a number, which is then increased by the product of -11 and a number

69. The quotient of -20 and a number increased by the quotient of 3 and a number

70. The quotient of -15 and a number increased by the quotient of 4 and a number

Application Exercises

Solve Exercises 71–78 by writing a sum of signed numbers and adding.

71. The greatest temperature variation recorded in a day is 100 degrees in Browning, Montana, on January 23, 1916. The low temperature was $-56°F$. What was the high temperature?

72. In Spearfish, South Dakota, on January 22, 1943, the temperature rose 49 degrees in two minutes. If the initial temperature was $-4°F$, what was the temperature two minutes later?

73. The Dead Sea is the lowest elevation on earth, 1312 feet below sea level. What is the elevation of a person standing 712 feet above the Dead Sea?

74. Lake Assal in Africa is 512 feet below sea level. What is the elevation of a person standing 642 feet above Lake Assal?

75. The temperature at 8:00 A.M. was $-7°F$. By noon it had risen 15°F, but by 4:00 P.M. it had fallen 5°F. What was the temperature at 4:00 P.M.?

76. On three successive plays, a football team lost 15 yards, gained 13 yards, and then lost 4 yards. What was the team's total gain or loss for the three plays?

77. A football team started with the football at the 27-yard line, advancing toward the center of the field (the 50-yard line). Four successive plays resulted in a 4-yard gain, a 2-yard loss, an 8-yard gain, and a 12-yard loss. What was the location of the football at the end of the fourth play?

78. The water level of a reservoir is measured over a five-month period. At the beginning, the level is 20 feet. During this time, the level rose 3 feet, then fell 2 feet, then fell 1 foot, then fell 4 feet, and then rose 2 feet. What is the reservoir's water level at the end of the five months?

The bar graph shows that in 2000 and 2001, the U.S. government collected more in taxes than it spent, so there was a budget surplus for each of these years. By contrast, in 2002 through 2005, the government spent more than it collected, resulting in budget deficits. Exercises 79–80 involve these deficits.

United States Government Budget Surplus/Deficit

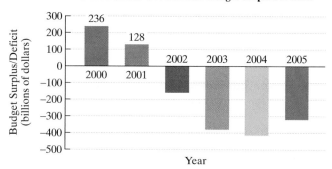

Source: Budget of the U.S. Government

79. a. In 2004, the government collected $1880 billion and spent $2293 billion. Find $1880 + (-2293)$ and determine the deficit, in billions of dollars, for 2004.

b. In 2005, the government collected $2154 billion and spent $2472 billion. Find the deficit, in billions of dollars, for 2005.

c. Use your answers from parts (a) and (b) to determine the combined deficit, in billions of dollars, for 2004 and 2005.

80. a. In 2002, the government collected $1853 billion and spent $2011 billion. Find $1853 + (-2011)$ and determine the deficit, in billions of dollars, for 2002.

b. In 2003, the government collected $1782 billion and spent $2160 billion. Find the deficit, in billions of dollars, for 2003.

c. Use your answers from parts (a) and (b) to determine the combined deficit, in billions of dollars, for 2002 and 2003.

Writing in Mathematics

81. Explain how to add two numbers with a number line. Provide an example with your explanation.

82. What are additive inverses?

83. Describe how the inverse property of addition

$$a + (-a) = 0$$

can be shown on a number line.

84. Without using a number line, describe how to add two numbers with the same sign. Give an example.

85. Without using a number line, describe how to add two numbers with different signs. Give an example.

86. Write a problem that can be solved by finding the sum of at least three numbers, some positive and some negative. Then explain how to solve the problem.

87. Without a calculator, you can add numbers using a number line, using absolute value, or using gains and losses. Which method do you find most helpful? Why is this so?

Critical Thinking Exercises

Make Sense?   *In Exercises 88–91, determine whether each statement "makes sense" or "does not make sense" and explain your reasoning.*

88. It takes me too much time to add real numbers with a number line.

89. I found the sum of -13 and 4 by thinking of temperatures above and below zero: If it's 13 below zero and the temperature rises 4 degrees, the new temperature will be 9 below zero, so $-13 + 4 = -9$.

90. I added two negative numbers and obtained a positive sum.

91. Without adding numbers, I can see that the sum of -227 and 319 is greater than the sum of 227 and -319.

In Exercises 92–95, determine whether each statement is true or false. If the statement is false, make the necessary change(s) to produce a true statement.

92. $\frac{3}{4} + \left(-\frac{3}{5}\right) = -\frac{3}{20}$

93. The sum of zero and a negative number is always a negative number.

94. If one number is positive and the other negative, then the absolute value of their sum equals the sum of their absolute values.

95. The sum of a positive number and a negative number is always a positive number.

In Exercises 96–99, use the number line to determine whether each expression is positive or negative.

96. $a + b$ **97.** $a + c$

98. $b + c$ **99.** $|a + c|$

Technology Exercises

100. Use a calculator to verify any five of the sums that you found in Exercises 9–46.

101. Use a calculator to verify any three of the answers that you obtained in Application Exercises 71–78.

Review Exercises

From here on, each exercise set will contain three review exercises. It is essential to review previously covered topics to improve your understanding of the topics and to help maintain your mastery of the material. If you are not certain how to solve a review exercise, turn to the section and the worked example given in parentheses at the end of each exercise.

102. Consider the set

$$\{-6, -\pi, 0, 0.\overline{7}, \sqrt{3}, \sqrt{4}\}.$$

List all numbers from the set that are **a.** natural numbers, **b.** whole numbers, **c.** integers, **d.** rational numbers, **e.** irrational numbers, **f.** real numbers. (Section 1.3, Example 5)

103. Determine whether this inequality is true or false: $19 \geq -18$. (Section 1.3, Example 7)

104. Determine whether 18 is a solution of $16 = 2(x - 1) - x$. (Section 1.1, Example 4)

Preview Exercises

Exercises 105–107 will help you prepare for the material covered in the next section. In each exercise, a subtraction is expressed as addition of an opposite. Find this sum, indicated by a question mark.

105. $7 - 10 = 7 + (-10) = ?$

106. $-8 - 13 = -8 + (-13) = ?$

107. $-8 - (-13) = -8 + 13 = ?$

Subtraction of Real Numbers

Objectives

1 Subtract real numbers.

2 Simplify a series of additions and subtractions.

3 Use the definition of subtraction to identify terms.

4 Use the subtraction definition to simplify algebraic expressions.

5 Solve problems involving subtraction.

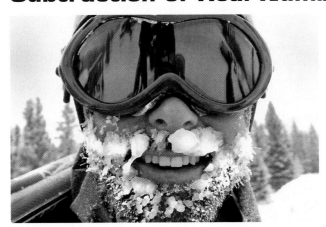

Everybody complains about the weather, but where is it really the worst? **Table 1.5** shows that in the United States, Alaska and Montana are the record-holders for the coldest temperatures.

Table 1.5	Lowest Recorded U.S. Temperatures	
State	**Year**	**Temperature (°F)**
Alaska	1971	−80
Montana	1954	−70

Source: National Climatic Data Center

The table shows that Montana's record low temperature, −70, decreased by 10 degrees gives Alaska's record low temperature, −80. This can be expressed in two ways:

$$-70 - 10 = -80 \qquad \text{or} \qquad -70 + (-10) = -80.$$

This means that

$$-70 - 10 = -70 + (-10).$$

To subtract 10 from −70, we add −70 and the opposite, or additive inverse, of 10. Generalizing from this situation, we define subtraction as addition of an opposite.

> ### Definition of Subtraction
>
> For all real numbers a and b,
>
> $$a - b = a + (-b).$$
>
> In words: To subtract b from a, add the opposite, or additive inverse, of b to a. The result of subtraction is called the **difference**.

1 Subtract real numbers.

A Procedure for Subtracting Real Numbers

The definition of subtraction gives us a procedure for subtracting real numbers.

> ### Subtracting Real Numbers
>
> 1. Change the subtraction operation to addition.
> 2. Change the sign of the number being subtracted.
> 3. Add, using one of the rules for adding numbers with the same sign or different signs.

Using Technology

You can use a calculator to subtract signed numbers. Here are the keystrokes for finding $5 - (-6)$:

Scientific Calculator

5 [−] 6 [+/−] [=]

Graphing Calculator

5 [−] [(−)] 6 [ENTER]

Here are the keystrokes for finding $-9 - (-3)$:

Scientific Calculator

9 [+/−] [−] 3 [+/−] [=]

Graphing Calculator

[(−)] 9 [−] [(−)] 3 [ENTER]

Don't confuse the subtraction key on a graphing calculator, [−], with the sign change or additive inverse key, [(−)]. What happens if you do?

EXAMPLE 1   Using the Definition of Subtraction

Subtract:

a. $7 - 10$ **b.** $5 - (-6)$ **c.** $-9 - (-3)$.

Solution

a. $7 - 10 = 7 + (-10) = -3$

Change the subtraction to addition.   Replace 10 with its opposite.

b. $5 - (-6) = 5 + 6 = 11$

Change the subtraction to addition.   Replace −6 with its opposite.

c. $-9 - (-3) = -9 + 3 = -6$

Change the subtraction to addition.   Replace −3 with its opposite.

☑ **CHECK POINT 1**   Subtract:

a. $3 - 11$ **b.** $4 - (-5)$ **c.** $-7 - (-2)$.

The definition of subtraction can be applied to real numbers that are not integers.

EXAMPLE 2   Using the Definition of Subtraction

Subtract:

a. $-5.2 - (-11.4)$ **b.** $-\dfrac{3}{4} - \dfrac{2}{3}$ **c.** $4\pi - (-9\pi)$.

Solution

a. $-5.2 - (-11.4) = -5.2 + 11.4 = 6.2$

| Change the subtraction to addition. | Replace -11.4 with its opposite. |

b. $-\dfrac{3}{4} - \dfrac{2}{3} = -\dfrac{3}{4} + \left(-\dfrac{2}{3}\right) = -\dfrac{9}{12} + \left(-\dfrac{8}{12}\right) = -\dfrac{17}{12}$

| Change the subtraction to addition. | Replace $\frac{2}{3}$ with its opposite. |

c. $4\pi - (-9\pi) = 4\pi + 9\pi = (4 + 9)\pi = 13\pi$

| Change the subtraction to addition. | Replace -9π with its opposite. |

Reading the symbol "$-$" can be a bit tricky. The way you read it depends on where it appears. For example,

$$-5.2 - (-11.4)$$

is read "negative five point two minus negative eleven point four." Read parts (b) and (c) of Example 2 aloud. When is "$-$" read "negative" and when is it read "minus"?

☑ **CHECK POINT 2**   Subtract:

a. $-3.4 - (-12.6)$ **b.** $-\dfrac{3}{5} - \dfrac{1}{3}$ **c.** $5\pi - (-2\pi)$.

2   Simplify a series of additions and subtractions.

Problems Containing a Series of Additions and Subtractions

In some problems, several additions and subtractions occur together. We begin by converting all subtractions to additions of opposites, or additive inverses.

> ### Simplifying a Series of Additions and Subtractions
> **1.** Change all subtractions to additions of opposites.
> **2.** Group and then add all the positive numbers.
> **3.** Group and then add all the negative numbers.
> **4.** Add the results of steps 2 and 3.

EXAMPLE 3   Simplifying a Series of Additions and Subtractions

Simplify: $7 - (-5) - 11 - (-6) - 19$.

Solution

$$
\begin{aligned}
&7 - (-5) - 11 - (-6) - 19 \\
&= 7 + 5 + (-11) + 6 + (-19) \quad &&\text{Write subtractions as additions of opposites.} \\
&= (7 + 5 + 6) + [(-11) + (-19)] \quad &&\text{Group the positive numbers. Group the negative numbers.} \\
&= 18 + (-30) \quad &&\text{Add the positive numbers. Add the negative numbers.} \\
&= -12 \quad &&\text{Add the results.}
\end{aligned}
$$

☑ **CHECK POINT 3**   Simplify: $10 - (-12) - 4 - (-3) - 6$.

3 Use the definition of subtraction to identify terms.

Subtraction and Algebraic Expressions

We know that the terms of an algebraic expression are separated by addition signs. Let's use this idea to identify the terms of the following algebraic expression:

$$9x - 4y - 5.$$

Because terms are separated by addition, we rewrite the algebraic expression as additions of opposites. Thus,

$$9x - 4y - 5 = 9x + (-4y) + (-5).$$

The three terms of the algebraic expression are $9x$, $-4y$, and -5.

EXAMPLE 4   **Using the Definition of Subtraction to Identify Terms**

Identify the terms of the algebraic expression:

$$2xy - 13y - 6.$$

Solution   Rewrite the algebraic expression as additions of opposites.

$$2xy - 13y - 6 = 2xy + (-13y) + (-6)$$

| First term | Second term | Third term |

Because terms are separated by addition, the terms are $2xy$, $-13y$, and -6. ▬

☑ **CHECK POINT 4**   Identify the terms of the algebraic expression:

$$-6 + 4a - 7ab.$$

---

4 Use the subtraction definition to simplify algebraic expressions.

The procedure for subtracting real numbers can be used to simplify certain algebraic expressions that involve subtraction.

EXAMPLE 5   **Simplifying Algebraic Expressions**

Simplify:

a. $2 + 3x - 8x$ **b.** $-4x - 9y - 2x + 12y.$

Solution

a. $2 + 3x - 8x$ This is the given algebraic expression.

$= 2 + 3x + (-8x)$ Write the subtraction as the addition of an opposite.

$= 2 + [3 + (-8)]x$ Apply the distributive property.

$= 2 + (-5x)$ Add within the grouping symbols.

$= 2 - 5x$ Be concise and express as subtraction.

b. $-4x - 9y - 2x + 12y$ This is the given algebraic expression.

$= -4x + (-9y) + (-2x) + 12y$ Write the subtractions as the additions of opposites.

$= -4x + (-2x) + (-9y) + 12y$ Arrange like terms so that they are next to one another.

$= [-4 + (-2)]x + (-9 + 12)y$ Apply the distributive property.

$= -6x + 3y$ Add within the grouping symbols. ▬

Study Tip

You can think of gains and losses of money to work some of the steps in Example 5 mentally:

- $3x - 8x = -5x$ A gain of 3 dollars followed by a loss of 8 dollars is a net loss of 5 dollars.

- $-9y + 12y = 3y$ A loss of 9 dollars followed by a gain of 12 dollars is a net gain of 3 dollars.

✓ **CHECK POINT 5**   Simplify:

a. $4 + 2x - 9x$ b. $-3x - 10y - 6x + 14y$.

---

5 Solve problems involving subtraction.

Applications

Subtraction is used to solve problems in which the word *difference* appears. The difference between real numbers a and b is expressed as $a - b$.

| **EXAMPLE 6** | **An Application of Subtraction Using the Word *Difference*** |

Figure 1.12 shows the hottest and coldest temperatures ever recorded on Earth. The record high was 136°F, recorded in Libya in 1922. The record low was −129°F, at the Vostok Scientific Station in Antarctica, in 1983. What is the difference between these highest and lowest temperatures? (*Source: The World Almanac Book of Records*, 2006)

Solution

The difference	is	the record high	minus	the record low.

$$= \quad 136 \quad - \quad (-129)$$
$$= \quad 136 \quad + \quad 129 = 265$$

The difference between the two temperatures is 265°F. We can also say that the variation in temperature between these two world extremes is 265°F. ■

FIGURE 1.12

✓ **CHECK POINT 6**   The peak of Mount Everest is 8848 meters above sea level. The Marianas Trench, on the floor of the Pacific Ocean, is 10,915 meters below sea level. What is the difference in elevation between the peak of Mount Everest and the Marianas Trench?

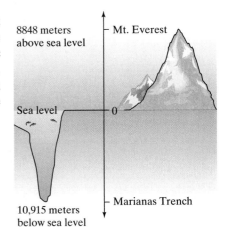

1.6 EXERCISE SET   MyMathLab

Practice Exercises

1. Consider the subtraction $5 - 12$.
   a. Find the opposite, or additive inverse, of 12.
   b. Rewrite the subtraction as the addition of the opposite of 12.

2. Consider the subtraction $4 - 10$.
   a. Find the opposite, or additive inverse, of 10.
   b. Rewrite the subtraction as the addition of the opposite of 10.

3. Consider the subtraction $5 - (-7)$.
   a. Find the opposite, or additive inverse, of -7.
   b. Rewrite the subtraction as the addition of the opposite of -7.

4. Consider the subtraction $2 - (-8)$.
   a. Find the opposite, or additive inverse, of -8.
   b. Rewrite the subtraction as the addition of the opposite of -8.

In Exercises 5–50, perform the indicated subtraction.

5. $14 - 8$
6. $15 - 2$
7. $8 - 14$
8. $2 - 15$
9. $3 - (-20)$
10. $5 - (-17)$
11. $-7 - (-18)$
12. $-5 - (-19)$
13. $-13 - (-2)$
14. $-21 - (-3)$
15. $-21 - 17$
16. $-29 - 21$
17. $-45 - (-45)$
18. $-65 - (-65)$
19. $23 - 23$
20. $26 - 26$
21. $13 - (-13)$
22. $15 - (-15)$
23. $0 - 13$
24. $0 - 15$
25. $0 - (-13)$
26. $0 - (-15)$
27. $\frac{3}{7} - \frac{5}{7}$
28. $\frac{4}{9} - \frac{7}{9}$
29. $\frac{1}{5} - \left(-\frac{3}{5}\right)$
30. $\frac{1}{7} - \left(-\frac{3}{7}\right)$
31. $-\frac{4}{5} - \frac{1}{5}$
32. $-\frac{4}{9} - \frac{1}{9}$
33. $-\frac{4}{5} - \left(-\frac{1}{5}\right)$
34. $-\frac{4}{9} - \left(-\frac{1}{9}\right)$
35. $\frac{1}{2} - \left(-\frac{1}{4}\right)$
36. $\frac{2}{5} - \left(-\frac{1}{10}\right)$
37. $\frac{1}{2} - \frac{1}{4}$
38. $\frac{2}{5} - \frac{1}{10}$
39. $9.8 - 2.2$
40. $5.7 - 3.3$
41. $-3.1 - (-1.1)$
42. $-4.6 - (-1.1)$
43. $1.3 - (-1.3)$
44. $1.4 - (-1.4)$
45. $-2.06 - (-2.06)$
46. $-3.47 - (-3.47)$
47. $5\pi - 2\pi$
48. $9\pi - 7\pi$
49. $3\pi - (-10\pi)$
50. $4\pi - (-12\pi)$

In Exercises 51–68, simplify each series of additions and subtractions.

51. $13 - 2 - (-8)$
52. $14 - 3 - (-7)$
53. $9 - 8 + 3 - 7$
54. $8 - 2 + 5 - 13$
55. $-6 - 2 + 3 - 10$
56. $-9 - 5 + 4 - 17$
57. $-10 - (-5) + 7 - 2$
58. $-6 - (-3) + 8 - 11$
59. $-23 - 11 - (-7) + (-25)$
60. $-19 - 8 - (-6) + (-21)$
61. $-823 - 146 - 50 - (-832)$
62. $-726 - 422 - 921 - (-816)$
63. $1 - \frac{2}{3} - \left(-\frac{5}{6}\right)$
64. $2 - \frac{3}{4} - \left(-\frac{7}{8}\right)$
65. $-0.16 - 5.2 - (-0.87)$
66. $-1.9 - 3 - (-0.26)$
67. $-\frac{3}{4} - \frac{1}{4} - \left(-\frac{5}{8}\right)$
68. $-\frac{1}{2} - \frac{2}{3} - \left(-\frac{1}{3}\right)$

In Exercises 69–72, identify the terms in each algebraic expression.

69. $-3x - 8y$
70. $-9a - 4b$
71. $12x - 5xy - 4$
72. $8a - 7ab - 13$

In Exercises 73–84, simplify each algebraic expression.

73. $3x - 9x$
74. $2x - 10x$
75. $4 + 7y - 17y$
76. $5 + 9y - 29y$
77. $2a + 5 - 9a$
78. $3a + 7 - 11a$
79. $4 - 6b - 8 - 3b$
80. $5 - 7b - 13 - 4b$
81. $13 - (-7x) + 4x - (-11)$
82. $15 - (-3x) + 8x - (-10)$
83. $-5x - 10y - 3x + 13y$
84. $-6x - 9y - 4x + 15y$

Practice PLUS

In Exercises 85–90, find the value of each expression.

85. $-|-9 - (-6)| - (-12)$
86. $-|-8 - (-2)| - (-6)$
87. $\frac{5}{8} - \left(\frac{1}{2} - \frac{3}{4}\right)$
88. $\frac{9}{10} - \left(\frac{1}{4} - \frac{7}{10}\right)$
89. $|-9 - (-3 + 7)| - |-17 - (-2)|$
90. $|24 - (-16)| - |-51 - (-31 + 2)|$

In Exercises 91–94, write each English phrase as an algebraic expression. Then simplify the expression. Let x represent the number.

91. The difference between 6 times a number and -5 times a number

92. The difference between 9 times a number and −4 times a number

93. The quotient of −2 and a number subtracted from the quotient of −5 and a number

94. The quotient of −7 and a number subtracted from the quotient of −12 and a number

Application Exercises

95. The peak of Mount Kilimanjaro, the highest point in Africa, is 19,321 feet above sea level. Qattara Depression, Egypt, one of the lowest points in Africa, is 436 feet below sea level. What is the difference in elevation between the peak of Mount Kilimanjaro and the Qattara Depression?

96. The peak of Mount Whitney is 14,494 feet above sea level. Mount Whitney can be seen directly above Death Valley, which is 282 feet below sea level. What is the difference in elevation between these geographic locations?

The bar graph shows the average daily low temperature for each month in Fairbanks, Alaska. Use the graph to solve Exercises 97–100.

Each Month's Average Daily Low Temperature in Fairbanks, Alaska

Source: The Weather Channel Enterprises, Inc.

97. What is the difference between the average daily low temperatures for March and February?

98. What is the difference between the average daily low temperatures for October and November?

99. How many degrees warmer is February's average low temperature than January's average low temperature?

100. How many degrees warmer is November's average low temperature than December's average low temperature?

The line graphs at the top of the next column show that from 1994 through 2004, the United States imported goods and services from other countries worth more than the value of its exports to those countries. As a result, we had a trade deficit for each year during this period. Exercises 101–102 are based on the data displayed by the graphs. Express all answers in billions of dollars.

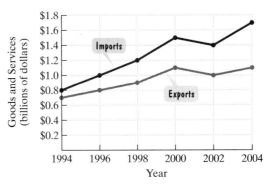

United States International Trade, in Billions of Dollars

Source: U.S. Department of Commerce

101. a. How much did we export to other countries in 2002?
   b. How much did we import from other countries in 2002?
   c. Find the trade deficit, the difference between exports and imports, for 2002.

102. a. How much did we export to other countries in 1998?
   b. How much did we import from other countries in 1998?
   c. Find the trade deficit, the difference between exports and imports, for 1998.

The data for U.S. international trade, shown by the line graphs in Exercises 101–102, can be approximated by the following mathematical models:

Exports, in billions of dollars $\quad E = 0.04n + 0.7$

Number of years after 1994

Imports, in billions of dollars $\quad I = 0.09n + 0.8.$

Use these formulas to solve Exercises 103–104. Express all answers in billions of dollars.

103. a. How much did we export to other countries in 2002?
   b. How much did we import from other countries in 2002?
   c. According to the models, what was the trade deficit for 2002?
   d. What is the difference between the actual trade deficit for 2002 that you found in Exercise 101(c) and the trade deficit given by the models?

104. a. How much did we export to other countries in 1998?
   b. How much did we import from other countries in 1998?
   c. According to the models, what was the trade deficit for 1998?
   d. What is the difference between the actual trade deficit for 1998 that you found in Exercise 102(c) and the trade deficit given by the models?

Writing in Mathematics

105. Explain how to subtract real numbers.

106. How is $4 - (-2)$ read?

107. Explain how to simplify a series of additions and subtractions. Provide an example with your explanation.

108. Explain how to find the terms of the algebraic expression $5x - 2y - 7$.

109. Write a problem that can be solved by finding the difference between two numbers. At least one of the numbers should be negative. Then explain how to solve the problem.

Critical Thinking Exercises

Make Sense?   *In Exercises 110–113, determine whether each statement "makes sense" or "does not make sense" and explain your reasoning.*

110. I already knew how to add positive and negative numbers, so there was not that much new to learn when it came to subtracting them.

111. I can find the closing price of stock PQR on Wednesday by subtracting the change in price, -1.23, from the closing price on Thursday, 47.19.

	C3	
STOCK PRICES		
Thursday		
Stock	**Close**	**Change**
ABC	31.54	0.47
XYZ	16.23	−0.87
PQR	47.19	−1.23
DEF	21.54	0.21

112. I found the variation in elevation between two heights by taking the difference between the high point and the low point.

113. I found the variation in U.S. temperature by subtracting the record low temperature, a negative number, from the record high temperature, a positive number.

In Exercises 114–117, determine whether each statement is true or false. If the statement is false, make the necessary change(s) to produce a true statement.

114. If a and b are negative numbers, then $a - b$ is sometimes a negative number.

115. $7 - (-2) = 5$

116. The difference between 0 and a negative number is always a positive number.

117. $|a - b| = |b - a|$

In Exercises 118–121, use the number line to determine whether each difference is positive or negative.

118. $c - a$ **119.** $a - b$

120. $b - c$ **121.** $0 - b$

122. Order the expressions $|x - y|, |x| - |y|$, and $|x + y|$ from least to greatest for $x = -6$ and $y = -8$.

Technology Exercises

123. Use a calculator to verify any five of the differences that you found in Exercises 5–46.

124. Use a calculator to verify any three of the answers that you found in Exercises 51–68.

Review Exercises

125. Determine whether 2 is a solution of $13x + 3 = 3(5x - 1)$. (Section 1.1, Example 4)

126. Simplify: $5(3x + 2y) + 6(5y)$. (Section 1.4, Example 11)

127. Give an example of an integer that is not a natural number. (Section 1.3; Example 5)

Preview Exercises

Exercises 128–130 will help you prepare for the material covered in the next section. In Exercises 128–129, a multiplication is expressed as a repeated addition. Find this sum, indicated by a question mark.

128. $4(-3) = (-3) + (-3) + (-3) + (-3) = ?$

129. $3(-3) = (-3) + (-3) + (-3) = ?$

130. The list shows a pattern for various products.

$$2(-3) = (-3) + (-3) = -6$$

$$2(-3) = -6$$
$$1(-3) = -3$$
$$0(-3) = 0$$
$$-1(-3) = 3$$
$$-2(-3) = 6$$
$$-3(-3) = 9$$
$$-4(-3) = ?$$

Reading down the list, products keep increasing by 3.

Use this pattern to find $-4(-3)$.

Multiplication and Division of Real Numbers

Objectives

1 Multiply real numbers.

2 Multiply more than two real numbers.

3 Find multiplicative inverses.

4 Use the definition of division.

5 Divide real numbers.

6 Simplify algebraic expressions involving multiplication.

7 Determine whether a number is a solution of an equation.

8 Use mathematical models involving multiplication and division.

Where the Boys Are, released in 1960, was the first college spring-break movie. Today, visitors to college campuses can't help but ask: Where are the boys?

In 2007, 135 women received bachelor's degrees for every 100 men. In this section, we use data from a report by the U.S. Department of Education to develop mathematical models showing that this gender imbalance will widen in the coming years. Working with these models requires that we know how to multiply and divide real numbers.

Multiplying Real Numbers

Multiplication is repeated addition. For example, $5(-3)$ means that -3 is added five times:

$$5(-3) = (-3) + (-3) + (-3) + (-3) + (-3) = -15.$$

The result of the multiplication, -15, is called the **product** of 5 and -3. The numbers being multiplied, 5 and -3, are called the **factors** of the product.

Rules for multiplying real numbers are described in terms of absolute value. For example, $5(-3) = -15$ illustrates that the product of numbers with different signs is found by multiplying their absolute values. The product is negative.

$$5(-3) = -15$$

Multiply absolute values:
$|5| \cdot |-3| = 5 \cdot 3 = 15.$

Factors have different signs and the product is negative.

The following rules are used to determine the sign of the product of two numbers:

1 Multiply real numbers.

The Product of Two Real Numbers

- The product of two real numbers with **different signs** is found by multiplying their absolute values. The product is **negative**.
- The product of two real numbers with the **same sign** is found by multiplying their absolute values. The product is **positive**.
- The product of 0 and any real number is 0. Thus, for any real number a,

$$a \cdot 0 = 0 \quad \text{and} \quad 0 \cdot a = 0.$$

EXAMPLE 1   Multiplying Real Numbers

Multiply:

a. $6(-3)$ **b.** $-\dfrac{1}{5} \cdot \dfrac{2}{3}$ **c.** $(-9)(-10)$ **d.** $(-1.4)(-2)$ **e.** $(-372)(0)$.

Solution

a. $6(-3) = -18$ Multiply absolute values: $6 \cdot 3 = 18$.

Different signs: negative product

b. $-\dfrac{1}{5} \cdot \dfrac{2}{3} = -\dfrac{2}{15}$ Multiply absolute values: $\frac{1}{5} \cdot \frac{2}{3} = \frac{1 \cdot 2}{5 \cdot 3} = \frac{2}{15}$.

Different signs: negative product

c. $(-9)(-10) = 90$ Multiply absolute values: $9 \cdot 10 = 90$.

Same sign: positive product

d. $(-1.4)(-2) = 2.8$ Multiply absolute values: $(1.4)(2) = 2.8$.

Same sign: positive product

e. $(-372)(0) = 0$ The product of 0 and any real number is 0: $a \cdot 0 = 0$.

☑ **CHECK POINT 1**   Multiply:

a. $8(-5)$ **b.** $-\dfrac{1}{3} \cdot \dfrac{4}{7}$ **c.** $(-12)(-3)$

d. $(-1.1)(-5)$ **e.** $(-543)(0)$.

---

2 Multiply more than two real numbers.

Multiplying More Than Two Numbers

How do we perform more than one multiplication, such as

$$-4(-3)(-2)?$$

Because of the associative and commutative properties, we can order and group the numbers in any manner. Each pair of negative numbers will produce a positive product. Thus, the product of an even number of negative numbers is always positive. By contrast, the product of an odd number of negative numbers is always negative.

$$-4(-3)(-2) = -24$$ Multiply absolute values: $4 \cdot 3 \cdot 2 = 24$.

Odd number of negative numbers (three): negative product

Multiplying More Than Two Numbers

1. Assuming that no factor is zero,
   - The product of an **even** number of **negative numbers** is **positive**.
   - The product of an **odd** number of **negative numbers** is **negative**.

   The multiplication is performed by multiplying the absolute values of the given numbers.

2. If any factor is 0, the product is 0.

EXAMPLE 2 Multiplying More Than Two Numbers

Multiply:

    a. $(-3)(-1)(2)(-2)$ **b.** $(-1)(-2)(-2)(3)(-4)$.

Solution

    a. $(-3)(-1)(2)(-2) = -12$ Multiply absolute values: $3 \cdot 1 \cdot 2 \cdot 2 = 12.$

Odd number of negative numbers (three): negative product

    b. $(-1)(-2)(-2)(3)(-4) = 48$ Multiply absolute values: $1 \cdot 2 \cdot 2 \cdot 3 \cdot 4 = 48.$

Even number of negative numbers (four): positive product

■

☑ **CHECK POINT 2** Multiply:

    a. $(-2)(3)(-1)(4)$ **b.** $(-1)(-3)(2)(-1)(5)$.

Is it always necessary to count the number of negative factors when multiplying more than two numbers? No. If any factor is 0, you can immediately write 0 for the product. For example,

$$(-37)(423)(0)(-55)(-3.7) = 0.$$

If any factor is 0, the product is 0.

3 Find multiplicative inverses.

The Meaning of Division

The result of dividing the real number a by the nonzero real number b is called the **quotient** of a and b. We can write this quotient as $a \div b$ or $\dfrac{a}{b}$.

We know that subtraction is defined in terms of addition of an additive inverse, or opposite:

$$a - b = a + (-b).$$

In a similar way, we can define division in terms of multiplication. For example, the quotient of 8 and 2 can be written as multiplication:

$$8 \div 2 = 8 \cdot \frac{1}{2}.$$

We call $\frac{1}{2}$ the *multiplicative inverse*, or *reciprocal*, of 2. Two numbers whose product is 1 are called **multiplicative inverses** or **reciprocals** of each other. Thus, the multiplicative inverse of 2 is $\frac{1}{2}$ and the multiplicative inverse of $\frac{1}{2}$ is 2 because $2 \cdot \frac{1}{2} = 1$.

EXAMPLE 3 Finding Multiplicative Inverses

Find the multiplicative inverse of each number:

    a. 5   **b.** $\dfrac{1}{3}$ **c.** -4 **d.** $-\dfrac{4}{5}$.

Solution

a. The multiplicative inverse of 5 is $\frac{1}{5}$ because $5 \cdot \frac{1}{5} = 1$.

b. The multiplicative inverse of $\frac{1}{3}$ is 3 because $\frac{1}{3} \cdot 3 = 1$.

c. The multiplicative inverse of -4 is $-\frac{1}{4}$ because $(-4)\left(-\frac{1}{4}\right) = 1$.

d. The multiplicative inverse of $-\frac{4}{5}$ is $-\frac{5}{4}$ because $\left(-\frac{4}{5}\right)\left(-\frac{5}{4}\right) = 1$.   ▄

☑ **CHECK POINT 3**   Find the multiplicative inverse of each number:

a. 7   b. $\frac{1}{8}$ c. -6 d. $-\frac{7}{13}$.

Can you think of a real number that has no multiplicative inverse? The number **0 has no multiplicative inverse** because 0 multiplied by any number is never 1, but always 0.

We now define division in terms of multiplication by a multiplicative inverse.

4 Use the definition of division.

Definition of Division

If a and b are real numbers and b is not 0, then the quotient of a and b is defined as

$$a \div b = a \cdot \frac{1}{b}.$$

In words: The quotient of two real numbers is the product of the first number and the multiplicative inverse of the second number.

Using Technology

You can use a calculator to multiply and divide signed numbers. Here are the keystrokes for finding

$$(-173)(-256):$$

Scientific Calculator

173 $\boxed{+/_}$ $\boxed{\times}$ 256 $\boxed{+/_}$ $\boxed{=}$

Graphing Calculator

$\boxed{(-)}$ 173 $\boxed{\times}$ $\boxed{(-)}$ 256 $\boxed{\text{ENTER}}$

The number 44288 should be displayed.
Division is performed in the same manner, using $\boxed{\div}$ instead of $\boxed{\times}$. What happens when you divide by 0? Try entering

$$8 \boxed{\div} 0$$

and pressing $\boxed{=}$ or $\boxed{\text{ENTER}}$.

EXAMPLE 4   **Using the Definition of Division**

Use the definition of division to find each quotient:

a. $-15 \div 3$ b. $\dfrac{-20}{-4}$.

Solution

a. $-15 \div 3 = -15 \cdot \dfrac{1}{3} = -5$

> Change the division to multiplication.   Replace 3 with its multiplicative inverse.

b. $\dfrac{-20}{-4} = -20 \cdot \left(-\dfrac{1}{4}\right) = 5$

> Change the division to multiplication.   Replace −4 with its multiplicative inverse.

☑ **CHECK POINT 4**   Use the definition of division to find each quotient:

a. $-28 \div 7$ b. $\dfrac{-16}{-2}$.

5 Divide real numbers.

A Procedure for Dividing Real Numbers

Because the quotient $a \div b$ is defined as the product $a \cdot \frac{1}{b}$, the sign rules for dividing numbers are the same as the sign rules for multiplying them.

> ### The Quotient of Two Real Numbers
> - The quotient of two real numbers with **different signs** is found by dividing their absolute values. The quotient is **negative**.
> - The quotient of two real numbers with the **same sign** is found by dividing their absolute values. The quotient is **positive**.
> - Division of any number by zero is undefined.
> - Any nonzero number divided into 0 is 0.

EXAMPLE 5 Dividing Real Numbers

Divide:

a. $\dfrac{8}{-2}$ **b.** $-\dfrac{3}{4} \div \left(-\dfrac{5}{9}\right)$ **c.** $\dfrac{-20.8}{4}$ **d.** $\dfrac{0}{-7}$.

Solution

a. $\dfrac{8}{-2} = -4$ Divide absolute values: $\frac{8}{2} = 4$.

Different signs: negative quotient

b. $-\dfrac{3}{4} \div \left(-\dfrac{5}{9}\right) = \dfrac{27}{20}$ Divide absolute values: $\frac{3}{4} \div \frac{5}{9} = \frac{3}{4} \cdot \frac{9}{5} = \frac{27}{20}$.

Same sign: positive quotient

c. $\dfrac{-20.8}{4} = -5.2$ Divide absolute values: $4\overline{)20.8}$ $= 5.2$.

Different signs: negative quotient

d. $\dfrac{0}{-7} = 0$ Any nonzero number divided into 0 is 0.

Can you see why $\frac{0}{-7}$ must be 0? The definition of division tells us that

$$\frac{0}{-7} = 0 \cdot \left(-\frac{1}{7}\right)$$

and the product of 0 and any real number is 0. By contrast, the definition of division does not allow for division by 0 because 0 does not have a multiplicative inverse. It is incorrect to write

$$\frac{-7}{0} = -7 \cdot \frac{1}{0}.$$

0 does not have a multiplicative inverse.

Division by zero is not allowed, or not defined. Thus, $\frac{-7}{0}$ does not represent a real number. A real number can never have a denominator of 0.

Study Tip

Zero Under the Line Is Undefined

Here's another reason why division by zero is undefined. We know that

$$\frac{12}{4} = 3 \text{ because } 3 \cdot 4 = 12.$$

Now think about $\frac{-7}{0}$. If

$$\frac{-7}{0} = ? \text{ then } ? \cdot 0 = -7.$$

However, any real number multiplied by 0 is 0 and not -7. No matter what number we replace by ? in $? \cdot 0 = -7$, we can never obtain -7.

✓ **CHECK POINT 5** Divide:

a. $\dfrac{-32}{-4}$ **b.** $-\dfrac{2}{3} \div \dfrac{5}{4}$ **c.** $\dfrac{21.9}{-3}$ **d.** $\dfrac{0}{-5}$.

6 Simplify algebraic expressions involving multiplication.

Multiplication and Algebraic Expressions

In Section 1.4, we discussed the commutative and associative properties of multiplication. We also know that multiplication distributes over addition and subtraction. We now add some additional properties to our previous list (**Table 1.6**). These properties are frequently helpful in simplifying algebraic expressions.

Table 1.6	Additional Properties of Multiplication	
Let a be a real number, a variable, or an algebraic expression.		
Property	**Meaning**	**Examples**
Identity Property of Multiplication	1 can be deleted from a product. $a \cdot 1 = a$ $1 \cdot a = a$	• $\sqrt{3} \cdot 1 = \sqrt{3}$ • $1x = x$ • $1(2x + 3) = 2x + 3$
Inverse Property of Multiplication	If a is not 0: $a \cdot \dfrac{1}{a} = 1$ $\dfrac{1}{a} \cdot a = 1$ The product of a nonzero number and its multiplicative inverse, or reciprocal, gives 1, the multiplicative identity.	• $6 \cdot \dfrac{1}{6} = 1$ • $3x \cdot \dfrac{1}{3x} = 1$ (x is not 0.) • $\dfrac{1}{(y - 2)} \cdot (y - 2) = 1$ (y is not 2.)
Multiplication Property of −1	Negative 1 times a is the opposite, or additive inverse, of a. $-1 \cdot a = -a$ $a(-1) = -a$	• $-1 \cdot \sqrt{3} = -\sqrt{3}$ • $-1\left(-\frac{3}{4}\right) = \frac{3}{4}$ • $-1x = -x$ • $-(x + 4) = -1(x + 4)$ $\quad = -x - 4$
Double Negative Property	The opposite of $-a$ is a. $-(-a) = a$	• $-(-4) = 4$ • $-(-6y) = 6y$

In the preceding table, we used two steps to remove the parentheses from $-(x + 4)$. First, we used the multiplication property of -1.

$$-(x + 4) = -1(x + 4)$$

Then we used the distributive property, distributing -1 to each term in parentheses.

$$-1(x + 4) = (-1)x + (-1)4 = -x + (-4) = -x - 4$$

There is a fast way to obtain $-(x + 4) = -x - 4$ in just one step.

Negative Signs and Parentheses

If a negative sign precedes parentheses, remove the parentheses and change the sign of every term within the parentheses.

Here are some examples that illustrate this method.

$$-(11x + 5) = -11x - 5$$
$$-(11x - 5) = -11x + 5$$
$$-(-11x + 5) = 11x - 5$$
$$-(-11x - 5) = 11x + 5$$

EXAMPLE 6   Simplifying Algebraic Expressions

Simplify:

　　a. $-2(3x)$　　**b.** $6x + x$　　**c.** $8a - 9a$　　**d.** $-3(2x - 5)$　　**e.** $-(3y - 8)$.

Discover for Yourself

Verify each simplification in Example 6 by substituting 5 for the variable. The value of the given expression should be the same as the value of the simplified expression. Which expression is easier to evaluate?

Solution   We will show all steps in the solution process. However, you probably are working many of these steps mentally.

a. $-2(3x)$ 　　　　　This is the given algebraic expression.

　　$= (-2 \cdot 3)x$ 　　Use the associative property and group the first two numbers.

　　$= -6x$ 　　　　　Numbers with opposite signs have a negative product.

b. $6x + x$ 　　　　This is the given algebraic expression.

　　$= 6x + 1x$ 　　Use the multiplication property of 1.

　　$= (6 + 1)x$ 　　Apply the distributive property.

　　$= 7x$ 　　　　Add within parentheses.

c. $8a - 9a$ 　　　This is the given algebraic expression.

　　$= (8 - 9)a$ 　Apply the distributive property.

　　$= -1a$ 　　　Subtract within parentheses: $8 - 9 = 8 + (-9) = -1$.

　　$= -a$ 　　　Apply the multiplication property of -1.

d. $-3(2x - 5)$ 　　　　　　　This is the given algebraic expression.

　　$= -3(2x) - (-3) \cdot (5)$ 　Apply the distributive property.

　　$= -6x - (-15)$ 　　　Multiply.

　　$= -6x + 15$ 　　　　Subtraction is the addition of an opposite.

e. $-(3y - 8)$ 　　　　This is the given algebraic expression.

　　$= -3y + 8$ 　　　Remove parentheses by changing the sign of every term inside the parentheses.

☑ **CHECK POINT 6**   Simplify:

　　a. $-4(5x)$　　　　　　　　**b.** $9x + x$　　　　　　　**c.** $13b - 14b$

　　d. $-7(3x - 4)$　　　　　　**e.** $-(7y - 6)$.

EXAMPLE 7   Simplifying an Algebraic Expression

Simplify: $5(2y - 9) - (9y - 8)$.

Solution

$$5(2y - 9) - (9y - 8)$$ This is the given algebraic expression.

$$= 5 \cdot 2y - 5 \cdot 9 - (9y - 8)$$ Apply the distributive property over the first parentheses.

$$= 10y - 45 - (9y - 8)$$ Multiply.

$$= 10y - 45 - 9y + 8$$ Remove the remaining parentheses by changing the sign of each term within parentheses.

$$= (10y - 9y) + (-45 + 8)$$ Group like terms.

$$= 1y + (-37)$$ Combine like terms. For the variable terms, $10y - 9y = 10y + (-9y) = [10 + (-9)]y = 1y.$

$$= y + (-37)$$ Use the multiplication property of 1: $1y = y.$

$$= y - 37$$ Express addition of an opposite as subtraction. ■

☑ **CHECK POINT 7**   Simplify: $4(3y - 7) - (13y - 2).$

A Summary of Operations with Real Numbers

Operations with real numbers are summarized in **Table 1.7**.

Table 1.7   **Summary of Operations with Real Numbers**

Signs of Numbers	Addition	Subtraction	Multiplication	Division
Both numbers are positive.	Sum is always positive.	Difference may be either positive or negative.	Product is always positive.	Quotient is always positive.
Examples				
8 and 2	$8 + 2 = 10$	$8 - 2 = 6$	$8 \cdot 2 = 16$	$8 \div 2 = 4$
2 and 8	$2 + 8 = 10$	$2 - 8 = -6$	$2 \cdot 8 = 16$	$2 \div 8 = \frac{1}{4}$
One number is positive and the other number is negative.	Sum may be either positive or negative.	Difference may be either positive or negative.	Product is always negative.	Quotient is always negative.
Examples				
8 and -2	$8 + (-2) = 6$	$8 - (-2) = 10$	$8(-2) = -16$	$8 \div (-2) = -4$
-8 and 2	$-8 + 2 = -6$	$-8 - 2 = -10$	$-8(2) = -16$	$-8 \div 2 = -4$
Both numbers are negative.	Sum is always negative.	Difference may be either positive or negative.	Product is always positive.	Quotient is always positive.
Examples				
-8 and -2	$-8 + (-2) = -10$	$-8 - (-2) = -6$	$-8(-2) = 16$	$-8 \div (-2) = 4$
-2 and -8	$-2 + (-8) = -10$	$-2 - (-8) = 6$	$-2(-8) = 16$	$-2 \div (-8) = \frac{1}{4}$

7   Determine whether a number is a solution of an equation.

To determine if a number is a solution of an equation, it is often necessary to perform more than one operation with real numbers.

EXAMPLE 8   **Determining Whether a Number Is a Solution of an Equation**

Determine whether -3 is a solution of

$$6x + 14 = -7 - x.$$

Solution

$$6x + 14 = -7 - x$$ This is the given equation.

Is −3 a solution?

$$6(-3) + 14 \stackrel{?}{=} -7 - (-3)$$ Substitute −3 for x.

$$-18 + 14 \stackrel{?}{=} -7 - (-3)$$ Multiply: 6(−3) = −18.

$$-18 + 14 \stackrel{?}{=} -7 + 3$$ Rewrite subtraction as addition of an opposite.

This statement is true $$-4 = -4$$ Add: −18 + 14 = −4 and −7 + 3 = −4.

Because the values on both sides of the equation are the same, −3 is a solution of the equation.

✓ **CHECK POINT 8**   Determine whether −3 is a solution of $2x - 5 = 8x + 7$.

---

8 Use mathematical models involving multiplication and division.

Applications

Mathematical models frequently involve multiplication and division.

EXAMPLE 9   Big (Lack of) Men on Campus

In 2007, 135 women received bachelor's degrees for every 100 men. According to the U.S. Department of Education, that gender imbalance will widen in the coming years, as shown by the bar graphs in **Figure 1.13**.

Percentage of College Degrees Awarded to United States Men and Women

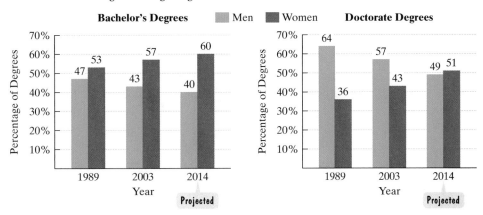

FIGURE 1.13

Source: U.S. Department of Education

The data for bachelor's degrees can be described by the following mathematical models:

Percentage of bachelor's degrees awarded to men $$M = -0.28n + 47$$

Number of years after 1989

Percentage of bachelor's degrees awarded to women $$W = 0.28n + 53.$$

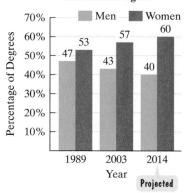

Percentage of College Degrees Awarded to United States Men and Women

Bachelor's Degrees

FIGURE 1.13 (repeated)

a. According to the first formula, $M = -0.28n + 47$, what percentage of bachelor's degrees were awarded to men in 2003? Does this underestimate or overestimate the actual percent shown in **Figure 1.13** for 2003? By how much?

b. A **ratio** is a comparison by division. Write a formula that describes the ratio of the percentage of bachelor's degrees received by men, $M = -0.28n + 47$, to the percentage of bachelor's degrees received by women, $W = 0.28n + 53$. Name this new mathematical model R, for ratio.

c. Use the formula for R to find the projected ratio of bachelor's degrees received by men to degrees received by women in 2014.

Solution

a. We are interested in the percentage of bachelor's degrees awarded to men in 2003. The year 2003 is 14 years after 1989: $2003 - 1989 = 14$. We substitute 14 for n in the formula for men and evaluate the formula.

$$M = -0.28n + 47$$ *This is the model for the percentage of degrees awarded to men.*

$$M = -0.28(14) + 47$$ *Replace n with 14.*

$$M = -3.92 + 47$$ *Multiply: $-0.28(14) = -3.92$.*

$$M = 43.08$$ *Add. Use the sign of 47 and subtract absolute values:*

$$\begin{array}{r} \overset{6\ \ 9}{4\cancel{7}.\cancel{0}0} \\ -\ \ 3.92 \\ \hline 43.08 \end{array}$$

The model indicates that 43.08% of bachelor's degrees were awarded to men in 2003. Because the number displayed in **Figure 1.13** is 43%, the formula overestimates the actual percent by 0.08%.

b. Now we use division to write a model comparing the ratio, R, of bachelor's degrees received by men to degrees received by women.

$$R = \frac{M}{W} = \frac{-0.28n + 47}{0.28n + 53}$$

Formula for percentage of degrees awarded to men

Formula for percentage of degrees awarded to women

c. Let's see what happens to the ratio, R, in 2014. The year 2014 is 25 years after 1989 ($2014 - 1989 = 25$), so substitute 25 for n.

$$R = \frac{-0.28n + 47}{0.28n + 53}$$ *This is the model for the ratio of the percentage of degrees awarded to men and to women.*

$$R = \frac{-0.28(25) + 47}{0.28(25) + 53}$$ *Replace each occurrence of n with 25.*

$$R = \frac{-7 + 47}{7 + 53}$$ *Multiply: $-0.28(25) = -7$ and $0.28(25) = 7$.*

*The models for M and W give the projected percents for 2014 in **Figure 1.13**.*

$$R = \frac{40}{60}$$ *Add: $-7 + 47 = 40$ and $7 + 53 = 60$.*

$$R = \frac{2}{3}$$ *Reduce: $\dfrac{40}{60} = \dfrac{2 \cdot \cancel{20}}{3 \cdot \cancel{20}} = \dfrac{2}{3}$.*

Our model for R, consistent with the actual data, shows that in 2014, the ratio of bachelor's degrees received by men to degrees received by women is projected to be 2 to 3. If model breakdown does not occur, three women will receive bachelor's degrees for every two men.

✓ **CHECK POINT 9** The data for doctorate degrees shown in **Figure 1.13** on page 77 can be described by $M = -0.6n + 64.4$, where M is the percentage of doctorate degrees awarded to men n years after 1989. According to this mathematical model, what percentage of doctorate degrees are projected to be received by men in 2014? Does this underestimate or overestimate the projection shown in **Figure 1.13**? By how much?

1.7 EXERCISE SET MyMathLab

 Math XL PRACTICE   WATCH   DOWNLOAD READ   REVIEW

Practice Exercises

In Exercises 1–34, perform the indicated multiplication.

1. $5(-9)$ **2.** $10(-7)$

3. $(-8)(-3)$ **4.** $(-9)(-5)$

5. $(-3)(7)$ **6.** $(-4)(8)$

7. $(-19)(-1)$ **8.** $(-11)(-1)$

9. $0(-19)$ **10.** $0(-11)$

11. $\frac{1}{2}(-24)$ **12.** $\frac{1}{3}(-21)$

13. $\left(-\frac{3}{4}\right)(-12)$ **14.** $\left(-\frac{4}{5}\right)(-30)$

15. $-\frac{3}{5} \cdot \left(-\frac{4}{7}\right)$ **16.** $-\frac{5}{7} \cdot \left(-\frac{3}{8}\right)$

17. $-\frac{7}{9} \cdot \frac{2}{3}$ **18.** $-\frac{5}{11} \cdot \frac{2}{7}$

19. $3(-1.2)$ **20.** $4(-1.2)$

21. $-0.2(-0.6)$ **22.** $-0.3(-0.7)$

23. $(-5)(-2)(3)$ **24.** $(-6)(-3)(10)$

25. $(-4)(-3)(-1)(6)$ **26.** $(-2)(-7)(-1)(3)$

27. $-2(-3)(-4)(-1)$ **28.** $-3(-2)(-5)(-1)$

29. $(-3)(-3)(-3)$ **30.** $(-4)(-4)(-4)$

31. $5(-3)(-1)(2)(3)$ **32.** $2(-5)(-2)(3)(1)$

33. $(-8)(-4)(0)(-17)(-6)$

34. $(-9)(-12)(-18)(0)(-3)$

In Exercises 35–42, find the multiplicative inverse of each number.

35. 4 **36.** 3 **37.** $\frac{1}{5}$

38. $\frac{1}{7}$ **39.** -10 **40.** -12

41. $-\frac{2}{5}$ **42.** $-\frac{4}{9}$

In Exercises 43–46,

 a. *Rewrite the division as multiplication involving a multiplicative inverse.*

 b. *Use the multiplication from part (a) to find the given quotient.*

43. $-32 \div 4$

44. $-18 \div 6$

45. $\frac{-60}{-5}$

46. $\frac{-30}{-5}$

In Exercises 47–76, perform the indicated division or state that the expression is undefined.

47. $\frac{12}{-4}$ **48.** $\frac{40}{-5}$ **49.** $\frac{-21}{3}$

50. $\frac{-60}{6}$ **51.** $\frac{-90}{-3}$ **52.** $\frac{-66}{-6}$

53. $\frac{0}{-7}$ **54.** $\frac{0}{-8}$ **55.** $\frac{7}{0}$

56. $\frac{-8}{0}$ **57.** $-15 \div 3$ **58.** $-80 \div 8$

59. $120 \div (-10)$ **60.** $130 \div (-10)$

61. $(-180) \div (-30)$ **62.** $(-150) \div (-25)$

63. $0 \div (-4)$ **64.** $0 \div (-10)$

65. $-4 \div 0$ **66.** $-10 \div 0$

67. $\frac{-12.9}{3}$ **68.** $\frac{-21.6}{3}$

69. $-\frac{1}{2} \div \left(-\frac{3}{5}\right)$ **70.** $-\frac{1}{2} \div \left(-\frac{7}{9}\right)$

71. $-\frac{14}{9} \div \frac{7}{8}$ **72.** $-\frac{5}{16} \div \frac{25}{8}$

73. $\frac{1}{3} \div \left(-\frac{1}{3}\right)$ **74.** $\frac{1}{5} \div \left(-\frac{1}{5}\right)$

75. $6 \div \left(-\frac{2}{5}\right)$ **76.** $8 \div \left(-\frac{2}{9}\right)$

In Exercises 77–96, simplify each algebraic expression.

77. $-5(2x)$ **78.** $-9(3x)$

79. $-4\left(-\frac{3}{4}y\right)$ **80.** $-5\left(-\frac{3}{5}y\right)$

81. $8x + x$ **82.** $12x + x$

83. $-5x + x$ **84.** $-6x + x$

85. $6b - 7b$ **86.** $12b - 13b$

87. $-y + 4y$

88. $-y + 9y$

89. $-4(2x - 3)$

90. $-3(4x - 5)$

91. $-3(-2x + 4)$

92. $-4(-3x + 2)$

93. $-(2y - 5)$

94. $-(3y - 1)$

95. $4(2y - 3) - (7y + 2)$

96. $5(3y - 1) - (14y - 2)$

In Exercises 97–108, determine whether the given number is a solution of the equation.

97. $4x = 2x - 10; -5$

98. $5x = 3x - 6; -3$

99. $-7y + 18 = -10y + 6; -4$

100. $-4y + 21 = -7y + 15; -2$

101. $5(w + 3) = 2w - 21; -10$

102. $6(w + 2) = 4w - 10; -9$

103. $4(6 - z) + 7z = 0; -8$

104. $5(7 - z) + 12z = 0; -5$

105. $14 - 2x = -4x + 7; -2\frac{1}{2}$

106. $16 - 4x = -2x + 21; -3\frac{1}{2}$

107. $\dfrac{5m - 1}{6} = \dfrac{3m - 2}{4}; -4$

108. $\dfrac{6m - 5}{11} = \dfrac{3m - 2}{5}; -1$

Practice PLUS

In Exercises 109–116, write a numerical expression for each phrase. Then simplify the numerical expression by performing the given operations.

109. 8 added to the product of 4 and -10

110. 14 added to the product of 3 and -15

111. The product of -9 and -3, decreased by -2

112. The product of -6 and -4, decreased by -5

113. The quotient of -18 and the sum of -15 and 12

114. The quotient of -25 and the sum of -21 and 16

115. The difference between -6 and the quotient of 12 and -4

116. The difference between -11 and the quotient of 20 and -5

Application Exercises

In Exercises 117–118, use the formula $C = \frac{5}{9}(F - 32)$ to express each Fahrenheit temperature, F, as its equivalent Celsius temperature, C.

117. $-22°F$

118. $-31°F$

The graph shows the percentage of high school seniors who used alcohol or marijuana during the 30 days prior to being surveyed for the University of Michigan's Monitoring the Future *study.*

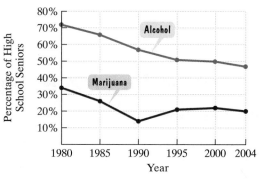

Alcohol and Marijuana Use by United States High School Seniors

Source: U.S. Department of Health and Human Services

The data can be described by the following mathematical models:

Percentage of seniors using alcohol → $A = -n + 70$

Number of years after 1980

Percentage of seniors using marijuana → $M = -0.5n + 28.$

Use this information to solve Exercises 119–120.

119. a. Use the appropriate line graph to determine the percentage of seniors who used alcohol in 2000.

b. Use the appropriate formula to determine the percentage of seniors who used alcohol in 2000. What do you observe?

c. Use the appropriate line graph to estimate the percentage of seniors who used marijuana in 2000.

d. Use the appropriate formula to determine the percentage of seniors who used marijuana in 2000. How does this compare with your estimate in part (c)?

e. Write a formula that describes the ratio of the percentage of seniors who used marijuana to the percentage who used alcohol. Name this new mathematical model R, for ratio.

120. a. Use the appropriate line graph to estimate the percentage of seniors who used alcohol in 2004.

b. Use the appropriate formula to determine the percentage of seniors who used alcohol in 2004. How does this compare with your estimate in part (a)?

c. Use the appropriate line graph to determine the percentage of seniors who used marijuana in 2004.

d. Use the appropriate formula to determine the percentage of seniors who used marijuana in 2004.

Does this underestimate or overestimate the percent that you found in part (c)? By how much?

e. Write a formula that describes the ratio of the percentage of seniors who used alcohol to the percentage who used marijuana. Name this new mathematical model R, for ratio.

121. The graph shows the sources of income for the federal government in 1960 and 2005.

Sources of Income for the United States Government

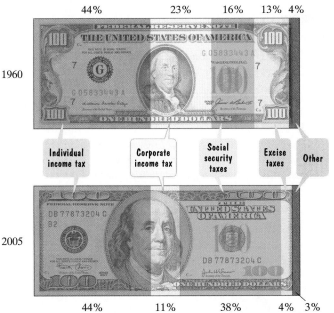

Source: Teller-Elsberg et al., *Field Guide to the U.S. Economy*, The New Press, 2006

The graph indicates that the percentage of money collected from corporations has decreased and the percentage coming from income taxes has not changed.

a. Use the data in the graph to write a simplified fraction that represents the 2005 ratio of percentage of income taxes from corporations to percentage from individual income taxes. According to this rational number, for every $10,000 collected in individual income taxes, how much was collected from corporate income taxes?

b. Write the rational number in part (a) in decimal form.

c. Here is a mathematical model that describes the diminishing share, D, of the tax burden by corporations n years after 1960:

Number of years after 1960

Ratio of percentage of taxes from corporations to percentage from income taxes

$$D = \frac{-0.27n + 23}{44}.$$

Use the model to write a quotient of two numbers that represents the 2005 ratio of percentage of corporate income taxes to percentage of individual income taxes.

d. Express the quotient of the two numbers in part (c) in decimal form, correct to two decimal places. How well does this rounded decimal describe the value obtained from the data in part (b)?

Writing in Mathematics

122. Explain how to multiply two real numbers. Provide examples with your explanation.

123. Explain how to determine the sign of a product that involves more than two numbers.

124. Explain how to find the multiplicative inverse of a number.

125. Why is it that 0 has no multiplicative inverse?

126. Explain how to divide real numbers.

127. Why is division by zero undefined?

128. Explain how to simplify an algebraic expression in which a negative sign procedes parentheses.

129. Do you believe that the trend in the graphs showing a decline in male college attendance (**Figure 1.13** on page 77) should be reversed by providing admissions preferences for men? Explain your position on this issue.

Critical Thinking Exercises

Make Sense? *In Exercises 130–133, determine whether each statement "makes sense" or "does not make sense" and explain your reasoning.*

130. I've noticed that the sign rules for dividing real numbers are slightly different than the sign rules for multiplying real numbers.

131. Just as two negative factors give a positive product, I've seen the same thing occur with double negatives in English:

That rumor is not unknown   means   That rumor is known.

132. This pattern suggests that multiplying two negative numbers results in a positive answer:

Decreasing by 1 from left to right

$2(-3) = -6; 1(-3) = -3; 0(-3) = 0; -1(-3) = 3; -2(-3) = 6.$

Increasing by 3 from left to right

133. When I used

$$R = \frac{M}{W} = \frac{-0.28n + 47}{0.28n + 53}$$

Number of years after 1989

to project the ratio of bachelor's degrees received by men to degrees received by women in 2020, I had to find the quotient of two negative numbers.

In Exercises 134–137, determine whether each statement is true or false. If the statement is false, make the necessary change(s) to produce a true statement.

134. Both the addition and the multiplication of two negative numbers result in a positive number.

135. Multiplying a negative number by a nonnegative number will always give a negative number.

136. $0 \div \left(-\sqrt{2}\right)$ is undefined.

137. If a is negative, b is positive, and c is negative, then $\frac{a}{bc}$ is positive.

In Exercises 138–141, write an algebraic expression for the given English phrase.

138. The value, in cents, of x nickels

139. The distance covered by a car traveling at 50 miles per hour for x hours

140. The monthly salary, in dollars, for a person earning x dollars per year

141. The fraction of people in a room who are women if there are 40 women and x men in the room

Technology Exercises

142. Use a calculator to verify any five of the products that you found in Exercises 1–34.

143. Use a calculator to verify any five of the quotients that you found in Exercises 47–76.

144. Simplify using a calculator:

$$0.3(4.7x - 5.9) - 0.07(3.8x - 61).$$

145. Use your calculator to attempt to find the quotient of -3 and 0. Describe what happens. Does the same thing occur when finding the quotient of 0 and -3? Explain the difference. Finally, what happens when you enter the quotient of 0 and itself?

Review Exercises

In Exercises 146–148, perform the indicated operation.

146. $-6 + (-3)$ (Section 1.5, Example 3)

147. $-6 - (-3)$ (Section 1.6, Example 1)

148. $-6 \div (-3)$ (Section 1.7, Example 4)

Preview Exercises

Exercises 149–151 will help you prepare for the material covered in the next section. In each exercise, an expression with an exponent is written as a repeated multiplication. Find this product, indicated by a question mark.

149. $(-6)^2 = (-6)(-6) = \,?$

150. $(-5)^3 = (-5)(-5)(-5) = \,?$

151. $(-2)^4 = (-2)(-2)(-2)(-2) = \,?$

SECTION 1.8

Exponents and Order of Operations

Objectives

1. Evaluate exponential expressions.

2. Simplify algebraic expressions with exponents.

3. Use the order of operations agreement.

4. Evaluate mathematical models.

Traffic is really backed up on the highway. Finally, you see the source of the traffic jam—a minor fender-bender. Still stuck in traffic, you notice that the driver appears to be quite young. This might seem like a strange observation. After all, what does a driver's age have to do with his or her chance of getting into an accident? In this section, we continue to see how mathematical models describe your world, including a formula that provides a relationship between age and number of car accidents.

1. Evaluate exponential expressions.

Natural Number Exponents

Although people do a great deal of talking, the total output since the beginning of gabble to the present day, including all baby talk, love songs, and congressional debates, only amounts to about 10 million billion words. This can be expressed as 16 factors of 10, or 10^{16} words.

Exponents such as 2, 3, 4, and so on are used to indicate repeated multiplication. For example,

$$10^2 = 10 \cdot 10 = 100,$$
$$10^3 = 10 \cdot 10 \cdot 10 = 1000, \quad 10^4 = 10 \cdot 10 \cdot 10 \cdot 10 = 10,000.$$

The 10 that is repeated when multiplying is called the **base**. The small numbers above and to the right of the base are called **exponents** or **powers**. The exponent tells the number of times the base is to be used when multiplying. In 10^3, the base is 10 and the exponent is 3.

Any number with an exponent of 1 is the number itself. Thus, $10^1 = 10$.

Multiplications that are expressed in exponential notation are read as follows:

10^1: "ten to the first power"
10^2: "ten to the second power" or "ten squared"
10^3: "ten to the third power" or "ten cubed"
10^4: "ten to the fourth power"
10^5: "ten to the fifth power"
etc.

Any real number can be used as the base. Thus,

$$7^2 = 7 \cdot 7 = 49 \quad \text{and} \quad (-3)^4 = (-3)(-3)(-3)(-3) = 81.$$

The bases are 7 and −3, respectively. Do not confuse $(-3)^4$ and -3^4.

$$-3^4 = -(3 \cdot 3 \cdot 3 \cdot 3) = -81$$

> The negative is not taken to the power because it is not inside parentheses.

An exponent applies only to a base. A negative sign is not part of a base unless it appears in parentheses.

Using Technology

You can use a calculator to evaluate exponential expressions. For example, to evaluate 5^3, press the following keys:

Many Scientific Calculators

5 y^x 3 $=$

Many Graphing Calculators

5 $\wedge$ 3 ENTER

Although calculators have special keys to evaluate powers of ten and squaring bases, you can always use one of the sequences shown here.

EXAMPLE 1 Evaluating Exponential Expressions

Evaluate:

a. 4^2 **b.** $(-5)^3$ **c.** $(-2)^4$ **d.** -2^4.

Solution

> Exponent is 2.

a. $4^2 = 4 \cdot 4 = 16$ The exponent indicates that the base is used as a factor two times.

> Base is 4.

We read $4^2 = 16$ as "4 to the second power is 16" or "4 squared is 16."

> Exponent is 3.

b. $(-5)^3 = (-5)(-5)(-5)$ The exponent indicates that the base is used as a factor three times.

> Base is −5.

$= -125$ An odd number of negative factors yields a negative product.

We read $(-5)^3 = -125$ as "the number negative 5 to the third power is negative 125" or "negative 5 cubed is negative 125."

Exponent is 4.

c. $(-2)^4 = (-2)(-2)(-2)(-2)$ The exponent indicates that the base is used as a factor four times.

Base is −2.

$= 16$ An even number of negative factors yields a positive product.

We read $(-2)^4 = 16$ as "the number negative 2 to the fourth power is 16."

Exponent is 4.

d. $-2^4 = -(2 \cdot 2 \cdot 2 \cdot 2)$ The negative is not inside parentheses and is not taken to the fourth power.

Base is 2.

$= -16$ Multiply the twos and copy the negative.

We read $-2^4 = -16$ as "the negative of 2 raised to the fourth power is negative 16" or "the opposite, or additive inverse, of 2 raised to the fourth power is negative 16."

✓ **CHECK POINT 1**   Evaluate:

   a. 6^2 **b.** $(-4)^3$ **c.** $(-1)^4$ **d.** -1^4.

The formal algebraic definition of a natural number exponent summarizes our discussion:

Definition of a Natural Number Exponent

If b is a real number and n is a natural number,

Exponent

$$b^n = \underbrace{b \cdot b \cdot b \cdots \cdot b}_{\substack{b \text{ appears as a} \\ \text{factor } n \text{ times.}}}$$

Base

b^n is read "the nth power of b" or "b to the nth power." Thus, the nth power of b is defined as the product of n factors of b. The expression b^n is called an **exponential expression**.

    Furthermore, $b^1 = b$.

---

2 Simplify algebraic expressions with exponents.

Exponents and Algebraic Expressions

The distributive property can be used to simplify certain algebraic expressions that contain exponents. For example, we can use the distributive property to combine like terms in the algebraic expression $4x^2 + 6x^2$:

$$4x^2 + 6x^2 \quad = \quad (4 + 6)x^2 = 10x^2.$$

First term with variable factor x^2 Second term with variable factor x^2 The common variable factor is x^2.

Study Tip

When adding algebraic expressions, if you have like terms you add only the numerical coefficients—not the exponents. **Exponents are never added when the operation is addition.** Avoid these common errors.

Incorrect

- $7x^3 + 2x^3 = 9x^6$
- $5x^2 + x^2 = 6x^4$
- $3x^2 + 4x^3 = 7x^5$

EXAMPLE 2 **Simplifying Algebraic Expressions**

Simplify, if possible:

a. $7x^3 + 2x^3$ **b.** $5x^2 + x^2$ **c.** $3x^2 + 4x^3$.

Solution

a. $7x^3 + 2x^3$ There are two like terms with the same variable factor, namely x^3.

$= (7 + 2)x^3$ Apply the distributive property.

$= 9x^3$ Add within parentheses.

b. $5x^2 + x^2$ There are two like terms with the same variable factor, namely x^2.

$= 5x^2 + 1x^2$ Use the multiplication property of 1.

$= (5 + 1)x^2$ Apply the distributive property.

$= 6x^2$ Add within parentheses.

c. $3x^2 + 4x^3$ cannot be simplified. The terms $3x^2$ and $4x^3$ are not like terms because they have different variable factors, namely x^2 and x^3. ■

☑ **CHECK POINT 2** Simplify, if possible:

a. $16x^2 + 5x^2$ **b.** $7x^3 + x^3$ **c.** $10x^2 + 8x^3$.

---

3 Use the order of operations agreement.

Order of Operations

Suppose that you want to find the value of $3 + 7 \cdot 5$. Which procedure shown is correct?

$$3 + 7 \cdot 5 = 3 + 35 = 38 \quad \text{or} \quad 3 + 7 \cdot 5 = 10 \cdot 5 = 50$$

You know the answer because we introduced certain rules, called the **order of operations**, at the beginning of the chapter. One of these rules stated that if a problem contains no parentheses or other grouping symbols, perform multiplication before addition. Thus, the procedure on the left is correct because the multiplication of 7 and 5 is done first. Then the addition is performed. The correct answer is 38.

Some problems contain grouping symbols, such as parentheses, ( ); brackets, [ ]; braces, { }; absolute value symbols, | |; or fraction bars. These grouping symbols tell us what to do first. Here are two examples:

- $$(3 + 7) \cdot 5 = 10 \cdot 5 = 50$$

  First, perform operations in grouping symbols.

- $$8|6 - 16| = 8|-10| = 8 \cdot 10 = 80.$$

Here are the rules for determining the order in which operations should be performed:

Study Tip

Here's a sentence to help remember the order of operations: "Please excuse my dear Aunt Sally."

Please	**Parentheses**
Excuse	**Exponents**
{ **My**	{ **Multiplication**
{ **Dear**	{ **Division**
{ **Aunt**	{ **Addition**
{ **Sally**	{ **Subtraction**

Order of Operations

1. Perform all operations within grouping symbols.

2. Evaluate all exponential expressions.

3. Do all multiplications and divisions in the order in which they occur, working from left to right.

4. Finally, do all additions and subtractions using one of the following procedures:

   a. Work from left to right and do additions and subtractions in the order in which they occur.

   or

   b. Rewrite subtractions as additions of opposites. Combine positive and negative numbers separately, and then add these results.

The last step in the order of operations indicates that you have a choice when working with additions and subtractions, although in this section we will perform these operations from left to right. However, when working with multiplications and divisions, you must perform these operations *as they occur* from left to right. For example,

$$8 \div 4 \cdot 2 = 2 \cdot 2 = 4 \qquad \text{Do the division first because it occurs first.}$$

$$8 \cdot 4 \div 2 = 32 \div 2 = 16. \qquad \text{Do the multiplication first because it occurs first.}$$

EXAMPLE 3   Using the Order of Operations

Simplify: $18 + 2 \cdot 3 - 10$.

Solution   There are no grouping symbols or exponential expressions. In cases like this, we multiply and divide before adding and subtracting.

$$
\begin{aligned}
18 + 2 \cdot 3 - 10 &= 18 + 6 - 10 \qquad && \text{Multiply: } 2 \cdot 3 = 6.\\
&= 24 - 10 && \text{Add and subtract from left to right:}\\
& && 18 + 6 = 24.\\
&= 14 && \text{Subtract: } 24 - 10 = 14. \quad \blacksquare
\end{aligned}
$$

☑ **CHECK POINT 3**   Simplify: $20 + 4 \cdot 3 - 17$.

EXAMPLE 4   Using the Order of Operations

Simplify: $6^2 - 24 \div 2^2 \cdot 3 - 1$.

Solution   There are no grouping symbols. Thus, we begin by evaluating exponential expressions. Then we multiply or divide. Finally, we add or subtract.

$$
\begin{aligned}
6^2 &- 24 \div 2^2 \cdot 3 - 1\\
&= 36 - 24 \div 4 \cdot 3 - 1 \qquad && \text{Evaluate exponential expressions:}\\
& && 6^2 = 6 \cdot 6 = 36 \text{ and } 2^2 = 2 \cdot 2 = 4.\\
&= 36 - 6 \cdot 3 - 1 && \text{Perform the multiplications and divisions}\\
& && \text{from left to right. Start with } 24 \div 4 = 6.\\
&= 36 - 18 - 1 && \text{Now do the multiplication: } 6 \cdot 3 = 18.\\
&= 18 - 1 && \text{Finally, perform the subtraction from left to}\\
& && \text{right: } 36 - 18 = 18.\\
&= 17 && \text{Complete the subtraction: } 18 - 1 = 17. \quad \blacksquare
\end{aligned}
$$

☑ **CHECK POINT 4**   Simplify: $7^2 - 48 \div 4^2 \cdot 5 - 2$.

EXAMPLE 5   Using the Order of Operations

Simplify:

   a. $(2 \cdot 5)^2$ **b.** $2 \cdot 5^2$.

Solution

   a. Because $(2 \cdot 5)^2$ contains grouping symbols, namely parentheses, we perform the operation within parentheses first.

$$
\begin{aligned}
(2 \cdot 5)^2 &= 10^2 \qquad && \text{Multiply within parentheses: } 2 \cdot 5 = 10.\\
&= 100 && \text{Evaluate the exponential expression: } 10^2 = 10 \cdot 10 = 100.
\end{aligned}
$$

b. Because $2 \cdot 5^2$ does not contain grouping symbols, we begin by evaluating the exponential expression.

$$2 \cdot 5^2 = 2 \cdot 25 \qquad \text{Evaluate the exponential expression: } 5^2 = 5 \cdot 5 = 25.$$
$$= 50 \qquad \text{Now do the multiplication: } 2 \cdot 25 = 50. \qquad \blacksquare$$

✓ **CHECK POINT 5**   Simplify:

a. $(3 \cdot 2)^2$ **b.** $3 \cdot 2^2$.

EXAMPLE 6   **Using the Order of Operations**

Simplify: $\left(\dfrac{1}{2}\right)^3 - \left(\dfrac{1}{2} - \dfrac{3}{4}\right)^2 (-4)$.

Solution   Because grouping symbols appear, we perform the operation within parentheses first.

$$\left(\frac{1}{2}\right)^3 - \left(\frac{1}{2} - \frac{3}{4}\right)^2 (-4)$$

$$= \left(\frac{1}{2}\right)^3 - \left(-\frac{1}{4}\right)^2 (-4) \qquad \text{Work inside parentheses first:}$$
$$\frac{1}{2} - \frac{3}{4} = \frac{2}{4} - \frac{3}{4} = \frac{2}{4} + \left(-\frac{3}{4}\right) = -\frac{1}{4}.$$

$$= \frac{1}{8} - \frac{1}{16}(-4) \qquad \text{Evaluate exponential expressions:}$$
$$\left(\frac{1}{2}\right)^3 = \frac{1}{2} \cdot \frac{1}{2} \cdot \frac{1}{2} = \frac{1}{8} \text{ and } \left(-\frac{1}{4}\right)^2 = \left(-\frac{1}{4}\right)\left(-\frac{1}{4}\right) = \frac{1}{16}.$$

$$= \frac{1}{8} - \left(-\frac{1}{4}\right) \qquad \text{Multiply: } \frac{1}{16} \cdot \left(\frac{-4}{1}\right) = -\frac{4}{16} = -\frac{1}{4}.$$

$$= \frac{3}{8} \qquad \text{Subtract: } \frac{1}{8} - \left(-\frac{1}{4}\right) = \frac{1}{8} + \frac{1}{4} = \frac{1}{8} + \frac{2}{8} = \frac{3}{8}. \qquad \blacksquare$$

✓ **CHECK POINT 6**   Simplify: $\left(-\dfrac{1}{2}\right)^2 - \left(\dfrac{7}{10} - \dfrac{8}{15}\right)^2 (-18)$.

Some expressions contain many grouping symbols. An example of such an expression is $2[5(4 - 7) + 9]$. The grouping symbols are the parentheses and the brackets.

> The parentheses, the innermost grouping symbols, group $4 - 7$.

$$2[5(4 - 7) + 9]$$

> The brackets, the outermost grouping symbols, group $5(4 - 7) + 9$.

When combinations of grouping symbols appear, **perform operations within the innermost grouping symbols first**. Then work to the outside, performing operations within the outermost grouping symbols.

EXAMPLE 7 **Using the Order of Operations**

Simplify: $2[5(4 - 7) + 9]$.

Solution

$$2[5(4 - 7) + 9]$$

$= 2[5(-3) + 9]$ Work inside parentheses first:
$4 - 7 = 4 + (-7) = -3$.

$= 2[-15 + 9]$ Work inside brackets and multiply:
$5(-3) = -15$.

$= 2[-6]$ Add inside brackets: $-15 + 9 = -6$. The
resulting problem can also be expressed as $2(-6)$.

$= -12$ Multiply: $2[-6] = -12$. ■

Parentheses can be used for both innermost and outermost grouping symbols. For example, the expression $2[5(4 - 7) + 9]$ can also be written $2(5(4 - 7) + 9)$. However, too many parentheses can be confusing. The use of both parentheses and brackets makes it easier to identify inner and outer groupings.

☑ **CHECK POINT 7** Simplify: $4[3(6 - 11) + 5]$.

EXAMPLE 8 **Using the Order of Operations**

Simplify: $18 \div 6 + 4\left[5 + 2(8 - 10)^3\right]$.

Solution

$$18 \div 6 + 4\left[5 + 2(8 - 10)^3\right]$$

$= 18 \div 6 + 4\left[5 + 2(-2)^3\right]$ Work inside parentheses first:
$8 - 10 = 8 + (-10) = -2$.

$= 18 \div 6 + 4[5 + 2(-8)]$ Work inside brackets and evaluate the
exponential expression:
$(-2)^3 = (-2)(-2)(-2) = -8$.

$= 18 \div 6 + 4[5 + (-16)]$ Work inside brackets and multiply:
$2(-8) = -16$.

$= 18 \div 6 + 4[-11]$ Work inside brackets and add:
$5 + (-16) = -11$.

$= 3 + 4[-11]$ Perform the multiplications and divisions
from left to right. Start with $18 \div 6 = 3$.

$= 3 + (-44)$ Now do the multiplication: $4(-11) = -44$.

$= -41$ Finally, perform the addition:
$3 + (-44) = -41$. ■

☑ **CHECK POINT 8** Simplify: $25 \div 5 + 3\left[4 + 2(7 - 9)^3\right]$.

Fraction bars are grouping symbols that separate expressions into two parts, the numerator and the denominator. Consider, for example,

The fraction bar is the grouping symbol.

The numerator is one part of the expression.

$$\frac{2(3 - 12) + 6 \cdot 4}{2^4 + 1}.$$

The denominator is the other part of the expression.

We can use brackets instead of the fraction bar. An equivalent expression is

$$\left[2(3 - 12) + 6 \cdot 4\right] \div \left[2^4 + 1\right].$$

The grouping suggests a method for simplifying expressions with fraction bars as grouping symbols:

- Simplify the numerator.
- Simplify the denominator.
- If possible, simplify the fraction.

EXAMPLE 9   **Using the Order of Operations**

Simplify: $\dfrac{2(3 - 12) + 6 \cdot 4}{2^4 + 1}$.

Solution

$$\dfrac{2(3 - 12) + 6 \cdot 4}{2^4 + 1}$$

$$= \dfrac{2(-9) + 6 \cdot 4}{16 + 1}$$

Work inside parentheses in the numerator: $3 - 12 = 3 + (-12) = -9$. Evaluate the exponential expression in the denominator: $2^4 = 2 \cdot 2 \cdot 2 \cdot 2 = 16$.

$$= \dfrac{-18 + 24}{16 + 1}$$

Multiply in the numerator: $2(-9) = -18$ and $6 \cdot 4 = 24$.

$$= \dfrac{6}{17}$$

Perform the addition in the numerator and in the denominator. ∎

☑ **CHECK POINT 9**   Simplify: $\dfrac{5(4 - 9) + 10 \cdot 3}{2^3 - 1}$.

EXAMPLE 10   **Using the Order of Operations**

Evaluate: $-x^2 - 7x$ for $x = -2$.

Solution   We begin by substituting -2 for each occurrence of x in the algebraic expression. Then we use the order of operations to evaluate the expression.

$$-x^2 - 7x$$

Replace x with -2. Place parentheses around -2.

$$= -(-2)^2 - 7(-2)$$

$$= -4 - 7(-2)$$

Evaluate the exponential expression: $(-2)^2 = (-2)(-2) = 4$.

$$= -4 - (-14)$$

Multiply: $7(-2) = -14$.

$$= 10$$

Subtract: $-4 - (-14) = -4 + 14 = 10$. ∎

☑ **CHECK POINT 10**   Evaluate: $-x^2 - 4x$ for $x = -5$.

Some algebraic expressions contain two sets of grouping symbols. Using the order of operations, grouping symbols are removed from innermost (parentheses) to outermost (brackets).

EXAMPLE 11 **Simplifying an Algebraic Expression**

Simplify: $18x^2 + 4 - [6(x^2 - 2) + 5]$.

Solution

$$18x^2 + 4 - [6(x^2 - 2) + 5]$$
$$= 18x^2 + 4 - [6x^2 - 12 + 5] \qquad \text{Use the distributive property to remove parentheses:}$$
$$6(x^2 - 2) = 6x^2 - 6 \cdot 2 = 6x^2 - 12.$$
$$= 18x^2 + 4 - [6x^2 - 7] \qquad \text{Add inside brackets: } -12 + 5 = -7.$$
$$= 18x^2 + 4 - 6x^2 + 7 \qquad \text{Remove brackets by changing the sign of each term within brackets.}$$
$$= (18x^2 - 6x^2) + 4 + 7 \qquad \text{Group like terms.}$$
$$= 12x^2 + 11 \qquad \text{Combine like terms.}$$

✓ **CHECK POINT 11** Simplify: $14x^2 + 5 - [7(x^2 - 2) + 4]$.

4 Evaluate mathematical models.

Applications

In Examples 12 and 13, we use the order of operations to evaluate mathematical models.

EXAMPLE 12 **Car Accidents and Age**

The mathematical model

$$N = 0.4x^2 - 36x + 1000$$

approximates the number of accidents, N, per 50 million miles driven, for drivers who are x years old. The formula applies to drivers ages 16 through 74, inclusive. How many accidents, per 50 million miles driven, are there for 20-year-old drivers?

Solution In the mathematical model, x represents the age of the driver. We are interested in 20-year-old drivers. Thus, we substitute 20 for each occurrence of x. Then we use the order of operations to find N, the number of accidents per 50 million miles driven.

Study Tip

To find

$$160 - 720 + 1000$$
$$= 160 + (-720) + 1000,$$

you can also first add the positive numbers

$$160 + 1000 = 1160$$

and then add the negative number to that result:

$$1160 + (-720) = 440.$$

$$N = 0.4x^2 - 36x + 1000 \qquad \text{This is the given mathematical model.}$$
$$N = 0.4(20)^2 - 36(20) + 1000 \qquad \text{Replace each occurrence of } x \text{ with 20.}$$
$$N = 0.4(400) - 36(20) + 1000 \qquad \text{Evaluate the exponential expression:} \\ (20)^2 = (20)(20) = 400.$$
$$N = 160 - 720 + 1000 \qquad \text{Multiply from left to right:} \\ 0.4(400) = 160 \text{ and } 36(20) = 720.$$
$$N = -560 + 1000 \qquad \text{Perform additions and subtractions from left to right. Subtract:} \\ 160 - 720 = 160 + (-720) = -560.$$
$$N = 440 \qquad \text{Add: } -560 + 1000 = 440.$$

Thus, 20-year-old drivers have 440 accidents per 50 million miles driven.

How do the number of accidents for 20-year-old drivers compare to, say, the number for 45-year-old drivers? The answer is given on the next page by the bar graph in **Figure 1.14**. We used the evaluation from Example 12 to construct the first bar: 20-year-old drivers have 440 accidents per 50 million miles driven. We then evaluated the model for $x = 30, 40, 45, 50, 60,$ and 70 to draw the other six bars.

Car Accidents and Age

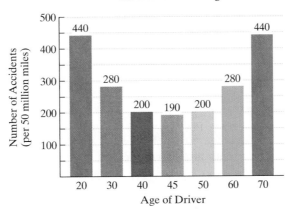

FIGURE 1.14

Source: Government Accountability Office

Can you see that the bar for 45-year-olds is lower than the other six bars? In practical terms, this indicates that 45-year-olds have fewer car accidents, 190 per 50 million miles driven, than do the younger and older drivers represented in **Figure 1.14**.

✓ **CHECK POINT 12** The mathematical model

$$P = -0.05x^2 + 3.6x - 15$$

describes the percentage of U.S. adults, P, tested for HIV at age x. According to the formula, what percentage of U.S. adults who are 20 years old have been tested for HIV?

EXAMPLE 13 Shrinky-Dink Size without Rinky-Dink Style

A company has decided to jump into the pedal game with foldable, ultraportable bikes. Its fixed monthly costs are $500,000 and it will cost $400 to manufacture each bike. The average cost per bike, $\overline{C}$, for the company to manufacture x foldable bikes per month is modeled by

$$\overline{C} = \frac{400x + 500,000}{x}.$$

Find the average cost per bike for the company to manufacture

a. 10,000 bikes per month.
b. 50,000 bikes per month.
c. 100,000 bikes per month.

What happens to the average cost per bike as the production level increases?

Solution

a. We are interested in the average cost per bike for the company if 10,000 bikes are manufactured per month. Because x represents the number of bikes manufactured per month, we substitute 10,000 for x in the given mathematical model.

$$\overline{C} = \frac{400x + 500,000}{x} = \frac{400(10,000) + 500,000}{10,000} = \frac{4,000,000 + 500,000}{10,000}$$

$$= \frac{4,500,000}{10,000} = 450$$

The average cost per bike of producing 10,000 bikes per month is $450.

b. Now, 50,000 bikes are manufactured per month. We find the average cost per bike by substituting 50,000 for x in the given mathematical model.

$$\overline{C} = \frac{400x + 500,000}{x} = \frac{400(50,000) + 500,000}{50,000} = \frac{20,000,000 + 500,000}{50,000}$$

$$= \frac{20,500,000}{50,000} = 410$$

The average cost per bike of producing 50,000 bikes per month is $410.

c. Finally, the production level has increased to 100,000 bikes per month. We find the average cost per bike for the company by substituting 100,000 for x in the given mathematical model.

$$\overline{C} = \frac{400x + 500,000}{x} = \frac{400(100,000) + 500,000}{100,000} = \frac{40,000,000 + 500,000}{100,000}$$

$$= \frac{40,500,000}{100,000} = 405$$

The average cost per bike of producing 100,000 bikes per month is $405.

As the production level increases, the average cost of producing each foldable bike decreases. This illustrates the difficulty with small businesses. It is nearly impossible to have competitively low prices when production levels are low. ■

The graph in **Figure 1.15** shows the relationship between production level and cost. The three points with the voice balloons illustrate our evaluations in Example 13. The other unlabeled points along the smooth blue curve represent the company's average costs for various production levels. The symbol $\dagger$ on the vertical axis indicates that there is a break in the values between 0 and 400. Thus, values for the average cost per bike begin at $400.

FIGURE 1.15

The points that lie along the blue curve in **Figure 1.15** are falling from left to right. Can you see how this shows that the company's cost per foldable bike is decreasing as the production level increases?

☑ **CHECK POINT 13** A company that manufactures running shoes has weekly fixed costs of $300,000 and it costs $30 to manufacture each pair of running shoes. The average cost per pair of running shoes, $\overline{C}$, for the company to manufacture x pairs per week is modeled by

$$\overline{C} = \frac{30x + 300,000}{x}.$$

Find the average cost per pair of running shoes for the company to manufacture
a. 1000 pairs per week.   **b.** 10,000 pairs per week.
c. 100,000 pairs per week.

Practice Exercises

In Exercises 1–14, evaluate each exponential expression.

1. 9^2 **2.** 3^2 **3.** 4^3

4. 6^3 **5.** $(-4)^2$ **6.** $(-10)^2$

7. $(-4)^3$ **8.** $(-10)^3$ **9.** $(-5)^4$

10. $(-1)^6$ **11.** -5^4 **12.** -1^6

13. -10^2 **14.** -8^2

In Exercises 15–28, simplify each algebraic expression, or explain why the expression cannot be simplified.

15. $7x^2 + 12x^2$ **16.** $6x^2 + 18x^2$

17. $10x^3 + 5x^3$ **18.** $14x^3 + 8x^3$

19. $8x^4 + x^4$ **20.** $14x^4 + x^4$

21. $26x^2 - 27x^2$ **22.** $29x^2 - 30x^2$

23. $27x^3 - 26x^3$ **24.** $30x^3 - 29x^3$

25. $5x^2 + 5x^3$

26. $8x^2 + 8x^3$

27. $16x^2 - 16x^2$ **28.** $34x^2 - x^2$

In Exercises 29–72, use the order of operations to simplify each expression.

29. $7 + 6 \cdot 3$ **30.** $3 + 4 \cdot 5$

31. $45 \div 5 \cdot 3$ **32.** $40 \div 4 \cdot 2$

33. $6 \cdot 8 \div 4$ **34.** $8 \cdot 6 \div 2$

35. $14 - 2 \cdot 6 + 3$ **36.** $36 - 12 \div 4 + 2$

37. $8^2 - 16 \div 2^2 \cdot 4 - 3$ **38.** $10^2 - 100 \div 5^2 \cdot 2 - 1$

39. $3(-2)^2 - 4(-3)^2$ **40.** $5(-3)^2 - 2(-4)^2$

41. $(4 \cdot 5)^2 - 4 \cdot 5^2$ **42.** $(3 \cdot 5)^2 - 3 \cdot 5^2$

43. $(2 - 6)^2 - (3 - 7)^2$ **44.** $(4 - 6)^2 - (5 - 9)^2$

45. $6(3 - 5)^3 - 2(1 - 3)^3$

46. $-3(-6 + 8)^3 - 5(-3 + 5)^3$

47. $[2(6 - 2)]^2$ **48.** $[3(4 - 6)]^3$

49. $2[5 + 2(9 - 4)]$ **50.** $3[4 + 3(10 - 8)]$

51. $[7 + 3(2^3 - 1)] \div 21$ **52.** $[11 - 4(2 - 3^3)] \div 37$

53. $\dfrac{10 + 8}{5^2 - 4^2}$ **54.** $\dfrac{6^2 - 4^2}{2 - (-8)}$

55. $\dfrac{37 + 15 \div (-3)}{2^4}$ **56.** $\dfrac{22 + 20 \div (-5)}{3^2}$

57. $\dfrac{(-11)(-4) + 2(-7)}{7 - (-3)}$ **58.** $\dfrac{-5(7 - 2) - 3(4 - 7)}{-13 - (-5)}$

59. $4|10 - (8 - 20)|$ **60.** $6|7 - 4 \cdot 3|$

61. $8(-10) + |4(-5)|$

62. $4(-15) + |3(-10)|$

63. $-2^2 + 4[16 \div (3 - 5)]$

64. $-3^2 + 2[20 \div (7 - 11)]$

65. $24 \div \dfrac{3^2}{8 - 5} - (-6)$

66. $30 \div \dfrac{5^2}{7 - 12} - (-9)$

67. $\dfrac{\dfrac{1}{4} - \dfrac{1}{2}}{\dfrac{1}{3}}$ **68.** $\dfrac{\dfrac{3}{5} - \dfrac{7}{10}}{\dfrac{1}{2}}$

69. $-\dfrac{9}{4}\left(\dfrac{1}{2}\right) + \dfrac{3}{4} \div \dfrac{5}{6}$

70. $\left[-\dfrac{4}{7} - \left(-\dfrac{2}{5}\right)\right]\left[-\dfrac{3}{8} + \left(-\dfrac{1}{9}\right)\right]$

71. $\dfrac{\dfrac{7}{9} - 3}{\dfrac{5}{6}} \div \dfrac{3}{2} + \dfrac{3}{4}$

72. $\dfrac{\dfrac{17}{25}}{\dfrac{3}{5} - 4} \div \dfrac{1}{5} + \dfrac{1}{2}$

In Exercises 73–80, evaluate each algebraic expression for the given value of the variable.

73. $x^2 + 5x; x = 3$ **74.** $x^2 - 2x; x = 6$

75. $3x^2 - 8x; x = -2$ **76.** $4x^2 - 2x; x = -3$

77. $-x^2 - 10x; x = -1$ **78.** $-x^2 - 14x; x = -1$

79. $\dfrac{6y - 4y^2}{y^2 - 15}; y = 5$ **80.** $\dfrac{3y - 2y^2}{y(y - 2)}; y = 5$

In Exercises 81–88, simplify each algebraic expression by removing parentheses and brackets.

81. $3[5(x - 2) + 1]$

82. $4[6(x - 3) + 1]$

83. $3[6 - (y + 1)]$

84. $5[2 - (y + 3)]$

85. $7 - 4[3 - (4y - 5)]$

86. $6 - 5[8 - (2y - 4)]$

87. $2(3x^2 - 5) - [4(2x^2 - 1) + 3]$

88. $4(6x^2 - 3) - [2(5x^2 - 1) + 1]$

Practice PLUS

In Exercises 89–92, express each sentence as a single numerical expression. Then use the order of operations to simplify the expression.

89. Cube -2. Subtract this exponential expression from -10.

90. Cube -5. Subtract this exponential expression from -100.

91. Subtract 10 from 7. Multiply this difference by 2. Square this product.

92. Subtract 11 from 9. Multiply this difference by 2. Raise this product to the fourth power.

In Exercises 93–96, let x represent the number. Express each sentence as a single algebraic expression. Then simplify the expression.

93. Multiply a number by 5. Add 8 to this product. Subtract this sum from the number.

94. Multiply a number by 3. Add 9 to this product. Subtract this sum from the number.

95. Cube a number. Subtract 4 from this exponential expression. Multiply this difference by 5.

96. Cube a number. Subtract 6 from this exponential expression. Multiply this difference by 4.

Application Exercises

The United States has more people in prison, as well as more people in prison per capita, than any other western industrialized nation. The bar graph shows the number of inmates in U.S. state and federal prisons for five selected years from 1980 through 2004.

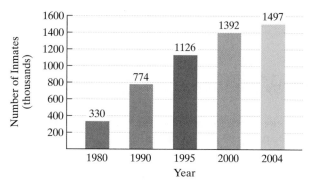

Number of Inmates in United States State and Federal Prisons

Source: U.S. Justice Department

The mathematical model

$$I = -0.09n^2 + 53n + 315$$

describes the number of inmates, I, in thousands, n years after 1980. Use this information to solve Exercises 97–98.

97. Use the formula to find the number of inmates, in thousands, in 1990. Does this underestimate or overestimate the actual number shown by the graph? By how much?

98. Use the formula to find the number of inmates, in thousands, in 2000. Does this underestimate or overestimate the actual number shown by the graph? By how much?

The line graphs show the percentage of people who used the Internet for four selected years, by level of education. Exercises 99–101 involve mathematical models for the data.

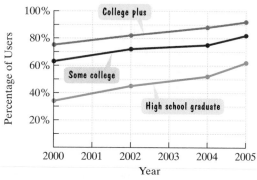

United States Internet Users, by Level of Education

Source: Pew Internet and American Life Project

99. a. Use the appropriate line graph to estimate the percentage of people at the college plus level of education who used the Internet in 2005.

b. The mathematical model

$$C = 0.006n^2 + 3.3n + 75$$

describes the percentage of people in the college plus group, C, who used the Internet n years after 2000. Use the formula to find the percentage of people at this education level who used the Internet in 2005. How does this value compare to your estimate in part (a)?

100. a. Use the appropriate line graph to estimate the percentage of people with some college who used the Internet in 2004.

b. The mathematical model

$$S = -0.01n^2 + 3.5n + 63.5$$

describes the percentage of people with some college, S, who used the Internet n years after 2000. Use the formula to find the percentage of people at this education level who used the Internet in 2004. How does this value compare to your estimate in part (a)?

101. a. Use the appropriate line graph to estimate the percentage of people at the high school graduate level of education who used the Internet for each of the four years shown.

b. Based on your estimates in part (a), select the correct mathematical model, model 1 or model 2, that describes the percentage of people in the high school graduate group, H, who used the Internet n years after 2000.

Model 1 $H = 0.41n^2 + 2.1n + 17$

Model 2 $H = 0.26n^2 + 3.9n + 34.5$

In Palo Alto, California, a government agency ordered computer-related companies to contribute to a pool of money to clean up underground water supplies. (The companies had stored toxic chemicals in leaking underground containers.) The mathematical model

$$C = \frac{200x}{100 - x}$$

describes the cost, C, in tens of thousands of dollars, for removing x percent of the contaminants. Use this formula to solve Exercises 102–103.

102. a. Find the cost, in tens of thousands of dollars, for removing 50% of the contaminants.

b. Find the cost, in tens of thousands of dollars, for removing 80% of the contaminants.

c. Describe what is happening to the cost of the cleanup as the percentage of contaminant removed increases.

103. a. Find the cost, in tens of thousands of dollars, for removing 60% of the contaminants.

b. Find the cost, in tens of thousands of dollars, for removing 90% of the contaminants.

c. Describe what is happening to the cost of the cleanup as the percentage of contaminants removed increases.

Writing in Mathematics

104. Describe what it means to raise a number to a power. In your description, include a discussion of the difference between -5^2 and $(-5)^2$.

105. Explain how to simplify $4x^2 + 6x^2$. Why is the sum not equal to $10x^4$?

106. Why is the order of operations agreement needed?

Critical Thinking Exercises

Make Sense? *In Exercises 107–110, determine whether each statement "makes sense" or "does not make sense" and explain your reasoning.*

107. Without parentheses, an exponent has only the number next to it as its base.

108. I read that a certain star is 10^4 light-years from Earth, which means 100,000 light-years.

109. When I evaluated $(-1)^n$, I obtained positive numbers when n was even and negative numbers when n was odd.

110. The rules for the order of operations avoid the confusion of obtaining different results when I simplify the same expression.

In Exercises 111–114, determine whether each statement is true or false. If the statement is false, make the necessary change(s) to produce a true statement.

111. If x is -3, then the value of $-3x - 9$ is -18.

112. The algebraic expression $\dfrac{6x + 6}{x + 1}$ cannot have the same value when two different replacements are made for x such as $x = -3$ and $x = 2$.

113. The value of $\dfrac{|3 - 7| - 2^3}{(-2)(-3)}$ is the fraction that results when $\dfrac{1}{3}$ is subtracted from $-\dfrac{1}{3}$.

114. $-2(6 - 4^2)^3 = -2(6 - 16)^3$
$= -2(-10)^3 = (-20)^3 = -8000$

115. Simplify: $\dfrac{1}{4} - 6(2 + 8) \div \left(-\dfrac{1}{3}\right)\left(-\dfrac{1}{9}\right).$

In Exercises 116–117, insert parentheses in each expression so that the resulting value is 45.

116. $2 \cdot 3 + 3 \cdot 5$

117. $2 \cdot 5 - \dfrac{1}{2} \cdot 10 \cdot 9$

Review Exercises

118. Simplify: $-8 - 2 - (-5) + 11$. (Section 1.6, Example 3)

119. Multiply: $-4(-1)(-3)(2)$. (Section 1.7, Example 2)

120. Give an example of a real number that is not an irrational number. (Section 1.3, Example 5).

Preview Exercises

Exercises 121–123 will help you prepare for the material covered in the first section of the next chapter. In each exercise, determine whether the given number is a solution of the equation.

121. $-\dfrac{1}{2} = x - \dfrac{2}{3}; \dfrac{1}{6}$

122. $5y + 3 - 4y - 8 = 15; 20$

123. $4x + 2 = 3(x - 6) + 8; -11$

GROUP PROJECT

CHAPTER 1

One measure of physical fitness is your *resting heart rate*. Generally speaking, the more fit you are, the lower your resting heart rate. The best time to take this measurement is when you first awaken in the morning, before you get out of bed. Lie on your back with no body parts crossed and take your pulse in your neck or wrist. Use your index and second fingers and count your pulse beat for one full minute to get your resting heart rate. A resting heart rate under 48 to 57 indicates high fitness, 58 to 62, above average fitness, 63 to 70, average fitness, 71 to 82, below average fitness, and 83 or more, low fitness.

(The project continues on the next page.)

Another measure of physical fitness is your percentage of body fat. You can estimate your body fat using the following formulas:

For men: Body fat $= -98.42 + 4.15w - 0.082b$

For women: Body fat $= -76.76 + 4.15w - 0.082b$

where w = waist measurement, in inches, and b = total body weight, in pounds. Then divide your body fat by your total weight to get your body fat percentage. For men, less than 15% is considered athletic, 25% about average. For women, less than 22% is considered athletic, 30% about average.

Each group member should bring his or her age, resting heart rate, and body fat percentage to the group. Using the data, the group should create three graphs.

a. Create a graph that shows age and resting heart rate for group members.

b. Create a graph that shows age and body fat percentage for group members.

c. Create a graph that shows resting heart rate and body fat percentage for group members.

For each graph, select the style (line or bar) that is most appropriate.

Chapter 1 Summary

Definitions and Concepts	Examples

Section 1.1 Introduction to Algebra: Variables and Mathematical Models

A letter that represents a variety of different numbers is called a variable. An algebraic expression is a combination of variables, numbers, and operation symbols. To evaluate an algebraic expression, substitute a given number for the variable and simplify:

1. Perform calculations within parentheses first.
2. Perform multiplication before addition or subtraction.

(A more detailed order of operations is given in Section 1.8.)

Evaluate $6(x - 3) + 4x$ for $x = 5$.

Replace x with 5.

$$6(5 - 3) + 4 \cdot 5$$
$$= 6(2) + 4 \cdot 5$$
$$= 12 + 20$$
$$= 32$$

Here are some key words for translating into algebraic expressions:

- Addition: plus, sum, more than, increased by
- Subtraction: minus, difference, less than, decreased by
- Multiplication: times, product, twice, multiplied by
- Division: divide, quotient, ratio, divided by

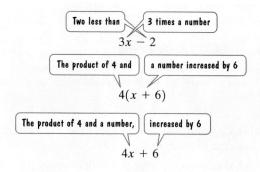

Two less than / 3 times a number

$$3x - 2$$

The product of 4 and / a number increased by 6

$$4(x + 6)$$

The product of 4 and a number, / increased by 6

$$4x + 6$$

An equation is a statement that two algebraic expressions are equal. Solutions of an equation are values of the variable that make the equation a true statement. To determine whether a number is a solution, substitute that number for the variable and evaluate each side of the equation. If the values on both sides of the equation are the same, the number is a solution.

Is 5 a solution of

$$3(3x + 5) = 10x + 10?$$

Substitute 5 for x.

$$3(3 \cdot 5 + 5) \stackrel{?}{=} 10 \cdot 5 + 10$$
$$3(15 + 5) \stackrel{?}{=} 50 + 10$$
$$3(20) \stackrel{?}{=} 60$$
$$60 = 60, \quad \text{true}$$

The true statement indicates that 5 is a solution.

Definitions and Concepts	**Examples**

Section 1.1 Introduction to Algebra: Variables and Mathematical Models (continued)

A formula is an equation that expresses a relationship between two or more variables. The process of finding formulas to describe real-world phenomena is called mathematical modeling. Such formulas, together with the meaning assigned to the variables, are called mathematical models. These formulas model, or describe, the relationship among the variables.

The formulas
$$M = 0.15n + 12.5$$
and
$$W = -0.44n + 39.6$$
model the time each week that men, M, and women, W, devoted to housework n years after 1965.
(*Source*: Robinson and Godbey, *Time for Life: Surprising Ways Americans Use Their Time*, Pennsylvania State University Press)

Section 1.2 Fractions in Algebra

Mixed Numbers and Improper Fractions

A mixed number consists of the addition of a natural number $(1, 2, 3, \dots)$ and a fraction, expressed without the use of an addition sign. An improper fraction has a numerator that is greater than its denominator. To convert a mixed number to an improper fraction, multiply the denominator by the natural number and add the numerator. Then place this result over the original denominator.

To convert an improper fraction to a mixed number, divide the denominator into the numerator and write the mixed number using

$$\text{quotient} \frac{\text{remainder}}{\text{original denominator}}.$$

Convert $5\frac{3}{7}$ to an improper fraction.
$$5\frac{3}{7} = \frac{7 \cdot 5 + 3}{7} = \frac{35 + 3}{7} = \frac{38}{7}$$

Convert $\frac{14}{3}$ to a mixed number.

$$\frac{14}{3} = 4\frac{2}{3} \qquad \begin{array}{r} 4 \\ 3\overline{)14} \\ \underline{12} \\ 2 \end{array}$$

A prime number is a natural number greater than 1 that has only itself and 1 as factors. A composite number is a natural number greater than 1 that is not a prime number. The prime factorization of a composite number is the expression of the composite number as the product of prime numbers.

Find the prime factorization:
$$\begin{aligned} 60 &= 6 \cdot 10 \\ &= 2 \cdot 3 \cdot 2 \cdot 5 \end{aligned}$$

A fraction is reduced to its lowest terms when the numerator and denominator have no common factors other than 1. To reduce a fraction to its lowest terms, divide both the numerator and the denominator by their greatest common factor. The greatest common factor can be found by inspection or prime factorizations of the numerator and the denominator.

Reduce to lowest terms:
$$\frac{8}{14} = \frac{\cancel{2} \cdot 4}{\cancel{2} \cdot 7} = \frac{4}{7}$$

Multiplying Fractions

The product of two or more fractions is the product of their numerators divided by the product of their denominators.

Multiply:
$$\frac{2}{7} \cdot \frac{5}{9} = \frac{2 \cdot 5}{7 \cdot 9} = \frac{10}{63}$$

Dividing Fractions

The quotient of two fractions is the first multiplied by the reciprocal (or multiplicative inverse) of the second.

Divide:
$$\frac{4}{9} \div \frac{3}{7} = \frac{4}{9} \cdot \frac{7}{3} = \frac{4 \cdot 7}{9 \cdot 3} = \frac{28}{27}$$

Adding and Subtracting Fractions with Identical Denominators

Add or subtract numerators. Put this result over the common denominator.

Subtract:
$$\frac{5}{8} - \frac{3}{8} = \frac{5 - 3}{8} = \frac{2}{8} = \frac{\cancel{2} \cdot 1}{\cancel{2} \cdot 4} = \frac{1}{4}$$

Definitions and Concepts	**Examples**

Section 1.2 Fractions in Algebra (continued)

Adding and Subtracting Fractions with Unlike Denominators

Rewrite the fractions as equivalent fractions with the least common denominator. Then add or subtract numerators, putting this result over the common denominator.

Add:

$$\frac{3}{8} + \frac{5}{12} = \frac{3 \cdot 3}{8 \cdot 3} + \frac{5 \cdot 2}{12 \cdot 2}$$

The LCD is 24. $= \dfrac{9}{24} + \dfrac{10}{24} = \dfrac{19}{24}$

Fractions appear throughout algebra. Many formulas and mathematical models contain fractions. Operations with fractions can be used to determine whether a particular fraction is a solution of an equation. When used with fractions, the word *of* represents multiplication.

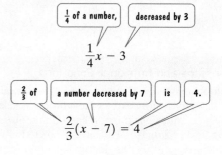

$\frac{1}{4}$ of a number, decreased by 3

$$\frac{1}{4}x - 3$$

$\frac{2}{3}$ of a number decreased by 7 is 4.

$$\frac{2}{3}(x - 7) = 4$$

Section 1.3 The Real Numbers

A set is a collection of objects, called elements, whose contents can be clearly determined.

$$\{a, b, c\}$$

A line used to visualize numbers is called a number line.

<number line from −4 to 4>
−4 −3 −2 −1 0 1 2 3 4

Real Numbers: the set of all numbers that can be represented by points on the number line

The Sets That Make Up the Real Numbers
- Natural Numbers: $\{1, 2, 3, 4, \dots\}$
- Whole Numbers: $\{0, 1, 2, 3, 4, \dots\}$
- Integers: $\{\dots, -3, -2, -1, 0, 1, 2, 3, \dots\}$
- Rational Numbers: the set of numbers that can be expressed as the quotient of an integer and a nonzero integer; can be expressed as terminating or repeating decimals
- Irrational Numbers: the set of numbers that cannot be expressed as the quotient of integers; decimal representations neither terminate nor repeat.

Given the set

$$\left\{-1.4, 0, 0.\overline{7}, \frac{9}{10}, \sqrt{2}, \sqrt{4}\right\}$$

list the

- natural numbers: $\sqrt{4}$, or 2
- whole numbers: $0, \sqrt{4}$
- integers: $0, \sqrt{4}$
- rational numbers: $-1.4, 0, 0.\overline{7}, \frac{9}{10}, \sqrt{4}$
- irrational numbers: $\sqrt{2}$
- real numbers:

$$-1.4, 0, 0.\overline{7}, \frac{9}{10}, \sqrt{2}, \sqrt{4}$$

For any two real numbers, a and b, a is less than b if a is to the left of b on the number line.

Inequality Symbols
$<$: is less than $>$: is greater than
$\leq$: is less than or equal to $\geq$: is greater than or equal to

<number line from −4 to 4>
−4 −3 −2 −1 0 1 2 3 4

$-2 < 0$ $0 > -2$
$0 < 2.5$ $2.5 > 0$

The absolute value of a, written $|a|$, is the distance from 0 to a on the number line.

$|4| = 4$ $|0| = 0$ $|-6| = 6$

Definitions and Concepts

Examples

Section 1.4 Basic Rules of Algebra

Terms of an algebraic expression are separated by addition. The parts of each term that are multiplied are its factors. The numerical part of a term is its coefficient. Like terms have the same variable factors raised to the same powers.

- An expression with three terms:

The like terms are $8x$ and $6x$.

Properties of Real Numbers and Algebraic Expressions

- Commutative Properties:
$$a + b = b + a$$
$$ab = ba$$

- Associative Properties:
$$(a + b) + c = a + (b + c)$$
$$(ab)c = a(bc)$$

- Distributive Properties:
$$a(b + c) = ab + ac$$
$$(b + c)a = ba + ca$$
$$a(b - c) = ab - ac$$
$$(b - c)a = ba - ca$$
$$a(b + c + d) = ab + ac + ad$$

Commutative of Addition:
$$5x + 4 = 4 + 5x$$

Commutative of Multiplication:
$$5x + 4 = x5 + 4$$

Associative of Addition:
$$6 + (4 + x) = (6 + 4) + x = 10 + x$$

Associative of Multiplication:
$$7(10x) = (7 \cdot 10)x = 70x$$

Distributive:
$$8(x + 5 + 4y) = 8x + 40 + 32y$$

Distributive to Combine Like Terms:
$$8x + 12x = (8 + 12)x = 20x$$

Simplifying Algebraic Expressions

Use the distributive property to remove grouping symbols. Then combine like terms.

$$4(5x + 7) + 13x$$
$$= 20x + 28 + 13x$$
$$= (20x + 13x) + 28$$
$$= 33x + 28$$

Section 1.5 Addition of Real Numbers

Sums on a Number Line

To find $a + b$, the sum of a and b, on a number line, start at a. If b is positive, move b units to the right. If b is negative, move b units to the left. If b is 0, stay at a. The number where we finish on the number line represents $a + b$.

$$-7 + 5 = -2$$

Start at −7. Move 5 units to the right.

$$\begin{array}{ccccccccccc} & & & & & & & & & & \\ -10 & -9 & -8 & -7 & -6 & -5 & -4 & -3 & -2 & -1 & 0 \end{array}$$

Additive inverses, or opposites, are pairs of real numbers that are the same number of units from zero on the number line, but on opposite sides of zero.

- Identity Property of Addition:
$$a + 0 = 0 \qquad 0 + a = a$$

- Inverse Property of Addition:
$$a + (-a) = 0 \qquad (-a) + a = 0$$

The additive inverse (or opposite) of 4 is -4. The additive inverse of -1.7 is 1.7.

Identity Property of Addition:
$$4x + 0 = 4x$$

Inverse Property of Addition:
$$4x + (-4x) = 0$$

Definitions and Concepts	**Examples**

Section 1.5 Addition of Real Numbers (continued)

Addition without a Number Line

To add two numbers with the same sign, add their absolute values and use their common sign. To add two numbers with different signs, subtract the smaller absolute value from the greater absolute value and use the sign of the number with the greater absolute value.

Add:

$$10 + 4 = 14$$
$$-4 + (-6) = -10$$
$$-30 + 5 = -25$$
$$12 + (-8) = 4$$

To add a series of positive and negative numbers, add all the positive numbers and add all the negative numbers. Then add the resulting positive and negative sums.

$$5 + (-3) + (-7) + 2$$
$$= (5 + 2) + [(-3) + (-7)]$$
$$= 7 + (-10)$$
$$= -3$$

Section 1.6 Subtraction of Real Numbers

To subtract b from a, add the opposite, or additive inverse, of b to a:

$$a - b = a + (-b).$$

The result is called the difference between a and b.

Subtract:

$$-7 - (-5) = -7 + 5 = -2$$
$$-\frac{3}{4} - \frac{1}{2} = -\frac{3}{4} + \left(-\frac{1}{2}\right)$$
$$= -\frac{3}{4} + \left(-\frac{2}{4}\right) = -\frac{5}{4}$$

To simplify a series of additions and subtractions, change all subtractions to additions of opposites. Then use the procedure for adding a series of positive and negative numbers.

Simplify:

$$-6 - 2 - (-3) + 10$$
$$= -6 + (-2) + 3 + 10$$
$$= -8 + 13$$
$$= 5$$

Section 1.7 Multiplication and Division of Real Numbers

The result of multiplying a and b, ab, is called the product of a and b. If the two numbers have different signs, the product is negative. If the two numbers have the same sign, the product is positive. If either number is 0, the product is 0.

Multiply:

$$-5(-10) = 50$$
$$\frac{3}{4}\left(-\frac{5}{7}\right) = -\frac{3}{4}\cdot\frac{5}{7} = -\frac{15}{28}$$

Assuming that no number is 0, the product of an even number of negative numbers is positive. The product of an odd number of negative numbers is negative. If any number is 0, the product is 0.

Multiply:

$$(-3)(-2)(-1)(-4) = 24$$
$$(-3)(2)(-1)(-4) = -24$$

The result of dividing the real number a by the nonzero real number b is called the quotient of a and b. If two numbers have different signs, their quotient is negative. If two numbers have the same sign, their quotient is positive. Division by zero is undefined.

Divide:

$$\frac{21}{-3} = -7$$
$$-\frac{1}{3} \div (-3) = \frac{1}{3}\cdot\frac{1}{3} = \frac{1}{9}$$

Definitions and Concepts	**Examples**

Section 1.7 Multiplication and Division of Real Numbers (continued)

Two numbers whose product is 1 are called multiplicative inverses or reciprocals of each other. The number 0 has no multiplicative inverse.	The multiplicative inverse of 4 is $\frac{1}{4}$.
	The multiplicative inverse of $-\frac{1}{3}$ is -3.

- Identity Property of Multiplication

$$a \cdot 1 = a \qquad 1 \cdot a = a$$

- Inverse Property of Multiplication
  If a is not 0:

$$a \cdot \frac{1}{a} = 1 \qquad \frac{1}{a} \cdot a = 1$$

- Multiplication Property of -1

$$-1a = -a \qquad a(-1) = -a$$

- Double Negative Property

$$-(-a) = a$$

Simplify:

$$1x = x$$

$$7x \cdot \frac{1}{7x} = 1, \quad x \neq 0$$

$$4x - 5x = -1x = -x$$

$$-(-7y) = 7y$$

If a negative sign precedes parentheses, remove parentheses and change the sign of every term within parentheses.

Simplify:

$$-(7x - 3y + 2) = -7x + 3y - 2$$

Section 1.8 Exponents and Order of Operations

If b is a real number and n is a natural number, b^n, the nth power of b, is the product of n factors of b. Furthermore, $b^1 = b$.

Evaluate:

$$8^2 = 8 \cdot 8 = 64$$
$$(-5)^3 = (-5)(-5)(-5) = -125$$

Order of Operations

1. Perform operations within grouping symbols, starting with the innermost grouping symbols. Grouping symbols include parentheses, brackets, fraction bars, absolute value symbols, and square root signs.
2. Evaluate exponential expressions.
3. Multiply and divide, from left to right.
4. Add and subtract. In this step, you have a choice. You can add and subtract in order from left to right. You can also rewrite subtractions as additions of opposites, combine positive and negative numbers separately, and then add these results.

Simplify:

$$5(4 - 6)^2 - 2(1 - 3)^3$$
$$= 5(-2)^2 - 2(-2)^3$$
$$= 5(4) - 2(-8)$$
$$= 20 - (-16)$$
$$= 20 + 16 = 36$$

Some algebraic expressions contain two sets of grouping symbols: parentheses, the inner grouping symbols, and brackets, the outer grouping symbols. To simplify such expressions, use the order of operations and remove grouping symbols from innermost (parentheses) to outermost (brackets).

Simplify:

$$5 - 3[2(x + 1) - 7]$$
$$= 5 - 3[2x + 2 - 7]$$
$$= 5 - 3[2x - 5]$$
$$= 5 - 6x + 15$$
$$= -6x + 20$$

CHAPTER 1 REVIEW EXERCISES

1.1   *In Exercises 1–2, evaluate each expression for x = 6.*

1. $10 + 5x$

2. $8(x - 2) + 3x$

In Exercises 3–4, evaluate each expression for x = 8 and y = 10.

3. $\dfrac{40}{x} - \dfrac{y}{5}.$

4. $3(2y + x)$

In Exercises 5–8, translate from English to an algebraic expression or equation, whichever is appropriate. Let the variable x represent the number.

5. Six subtracted from the product of 7 and a number

6. The quotient of a number and 5, decreased by 2, is 18.

7. Nine less twice a number is 14.

8. The product of 3 and 7 more than a number

In Exercises 9–11, determine whether the given number is a solution of the equation.

9. $4x + 5 = 13; 3$

10. $2y + 7 = 4y - 5; 6$

11. $3(w + 1) + 11 = 2(w + 8); 2$

Exercises 12–13 involve an experiment on memory. Students in a language class are asked to memorize 40 vocabulary words in Latin, a language with which they are not familiar. After studying the words for one day, students are tested each day after to see how many words they remember. The class average is taken and the results are shown as ten points on a line graph.

Average Number of Words Remembered Over Time

12. Use the line graph to estimate the average number of Latin words the class remembered after 5 days.

13. The mathematical model

$$L = \frac{5n + 30}{n}$$

describes the average number of Latin words remembered by the students, L, after n days. Use the formula to find the average number of words remembered after 5 days. How well does this compare with your estimate from Exercise 12?

14. Great expectations for sophomores: The bar graph shows an increase in the percentage of U.S. high school sophomores who expect to earn a college degree.

Percentage of United States High School Sophomores Expecting to Earn a College Degree

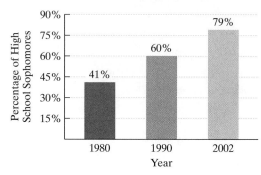

Source: U.S. Department of Education

Here is a mathematical model that approximates the data displayed by the bar graph:

$$p = 1.7n + 42.$$

Number of years after 1980

Percentage of sophomores expecting to earn a college degree

a. Use the formula to find the percentage of sophomores who expected to earn a college degree in 1990. Does the mathematical model underestimate or overestimate the actual percent shown by the bar graph? By how much?

b. Use the formula to project the percentage of sophomores who will expect to earn a college degree in 2010.

1.2   *In Exercises 15–16, convert each mixed number to an improper fraction.*

15. $3\dfrac{2}{7}$

16. $5\dfrac{9}{11}$

In Exercises 17–18, convert each improper fraction to a mixed number.

17. $\dfrac{17}{9}$

18. $\dfrac{27}{5}$

In Exercises 19–21, identify each natural number as prime or composite. If the number is composite, find its prime factorization.

19. 60

20. 63

21. 67

In Exercises 22–23, simplify each fraction by reducing it to its lowest terms.

22. $\dfrac{15}{33}$

23. $\dfrac{40}{75}$

In Exercises 24–29, perform the indicated operation. Where possible, reduce the answer to its lowest terms.

24. $\dfrac{3}{5} \cdot \dfrac{7}{10}$ **25.** $\dfrac{4}{5} \div \dfrac{3}{10}$ **26.** $1\dfrac{2}{3} \div 6\dfrac{2}{3}$

27. $\dfrac{2}{9} + \dfrac{4}{9}$ **28.** $\dfrac{5}{6} + \dfrac{7}{9}$ **29.** $\dfrac{3}{4} - \dfrac{2}{15}$

In Exercises 30–31, determine whether the given number is a solution of the equation.

30. $x - \dfrac{3}{4} = \dfrac{7}{4}; \; 2\dfrac{1}{2}$

31. $\dfrac{2}{3}w = \dfrac{1}{15}w + \dfrac{3}{5}; \; 2$

In Exercises 32–33, translate from English to an algebraic expression or equation, whichever is appropriate. Let the variable x represent the number.

32. Two decreased by half of a number is $\frac{1}{4}$ of the number.

33. $\frac{3}{5}$ of a number increased by 6

34. Suppose that the target heart rate, H, in beats per minute, for your exercise goal is given by

$$H = \frac{4}{5}(220 - a),$$

where a is your age. If you are 30 years old, what is your target heart rate?

1.3 *In Exercises 35–36, graph each real number on a number line.*

35. -2.5 **36.** $4\dfrac{3}{4}$

In Exercises 37–38, express each rational number as a decimal.

37. $\dfrac{5}{8}$ **38.** $\dfrac{3}{11}$

39. Consider the set

$$\left\{ -17, -\frac{9}{13}, 0, 0.75, \sqrt{2}, \pi, \sqrt{81} \right\}.$$

List all numbers from the set that are: **a.** natural numbers, **b.** whole numbers, **c.** integers, **d.** rational numbers, **e.** irrational numbers, **f.** real numbers.

40. Give an example of an integer that is not a natural number.

41. Give an example of a rational number that is not an integer.

42. Give an example of a real number that is not a rational number.

In Exercises 43–46, insert either $<$ or $>$ in the shaded area between each pair of numbers to make a true statement.

43. -93 ▨ 17 **44.** -2 ▨ -200

45. 0 ▨ $-\dfrac{1}{3}$ **46.** $-\dfrac{1}{4}$ ▨ $-\dfrac{1}{5}$

In Exercises 47–48, determine whether each inequality is true or false.

47. $-13 \geq -11$ **48.** $-126 \leq -126$

In Exercises 49–50, find each absolute value.

49. $|-58|$ **50.** $|2.75|$

1.4

51. Use the commutative property of addition to write an equivalent algebraic expression: $7 + 13y$.

52. Use the commutative property of multiplication to write an equivalent algebraic expression: $9(x + 7)$.

In Exercises 53–54, use an associative property to rewrite each algebraic expression. Then simplify the resulting algebraic expression.

53. $6 + (4 + y)$

54. $7(10x)$

55. Use the distributive property to rewrite without parentheses: $6(4x - 2 + 5y)$.

In Exercises 56–57, simplify each algebraic expression.

56. $4a + 9 + 3a - 7$

57. $6(3x + 4) + 5(2x - 1)$

1.5

58. Use a number line to find the sum: $-6 + 8$.

In Exercises 59–61, find each sum without the use of a number line.

59. $8 + (-11)$ **60.** $-\dfrac{3}{4} + \dfrac{1}{5}$

61. $7 + (-5) + (-13) + 4$

In Exercises 62–63, simplify each algebraic expression.

62. $8x + (-6y) + (-12x) + 11y$

63. $10(3y + 4) + (-40y)$

64. The Dead Sea is the lowest elevation on Earth, 1312 feet below sea level. If a person is standing 512 feet above the Dead Sea, what is that person's elevation?

65. The water level of a reservoir is measured over a five-month period. At the beginning, the level is 25 feet. During this time, the level fell 3 feet, then rose 2 feet, then rose 1 foot, then fell 4 feet, and then rose 2 feet. What is the reservoir's water level at the end of the five months?

1.6

66. Rewrite $9 - 13$ as the addition of an opposite.

In Exercises 67–69, perform the indicated subtraction.

67. $-9 - (-13)$ **68.** $-\dfrac{7}{10} - \dfrac{1}{2}$

69. $-3.6 - (-2.1)$

In Exercises 70–71, simplify each series of additions and subtractions.

70. $-7 - (-5) + 11 - 16$

71. $-25 - 4 - (-10) + 16$

72. Simplify: $3 - 6a - 8 - 2a$.

73. What is the difference in elevation between a plane flying 26,500 feet above sea level and a submarine traveling 650 feet below sea level?

1.7 *In Exercises 74–76, perform the indicated multiplication.*

74. $-7(-12)$

75. $\frac{3}{5}\left(-\frac{5}{11}\right)$

76. $5(-3)(-2)(-4)$

In Exercises 77–79, perform the indicated division or state that the expression is undefined.

77. $\frac{45}{-5}$

78. $-17 \div 0$

79. $-\frac{4}{5} \div \left(-\frac{2}{5}\right)$

In Exercises 80–81, simplify each algebraic expression.

80. $-4\left(-\frac{3}{4}x\right)$

81. $-3(2x - 1) - (4 - 5x)$

In Exercises 82–83, determine whether the given number is a solution of the equation.

82. $5x + 16 = -8 - x;\ -6$

83. $2(x + 3) - 18 = 5x;\ -4$

84. The bar graph shows fewer adolescents doing time.

Number of Inmates Younger Than 18 Held in United States State Prisons

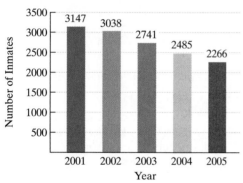

Source: U.S. Department of Justice

The number of inmates younger than 18 held in U.S. state prisons, I, can be described by the mathematical model

$$I = -232n + 3430,$$

where n is the number of years after 2000.

a. Use the formula to find the number of inmates younger than 18 in 2003. Does the mathematical model underestimate or overestimate the actual number of inmates shown by the bar graph? By how much?

b. Use the formula to project the number of inmates younger than 18 in 2010.

1.8 *In Exercises 85–87, evaluate each exponential expression.*

85. $(-6)^2$

86. -6^2

87. $(-2)^5$

In Exercises 88–89, simplify each algebraic expression, or explain why the expression cannot be simplified.

88. $4x^3 + 2x^3$

89. $4x^3 + 4x^2$

In Exercises 90–98, use the order of operations to simplify each expression.

90. $-40 \div 5 \cdot 2$

91. $-6 + (-2) \cdot 5$

92. $6 - 4(-3 + 2)$

93. $28 \div (2 - 4^2)$

94. $36 - 24 \div 4 \cdot 3 - 1$

95. $-8[-4 - 5(-3)]$

96. $\frac{6(-10 + 3)}{2(-15) - 9(-3)}$

97. $\left(\frac{1}{2} + \frac{1}{3}\right) \div \left(\frac{1}{4} - \frac{3}{8}\right)$

98. $\frac{1}{2} - \frac{2}{3} \div \frac{5}{9} + \frac{3}{10}$

In Exercises 99–100, evaluate each algebraic expression for the given value of the variable.

99. $x^2 - 2x + 3;\ x = -1$

100. $-x^2 - 7x;\ x = -2$

In Exercises 101–102, simplify each algebraic expression.

101. $4[7(a - 1) + 2]$

102. $-6[4 - (y + 2)]$

103. The graph shows a drop in volunteerism.

Number of Americans Who Volunteered at Least Once

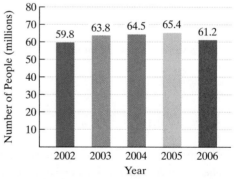

Source: U.S. Bureau of Labor Statistics

The number of Americans who volunteered at least once, V, in millions, can be described by the mathematical model

$$V = -1.2n^2 + 10n + 45,$$

where n is the number of years after 2000. Use the formula to find the number of volunteers in 2005. Does the mathematical model underestimate or overestimate the actual number of volunteers shown by the bar graph? By how much?

In Exercises 1–10, perform the indicated operation or operations.

1. $1.4 - (-2.6)$

2. $-9 + 3 + (-11) + 6$

3. $3(-17)$

4. $\left(-\dfrac{3}{7}\right) \div \left(-\dfrac{15}{7}\right)$

5. $\left(3\dfrac{1}{3}\right)\left(-1\dfrac{3}{4}\right)$

6. $-50 \div 10$

7. $-6 - (5 - 12)$

8. $(-3)(-4) \div (7 - 10)$

9. $(6 - 8)^2(5 - 7)^3$

10. $\dfrac{3(-2) - 2(2)}{-2(8 - 3)}$

In Exercises 11–13, simplify each algebraic expression.

11. $11x - (7x - 4)$

12. $5(3x - 4y) - (2x - y)$

13. $6 - 2[3(x + 1) - 5]$

14. List all the rational numbers in this set.

$$\left\{-7, -\dfrac{4}{5}, 0, 0.25, \sqrt{3}, \sqrt{4}, \dfrac{22}{7}, \pi\right\}$$

15. Insert either $<$ or $>$ in the shaded area to make a true statement: $-1 \;\rule{1.2em}{0.8em}\; -100$.

16. Find the absolute value: $|-12.8|$.

In Exercises 17–18, evaluate each algebraic expression for the given value of the variable.

17. $5(x - 7); x = 4$

18. $x^2 - 5x; x = -10$

19. Use the commutative property of addition to write an equivalent algebraic expression: $2(x + 3)$.

20. Use the associative property of multiplication to rewrite $-6(4x)$. Then simplify the expression.

21. Use the distributive property to rewrite without parentheses: $7(5x - 1 + 2y)$.

22. What is the difference in elevation between a plane flying 16,200 feet above sea level and a submarine traveling 830 feet below sea level?

In Exercises 23–24, determine whether the given number is a solution of the equation.

23. $\dfrac{1}{5}(x + 2) = \dfrac{1}{10}x + \dfrac{3}{5}; 3$

24. $3(x + 2) - 15 = 4x; -9$

In Exercises 25–26, translate from English to an algebraic expression or equation, whichever is appropriate. Let the variable x represent the number.

25. $\frac{1}{4}$ of a number, decreased by 5, is 32.

26. Seven subtracted from the product of 5 and 4 more than a number

27. *The Sopranos* saga showed just how complex and involving TV storytelling could be. The bar graph shows the number of viewers in the series opening episodes.

Number of Viewers of Opening Episodes of *The Sopranos*

Source: Nielsen Media Research

The number of viewers of the opening episodes of *The Sopranos*, V, in millions, can be described by the mathematical model

$$V = -n^2 + 8n - 4,$$

where n is the season number. Use the formula to find the number of viewers of the opening episode of season 3. Does the mathematical model underestimate or overestimate the actual number of viewers shown by the bar graph? By how many million?

28. Electrocardiograms are used in exercise stress tests to determine a person's fitness for strenuous exercise. The target heart rate, in beats per minute, for such tests depends on a person's age. The line graph shows target heart rates for stress tests for people of various ages.

Target Heart Rates for Stress Tests

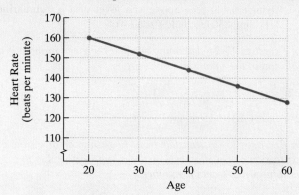

Use the graph to estimate the target heart rate for a 40-year-old taking a stress test.

29. The formula $H = \frac{4}{5}(220 - a)$ gives the target heart rate, H, in beats per minute, on a stress test for a person of age a. Use the formula to find the target heart rate for a 40-year-old. How does this compare with your estimate from Example 28?

Linear Equations and Inequalities in One Variable

"In high school, I could not pass a math test.
I couldn't pass a drug test either.
There may be a correlation."
–Lynda Montgomery

The belief that humor and laughter can have positive effects on our lives is not new. The Bible tells us, "A merry heart doeth good like a medicine, but a broken spirit drieth the bones." (Proverbs 17:22)

Some random humor factoids: • The average adult laughs 15 times each day. (Newhouse News Service) • Forty-six percent of people who are telling a joke laugh more than the people they are telling it to. (*U.S. News and World Report*) • Jews tell the most jokes about themselves. Japanese tell the fewest. (Davies, *The Mirth of Nations*) • Algebra can be used to model the influence that humor plays in our responses to negative life events. (Bob Blitzer, *Introductory Algebra*)

"As long as there is algebra,
there will be prayer in school."
–Larry Miller

That last tidbit that your author threw into the list is true. Based on our sense of humor, there is actually a formula that predicts how we will respond to difficult life events.

Formulas can be used to explain what is happening in the present and to make predictions about what might occur in the future. In this chapter, you will learn to use formulas in new ways that will help you to recognize patterns, logic, and order in a world that can appear chaotic to the untrained eye.

A mathematical model that includes sense of humor as a variable is developed in Example 7 of Section 2.3.

2.1

Objectives

1 Identify linear equations in one variable.

2 Use the addition property of equality to solve equations.

3 Solve applied problems using formulas.

The Addition Property of Equality

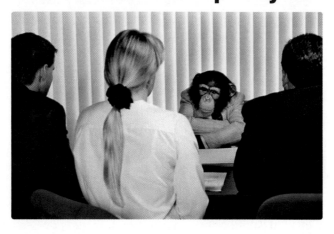

"I had the meanest boss in the world, so I would call in sick a lot. I would say I had 'female problems.' My boss didn't know I meant her."
—*Comedian Wendy Liebman*

Job satisfaction, or a positive feeling toward one's job, is linked to productivity and absenteeism. In this section's exercise set, you will work with a mathematical model that shows a nationwide decline in job satisfaction from 1987 through 2006. Your work with this model will involve *solving linear equations.*

Linear Equations in One Variable

1 Identify linear equations in one variable.

In Chapter 1, we learned that an equation is a statement that two algebraic expressions are equal. We determined whether a given number is an equation's solution by substituting that number for each occurrence of the variable. When the substitution resulted in a true statement, that number was a solution. When the substitution resulted in a false statement, that number was not a solution.

In the next three sections, we will study how to solve equations in one variable. **Solving an equation** is the process of finding the number (or numbers) that make the equation a true statement. These numbers are called the **solutions**, or **roots**, of the equation, and we say that they **satisfy** the equation.

In this chapter, you will learn to solve the simplest type of equation, called a *linear equation in one variable.*

Definition of a Linear Equation in One Variable

A **linear equation in one variable** x is an equation that can be written in the form

$$ax + b = c,$$

where a, b, and c are real numbers, and $a \neq 0$ (a is not equal to 0).

Linear Equations in One Variable (x)

$3x + 7 = 9$	$-15x = 45$	$x = 6.8$
$ax + b = c$, **with** $a = 3, b = 7,$ **and** $c = 9$	$ax + b = c$, **with** $a = -15, b = 0,$ **and** $c = 45$	$ax + b = c$ ($1x + 0 = 6.8$), **with** $a = 1, b = 0,$ **and** $c = 6.8$

Nonlinear Equations in One Variable (x)

$3x^2 + 7 = 9$	$-\dfrac{15}{x} = 45$	$\lvert x \rvert = 6.8$
Nonlinear because x is squared	Nonlinear because x is in the denominator	Nonlinear because of the absolute value bars around x

The adjective *linear* contains the word *line*. As we shall see, linear equations are related to graphs whose points lie along a straight line.

2 Use the addition property of equality to solve equations.

Using the Addition Property of Equality to Solve Equations

Consider the equation

$$x = 11.$$

By inspection, we can see that the solution to this equation is 11. If we substitute 11 for x, we obtain the true statement $11 = 11$.

Now consider the equation

$$x - 3 = 8.$$

If we substitute 11 for x, we obtain $11 - 3 \overset{?}{=} 8$. Subtracting on the left side, we get the true statement $8 = 8$.

The equations $x - 3 = 8$ and $x = 11$ both have the same solution, namely 11, and are called *equivalent equations*. **Equivalent equations** are equations that have the same solution.

The idea in solving a linear equation is to get an equivalent equation with the variable (the letter) by itself on one side of the equal sign and a number by itself on the other side. For example, consider the equation $x - 3 = 8$. To get x by itself on the left side, add 3 to the left side, because $x - 3 + 3$ gives $x + 0$, or just x. You must then add 3 to the right side also. By doing this, we are using the **addition property of equality**.

The Addition Property of Equality

The same real number (or algebraic expression) may be added to both sides of an equation without changing the equation's solution. This can be expressed symbolically as follows:

$$\text{If } a = b, \text{ then } a + c = b + c.$$

EXAMPLE 1 **Solving an Equation Using the Addition Property**

Solve the equation: $x - 3 = 8$.

Study Tip

Some people prefer to show the number added to both sides below the equation:

$$\begin{array}{r} x - 3 = 8 \\ +3 +3 \\ \hline x = 11. \end{array}$$

Solution We can isolate the variable, x, by adding 3 to both sides of the equation.

$x - 3 = 8$	This is the given equation.
$x - 3 + 3 = 8 + 3$	Add 3 to both sides.
$x + 0 = 11$	This step is often done mentally and not listed.
$x = 11$	

By inspection, we can see that the solution to $x = 11$ is 11. To check this proposed solution, replace x with 11 in the original equation.

Check		
	$x - 3 = 8$	This is the original equation.
	$11 - 3 \overset{?}{=} 8$	Substitute 11 for x.
This statement is true.	$8 = 8$	Subtract: $11 - 3 = 8$.

Because the check results in a true statement, we conclude that the solution to the given equation is 11.

Study Tip

You can think of an equation as a balanced scale—balanced because its two sides are equal. To maintain this balance, whatever you do to one side must also be done to the other side.

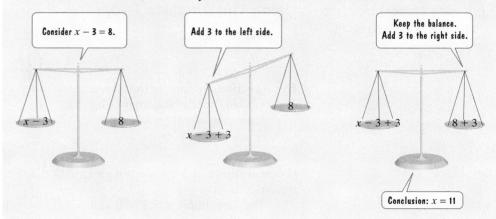

The set of an equation's solutions is called its **solution set**. Thus, the solution set of the equation in Example 1, $x - 3 = 8$, is $\{11\}$. The solution can be expressed as 11 or, using set notation, $\{11\}$.

✓ **CHECK POINT 1**   Solve the equation and check your proposed solution:

$$x - 5 = 12.$$

When we use the addition property of equality, we add the same number to both sides of an equation. We know that subtraction is the addition of an opposite, or additive inverse. Thus, the addition property also lets us subtract the same number from both sides of an equation without changing the equation's solution.

EXAMPLE 2   Subtracting the Same Number from Both Sides

Solve and check: $z + 1.4 = 2.06$.

Solution

$$z + 1.4 = 2.06 \qquad \text{This is the given equation.}$$
$$z + 1.4 - 1.4 = 2.06 - 1.4 \qquad \text{Subtract 1.4 from both sides. This is equivalent to adding} -1.4 \text{ to both sides.}$$
$$z = 0.66 \qquad \text{Subtracting 1.4 from both sides eliminates 1.4 on the left.}$$

Can you see that the solution to $z = 0.66$ is 0.66? To check this proposed solution, replace z with 0.66 in the original equation.

Check $z + 1.4 = 2.06$ This is the original equation.

$$0.66 + 1.4 \overset{?}{=} 2.06 \qquad \text{Substitute 0.66 for } z.$$
$$2.06 = 2.06 \qquad \text{This statement is true.}$$

This true statement indicates that the solution is 0.66, or the solution set is $\{0.66\}$. ■

✓ **CHECK POINT 2**   Solve and check: $z + 2.8 = 5.09$.

When isolating the variable, we can isolate it on either the left side or the right side of an equation.

EXAMPLE 3 Isolating the Variable on the Right

Solve and check: $-\dfrac{1}{2} = x - \dfrac{2}{3}$.

Solution We can isolate the variable, x, on the right side by adding $\frac{2}{3}$ to both sides of the equation.

$$-\frac{1}{2} = x - \frac{2}{3} \qquad \text{This is the given equation.}$$

$$-\frac{1}{2} + \frac{2}{3} = x - \frac{2}{3} + \frac{2}{3} \qquad \text{Add } \tfrac{2}{3} \text{ to both sides, isolating } x \text{ on the right.}$$

$$-\frac{3}{6} + \frac{4}{6} = x \qquad \text{Rewrite each fraction as an equivalent fraction with a}$$

denominator of 6: $-\dfrac{1}{2} + \dfrac{2}{3} = -\dfrac{1}{2}\cdot\dfrac{3}{3} + \dfrac{2}{3}\cdot\dfrac{2}{2} = -\dfrac{3}{6} + \dfrac{4}{6}.$

$$\frac{1}{6} = x \qquad \text{Add on the left side: } -\dfrac{3}{6} + \dfrac{4}{6} = \dfrac{-3+4}{6} = \dfrac{1}{6}.$$

Take a moment to check the proposed solution, $\frac{1}{6}$. Substitute $\frac{1}{6}$ for x in the original equation. You should obtain $-\frac{1}{2} = -\frac{1}{2}$. This true statement indicates that the solution is $\frac{1}{6}$, or the solution set is $\left\{\frac{1}{6}\right\}$. ∎

☑ CHECK POINT 3 Solve and check: $-\dfrac{1}{2} = x - \dfrac{3}{4}$.

In Example 4, we combine like terms before using the addition property.

EXAMPLE 4 Combining Like Terms before Using the Addition Property

Solve and check: $5y + 3 - 4y - 8 = 6 + 9$.

Solution

$$5y + 3 - 4y - 8 = 6 + 9 \qquad \text{This is the given equation.}$$

$$y - 5 = 15 \qquad \begin{array}{l}\text{Combine like terms:}\\ 5y - 4y = y, 3 - 8 = -5, \text{ and } 6 + 9 = 15.\end{array}$$

$$y - 5 + 5 = 15 + 5 \qquad \text{Add 5 to both sides.}$$

$$y = 20 \qquad \text{Simplify.}$$

To check the proposed solution, 20, replace y with 20 in the original equation.

Check $\quad 5y + 3 - 4y - 8 = 6 + 9$ Be sure to use the original equation and not the simplified form from the second step above. (Why?)

$$5(20) + 3 - 4(20) - 8 \overset{?}{=} 6 + 9 \qquad \text{Substitute 20 for } y.$$

$$100 + 3 - 80 - 8 \overset{?}{=} 6 + 9 \qquad \text{Multiply on the left.}$$

$$103 - 88 \overset{?}{=} 6 + 9 \qquad \begin{array}{l}\text{Combine positive numbers and combine}\\ \text{negative numbers on the left.}\end{array}$$

$$15 = 15 \qquad \text{This statement is true.}$$

This true statement verifies that the solution is 20, or the solution set is $\{20\}$. ∎

☑ CHECK POINT 4 Solve and check: $8y + 7 - 7y - 10 = 6 + 4$.

Adding and Subtracting Variable Terms on Both Sides of an Equation

In some equations, variable terms appear on both sides. Here is an example:

$$4x = 7 + 3x.$$

| A variable term, $4x$, is on the left side. | A variable term, $3x$, is on the right side. |

Our goal is to isolate all the variable terms on one side of the equation. We can use the addition property of equality to do this. The property allows us to add or subtract the same variable term on both sides of an equation without changing the solution. Let's see how we can use this idea to solve $4x = 7 + 3x$.

EXAMPLE 5 **Using the Addition Property to Isolate Variable Terms**

Solve and check: $4x = 7 + 3x$.

Solution In the given equation, variable terms appear on both sides. We can isolate them on one side by subtracting $3x$ from both sides of the equation.

$4x = 7 + 3x$	This is the given equation.
$4x - 3x = 7 + 3x - 3x$	Subtract 3x from both sides and isolate variable terms on the left.
$x = 7$	Subtracting 3x from both sides eliminates 3x on the right. On the left, 4x − 3x = 1x = x.

To check the proposed solution, 7, replace x with 7 in the original equation.

Check	$4x = 7 + 3x$	Use the original equation.
	$4(7) \stackrel{?}{=} 7 + 3(7)$	Substitute 7 for x.
	$28 \stackrel{?}{=} 7 + 21$	Multiply: 4(7) = 28 and 3(7) = 21.
	$28 = 28$	This statement is true.

This true statement verifies that the solution is 7, or the solution set is $\{7\}$. ▬

✓ **CHECK POINT 5** Solve and check: $7x = 12 + 6x$.

EXAMPLE 6 **Solving an Equation by Isolating the Variable**

Solve and check: $3y - 9 = 2y + 6$.

Solution Our goal is to isolate variable terms on one side and constant terms on the other side. Let's begin by isolating the variable on the left.

$3y - 9 = 2y + 6$	This is the given equation.
$3y - 2y - 9 = 2y - 2y + 6$	Isolate the variable terms on the left by subtracting 2y from both sides.
$y - 9 = 6$	Subtracting 2y from both sides eliminates 2y on the right. On the left, 3y − 2y = 1y = y.

Now we isolate the constant terms on the right by adding 9 to both sides.

$y - 9 + 9 = 6 + 9$	Add 9 to both sides.
$y = 15$	Simplify.

Check
$$3y - 9 = 2y + 6 \qquad \text{Use the original equation.}$$
$$3(15) - 9 \stackrel{?}{=} 2(15) + 6 \qquad \text{Substitute 15 for } y.$$
$$45 - 9 \stackrel{?}{=} 30 + 6 \qquad \text{Multiply: } 3(15) = 45 \text{ and } 2(15) = 30.$$
$$36 = 36 \qquad \text{This statement is true.}$$

The solution is 15, or the solution set is $\{15\}$.

☑ **CHECK POINT 6**   Solve and check: $3x - 6 = 2x + 5$.

---

3 Solve applied problems using formulas.

Applications

Our next example shows how the addition property of equality can be used to find the value of a variable in a mathematical model.

EXAMPLE 7   An Application: Vocabulary and Age

There is a relationship between the number of words in a child's vocabulary, V, and the child's age, A, in months, for ages between 15 and 50 months, inclusive. This relationship can be modeled by the formula

$$V + 900 = 60A.$$

Use the formula to find the number of words in a child's vocabulary at the age of 30 months.

Solution   In the formula, A represents the child's age, in months. We are interested in a 30-month-old child. Thus, we substitute 30 for A. Then we use the addition property of equality to find V, the number of words in the child's vocabulary.

$$V + 900 = 60A \qquad \text{This is the given formula.}$$
$$V + 900 = 60(30) \qquad \text{Substitute 30 for } A.$$
$$V + 900 = 1800 \qquad \text{Multiply: } 60(30) = 1800.$$
$$V + 900 - 900 = 1800 - 900 \qquad \text{Subtract 900 from both sides and solve for } V.$$
$$V = 900$$

At the age of 30 months, a child has a vocabulary of 900 words.

☑ **CHECK POINT 7**   Use the formula $V + 900 = 60A$ to find the number of words in a child's vocabulary at the age of 50 months.

The line graph in **Figure 2.1** allows us to "see" the formula $V + 900 = 60A$. The two points labeled with voice balloons illustrate what we learned about vocabulary and age by solving equations in Example 7 and Check Point 7.

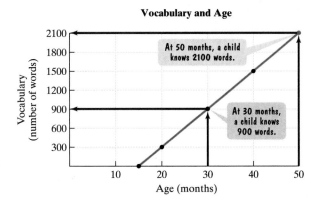

Vocabulary and Age

At 50 months, a child knows 2100 words.

At 30 months, a child knows 900 words.

Vocabulary (number of words)

Age (months)

FIGURE 2.1

Later in the book, you will learn to create graphs for mathematical models and formulas. Linear equations are related to models whose graphs are straight lines, like the one in **Figure 2.1**. For this reason, these equations are called *linear* equations.

2.1 EXERCISE SET *MyMathLab* Math XL PRACTICE WATCH DOWNLOAD READ REVIEW

Practice Exercises

In Exercises 1–10, identify the linear equations in one variable.

1. $x - 9 = 13$

2. $x - 15 = 20$

3. $x^2 - 9 = 13$

4. $x^2 - 15 = 20$

5. $\dfrac{9}{x} = 13$

6. $\dfrac{15}{x} = 20$

7. $\sqrt{2}x + \pi = 0.\overline{3}$

8. $\sqrt{3}x + \pi = 0.\overline{6}$

9. $|x + 2| = 5$

10. $|x + 5| = 8$

Solve each equation in Exercises 11–54 using the addition property of equality. Be sure to check your proposed solutions.

11. $x - 4 = 19$

12. $y - 5 = -18$

13. $z + 8 = -12$

14. $z + 13 = -15$

15. $-2 = x + 14$

16. $-13 = x + 11$

17. $-17 = y - 5$

18. $-21 = y - 4$

19. $7 + z = 11$

20. $18 + z = 14$

21. $-6 + y = -17$

22. $-8 + y = -29$

23. $x + \dfrac{1}{3} = \dfrac{7}{3}$

24. $x + \dfrac{7}{8} = \dfrac{9}{8}$

25. $t + \dfrac{5}{6} = -\dfrac{7}{12}$

26. $t + \dfrac{2}{3} = -\dfrac{7}{6}$

27. $x - \dfrac{3}{4} = \dfrac{9}{2}$

28. $x - \dfrac{3}{5} = \dfrac{7}{10}$

29. $-\dfrac{1}{5} + y = -\dfrac{3}{4}$

30. $-\dfrac{1}{8} + y = -\dfrac{1}{4}$

31. $3.2 + x = 7.5$

32. $-2.7 + w = -5.3$

33. $x + \dfrac{3}{4} = -\dfrac{9}{2}$

34. $r + \dfrac{3}{5} = -\dfrac{7}{10}$

35. $5 = -13 + y$

36. $-11 = 8 + x$

37. $-\dfrac{3}{5} = -\dfrac{3}{2} + s$

38. $\dfrac{7}{3} = -\dfrac{5}{2} + z$

39. $830 + y = 520$

40. $-90 + t = -35$

41. $r + 3.7 = 8$

42. $x + 10.6 = -9$

43. $-3.7 + m = -3.7$

44. $y + \dfrac{7}{11} = \dfrac{7}{11}$

45. $6y + 3 - 5y = 14$

46. $-3x - 5 + 4x = 9$

47. $7 - 5x + 8 + 2x + 4x - 3 = 2 + 3 \cdot 5$

48. $13 - 3r + 2 + 6r - 2r - 1 = 3 + 2 \cdot 9$

49. $7y + 4 = 6y - 9$

50. $4r - 3 = 5 + 3r$

51. $12 - 6x = 18 - 7x$

52. $20 - 7s = 26 - 8s$

53. $4x + 2 = 3(x - 6) + 8$

54. $7x + 3 = 6(x - 1) + 9$

Practice PLUS

The equations in Exercises 55–58 contain small geometric figures that represent real numbers. Use the addition property of equality to isolate x on one side of the equation and the geometric figures on the other side.

55. $x - \square = \triangle$

56. $x + \square = \triangle$

57. $2x + \triangle = 3x + \square$

58. $6x - \triangle = 7x - \square$

In Exercises 59–62, use the given information to write an equation. Let x represent the number described in each exercise. Then solve the equation and find the number.

59. If 12 is subtracted from a number, the result is -2. Find the number.

60. If 23 is subtracted from a number, the result is -8. Find the number.

61. The difference between $\dfrac{2}{5}$ of a number and 8 is $\dfrac{7}{5}$ of that number. Find the number.

62. The difference between 3 and $\dfrac{2}{7}$ of a number is $\dfrac{5}{7}$ of that number. Find the number.

Application Exercises

Formulas frequently appear in the business world. For example, the cost, C, of an item (the price paid by a retailer) plus the markup, M, on that item (the retailer's profit) equals the selling price, S, of the item. The formula is

$$C + M = S.$$

Use the formula to solve Exercises 63–64.

63. The selling price of a computer is $1850. If the markup on the computer is $150, find the cost to the retailer for the computer.

64. The selling price of a television is $650. If the cost to the retailer for the television is $520, find the markup.

We appear to be at home with more TV channels.

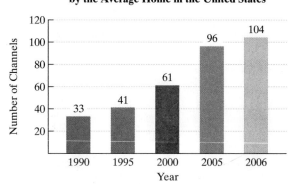

Number of Television Channels Received by the Average Home in the United States

Source: Nielsen Media Research

The data displayed by the bar graph on the previous page can be described by the mathematical model

$$C - 4.6x = 25,$$

where C is the number of television channels received by the average U.S. home x years after 1990. Use this information to solve Exercises 65–66.

65. According to the formula, how many television channels were received by the average U.S. home in 2005? Does this underestimate or overestimate the number displayed by the bar graph on the previous page? By how much?

66. According to the formula, how many television channels were received by the average U.S. home in 1995? Does this underestimate or overestimate the number displayed by the bar graph on the previous page? By how much?

The line graph shows the decline in U.S. job satisfaction for four years in the period from 1987 through 2006.

Percentage of United States Workers Satisfied with Their Jobs

Source: The Conference Board

The graph was obtained from the mathematical model

$$S + 0.8x = 62,$$

where S is the percentage of workers satisfied with their jobs x years after 1987. Use this information to solve Exercises 67–68.

67. a. Use the line graph to estimate the percentage of U.S. workers who were satisfied with their jobs in 2006.

b. Use the formula to determine the percentage of U.S. workers who were satisfied with their jobs in 2006. How does this compare with your graphical estimate from part (a)?

68. a. Use the line graph to estimate the percentage of U.S. workers who were satisfied with their jobs in 2000.

b. Use the formula to determine the percentage of U.S. workers who were satisfied with their jobs in 2000. How does this compare with your graphical estimate from part (a)?

Writing in Mathematics

69. State the addition property of equality and give an example.

70. Explain why $x + 2 = 9$ and $x + 2 = -6$ are not equivalent equations.

71. What is the difference between solving an equation such as

$$5y + 3 - 4y - 8 = 6 + 9$$

and simplifying an algebraic expression such as

$$5y + 3 - 4y - 8?$$

If there is a difference, which topic should be taught first? Why?

72. Look, again, at the graph for Exercises 67–68 that shows the percentage of U.S. workers satisfied with their jobs over time. You used a *linear* equation to solve Exercise 67 or 68. What does the adjective *linear* have to do with the relationship among the four data points in the graph?

Critical Thinking Exercises

Make Sense? *In Exercises 73–76, determine whether each statement "makes sense" or "does not make sense" and explain your reasoning.*

73. The book is teaching me totally different things than my instructor: The book adds the number to both sides beside the equation, but my instructor adds the number underneath.

74. There are times that I prefer to check an equation's solution in my head and not show the check.

75. Solving an equation reminds me of keeping a barbell balanced: If I add weight to or subtract weight from one side of the bar, I must do the same thing to the other side.

76. I used a linear equation to explore data points lying on the same line.

In Exercises 77–80, determine whether each statement is true or false. If the statement is false, make the necessary change(s) to produce a true statement.

77. If $y - a = -b$, then $y = a + b$.

78. If $y + 7 = 0$, then $y = 7$.

79. If $2x - 5 = 3x$, then $x = -5$.

80. If $3x = 18$, then $x = 18 - 3$.

81. Write an equation with a negative solution that can be solved by adding 100 to both sides.

Technology Exercises

Use a calculator to solve each equation in Exercises 82–83.

82. $x - 7.0463 = -9.2714$

83. $6.9825 = 4.2296 + y$

Review Exercises

84. Write as an algebraic expression in which x represents the number: the quotient of 9 and a number, decreased by 4 times the number. (Section 1.1, Example 3)

85. Simplify: $-16 - 8 \div 4 \cdot (-2)$. (Section 1.8, Example 4)

86. Simplify: $3[7x - 2(5x - 1)]$. (Section 1.8, Example 11)

Preview Exercises

Exercises 87–89 will help you prepare for the material covered in the next section.

87. Multiply and simplify: $5 \cdot \dfrac{x}{5}$.

88. Divide and simplify: $\dfrac{-7y}{-7}$.

89. Is 4 a solution of $3x - 14 = -2x + 6$?

SECTION **2.2**

Objectives

1. Use the multiplication property of equality to solve equations.
2. Solve equations in the form $-x = c$.
3. Use the addition and multiplication properties to solve equations.
4. Solve applied problems using formulas.

1. Use the multiplication property of equality to solve equations.

The Multiplication Property of Equality

In 2006, more than three-fourths of the United States population had cellphones. As we go wireless, what does this mean for the future of pay phones? In this section, we introduce a new property for solving equations to answer the question.

Using the Multiplication Property of Equality to Solve Equations

Can the addition property of equality be used to solve every linear equation in one variable? No. For example, consider the equation

$$\frac{x}{5} = 9.$$

We cannot isolate the variable x by adding or subtracting 5 on both sides. To get x by itself on the left side, multiply the left side by 5:

$$5 \cdot \frac{x}{5} = \left(5 \cdot \frac{1}{5}\right)x = 1x = x.$$

5 is the multiplicative inverse of $\frac{1}{5}$.

You must then multiply the right side by 5 also. By doing this, we are using the **multiplication property of equality**.

The Multiplication Property of Equality

The same nonzero real number (or algebraic expression) may multiply both sides of an equation without changing the solution. This can be expressed symbolically as follows:

If $a = b$ and $c \neq 0$, then $ac = bc$.

EXAMPLE 1 Solving an Equation Using the Multiplication Property

Solve the equation: $\dfrac{x}{5} = 9$.

Solution We can isolate the variable, x, by multiplying both sides of the equation by 5.

$$\dfrac{x}{5} = 9 \qquad \text{This is the given equation.}$$

$$5 \cdot \dfrac{x}{5} = 5 \cdot 9 \qquad \text{Multiply both sides by 5.}$$

$$1x = 45 \qquad \text{Simplify.}$$

$$x = 45 \qquad \text{1x = x}$$

By substituting 45 for x in the original equation, we obtain the true statement $9 = 9$. This verifies that the solution is 45, or the solution set is $\{45\}$. ◼

 ✓ **CHECK POINT 1** Solve the equation: $\dfrac{x}{3} = 12$.

When we use the multiplication property of equality, we multiply both sides of an equation by the same nonzero number. We know that division is multiplication by a multiplicative inverse. Thus, the multiplication property also lets us divide both sides of an equation by a nonzero number without changing the solution.

EXAMPLE 2 Dividing Both Sides by the Same Nonzero Number

Solve: **a.** $6x = 30$ **b.** $-7y = 56$ **c.** $-18.9 = 3z$.

Solution In each equation, the variable is multiplied by a number. We can isolate the variable by dividing both sides of the equation by that number.

a.
$$6x = 30 \qquad \text{This is the given equation.}$$
$$\dfrac{6x}{6} = \dfrac{30}{6} \qquad \text{Divide both sides by 6.}$$
$$1x = 5 \qquad \text{Simplify.}$$
$$x = 5 \qquad \text{1x = x}$$

By substituting 5 for x in the original equation, we obtain the true statement $30 = 30$. The solution is 5, or the solution set is $\{5\}$.

b.
$$-7y = 56 \qquad \text{This is the given equation.}$$
$$\dfrac{-7y}{-7} = \dfrac{56}{-7} \qquad \text{Divide both sides by } -7.$$
$$1y = -8 \qquad \text{Simplify.}$$
$$y = -8 \qquad \text{1y = y}$$

By substituting -8 for y in the original equation, we obtain the true statement $56 = 56$. The solution is -8, or the solution set is $\{-8\}$.

c.
$$-18.9 = 3z \qquad \text{This is the given equation.}$$
$$\dfrac{-18.9}{3} = \dfrac{3z}{3} \qquad \text{Divide both sides by 3.}$$
$$-6.3 = 1z \qquad \text{Simplify.}$$
$$-6.3 = z \qquad \text{1z = z}$$

By substituting -6.3 for z in the original equation, we obtain the true statement $-18.9 = -18.9$. The solution is -6.3, or the solution set is $\{-6.3\}$. ◼

☑ **CHECK POINT 2** Solve:

a. $4x = 84$
b. $-11y = 44$
c. $-15.5 = 5z.$

Some equations have a variable term with a fractional coefficient. Here is an example:

$$\frac{3}{4}y = 12.$$

> The coefficient of the term $\frac{3}{4}y$ is $\frac{3}{4}$.

To isolate the variable, multiply both sides of the equation by the multiplicative inverse of the fraction. For the equation $\frac{3}{4}y = 12$, the multiplicative inverse of $\frac{3}{4}$ is $\frac{4}{3}$. Thus, we solve $\frac{3}{4}y = 12$ by multiplying both sides by $\frac{4}{3}$.

EXAMPLE 3 **Using the Multiplication Property to Eliminate a Fractional Coefficient**

Solve: **a.** $\frac{3}{4}y = 12$ **b.** $9 = -\frac{3}{5}x.$

Solution

a. $\frac{3}{4}y = 12$ This is the given equation.

$\frac{4}{3}\left(\frac{3}{4}y\right) = \frac{4}{3} \cdot 12$ Multiply both sides by $\frac{4}{3}$, the multiplicative inverse of $\frac{3}{4}$.

$1y = 16$ On the left, $\frac{4}{3}\left(\frac{3}{4}y\right) = \left(\frac{4}{3} \cdot \frac{3}{4}\right)y = 1y.$
              On the right, $\frac{4}{3} \cdot \frac{12}{1} = \frac{48}{3} = 16.$

$y = 16$ $1y = y$

By substituting 16 for y in the original equation, we obtain the true statement $12 = 12$. The solution is 16, or the solution set is $\{16\}$.

b. $9 = -\frac{3}{5}x$ This is the given equation.

$-\frac{5}{3} \cdot 9 = -\frac{5}{3}\left(-\frac{3}{5}x\right)$ Multiply both sides by $-\frac{5}{3}$, the multiplicative inverse of $-\frac{3}{5}$.

$-15 = 1x$ Simplify.

$-15 = x$ $1x = x$

By substituting -15 for x in the original equation, we obtain the true statement $9 = 9$. The solution is -15, or the solution set is $\{-15\}$. ■

☑ **CHECK POINT 3** Solve:

a. $\frac{2}{3}y = 16$ b. $28 = -\frac{7}{4}x.$

---

2 Solve equations in the form $-x = c.$

Equations and Coefficients of −1

How do we solve an equation in the form $-x = c$, such as $-x = 4$? Because the equation means $-1x = 4$, we have not yet obtained a solution. The solution of an equation is obtained from the form $x =$ some number. The equation $-x = 4$ is not yet in this form. We still need to isolate x. We can do this by multiplying or dividing both sides of the equation by -1. We will multiply by -1.

EXAMPLE 4   Solving Equations in the Form $-x = c$

Solve: **a.** $-x = 4$ **b.** $-x = -7$.

Solution   We multiply both sides of each equation by -1. This will isolate x on the left side.

a.

$-x = 4$	This is the given equation.
$-1x = 4$	Rewrite $-x$ as $-1x$.
$(-1)(-1x) = (-1)(4)$	Multiply both sides by -1.
$1x = -4$	On the left, $(-1)(-1) = 1$. On the right, $(-1)(4) = -4$.
$x = -4$	$1x = x$

Check

$-x = 4$	This is the original equation.
$-(-4) \overset{?}{=} 4$	Substitute -4 for x.
$4 = 4$	$-(-a) = a$, so $-(-4) = 4$.

This true statement indicates that the solution is -4, or the solution set is $\{-4\}$.

b.

$-x = -7$	This is the given equation.
$-1x = -7$	Rewrite $-x$ as $-1x$.
$(-1)(-1x) = (-1)(-7)$	Multiply both sides by -1.
$1x = 7$	$(-1)(-1) = 1$ and $(-1)(-7) = 7$.
$x = 7$	$1x = x$

By substituting 7 for x in the original equation, we obtain the true statement $-7 = -7$. The solution is 7, or the solution set is $\{7\}$. ■

 CHECK POINT 4   Solve: **a.** $-x = 5$ **b.** $-x = -3$.

Study Tip

If $-x = c$, then the equation's solution is the opposite, or additive inverse, of c. For example, the solution of $-x = -7$ is the opposite of -7, which is 7.

3 Use the addition and multiplication properties to solve equations.

Equations Requiring Both the Addition and Multiplication Properties

When an equation does not contain fractions, we will often use the addition property of equality before the multiplication property of equality. Our overall goal is to isolate the variable with a coefficient of 1 on either the left or right side of the equation.

Here is the procedure that we will be using to solve the equations in the next three examples:

- Use the addition property of equality to isolate the variable term.
- Use the multiplication property of equality to isolate the variable.

EXAMPLE 5   Using Both the Addition and Multiplication Properties

Solve: $3x + 1 = 7$.

Solution   We first isolate the variable term, $3x$, by subtracting 1 from both sides. Then we isolate the variable, x, by dividing both sides by 3.

- **Use the addition property of equality to isolate the variable term.**

$3x + 1 = 7$	This is the given equation.
$3x + 1 - 1 = 7 - 1$	Use the addition property, subtracting 1 from both sides.
$3x = 6$	Simplify.

- **Use the multiplication property of equality to isolate the variable.**

$$\frac{3x}{3} = \frac{6}{3}$$ Divide both sides of 3x = 6 by 3.

$$x = 2$$ Simplify.

By substituting 2 for x in the original equation, $3x + 1 = 7$, we obtain the true statement $7 = 7$. The solution is 2, or the solution set is $\{2\}$.

☑ **CHECK POINT 5** Solve: $4x + 3 = 27$.

EXAMPLE 6 **Using Both the Addition and Multiplication Properties**

Solve: $-2y - 28 = 4$.

Solution We first isolate the variable term, $-2y$, by adding 28 to both sides. Then we isolate the variable, y, by dividing both sides by -2.

- **Use the addition property of equality to isolate the variable term.**

$$-2y - 28 = 4$$ This is the given equation.

$$-2y - 28 + 28 = 4 + 28$$ Use the addition property, adding 28 to both sides.

$$-2y = 32$$ Simplify.

- **Use the multiplication property of equality to isolate the variable.**

$$\frac{-2y}{-2} = \frac{32}{-2}$$ Divide both sides by -2.

$$y = -16$$ Simplify.

Take a moment to substitute -16 for y in the given equation. Do you obtain the true statement $4 = 4$? The solution is -16, or the solution set is $\{-16\}$.

☑ **CHECK POINT 6** Solve: $-4y - 15 = 25$.

EXAMPLE 7 **Using Both the Addition and Multiplication Properties**

Solve: $3x - 14 = -2x + 6$.

Solution We will use the addition property to collect all terms involving x on the left and all numerical terms on the right. Then we will isolate the variable, x, by dividing both sides by its coefficient.

- **Use the addition property of equality to isolate the variable term.**

$$3x - 14 = -2x + 6$$ This is the given equation.

$$3x + 2x - 14 = -2x + 2x + 6$$ Add 2x to both sides.

$$5x - 14 = 6$$ Simplify.

$$5x - 14 + 14 = 6 + 14$$ Add 14 to both sides.

$$5x = 20$$ Simplify. The variable term, 5x, is isolated on the left. The numerical term, 20, is isolated on the right.

- **Use the multiplication property of equality to isolate the variable.**

$$\frac{5x}{5} = \frac{20}{5}$$ Divide both sides by 5.

$$x = 4$$ Simplify.

Check

$$3x - 14 = -2x + 6 \qquad \text{Use the original equation.}$$
$$3(4) - 14 \overset{?}{=} -2(4) + 6 \qquad \text{Substitute the proposed solution, 4, for x.}$$
$$12 - 14 \overset{?}{=} -8 + 6 \qquad \text{Multiply.}$$
$$-2 = -2 \qquad \text{Simplify.}$$

The true statement $-2 = -2$ verifies that the solution is 4, or the solution set is $\{4\}$.

☑ **CHECK POINT 7**   Solve: $2x - 15 = -4x + 21$.

4 Solve applied problems using formulas.

Applications

Our next example shows how solving linear equations can be used to explore the behavior of a mathematical model.

> **EXAMPLE 8**   The Disappearing Pay Phone

The bar graph in **Figure 2.2** shows the number of pay phones in the United States from 2000 through 2006.

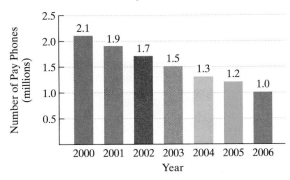

Number of Pay Phones in the United States

FIGURE 2.2

Source: Federal Communications Commission

The data can be described by the mathematical model

$$P = -0.18n + 2.1,$$

where P is the number of pay phones, in millions, n years after 2000.

    a. Does the formula underestimate or overestimate the number of pay phones in 2006? By how much?

    b. If trends shown by the formula continue, when will there be no pay phones in the United States? Round to the nearest year.

Solution

    a. **Figure 2.2** indicates that there were 1.0 million pay phones in 2006. Let's see if the formula underestimates or overestimates this value. The year 2006 is 6 years after 2000, so we substitute 6 for n.

$$P = -0.18n + 2.1 \qquad \text{This is the given formula.}$$
$$P = -0.18(6) + 2.1 \qquad \text{Replace n with 6.}$$
$$P = -1.08 + 2.1 \qquad \text{Multiply: } -0.18(6) = -1.08.$$
$$P = 1.02 \qquad \text{Add: } -1.08 + 2.1 = 1.02.$$

The model indicates that there were 1.02 million pay phones in 2006. This overestimates the number displayed by the bar graph, 1.0 million, by 0.02 million phones.

b. We are interested in when there will be no pay phones. We substitute 0 for the number of pay phones, P, in the given formula. Then we solve for n, the number of years after 2000.

$$P = -0.18n + 2.1$$ This is the given formula.

$$0 = -0.18n + 2.1$$ Replace P with 0.

Our goal is to isolate n.

$$0 - 2.1 = -0.18n + 2.1 - 2.1$$ Isolate the term containing n by subtracting 2.1 from both sides.

$$-2.1 = -0.18n$$ Simplify.

$$\frac{-2.1}{-0.18} = \frac{-0.18n}{-0.18}$$ Divide both sides by -0.18.

$$12 \approx n$$ Simplify: $\dfrac{11.6 \approx 12.}{0.18\,)\overline{210.0}}$

The formula indicates that approximately 12 years after 2000, or in 2012, there will be no pay phones in the United States. ∎

Study Tip

Notice the two different ways that we used the formula $P = -0.18n + 2.1$. In Example 8(a), we substituted a value for n and used the order of operations to determine P. In Example 8(b), we substituted a value for P and solved an equation to determine n.

✓ **CHECK POINT 8**   Use the mathematical model $P = -0.18n + 2.1$, described in Example 8, to solve this exercise.

a. Does the formula underestimate or overestimate the number of pay phones shown in **Figure 2.2** on page 121 for 2004? By how much?

b. If trends shown by the formula continue, when will there be only 0.3 million pay phones in the United States?

2.2 EXERCISE SET   *MyMathLab*   PRACTICE   WATCH   DOWNLOAD   READ   REVIEW

Practice Exercises

Solve each equation in Exercises 1–28 using the multiplication property of equality. Be sure to check your proposed solutions.

1. $\dfrac{x}{6} = 5$ **2.** $\dfrac{x}{7} = 4$

3. $\dfrac{x}{-3} = 11$ **4.** $\dfrac{x}{-5} = 8$

5. $5y = 35$ **6.** $6y = 42$

7. $-7y = 63$ **8.** $-4y = 32$

9. $-28 = 8z$ **10.** $-36 = 8z$

11. $-18 = -3z$ **12.** $-54 = -9z$

13. $-8x = 6$ **14.** $-8x = 4$

15. $17y = 0$ **16.** $-16y = 0$

17. $\dfrac{2}{3}y = 12$ **18.** $\dfrac{3}{4}y = 15$

19. $28 = -\dfrac{7}{2}x$ **20.** $20 = -\dfrac{5}{8}x$

21. $-x = 17$ **22.** $-x = 23$

23. $-47 = -y$ **24.** $-51 = -y$

25. $-\dfrac{x}{5} = -9$

26. $-\dfrac{x}{5} = -1$

27. $2x - 12x = 50$

28. $8x - 3x = -45$

Solve each equation in Exercises 29–54 using both the addition and multiplication properties of equality. Check proposed solutions.

29. $2x + 1 = 11$

30. $2x + 5 = 13$

31. $2x - 3 = 9$

32. $3x - 2 = 9$

33. $-2y + 5 = 7$

34. $-3y + 4 = 13$

35. $-3y - 7 = -1$

36. $-2y - 5 = 7$

37. $12 = 4z + 3$

38. $14 = 5z - 21$

39. $-x - 3 = 3$

40. $-x - 5 = 5$

41. $6y = 2y - 12$

42. $8y = 3y - 10$

43. $3z = -2z - 15$

44. $2z = -4z + 18$

45. $-5x = -2x - 12$

46. $-7x = -3x - 8$

47. $8y + 4 = 2y - 5$

48. $5y + 6 = 3y - 6$

49. $6z - 5 = z + 5$

50. $6z - 3 = z + 2$

51. $6x + 14 = 2x - 2$

52. $9x + 2 = 6x - 4$

53. $-3y - 1 = 5 - 2y$

54. $-3y - 2 = -5 - 4y$

Practice PLUS

The equations in Exercises 55–58 contain small geometric figures that represent real numbers. Use the multiplication property of equality to isolate x on one side of the equation and the geometric figures on the other side.

55. $\dfrac{x}{\square} = \triangle$

56. $\triangle = \square\, x$

57. $\triangle = -x$

58. $\dfrac{-x}{\square} = \triangle$

In Exercises 59–66, use the given information to write an equation. Let x represent the number described in each exercise. Then solve the equation and find the number.

59. If a number is multiplied by 6, the result is 10. Find the number.

60. If a number is multiplied by −6, the result is 20. Find the number.

61. If a number is divided by −9, the result is 5. Find the number.

62. If a number is divided by −7, the result is 8. Find the number.

63. Eight subtracted from the product of 4 and a number is 56.

64. Ten subtracted from the product of 3 and a number is 23.

65. Negative three times a number, increased by 15, is −6.

66. Negative five times a number, increased by 11, is −29.

Application Exercises

The formula

$$M = \dfrac{n}{5}$$

models your distance, M, in miles, from a lightning strike in a thunderstorm if it takes n seconds to hear thunder after seeing the lightning. Use this formula to solve Exercises 67–68.

67. If you are 2 miles away from the lightning flash, how long will it take the sound of thunder to reach you?

68. If you are 3 miles away from the lightning flash, how long will it take the sound of thunder to reach you?

The Mach number is a measurement of speed, named after the man who suggested it, Ernst Mach (1838–1916). The formula

$$M = \dfrac{A}{740}$$

indicates that the speed of an aircraft, A, in miles per hour, divided by the speed of sound, approximately 740 miles per hour, results in the Mach number, M. Use the formula to determine the speed, in miles per hour, of the aircrafts in Exercises 69–70. (Note: When an aircraft's speed increases beyond Mach 1, it is said to have broken the sound barrier.)

69.

70.

Concorde
Mach 2.03
Speed = ?

SR-71 Blackbird
Mach 3.3
Speed = ?

In 2006, the United States Senate voted to raise the amount of money the federal government was allowed to borrow, called the debt limit, to $8.96 trillion. That amounts to approximately $30,200 per U.S. citizen. Yearly budget deficits had pushed the nation's debt into the trillions of dollars and without more borrowing, the government would run short of money to pay its bills. The bar graph shows the federal debt limit, in trillions of dollars, and what this means per U.S. citizen.

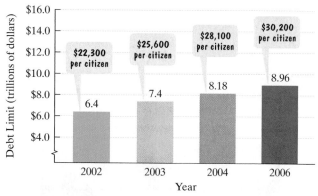

United States Government Debt Limit

Source: White House Office of Management and Budget

The federal debt limit per citizen, D, in dollars, can be described by the mathematical model

$$D = 1914n + 19,371,$$

where n is the number of years after 2000. Use this information to solve Exercises 71–72.

71. a. Does the formula underestimate or overestimate the debt limit per citizen in 2006? By how much?

  b. If trends shown by the formula continue, when will the debt limit per citizen be $40,425?

72. a. Does the formula underestimate or overestimate the debt limit per citizen in 2004? By how much?

  b. If trends shown by the formula continue, when will the debt limit per citizen be $44,253?

Writing in Mathematics

73. State the multiplication property of equality and give an example.

74. Explain how to solve the equation $-x = -50$.

75. Explain how to solve the equation $2x + 8 = 5x - 3$.

Critical Thinking Exercises

Make Sense?   In Exercises 76–79, determine whether each statement "makes sense" or "does not make sense" and explain your reasoning.

76. I used the addition and multiplication properties of equality to solve $3x = 20 + 4$.

77. I used the addition and multiplication properties of equality to solve $12 - 3x = 15$ as follows:

$$12 - 3x = 15$$
$$3x = 3$$
$$x = 1.$$

78. When I use the addition and multiplication properties to solve $2x + 5 = 17$, I undo the operations in the opposite order in which they are performed.

79. The model $P = -0.18n + 2.1$ describes the number of pay phones, P, in millions, n years after 2000, so I have to solve a linear equation to determine the number of pay phones in 2006.

In Exercises 80–83, determine whether each statement is true or false. If the statement is false, make the necessary change(s) to produce a true statement.

80. If $7x = 21$, then $x = 21 - 7$.

81. If $3x - 4 = 16$, then $3x = 12$.

82. If $3x + 7 = 0$, then $x = \dfrac{7}{3}$.

83. The solution of $6x = 0$ is not a natural number.

In Exercises 84–85, write an equation with the given characteristics.

84. The solution is a positive integer and the equation can be solved by dividing both sides by -60.

85. The solution is a negative integer and the equation can be solved by multiplying both sides by $\dfrac{4}{5}$.

Technology Exercises

Solve each equation in Exercises 86–87. Use a calculator to help with the arithmetic. Check your solution using the calculator.

86. $3.7x - 19.46 = -9.988$

87. $-72.8y - 14.6 = -455.43 - 4.98y$

Review Exercises

88. Evaluate: $(-10)^2$. (Section 1.8, Example 1)

89. Evaluate: -10^2. (Section 1.8, Example 1)

90. Evaluate $x^3 - 4x$ for $x = -1$. (Section 1.8, Example 10)

Preview Exercises

Exercises 91–93 will help you prepare for the material covered in the next section.

91. Simplify: $13 - 3(x + 2)$.

92. Is 6 a solution of $2(x - 3) - 17 = 13 - 3(x + 2)$?

93. Multiply and simplify: $10\left(\dfrac{x}{5} - \dfrac{39}{5}\right)$.

Solving Linear Equations

Sense of Humor and Depression

FIGURE 2.3

Source: Steven Davis and Joseph Palladino. *Psychology*, 5th Edition, Prentice Hall, 2007

Objectives

1 Solve linear equations.

2 Solve linear equations containing fractions.

3 Identify equations with no solution or infinitely many solutions.

4 Solve applied problems using formulas.

The belief that humor and laughter can have positive benefits on our lives is not new. The graphs in **Figure 2.3** indicate that persons with a low sense of humor have higher levels of depression in response to negative life events than those with a high sense of humor. In this section, we will see how algebra models these relationships. To use these mathematical models, it would be helpful to have a systematic procedure for solving linear equations. We open this section with such a procedure.

1 Solve linear equations.

A Step-by-Step Procedure for Solving Linear Equations

Here is a step-by-step procedure for solving a linear equation in one variable. Not all of these steps are necessary to solve every equation.

> ### Solving A Linear Equation
> **1.** Simplify the algebraic expression on each side.
> **2.** Collect all the variable terms on one side and all the constant terms on the other side.
> **3.** Isolate the variable and solve.
> **4.** Check the proposed solution in the original equation.

EXAMPLE 1   Solving a Linear Equation

Solve and check: $2x - 8x + 40 = 13 - 3x - 3$.

Solution

Step 1. **Simplify the algebraic expression on each side.**

$$2x - 8x + 40 = 13 - 3x - 3 \qquad \text{This is the given equation.}$$
$$-6x + 40 = 10 - 3x \qquad \text{Combine like terms: } 2x - 8x = -6x \text{ and } 13 - 3 = 10.$$

Step 2. Collect variable terms on one side and constant terms on the other side. The simplified equation is $-6x + 40 = 10 - 3x$. We will collect variable terms on the left by adding $3x$ to both sides. We will collect the numbers on the right by subtracting 40 from both sides.

$$-6x + 40 + 3x = 10 - 3x + 3x \qquad \text{Add 3x to both sides.}$$
$$-3x + 40 = 10 \qquad \text{Simplify: } -6x + 3x = -3x.$$
$$-3x + 40 - 40 = 10 - 40 \qquad \text{Subtract 40 from both sides.}$$
$$-3x = -30 \qquad \text{Simplify.}$$

Step 3. Isolate the variable and solve. We isolate the variable, x, by dividing both sides by -3.

$$\frac{-3x}{-3} = \frac{-30}{-3} \qquad \text{Divide both sides by -3.}$$
$$x = 10 \qquad \text{Simplify.}$$

Step 4. Check the proposed solution in the original equation. Substitute 10 for x in the original equation.

$$2x - 8x + 40 = 13 - 3x - 3 \qquad \text{This is the original equation.}$$
$$2 \cdot 10 - 8 \cdot 10 + 40 \stackrel{?}{=} 13 - 3 \cdot 10 - 3 \qquad \text{Substitute 10 for x.}$$
$$20 - 80 + 40 \stackrel{?}{=} 13 - 30 - 3 \qquad \text{Perform the indicated multiplications.}$$
$$-60 + 40 \stackrel{?}{=} -17 - 3 \qquad \text{Subtract: } 20 - 80 = -60 \text{ and } 13 - 30 = -17.$$
$$-20 = -20 \qquad \text{Simplify.}$$

By substituting 10 for x in the original equation, we obtain the true statement $-20 = -20$. This verifies that the solution is 10, or the solution set is $\{10\}$. ■

☑ **CHECK POINT 1** Solve and check: $-7x + 25 + 3x = 16 - 2x - 3$.

EXAMPLE 2   Solving a Linear Equation

Solve and check: $5x = 8(x + 3)$.

Solution

Step 1. Simplify the algebraic expression on each side. Use the distributive property to remove parentheses on the right.

$$5x = 8(x + 3) \qquad \text{This is the given equation.}$$
$$5x = 8x + 24 \qquad \text{Use the distributive property.}$$

Step 2. Collect variable terms on one side and constant terms on the other side. We will work with $5x = 8x + 24$ and collect variable terms on the left by subtracting $8x$ from both sides. The only constant term, 24, is already on the right.

$$5x - 8x = 8x + 24 - 8x \qquad \text{Subtract 8x from both sides.}$$
$$-3x = 24 \qquad \text{Simplify: } 5x - 8x = -3x.$$

Step 3. Isolate the variable and solve. We isolate the variable, x, by dividing both sides by -3.

$$\frac{-3x}{-3} = \frac{24}{-3} \qquad \text{Divide both sides by -3.}$$
$$x = -8 \qquad \text{Simplify.}$$

Step 4. Check the proposed solution in the original equation. Substitute -8 for x in the original equation.

$$5x = 8(x + 3) \qquad \text{This is the original equation.}$$
$$5(-8) \stackrel{?}{=} 8(-8 + 3) \qquad \text{Substitute } -8 \text{ for } x.$$
$$5(-8) \stackrel{?}{=} 8(-5) \qquad \text{Perform the addition in parentheses: } -8 + 3 = -5.$$
$$-40 = -40 \qquad \text{Multiply.}$$

The true statement $-40 = -40$ verifies that -8 is the solution, or the solution set is $\{-8\}$. ■

☑ **CHECK POINT 2**   Solve and check: $8x = 2(x + 6)$.

EXAMPLE 3   **Solving a Linear Equation**

Solve and check: $2(x - 3) - 17 = 13 - 3(x + 2)$.

Solution

Step 1. Simplify the algebraic expression on each side.

> Do not begin with $13 - 3$. Multiplication (the distributive property) is applied before subtraction.

$$2(x - 3) - 17 = 13 - 3(x + 2) \qquad \text{This is the given equation.}$$
$$2x - 6 - 17 = 13 - 3x - 6 \qquad \text{Use the distributive property.}$$
$$2x - 23 = -3x + 7 \qquad \text{Combine like terms.}$$

Step 2. Collect variable terms on one side and constant terms on the other side. We will collect variable terms on the left by adding $3x$ to both sides. We will collect the numbers on the right by adding 23 to both sides.

$$2x - 23 + 3x = -3x + 7 + 3x \qquad \text{Add } 3x \text{ to both sides.}$$
$$5x - 23 = 7 \qquad \text{Simplify: } 2x + 3x = 5x.$$
$$5x - 23 + 23 = 7 + 23 \qquad \text{Add 23 to both sides.}$$
$$5x = 30 \qquad \text{Simplify.}$$

Step 3. Isolate the variable and solve. We isolate the variable, x, by dividing both sides by 5.

$$\frac{5x}{5} = \frac{30}{5} \qquad \text{Divide both sides by 5.}$$
$$x = 6 \qquad \text{Simplify.}$$

Step 4. Check the proposed solution in the original equation. Substitute 6 for x in the original equation.

$$2(x - 3) - 17 = 13 - 3(x + 2) \qquad \text{This is the original equation.}$$
$$2(6 - 3) - 17 \stackrel{?}{=} 13 - 3(6 + 2) \qquad \text{Substitute 6 for } x.$$
$$2(3) - 17 \stackrel{?}{=} 13 - 3(8) \qquad \text{Simplify inside parentheses.}$$
$$6 - 17 \stackrel{?}{=} 13 - 24 \qquad \text{Multiply.}$$
$$-11 = -11 \qquad \text{Subtract.}$$

The true statement $-11 = -11$ verifies that 6 is the solution, or the solution set is $\{6\}$. ■

☑ **CHECK POINT 3**   Solve and check: $4(2x + 1) - 29 = 3(2x - 5)$.

2 Solve linear equations containing fractions.

Linear Equations with Fractions

Equations are easier to solve when they do not contain fractions. How do we remove fractions from an equation? We begin by multiplying both sides of the equation by the least common denominator of any fractions in the equation. The least common denominator is the smallest number that all denominators will divide into. Multiplying every term on both sides of the equation by the least common denominator will eliminate the fractions in the equation. Example 4 shows how we "clear an equation of fractions."

EXAMPLE 4 **Solving a Linear Equation Involving Fractions**

Solve and check: $\dfrac{3x}{2} = \dfrac{x}{5} - \dfrac{39}{5}$.

Solution The denominators are 2, 5, and 5. The smallest number that is divisible by 2, 5, and 5 is 10. We begin by multiplying both sides of the equation by 10, the least common denominator.

$$\frac{3x}{2} = \frac{x}{5} - \frac{39}{5}$$ This is the given equation.

$$10 \cdot \frac{3x}{2} = 10\left(\frac{x}{5} - \frac{39}{5}\right)$$ Multiply both sides by 10.

$$10 \cdot \frac{3x}{2} = 10 \cdot \frac{x}{5} - 10 \cdot \frac{39}{5}$$ Use the distributive property. Be sure to multiply all terms by 10.

$$\overset{5}{\cancel{10}} \cdot \frac{3x}{2}_{1} = \overset{2}{\cancel{10}} \cdot \frac{x}{\cancel{5}}_{1} - \overset{2}{\cancel{10}} \cdot \frac{39}{\cancel{5}}_{1}$$ Divide out common factors in the multiplications.

$$15x = 2x - 78$$ Complete the multiplications. The fractions are now cleared.

At this point, we have an equation similar to those we previously have solved. Collect the variable terms on one side and the constant terms on the other side.

$$15x - 2x = 2x - 2x - 78$$ Subtract 2x to get the variable terms on the left.

$$13x = -78$$ Simplify.

Isolate x by dividing both sides by 13.

$$\frac{13x}{13} = \frac{-78}{13}$$ Divide both sides by 13.

$$x = -6$$ Simplify.

Check

$$\frac{3x}{2} = \frac{x}{5} - \frac{39}{5}$$ This is the original equation.

$$\frac{3(-6)}{2} \overset{?}{=} \frac{-6}{5} - \frac{39}{5}$$ Substitute −6 for x.

$$-9 \overset{?}{=} \frac{-6}{5} - \frac{39}{5}$$ Simplify the left side: $\dfrac{3(-6)}{2} = \dfrac{-18}{2} = -9.$

$$-9 \overset{?}{=} \frac{-45}{5}$$ Subtract on the right side:

$$\frac{-6}{5} - \frac{39}{5} = \frac{-6}{5} + \left(\frac{-39}{5}\right) = \frac{-45}{5}.$$

$$-9 = -9$$ Simplify: $\dfrac{-45}{5} = -9.$

The true statement $-9 = -9$ verifies that -6 is the solution, or the solution set is $\{-6\}$.

☑ **CHECK POINT 4**   Solve and check: $\dfrac{x}{4} = \dfrac{2x}{3} + \dfrac{5}{6}$.

---

3   Identify equations with no solution or infinitely many solutions.

Equations with No Solution or Infinitely Many Solutions

Thus far, each equation that we have solved has had a single solution. However, some equations are not true for even one real number. Such an equation is called an **inconsistent equation**. Here is an example of such an equation:

$$x = x + 4.$$

There is no number that is equal to itself plus 4. This equation has no solution.

You can express the fact that an equation has no solution using words or set notation.

- Use the phrase "no solution."
- Use set notation: $\varnothing$.

> This symbol stands for the empty set, a set with no elements.

An equation that is true for all real numbers is called an **identity**. An example of an identity is

$$x + 3 = x + 2 + 1.$$

Every number plus 3 is equal to that number plus 2 plus 1. Every real number is a solution to this equation.

You can express the fact that every real number is a solution of an equation using words or set notation.

- Use the phrase "all real numbers."
- Use set notation: $\{x \mid x \text{ is a real number}\}$.

> The set of | all x | such that | x is a real number

Recognizing Inconsistent Equations and Identities

If you attempt to solve an equation with no solution or one that is true for every real number, you will eliminate the variable.

- An inconsistent equation with no solution results in a false statement, such as $2 = 5$.
- An identity that is true for every real number results in a true statement, such as $4 = 4$.

EXAMPLE 5 Solving an Equation

Solve: $2x + 6 = 2(x + 4)$.

Solution

$2x + 6 = 2(x + 4)$	This is the given equation.
$2x + 6 = 2x + 8$	Use the distributive property.
$2x + 6 - 2x = 2x + 8 - 2x$	Subtract 2x from both sides.
$6 = 8$	Simplify.

Keep reading. 6 = 8 is not the solution.

The original equation is equivalent to the false statement $6 = 8$, which is false for every value of x. The equation is inconsistent and has no solution. You can express this by writing "no solution" or using the symbol for the empty set, $\varnothing$.

✓ **CHECK POINT 5** Solve: $3x + 7 = 3(x + 1)$.

EXAMPLE 6 Solving an Equation

Solve: $-3x + 5 + 5x = 4x - 2x + 5$.

Solution

$-3x + 5 + 5x = 4x - 2x + 5$	This is the given equation.
$2x + 5 = 2x + 5$	Combine like terms: $-3x + 5x = 2x$ and $4x - 2x = 2x$.
$2x + 5 - 2x = 2x + 5 - 2x$	Subtract 2x from both sides.
$5 = 5$	Simplify.

Keep reading. 5 = 5 is not the solution.

The original equation is equivalent to the statement $5 = 5$, which is true for every value of x. The equation is an identity and all real numbers are solutions. You can express this by writing "all real numbers" or using set notation: $\{x \mid x \text{ is a real number}\}$.

✓ **CHECK POINT 6** Solve: $3(x - 1) + 9 = 8x + 6 - 5x$.

---

4 Solve applied problems using formulas.

Applications

The next example shows how our procedure for solving equations with fractions can be used to find the value of a variable in a mathematical model.

EXAMPLE 7 An Application: Responding to Negative Life Events

In the section opener, we introduced line graphs, repeated in **Figure 2.3** on the next page, indicating that persons with a low sense of humor have higher levels of depression in response to negative life events than those with a high sense of humor. These graphs can be modeled by the following formulas:

Low-Humor Group   High-Humor Group

$$D = \frac{10}{9}x + \frac{53}{9} \qquad\qquad D = \frac{1}{9}x + \frac{26}{9}.$$

Sense of Humor and Depression

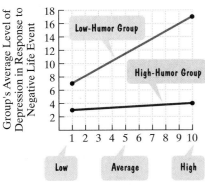

FIGURE 2.3 (repeated)

In each formula, x represents the intensity of a negative life event (from 1, low, to 10, high) and D is the level of depression in response to that event. If the high-humor group averages a level of depression of 3.5, or $\frac{7}{2}$, in response to a negative life event, what is the intensity of that event? How is the solution shown on the red line graph in **Figure 2.3**?

Solution We are interested in the intensity of a negative life event with an average level of depression of $\frac{7}{2}$ for the high-humor group. We substitute $\frac{7}{2}$ for D in the high-humor model and solve for x, the intensity of the negative life event.

$$D = \frac{1}{9}x + \frac{26}{9}$$ This is the given formula for the high-humor group.

$$\frac{7}{2} = \frac{1}{9}x + \frac{26}{9}$$ Replace D with $\frac{7}{2}$.

$$18 \cdot \frac{7}{2} = 18\left(\frac{1}{9}x + \frac{26}{9}\right)$$ Multiply both sides by 18, the least common denominator.

$$18 \cdot \frac{7}{2} = 18 \cdot \frac{1}{9}x + 18 \cdot \frac{26}{9}$$ Use the distributive property.

$$\overset{9}{\cancel{18}} \cdot \frac{7}{\underset{1}{2}} = \overset{2}{\cancel{18}} \cdot \frac{1}{\underset{1}{9}}x + \overset{2}{\cancel{18}} \cdot \frac{26}{\underset{1}{9}}$$ Divide out common factors in the multiplications.

$$63 = 2x + 52$$ Complete the multiplications. The fractions are now cleared.

$$63 - 52 = 2x + 52 - 52$$ Subtract 52 from both sides to get constants on the left.

$$11 = 2x$$ Simplify.

$$\frac{11}{2} = \frac{2x}{2}$$ Divide both sides by 2.

$$\frac{11}{2} = x$$ Simplify.

The formula indicates that if the high-humor group averages a level of depression of 3.5 in response to a negative life event, the intensity of that event is $\frac{11}{2}$, or 5.5. This is illustrated on the line graph for the high-humor group in **Figure 2.4**.

FIGURE 2.4

☑ **CHECK POINT 7** Use the model for the low-humor group given in Example 7 to solve this problem. If the low-humor group averages a level of depression of 10 in response to a negative life event, what is the intensity of that event? How is the solution shown on the blue line graph in **Figure 2.3**?

2.3 EXERCISE SET *MyMathLab* MathXL PRACTICE WATCH DOWNLOAD READ REVIEW

Practice Exercises

In Exercises 1–30, solve each equation. Be sure to check your proposed solution by substituting it for the variable in the original equation.

1. $5x + 3x - 4x = 10 + 2$

2. $4x + 8x - 2x = 20 - 15$

3. $4x - 9x + 22 = 3x + 30$

4. $3x + 2x + 64 = 40 - 7x$

5. $3x + 6 - x = 8 + 3x - 6$

6. $3x + 2 - x = 6 + 3x - 8$

7. $4(x + 1) = 20$

8. $3(x - 2) = -6$

9. $7(2x - 1) = 42$

10. $4(2x - 3) = 32$

11. $38 = 30 - 2(x - 1)$

12. $20 = 44 - 8(2 - x)$

13. $2(4z + 3) - 8 = 46$

14. $3(3z + 5) - 7 = 89$

15. $6x - (3x + 10) = 14$

16. $5x - (2x + 14) = 10$

17. $5(2x + 1) = 12x - 3$

18. $3(x + 2) = x + 30$

19. $3(5 - x) = 4(2x + 1)$

20. $3(3x - 1) = 4(3 + 3x)$

21. $8(y + 2) = 2(3y + 4)$

22. $8(y + 3) = 3(2y + 12)$

23. $3(x + 1) = 7(x - 2) - 3$

24. $5x - 4(x + 9) = 2x - 3$

25. $5(2x - 8) - 2 = 5(x - 3) + 3$

26. $7(3x - 2) + 5 = 6(2x - 1) + 24$

27. $6 = -4(1 - x) + 3(x + 1)$

28. $100 = -(x - 1) + 4(x - 6)$

29. $10(z + 4) - 4(z - 2) = 3(z - 1) + 2(z - 3)$

30. $-2(z - 4) - (3z - 2) = -2 - (6z - 2)$

Solve and check each equation in Exercises 31–46. Begin your work by rewriting each equation without fractions.

31. $\dfrac{x}{5} - 4 = -6$

32. $\dfrac{x}{2} + 13 = -22$

33. $\dfrac{2x}{3} - 5 = 7$

34. $\dfrac{3x}{4} - 9 = -6$

35. $\dfrac{2y}{3} - \dfrac{3}{4} = \dfrac{5}{12}$

36. $\dfrac{3y}{4} - \dfrac{2}{3} = \dfrac{7}{12}$

37. $\dfrac{x}{3} + \dfrac{x}{2} = \dfrac{5}{6}$

38. $\dfrac{x}{4} - \dfrac{x}{5} = 1$

39. $20 - \dfrac{z}{3} = \dfrac{z}{2}$

40. $\dfrac{z}{5} - \dfrac{1}{2} = \dfrac{z}{6}$

41. $\dfrac{y}{3} + \dfrac{2}{5} = \dfrac{y}{5} - \dfrac{2}{5}$

42. $\dfrac{y}{12} + \dfrac{1}{6} = \dfrac{y}{2} - \dfrac{1}{4}$

43. $\dfrac{3x}{4} - 3 = \dfrac{x}{2} + 2$

44. $\dfrac{3x}{5} - \dfrac{2}{5} = \dfrac{x}{3} + \dfrac{2}{5}$

45. $\dfrac{x - 3}{5} - 1 = \dfrac{x - 5}{4}$

46. $\dfrac{x - 2}{3} - 4 = \dfrac{x + 1}{4}$

In Exercises 47–64, solve each equation. Use words or set notation to identify equations that have no solution, or equations that are true for all real numbers.

47. $3x - 7 = 3(x + 1)$

48. $2(x - 5) = 2x + 10$

49. $2(x + 4) = 4x + 5 - 2x + 3$

50. $3(x - 1) = 8x + 6 - 5x - 9$

51. $7 + 2(3x - 5) = 8 - 3(2x + 1)$

52. $2 + 3(2x - 7) = 9 - 4(3x + 1)$

53. $4x + 1 - 5x = 5 - (x + 4)$

54. $5x - 5 = 3x - 7 + 2(x + 1)$

55. $4(x + 2) + 1 = 7x - 3(x - 2)$

56. $5x - 3(x + 1) = 2(x + 3) - 5$

57. $3 - x = 2x + 3$

58. $5 - x = 4x + 5$

59. $\dfrac{x}{3} + 2 = \dfrac{x}{3}$

60. $\dfrac{x}{4} + 3 = \dfrac{x}{4}$

61. $\dfrac{x}{2} - \dfrac{x}{4} + 4 = x + 4$

62. $\dfrac{x}{2} + \dfrac{2x}{3} + 3 = x + 3$

63. $\dfrac{2}{3}x = 2 - \dfrac{5}{6}x$

64. $\dfrac{2}{3}x = \dfrac{1}{4}x - 8$

Practice PLUS

The equations in Exercises 65–66 contain small figures ($\square$, $\triangle$, and $\$$) that represent real numbers. Use properties of equality to isolate x on one side of the equation and the small figures on the other side.

65. $\dfrac{x}{\square} + \triangle = \$$

66. $\dfrac{x}{\square} - \triangle = -\$$

67. If $\dfrac{x}{5} - 2 = \dfrac{x}{3}$, evaluate $x^2 - x$.

68. If $\dfrac{3x}{2} + \dfrac{3x}{4} = \dfrac{x}{4} - 4$, evaluate $x^2 - x$.

In Exercises 69–72, use the given information to write an equation. Let x represent the number described in each exercise. Then solve the equation and find the number.

69. When one-third of a number is added to one-fifth of the number, the sum is 16. What is the number?

70. When two-fifths of a number is added to one-fourth of the number, the sum is 13. What is the number?

71. When 3 is subtracted from three-fourths of a number, the result is equal to one-half of the number. What is the number?

72. When 30 is subtracted from seven-eights of a number, the result is equal to one-half of the number. What is the number?

Application Exercises

In Massachusetts, speeding fines are determined by the formula

$$F = 10(x - 65) + 50,$$

where F is the cost, in dollars, of the fine if a person is caught driving x miles per hour. Use this formula to solve Exercises 73–74.

73. If a fine comes to $250, how fast was that person driving?

74. If a fine comes to $400, how fast was that person driving?

The latest guidelines, which apply to both men and women, give healthy weight ranges, rather than specific weights, for your height. The further you are above the upper limit of your range, the greater are the risks of developing weight-related health problems. The bar graph shows these ranges for various heights for people between the ages of 19 and 34, inclusive.

Healthy Weight Ranges for Men and Women, Ages 19 to 34

■ Lower end of range   ■ Upper end of range

Source: U.S. Department of Health and Human Services

The mathematical model

$$\frac{W}{2} - 3H = 53$$

describes a weight, W, in pounds, that lies within the healthy weight range for a person whose height is H inches over 5 feet. Use this information to solve Exercises 75–76.

75. Use the formula to find a healthy weight for a person whose height is 5′6″. (*Hint:* H = 6 because this person's height is 6 inches over 5 feet.) How many pounds is this healthy weight below the upper end of the range shown by the bar graph?

76. Use the formula to find a healthy weight for a person whose height is 6′0″. (*Hint:* H = 12 because this person's height is 12 inches over 5 feet.) How many pounds is this healthy weight below the upper end of the range shown by the bar graph?

The formula

$$p = 15 + \frac{5d}{11}$$

describes the pressure of sea water, p, in pounds per square foot, at a depth of d feet below the surface. Use the formula to solve Exercises 77–78.

77. The record depth for breath-held diving, by Francisco Ferreras (Cuba) off Grand Bahama Island, on November 14, 1993, involved pressure of 201 pounds per square foot. To what depth did Ferreras descend on this ill-advised venture? (He was underwater for 2 minutes and 9 seconds!)

78. At what depth is the pressure 20 pounds per square foot?

Writing in Mathematics

79. In your own words, describe how to solve a linear equation.

80. Explain how to solve a linear equation containing fractions.

81. Suppose that you solve $\frac{x}{5} - \frac{x}{2} = 1$ by multiplying both sides by 20, rather than the least common denominator, 10. Describe what happens. If you get the correct solution, why do you think we clear the equation of fractions by multiplying by the *least* common denominator?

82. Suppose you are an algebra teacher grading the following solution on an examination:

Solve: $-3(x - 6) = 2 - x$

Solution: $-3x - 18 = 2 - x$

$-2x - 18 = 2$

$-2x = -16$

$x = 8.$

You should note that 8 checks, and the solution is 8. The student who worked the problem therefore wants full credit. Can you find any errors in the solution? If full credit is 10 points, how many points should you give the student? Justify your position.

Critical Thinking Exercises

Make Sense?   *In Exercises 83–86, determine whether each statement "makes sense" or "does not make sense" and explain your reasoning.*

83. Although I can solve $3x + \frac{1}{5} = \frac{1}{4}$ by first subtracting $\frac{1}{5}$ from both sides, I find it easier to begin by multiplying both sides by 20, the least common denominator.

84. I can use any common denominator to clear an equation of fractions, but using the least common denominator makes the arithmetic easier.

85. When I substituted 5 for x in the equation

$$4x + 6 = 6(x + 1) - 2x$$

I obtained a true statement, so the equation's solution is 5.

86. Because I know how to clear an equation of fractions, I decided to clear the equation $0.5x + 8.3 = 12.4$ of decimals by multiplying both sides by 10.

In Exercises 87–90, determine whether each statement is true or false. If the statement is false, make the necessary change(s) to produce a true statement.

87. The equation $3(x + 4) = 3(4 + x)$ has precisely one solution.

88. The equation $2y + 5 = 0$ is equivalent to $2y = 5$.

89. If $2 - 3y = 11$ and the solution to the equation is substituted into $y^2 + 2y - 3$, a number results that is neither positive nor negative.

90. The equation $x + \frac{1}{3} = \frac{1}{2}$ is equivalent to $x + 2 = 3$.

91. A woman's height, h, is related to the length of her femur, f (the bone from the knee to the hip socket), by the formula $f = 0.432h - 10.44$. Both h and f are measured in inches. A partial skeleton is found of a woman in which the femur is 16 inches long. Police find the skeleton in an area where a woman slightly over 5 feet tall has been missing for over a year. Can the partial skeleton be that of the missing woman? Explain.

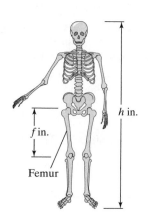

h in.

f in.

Femur

Solve each equation in Exercises 92–93.

92. $\dfrac{2x - 3}{9} + \dfrac{x - 3}{2} = \dfrac{x + 5}{6} - 1$

93. $2(3x + 4) = 3x + 2[3(x - 1) + 2]$

Technology Exercises

Solve each equation in Exercises 94–95. Use a calculator to help with the arithmetic. Check your solution using the calculator.

94. $2.24y - 9.28 = 5.74y + 5.42$

95. $4.8y + 32.5 = 124.8 - 9.4y$

Review Exercises

In Exercises 96–97, insert either $<$ or $>$ in the shaded area between each pair of numbers to make a true statement.

96. -24 ▢ -20 (Section 1.3, Example 6)

97. $-\dfrac{1}{3}$ ▢ $-\dfrac{1}{5}$ (Section 1.3, Example 6)

98. Simplify: $-9 - 11 + 7 - (-3)$. (Section 1.6, Example 3)

Preview Exercises

Exercises 99–101 will help you prepare for the material covered in the next section.

99. Consider the formula

$$T = D + pm.$$

a. Subtract D from both sides and write the resulting formula.

b. Divide both sides of your formula from part (a) by p and write the resulting formula.

100. Solve: $4 = 0.25B$.

101. Solve: $1.3 = P \cdot 26$.

Formulas and Percents

Objectives

1. Solve a formula for a variable.
2. Express a percent as a decimal.
3. Express a decimal as a percent.
4. Use the percent formula.
5. Solve applied problems involving percent change.

"And, if elected, it is my solemn pledge to cut your taxes by 10% for each of my first three years in office, for a total cut of 30%."

Did you know that one of the most common ways that you are given numerical information is with percents? In this section, you will learn to use a formula that will help you to understand percents from an algebraic perspective, enabling you to make sense of the politician's promise.

1. Solve a formula for a variable.

Solving a Formula for One of Its Variables

We know that solving an equation is the process of finding the number (or numbers) that make the equation a true statement. All of the equations we have solved contained only one letter, such as x or y.

By contrast, the formulas we have seen contain two or more letters, representing two or more variables. Here is an example:

$$C + M = S.$$

| A retailer's cost of an item | plus | the retailer's markup (profit) | equals | the items's selling price. |

We say that this formula is solved for the variable S because S is alone on one side of the equation and the other side does not contain an S.

Solving a formula for a variable means rewriting the formula so that the variable is isolated on one side of the equation. It does not mean obtaining a numerical value for that variable.

The addition and multiplication properties of equality are used to solve a formula for one of its variables. Consider the retailer's formula, $C + M = S$. How do we solve this formula for C? Use the addition property to isolate C by subtracting M from both sides:

We need to isolate C.

$$C + M = S \qquad \text{This is the given formula.}$$
$$C + M - M = S - M \qquad \text{Subtract M from both sides.}$$
$$C = S - M \qquad \text{Simplify.}$$

Solved for C, the formula $C = S - M$ tells us that the cost of an item for a retailer is the item's selling price minus its markup.

To solve a formula for one of its variables, treat that variable as if it were the only variable in the equation. Think of the other variables as if they were numbers. Use the addition property of equality to isolate all terms with the specified variable on one side of the equation and all terms without the specified variable on the other side. Then use the multiplication property of equality to get the specified variable alone.

Our first example involves the formula for the area of a rectangle. The **area of a two-dimensional figure** is the number of square units it takes to fill the interior of the figure. A **square unit** is a square, each of whose sides is one unit in length, as illustrated in **Figure 2.5**. The figure shows that there are 12 square units contained within the rectangle. The area of the rectangle is 12 square units. Notice that the area can be determined in the following manner:

Square unit of measure

FIGURE 2.5

Across Down

4 units · 3 units = 4 · 3 units · units = 12 square units.

The area of a rectangle is the product of the distance across, its length, and the distance down, its width.

Area of a Rectangle

The area, A, of a rectangle with length l and width w is given by the formula

$$A = lw.$$

w

l

EXAMPLE 1   Solving a Formula for a Variable

Solve the formula $A = lw$ for w.

Solution   Our goal is to get w by itself on one side of the formula. There is only one term with w, lw, and it is already isolated on the right side. We isolate w on the right by using the multiplication property of equality and dividing both sides by l.

We need to isolate w.

$A = lw$ This is the given formula.

$\dfrac{A}{l} = \dfrac{lw}{l}$ Isolate w by dividing both sides by l.

$\dfrac{A}{l} = w$ Simplify: $\dfrac{lw}{l} = 1w = w$.

The formula solved for w is $\dfrac{A}{l} = w$ or $w = \dfrac{A}{l}$. Thus, the area of a rectangle divided by its length is equal to its width.

☑ **CHECK POINT 1**   Solve the formula $A = lw$ for l.

The **perimeter of a two-dimensional figure** is the sum of the lengths of its sides. Perimeter is measured in linear units, such as inches, feet, yards, meters, or kilometers.

Example 2 involves the perimeter, P, of a rectangle. Because perimeter is the sum of the lengths of the sides, the perimeter of the rectangle shown in **Figure 2.6** is $l + w + l + w$. This can be expressed as

$$P = 2l + 2w.$$

FIGURE 2.6
A rectangle with length l and width w

Perimeter of a Rectangle

The perimeter, P, of a rectangle with length l and width w is given by the formula

$$P = 2l + 2w.$$

The perimeter of a rectangle is the sum of twice the length and twice the width.

EXAMPLE 2   Solving a Formula for a Variable

Solve the formula $2l + 2w = P$ for w.

Solution   First, isolate $2w$ on the left by subtracting $2l$ from both sides. Then solve for w by dividing both sides by 2.

> We need to isolate w.

$2l + 2w = P$	This is the given formula.
$2l - 2l + 2w = P - 2l$	Isolate 2w by subtracting 2l from both sides.
$2w = P - 2l$	Simplify.
$\dfrac{2w}{2} = \dfrac{P - 2l}{2}$	Isolate w by dividing both sides by 2.
$w = \dfrac{P - 2l}{2}$	Simplify.

☑ **CHECK POINT 2**   Solve the formula $2l + 2w = P$ for l.

EXAMPLE 3   Solving a Formula for a Variable

The total price of an article purchased on a monthly deferred payment plan is described by the following formula:

$$T = D + pm.$$

In this formula, T is the total price, D is the down payment, p is the amount of the monthly payment, and m is the number of payments. Solve the formula for p.

Solution   First, isolate pm on the right by subtracting D from both sides. Then isolate p from pm by dividing both sides of the formula by m.

> We need to isolate p.

$T = D + pm$	This is the given formula. We want p alone.
$T - D = D - D + pm$	Isolate pm by subtracting D from both sides.
$T - D = pm$	Simplify.
$\dfrac{T - D}{m} = \dfrac{pm}{m}$	Now isolate p by dividing both sides by m.
$\dfrac{T - D}{m} = p$	Simplify: $\dfrac{pm}{m} = \dfrac{p\overset{1}{\cancel{m}}}{\underset{1}{\cancel{m}}} = p \cdot 1 = p.$

☑ **CHECK POINT 3**   Solve the formula $T = D + pm$ for m.

The next example has a formula that contains a fraction. To solve for a variable in a formula involving fractions, we begin by multiplying both sides by the least common denominator of all fractions in the formula. This will eliminate the fractions. Then we solve for the specified variable.

EXAMPLE 4   Solving a Formula Containing a Fraction for a Variable

Solve the formula $\dfrac{W}{2} - 3H = 53$ for W.

Solution   Do you remember seeing this formula in the last section's exercise set? It models a person's healthy weight, W, where H represents that person's height, in inches, over 5 feet. We begin by multiplying both sides of the formula by 2 to eliminate the fraction. Then we isolate the variable W.

$$\frac{W}{2} - 3H = 53 \qquad \text{This is the given formula.}$$

$$2\left(\frac{W}{2} - 3H\right) = 2 \cdot 53 \qquad \text{Multiply both sides by 2.}$$

$$2 \cdot \frac{W}{2} - 2 \cdot 3H = 2 \cdot 53 \qquad \text{Use the distributive property.}$$

We need to isolate W.

$$W - 6H = 106 \qquad \text{Simplify.}$$

$$W - 6H + 6H = 106 + 6H \qquad \text{Isolate } W \text{ by adding 6H to both sides.}$$

$$W = 106 + 6H \qquad \text{Simplify.}$$

This form of the formula makes it easy to find a person's healthy weight, W, if we know that person's height, H, in inches, over 5 feet.   ∎

☑ **CHECK POINT 4**   Solve for x: $\dfrac{x}{3} - 4y = 5$.

Before turning to a formula involving percent, let's review some of the basics of percent.

Basics of Percent

We have frequently worked with mathematical models that described data displayed in percent notation. **Percents** are the result of expressing numbers as a part of 100. The word *percent* means *per hundred*.

Percent Notation

$$n\% \quad \text{means} \quad \frac{n}{100}.$$

Study Tip

Dictionaries indicate that the word *percentage* has the same meaning as the word *percent*. Use the word that sounds best in the circumstance.

Figure 2.7 shows the percentage of college freshmen and college seniors who often ask questions in class. The data indicate that 62 out of every 100 freshmen and 75 out of every 100 seniors often ask questions in class.

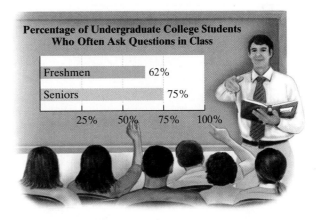

FIGURE 2.7

Source: National Survey of Student Engagement

By definition, $62\% = \frac{62}{100}$ and $75\% = \frac{75}{100}$. Observe that we can express the resulting fractions in decimal notation:

$$62\% = \frac{62}{100} = 100\overline{)62.00}^{\,0.62} \qquad \boxed{62\% = 0.62}$$

$$75\% = \frac{75}{100} = 100\overline{)75.00}^{\,0.75} \qquad \boxed{75\% = 0.75}$$

We can generalize the observations in the voice balloons and quickly express any percent as a decimal number.

2 Express a percent as a decimal.

Expressing a Percent as a Decimal Number

1. Move the decimal point two places to the left.
2. Remove the percent sign.

EXAMPLE 5   Expressing Percents as Decimals

Express each percent as a decimal:

a. 19%   **b.** 180%.

Solution   Use the two steps in the box.

a.

$$19\% = 19.\% = 0.19\%$$

The percent sign is removed.

The decimal point starts at the far right.

The decimal point is moved two places to the left.

Thus, $19\% = 0.19$.

b. $180\% = 1.80\% = 1.80$ or 1.8.

✓ **CHECK POINT 5**   Express each percent as a decimal:

a. 67%   **b.** 250%.

3 Express a decimal as a percent.

We reverse the procedure of Example 5 to express a decimal number as a percent.

> **Expressing a Decimal Number as a Percent**
> 1. Move the decimal point two places to the right.
> 2. Attach a percent sign.

EXAMPLE 6 Expressing a Decimal as a Percent

Express 0.47 as a percent.

Solution

> Move the decimal point two places to the right.

$$047. \quad \% \quad \text{Attach a percent sign.}$$

Thus, $0.47 = 47\%$.

✓ **CHECK POINT 6** Express 0.023 as a percent.

4 Use the percent formula.

A Formula Involving Percent

Percents are useful in comparing two numbers. To compare the number A to the number B using a percent P, the following formula is used:

> A is P percent of B.
>
> $A = P \cdot B$.

In the formula

$$A = PB,$$

B = the base number, P = the percent (in decimal form), and A = the number being compared to B.

There are three basic types of percent problems that can be solved using the percent formula

$$A = PB. \quad \text{A is P percent of B.}$$

Question	Given	Percent Formula
What is P percent of B?	P and B	Solve for A.
A is P percent of what?	A and P	Solve for B.
A is what percent of B?	A and B	Solve for P.

Let's look at an example of each type of problem.

EXAMPLE 7 Using the Percent Formula:
What Is P Percent of B?

What is 8% of 20?

Solution We use the formula $A = PB$: A is P percent of B. We are interested in finding the quantity A in this formula.

What | is | 8% | of | 20?

$$A \quad = \quad 0.08 \quad \cdot \quad 20 \qquad \text{Express 8\% as 0.08.}$$

$$A \quad = \quad 1.6 \qquad\qquad \text{Multiply:} \quad \begin{array}{r} .08 \\ \times\ 20 \\ \hline 1.60 \end{array}$$

Thus, 1.6 is 8% of 20. The answer is 1.6.

✓ **CHECK POINT 7** What is 9% of 50?

EXAMPLE 8 Using the Percent Formula:
A Is *P* Percent of What?

4 is 25% of what?

Solution We use the formula $A = PB$: A is P percent of B. We are interested in finding the quantity B in this formula.

4 | is | 25% | of | what?

$$4 \quad = \quad 0.25 \quad \cdot \quad B \qquad \text{Express 25\% as 0.25.}$$

$$\frac{4}{0.25} \quad = \quad \frac{0.25B}{0.25} \qquad \text{Divide both sides by 0.25.}$$

$$16 \quad = \quad B \qquad \text{Simplify:} \quad 0.25)\overline{4.00}^{\,16.}$$

Thus, 4 is 25% of 16. The answer is 16.

✓ **CHECK POINT 8** 9 is 60% of what?

EXAMPLE 9 Using the Percent Formula:
A Is What Percent of *B*?

1.3 is what percent of 26?

Solution We use the formula $A = PB$: A is P percent of B. We are interested in finding the quantity P in this formula.

1.3 | is | what percent | of | 26?

$$1.3 \quad = \quad P \quad \cdot \quad 26$$

$$\frac{1.3}{26} = \frac{P \cdot 26}{26} \qquad \text{Divide both sides by 26.}$$

$$0.05 = P \qquad \text{Simplify:} \quad 26)\overline{1.30}^{\,0.05}$$

We change 0.05 to a percent by moving the decimal point two places to the right and adding a percent sign: 0.05 = 5%. Thus, 1.3 is 5% of 26. The answer is 5%.

✓ **CHECK POINT 9** 18 is what percent of 50?

Solve applied problems involving percent change.

Applications

Percents are used for comparing changes, such as increases or decreases in sales, population, prices, and production. If a quantity changes, its percent increase or percent decrease can be determined by asking the following question:

The change is what percent of the original amount?

The question is answered using the percent formula as follows:

EXAMPLE 10   Finding Percent Decrease

A jacket regularly sells for $135.00 The sale price is $60.75. Find the percent decrease in the jacket's price.

Solution   The percent decrease in price can be determined by asking the following question:

The price decrease is what percent of the original price ($135.00)?

The price decrease is the difference between the original price and the sale price ($60.75):

$$\$135.00 - \$60.75 = \$74.25.$$

Now we use the percent formula to find the percent decrease.

$$74.25 = P \cdot 135$$

$$\frac{74.25}{135} = \frac{P \cdot 135}{135} \qquad \text{Divide both sides by 135.}$$

$$0.55 = P \qquad \text{Simplify: } 135\overline{)74.25}^{\,0.55}$$

We change 0.55 to a percent by moving the decimal point two places to the right and adding a percent sign: 0.55 = 55%. Thus, the percent decrease in the jacket's price is 55%.   ■

✓ **CHECK POINT 10**   A television regularly sells for $940. The sale price is $611. Find the percent decrease in the television's price.

In our next example, we look at one of the many ways that percent can be used incorrectly.

EXAMPLE 11   Promises of a Politician

A politician states, "If you elect me to office, I promise to cut your taxes for each of my first three years in office by 10% each year, for a total reduction of 30%." Evaluate the accuracy of the politician's statement.

Solution   To make things simple, let's assume that a taxpayer paid $100 in taxes in the year previous to the politician's election. A 10% reduction during year 1 is 10% of $100.

$$10\% \text{ of previous year's tax} = 10\% \text{ of } \$100 = 0.10 \cdot \$100 = \$10$$

With a 10% reduction the first year, the taxpayer will pay only $100 − $10, or $90, in taxes during the politician's first year in office.

The table below shows how we calculate the new, reduced tax for each of the first three years in office.

Year	Tax Paid the Year Before	10% Reduction	Taxes Paid This Year
1	$100	$0.10 \cdot \$100 = \$10$	$\$100 - \$10 = \$90$
2	$90	$0.10 \cdot \$90 = \$9$	$\$90 - \$9 = \$81$
3	$81	$0.10 \cdot \$81 = \$8.10$	$\$81 - \$8.10 = \$72.90$

The tax reduction is the amount originally paid, $100.00, minus the amount paid during the politician's third year in office, $72.90:

$$\$100.00 - \$72.90 = \$27.10.$$

Now we use the percent formula to determine the percent decrease in taxes over the three years.

$$27.1 = P \cdot 100$$

$$\frac{27.1}{100} = \frac{P \cdot 100}{100} \qquad \text{Divide both sides by 100.}$$

$$0.271 = P \qquad \text{Simplify.}$$

Change 0.271 to a percent: 0.271 = 27.1%. The percent decrease is 27.1%. The taxes decline by 27.1%, not by 30%. The politician is ill-informed in saying that three consecutive 10% cuts add up to a total tax cut of 30%. In our calculation, which serves as a counterexample to the promise, the total tax cut is only 27.1%. ∎

✓ **CHECK POINT 11** Suppose you paid $1200 in taxes. During year 1, taxes decrease by 20%. During year 2, taxes increase by 20%.

  a. What do you pay in taxes for year 2?

  b. How do your taxes for year 2 compare with what you originally paid, namely $1200? If the taxes are not the same, find the percent increase or decrease.

2.4 EXERCISE SET **MyMathLab**   PRACTICE   WATCH   DOWNLOAD   READ   REVIEW

Practice Exercises

In Exercises 1–26, solve each formula for the specified variable. Do you recognize the formula? If so, what does it describe?

1. $d = rt$ for r

2. $d = rt$ for t

3. $I = Prt$ for P

4. $I = Prt$ for r

5. $C = 2\pi r$ for r

6. $C = \pi d$ for d

7. $E = mc^2$ for m

8. $V = \pi r^2 h$ for h

9. $y = mx + b$ for m

10. $y = mx + b$ for x

11. $T = D + pm$ for D

12. $P = C + MC$ for M

13. $A = \frac{1}{2}bh$ for b

14. $A = \frac{1}{2}bh$ for h

15. $M = \frac{n}{5}$ for n

16. $M = \frac{A}{740}$ for A

17. $\frac{c}{2} + 80 = 2F$ for c

18. $p = 15 + \frac{5d}{11}$ for d

19. $A = \frac{1}{2}(a + b)$ for a

20. $A = \frac{1}{2}(a + b)$ for b

21. $S = P + Prt$ for r

22. $S = P + Prt$ for t

23. $A = \frac{1}{2}h(a + b)$ for b

24. $A = \frac{1}{2}h(a + b)$ for a

25. $Ax + By = C$ for x

26. $Ax + By = C$ for y

In Exercises 27–36, express each percent as a decimal.

27. 27%   **28.** 83%   **29.** 63.4%

30. 2.15%   **31.** 170%   **32.** 360%

33. 3%   **34.** 8%   **35.** $\frac{1}{2}$%

36. $\frac{1}{4}$%

In Exercises 37–44, express each decimal as a percent.

37. 0.89   **38.** 0.16   **39.** 0.002

40. 0.008   **41.** 4.78   **42.** 5.38

43. 100   **44.** 85

Use the percent formula, $A = PB$: A is P percent of B, to solve Exercises 45–56.

45. What is 3% of 200?   **46.** What is 8% of 300?

47. What is 18% of 40?   **48.** What is 16% of 90?

49. 3 is 60% of what?   **50.** 8 is 40% of what?

51. 24% of what number is 40.8?

52. 32% of what number is 51.2?

53. 3 is what percent of 15?

54. 18 is what percent of 90?

55. What percent of 2.5 is 0.3?

56. What percent of 7.5 is 0.6?

57. If 5 is increased to 8, the increase is what percent of the original number?

58. If 5 is increased to 9, the increase is what percent of the original number?

59. If 4 is decreased to 1, the decrease is what percent of the original number?

60. If 8 is decreased to 6, the decrease is what percent of the original number?

Practice PLUS

In Exercises 61–68, solve each equation for x.

61. $y = (a + b)x$

62. $y = (a - b)x$

63. $y = (a - b)x + 5$

64. $y = (a + b)x - 8$

65. $y = cx + dx$

66. $y = cx - dx$

67. $y = Ax - Bx - C$

68. $y = Ax + Bx + C$

Application Exercises

69. The average, or mean, A, of three exam grades, x, y, and z, is given by the formula

$$A = \frac{x + y + z}{3}.$$

a. Solve the formula for z.

b. Use the formula in part (a) to solve this problem. On your first two exams, your grades are 86% and 88%: $x = 86$ and $y = 88$. What must you get on the third exam to have an average of 90%?

70. The average, or mean, A, of four exam grades, x, y, z, and w, is given by the formula

$$A = \frac{x + y + z + w}{4}.$$

a. Solve the formula for w.

b. Use the formula in part (a) to solve this problem. On your first three exams, your grades are 76%, 78%, and 79%: $x = 76$, $y = 78$, and $z = 79$. What must you get on the fourth exam to have an average of 80%?

71. If you are traveling in your car at an average rate of r miles per hour for t hours, then the distance, d, in miles, that you travel is described by the formula $d = rt$: distance equals rate times time.

a. Solve the formula for t.

b. Use the formula in part (a) to find the time that you travel if you cover a distance of 100 miles at an average rate of 40 miles per hour.

72. The formula $F = \frac{9}{5}C + 32$ expresses the relationship between Celsius temperature, C, and Fahrenheit temperature, F.

a. Solve the formula for C.

b. Use the formula from part (a) to find the equivalent Celsius temperature for a Fahrenheit temperature of 59°.

A Time/CNN telephone poll included never-married single women between the ages of 18 and 49 and never-married single men between the ages of 18 and 49. The circle graphs show the results for one of the questions in the poll. Use this information to solve Exercises 73–74.

If You Couldn't Find the Perfect Mate, Would You Marry Someone Else?

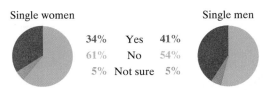

Single women			Single men
34%	Yes	**41%**	
61%	No	54%	
5%	Not sure	5%	

Source: Time

73. There were 1200 single women who participated in the poll. How many stated they would marry someone other than the perfect mate?

74. There were 1200 single men who participated in the poll. How many stated they would marry someone other than the perfect mate?

The graph shows the composition of a typical American community's trash. Use this information to solve Exercises 75–76.

Types of Trash in an American Community by Percentage of Total Weight

Paper 35%
Yard waste 12%
Food waste 12%
Plastic 11%
Metal 8%
Glass 5%
Other 17%

Source: U.S. Environmental Protection Agency

75. Across the United States, people generate approximately 175 billion pounds of trash in the form of paper each year. (That's approximately 580 pounds per person per year.) How many billions of pounds of trash do we throw away each year?

76. Across the United States, people generate approximately 55 billion pounds of trash in the form of plastic each year. (That's approximately 180 pounds per person per year.) How many billions of pounds of trash do we throw away each year?

The circle graphs show the number of free, partly free, and not free countries in 1974 and 2004. Use this information to solve Exercises 77–78. Round answers to the nearest percent.

Global Trends in Freedom

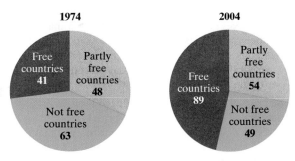

1974	2004
Free countries 41	Free countries 89
Partly free countries 48	Partly free countries 54
Not free countries 63	Not free countries 49

Source: Freedom House

77. a. What percentage of the world's countries were free in 1974?

b. What percentage of the world's countries were free in 2004?

c. Find the percent increase in the number of free countries from 1974 to 2004.

78. a. What percentage of the world's countries were not free in 1974?

b. What percentage of the world's countries were not free in 2004?

c. Find the percent decrease in the number of not free countries from 1974 to 2004.

79. A charity has raised $7500, with a goal of raising $60,000. What percent of the goal has been raised?

80. A charity has raised $225,000, with a goal of raising $500,000. What percent of the goal has been raised?

81. A restaurant bill came to $60. If 15% of this amount was left as a tip, how much was the tip?

82. If income tax is $3502 plus 28% of taxable income over $23,000, how much is the income tax on a taxable income of $35,000?

83. Suppose that the local sales tax rate is 6% and you buy a car for $16,800.

a. How much tax is due?

b. What is the car's total cost?

84. Suppose that the local sales tax rate is 7% and you buy a graphing calculator for $96.

a. How much tax is due?

b. What is the calculator's total cost?

85. An exercise machine with an original price of $860 is on sale at 12% off.

a. What is the discount amount?

b. What is the exercise machine's sale price?

86. A dictionary that normally sells for $16.50 is on sale at 40% off.

  a. What is the discount amount?

  b. What is the dictionary's sale price?

87. A sofa regularly sells for $840. The sale price is $714. Find the percent decrease in the sofa's price.

88. A fax machine regularly sells for $380. The sale price is $266. Find the percent decrease in the machine's price.

89. Suppose that you put $10,000 in a rather risky investment recommended by your financial advisor. During the first year, your investment decreases by 30% of its original value. During the second year, your investment increases by 40% of its first-year value. Your advisor tells you that there must have been a 10% overall increase of your original $10,000 investment. Is your financial advisor using percentages properly? If not, what is the actual percent gain or loss on your original $10,000 investment?

90. The price of a color printer is reduced by 30% of its original price. When it still does not sell, its price is reduced by 20% of the reduced price. The salesperson informs you that there has been a total reduction of 50%. Is the salesperson using percentages properly? If not, what is the actual percent reduction from the original price?

Writing in Mathematics

91. Explain what it means to solve a formula for a variable.

92. What is a percent?

93. Describe how to express a decimal number as a percent and give an example.

94. Describe how to express a percent as a decimal number and give an example.

95. What does the percent formula, $A = PB$, describe? Give an example of how the formula is used.

Critical Thinking Exercises

Make Sense? *In Exercises 96–99, determine whether each statement "makes sense" or "does not make sense" and explain your reasoning.*

96. To help me get started, I circle the variable that I need to solve for in a formula.

97. By solving a formula for one of its variables, I find a numerical value for that variable.

98. I have $100 and my restaurant bill comes to $80, which is not enough to leave a 20% tip.

99. I found the percent decrease in a jacket's price to be 120%.

In Exercises 100–103, determine whether each statement is true or false. If the statement is false, make the necessary change(s) to produce a true statement.

100. If $ax + b = 0$, then $x = \dfrac{b}{a}$.

101. If $A = lw$, then $w = \dfrac{l}{A}$.

102. If $A = \dfrac{1}{2}bh$, then $b = \dfrac{A}{2h}$.

103. Solving $x - y = -7$ for y gives $y = x + 7$.

104. In psychology, an intelligence quotient, Q, also called IQ, is measured by the formula

$$Q = \frac{100M}{C},$$

where M = mental age and C = chronological age. Solve the formula for C.

Review Exercises

105. Solve and check: $5x + 20 = 8x - 16$. (Section 2.2, Example 7)

106. Solve and check: $5(2y - 3) - 1 = 4(6 + 2y)$. (Section 2.3, Example 3)

107. Simplify: $x - 0.3x$. (Section 1.4, Example 8)

Preview Exercises

Exercises 108–110 will help you prepare for the material covered in the next section. In each exercise, let x represent the number and write the phrase as an algebraic expression.

108. The quotient of 13 and a number, decreased by 7 times the number

109. Eight times the sum of a number and 14

110. Nine times the difference of a number and 5

| **MID-CHAPTER CHECK POINT** | Section 2.1–Section 2.4 |

✓ **What You Know:** We learned a step-by-step procedure for solving linear equations, including equations with fractions. We saw that some equations have no solution, whereas others have all real numbers as solutions. We used the addition and multiplication properties of equality to solve formulas for variables. Finally, we worked with the percent formula $A = PB$: A is P percent of B.

1. Solve: $\dfrac{x}{2} = 12 - \dfrac{x}{4}$.

2. Solve: $5x - 42 = -57$.

3. Solve for C: $H = \dfrac{EC}{825}$.

4. What is 6% of 140?

5. Solve: $\dfrac{-x}{10} = -3$.

6. Solve: $1 - 3(y - 5) = 4(2 - 3y)$.

7. Solve for r: $S = 2\pi rh$.

8. 12 is 30% of what?

9. Solve: $\dfrac{3y}{5} + \dfrac{y}{2} = \dfrac{5y}{4} - 3$.

10. Solve: $5z + 7 = 6(z - 2) - 4(2z - 3)$.

11. Solve for x: $Ax - By = C$.

12. Solve: $6y + 7 + 3y = 3(3y - 1)$.

13. Solve: $10\left(\dfrac{1}{2}x + 3\right) = 10\left(\dfrac{3}{5}x - 1\right)$.

14. 50 is what percent of 400?

15. Solve: $\dfrac{3(m + 2)}{4} = 2m + 3$.

16. If 40 is increased to 50, the increase is what percent of the original number?

17. Solve: $12x - 4 + 8x - 4 = 4(5x - 2)$.

18. The bar graph shows gun ownership in the United States for selected years from 1980 through 2004.

Percentage of Americans with Guns in Their Homes

A bar graph with the vertical axis labeled "Percentage with Guns" ranging from 0 to 70% and the horizontal axis labeled "Year." Bar values: 1980: 47.7, 1985: 44.3, 1990: 42.7, 1996: 40.2, 2004: 36.5.

Source: The General Social Survey

The data can be described by the mathematical model

$$G = -\frac{1}{2}n + 47,$$

where G is the percentage of Americans with guns in their homes n years after 1980.

a. Does the formula underestimate or overestimate the percentage with guns in 1990? By how much?

b. If trends shown by the formula continue, when will the percentage of Americans with guns in their homes decline to 30%?

SECTION

2.5

An Introduction to Problem Solving

Objectives

1 Translate English phrases into algebraic expressions.

2 Solve algebraic word problems using linear equations.

How Long It Takes to Earn $1000

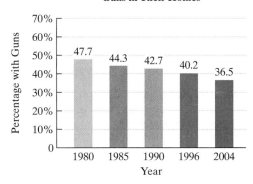

Howard Stern Radio host 24 sec.	**Dr. Phil McGraw** Television host 2 min. 24 sec.	**Brad Pitt** Actor 4 min. 48 sec.	**Kobe Bryant** Basketball player 5 min. 30 sec.
Chief executive U.S. average 2 hr. 55 min.	**Doctor, G.P.** U.S. average 13 hr. 5 min.	**High school teacher** U.S. average 43 hours	**Janitor** U.S. average 103 hours

Source: *Time*

In this section, you'll see examples and exercises focused on how much money Americans earn. These situations illustrate a step-by-step strategy for solving problems. As you become familiar with this strategy, you will learn to solve a wide variety of problems.

1 Translate English phrases into algebraic expressions.

Algebraic Expressions for English Phrases

The hardest thing about word problems is writing an equation that translates, or models, the problem's conditions. Throughout Chapter 1, you wrote algebraic expressions and equations for conditions about numbers. **Table 2.1** summarizes many of the algebraic expressions that you wrote for these conditions. We are using x to represent the variable, but we can use any letter.

Study Tip

Cover the right column in **Table 2.1** with a sheet of paper and attempt to formulate the algebraic expression for the English phrase in the left column on your own. Then slide the paper down and check your answer. Work through the entire table in this manner.

Table 2.1 **Algebraic Translations of English Phrases**

English Phrase	Algebraic Expression
Addition	
The sum of a number and 7	$x + 7$
Five more than a number; a number plus 5	$x + 5$
A number increased by 6; 6 added to a number	$x + 6$
Subtraction	
A number minus 4	$x - 4$
A number decreased by 5	$x - 5$
A number subtracted from 8	$8 - x$
The difference between a number and 6	$x - 6$
The difference between 6 and a number	$6 - x$
Seven less than a number	$x - 7$
Seven minus a number	$7 - x$
Nine fewer than a number	$x - 9$
Multiplication	
Five times a number	$5x$
The product of 3 and a number	$3x$
Two-thirds of a number (used with fractions)	$\frac{2}{3}x$
Seventy-five percent of a number (used with decimals)	$0.75x$
Thirteen multiplied by a number	$13x$
A number multiplied by 13	$13x$
Twice a number	$2x$
Division	
A number divided by 3	$\dfrac{x}{3}$
The quotient of 7 and a number	$\dfrac{7}{x}$
The quotient of a number and 7	$\dfrac{x}{7}$
The reciprocal of a number	$\dfrac{1}{x}$
More than one operation	
The sum of twice a number and 7	$2x + 7$
Twice the sum of a number and 7	$2(x + 7)$
Three times the sum of 1 and twice a number	$3(1 + 2x)$
Nine subtracted from 8 times a number	$8x - 9$
Twenty-five percent of the sum of 3 times a number and 14	$0.25(3x + 14)$
Seven times a number, increased by 24	$7x + 24$
Seven times the sum of a number and 24	$7(x + 24)$

A Strategy for Solving Word Problems Using Equations

Here are some general steps we will follow in solving word problems:

Strategy for Solving Word Problems

Step 1. Read the problem carefully. Attempt to state the problem in your own words and state what the problem is looking for. Let x (or any variable) represent one of the unknown quantities in the problem.

Step 2. If necessary, write expressions for any other unknown quantities in the problem in terms of x.

Step 3. Write an equation in x that translates, or models, the conditions of the problem.

Step 4. Solve the equation and answer the problem's question.

Step 5. Check the solution *in the original wording* of the problem, not in the equation obtained from the words.

Study Tip

There is great value in reasoning through the steps for solving a word problem. This value comes from the problem-solving skills that you will attain and is often more important than the specific problem or its solution.

2 Solve algebraic word problems using linear equations.

Applying the Strategy for Solving Word Problems

Now that you've read the study tip, let's apply our five-step strategy for solving word problems.

EXAMPLE 1   Solving a Word Problem

Nine subtracted from eight times a number is 39. Find the number.

Solution

Step 1. Let x represent one of the quantities. Because we are asked to find a number, let

$$x = \text{the number.}$$

Step 2. Represent other unknown quantities in terms of x. There are no other unknown quantities to find, so we can skip this step.

Step 3. Write an equation in x that models the conditions.

Nine subtracted from	eight times a number	is	39.
$8x$	$- 9$	$=$	39

Step 4. Solve the equation and answer the question.

$$8x - 9 = 39 \qquad \text{This is the equation that models the problem's conditions.}$$
$$8x - 9 + 9 = 39 + 9 \qquad \text{Add 9 to both sides.}$$
$$8x = 48 \qquad \text{Simplify.}$$
$$\frac{8x}{8} = \frac{48}{8} \qquad \text{Divide both sides by 8.}$$
$$x = 6 \qquad \text{Simplify.}$$

The number is 6.

Step 5. Check the proposed solution in the original wording of the problem. "Nine subtracted from eight times a number is 39." The proposed number is 6. Eight times 6 is $8 \cdot 6$, or 48. Nine subtracted from 48 is $48 - 9$, or 39. The proposed solution checks in the problem's wording, verifying that the number is 6.

☑ **CHECK POINT 1**   Four subtracted from six times a number is 68. Find the number.

EXAMPLE 2   Academic Salary Gap

According to the U.S. Department of Education (2007 data), there is a gap between teaching salaries for men and women at private colleges and universities. The average salary for men exceeds the average salary for women by $14,037. Combined, their average salaries are $130,015. Determine the average teaching salaries at private colleges for women and for men.

Solution

Step 1. Let x represent one of the quantities. We know something about the average salary for men: It exceeds the average salary for women by $14,037. This means that men earn $14,037 more than women. We will let

$$x = \text{the average teaching salary for women.}$$

Step 2. Represent other unknown quantities in terms of x. The other unknown quantity is the average teaching salary for men. Because it is $14,037 more than women, let

$$x + 14{,}037 = \text{the average teaching salary for men.}$$

Step 3. Write an equation in x that models the conditions. Combined, the average salaries for women and men is $130,015.

The average salary for women	plus	the average salary for men	is	$130,015.
x	$+$	$(x + 14{,}037)$	$=$	$130{,}015$

Step 4. Solve the equation and answer the question.

$$x + (x + 14{,}037) = 130{,}015 \qquad \text{This is the equation that models the problem's conditions.}$$

$$2x + 14{,}037 = 130{,}015 \qquad \text{Regroup and combine like terms on the left side.}$$

$$2x + 14{,}037 - 14{,}037 = 130{,}015 - 14{,}037 \qquad \text{Subtract 14,037 from both sides.}$$

$$2x = 115{,}978 \qquad \text{Simplify.}$$

$$\frac{2x}{2} = \frac{115{,}978}{2} \qquad \text{Divide both sides by 2.}$$

$$x = 57{,}989 \qquad \text{Simplify.}$$

Because x represents the average teaching salary for women, we see that women earn an average of $57,989 at private colleges. Because $x + 14,037$ represents the average teaching salary for men, they earn $57,989 + 14,037$, or $72,026 at private colleges.

Step 5. Check the proposed solution in the original wording of the problem. The problem states that combined, the average teaching salaries are $130,015. By adding $57,989, the women's average salary, and $72,026, the men's average salary, we do, indeed, obtain a sum of $130,015.   ▬

Study Tip

Modeling with the word "exceeds" can be a bit tricky. It's helpful to identify the smaller quantity. Then add to this quantity to represent the larger quantity. For example, suppose that Tim's height exceeds Tom's height by a inches. Tom is the shorter person. If Tom's height is represented by x, then Tim's height is represented by $x + a$.

✓ **CHECK POINT 2**   The average salary for lawyers exceeds the average salary for elementary school teachers by \$54,890. (*Source*: 2007 data from salary.com) Combined, the average salaries for both professions are \$142,970. Determine the average salaries for elementary school teachers and for lawyers.

EXAMPLE 3   Consecutive Integers and the Super Bowl

Only once, in 1991, were the winning and losing scores in the Super Bowl consecutive integers. The New York Giants beat the Buffalo Bills in a nearly error-free game. If the sum of the scores was 39, determine the points scored by the losing team, the Bills, and the winning team, the Giants.

Solution

Step 1. Let x represent one of the quantities. We will let

$$x = \text{points scored by the losing team, the Bills.}$$

Step 2. Represent other unknown quantities in terms of x. The other unknown quantity involves points scored by the winning team, the Giants. Because the scores were consecutive integers, the winning team scored one point more than the losing team. Thus,

$$x + 1 = \text{points scored by the winning team, the Giants.}$$

Step 3. Write an equation in x that models the conditions. We are told that the sum of the scores was 39.

Points scored by the losing team	plus	points scored by the winning team	result in	39 points.
x	$+$	$(x + 1)$	$=$	39

Step 4. Solve the equation and answer the question.

$$x + (x + 1) = 39 \qquad \text{This is the equation that models the problem's conditions.}$$
$$2x + 1 = 39 \qquad \text{Regroup and combine like terms.}$$
$$2x + 1 - 1 = 39 - 1 \qquad \text{Subtract 1 from both sides.}$$
$$2x = 38 \qquad \text{Simplify.}$$
$$\frac{2x}{2} = \frac{38}{2} \qquad \text{Divide both sides by 2.}$$
$$x = 19 \qquad \text{Simplify.}$$

Thus,

$$\text{points scored by the losing team, the Bills } = x = 19$$

and

$$\text{points scored by the winning team, the Giants } = x + 1 = 20.$$

With the closest final score in Super Bowl history, the Giants scored 20 points and the Bills scored 19 points.

Step 5. Check the proposed solution in the original wording of the problem. The problem states that the sum of the scores was 39. With a final score of 20 to 19, we see that the sum of these numbers is, indeed, 39.

✓ **CHECK POINT 3** Page numbers on facing pages of a book are consecutive integers. Two pages that face each other have 145 as the sum of their page numbers. What are the page numbers?

Example 3 and Check Point 3 involved consecutive integers. By contrast, some word problems involve consecutive odd integers, such as 5, 7, and 9. Other word problems involve consecutive even integers, such as 6, 8, and 10. When working with consecutive even or consecutive odd integers, we must continually add 2 to move from one integer to the next successive integer in the list.

Table 2.2 should be helpful in solving consecutive integer problems.

Table 2.2	Consecutive Integers	
English Phrase	**Algebraic Expressions**	**Example**
Two consecutive integers	$x, x + 1$	13, 14
Three consecutive integers	$x, x + 1, x + 2$	$-8, -7, -6$
Two consecutive even integers	$x, x + 2$	40, 42
Two consecutive odd integers	$x, x + 2$	$-37, -35$
Three consecutive even integers	$x, x + 2, x + 4$	30, 32, 34
Three consecutive odd integers	$x, x + 2, x + 4$	9, 11, 13

EXAMPLE 4 Renting a Car

Rent-a-Heap Agency charges $125 per week plus $0.20 per mile to rent a small car. How many miles can you travel for $335?

Solution

Step 1. Let x represent one of the quantities. Because we are asked to find the number of miles we can travel for $335, let

$$x = \text{the number of miles.}$$

Step 2. Represent other unknown quantities in terms of x. There are no other unknown quantities to find, so we can skip this step.

Step 3. Write an equation in x that models the conditions. Before writing the equation, let us consider a few specific values for the number of miles traveled. The rental charge is $125 plus $0.20 for each mile.

3 miles: The rental charge is $125 + \$0.20(3)$.

30 miles: The rental charge is $125 + \$0.20(30)$.

100 miles: The rental charge is $125 + \$0.20(100)$.

x miles: The rental charge is $125 + \$0.20x$.

The weekly charge of \$125	plus	the charge of \$0.20 per mile for x miles	equals	the total \$335 rental charge.
125	+	$0.20x$	=	335

Step 4. Solve the equation and answer the question.

$$125 + 0.20x = 335 \qquad \text{This is the equation that models the conditions of the problem.}$$

$$125 + 0.20x - 125 = 335 - 125 \qquad \text{Subtract 125 from both sides.}$$

$$0.20x = 210 \qquad \text{Simplify.}$$

$$\frac{0.20x}{0.20} = \frac{210}{0.20} \qquad \text{Divide both sides by 0.20.}$$

$$x = 1050 \qquad \text{Simplify.}$$

You can travel 1050 miles for $335.

Step 5. Check the proposed solution in the original wording of the problem. Traveling 1050 miles should result in a total rental charge of $335. The mileage charge of $0.20 per mile is

$$\$0.20(1050) = \$210.$$

Adding this to the $125 weekly charge gives a total rental charge of

$$\$125 + \$210 = \$335.$$

Because this results in the given rental charge of $335, this verifies that you can travel 1050 miles. ▬

☑ **CHECK POINT 4**   A taxi charges $2.00 to turn on the meter plus $0.25 for each eighth of a mile. If you have $10.00, how many eighths of a mile can you go? How many miles is that?

We will be using the formula for the perimeter of a rectangle, $P = 2l + 2w$, in our next example. Twice the rectangle's length plus twice the rectangle's width is its perimeter.

EXAMPLE 5   Finding the Dimensions of a Soccer Field

A rectangular soccer field is twice as long as it is wide. If the perimeter of a soccer field is 300 yards, what are the field's dimensions?

Solution

Step 1. Let x represent one of the quantities. We know something about the length; the field is twice as long as it is wide. We will let

$$x = \text{the width.}$$

Step 2. Represent other unknown quantities in terms of x. Because the field is twice as long as it is wide, let

$$2x = \text{the length.}$$

Figure 2.8 illustrates the soccer field and its dimensions.

Study Tip

When solving word problems, particularly problems involving geometric figures, drawing a picture of the situation is often helpful. Label x on your drawing and, where appropriate, label other parts of the drawing in terms of x.

FIGURE 2.8

FIGURE 2.8 (repeated)

Step 3. Write an equation in *x* that models the conditions. Because the perimeter of a soccer field is 300 yards,

Twice the length	plus	twice the width	is	the perimeter.
$2 \cdot 2x$	$+$	$2 \cdot x$	$=$	$300.$

Step 4. Solve the equation and answer the question.

$$2 \cdot 2x + 2 \cdot x = 300 \qquad \text{This is the equation that models the problem's conditions.}$$

$$4x + 2x = 300 \qquad \text{Multiply.}$$

$$6x = 300 \qquad \text{Combine like terms.}$$

$$\frac{6x}{6} = \frac{300}{6} \qquad \text{Divide both sides by 6.}$$

$$x = 50 \qquad \text{Simplify.}$$

Thus,
$$\text{width} = x = 50$$
$$\text{length} = 2x = 2(50) = 100.$$

The dimensions of a soccer field are 50 yards by 100 yards.

Step 5. Check the proposed solution in the original wording of the problem. The perimeter of the soccer field, using the dimensions that we found, is 2(50 yards) + 2(100 yards) = 100 yards + 200 yards, or 300 yards. Because the problem's wording tells us that the perimeter is 300 yards, our dimensions are correct. ▪

✓ **CHECK POINT 5**   A rectangular swimming pool is three times as long as it is wide. If the perimeter of the pool is 320 feet, what are the pool's dimensions?

EXAMPLE 6   A Price Reduction

Your local computer store is having a sale. After a 30% price reduction, you purchase a new computer for $980. What was the computer's price before the reduction?

Solution

Step 1. Let *x* represent one of the quantities. We will let
$$x = \text{the original price of the computer.}$$

Step 2. Represent other unknown quantities in terms of *x*. There are no other unknown quantities to find, so we can skip this step.

Step 3. Write an equation in *x* that models the conditions. The computer's original price minus the 30% reduction is the reduced price, $980.

Original price	minus	the reduction (30% of the original price)	is	the reduced price, $980.
x	$-$	$0.3x$	$=$	980

Step 4. Solve the equation and answer the question.

$$x - 0.3x = 980 \qquad \text{This is the equation that models the problem's conditions.}$$

$$0.7x = 980 \qquad \text{Combine like terms: } x - 0.3x = 1x - 0.3x = 0.7x.$$

$$\frac{0.7x}{0.7} = \frac{980}{0.7} \qquad \text{Divide both sides by 0.7.}$$

$$x = 1400 \qquad \text{Simplify: } 0.7{\overline{)980.0}}\ ^{1400}$$

The computer's price before the reduction was $1400.

Step 5. Check the proposed solution in the original wording of the problem. The price before the reduction, $1400, minus the reduction in price should equal the reduced price given in the original wording, $980. The reduction in price is equal to 30% of the price before the reduction, $1400. To find the reduction, we multiply the decimal equivalent of 30%, 0.30 or 0.3, by the original price, $1400:

$$30\% \text{ of } \$1400 = (0.3)(\$1400) = \$420.$$

Now we can determine whether the calculation for the price before the reduction, $1400, minus the reduction, $420, is equal to the reduced price given in the problem, $980. We subtract:

$$\$1400 - \$420 = \$980.$$

This verifies that the price of the computer before the reduction was $1400.

☑ **CHECK POINT 6** After a 40% price reduction, an exercise machine sold for $564. What was the exercise machine's price before this reduction?

2.5 EXERCISE SET **MyMathLab**
PRACTICE   WATCH   DOWNLOAD   READ   REVIEW

Study Tip

Do not expect to solve every word problem immediately. As you read each problem, underline the important parts and don't quit. It's a good idea to read the problem at least twice. Be persistent. Avoid these false beliefs:

Challenging Your Negative Beliefs

If I can't solve the problem immediately, I never will. — **FALSE** — If I can't complete this homework, I will never be able to get algebra.

Practice Exercises

In Exercises 1–20, let x represent the number. Use the given conditions to write an equation. Solve the equation and find the number.

1. A number increased by 60 is equal to 410. Find the number.

2. The sum of a number and 43 is 107. Find the number.

3. A number decreased by 23 is equal to 214. Find the number.

4. The difference between a number and 17 is 96. Find the number.

5. The product of 7 and a number is 126. Find the number.

6. The product of 8 and a number is 272. Find the number.

7. The quotient of a number and 19 is 5. Find the number.

8. The quotient of a number and 14 is 8. Find the number.

9. The sum of four and twice a number is 56. Find the number.

10. The sum of five and three times a number is 59. Find the number.

11. Seven subtracted from five times a number is 178. Find the number.

12. Eight subtracted from six times a number is 298. Find the number.

13. A number increased by 5 is two times the number. Find the number.

14. A number increased by 12 is four times the number. Find the number.

15. Twice the sum of four and a number is 36. Find the number.

16. Three times the sum of five and a number is 48. Find the number.

17. Nine times a number is 30 more than three times that number. Find the number.

18. Five more than four times a number is that number increased by 35. Find the number.

19. If the quotient of three times a number and five is increased by four, the result is 34. Find the number.

20. If the quotient of three times a number and four is decreased by three, the result is nine. Find the number.

Application Exercises

In Exercises 21–46, use the five-step strategy to solve each problem.

21. The bar graph shows the time Americans spent using various media in 2007.

Average Number of Hours Americans Used Various Media in 2007

Source: Communications Industry Forecast and Report

Time spent watching TV exceeded time spent listening to radio by 581 hours. The combined time devoted to these two media was 2529 hours. In 2007, how many hours did Americans spend listening to radio and how many hours were spent watching TV?

22. Compared with Europeans, American employees use less vacation time.

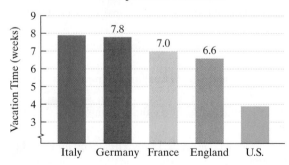

Average Vacation Time for Europeans and Americans

Source: The State of Working America 2006/2007

The average time Italians spend on vacation exceeds the average American vacation time by 4 weeks. The combined average vacation time for Americans and Italians is 11.8 weeks. On average, how many weeks do Americans spend on vacation and how many weeks do Italians spend on vacation?

Exercises 23–24 involve the average salaries represented by the bar graph.

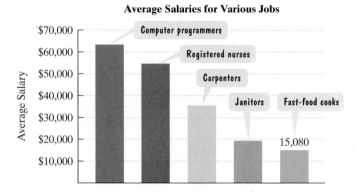

Average Salaries for Various Jobs

Source: 2007 data from salary.com

23. The average salary for computer programmers is $7740 less than twice the average salary for carpenters. Combined, their average salaries are $99,000. Determine the average salary for each of these jobs.

24. The average salary for registered nurses is $3500 less than three times the average salary for janitors. Combined, their average salaries are $74,060. Determine the average salary for each of these jobs.

In Exercises 25–26, use the fact that page numbers on facing pages of a book are consecutive integers.

25. The sum of the page numbers on the facing pages of a book is 629. What are the page numbers?

26. The sum of the page numbers on the facing pages of a book is 525. What are the page numbers?

27. Roger Maris and Babe Ruth are among the ten baseball players with the most home runs in a major league baseball season.

The number of home runs by these players for the seasons shown are consecutive odd integers whose sum is 120. Determine the number of homers hit by Ruth and by Maris.

28. Babe Ruth and Hank Greenberg are among the ten baseball players with the most home runs in a major league baseball season.

The number of home runs by these players for the seasons shown are consecutive even integers whose sum is 118. Determine the number of homers hit by Greenberg and by Ruth.

29. A car rental agency charges $200 per week plus $0.15 per mile to rent a car. How many miles can you travel in one week for $320?

30. A car rental agency charges $180 per week plus $0.25 per mile to rent a car. How many miles can you travel in one week for $395?

The bar graph shows that average rent and mortgage payments in the United States have increased since 1975, even after taking inflation into account. Exercises 31–32 are based on the information displayed by the graph.

Source: *The State of the Nation's Housing,* Harvard University, 2005

31. In 2004, mortgage payments averaged $960 per month. For the period shown, monthly mortgage payments increased by approximately $11.40 per year. If this trend continues, how many years after 2004 will mortgage payments average $1074? In which year will this occur?

32. In 2004, rent payments averaged $630 per month. For the period shown, monthly rent payments increased by approximately $1.40 per year. If this trend continues, how many years after 2004 will rent payments average $651? In which year will this occur?

33. A rectangular field is four times as long as it is wide. If the perimeter of the field is 500 yards, what are the field's dimensions?

34. A rectangular field is five times as long as it is wide. If the perimeter of the field is 288 yards, what are the field's dimensions?

35. An American football field is a rectangle with a perimeter of 1040 feet. The length is 200 feet more than the width. Find the width and length of the rectangular field.

36. A basketball court is a rectangle with a perimeter of 86 meters. The length is 13 meters more than the width. Find the width and length of the basketball court.

37. A bookcase is to be constructed as shown in the figure. The length is to be 3 times the height. If 60 feet of lumber is available for the entire unit, find the length and height of the bookcase.

38. The height of the bookcase in the figure is 3 feet longer than the length of a shelf. If 18 feet of lumber is available for the entire unit, find the length and height of the unit.

39. After a 20% reduction, you purchase a television for $320. What was the television's price before the reduction?

40. After a 30% reduction, you purchase a DVD player for $98. What was the price before the reduction?

41. This year's salary, $50,220, is an 8% increase over last year's salary. What was last year's salary?

42. This year's salary, $42,074, is a 9% increase over last year's salary. What was last year's salary?

43. Including 6% sales tax, a car sold for $23,850. Find the price of the car before the tax was added.

44. Including 8% sales tax, a bed-and-breakfast inn charges $172.80 per night. Find the inn's nightly cost before the tax is added.

45. An automobile repair shop charged a customer $448, listing $63 for parts and the remainder for labor. If the cost of labor is $35 per hour, how many hours of labor did it take to repair the car?

46. A repair bill on a sailboat came to $1603, including $532 for parts and the remainder for labor. If the cost of labor is $63 per hour, how many hours of labor did it take to repair the sailboat?

Writing in Mathematics

47. In your own words, describe a step-by-step approach for solving algebraic word problems.

48. Many students find solving linear equations much easier than solving algebraic word problems. Discuss some of the reasons why this is the case.

49. Did you have some difficulties solving some of the problems that were assigned in this exercise set? Discuss what you did if this happened to you. Did your course of action enhance your ability to solve algebraic word problems?

50. Write an original word problem that can be solved using a linear equation. Then solve the problem.

Critical Thinking Exercises

Make Sense? *In Exercises 51–54, determine whether each statement "makes sense" or "does not make sense" and explain your reasoning.*

51. Rather than struggling with the assigned word problems, I'll ask my instructor to solve them all in class and then study the solutions.

52. By reasoning through word problems, I can increase my problem-solving skills in general.

53. I find the hardest part in solving a word problem is writing the equation that models the verbal conditions.

54. I made a mistake when I used x and $x + 2$ to represent two consecutive odd integers, because 2 is even.

In Exercises 55–58, determine whether each statement is true or false. If the statement is false, make the necessary change(s) to produce a true statement.

55. Ten pounds less than Bill's weight, x, equals 160 pounds is modeled by $10 - x = 160$.

56. After a 35% reduction, a computer's price is $780, so its original price, x, can be found by solving $x - 0.35 = 780$.

57. If the length of a rectangle is 6 inches more than its width, and its perimeter is 24 inches, the distributive property must be used to solve the equation that determines the length.

58. On a number line, consecutive integers do not have any other integers between them.

59. An HMO pamphlet contains the following recommended weight for women: "Give yourself 100 pounds for the first 5 feet plus 5 pounds for every inch over 5 feet tall." Using this description, which height corresponds to an ideal weight of 135 pounds?

60. The rate for a particular international telephone call is $0.55 for the first minute and $0.40 for each additional minute. Determine the length of a call that costs $6.95.

61. In a film, the actor Charles Coburn played an elderly "uncle" character criticized for marrying a woman when he is 3 times her age. He wittily replies, "Ah, but in 20 years time I shall only be twice her age." How old is the "uncle" and the woman?

62. Answer the question in the following *Peanuts* cartoon strip. (*Note:* You may not use the answer given in the cartoon!)

PEANUTS © United Feature Syndicate, Inc.

Review Exercises

63. Solve and check: $\dfrac{4}{5}x = -16$. (Section 2.2, Example 3)

64. Solve and check: $6(y - 1) + 7 = 9y - y + 1$. (Section 2.3, Example 3)

65. Solve for w: $V = \dfrac{1}{3}lwh$. (Section 2.4, Example 4)

Preview Exercises

Exercises 66–68 will help you prepare for the material covered in the next section.

66. Is 2 a solution of $x + 3 < 8$?

67. Is 6 a solution of $4y - 7 \geq 5$?

68. Solve: $2(x - 3) + 5x = 8(x - 1)$.

Objectives

1 Graph the solutions of an inequality on a number line.

2 Use interval notation.

3 Understand properties used to solve linear inequalities.

4 Solve linear inequalities.

5 Identify inequalities with no solution or infinitely many solutions.

6 Solve problems using linear inequalities.

Solving Linear Inequalities

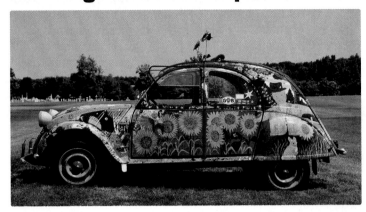

Do you remember Rent-a-Heap, the car rental company that charged $125 per week plus $0.20 per mile to rent a small car? In Example 4 on page 152, we asked the question: How many miles can you travel for $335? We let x represent the number of miles and set up a linear equation as follows:

The weekly charge of $125	plus	the charge of $0.20 per mile for x miles	equals	the total $335 rental charge.
125	+	0.20x	=	335.

Because we are limited by how much money we can spend on everything from buying clothing to renting a car, it is also possible to ask: How many miles can you travel if you can spend *at most* $335? We again let x represent the number of miles. Spending *at most* $335 means that the amount spent on the weekly rental must be *less than or equal to* $335:

The weekly charge of $125	plus	the charge of $0.20 per mile for x miles	must be less than or equal to	$335.
125	+	0.20x	≤	335.

Using the commutative property of addition, we can express this inequality as

$$0.20x + 125 \le 335.$$

The form of this inequality is $ax + b \le c$, with $a = 0.20$, $b = 125$, and $c = 335$. Any inequality in this form is called a **linear inequality in one variable**. The symbol between $ax + b$ and c can be $\le$ (is less than or equal to), $<$ (is less than), $\ge$ (is greater than or equal to), or $>$ (is greater than). The greatest exponent on the variable in such an inequality is 1.

In this section, we will study how to solve linear inequalities such as $0.20x + 125 \le 335$. **Solving an inequality** is the process of finding the set of numbers that will make the inequality a true statement. These numbers are called the **solutions** of the inequality, and we say that they **satisfy** the inequality. The set of all solutions is called the **solution set** of the inequality. We begin by discussing how to graph and how to represent these solution sets.

1 Graph the solutions of an inequality on a number line.

Graphs of Inequalities

There are infinitely many solutions to the inequality $x < 3$, namely, all real numbers that are less than 3. Although we cannot list all the solutions, we can make a drawing on a number line that represents these solutions. Such a drawing is called the **graph of the inequality**.

Graphs of solutions to linear inequalities are shown on a number line by shading all points representing numbers that are solutions. **Square brackets, [ ], indicate endpoints that are solutions. Parentheses, ( ), indicate endpoints that are not solutions.**

Study Tip

Because an inequality symbol points to the smaller number, $x < 3$ (x is less than 3) may be expressed as $3 > x$ (3 is greater than x).

EXAMPLE 1   Graphing Inequalities

Graph the solutions of each inequality:

    a. $x < 3$ **b.** $x \geq -1$ **c.** $-1 < x \leq 3$.

Solution

    a. The solutions of $x < 3$ are all real numbers that are less than 3. They are graphed on a number line by shading all points to the left of 3. The parenthesis at 3 indicates that 3 is not a solution, but numbers such as 2.9999 and 2.6 are. The arrow shows that the graph extends indefinitely to the left.

    b. The solutions of $x \geq -1$ are all real numbers that are greater than or equal to -1. We shade all points to the right of -1 and the point for -1 itself. The square bracket at -1 shows that -1 is a solution of the given inequality. The arrow shows that the graph extends indefinitely to the right.

    c. The inequality $-1 < x \leq 3$ is read "-1 is less than x *and* x is less than or equal to 3," or "x is greater than -1 *and* less than or equal to 3." The solutions of $-1 < x \leq 3$ are all real numbers between -1 and 3, not including -1 but including 3. The parenthesis at -1 indicates that -1 is not a solution. The square bracket at 3 shows that 3 is a solution. Shading indicates the other solutions.

    ☑ **CHECK POINT 1**   Graph the solutions of each inequality:

    a. $x < 4$ **b.** $x \geq -2$ **c.** $-4 \leq x < 1$.

---

2 Use interval notation.

Interval Notation

The solutions of $x < 3$ are all real numbers that are less than 3:

These numbers form an *interval* on the number line. The solution set of $x < 3$ can be expressed in **interval notation** as

$$(-\infty, 3).$$

The negative infinity symbol indicates that the interval extends indefinitely to the left.

The parenthesis indicates that 3 is excluded from the interval.

The solution set of $x < 3$ can also be expressed in set-builder notation as

$$\{ \ x \ | \ x \ < \ 3 \ \}.$$

| The set of | all x | such that | x is less than 3. |

Table 2.3 shows four inequalities, their solution sets using interval and set-builder notations, and graphs of the solution sets.

Table 2.3   Solution Sets of Inequalities

Let a and b be real numbers.

Inequality	Interval Notation	Set-Builder Notation	Graph
$x > a$	(a, ∞)	$\{x \mid x > a\}$	
$x \geq a$	$[a, \infty)$	$\{x \mid x \geq a\}$	
$x < b$	$(-\infty, b)$	$\{x \mid x < b\}$	
$x \leq b$	$(-\infty, b]$	$\{x \mid x \leq b\}$	

Parentheses and Brackets in Interval Notation

Parentheses indicate endpoints that are not included in an interval. Square brackets indicate endpoints that are included in an interval. Parentheses are always used with ∞ or $-\infty$.

EXAMPLE 2   Using Interval Notation

Express the solution set of each inequality in interval notation and graph the interval:

a. $x \leq -1$ b. $x > 2$.

Solution

Inequality	Interval Notation	Graph
a. $x \leq -1$	$(-\infty, -1]$	
b. $x > 2$	$(2, \infty)$	

a. x is less than or equal to -1. | The interval extends indefinitely to the left. | The square bracket on the graph and in interval notation shows -1 is included in the interval.

b. x is greater than 2. | The interval extends indefinitely to the right. | The parenthesis on the graph and in interval notation shows 2 is excluded from the interval.

☑ **CHECK POINT 2** Express the solution set of each inequality in interval notation and graph the interval:

a. $x \geq 0$ **b.** $x < 5$.

---

3 Understand properties used to solve linear inequalities.

Properties Used to Solve Linear Inequalities

Back to our question that opened this section: How many miles can you drive your Rent-a-Heap car if you can spend at most $335 per week? We answer the question by solving

$$0.20x + 125 \leq 335$$

for x. The solution procedure is nearly identical to that for solving

$$0.20x + 125 = 335.$$

Our goal is to get x by itself on the left side. We do this by first subtracting 125 from both sides to isolate $0.20x$:

$0.20x + 125 \leq 335$	This is the given inequality.
$0.20x + 125 - 125 \leq 335 - 125$	Subtract 125 from both sides.
$0.20x \leq 210.$	Simplify.

Finally, we isolate x from $0.20x$ by dividing both sides of the inequality by 0.20:

$\dfrac{0.20x}{0.20} \leq \dfrac{210}{0.20}$	Divide both sides by 0.20.
$x \leq 1050.$	Simplify.

With at most $335 per week to spend, you can travel at most 1050 miles.

We started with the inequality $0.20x + 125 \leq 335$ and obtained the inequality $x \leq 1050$ in the final step. Both of these inequalities have the same solution set, namely $\{x \mid x \leq 1050\}$. Inequalities such as these, with the same solution set, are said to be **equivalent**.

We isolated x from $0.20x$ by dividing both sides of $0.20x \leq 210$ by 0.20, a positive number. Let's see what happens if we divide both sides of an inequality by a negative number. Consider the inequality $10 < 14$. Divide both 10 and 14 by -2:

$$\frac{10}{-2} = -5 \quad \text{and} \quad \frac{14}{-2} = -7.$$

Because -5 lies to the right of -7 on the number line, -5 is greater than -7:

$$-5 > -7.$$

Notice that the direction of the inequality symbol is reversed:

$$10 < 14$$

Dividing by -2 changes the direction of the inequality symbol.

$$-5 > -7.$$

In general, **when we multiply or divide both sides of an inequality by a negative number, the direction of the inequality symbol is reversed**. When we reverse the direction of the inequality symbol, we say that we change the *sense* of the inequality.

Study Tip

English phrases such as "at least" and "at most" can be represented by inequalities.

English Sentence	Inequality
x is at least 5.	$x \geq 5$
x is at most 5.	$x \leq 5$
x is no more than 5.	$x \leq 5$
x is no less than 5.	$x \geq 5$

We can isolate a variable in a linear inequality the same way we can isolate a variable in a linear equation. The following properties are used to create equivalent inequalities:

Properties of Inequalities

Property	The Property in Words	Example
The Addition Property of Inequality If $a < b$, then $a + c < b + c$. If $a < b$, then $a - c < b - c$.	If the same quantity is added to or subtracted from both sides of an inequality, the resulting inequality is equivalent to the original one.	$2x + 3 < 7$ Subtract 3: $2x + 3 - 3 < 7 - 3$. Simplify: $2x < 4$.
The Positive Multiplication Property of Inequality If $a < b$ and c is positive, then $ac < bc$. If $a < b$ and c is positive, then $\dfrac{a}{c} < \dfrac{b}{c}$.	If we multiply or divide both sides of an inequality by the same positive quantity, the resulting inequality is equivalent to the original one.	$2x < 4$ Divide by 2: $\dfrac{2x}{2} < \dfrac{4}{2}$. Simplify: $x < 2$.
The Negative Multiplication Property of Inequality If $a < b$ and c is negative, then $ac > bc$. If $a < b$ and c is negative, then $\dfrac{a}{c} > \dfrac{b}{c}$.	If we multiply or divide both sides of an inequality by the same negative quantity and reverse the direction of the inequality symbol, the resulting inequality is equivalent to the original one.	$-4x < 20$ Divide by -4 and reverse the direction of the inequality symbol: $\dfrac{-4x}{-4} > \dfrac{20}{-4}$. Simplify: $x > -5$.

4 Solve linear inequalities.

Solving Linear Inequalities Involving Only One Property of Inequality

If you can solve a linear equation, it is likely that you can solve a linear inequality. Why? The procedure for solving linear inequalities is nearly the same as the procedure for solving linear equations, with one important exception: **When multiplying or dividing by a negative number, reverse the direction of the inequality symbol, changing the sense of the inequality**.

EXAMPLE 3   Solving a Linear Inequality

Solve and graph the solution set on a number line:

$$x + 3 < 8.$$

Solution   Our goal is to isolate x. We can do this by using the addition property, subtracting 3 from both sides.

$$x + 3 < 8 \qquad \text{This is the given inequality.}$$
$$x + 3 - 3 < 8 - 3 \qquad \text{Subtract 3 from both sides.}$$
$$x < 5 \qquad \text{Simplify.}$$

The solution set consists of all real numbers that are less than 5. We express this in interval notation as $(-\infty, 5)$, or in set-builder notation as $\{x \mid x < 5\}$. The graph of the solution set is shown as follows:

Discover for Yourself

Can you check all solutions to Example 3 in the given inequality? Is a partial check possible? Select a real number that is less than 5 and show that it satisfies $x + 3 < 8$.

☑ **CHECK POINT 3** Solve and graph the solution set on a number line:

$$x + 6 < 9.$$

EXAMPLE 4 Solving a Linear Inequality

Solve and graph the solution set on a number line:

$$4x - 1 \geq 3x - 6.$$

Solution Our goal is to isolate all terms involving x on one side and all numerical terms on the other side, exactly as we did when solving equations. Let's begin by using the addition property to isolate variable terms on the left.

$4x - 1 \geq 3x - 6$	This is the given inequality.
$4x - 3x - 1 \geq 3x - 3x - 6$	Subtract 3x from both sides.
$x - 1 \geq -6$	Simplify.

Now we isolate the numerical terms on the right. Use the addition property and add 1 to both sides.

$x - 1 + 1 \geq -6 + 1$	Add 1 to both sides.
$x \geq -5$	Simplify.

The solution set consists of all real numbers that are greater than or equal to -5. We express this in interval notation as $[-5, \infty)$, or in set-builder notation as $\{x \mid x \geq -5\}$.

The graph of the solution set is shown as follows:

☑ **CHECK POINT 4** Solve and graph the solution set on a number line:

$$8x - 2 \geq 7x - 4.$$

We solved the inequalities in Examples 3 and 4 using the addition property of inequality. Now let's practice using the multiplication property of inequality. Do not forget to reverse the direction of the inequality symbol when multiplying or dividing both sides by a negative number.

EXAMPLE 5 Solving Linear Inequalities

Solve and graph the solution set on a number line:

a. $\frac{1}{3}x < 5$ **b.** $-3x < 21$.

Solution In each case, our goal is to isolate x. In the first inequality, this is accomplished by multiplying both sides by 3. In the second inequality, we can do this by dividing both sides by -3.

a. $\frac{1}{3}x < 5$ This is the given inequality.

$3 \cdot \frac{1}{3}x < 3 \cdot 5$ Isolate x by multiplying by 3 on both sides.

            The symbol $<$ stays the same because we are multiplying both sides by a positive number.

$x < 15$ Simplify.

The solution set consists of all real numbers that are less than 15. We express this in interval notation as $(-\infty, 15)$, or in set-builder notation as $\{x \mid x < 15\}$. The graph of the solution set is shown as follows:

b. $-3x < 21$ *This is the given inequality.*

$$\frac{-3x}{-3} > \frac{21}{-3}$$ *Isolate x by dividing by −3 on both sides.*
    The symbol < must be reversed because we are dividing both sides by a negative number.

$\qquad x > -7$ *Simplify.*

The solution set consists of all real numbers that are greater than -7. We express this in interval notation as $(-7, \infty)$, or in set-builder notation as $\{x \mid x > -7\}$. The graph of the solution set is shown as follows:

✓ **CHECK POINT 5**   Solve and graph the solution set on a number line:

a. $\dfrac{1}{4}x < 2$

b. $-6x < 18.$

Inequalities Requiring Both the Addition and Multiplication Properties

If a linear inequality does not contain fractions, it can be solved using the following procedure. Notice, again, how similar this procedure is to the procedure for solving an equation.

Solving a Linear Inequality

1. Simplify the algebraic expression on each side.
2. Use the addition property of inequality to collect all the variable terms on one side and all the constant terms on the other side.
3. Use the multiplication property of inequality to isolate the variable and solve. Change the sense of the inequality when multiplying or dividing both sides by a negative number.
4. Express the solution set in interval or set-builder notation, and graph the solution set on a number line.

EXAMPLE 6   Solving a Linear Inequality

Solve and graph the solution set on a number line:

$$4y - 7 \geq 5.$$

Solution

Step 1. Simplify each side. Because each side is already simplified, we can skip this step.

Step 2. Collect variable terms on one side and constant terms on the other side. The only variable term, $4y$, is already on the left side of $4y - 7 \geq 5$. We will collect constant terms on the right by adding 7 to both sides.

$$4y - 7 \geq 5 \qquad \text{\small This is the given inequality.}$$
$$4y - 7 + 7 \geq 5 + 7 \qquad \text{\small Add 7 to both sides.}$$
$$4y \geq 12 \qquad \text{\small Simplify.}$$

Step 3. Isolate the variable and solve. We isolate the variable, y, by dividing both sides by 4. Because we are dividing by a positive number, we do not reverse the inequality symbol.

$$\frac{4y}{4} \geq \frac{12}{4} \qquad \text{\small Divide both sides by 4.}$$
$$y \geq 3 \qquad \text{\small Simplify.}$$

Step 4. Express the solution set in interval or set-builder notation, and graph the set on a number line. The solution set consists of all real numbers that are greater than or equal to 3, expressed in interval notation as $[3, \infty)$, or in set-builder notation as $\{y | y \geq 3\}$. The graph of the solution set is shown as follows:

✓ **CHECK POINT 6** Solve and graph the solution set on a number line:

$$5y - 3 \geq 17.$$

EXAMPLE 7 Solving a Linear Inequality

Solve and graph the solution set on a number line:

$$7x + 15 \geq 13x + 51.$$

Solution

Step 1. Simplify each side. Because each side is already simplified, we can skip this step.

Step 2. Collect variable terms on one side and constant terms on the other side. We will collect variable terms on the left and constant terms on the right.

$$7x + 15 \geq 13x + 51 \qquad \text{\small This is the given inequality.}$$
$$7x + 15 - 13x \geq 13x + 51 - 13x \qquad \text{\small Subtract 13x from both sides.}$$
$$-6x + 15 \geq 51 \qquad \text{\small Simplify.}$$
$$-6x + 15 - 15 \geq 51 - 15 \qquad \text{\small Subtract 15 from both sides.}$$
$$-6x \geq 36 \qquad \text{\small Simplify.}$$

Step 3. Isolate the variable and solve. We isolate the variable, x, by dividing both sides by -6. Because we are dividing by a negative number, we must reverse the inequality symbol.

$$\frac{-6x}{-6} \leq \frac{36}{-6} \qquad \text{\small Divide both sides by -6 and change the sense of the inequality.}$$
$$x \leq -6 \qquad \text{\small Simplify.}$$

Step 4. Express the solution set in interval or set-builder notation, and graph the set on a number line. The solution set consists of all real numbers that are less than or equal to -6, expressed in interval notation as $(-\infty, -6]$, or in set-builder notation as $\{x | x \leq -6\}$. The graph of the solution set is shown as follows:

Study Tip

It is possible to perform a partial check for an inequality. Select one number from the solution set. Substitute that number into the original inequality and perform the resulting computations. You should obtain a true statement.

Study Tip

You can solve

$$7x + 15 \geq 13x + 51$$

by isolating x on the right side. Subtract $7x$ from both sides:

$$7x + 15 - 7x$$
$$\geq 13x + 51 - 7x$$
$$15 \geq 6x + 51.$$

Now subtract 51 from both sides:

$$15 - 51 \geq 6x + 51 - 51$$
$$-36 \geq 6x.$$

Finally, divide both sides by 6:

$$\frac{-36}{6} \geq \frac{6x}{6}$$
$$-6 \geq x.$$

This last inequality means the same thing as

$$x \leq -6.$$

☑ **CHECK POINT 7**   Solve and graph the solution set: $6 - 3x \leq 5x - 2$.

EXAMPLE 8   Solving a Linear Inequality

Solve and graph the solution set on a number line:

$$2(x - 3) + 5x \leq 8(x - 1).$$

Solution

Step 1. Simplify each side. We use the distributive property to remove parentheses. Then we combine like terms.

$2(x - 3) + 5x \leq 8(x - 1)$	This is the given inequality.
$2x - 6 + 5x \leq 8x - 8$	Use the distributive property.
$7x - 6 \leq 8x - 8$	Add like terms on the left.

Step 2. Collect variable terms on one side and constant terms on the other side. We will collect variable terms on the left and constant terms on the right.

$7x - 8x - 6 \leq 8x - 8x - 8$	Subtract 8x from both sides.
$-x - 6 \leq -8$	Simplify.
$-x - 6 + 6 \leq -8 + 6$	Add 6 to both sides.
$-x \leq -2$	Simplify.

Step 3. Isolate the variable and solve. To isolate x in $-x \leq -2$, we must eliminate the negative sign in front of the x. Because $-x$ means $-1x$, we can do this by multiplying (or dividing) both sides of the inequality by -1. We are multiplying by a negative number. Thus, we must reverse the inequality symbol.

$(-1)(-x) \geq (-1)(-2)$	Multiply both sides of $-x \leq -2$ by -1 and change the sense of the inequality.
$x \geq 2$	Simplify.

Step 4. Express the solution set in interval or set-builder notation, and graph the set on a number line. The solution set consists of all real numbers that are greater than or equal to 2, expressed in interval notation as $[2, \infty)$, or in set-builder notation as $\{x \mid x \geq 2\}$. The graph of the solution set is shown as follows:

☑ **CHECK POINT 8**   Solve and graph the solution set:

$$2(x - 3) - 1 \leq 3(x + 2) - 14.$$

---

5 Identify inequalities with no solution or infinitely many solutions.

Inequalities with Unusual Solution Sets

We have seen that some equations have no solution. This is also true for some inequalities. An example of such an inequality is

$$x > x + 1.$$

There is no number that is greater than itself plus 1. This inequality has no solution. Its solution set is $\varnothing$, the empty set.

By contrast, some inequalities are true for all real numbers. An example of such an inequality is

$$x < x + 1.$$

Every real number is less than itself plus 1. The solution set is expressed in interval notation as $(-\infty, \infty)$, or in set-builder notation as $\{x | x \text{ is a real number}\}$.

Recognizing Inequalities with No Solution or Infinitely Many Solutions

If you attempt to solve an inequality with no solution or one that is true for every real number, you will eliminate the variable.

- An inequality with no solution results in a false statement, such as $0 > 1$. The solution set is $\varnothing$, the empty set.

- An inequality that is true for every real number results in a true statement, such as $0 < 1$. The solution set is $(-\infty, \infty)$ or $\{x | x \text{ is a real number}\}$.

EXAMPLE 9 Solving a Linear Inequality

Solve: $3(x + 1) > 3x + 5$.

Solution

$$3(x + 1) > 3x + 5 \qquad \text{This is the given inequality.}$$
$$3x + 3 > 3x + 5 \qquad \text{Apply the distributive property.}$$
$$3x + 3 - 3x > 3x + 5 - 3x \qquad \text{Subtract 3x from both sides.}$$

Keep reading. $3 > 5$ is not the solution. $\qquad 3 > 5 \qquad \text{Simplify.}$

The original inequality is equivalent to the statement $3 > 5$, which is false for every value of x. The inequality has no solution. The solution set is $\varnothing$, the empty set.

☑ **CHECK POINT 9** Solve: $4(x + 2) > 4x + 15$.

EXAMPLE 10 Solving a Linear Inequality

Solve: $2(x + 5) \leq 5x - 3x + 14$.

Solution

$$2(x + 5) \leq 5x - 3x + 14 \qquad \text{This is the given inequality.}$$
$$2x + 10 \leq 5x - 3x + 14 \qquad \text{Apply the distributive property.}$$
$$2x + 10 \leq 2x + 14 \qquad \text{Combine like terms.}$$
$$2x + 10 - 2x \leq 2x + 14 - 2x \qquad \text{Subtract 2x from both sides.}$$

Keep reading. $10 \leq 14$ is not the solution. $\qquad 10 \leq 14 \qquad \text{Simplify.}$

The original inequality is equivalent to the statement $10 \leq 14$, which is true for every value of x. The solution is the set of all real numbers, expressed in interval notation as $(-\infty, \infty)$, or in set-builder notation as $\{x | x \text{ is a real number}\}$.

☑ **CHECK POINT 10** Solve: $3(x + 1) \geq 2x + 1 + x$.

6 Solve problems using linear inequalities.

Applications

As you know, different professors may use different grading systems to determine your final course grade. Some professors require a final examination; others do not. In our next example, a final exam is required *and* it also counts as two grades.

EXAMPLE 11 An Application: Final Course Grade

To earn an A in a course, you must have a final average of at least 90%. On the first four examinations, you have grades of 86%, 88%, 92%, and 84%. If the final examination counts as two grades, what must you get on the final to earn an A in the course?

Solution We will use our five-step strategy for solving algebraic word problems.

Steps 1 and 2. Represent unknown quantities in terms of x. Let

$$x = \text{your grade on the final examination.}$$

Step 3. Write an inequality in x that models the conditions. The average of the six grades is found by adding the grades and dividing the sum by 6.

$$\text{Average} = \frac{86 + 88 + 92 + 84 + x + x}{6}$$

Because the final counts as two grades, the x (your grade on the final examination) is added twice. This is also why the sum is divided by 6.

To get an A, your average must be at least 90. This means that your average must be greater than or equal to 90.

| Your average | must be greater than or equal to | 90. |

$$\frac{86 + 88 + 92 + 84 + x + x}{6} \geq 90$$

Step 4. Solve the inequality and answer the problem's question.

$$\frac{86 + 88 + 92 + 84 + x + x}{6} \geq 90 \qquad \text{This is the inequality that models the given conditions.}$$

$$\frac{350 + 2x}{6} \geq 90 \qquad \text{Combine like terms in the numerator.}$$

$$6\left(\frac{350 + 2x}{6}\right) \geq 6(90) \qquad \text{Multiply both sides by 6, clearing the fraction.}$$

$$350 + 2x \geq 540 \qquad \text{Multiply.}$$

$$350 + 2x - 350 \geq 540 - 350 \qquad \text{Subtract 350 from both sides.}$$

$$2x \geq 190 \qquad \text{Simplify.}$$

$$\frac{2x}{2} \geq \frac{190}{2} \qquad \text{Divide both sides by 2.}$$

$$x \geq 95 \qquad \text{Simplify.}$$

You must get at least 95% on the final examination to earn an A in the course.

Step 5. Check. We can perform a partial check by computing the average with any grade that is at least 95. We will use 96. If you get 96% on the final examination, your average is

$$\frac{86 + 88 + 92 + 84 + 96 + 96}{6} = \frac{542}{6} = 90\frac{1}{3}.$$

Because $90\frac{1}{3} > 90$, you earn an A in the course.

✓ **CHECK POINT 11** To earn a B in a course, you must have a final average of at least 80%. On the first three examinations, you have grades of 82%, 74%, and 78%. If the final examination counts as two grades, what must you get on the final to earn a B in the course?

2.6 EXERCISE SET   *MyMathLab*

Practice Exercises

In Exercises 1–12, graph the solutions of each inequality on a number line.

1. $x > 5$　　　　**2.** $x > -3$
3. $x < -2$　　　　**4.** $x < 0$
5. $x \geq -4$　　　　**6.** $x \geq -6$
7. $x \leq 4.5$　　　　**8.** $x \leq 7.5$
9. $-2 < x \leq 6$　　**10.** $-3 \leq x < 6$
11. $-1 < x < 3$　　**12.** $-2 \leq x \leq 0$

In Exercises 13–20, express the solution set of each inequality in interval notation and graph the interval.

13. $x \leq 3$　　　　**14.** $x \leq 5$
15. $x > \dfrac{5}{2}$　　　**16.** $x > \dfrac{7}{2}$
17. $x \leq 0$　　　　**18.** $x \leq 1$
19. $x < 4$　　　　**20.** $x < 5$

Use the addition property of inequality to solve each inequality in Exercises 21–38 and graph the solution set on a number line.

21. $x - 3 > 4$
22. $x + 1 < 6$
23. $x + 4 \leq 10$
24. $x - 5 \geq 2$
25. $y - 2 < 0$
26. $y + 3 \geq 0$
27. $3x + 4 \leq 2x + 7$
28. $2x + 9 \leq x + 2$
29. $5x - 9 < 4x + 7$
30. $3x - 8 < 2x + 11$
31. $7x - 7 > 6x - 3$
32. $8x - 9 > 7x - 3$
33. $x - \dfrac{2}{3} > \dfrac{1}{2}$
34. $x - \dfrac{1}{3} \geq \dfrac{5}{6}$
35. $y + \dfrac{7}{8} \leq \dfrac{1}{2}$
36. $y + \dfrac{1}{3} \leq \dfrac{3}{4}$

37. $-15y + 13 > 13 - 16y$
38. $-12y + 17 > 20 - 13y$

Use the multiplication property of inequality to solve each inequality in Exercises 39–56 and graph the solution set on a number line.

39. $\dfrac{1}{2}x < 4$　　　**40.** $\dfrac{1}{2}x > 3$
41. $\dfrac{x}{3} > -2$　　　**42.** $\dfrac{x}{4} < -1$
43. $4x < 20$　　　　**44.** $6x < 18$
45. $3x \geq -21$　　　**46.** $7x \geq -56$
47. $-3x < 15$　　　**48.** $-7x > 21$
49. $-3x \geq 15$　　　**50.** $-7x \leq 21$
51. $-16x > -48$　　**52.** $-20x > -140$
53. $-4y \leq \dfrac{1}{2}$　　　**54.** $-2y \leq \dfrac{1}{2}$
55. $-x < 4$　　　　**56.** $-x > -3$

Use both the addition and multiplication properties of inequality to solve each inequality in Exercises 57–80 and graph the solution set on a number line.

57. $2x - 3 > 7$
58. $3x + 2 \leq 14$
59. $3x + 3 < 18$
60. $8x - 4 > 12$
61. $3 - 7x \leq 17$
62. $5 - 3x \geq 20$
63. $-2x - 3 < 3$
64. $-3x + 14 < 5$
65. $5 - x \leq 1$
66. $3 - x \geq -3$
67. $2x - 5 > -x + 6$
68. $6x - 2 \geq 4x + 6$
69. $2y - 5 < 5y - 11$
70. $4y - 7 > 9y - 2$
71. $3(2y - 1) < 9$
72. $4(2y - 1) > 12$
73. $3(x + 1) - 5 < 2x + 1$
74. $4(x + 1) + 2 \geq 3x + 6$

75. $8x + 3 > 3(2x + 1) - x + 5$

76. $7 - 2(x - 4) < 5(1 - 2x)$

77. $\dfrac{x}{3} - 2 \geq 1$

78. $\dfrac{x}{4} - 3 \geq 1$

79. $1 - \dfrac{x}{2} > 4$

80. $1 - \dfrac{x}{2} < 5$

In Exercises 81–90, solve each inequality.

81. $4x - 4 < 4(x - 5)$

82. $3x - 5 < 3(x - 2)$

83. $x + 3 < x + 7$

84. $x + 4 < x + 10$

85. $7x \leq 7(x - 2)$

86. $3x + 1 \leq 3(x - 2)$

87. $2(x + 3) > 2x + 1$

88. $5(x + 4) > 5x + 10$

89. $5x - 4 \leq 4(x - 1)$

90. $6x - 3 \leq 3(x - 1)$

Practice PLUS

In Exercises 91–94, use properties of inequality to rewrite each inequality so that x is isolated on one side.

91. $3x + a > b$

92. $-2x - a \leq b$

93. $y \leq mx + b$ and $m < 0$

94. $y > mx + b$ and $m > 0$

We know that $|x|$ represents the distance from 0 to x on a number line. In Exercises 95–98, use each sentence to describe all possible locations of x on a number line. Then rewrite the given sentence as an inequality involving $|x|$.

95. The distance from 0 to x on a number line is less than 2.

96. The distance from 0 to x on a number line is less than 3.

97. The distance from 0 to x on a number line is greater than 2.

98. The distance from 0 to x on a number line is greater than 3.

Application Exercises

The bar graph at the top of the next column shows the percentage of wages full-time workers pay in income tax in eight selected countries. (The percents shown are averages for single earners without children.) Let x represent the percentage of wages workers

pay in income tax. In Exercises 99–104, name the country or countries described by the given inequality.

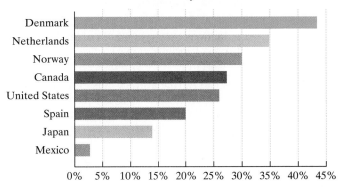

Percentage of Wages Full-Time Workers Pay in Income Tax

Source: The Washington Post

99. $x \geq 30\%$

100. $x > 30\%$

101. $x < 20\%$

102. $x \leq 20\%$

103. $25\% \leq x < 40\%$

104. $5\% < x \leq 25\%$

The line graph shows the declining consumption of cigarettes in the United States. The data shown by the graph can be modeled by

$$N = 550 - 9x,$$

where N is the number of cigarettes consumed, in billions, x years after 1988. Use this formula to solve Exercises 105–106.

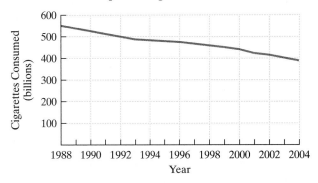

Consumption of Cigarettes in the United States

Source: Economic Research Service, USDA

105. Describe how many years after 1988 cigarette consumption will be less than 370 billion cigarettes each year. Which years are included in your description?

106. Describe how many years after 1988 cigarette consumption will be less than 325 billion cigarettes each year. Which years are included in your description?

107. On two examinations, you have grades of 86 and 88. There is an optional final examination, which counts as one grade. You decide to take the final in order to get a course grade of A, meaning a final average of at least 90.

    a. What must you get on the final to earn an A in the course?

b. By taking the final, if you do poorly, you might risk the B that you have in the course based on the first two exam grades. If your final average is less than 80, you will lose your B in the course. Describe the grades on the final that will cause this to happen.

108. On three examinations, you have grades of 88, 78, and 86. There is still a final examination, which counts as one grade.

a. In order to get an A, your average must be at least 90. If you get 100 on the final, compute your average and determine if an A in the course is possible.

b. To earn a B in the course, you must have a final average of at least 80. What must you get on the final to earn a B in the course?

109. A car can be rented from Continental Rental for $80 per week plus 25 cents for each mile driven. How many miles can you travel if you can spend at most $400 for the week?

110. A car can be rented from Basic Rental for $60 per week plus 50 cents for each mile driven. How many miles can you travel if you can spend at most $600 for the week?

111. An elevator at a construction site has a maximum capacity of 3000 pounds. If the elevator operator weighs 245 pounds and each cement bag weighs 95 pounds, how many bags of cement can be safely lifted on the elevator in one trip?

112. An elevator at a construction site has a maximum capacity of 2800 pounds. If the elevator operator weighs 265 pounds and each cement bag weighs 65 pounds, how many bags of cement can be safely lifted on the elevator in one trip?

Writing in Mathematics

113. When graphing the solutions of an inequality, what is the difference between a parenthesis and a bracket?

114. When solving an inequality, when is it necessary to change the direction of the inequality symbol? Give an example.

115. Describe ways in which solving a linear inequality is similar to solving a linear equation.

116. Describe ways in which solving a linear inequality is different from solving a linear equation.

Critical Thinking Exercises

Make Sense?   *In Exercises 117–120, determine whether each statement "makes sense" or "does not make sense" and explain your reasoning.*

117. I prefer interval notation over set-builder notation because it takes less space to write solution sets.

118. I can check inequalities by substituting 0 for the variable: When 0 belongs to the solution set, I should obtain a true statement, and when 0 does not belong to the solution set, I should obtain a false statement.

119. In an inequality such as $5x + 4 < 8x - 5$, I can avoid division by a negative number depending on which side I collect the variable terms and on which side I collect the constant terms.

120. I solved $-2x + 5 \geq 13$ and concluded that -4 is the greatest integer in the solution set.

In Exercises 121–124, determine whether each statement is true or false. If the statement is false, make the necessary change(s) to produce a true statement.

121. The inequality $x - 3 > 0$ is equivalent to $x < 3$.

122. The statement "x is at most 5" is written $x < 5$.

123. The inequality $-4x < -20$ is equivalent to $x > -5$.

124. The statement "the sum of x and 6% of x is at least 80" is modeled by $x + 0.06x \geq 80$.

125. A car can be rented from Basic Rental for $260 per week with no extra charge for mileage. Continental charges $80 per week plus 25 cents for each mile driven to rent the same car. How many miles should be driven in a week to make the rental cost for Basic Rental a better deal than Continental's?

126. Membership in a fitness club costs $500 yearly plus $1 per hour spent working out. A competing club charges $440 yearly plus $1.75 per hour for use of their equipment. How many hours must a person work out yearly to make membership in the first club cheaper than membership in the second club?

Technology Exercises

Solve each inequality in Exercises 127–128. Use a calculator to help with the arithmetic.

127. $1.45 - 7.23x > -1.442$

128. $126.8 - 9.4y \leq 4.8y + 34.5$

Review Exercises

129. 8 is 40% of what number? (Section 2.4, Example 8)

130. The length of a rectangle exceeds the width by 5 inches. The perimeter is 34 inches. What are the rectangle's dimensions? (Section 2.5, Example 5)

131. Solve and check: $5x + 16 = 3(x + 8)$. (Section 2.3, Example 2)

Preview Exercises

Exercises 132–134 will help you prepare for the material covered in the first section of the next chapter.

132. Evaluate $I = Prt$ for $P = 2000$, $r = 0.06$, and $t = 2$.

133. Solve: $0.06x + 0.08(16,000 - x) = 1180$.

134. Solve: $0.18x + 0.45(12 - x) = 0.36(12)$.

GROUP PROJECT

CHAPTER 2

One of the best ways to learn how to *solve* a word problem in algebra is to *design* word problems of your own. Creating a word problem makes you very aware of precisely how much information is needed to solve the problem. You must also focus on the best way to present information to a reader and on how much information to give. As you write your problem, you gain skills that will help you solve problems created by others.

The group should design five different word problems that can be solved using an algebraic equation. All of the problems should be on different topics. For example, the group should not have more than one problem on finding a number. The group should turn in both the problems and their algebraic solutions.

Chapter 2 Summary

Definitions and Concepts	Examples

Section 2.1 The Addition Property of Equality

A linear equation in one variable can be written in the form $ax + b = c$, where a is not zero.

$3x + 7 = 9$ is a linear equation.

Equivalent equations have the same solution.

$2x - 4 = 6$, $2x = 10$, and $x = 5$ are equivalent equations.

The Addition Property of Equality

Adding the same number (or algebraic expression) to both sides of an equation or subtracting the same number (or algebraic expression) from both sides of an equation does not change its solution.

$$\begin{aligned} x - 3 &= 8 \\ x - 3 + 3 &= 8 + 3 \\ x &= 11 \end{aligned}$$

$$\begin{aligned} x + 4 &= 10 \\ x + 4 - 4 &= 10 - 4 \\ x &= 6 \end{aligned}$$

Section 2.2 The Multiplication Property of Equality

The Multiplication Property of Equality

Multiplying both sides of an equation or dividing both sides of an equation by the same nonzero real number (or algebraic expression) does not change the solution.

$$\begin{aligned} \frac{x}{-5} &= 6 \\ -5\left(\frac{x}{-5}\right) &= -5(6) \\ x &= -30 \end{aligned}$$

$$\begin{aligned} -50 &= -5y \\ \frac{-50}{-5} &= \frac{-5y}{-5} \\ 10 &= y \end{aligned}$$

Equations and Coefficients of -1

If $-x = c$, multiply both sides by -1 to solve for x. The solution is the opposite, or additive inverse, of c.

$$\begin{aligned} -x &= -12 \\ (-1)(-x) &= (-1)(-12) \\ x &= 12 \end{aligned}$$

Using the Addition and Multiplication Properties

If an equation does not contain fractions,

- Use the addition property to isolate the variable term.
- Use the multiplication property to isolate the variable.

$$\begin{aligned} -2x - 5 &= 11 \\ -2x - 5 + 5 &= 11 + 5 \\ -2x &= 16 \\ \frac{-2x}{-2} &= \frac{16}{-2} \\ x &= -8 \end{aligned}$$

Definitions and Concepts	**Examples**

Section 2.3 Solving Linear Equations

Solving a Linear Equation

1. Simplify each side.

Solve:
$$7 - 4(x - 1) = x + 1.$$
$$7 - 4x + 4 = x + 1$$
$$-4x + 11 = x + 1$$

2. Collect all the variable terms on one side and all the constant terms on the other side.

$$-4x - x + 11 = x - x + 1$$
$$-5x + 11 = 1$$
$$-5x + 11 - 11 = 1 - 11$$
$$-5x = -10$$

3. Isolate the variable and solve. (If the variable is eliminated and a false statement results, the inconsistent equation has no solution. If a true statement results, all real numbers are solutions of the identity.)

$$\frac{-5x}{-5} = \frac{-10}{-5}$$
$$x = 2$$

4. Check the proposed solution in the original equation.

$$7 - 4(x - 1) = x + 1$$
$$7 - 4(2 - 1) \stackrel{?}{=} 2 + 1$$
$$7 - 4(1) \stackrel{?}{=} 2 + 1$$
$$7 - 4 \stackrel{?}{=} 2 + 1$$
$$3 = 3, \text{ true}$$

The solution is 2, or the solution set is $\{2\}$.

Equations Containing Fractions

Multiply both sides (all terms) by the least common denominator. This clears the equation of fractions.

$$\frac{x}{5} + \frac{1}{2} = \frac{x}{2} - 1$$
$$10\left(\frac{x}{5} + \frac{1}{2}\right) = 10\left(\frac{x}{2} - 1\right)$$
$$10 \cdot \frac{x}{5} + 10 \cdot \frac{1}{2} = 10 \cdot \frac{x}{2} - 10 \cdot 1$$
$$2x + 5 = 5x - 10$$
$$-3x = -15$$
$$x = 5$$

The solution is 5, or the solution set is $\{5\}$.

Inconsistent equations with no solution (solution set: $\varnothing$) result in false statements in the solution process. Identities, true for every real number (solution set: $\{x | x \text{ is a real number}\}$) result in true statements in the solution process.

Solve:
$$3x + 2 = 3(x + 5).$$
$$3x + 2 = 3x + 15$$
$$3x + 2 - 3x = 3x + 15 - 3x$$
$$2 = 15 \quad \text{(false)}$$
No solution: $\varnothing$

Solve:
$$2(x + 4) = x + x + 8.$$
$$2x + 8 = 2x + 8$$
$$2x + 8 - 2x = 2x + 8 - 2x$$
$$8 = 8 \quad \text{(true)}$$
$$\{x | x \text{ is a real number}\}$$

Definitions and Concepts	**Examples**

Section 2.4 Formulas and Percents

To solve a formula for one of its variables, use the steps for solving a linear equation and isolate the specified variable on one side of the equation.

Solve for l:

$$w = \frac{P - 2l}{2}.$$

$$2w = 2\left(\frac{P - 2l}{2}\right)$$

$$2w = P - 2l$$

$$2w - P = P - P - 2l$$

$$2w - P = -2l$$

$$\frac{2w - P}{-2} = \frac{-2l}{-2}$$

$$\frac{2w - P}{-2} = l$$

The word *percent* means *per hundred*. The symbol % denotes percent.

$$47\% = \frac{47}{100} \qquad 3\% = \frac{3}{100}$$

To express a percent as a decimal, move the decimal point two places to the left and remove the percent sign.

$$250\% = 2\underset{\smile}{50\%} = 2.5$$

$$4\% = 0\underset{\smile}{4.\%} = 0.04$$

To express a decimal as a percent, move the decimal point two places to the right and add a percent sign.

$$0.37 = 37\%$$

$$0.006 = 0.6\%$$

A Formula Involving Percent

$$\boxed{A} \quad \boxed{is} \quad \boxed{P \text{ percent}} \quad \boxed{of} \quad \boxed{B.}$$

$$A \quad = \quad P \quad \cdot \quad B$$

In the formula $A = PB$, P is expressed as a decimal.

• $\boxed{What} \quad \boxed{is} \quad \boxed{5\%} \quad \boxed{of} \quad \boxed{20?}$

$$A \quad = \quad 0.05 \quad \cdot \quad 20$$

$$A = 1$$

Thus, 1 is 5% of 20.

• $\boxed{6} \quad \boxed{is} \quad \boxed{30\%} \quad \boxed{of} \quad \boxed{what?}$

$$6 \quad = \quad 0.3 \quad \cdot \quad B$$

$$\frac{6}{0.3} = B$$

$$20 = B$$

Thus, 6 is 30% of 20.

• $\boxed{33} \quad \boxed{is} \quad \boxed{what\ percent} \quad \boxed{of} \quad \boxed{75?}$

$$33 \quad = \quad P \quad \cdot \quad 75$$

$$\frac{33}{75} = P$$

$$P = 0.44 = 44\%$$

Thus, 33 is 44% of 75.

Definitions and Concepts	**Examples**

Section 2.5 An Introduction to Problem Solving

Strategy for Solving Word Problems

Step 1. Let x represent one of the quantities.

Step 2. Represent other unknown quantities in terms of x.

Step 3. Write an equation that models the conditions.

Step 4. Solve the equation and answer the question.

Step 5. Check the proposed solution in the original wording of the problem.

The length of a rectangle exceeds the width by 3 inches. The perimeter is 26 inches. What are the rectangle's dimensions?

Let x = the width.

$x + 3$ = the length

$$2(x + 3) \quad + \quad 2x \quad = \quad 26$$

$$2(x + 3) + 2x = 26$$
$$2x + 6 + 2x = 26$$
$$4x + 6 = 26$$
$$4x = 20$$
$$x = 5$$

The width (x) is 5 inches and the length $(x + 3)$ is $5 + 3$, or 8 inches.

$$\text{Perimeter} = 2(5 \text{ in.}) + 2(8 \text{ in.})$$
$$= 10 \text{ in.} + 16 \text{ in.} = 26 \text{ in.}$$

This checks with the given perimeter.

Section 2.6 Solving Linear Inequalities

A linear inequality in one variable can be written in one of these forms:

$$ax + b < c \qquad ax + b \le c$$
$$ax + b > c \qquad ax + b \ge c.$$

$3x + 6 > 12$ is a linear inequality.

Graphs of solutions to linear inequalities are shown on a number line by shading all points representing numbers that are solutions. Square brackets, [ ], indicate endpoints that are solutions. Parentheses, ( ), indicate endpoints that are not solutions.

- Graph the solutions of $x < 4$.

  −3 −2 −1 0 1 2 3 4 5

- Graph the solutions of $-2 < x \le 1$.

  −4 −3 −2 −1 0 1 2 3 4

Solutions of inequalities can be expressed in interval notation or set-builder notation.

Inequality	Interval Notation	Set-Builder Notation	Graph
$x > b$	(b, ∞)	$\{x \mid x > b\}$	$a \quad b$
$x \le a$	$(-\infty, a]$	$\{x \mid x \le a\}$	$a \quad b$

Express the solution set in interval notation and graph the interval:

- $x \le 1$
  $(-\infty, 1]$

  −4 −3 −2 −1 0 1 2 3 4

- $x > -2$
  $(-2, \infty)$
  −4 −3 −2 −1 0 1 2 3 4

Definitions and Concepts	Examples

Section 2.6 Solving Linear Inequalities (continued)

The Addition Property of Inequality

Adding the same number to both sides of an inequality or subtracting the same number from both sides of an inequality does not change the solutions.

$$x + 3 < 8$$
$$x + 3 - 3 < 8 - 3$$
$$x < 5$$

The Positive Multiplication Property of Inequality

Multiplying or dividing both sides of an inequality by the same positive number does not change the solutions.

$$\frac{x}{6} \geq 5$$
$$6 \cdot \frac{x}{6} \geq 6 \cdot 5$$
$$x \geq 30$$

The Negative Multiplication Property of Inequality

Multiplying or dividing both sides of an inequality by the same negative number and reversing the direction of the inequality sign does not change the solutions.

$$-3x \leq 12$$
$$\frac{-3x}{-3} \geq \frac{12}{-3}$$
$$x \geq -4$$

Solving Linear Inequalities

Use the procedure for solving linear equations. When multiplying or dividing by a negative number, reverse the direction of the inequality symbol. Express the solution set in interval or set-builder notation, and graph the set on a number line. If the variable is eliminated and a false statement results, the inequality has no solution. The solution set is $\varnothing$, the empty set. If a true statement results, the solution is the set of all real numbers: $(-\infty, \infty)$ or $\{x \mid x \text{ is a real number}\}$.

Solve:
$$x + 4 \geq 6x - 16$$
$$x + 4 - 6x \geq 6x - 16 - 6x$$
$$-5x + 4 \geq -16$$
$$-5x + 4 - 4 \geq -16 - 4$$
$$-5x \geq -20$$
$$\frac{-5x}{-5} \leq \frac{-20}{-5}$$
$$x \leq 4$$

Solution set: $(-\infty, 4]$ or $\{x \mid x \leq 4\}$

$$-3 \ -2 \ -1 \ \ 0 \ \ 1 \ \ 2 \ \ 3 \ \ 4 \ \ 5$$

CHAPTER 2 REVIEW EXERCISES

2.1 *Solve each equation in Exercises 1–5 using the addition property of equality. Be sure to check proposed solutions.*

1. $x - 10 = 22$

2. $-14 = y + 8$

3. $7z - 3 = 6z + 9$

4. $4(x + 3) = 3x - 10$

5. $6x - 3x - 9 + 1 = -5x + 7x - 3$

2.2 *Solve each equation in Exercises 6–13 using the multiplication property of equality. Be sure to check proposed solutions.*

6. $\dfrac{x}{8} = 10$

7. $\dfrac{y}{-8} = 7$

8. $7z = 77$

9. $-36 = -9y$

10. $\dfrac{3}{5}x = -9$

11. $30 = -\dfrac{5}{2}y$

12. $-x = 25$

13. $\dfrac{-x}{10} = -1$

Solve each equation in Exercises 14–18 using both the addition and multiplication properties of equality. Check proposed solutions.

14. $4x + 9 = 33$

15. $-3y - 2 = 13$

16. $5z + 20 = 3z$

17. $5x - 3 = x + 5$

18. $3 - 2x = 9 - 8x$

19. The bar graph shows the average amount of oil imported monthly to the United States, in millions of barrels, for five selected years.

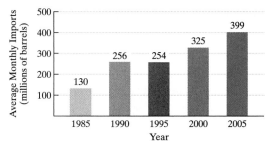

United States Oil Imports

Source: Energy Department (Monthly averages for first three months of each year)

The mathematical model

$$I = 12n + 151$$

describes the average amount of oil imported monthly to the United States, I, in millions of barrels, n years after 1985.

a. Does the formula underestimate or overestimate the monthly average of oil imports in 2005 shown by the bar graph on the previous page? By how many millions of barrels?

b. If trends shown by the formula continue, when will the United States import an average of 511 million barrels of oil per month?

2.3 *Solve and check each equation in Exercises 20–28.*

20. $5x + 9 - 7x + 6 = x + 18$

21. $3(x + 4) = 5x - 12$

22. $1 - 2(6 - y) = 3y + 2$

23. $2(x - 4) + 3(x + 5) = 2x - 2$

24. $-2(y - 4) - (3y - 2) = -2 - (6y - 2)$

25. $\dfrac{2x}{3} = \dfrac{x}{6} + 1$

26. $\dfrac{x}{2} - \dfrac{1}{10} = \dfrac{x}{5} + \dfrac{1}{2}$

27. $3(8x - 1) = 6(5 + 4x)$

28. $4(2x - 3) + 4 = 8x - 8$

29. The formula $H = 0.7(220 - a)$ can be used to determine target heart rate during exercise, H, in beats per minute, by a person of age a. If the target heart rate is 133 beats per minute, how old is that person?

2.4 *In Exercises 30–34, solve each formula for the specified variable.*

30. $I = Pr$ for r **31.** $V = \dfrac{1}{3}Bh$ for h

32. $P = 2l + 2w$ for w

33. $A = \dfrac{B + C}{2}$ for B

34. $T = D + pm$ for m

In Exercises 35–37, express each percent as a decimal.

35. 65%   **36.** 150%

37. 3%

In Exercises 38–39, express each decimal as a percent.

38. 0.72   **39.** 0.0035

40. What is 8% of 120?

41. 90 is 45% of what?

42. 36 is what percent of 75?

43. If 6 is increased to 12, the increase is what percent of the original number?

44. If 5 is decreased to 3, the decrease is what percent of the original number?

45. A college that had 40 students for each lecture course increased the number to 45 students. What is the percent increase in the number of students in a lecture course?

46. Consider the following statement:

My portfolio fell 10% last year, but then it rose 10% this year, so at least I recouped my losses.

Is this statement true? In particular, suppose you invested $10,000 in the stock market last year. How much money would be left in your portfolio with a 10% fall and then a 10% rise? If there is a loss, what is the percent decrease, to the nearest tenth of a percent, in your portfolio?

47. The radius is one of two bones that connect the elbow and the wrist. The formula $r = \dfrac{h}{7}$ models the length of a woman's radius, r, in inches, and her height, h, in inches.

a. Solve the formula for h.

b. Use the formula in part (a) to find a woman's height if her radius is 9 inches long.

48. Every day, the average U.S. household uses 91 gallons of water flushing toilets. The circle graph shows that this represents 26% of the total number of gallons of water used per day. How many gallons of water does the average U.S. household use per day?

Where United States Households Use Water

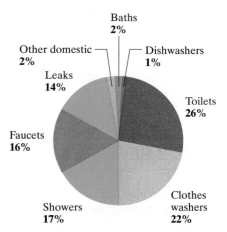

Source: American Water Works Association

2.5 *In Exercises 49–56, use the five-step strategy to solve each problem.*

49. Six times a number, decreased by 20, is four times the number. Find the number.

50. In 2006, the average amount that a dog owner spent on pet food exceeded the average amount that a cat owner spent by $29. Combined, $405 was spent on pet food for dogs and cats. Determine the average amount spent on pet food for cats and the average amount spent on pet food for dogs. (*Source*: American Pet Products Manufacturers Association)

51. Two pages that face each other in a book have 93 as the sum of their page numbers. What are the page numbers?

52. The graph shows the gender breakdown of the U.S. population at each end of the age spectrum.

Gender Breakdown of the American Population

☐ Male ■ Female

Younger than 20   70 and Older

Source: U.S. Census Bureau

For Americans under 20, the percentage of males and the percentage of females are consecutive odd integers. What percentage of Americans younger than 20 are females and what percentage are males?

53. In 2003, the average weekly salary for workers in the United States was $612. If this amount is increasing by $15 yearly, in how many years after 2003 will the average salary reach $747? In which year will that be?

54. A bank's total monthly charge for a checking account is $6 plus $0.05 per check. If your total monthly charge is $6.90, how many checks did you write during that month?

55. A rectangular field is three times as long as it is wide. If the perimeter of the field is 400 yards, what are the field's dimensions?

56. After a 25% reduction, you purchase a table for $180. What was the table's price before the reduction?

2.6 *In Exercises 57–58, graph the solution of each inequality on a number line.*

57. $x < -1$ **58.** $-2 < x \leq 4$

In Exercises 59–60, express the solution set of each inequality in interval notation and graph the interval.

59. $x \geq \dfrac{3}{2}$ **60.** $x < 0$

In Exercises 61–68, solve each inequality and graph the solution set on a number line. It is not necessary to provide graphs if the inequality has no solution or is true for all real numbers.

61. $2x - 5 < 3$

62. $\dfrac{x}{2} > -4$

63. $3 - 5x \leq 18$

64. $4x + 6 < 5x$

65. $6x - 10 \geq 2(x + 3)$

66. $4x + 3(2x - 7) \leq x - 3$

67. $2(2x + 4) > 4(x + 2) - 6$

68. $-2(x - 4) \leq 3x + 1 - 5x$

69. To pass a course, a student must have an average on three examinations of at least 60. If a student scores 42 and 74 on the first two tests, what must be earned on the third test to pass the course?

70. A long-distance telephone service charges 10¢ for the first minute and 5¢ for each minute thereafter. The cost, C, in cents, for a call lasting x minutes is modeled by the formula

$$C = 10 + 5(x - 1).$$

How many minutes can you talk on the phone if you do not want the cost to exceed $5, or 500¢?

CHAPTER 2 TEST

Remember to use your Chapter Test Prep Video CD to see the worked-out solutions to the test questions you want to review.

In Exercises 1–6, solve each equation.

1. $4x - 5 = 13$

2. $12x + 4 = 7x - 21$

3. $8 - 5(x - 2) = x + 26$

4. $3(2y - 4) = 9 - 3(y + 1)$

5. $\dfrac{3}{4}x = -15$

6. $\dfrac{x}{10} + \dfrac{1}{3} = \dfrac{x}{5} + \dfrac{1}{2}$

7. The formula $P = 2.4x + 180$ models U.S. population, P, in millions x years after 1960. How many years after 1960 is the U.S. population expected to reach 324 million? In which year is this expected to occur?

In Exercises 8–9, solve each formula for the specified variable.

8. $V = \pi r^2 h$ for h

9. $l = \dfrac{P - 2w}{2}$ for w

10. What is 6% of 140?

11. 120 is 80% of what?

12. 12 is what percent of 240?

In Exercises 13–17, solve each problem.

13. The product of 5 and a number, decreased by 9, is 306. What is the number?

14. The graph shows the number of people in the United States, ages 18 and older, who have had at least one major depressive episode in their lifetime, by income.

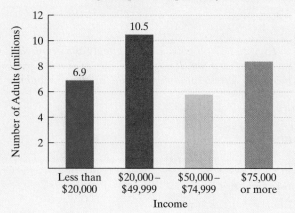

Number of Americans with At Least One Major Depressive Episode, by Income

Source: Department of Health and Human Services

The number of people with depressive episodes earning $75,000 or more exceeds the number in the $50,000–$74,999 range by 2.6 million. There are 14.2 million Americans in both these income groups combined who have had at least one major depressive episode in their lifetime. How many people, in millions, earning $50,000–$74,999 have had major depressive episodes and how many people earning $75,000 or more have had such episodes?

15. A long-distance telephone plan has a monthly fee of $15.00 and a rate of $0.05 per minute. How many minutes can you chat long distance in a month for a total cost, including the $15.00, of $45.00?

16. A rectangular field is twice as long as it is wide. If the perimeter of the field is 450 yards, what are the field's dimensions?

17. After a 20% reduction, you purchase a new Stephen King novel for $28. What was the book's price before the reduction?

In Exercises 18–19, express the solution set of each inequality in interval notation and graph the interval.

18. $x > -2$ **19.** $x \leq 3$

Solve each inequality in Exercises 20–22 and graph the solution set on a number line.

20. $\dfrac{x}{2} < -3$

21. $6 - 9x \geq 33$

22. $4x - 2 > 2(x + 6)$

23. A student has grades on three examinations of 76, 80, and 72. What must the student earn on a fourth examination to have an average of at least 80?

24. The length of a rectangle is 20 inches. For what widths is the perimeter greater than 56 inches?

CUMULATIVE REVIEW EXERCISES (CHAPTERS 1–2)

In Exercises 1–3, perform the indicated operation or operations.

1. $-8 - (12 - 16)$ **2.** $(-3)(-2) + (-2)(4)$

3. $(8 - 10)^3 (7 - 11)^2$

4. Simplify: $2 - 5[x + 3(x + 7)]$.

5. List all the rational numbers in this set:

$$\left\{-4, -\dfrac{1}{3}, 0, \sqrt{2}, \sqrt{4}, \dfrac{\pi}{2}, 1063\right\}.$$

6. Write as an algebraic expression, using x to represent the number:

The quotient of 5 and a number, decreased by 2 more than the number.

7. Insert either $<$ or $>$ in the shaded area to make a true statement:

$$-10,000 \quad \blacksquare \quad -2.$$

8. Use the distributive property to rewrite without parentheses:

$$6(4x - 1 - 5y).$$

The bar graph shows an increase in the percentage of American consumers looking for trans fats on food labels from 2004 through 2007.

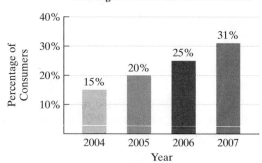

Percentage of American Consumers Looking for Trans Fats on Food Labels

Source: NPD Group

The mathematical model

$$T = 5.3n + 9.5$$

describes the percentage of consumers looking for trans fats on food labels, T, n years after 2003. Use this information to solve Exercises 9–10.

9. Does the formula underestimate or overestimate the percentage of consumers in 2007? By how much?

10. If trends shown by the formula continue, when will 89% of American consumers look for trans fats on food labels?

In Exercises 11–12, solve each equation.

11. $5 - 6(x + 2) = x - 14$

12. $\dfrac{x}{5} - 2 = \dfrac{x}{3}$

13. Solve for A: $V = \dfrac{1}{3}Ah$.

14. 48 is 30% of what number?

15. The length of a rectangular parking lot is 10 yards less than twice its width. If the perimeter of the lot is 400 yards, what are its dimensions?

16. A gas station owner makes a profit of 40 cents per gallon of gasoline sold. How many gallons of gasoline must be sold in a year to make a profit of $30,000 from gasoline sales?

17. Express the solution set of $x \le \frac{1}{2}$ in interval notation and graph the interval.

Solve each inequality in Exercises 18–19 and graph the solution set on a number line.

18. $3 - 3x > 12$

19. $5 - 2(3 - x) \le 2(2x + 5) + 1$

20. You take a summer job selling medical supplies. You are paid $600 per month plus 4% of the sales price of all the supplies you sell. If you want to earn more than $2500 per month, what value of medical supplies must you sell?

The beautiful actress America Ferrera won a Golden Globe for her portrayal of the fashion-challenged Betty Suarez on the popular TV sitcom Ugly Betty.

Problem Solving

I n a 1996 study, researchers found that employees described as plain had salaries below the median level and those described as attractive had salaries above the median level.

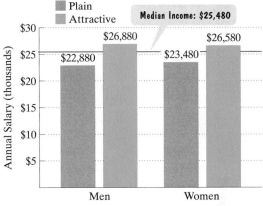

Annual Salaries of American Men and Women, by Physical Attractiveness

Source: Jeff Biddle and Daniel Hamermesh, CompPsych Survey

You can hardly pick up a newspaper without being bombarded with graphs that describe how we live, what we want, and even how being attractive pays off. Because the data presented by these graphs both record and influence our behavior (and provide an endless array of mathematical models!), it is important to identify possible deceptions in visual displays.

In this chapter, you will learn to be a better, more confident problem solver. We'll see how the thought and reasoning needed to solve geometry problems can be used to distinguish between good and bad methods for presenting data.

Deceptions in visual displays of data are discussed in the Blitzer Bonus on pages 207–208, Exercises 78 and 79 of Exercise Set 3.3, and the Group Project on page 215.

3.1

Objectives

1 Solve simple interest problems.

2 Solve mixture problems.

3 Solve motion problems.

Further Problem Solving

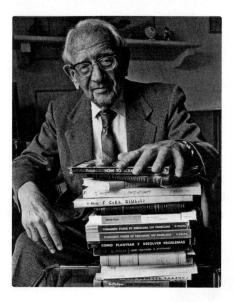

George Polya (1887–1985), author of How to Solve It

Understand the problem. Devise a plan for solving the problem. Carry out the plan and solve the problem. Look back and check the answer. These are the guidelines for problem solving offered by the charismatic teacher and mathematician George Polya in *How to Solve It*. First published in 1945, the book has sold more than one million copies and is available in 17 languages. Polya's book demonstrates how to think clearly in any field.

In this section, you will learn to solve problems involving simple interest, mixtures, and motion. To solve these problems, we will use the five-step strategy that was introduced in Chapter 2. We will also use three problem-solving tips from *How to Solve It*:

- Organize the given information in a table.
- Make a sketch to illustrate the problem.
- Relate the problem to similar problems that you have seen before.

1 Solve simple interest problems.

Problems Involving Simple Interest

Interest is the amount of money that we get paid for lending or investing money or that we pay for borrowing money. When we deposit money in a savings institution, the institution pays us interest for its use. When we borrow money, interest is the price we pay for the privilege of using the money until we repay it.

The amount of money that we deposit or borrow is called the **principal**. For example, if you deposit $2000 in a savings account, then $2000 is the principal. The amount of interest depends on the principal, the interest **rate**, which is given as a percent and varies from bank to bank, and the length of time for which the money is deposited. In this section, the rate is assumed to be annual (per year).

Simple interest involves interest calculated only on the principal. The following formula is used to find simple interest:

Calculating Simple Interest

$$\text{Interest} = \text{principal} \times \text{rate} \times \text{time}$$
$$I = Prt$$

The rate, r, is expressed as a decimal when calculating simple interest.

EXAMPLE 1   Calculating Simple Interest for a Year

You deposit $2000 in a savings account at Hometown Bank, which has a rate of 6%. Find the interest at the end of the first year.

Solution   To find the interest at the end of the first year, we use the simple interest formula.

$$I \;=\; Prt \;=\; (2000)\,(0.06)\,(1) \;=\; 120$$

| Principal, or amount deposited, is $2000. | Rate is 6% = 0.06. | Time is 1 year. |

At the end of the first year, the interest is $120. You can withdraw the $120 in interest, and you still have $2000 in the savings account.   ▬

☑ **CHECK POINT 1**   You deposit $3000 in a savings account at Yourtown Bank, which has a rate of 5%. Find the interest at the end of the first year.

Our next problem again involves simple interest earned for one year. Because the time is one year, we substitute 1 for t in the simple interest formula, $I = Prt$. The formula becomes $I = Pr \cdot 1$, or simply $I = Pr$. Simple interest earned in one year is the product of the principal and the rate.

EXAMPLE 2   Solving a Simple Interest Problem

You inherited $16,000 with the stipulation that for the first year the money must be invested in two accounts paying 6% and 8% annual interest. How much did you invest at each rate if the total interest earned for the year was $1180?

Solution

Step 1. Let x represent one of the quantities.

Let x = the amount invested at 6%.

Step 2. Represent other unknown quantities in terms of x. The other quantity that we seek is the amount invested at 8%. Because the total amount invested is $16,000, and we already used up x,

$16,000 - x$ = the amount invested at 8%.

Step 3. Write an equation in x that models the conditions. We can use a table to organize the information given in the problem.

Study Tip

Look at the expression in Step 2. Notice that when you add x and $16,000 - x$, you get 16,000, the total investment. In many word problems, a total amount is divided into two parts. If the total amount is T and x represents one part, the other part is represented by $T - x$. Furthermore, it does not matter which part x represents and which part $T - x$ represents.

	Principal	×	Rate	=	Interest
6% Investment	x		0.06		$0.06x$
8% Investment	$16,000 - x$		0.08		$0.08(16,000 - x)$

The interest for the two investments combined is $1180.

| Interest from 6% investment | plus | Interest from 8% investment | is | $1180. |

$$0.06x \;+\; 0.08(16,000 - x) \;=\; 1180$$

| rate times principal | rate times principal |

Step 4. Solve the equation and answer the question.

$$0.06x + 0.08(16{,}000 - x) = 1180 \qquad \text{This is the equation that models the problem's conditions.}$$

$$0.06x + 1280 - 0.08x = 1180 \qquad \text{Use the distributive property.}$$

$$-0.02x + 1280 = 1180 \qquad \text{Combine like terms.}$$

$$-0.02x + 1280 - 1280 = 1180 - 1280 \qquad \text{Subtract 1280 from both sides.}$$

$$-0.02x = -100 \qquad \text{Simplify.}$$

$$\frac{-0.02x}{-0.02} = \frac{-100}{-0.02} \qquad \text{Divide both sides by } -0.02.$$

$$x = 5000 \qquad \text{Simplify.}$$

Because x represents the amount invested at 6%, $5000 was invested at 6%. Because $16{,}000 - x$ represents the amount invested at 8%, $16{,}000 - \$5000$, or $11{,}000, was invested at 8%.

Step 5. Check the proposed solution in the original wording of the problem. The problem states that the total interest was $1180. The interest earned on $5000 at 6% is ($5000)(0.06), or $300. The interest earned on $11,000 at 8% is ($11,000)(0.08), or $880. The total interest is $300 + $880, or $1180, exactly the amount given in the problem. ■

✓ **CHECK POINT 2**   Suppose that you invested $25,000, part at 9% simple interest and the remainder at 12%. If the total yearly interest from these investments was $2550, find the amount invested at each rate.

---

2 Solve mixture problems.

Problems Involving Mixtures

Chemists and pharmacists often have to change the concentration of solutions and other mixtures. In these situations, the amount of a particular ingredient in the solution or mixture is expressed as a percentage of the total solution.

EXAMPLE 3   Finding the Amount of Acid in a Solution

A 40-milliliter solution of acid in water contains 35% acid. How much acid is in the solution?

Solution

| Amount of acid in the solution | is | 35% | of | total number of milliliters in the solution. |

$$= (0.35) \cdot (40)$$
$$= 14$$

There are 14 milliliters of acid in the solution. ■

✓ **CHECK POINT 3**   A 60-milliliter solution of acid in water contains 45% acid. How much acid is in the solution?

We can express our computation in Example 3 as follows:

$$(40)(0.35) = 14.$$

| total solution amount: 40 milliliters | rate of acid's concentration: 35% | amount of acid: 14 milliliters |

The total amount times the rate gives the amount. Can you see that this is the same idea we used to compute simple interest?

total money amount

rate of interest

amount of interest

Simple Interest Problems: principal · rate = interest

Percents are expressed as decimals in these equations.

Mixture Problems: solution · concentration = ingredient

total solution amount

percent of ingredient, or rate of concentration

amount of ingredient

In our simple interest problem of Example 2, we were mixing two investments. In the mixture problem that follows, we will be mixing two liquids. The equations in these problems are obtained from similar conditions:

Simple Interest Problems

interest from investment 1 + interest from investment 2 = amount of interest from mixed investments

Mixture Problems

ingredient amount in solution 1 + ingredient amount in solution 2 = amount of ingredient in mixture

We will continue using our five-step strategy to solve mixture problems. Organizing the given information in a table helps obtain an equation, which is usually the most difficult step in our strategy.

Study Tip

"Relate the problem to similar problems that you have seen before."

George Polya, *How to Solve It*

Being aware of the similarities between interest and mixture problems should make you a better problem solver in a variety of situations that involve mixtures. Mixing investments or liquids, combining two blends of coffee with different prices into a single blend, mixing coins or stamps of different values: These problems have nearly identical solutions. Only the nouns and the amounts change from situation to situation.

EXAMPLE 4   Solving a Solution Mixture Problem

A chemist needs to mix an 18% acid solution with a 45% acid solution to obtain 12 liters of a 36% acid solution. How many liters of each of the acid solutions must be used?

Solution

Step 1. Let x represent one of the quantities.

Let x = the number of liters of the 18% acid solution to be used in the mixture.

Step 2. Represent other unknown quantities in terms of x. The other quantity that we seek is the number of liters of the 45% acid solution to be used in the mixture. Because the total to be obtained is 12 liters, and we have already used up x liters,

$12 - x$ = the number of liters of the 45% acid solution to be used in the mixture.

Step 3. Write an equation in x that models the conditions. The situation is illustrated in **Figure 3.1**.

x liters

$12 - x$ liters

18% solution

45% solution

36% solution

12 liters

FIGURE 3.1   Obtaining 12 liters of a 36% acid mixture

The sum of the amounts of acid in the 18% solution and the 45% solution must equal the amount of acid in the 36%, 12-liter, mixture. We form a table that shows the amount of acid in each of the three solutions.

	Number of Liters	×	Percent of Acid	=	Amount of Acid
18% Acid Solution	x		18% = 0.18		$0.18x$
45% Acid Solution	$12 - x$		45% = 0.45		$0.45(12 - x)$
36% Acid Mixture	12		36% = 0.36		$0.36(12)$

Amount of acid in the 18% solution	plus	amount of acid in the 45% solution	equals	amount of acid in the 36% mixture.

$$0.18x \quad + \quad 0.45(12 - x) \quad = \quad 0.36(12)$$

Step 4. Solve the equation and answer the question.

$0.18x + 0.45(12 - x) = 0.36(12)$	This is the equation that models the problem's conditions.
$0.18x + 5.4 - 0.45x = 4.32$	Use the distributive property.
$-0.27x + 5.4 = 4.32$	Combine like terms.
$-0.27x = -1.08$	Subtract 5.4 from both sides.
$\dfrac{-0.27x}{-0.27} = \dfrac{-1.08}{-0.27}$	Divide both sides by −0.27.
$x = 4$	Simplify.

Because x represents the number of liters of the 18% solution, 4 liters of the 18% solution should be used in the mixture. Because $12 - x$ represents the number of liters of the 45% solution, $12 - 4$, or 8 liters of the 45% solution should be used in the mixture. Thus, the chemist should mix 4 liters of the 18% acid solution with 8 liters of the 45% acid solution.

Step 5. Check the proposed solution in the original wording of the problem. The problem states that the 12-liter mixture should be 36% acid. The amount of acid in this mixture is 0.36(12), or 4.32 liters of acid. The amount of acid in 4 liters of the 18% solution is 0.18(4), or 0.72 liter. The amount of acid in 8 liters of the 45% solution is 0.45(8), or 3.6 liters. The amount of acid in the two solutions used in the mixture is 0.72 liter + 3.6 liters, or 4.32 liters, exactly as it should be.

✓ **CHECK POINT 4**   A chemist needs to mix a 10% acid solution with a 60% acid solution to obtain 50 milliliters of a 30% acid solution. How many milliliters of each of the acid solutions must be used?

---

3 Solve motion problems.

Problems Involving Motion

Suppose that you ride your bike at an average speed of 12 miles per hour. What distance do you cover in 2 hours? Your distance is the product of your speed and the time that you travel:

$$\frac{12 \text{ miles}}{\text{hour}} \times 2 \text{ hours} = 24 \text{ miles}.$$

Your distance is 24 miles. Notice how the hour units cancel. The distance is expressed in miles.

In general, the distance covered by any moving body is the product of its average speed, or rate, and its time in motion:

> ### A Formula For Motion
>
> $$d = rt$$
>
> Distance equals rate times time.

EXAMPLE 5   Using the Motion Formula

Your nonstop flight from San Francisco to New York, a distance of about 2700 miles, takes 5 hours. What is the plane's average rate?

Solution   Use the formula $d = rt$. The distance, d, is 2700 miles. The time, t, is 5 hours. The problem asks us to find r, the plane's average rate.

$$d = rt \qquad \text{Use the motion formula.}$$

$$2700 = r \cdot 5 \qquad d = 2700 \text{ and } t = 5$$

$$\frac{2700}{5} = \frac{r \cdot 5}{5} \qquad \text{Divide both sides of the equation by 5.}$$

$$540 = r \qquad \text{Simplify.}$$

The plane's average rate, or speed, is 540 miles per hour.   ▬

✓ **CHECK POINT 5**   Riding your bike a distance of 27 miles takes 3 hours. What is your average rate?

The five-step strategy can be used to solve motion problems involving two rates. For example, a problem might involve two planes traveling at different speeds. Because there are two speeds, there are also two distances. The equations in these problems are obtained from the relationship between these distances. Depending on the problem's conditions, these distances are often either added or set equal to each other.

EXAMPLE 6 Solving a Problem Involving Motion

Two planes leave the same airport at the same time, flying in opposite directions. The rate of the faster plane is 300 miles per hour. The rate of the slower plane is 200 miles per hour. After how many hours will the planes be 1000 miles apart?

Solution

Step 1. Let x represent one of the quantities. Let $x =$ the time when the planes are 1000 miles apart.

Step 2. Represent other unknown quantities in terms of x. Because there are no other unknown quantities that we are asked to find, we can skip this step.

Step 3. Write an equation in x that models the conditions. We obtain this equation by finding a relationship between the distances of the two planes. Here is a table that summarizes the problem's information and includes these distances:

	Rate	×	Time	=	Distance
Faster Plane	300		x		$300x$
Slower Plane	200		x		$200x$

We are interested in finding x, the time when the planes are 1000 miles apart. At this time, the distance traveled by the faster plane plus the distance traveled by the slower plane equals 1000 miles.

Distance traveled by the faster plane	plus	distance traveled by the slower plane	equals	1000 miles.
$300x$	$+$	$200x$	$=$	1000

Step 4. Solve the equation and answer the question.

$$300x + 200x = 1000 \quad \text{This is the equation that models the problem's conditions.}$$
$$500x = 1000 \quad \text{Add like terms.}$$
$$\frac{500x}{500} = \frac{1000}{500} \quad \text{Divide both sides by 500.}$$
$$x = 2 \quad \text{Simplify.}$$

Because x represents the time when the planes are 1000 miles apart, we can conclude that they will be 1000 miles apart after 2 hours.

Step 5. Check the proposed solution in the original wording of the problem. The problem states that the planes are to be 1000 miles apart. After 2 hours, the distance covered by the plane averaging 300 miles per hour is $300 \cdot 2$, or 600 miles. The distance covered by the plane averaging 200 miles per hour is $200 \cdot 2$, or 400 miles. The distance between the planes is 600 miles + 400 miles, or 1000 miles, as specified by the problem's conditions.

☑ **CHECK POINT 6** Two cars leave from the same place, traveling in opposite directions. The rate of the faster car is 55 miles per hour. The rate of the slower car is 50 miles per hour. In how many hours will the cars be 420 miles apart?

Study Tip

In the section opener, we listed three problem-solving tips from *How to Solve It* that we used throughout this section. Here are some additional tips from Polya's book that you may find helpful in many problem-solving situations:

- Look for a pattern.
- Make a systematic list.
- Make an educated guess at the solution. Check the guess against the problem's conditions and work backward to determine the solution.
- Try expressing the problem more simply and solve a similar simpler problem.
- Use trial and error.
- Look for a "catch" if the answer seems too obvious. Perhaps the problem involves some sort of trick question deliberately intended to lead the problem solver in the wrong direction. (See the Blitzer Bonus that follows.)
- Use the given information to eliminate possibilities.
- Use common sense.

Blitzer Bonus

Trick Questions: Thinking Outside the Box

When people attempt to solve this problem, a mental set causes them to think of the dots as representing a box. As shown by the correct solution, it's necessary to draw the lines to extend out of the box of dots—literally "thinking outside the box." (*Source:* Saundra K. Ciccarelli and Glenn E. Meyer, *Psychology*, Prentice Hall, 2006)

Trick questions require you to think carefully before answering. Unless you think outside the box, you'll be duped by some sort of trick or catch.

Sample: Do they have a fourth of July in England?
Answer: Of course they do. However, there is no national holiday on that date!

See if you can answer the questions that follow without developing mental whiplash. The answers appear in the answer section.

1. A farmer had 17 sheep. All but 12 died. How many sheep does the farmer have left?
2. Some months have 30 days. Some have 31. How many months have 28 days?
3. A doctor had a brother, but this brother had no brothers. What was the relationship between doctor and brother?
4. If you had only one match and entered a log cabin in which there was a candle, a fireplace, and a woodburning stove, which should you light first?

3.1 EXERCISE SET   *MyMathLab*

PRACTICE   WATCH   DOWNLOAD   READ   REVIEW

Study Tip

Remember that it is the **thought** you put into solving a problem that will increase your problem-solving skills. Don't be afraid to make mistakes and to take risks. Maintain a sense of humor and have fun as you attempt these exercises. Make an effort to solve each assigned exercise, but don't let any single exercise create a negative mind set that will interfere with your brain's ability to work its way up the math mountain.

Practice and Application Exercises

Exercises 1–10 involve simple interest.

1. You deposit $4000 in a savings account, which has a rate of 4%. Find the interest at the end of the first year.

2. You deposit $9200 in a savings account, which has a rate of 4%. Find the interest at the end of the first year.

3. You invest $20,000 in two accounts paying 7% and 8% annual interest. How much should be invested at each rate if the total interest earned for the year is to be $1520? Begin by filling in the missing entry in the following table. Then use the fact that the interest for the two investments combined must be $1520.

	Principal	×	Rate	=	Interest
7% Investment	x		0.07		0.07x
8% Investment	20,000 − x		0.08		

4. You invested $20,000 in two accounts paying 7% and 9% annual interest. If the total interest earned for the year was $1550, how much was invested at each rate? Begin by filling in the missing entry in the following table. Then use the fact that the interest for the two investments combined was $1550.

	Principal	×	Rate	=	Interest
7% Investment	x		0.07		0.07x
9% Investment	20,000 − x		0.09		

5. A bank loaned out $120,000, part of it at the rate of 8% annual mortgage interest and the rest at the rate of 18% annual credit card interest. The interest received on both loans totaled $10,000. How much was loaned at each rate? Organize your work in the following table.

	Principal	×	Rate	=	Interest
8% Loan	x		0.08		
18% Loan			0.18		

6. A bank loaned out $250,000, part of it at the rate of 8% annual mortgage interest and the rest at the rate of 18% annual credit card interest. The interest received on both loans totaled $23,000. How much was loaned at each rate? Organize your work in the following table.

	Principal	×	Rate	=	Interest
8% Loan	x		0.08		
18% Loan			0.18		

7. You invest $6000 in two accounts paying 6% and 9% annual interest. At the end of the year, the accounts earn the same interest. How much was invested at each rate?

8. You invest $7200 in two accounts paying 8% and 10% annual interest. At the end of the year, the accounts earn the same interest. How much was invested at each rate?

9. Your grandmother needs your help. She has $50,000 to invest. Part of this money is to be invested in noninsured bonds paying 15% annual interest. The rest of this money is to be invested in a government-insured certificate of deposit paying 7% annual interest. She told you that she requires a total of $6000 per year in extra income from these investments. How much money should be placed in each investment?

10. Things did not go quite as planned. You invested part of $8000 in an account that paid 12% annual interest. However, the rest of the money suffered a 5% loss. If the total annual income from both investments was $620, how much was invested at each rate?

Exercises 11–18 involve mixtures.

11. A 20 milliliter solution of acid in water contains 30% acid. How much acid is in the solution?

12. A 70 milliliter solution of acid in water contains 60% acid. How much acid is in the solution?

13. A lab technician needs to mix a 5% fungicide solution with a 10% fungicide solution to obtain a 50-liter mixture consisting of 8% fungicide. How many liters of each of the fungicide solutions must be used? Begin by filling in the missing entries in the following table. Then use the fact that the amount of fungicide in the 5% solution plus the amount of fungicide in the 10% solution must equal the amount of fungicide in the 8% mixture.

	Number of Liters	×	Percent of Fungicide	=	Amount of Fungicide
5% Fungicide Solution	x		0.05		0.05x
10% Fungicide Solution	50 − x		0.10		
8% Fungicide Mixture	50		0.08		

14. A candy company needs to mix a 20% fat-content chocolate with a 15% fat-content chocolate to obtain 50 kilograms of a 16% fat-content chocolate. How many kilograms of each kind of chocolate must be used? Begin by filling in the missing entries in the following table. Then use the fact that the amount of fat in the 20% fat-content chocolate plus the amount of fat in the 15% fat-content chocolate must equal the amount of fat in the 16% mixture.

	Number of Kilograms	×	Percent of Fat	=	Amount of Fat
20% Fat Chocolate	x		0.20		0.20x
15% Fat Chocolate	50 − x		0.15		
16% Fat Mixture	50		0.16		

15. How many ounces of a 15% alcohol solution must be mixed with 4 ounces of a 20% alcohol solution to make a 17% alcohol solution?

16. How many ounces of a 50% alcohol solution must be mixed with 80 ounces of a 20% alcohol solution to make a 40% alcohol solution?

17. At the north campus of a performing arts school, 10% of the students are music majors. At the south campus, 90% of the students are music majors. The campuses are merged into one east campus. If 42% of the 1000 students at the east campus are music majors, how many students did each of the north and south campuses have before the merger?

18. At the north campus of a small liberal arts college, 10% of the students are women. At the south campus, 50% of the students are women. The campuses are merged into one east campus. If 40% of the 1200 students at the east campus are women, how many students did each of the north and south campuses have before the merger?

Exercises 19–26 involve motion.

19. Two cyclists, one averaging 10 miles per hour and the other 12 miles per hour, start from the same town at the same time. If they travel in opposite directions, after how long will they be 66 miles apart? Organize your work in the following table.

	Rate	×	Time	=	Distance
Slower Cyclist	10		x		
Faster Cyclist	12		x		

20. Two cars start from the same place at the same time. They travel in opposite directions. One averages 45 miles per hour and the other averages 30 miles per hour. After how long will they be 225 miles apart? Organize your work in the following table.

	Rate	×	Time	=	Distance
Faster Car	45		x		
Slower Car	30		x		

21. Two trucks leave a warehouse at the same time, traveling in opposite directions. The rate of the faster truck exceeds that of the slower truck by 5 miles per hour. After 5 hours, they are 600 miles apart. What are the rates of the trucks?

	Rate	×	Time	=	Distance
Faster Truck	$x + 5$		5		
Slower Truck	x		5		

22. Two buses leave a station at the same time, traveling in opposite directions. The rate of the faster bus exceeds that of the slower bus by 15 miles per hour. After 3 hours, they are 345 miles apart. What are the rates of the buses?

	Rate	×	Time	=	Distance
Faster Bus	$x + 15$		3		
Slower Bus	x		3		

23. New York City and Washington, D.C., are about 240 miles apart. A car leaves New York City traveling toward Washington, D.C., at 55 miles per hour. At the same time, a bus leaves Washington D.C., bound for New York City at 45 miles per hour. How long will it take before they meet?

24. Two cities are 315 miles apart. A car leaves one of the cities traveling toward the second city at 50 miles per hour. At the same time, a bus leaves the second city bound for the first city at 55 miles per hour. How long will it take for them to meet?

25. A bus traveled at an average rate of 50 miles per hour and then reduced its average rate to 40 miles per hour for the rest of the trip. If the 220-mile trip took 5 hours, determine how long the bus traveled at each rate.

26. A car traveled at an average rate of 40 miles per hour and then reduced its average rate to 30 miles per hour for the rest of the trip. If the 180-mile trip took 5 hours, determine how long the car traveled at each rate.

Practice and Application PLUS

In Exercises 27–36, you will use what you learned about solving interest and mixture problems to solve new kinds of mixture problems. Although tables may help you organize the information, use of tables is not necessary to obtain the equations.

Exercises 27–30 are called **value-mixture problems.** *In these problems, two ingredients that have different values are combined into a single blend. The value of each ingredient in the blend is determined as follows:*

amount of ingredient × cost per unit of the ingredient
= value of the ingredient.

Equations are obtained from the following condition:

| value of ingredient 1 | + | value of ingredient 2 | = | value of blend. |

Use this information to solve Exercises 27–30.

27. The manager of a Starbucks store plans to mix A-grade coffee that costs $9.50 per pound with B-grade coffee that costs $7.00 per pound to create a 20-pound blend that will sell for $8.50 per pound. How many pounds of each grade coffee are required?

28. The manager of a store selling tea plans to mix a more expensive tea that costs $5 per pound with a less expensive tea that costs $3 per pound to create a 100-pound blend that will sell for $4.50 per pound. How many pounds of each type of tea are required?

29. How many pounds of cashews selling for $7.40 per pound must be mixed with 6 pounds of peanuts selling for $2.60 per pound to create a mixture that sells for $5.80 per pound?

30. How many pounds of cashews selling for $8.96 per pound must be mixed with 12 pounds of chocolates selling for $4.48 per pound to create a mixture that sells for $7.28 per pound?

Exercises 31–36 involve mixtures of items with different monetary values, such as coins. If the collection contains coins, the total monetary value of each coin denomination is determined as follows:

number of coins × value of one coin, in cents
= total denomination value, in cents.

Equations are obtained from the following condition:

| total value of denomination 1 | + | total value of denomination 2 | = | total value of all the coins. |

Use this information to solve Exercises 31–36.

31. A coin purse contains a mixture of 32 coins in dimes and quarters. The coins have a total value of $5, or 500¢. Determine the number of dimes and the number of quarters in the purse.

32. A coin purse contains a mixture of 46 coins in dimes and quarters. The coins have a total value of $7, or 700¢. Determine the number of dimes and the number of quarters in the purse.

33. A bank teller cashed a check for $200 using $5 bills and $10 bills. The teller handed 25 more $5 bills than $10 bills to the customer. Determine how many bills of each denomination were handed to the customer.

34. A collection of stamps consists of 25¢ stamps and 40¢ stamps. The number of 40¢ stamps is 5 more than 3 times the number of 25¢ stamps. The total value of the stamps is $10.70. Determine how many stamps of each kind are in the collection.

35. Tickets to a community theater cost $25 for adults and $5 for children. At a recent performance, a total of 305 tickets were sold for $6025. Determine how many tickets of each kind were sold.

36. The manager of a men's clothing store sold out the entire stock of $60 jackets and $50 jackets for a total of $3080. If the manager sold 22 more $50 jackets than $60 jackets, how many jackets of each type were sold?

Writing in Mathematics

37. What is interest?

38. How is simple interest calculated?

39. Describe how to find the amount of an ingredient in a solution if the percentage of the ingredient in the solution is known.

40. Describe what the formula for motion, $d = rt$, means.

41. Is it necessary to use tables and charts to solve the problems discussed in this section? What advantage is there in using charts to solve interest, mixture, and motion problems?

42. Describe two similarities between interest and mixture problems.

43. Must the concentration of a mixture always be greater than the concentration of an ingredient in one of the solutions and less than the concentration of the ingredient in the other solution being mixed? Explain your answer.

Critical Thinking Exercises

Make Sense? *In Exercises 44–47, determine whether each statement "makes sense" or "does not make sense" and explain your reasoning.*

44. When I have trouble getting started on a word problem, it helps me to restate the problem in my own words.

45. When solving a problem involving a $20,000 investment, part at 5% and the remainder at 7%, it does not matter which investment amount is x and which is $20,000 − x$.

46. I organized the information in Exercise 29 in a table (be sure to reread the exercise) and it did not matter which row represented cashews and which represented peanuts.

47. I mixed a 50% acid solution with a 60% acid solution to obtain 12 liters of a 62% acid solution.

48. You have $70,000 to invest. Part of the money is to be placed in a certificate of deposit paying 8% per year. The rest is to be placed in corporate bonds paying 12% per year. If you wish to obtain an overall return of 9% per year, how much should you place in each investment?

49. How many milliliters of pure water should a pharmacist add to 50 milliliters of a 15% minoxidil solution (used to treat male pattern baldness) to make a 10% minoxidil solution?

50. Two thieves rob a jewelry store and make their getaway at 10:00 A.M., traveling at 85 miles per hour. At noon, the police leave the store, in hot pursuit of the robbers. If the police travel at 105 miles per hour, how long does it take for the cops to catch the robbers? At what time of the day does this occur? How far from the jewelry store is the arrest made?

Review Exercises

51. Simplify: $3^2 - 5[4 - 3(2 - 6)]$. (Section 1.8, Example 7)

52. List the integers in this set: $\{-100, -\frac{3}{4}, 0, \sqrt{3}, \sqrt{16}\}$. (Section 1.3, Example 5)

53. Solve: $3(x + 3) = 5(2x + 6)$. (Section 2.3, Example 3)

Preview Exercises

Exercises 54–56 will help you prepare for the material covered in the next section.

54. Reduce to lowest terms: $\frac{32}{200}$.

55. Is -12 a solution of $\frac{-8}{3} = \frac{32}{x}$?

56. Write as an equation, where x represents the number: The quotient of 63 and a number is equal to the quotient of 7 and 5.

Ratio and Proportion

SECTION 3.2

Objectives

1 Find ratios.

2 Solve proportions.

3 Solve problems using proportions.

The possibility of seeing a blue whale, the largest mammal ever to grace the earth, increases the excitement of gazing out over the ocean's swell of waves. Blue whales were hunted to near extinction in the last half of the nineteenth and the first half of the twentieth centuries. Using a method for estimating wildlife populations that we discuss in this section, by the mid-1960s it was determined that the world population of blue whales was less than 1000. This led the International Whaling Commission to ban the killing of blue whales to prevent their extinction. A dramatic increase in blue whale sightings indicates an ongoing increase in their population and the success of the killing ban.

1 Find ratios.

Ratios

A **ratio** compares quantities by division. For example, this year's entering class at a medical school contains 60 women and 30 men. The ratio of women to men is $\frac{60}{30}$. We can express this ratio as a fraction reduced to lowest terms:

$$\frac{60}{30} = \frac{\cancel{30} \cdot 2}{\cancel{30} \cdot 1} = \frac{2}{1}.$$

This ratio can be expressed as 2:1, or 2 to 1.

EXAMPLE 1 Finding Ratios

Imagine the entire global population as a village of precisely 200 people. **Figure 3.2** shows some numeric observations based on this scenario.

Earth's Population as a Village of 200 People

FIGURE 3.2

Source: Gary Rimmer, *Number Freaking*, The Disinformation Company Ltd., 2006

Use the numbers shown to find each of the following ratios:

a. The ratio of Asians to Americans

b. The ratio of people unable to read or write to all people in the village.

Solution

a. The ratio of Asians to Americans is determined as follows:

Refers to the U.S. and not all of North and South America

$$\frac{\text{number of Asians}}{\text{number of Americans}} = \frac{122}{9}.$$

The ratio of Asians to Americans is 122:9, or 122 to 9.

b. The ratio of people unable to read or write to all people in the village is determined as follows:

$$\frac{\text{number unable to read or write}}{\text{number of people in the village}} = \frac{32}{200} = \frac{8 \cdot 4}{8 \cdot 25} = \frac{4}{25}.$$

The ratio of people unable to read or write to all people in the village is 4:25, or 4 to 25. (Because this ratio reflects the entire global population and $\frac{4}{25} = 16\%$, this means that 16% of the world's population is illiterate.) ▬

☑ **CHECK POINT 1** Use the numbers shown in **Figure 3.2** to find each of the following ratios:

a. The ratio of people under age 15 to those over 65.

b. The ratio of Africans to all people in the village.

2 Solve proportions.

Proportions

A **proportion** is a statement that says that two ratios are equal. If the ratios are $\frac{a}{b}$ and $\frac{c}{d}$, then the proportion is

$$\frac{a}{b} = \frac{c}{d}.$$

We can clear this equation of fractions by multiplying both sides by bd, the least common denominator:

$$\frac{a}{b} = \frac{c}{d}$$ This is the given proportion.

$$bd \cdot \frac{a}{b} = bd \cdot \frac{c}{d}$$ Multiply both sides by bd ($b \neq 0$ and $d \neq 0$). Then simplify.

$$ad = bc$$ On the left, $\frac{bd}{1} \cdot \frac{a}{b} = da = ad$. On the right, $\frac{bd}{1} \cdot \frac{c}{d} = bc$.

We see that the following principle is true for any proportion:

The Cross-Products Principle for Proportions

If $\dfrac{a}{b} = \dfrac{c}{d}$, then $ad = bc$. ($b \neq 0$ and $d \neq 0$)

The cross products ad and bc are equal.

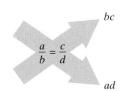

The cross-products principle:
$ad = bc$

For example, since $\frac{2}{3} = \frac{6}{9}$, we see that $2 \cdot 9 = 3 \cdot 6$, or $18 = 18$. We can also use $\frac{2}{3} = \frac{6}{9}$ and conclude that $3 \cdot 6 = 2 \cdot 9$. When using the cross-products principle, it does not matter on which side of the equation each product is placed.

If three of the numbers in a proportion are known, the value of the missing quantity can be found by using the cross-products principle. This idea is illustrated in Example 2.

EXAMPLE 2   Solving Proportions

Solve each proportion for x:

a. $\dfrac{63}{x} = \dfrac{7}{5}$ b. $\dfrac{-8}{3} = \dfrac{32}{x}$.

Solution

a.

$$\frac{63}{x} = \frac{7}{5}$$ This is the given proportion.

$$63 \cdot 5 = 7x$$ Apply the cross-products principle.

$$315 = 7x$$ Simplify.

$$\frac{315}{7} = \frac{7x}{7}$$ Divide both sides by 7.

$$45 = x$$ Simplify.

$$\frac{63}{x} = \frac{7}{5}$$

Cross products

This means that the ratio of 63 to 45 is the same as the ratio of 7 to 5. This can be checked by simplifying $\frac{63}{45}$, dividing the numerator and denominator by 9 as follows:

$$\frac{63}{45} = \frac{9 \cdot 7}{9 \cdot 5} = \frac{7}{5}.$$

The solution to the proportion is 45, or the solution set is $\{45\}$.

b.

$$\frac{-8}{3} = \frac{32}{x}$$ This is the given proportion.

$$-8x = 3 \cdot 32$$ Apply the cross-products principle.

$$-8x = 96$$ Simplify.

$$\frac{-8x}{-8} = \frac{96}{-8}$$ Divide both sides by -8.

$$x = -12$$ Simplify.

This means that the ratio of 32 to -12 is the same as the ratio of -8 to 3.

$$\frac{32}{-12} = \frac{4 \cdot 8}{-4 \cdot 3} = \frac{8}{-3} = \frac{-8}{3}$$

The solution to the proportion is -12, or the solution set is $\{-12\}$.

3 Solve problems using proportions.

Applications of Proportions

We now turn to practical application problems that can be solved using proportions. Here is a procedure for solving these problems:

> ### Solving Applied Problems Using Proportions
>
> 1. Read the problem and represent the unknown quantity by x (or any letter).
> 2. Set up a proportion by listing the given ratio on one side and the ratio with the unknown quantity on the other side. Each respective quantity should occupy the same corresponding position on each side of the proportion.
> 3. Drop units and apply the cross-products principle.
> 4. Solve for x and answer the question.

EXAMPLE 3 Applying Proportions: Calculating Taxes

The property tax on a house with an assessed value of $480,000 is $5760. Determine the property tax on a house with an assessed value of $600,000, assuming the same tax rate.

Solution

Step 1. Represent the unknown by x. Let $x =$ the tax on the $600,000 house.

Step 2. Set up a proportion. We will set up a proportion comparing taxes to assessed value.

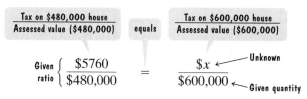

Step 3. Drop the units and apply the cross-products principle. We drop the dollar signs and begin to solve for x.

$$\frac{5760}{480,000} = \frac{x}{600,000} \qquad \text{This is the proportion that models the problem's conditions.}$$

$$480,000x = (5760)(600,000) \qquad \text{Apply the cross-products principle.}$$

$$480,000x = 3,456,000,000 \qquad \text{Multiply.}$$

Step 4. Solve for x and answer the question.

$$\frac{480,000x}{480,000} = \frac{3,456,000,000}{480,000} \qquad \text{Divide both sides by 480,000.}$$

$$x = 7200 \qquad \text{Simplify.}$$

The property tax on the $600,000 house is $7200.

☑ **CHECK POINT 3** The property tax on a house with an assessed value of $250,000 is $3500. Determine the property tax on a house with an assessed value of $420,000, assuming the same tax rate.

Study Tip

Here are three other correct proportions you can use in step 2:

- $\dfrac{\$480,000 \text{ value}}{\$5760 \text{ tax}} = \dfrac{\$600,000 \text{ value}}{\$x \text{ tax}}$

- $\dfrac{\$480,000 \text{ value}}{\$600,000 \text{ value}} = \dfrac{\$5760 \text{ tax}}{\$x \text{ tax}}$

- $\dfrac{\$600,000 \text{ value}}{\$480,000 \text{ value}} = \dfrac{\$x \text{ tax}}{\$5760 \text{ tax}}$.

Each proportion gives the same cross product obtained in step 3.

Sampling in Nature

The method that was used to estimate the blue whale population described in the section opener is called the **capture-recapture method**. Because it is impossible to count each individual animal within a population, wildlife biologists randomly catch and tag a given number of animals. Sometime later they recapture a second sample of animals and count the number of recaptured tagged animals. The total size of the wildlife population is then estimated using the following proportion:

Initially unknown (x) ⟶

$$\underbrace{\dfrac{\text{Original number of tagged animals}}{\text{Total number of animals in the population}} = \dfrac{\text{Number of recaptured tagged animals}}{\text{Number of animals in second sample}}}_{\text{Known ratio}}.$$

Although this is called the capture-recapture method, it is not necessary to recapture animals to observe whether or not they are tagged. This could be done from a distance, with binoculars for instance.

EXAMPLE 4   Applying Proportions: Estimating Wildlife Population

Wildlife biologists catch, tag, and then release 135 deer back into a wildlife refuge. Two weeks later they observe a sample of 140 deer, 30 of which are tagged. Assuming the ratio of tagged deer in the sample holds for all deer in the refuge, approximately how many deer are in the refuge?

Solution

Step 1. Represent the unknown by x. Let x = the total number of deer in the refuge.

Step 2. Set up a proportion.

Unknown ⟶ $\dfrac{\text{Original number of tagged deer}}{\text{Total number of deer}}$ equals $\underbrace{\dfrac{\text{Number of tagged deer in the observed sample}}{\text{Total number of deer in the observed sample}}}_{\text{Known ratio}}$

$$\dfrac{135}{x} = \dfrac{30}{140}$$

Steps 3 and 4. Apply the cross-products principle, solve, and answer the question.

$$\dfrac{135}{x} = \dfrac{30}{140} \qquad \text{This is the proportion that models the problem's conditions.}$$

$$(135)(140) = 30x \qquad \text{Apply the cross-products principle.}$$

$$18{,}900 = 30x \qquad \text{Multiply.}$$

$$\dfrac{18{,}900}{30} = \dfrac{30x}{30} \qquad \text{Divide both sides by 30.}$$

$$630 = x \qquad \text{Simplify.}$$

There are approximately 630 deer in the refuge. ▄

✓ **CHECK POINT 4**   Wildlife biologists catch, tag, and then release 120 deer back into a wildlife refuge. Two weeks later they observe a sample of 150 deer, 25 of which are tagged. Assuming the ratio of tagged deer in the sample holds for all deer in the refuge, approximately how many deer are in the refuge?

3.2 EXERCISE SET *MyMathLab*

PRACTICE   WATCH   DOWNLOAD   READ   REVIEW

Practice Exercises

In Exercises 1–6, express each ratio as a fraction reduced to lowest terms.

1. 24 to 48

2. 14 to 49

3. 48 to 20

4. 24 to 15

5. 27:36

6. 25:40

In a class, there are 20 men and 10 women. Find each ratio in Exercises 7–10. First express the ratio as a fraction reduced to lowest terms. Then rewrite the ratio using the reduced fraction and a colon.

7. Find the ratio of the number of men to the number of women.

8. Find the ratio of the number of women to the number of men.

9. Find the ratio of the number of women to the number of students in the class.

10. Find the ratio of the number of men to the number of students in the class.

Solve each proportion in Exercises 11–22.

11. $\dfrac{20}{x} = \dfrac{5}{3}$

12. $\dfrac{27}{x} = \dfrac{9}{4}$

13. $\dfrac{x}{3} = \dfrac{5}{2}$

14. $\dfrac{x}{6} = \dfrac{3}{2}$

15. $\dfrac{x}{12} = -\dfrac{3}{4}$

16. $\dfrac{x}{64} = -\dfrac{9}{16}$

17. $\dfrac{x-2}{12} = \dfrac{8}{3}$

18. $\dfrac{x-4}{10} = \dfrac{3}{5}$

19. $\dfrac{x}{7} = \dfrac{x+14}{5}$

20. $\dfrac{x}{5} = \dfrac{x-3}{2}$

21. $\dfrac{x+9}{5} = \dfrac{x-10}{11}$

22. $\dfrac{x+7}{-4} = \dfrac{x-12}{6}$

Practice PLUS

In Exercises 23–28, solve each proportion for x.

23. $\dfrac{x}{a} = \dfrac{b}{c}$

24. $\dfrac{a}{x} = \dfrac{b}{c}$

25. $\dfrac{a+b}{c} = \dfrac{x}{d}$

26. $\dfrac{a-b}{c} = \dfrac{x}{d}$

27. $\dfrac{x+a}{a} = \dfrac{b+c}{c}$

28. $\dfrac{ax-b}{b} = \dfrac{c-d}{d}$

Application Exercises

The bar graph shows drinks that contain caffeine and the number of milligrams in each serving.

Java Jolts: Drinks That Contain Caffeine

Source: The New York Times Practical Guide to Practically Everything

In Exercises 29–32, use the graph to find each ratio. First express the ratio as a fraction reduced to lowest terms. Then rewrite the ratio using the reduced fraction and a colon.

29. The number of milligrams of caffeine in a cup of strong coffee to that in a cup of weak coffee

30. The number of milligrams of caffeine in a cup of strong coffee to that in a cup of weak tea

31. The number of milligrams of caffeine in two glasses of cola to that in a cup of strong coffee

32. The number of milligrams of caffeine in three glasses of cola to that in a cup of strong coffee

Use a proportion to solve each problem in Exercises 33–42.

33. The tax on a property with an assessed value of $65,000 is $720. Find the tax on a property with an assessed value of $162,500.

34. The maintenance bill for a shopping center containing 180,000 square feet is $45,000. What is the bill for a store in the center that is 4800 square feet?

35. St. Paul Island in Alaska has 12 fur seal rookeries (breeding places). In 1961, to estimate the fur seal pup population in the Gorbath rookery, 4963 fur seal pups were tagged in early August. In late August, a sample of 900 pups was observed and 218 of these were found to have been previously tagged. Estimate the total number of fur seal pups in this rookery.

36. To estimate the number of bass in a lake, wildlife biologists tagged 50 bass and released them in the lake. Later they netted 108 bass and found that 27 of them were tagged. Approximately how many bass are in the lake?

37. In Example 1, we saw that in a global village of 200 people, 28 suffered from malnutrition. How many people of the world's 6.5 billion people (2006 population) suffer from malnutrition? Round to the nearest tenth of a billion.

38. According to the authors of *Number Freaking*, in a global village of 200 people, 9 get drunk every day. How many of the world's 6.5 billion people (2006 population) get drunk every day?

39. Height is proportional to foot length. A person whose foot length is 10 inches is 67 inches tall. In 1951, photos of large footprints were published. Some believed that these foot-prints were made by the "Abominable Snowman." Each footprint was 23 inches long. If indeed they belonged to the Abominable Snowman, how tall is the critter?

Roger Patterson comparing his foot with a plaster cast of a footprint of the purported "Bigfoot" that Mr. Patterson said he sighted in a California forest in 1967.

40. A person's hair length is proportional to the number of years it has been growing. After 2 years, a person's hair grows 8 inches. The longest moustache on record was grown by Kalyan Sain of India. Sain grew his moustache for 17 years. How long was each side of the moustache?

41. The formulas

$$C_{\text{Gas}} = 12{,}000 + 700x$$
$$C_{\text{Solar}} = 30{,}000 + 150x$$

model the total cost, in dollars, for gas and solar heating systems x years after installation. What is the ratio of the total cost for gas heating to the total cost for solar heating 5 years after installation? What is this ratio 40 years after installation?

42. The formulas

$$C_{\text{Electric}} = 5000 + 1100x$$
$$C_{\text{Solar}} = 30{,}000 + 150x$$

model the total cost, in dollars, for electric and solar heating systems x years after installation. What is the ratio of the total cost for electric heating to the total cost for solar heating 5 years after installation? What is this ratio 40 years after installation?

Writing in Mathematics

43. What is a ratio? Give an example with your description.

44. What is a proportion? Give an example with your description.

45. Explain the difference between a ratio and a proportion.

46. Explain how to solve a proportion. Illustrate your explanation with an example.

47. Explain the meaning of this statement: A company's monthly sales is proportional to its advertising budget.

Critical Thinking Exercises

Make Sense? *In Exercises 48–51, determine whether each statement "makes sense" or "does not make sense" and explain your reasoning.*

48. At the rate shown in the cartoon, five potatoes would cost $36.67.

If the people who own the shops at the airport owned other things.

Universal Press Syndicate

49. I can solve $\frac{x}{9} = \frac{4}{6}$ by using the cross-products principle or by multiplying both sides by 18, the least common denominator.

50. I made an error using the cross-products principle and put each product on the wrong side of the equation.

51. The capture-recapture method may be inaccurate if some of the animals are hiding.

In Exercises 52–55, determine whether each statement is true or false. If the statement is false, make the necessary change(s) to produce a true statement.

52. A statement that two proportions are equal is called a ratio.

53. The ratio of 1.25 to 2 is equal to the ratio 5:8.

54. The ratio of heights of a 12-inch plant to a 4-foot plant is 3:1.

55. If 30 people out of 70 are men, then the ratio of women to men is 40:100.

56. The front sprocket on a bicycle has 60 teeth and the rear sprocket has 20 teeth. For mountain biking, an owner needs a 5-to-1 front-to-rear ratio. If only one of the sprockets is to be replaced, describe the two ways in which this can be done.

57. A baseball player's batting average is the ratio of the number of hits to the number of times at bat. A baseball player has 10 hits out of 20 times at bat. How many consecutive times at bat must a hit be made to raise the player's batting average to 0.600?

Technology Exercises

Use a calculator to solve Exercises 58–59. Round your answer to two decimal places.

58. Solve: $\dfrac{7.32}{x} = \dfrac{-19.03}{28}$

59. On a map, 2 inches represent 13.47 miles. How many miles does a person plan to travel if the distance on the map is 9.85 inches?

Review Exercises

60. Solve and graph the solution set: $-6x + 2 \le 2(5 - x)$. (Section 2.6, Example 8)

61. 112 is 40% of what number? (Section 2.4, Example 8)

62. Simplify: $9 - 2[4(x - 3) + 7]$. (Section 1.8, Example 11)

Preview Exercises

Exercises 63–65 will help you prepare for the material covered in the next section.

63. Use $A = \frac{1}{2}bh$ to find h if $A = 30$ and $b = 12$.

64. Evaluate $A = \dfrac{1}{2}h(a + b)$ for $a = 10, b = 16$, and $h = 7$.

65. Solve: $x = 4(90 - x) - 40$.

MID-CHAPTER CHECK POINT   Section 3.1–Section 3.2

What You Know: We learned to solve problems involving simple interest, mixtures, and motion, using tables to organize the information. We used the cross-products principle to solve proportions and used proportions to solve a variety of practical problems.

1. If you can type 600 pages of manuscript in 21 days, how many days will it take to type 230 pages working at the same rate?

2. How many liters of a 30% alcohol solution must be mixed with a 70% alcohol solution to obtain 12 liters of a 60% alcohol solution?

3. Solve: $\dfrac{4x}{9} = \dfrac{5}{-15}$.

4. You invested $25,000 in two accounts paying 8% and 9% annual interest. At the end of the year, the total interest from these investments was $2135. How much was invested at each rate?

5. In a class, there are 25 men and 35 women. Find the ratio of the number of men to the number of students in the class. First express the ratio as a fraction reduced to lowest terms. Then rewrite the ratio using a second method.

6. Two planes leave an airport at the same time and fly in opposite directions. The rate of the faster plane exceeds that of the slower plane by 100 miles per hour. After 5 hours, they are 4250 miles apart. What are the rates of the planes?

7. To estimate the number of elk in a wildlife refuge, park rangers catch, tag, and then release 40 elk. One month later they observe a sample of 80 elk, 4 of which are tagged. Assuming the ratio of tagged elk in the sample holds for all elk in the refuge, approximately how many elk are in the refuge?

8. How many quarts of $\frac{1}{2}$%-fat milk should be added to 4 quarts of 2%-fat milk to obtain 1%-fat milk?

9. Solve: $\dfrac{2}{x + 1} = \dfrac{3}{2x + 3}$.

10. You invested $4000. On part of this investment, you earned 4% interest. On the remainder of the investment, you lost 3%. Combining earnings and losses, the annual income from these investments was $55. How much was invested at each rate?

11. To reach your vacation destination, you travel part of the distance by plane averaging 450 miles per hour and the remainder of the distance by car averaging 50 miles per hour. You spend 2 hours less traveling by car than by plane. If the entire trip is 1900 miles, how long do you travel by plane and how long do you travel by car?

12. The distance that a spring stretches is proportional to the weight attached to the spring. When a 60-pound weight is attached, it stretches the spring 4 inches. What distance will an 80-pound weight stretch the spring?

SECTION
3.3

Objectives

1 Solve problems using formulas for perimeter and area.

2 Solve problems using formulas for a circle's area and circumference.

3 Solve problems using formulas for volume.

4 Solve problems involving the angles of a triangle.

5 Solve problems involving complementary and supplementary angles.

Problem Solving in Geometry

A portion of the human lung magnified 160 times

Geometry is about the space you live in and the shapes that surround you. You're even made of it. The human lung consists of nearly 300 spherical air sacs, geometrically designed to provide the greatest surface area within the limited volume of our bodies. Viewed in this way, geometry becomes an intimate experience.

For thousands of years, people have studied geometry in some form to obtain a better understanding of the world in which they live. A study of the shape of your world will provide you with many practical applications that will help to increase your problem-solving skills.

1 Solve problems using formulas for perimeter and area.

Geometric Formulas for Perimeter and Area

Solving geometry problems often requires using basic geometric formulas. Formulas for perimeter and area are summarized in **Table 3.1**. Remember that perimeter is measured in linear units, such as feet or meters, and area is measured in square units, such as square feet, ft^2, or square meters, m^2.

| Table 3.1 | Common Formulas for Perimeter and Area |

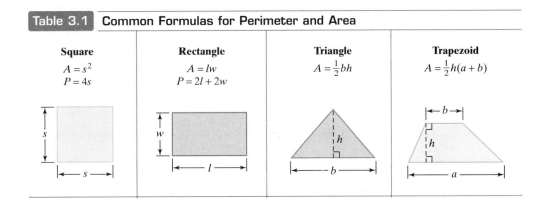

Square
$A = s^2$
$P = 4s$

Rectangle
$A = lw$
$P = 2l + 2w$

Triangle
$A = \frac{1}{2}bh$

Trapezoid
$A = \frac{1}{2}h(a + b)$

FIGURE 3.3 Finding the height of a triangular sail

EXAMPLE 1   Using the Formula for the Area of a Triangle

A sailboat has a triangular sail with an area of 30 square feet and a base that is 12 feet long. (See **Figure 3.3.**) Find the height of the sail.

Solution   We begin with the formula for the area of a triangle given on the previous page in **Table 3.1**.

$$A = \frac{1}{2}bh \qquad \text{The area of a triangle is } \tfrac{1}{2} \text{ the product of its base and height.}$$

$$30 = \frac{1}{2}(12)h \qquad \text{Substitute 30 for } A \text{ and 12 for } b.$$

$$30 = 6h \qquad \text{Simplify.}$$

$$\frac{30}{6} = \frac{6b}{6} \qquad \text{Divide both sides by 6.}$$

$$5 = h \qquad \text{Simplify.}$$

The height of the sail is 5 feet.

Check
The area is $A = \frac{1}{2}bh = \frac{1}{2}(12 \text{ feet})(5 \text{ feet}) = 30$ square feet.   ■

☑ **CHECK POINT 1**   A sailboat has a triangular sail with an area of 24 square feet and a base that is 4 feet long. Find the height of the sail.

---

2 Solve problems using formulas for a circle's area and circumference.

The point at which a pebble hits a flat surface of water becomes the center of a number of circular ripples.

Geometric Formulas for Circumference and Area of a Circle

It's a good idea to know your way around a circle. Clocks, angles, maps, and compasses are based on circles. Circles occur everywhere in nature: in ripples on water, patterns on a butterfly's wings, and cross sections of trees. Some consider the circle to be the most pleasing of all shapes.

A **circle** is a set of points in the plane equally distant from a given point, its center. **Figure 3.4** shows two circles. A **radius** (plural: radii), r, is a line segment from the center to any point on the circle. For a given circle, all radii have the same length. A **diameter**, d, is a line segment through the center whose endpoints both lie on the circle. For a given circle, all diameters have the same length. In any circle, **the length of a diameter is twice the length of a radius**.

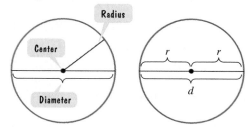

FIGURE 3.4

The words *radius* and *diameter* refer to both the line segments in **Figure 3.4** as well as their linear measures. The distance around a circle (its perimeter) is called its **circumference**. Formulas for the area and circumference of a circle are given in terms of π and appear in **Table 3.2**. We have seen that π is an irrational number and is approximately equal to 3.14.

Table 3.2	Formulas for Circles	
Circle	Area	Circumference
	$A = \pi r^2$	$C = 2\pi r$

When computing a circle's area or circumference by hand, round π to 3.14. When using a calculator, use the $\boxed{\pi}$ key, which gives the value of π rounded to approximately 11 decimal places. In either case, calculations involving π give approximate answers. These answers can vary slightly depending on how π is rounded. The symbol $\approx$ (is approximately equal to) will be written in these calculations.

EXAMPLE 2 Finding the Area and Circumference of a Circle

Find the area and circumference of a circle with a diameter measuring 20 inches.

Solution The radius is half the diameter, so $r = \frac{20}{2} = 10$ inches.

$$A = \pi r^2 \qquad C = 2\pi r \qquad \text{Use the formulas for area and circumference of a circle.}$$
$$A = \pi(10)^2 \qquad C = 2\pi(10) \quad \text{Substitute 10 for } r.$$
$$A = 100\pi \qquad C = 20\pi$$

The area of the circle is 100π square inches and the circumference is 20π inches. Using the fact that $\pi \approx 3.14$, the area is approximately 100(3.14), or 314 square inches. The circumference is approximately 20(3.14), or 62.8 inches. ■

☑ **CHECK POINT 2** The diameter of a circular landing pad for helicopters is 40 feet. Find the area and circumference of the landing pad. Express answers in terms of π. Then round answers to the nearest square foot and foot, respectively.

EXAMPLE 3 Problem Solving Using the Formula for a Circle's Area

Which one of the following is the better buy: a large pizza with a 16-inch diameter for $15.00 or a medium pizza with an 8-inch diameter for $7.50?

Solution The better buy is the pizza with the lower price per square inch. The radius of the large pizza is $\frac{1}{2} \cdot 16$ inches, or 8 inches, and the radius of the medium pizza is $\frac{1}{2} \cdot 8$ inches, or 4 inches. The area of the surface of each circular pizza is determined using the formula for the area of a circle.

$$\text{Large pizza:} \quad A = \pi r^2 = \pi(8 \text{ in.})^2 = 64\pi \text{ in.}^2 \approx 201 \text{ in.}^2$$
$$\text{Medium pizza:} \quad A = \pi r^2 = \pi(4 \text{ in.})^2 = 16\pi \text{ in.}^2 \approx 50 \text{ in.}^2$$

Using Technology

You can use your calculator to obtain the price per square inch for each pizza in Example 3. The price per square inch for the large pizza, $\frac{15}{64\pi}$, is approximated by one of the following keystrokes:

Many Scientific Calculators

$15 \boxed{\div} \boxed{(} 64 \boxed{\times} \boxed{\pi} \boxed{)} \boxed{=}$

Many Graphing Calculators

$15 \boxed{\div} \boxed{(} 64 \boxed{\pi} \boxed{)} \boxed{\text{ENTER}}$

For each pizza, the price per square inch is found by dividing the price by the area:

$$\text{Price per square inch for large pizza} = \frac{\$15.00}{64\pi \text{ in.}^2} \approx \frac{\$15.00}{201 \text{ in.}^2} \approx \frac{\$0.07}{\text{in.}^2}$$
$$\text{Price per square inch for medium pizza} = \frac{\$7.50}{16\pi \text{ in.}^2} \approx \frac{\$7.50}{50 \text{ in.}^2} = \frac{\$0.15}{\text{in.}^2}.$$

The large pizza costs approximately $0.07 per square inch and the medium pizza costs approximately $0.15 per square inch. Thus, the large pizza is the better buy. ■

In Example 3, did you at first think that the price per square inch would be the same for the large and the medium pizzas? After all, the radius of the large pizza is twice that of the medium pizza, and the cost of the large is twice that of the medium. However, the large pizza's area, 64π square inches, is *four times the area* of the medium pizza, 16π square inches. Doubling the radius of a circle increases its area by four times the original amount.

☑ **CHECK POINT 3**   Which one of the following is the better buy: a large pizza with an 18-inch diameter for $20.00 or a medium pizza with a 14-inch diameter for $14.00?

---

3 Solve problems using formulas for volume.

Geometric Formulas for Volume

A shoe box and a basketball are examples of three-dimensional figures. **Volume** refers to the amount of space occupied by such a figure. To measure this space, we begin by selecting a cubic unit. One such cubic unit, 1 cubic centimeter (cm³), is shown in **Figure 3.5**.

The edges of a cube all have the same length. Other cubic units used to measure volume include 1 cubic inch (in.³) and 1 cubic foot (ft³). The volume of a solid is the number of cubic units that can be contained in the solid.

Formulas for volumes of three-dimensional figures are given in **Table 3.3**.

FIGURE 3.5

Table 3.3	Common Formulas for Volume				
Cube	**Rectangular Solid**	**Circular Cylinder**	**Sphere**	**Cone**
$V = s^3$	$V = lwh$	$V = \pi r^2 h$	$V = \frac{4}{3}\pi r^3$	$V = \frac{1}{3}\pi r^2 h$

EXAMPLE 4   Using the Formula for the Volume of a Cylinder

A cylinder with a radius of 2 inches and a height of 6 inches has its radius doubled. (See **Figure 3.6**.) How many times greater is the volume of the larger cylinder than the volume of the smaller cylinder?

Solution   We begin with the formula for the volume of a cylinder given in **Table 3.3**. Find the volume of the smaller cylinder and the volume of the larger cylinder. To compare the volumes, divide the volume of the larger cylinder by the volume of the smaller cylinder.

Radius: 2 inches
Height: 6 inches

$$V = \pi r^2 h \qquad \text{Use the formula for the volume of a cylinder.}$$

Radius is doubled.

$$V_{\text{Smaller}} = \pi(2)^2(6) \quad V_{\text{Larger}} = \pi(4)^2(6) \quad \text{Substitute the given values.}$$

$$V_{\text{Smaller}} = \pi(4)(6) \quad V_{\text{Larger}} = \pi(16)(6)$$

$$V_{\text{Smaller}} = 24\pi \qquad V_{\text{Larger}} = 96\pi$$

The volume of the smaller cylinder is 24π cubic inches. The volume of the larger cylinder is 96π cubic inches. We use division to compare the volumes:

$$\frac{V_{\text{Larger}}}{V_{\text{Smaller}}} = \frac{96\pi}{24\pi} = \frac{4}{1}.$$

Radius: 4 inches
Height: 6 inches

FIGURE 3.6   Doubling a cylinder's radius

Thus, the volume of the larger cylinder is 4 times the volume of the smaller cylinder.   ■

✓ **CHECK POINT 4** A cylinder with a radius of 3 inches and a height of 5 inches has its height doubled. How many times greater is the volume of the larger cylinder than the volume of the smaller cylinder?

EXAMPLE 5 Applying Volume Formulas

An ice cream cone is 5 inches deep and has a radius of 1 inch. A spherical scoop of ice cream also has a radius of 1 inch. (See **Figure 3.7**.) If the ice cream melts into the cone, will it overflow?

Solution The ice cream will overflow if the volume of the ice cream, a sphere, is greater than the volume of the cone. Find the volume of each.

$$V_{\text{cone}} = \frac{1}{3}\pi r^2 h = \frac{1}{3}\pi (1 \text{ in.})^2 \cdot 5 \text{ in.} = \frac{5\pi}{3} \text{in.}^3 \approx 5 \text{ in.}^3$$

$$V_{\text{sphere}} = \frac{4}{3}\pi r^3 = \frac{4}{3}\pi (1 \text{ in.})^3 = \frac{4\pi}{3} \text{in.}^3 \approx 4 \text{ in.}^3$$

The volume of the spherical scoop of ice cream is less than the volume of the cone, so there will be no overflow.

1 in.

5 in.

FIGURE 3.7

✓ **CHECK POINT 5** A basketball has a radius of 4.5 inches. If 350 cubic inches of air are pumped into the ball, is this enough air to fill it completely?

Blitzer Bonus

Deceptions in Visual Displays of Data

Graphs can be used to distort data, making it difficult for the viewer to learn the truth. One potential source of misunderstanding involves geometric figures whose lengths are in the correct ratios for the displayed data, but whose areas or volumes are then varied to create a misimpression about how the data are changing over time. Here are two examples of misleading visual displays.

Graphic Display	Presentation Problems
Purchasing Power of the Diminishing Dollar $1.00 $1.00 63¢ 54¢ 48¢ 43¢ $0.50 0 1980 1990 1995 2000 2005 *Source*: Bureau of Labor Statistics	Although the length of each dollar bill is proportional to its spending power, the visual display varies both the length *and width* of the bills to show the diminishing power of the dollar over time. Because our eyes focus on the *areas* of the dollar-shaped bars, this creates the impression that the purchasing power of the dollar diminished even more than it really did. If the area of the dollar were drawn to reflect its purchasing power, the 2005 dollar would be approximately twice as large as the one shown in the graphic display.

(*continues on page 208*)

Graphic Display	Presentation Problems
Average Daily Price per Barrel of Oil 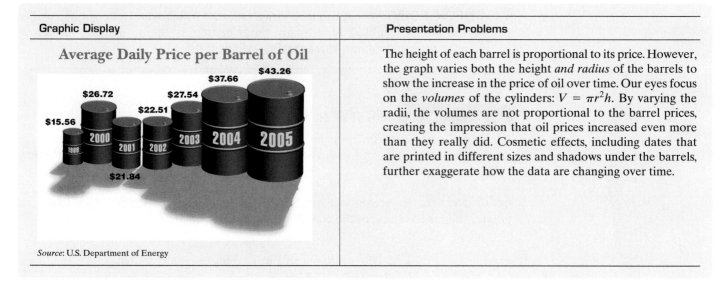 *Source*: U.S. Department of Energy	The height of each barrel is proportional to its price. However, the graph varies both the height *and radius* of the barrels to show the increase in the price of oil over time. Our eyes focus on the *volumes* of the cylinders: $V = \pi r^2 h$. By varying the radii, the volumes are not proportional to the barrel prices, creating the impression that oil prices increased even more than they really did. Cosmetic effects, including dates that are printed in different sizes and shadows under the barrels, further exaggerate how the data are changing over time.

4 Solve problems involving the angles of a triangle.

FIGURE 3.8 Clock with a ray rotating to form an angle

The Angles of a Triangle

The hour hand of a clock moves from 12 to 2. The hour hand suggests a **ray**, a part of a line that has only one endpoint and extends forever in the opposite direction. An *angle* is formed as the ray in **Figure 3.8** rotates from 12 to 2.

An **angle**, symbolized ∡, is made up of two rays that have a common endpoint. **Figure 3.9** shows an angle. The common endpoint, *B* in the figure, is called the **vertex** of the angle. The two rays that form the angle are called its **sides**. The four ways of naming the angle are shown to the right of **Figure 3.9**.

FIGURE 3.9 An angle: two rays with a common endpoint

One way to measure angles is in **degrees**, symbolized by a small, raised circle °. Think of the hour hand of a clock. From 12 noon to 12 midnight, the hour hand moves around in a complete circle. By definition, the ray has rotated through 360 degrees, or 360°. Using 360° as the amount of rotation of a ray back onto itself, **a degree, 1°, is** $\frac{1}{360}$ **of a complete rotation**.

Our next problem is based on the relationship among the three angles of any triangle.

The Angles of a Triangle

The sum of the measures of the three angles of any triangle is 180°.

EXAMPLE 6   Angles of a Triangle

In a triangle, the measure of the first angle is twice the measure of the second angle. The measure of the third angle is 20° less than the second angle. What is the measure of each angle?

Solution

Step 1. Let x represent one of the quantities. Let

$$x = \text{the measure of the second angle.}$$

Step 2. Represent other unknown quantities in terms of x. The measure of the first angle is twice the measure of the second angle. Thus, let

$$2x = \text{the measure of the first angle.}$$

The measure of the third angle is 20° less than the second angle. Thus, let

$$x - 20 = \text{the measure of the third angle.}$$

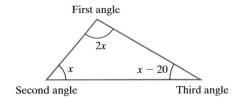

Step 3. Write an equation in x that models the conditions. Because we are working with a triangle, the sum of the measures of its three angles is 180°.

$$2x \quad + \quad x \quad + \quad (x - 20) \quad = \quad 180$$

Step 4. Solve the equation and answer the question.

$$
\begin{array}{ll}
2x + x + (x - 20) = 180 & \text{This is the equation that models the} \\
& \text{sum of the measures of the angles.} \\
4x - 20 = 180 & \text{Regroup and combine like terms.} \\
4x - 20 + 20 = 180 + 20 & \text{Add 20 to both sides.} \\
4x = 200 & \text{Simplify.} \\
\dfrac{4x}{4} = \dfrac{200}{4} & \text{Divide both sides by 4.} \\
x = 50 & \text{Simplify.}
\end{array}
$$

Measure of first angle $= 2x = 2 \cdot 50 = 100$

Measure of second angle $= x = 50$

Measure of third angle $= x - 20 = 50 - 20 = 30$

The angles measure 100°, 50°, and 30°.

Step 5. Check the proposed solution in the original wording of the problem. The problem tells us that we are working with a triangle's angles. Thus, the sum of the measures should be 180°. Adding the three measures, we obtain 100° + 50° + 30°, giving the required sum of 180°. ■

☑ **CHECK POINT 6**   In a triangle, the measure of the first angle is three times the measure of the second angle. The measure of the third angle is 20° less than the second angle. What is the measure of each angle?

---

5 Solve problems involving complementary and supplementary angles.

Complementary and Supplementary Angles

Two angles with measures having a sum of 90° are called **complementary angles**. For example, angles measuring 70° and 20° are complementary angles because 70° + 20° = 90°. For angles such as those measuring 70° and 20°, each angle is a **complement** of the other: The 70° angle is the complement of the 20° angle and the 20° angle is the complement of the 70° angle. The measure of the complement can be found by subtracting the angle's measure from 90°. For example, we can find the complement of a 25° angle by subtracting 25° from 90°: 90° − 25° = 65°. Thus, an angle measuring 65° is the complement of one measuring 25°.

Two angles with measures having a sum of 180° are called **supplementary angles**. For example, angles measuring 110° and 70° are supplementary angles because 110° + 70° = 180°. For angles such as those measuring 110° and 70°, each angle is a **supplement** of the other: The 110° angle is the supplement of the 70° angle and the 70° angle is the supplement of the 110° angle. The measure of the supplement can be found by subtracting the angle's measure from 180°. For example, we can find the supplement of a 25° angle by subtracting 25° from 180°: 180° − 25° = 155°. Thus, an angle measuring 155° is the supplement of one measuring 25°.

Algebraic Expressions for Complements and Supplements

Measure of an angle: x

Measure of the angle's complement: $90 - x$

Measure of the angle's supplement: $180 - x$

EXAMPLE 7   Angle Measures and Complements

The measure of an angle is 40° less than four times the measure of its complement. What is the angle's measure?

Solution

Step 1. Let x represent one of the quantities. Let

$$x = \text{the measure of the angle.}$$

Step 2. Represent other unknown quantities in terms of x. Because this problem involves an angle and its complement, let

$$90 - x = \text{the measure of the complement.}$$

Step 3. Write an equation in x that models the conditions.

$$x \quad = \quad 4(90 - x) \quad - \quad 40$$

Step 4. Solve the equation and answer the question.

$x = 4(90 - x) - 40$	This is the equation that models the problem's conditions.
$x = 360 - 4x - 40$	Use the distributive property.
$x = 320 - 4x$	Simplify: 360 − 40 = 320.
$x + 4x = 320 - 4x + 4x$	Add 4x to both sides.
$5x = 320$	Simplify.
$\dfrac{5x}{5} = \dfrac{320}{5}$	Divide both sides by 5.
$x = 64$	Simplify.

The angle measures 64°.

Step 5. Check the proposed solution in the original wording of the problem. The measure of the complement is 90° − 64° = 26°. Four times the measure of the complement is 4 · 26°, or 104°. The angle's measure, 64°, is 40° less than 104°: 104° − 40° = 64°. As specified by the problem's wording, the angle's measure is 40° less than four times the measure of its complement. ■

✓ **CHECK POINT 7**   The measure of an angle is twice the measure of its complement. What is the angle's measure?

3.3 EXERCISE SET   *MyMathLab*

Practice Exercises

*Use the formulas for perimeter and area in **Table 3.1** on page 203 to solve Exercises 1–12.*

In Exercises 1–2, find the perimeter and area of each rectangle.

1.

3 m
6 m

2.

3 ft
4 ft

In Exercises 3–4, find the area of each triangle.

3.
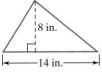
8 in.
14 in.

4.

33 m
30 m

In Exercises 5–6, find the area of each trapezoid.

5.

10 m
7 m
16 m

6.

26 m
18 m   18 m   21.1 m
37 m

7. A rectangular swimming pool has a width of 25 feet and an area of 1250 square feet. What is the pool's length?

8. A rectangular swimming pool has a width of 35 feet and an area of 2450 square feet. What is the pool's length?

9. A triangle has a base of 5 feet and an area of 20 square feet. Find the triangle's height.

10. A triangle has a base of 6 feet and an area of 30 square feet. Find the triangle's height.

11. A rectangle has a width of 44 centimeters and a perimeter of 188 centimeters. What is the rectangle's length?

12. A rectangle has a width of 46 centimeters and a perimeter of 208 centimeters. What is the rectangle's length?

*Use the formulas for the area and circumference of a circle in **Table 3.2** on page 204 to solve Exercises 13–18.*

In Exercises 13–16, find the area and circumference of each circle. Express answers in terms of π. Then round to the nearest whole number.

13.

4 cm

14.

9 m

15.

12 yd

16.

40 ft

17. The circumference of a circle is 14π inches. Find the circle's radius and diameter.

18. The circumference of a circle is 16π inches. Find the circle's radius and diameter.

*Use the formulas for volume in **Table 3.3** on page 206 to solve Exercises 19–30.*

In Exercises 19–26, find the volume of each figure. Where applicable, express answers in terms of π. Then round to the nearest whole number.

19.

4 in.
3 in.
3 in.

20.

3 cm
3 cm
5 cm

21.

5 cm
6 cm

22.

23.

24.

25.

26.

27. Solve the formula for the volume of a circular cylinder for *h*.

28. Solve the formula for the volume of a cone for *h*.

29. A cylinder with radius 3 inches and height 4 inches has its radius tripled. How many times greater is the volume of the larger cylinder than the smaller cylinder?

30. A cylinder with radius 2 inches and height 3 inches has its radius quadrupled. How many times greater is the volume of the larger cylinder than the smaller cylinder?

Use the relationship among the three angles of any triangle to solve Exercises 31–36.

31. Two angles of a triangle have the same measure and the third angle is 30° greater than the measure of the other two. Find the measure of each angle.

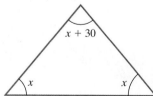

32. One angle of a triangle is three times as large as another. The measure of the third angle is 40° more than that of the smallest angle. Find the measure of each angle.

Find the measure of each angle whose degree measure is represented in terms of x in the triangles in Exercises 33–34.

33.

34.

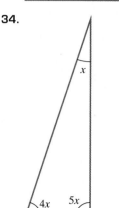

35. One angle of a triangle is twice as large as another. The measure of the third angle is 20° more than that of the smallest angle. Find the measure of each angle.

36. One angle of a triangle is three times as large as another. The measure of the third angle is 30° greater than that of the smallest angle. Find the measure of each angle.

In Exercises 37–40, find the measure of the complement of each angle.

37. 58°   **38.** 41°   **39.** 88°   **40.** 2°

In Exercises 41–44, find the measure of the supplement of each angle.

41. 132°   **42.** 93°

43. 90°   **44.** 179.5°

In Exercises 45–50, use the five-step problem-solving strategy to find the measure of the angle described.

45. The angle's measure is 60° more than that of its complement.

46. The angle's measure is 78° less than that of its complement.

47. The angle's measure is three times that of its supplement.

48. The angle's measure is 16° more than triple that of its supplement.

49. The measure of the angle's supplement is 10° more than three times that of its complement.

50. The measure of the angle's supplement is 52° more than twice that of its complement.

Practice PLUS

In Exercises 51–53, find the area of each figure.

51.

52.

53.

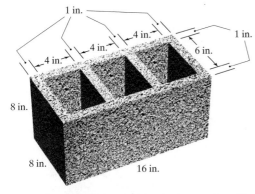

54. Find the area of the shaded region in terms of π.

55. Find the volume of the cement block in the figure shown.

56. Find the volume of the darkly shaded region. Express the answer in terms of π.

Application Exercises

*Use the formulas for perimeter and area in **Table 3.1** on page 203 to solve Exercises 57–58.*

57. Taxpayers with an office in their home may deduct a percentage of their home-related expenses. This percentage is based on the ratio of the office's area to the area of the home. A taxpayer with a 2200-square-foot home maintains a 20-foot by 16-foot office. If the yearly electricity bills for the home come to $4800, how much of this is deductible?

58. The lot in the figure shown, except for the house, shed, and driveway, is lawn. One bag of lawn fertilizer costs $25.00 and covers 4000 square feet.

   a. Determine the minimum number of bags of fertilizer needed for the lawn.

   b. Find the total cost of the fertilizer.

*Use the formulas for the area and the circumference of a circle in **Table 3.2** on page 204 to solve Exercises 59–64. Unless otherwise indicated, round all circumference and area calculations to the nearest whole number.*

59. Which one of the following is a better buy: a large pizza with a 14-inch diameter for $12.00 or a medium pizza with a 7-inch diameter for $5.00?

60. Which one of the following is a better buy: a large pizza with a 16-inch diameter for $12.00 or two small pizzas, each with a 10-inch diameter, for $12.00?

61. If asphalt pavement costs $0.80 per square foot, find the cost to pave the circular road in the figure shown. Round to the nearest dollar.

62. Hardwood flooring costs $10.00 per square foot. How much will it cost (to the nearest dollar) to cover the dance floor shown in the figure with hardwood flooring?

63. A glass window is to be placed in a house. The window consists of a rectangle, 6 feet high by 3 feet wide, with a semicircle at the top. Approximately how many feet of stripping, to the nearest tenth of a foot, will be needed to frame the window?

6 ft

3 ft

64. How many plants spaced every 6 inches are needed to surround a circular garden with a 30-foot radius?

Use the formulas for volume in **Table 3.3** *on page 206 to solve Exercises 65–69. When necessary, round all volume calculations to the nearest whole number.*

65. A water reservoir is shaped like a rectangular solid with a base that is 50 yards by 30 yards, and a vertical height of 20 yards. At the start of a three-month period of no rain, the reservoir was completely full. At the end of this period, the height of the water was down to 6 yards. How much water was used in the three-month period?

66. A building contractor is to dig a foundation 4 yards long, 3 yards wide, and 2 yards deep for a toll booth's foundation. The contractor pays $10 per load for trucks to remove the dirt. Each truck holds 6 cubic yards. What is the cost to the contractor to have all the dirt hauled away?

67. Two cylindrical cans of soup sell for the same price. One can has a diameter of 6 inches and a height of 5 inches. The other has a diameter of 5 inches and a height of 6 inches. Which can contains more soup and, therefore, is the better buy?

68. The tunnel under the English Channel that connects England and France is one of the world's longest tunnels. The Chunnel, as it is known, consists of three separate tunnels built side by side. Each is a half-cylinder that is 50,000 meters long and 4 meters high. How many cubic meters of dirt had to be removed to build the Chunnel?

69. You are about to sue your contractor who promised to install a water tank that holds 500 gallons of water. You know that 500 gallons is the capacity of a tank that holds 67 cubic feet. The cylindrical tank has a radius of 3 feet and a height of 2 feet 4 inches. Does the evidence indicate you can win the case against the contractor if it goes to court?

Writing in Mathematics

70. Using words only, describe how to find the area of a triangle.

71. Describe the difference between the following problems: How much fencing is needed to enclose a garden? How much fertilizer is needed for the garden?

72. Describe how volume is measured. Explain why linear or square units cannot be used.

73. What is an angle?

74. If the measures of two angles of a triangle are known, explain how to find the measure of the third angle.

75. Can a triangle contain two 90° angles? Explain your answer.

76. What are complementary angles? Describe how to find the measure of an angle's complement.

77. What are supplementary angles? Describe how to find the measure of an angle's supplement.

78. Describe what is misleading in this visual display of data.

Number of Square Feet in an Average U.S. Single-Family Home

1500 — 1970
2080 — 1990
2349 — 2004

Source: National Association of Home Builders

Critical Thinking Exercises

Make Sense? *In Exercises 79–82, determine whether each statement "makes sense" or "does not make sense" and explain your reasoning.*

79. There is nothing that is misleading in this visual display of data.

Book Title Output in the United States

114,487 — 2001
147,120 — 2002
171,061 — 2003
190,078 — 2004
172,000 — 2005

Source: R. R. Bowker

80. I solved a word problem and determined that a triangle had angles measuring 37°, 58°, and 86°.

81. I paid $10 for a pizza, so I would expect to pay approximately $20 for the same kind of pizza with twice the radius.

82. I find that my answers involving π can vary slightly depending on whether I round π mid-calculation or use the π key on my calculator and then round at the very end.

In Exercises 83–86, determine whether each statement is true or false. If the statement is false, make the necessary change(s) to produce a true statement.

83. It is possible to have a circle whose circumference is numerically equal to its area.

84. When the measure of a given angle is added to three times the measure of its complement, the sum equals the sum of the measures of the complement and supplement of the angle.

85. The complement of an angle that measures less than 90° is an angle that measures more than 90°.

86. Two complementary angles can be equal in measure.

87. Suppose you know the cost for building a rectangular deck measuring 8 feet by 10 feet. If you decide to increase the dimensions to 12 feet by 15 feet, by how many times will the cost increase?

88. A rectangular swimming pool measures 14 feet by 30 feet. The pool is surrounded on all four sides by a path that is 3 feet wide. If the cost to resurface the path is $2 per square foot, what is the total cost of resurfacing the path?

89. What happens to the volume of a sphere if its radius is doubled?

90. A scale model of a car is constructed so that its length, width, and height are each $\frac{1}{10}$ the length, width, and height of the actual car. By how many times does the volume of the car exceed its scale model?

91. Find the measure of the angle of inclination, denoted by x in the figure, for the road leading to the bridge.

Review Exercises

92. Solve for s: $P = 2s + b$. (Section 2.4, Example 3)

93. Solve for x: $\dfrac{x}{2} + 7 = 13 - \dfrac{x}{4}$ (Section 2.3, Example 4)

94. Simplify: $\left[3\left(12 \div 2^2 - 3\right)^2\right]^2$. (Section 1.8, Example 8)

Preview Exercises

Exercises 95–97 will help you prepare for the material covered in the first section of the next chapter.

95. Is $x - 4y = 14$ a true statement for $x = 2$ and $y = -3$?

96. Is $x - 4y = 14$ a true statement for $x = 12$ and $y = 1$?

97. If $y = \dfrac{2}{3}x + 1$, find the value of y for $x = -6$.

GROUP PROJECT

CHAPTER 3

In the Blitzer Bonus on pages 207–208, we saw two examples of misleading graphs. Here are some things to watch for in visual displays of data:

- Is there a title that explains what is being displayed?
- Do too many design and cosmetic effects draw attention from or distort the data?
- Are bar sizes in the correct ratios for the displayed data?
- Are numbers lined up with tick marks on the vertical axis that clearly indicate the scale?
- Has the scale been varied to create a more or less dramatic impression than shown by the actual data?
- Is there a source that indicates where the data in the display come from? Do the data come from an entire population or from a sample of the population?
- Who is presenting the visual display, and does that person or group have a special case to make for or against the trend shown by the graph?

With these questions in mind, each group member should find one example of a graph that presents data with integrity and one example of a graph that is misleading. Use newspapers, magazines, the Internet, books, and so forth. Once graphs have been collected, each member should share his or her graphs with the entire group. Be sure to explain why one graph depicts data in a forthright manner and how the other graph misleads the viewer.

Chapter 3 Summary

Definitions and Concepts	**Examples**

Section 3.1 Further Problem Solving

Simple interest involves interest calculated only on the principal. Information in these problems can be organized in a table.

	Principal	×	Rate	=	Interest
Investment 1	?				
Investment 2	?				

Represent the principal invested at each rate in terms of x. Set the sum of the interests from the final column equal to the given amount for the total interest from the two investments.

You invested $16,000 in two investments paying 6% and 8% annual interest. At the end of the year, total interest was $1180. How much was invested at each rate?

$$x = \text{amount invested at 6\%}$$
$$16,000 - x = \text{amount invested at 8\%}$$

	Principal	×	Rate	=	Interest
6% Investment	x		0.06		$0.06x$
8% Investment	$16,000 - x$		0.08		$0.08(16,000 - x)$

$$0.06x + 0.08(16,000 - x) = 1180$$
$$(\text{See Example 2 in Section 3.1.})$$
$$x = 5000$$
$$16,000 - x = 16,000 - 5000 = 11,000$$

Thus, $5000 was invested at 6% and $11,000 at 8%.

Mixture problems can be solved by organizing information in a table. The amount of the ingredient in solution 1 plus the amount of the ingredient in solution 2 equals the amount of the ingredient in the mixture.

How many liters of a 50% acid solution must be mixed with a 20% acid solution to obtain 12 liters of a 30% solution?

$$x = \text{number of liters of 50\% solution}$$
$$12 - x = \text{number of liters of 20\% solution}$$

	Number of Liters	×	Percent of Acid	=	Amount of Acid
50% Solution	x		0.5		$0.5x$
20% Solution	$12 - x$		0.2		$0.2(12 - x)$
30% Mixture	12		0.3		$0.3(12)$

$$0.5x + 0.2(12 - x) = 0.3(12)$$

Solving: $x = 4$.

Use 4 liters of the 50% solution and $12 - x = 12 - 4 = 8$ liters of the 20% solution.

Motion problems can be solved by organizing information in a table.

	Rate	×	Time	=	Distance
Moving object 1					
Moving object 2					

The algebraic expressions for distance in the final column are used to set up an equation based on the problem's conditions.

Two runners averaging 5 miles per hour and 6 miles per hour start in the same place and run in opposite directions. After how many hours will they be 22 miles apart?

$$x = \text{time when runners are 22 miles apart}$$

	Rate	×	Time	=	Distance
Runner 1	5		x		$5x$
Runner 2	6		x		$6x$

Definitions and Concepts	**Examples**

Section 3.1 Further Problem Solving (continued)

$$5x + 6x = 22$$
$$11x = 22$$
$$x = 2$$

The runners will be 22 miles apart after 2 hours.

Section 3.2 Ratio and Proportion

A ratio compares quantities by division. The ratio of a to b can be expressed as

$$\frac{a}{b} \quad \text{or} \quad a:b.$$

20 men, 30 women

Ratio of women to men

$$= \frac{30}{20} = \frac{3}{2} = 3:2 \text{ (3 to 2)}$$

Ratio of men to all people

$$= \frac{20}{50} = \frac{2}{5} = 2:5 \text{ (2 to 5)}$$

A proportion is a statement in the form $\frac{a}{b} = \frac{c}{d}$. The cross-products principal states that if $\frac{a}{b} = \frac{c}{d}$, then $ad = bc$ ($b \neq 0$ and $d \neq 0$).

Solve:

$$\frac{-3}{4} = \frac{x}{12}.$$
$$4x = -3(12)$$
$$4x = -36$$
$$x = -9$$

The solution is -9, or the solution set is $\{-9\}$.

Solving Applied Problems Using Proportions

1. Read the problem and represent the unknown quantity by x (or any letter).
2. Set up a proportion by listing the given ratio on one side and the ratio with the unknown quantity on the other side.
3. Drop units and apply the cross-products principle.
4. Solve for x and answer the question.

30 elk are tagged and released. Sometime later, a sample of 80 elk are observed and 10 are tagged. How many elk are there?

$$x = \text{number of elk}$$

$$\boxed{\text{Tagged}} \searrow \frac{30}{x} = \frac{10}{80}$$
$$\boxed{\text{Total}} \nearrow$$

$$10x = 30 \cdot 80$$
$$10x = 2400$$
$$x = 240$$

There are 240 elk.

Section 3.3 Problem Solving in Geometry

Solving geometry problems often requires using basic geometric formulas. Formulas for perimeter, area, circumference, and volume are given in **Table 3.1** (page 203), **Table 3.2** (page 204), and **Table 3.3** (page 206) in Section 3.3.

A sailboat's triangular sail has an area of 24 ft^2 and a base of 8 ft. Find its height.

$$A = \frac{1}{2}bh$$
$$24 = \frac{1}{2}(8)h$$
$$24 = 4h$$
$$6 = h$$

The sail's height is 6 ft.

Definitions and Concepts	**Examples**

Section 3.3 Problem Solving in Geometry (continued)

The sum of the measures of the three angles of any triangle is 180°.	In a triangle, the first angle measures 3 times the second and the third measures 40° less than the second. Find each angle's measure. $$\text{Second angle} = x$$ $$\text{First angle} = 3x$$ $$\text{Third angle} = x - 40$$ Sum of measures is 180°. $$x + 3x + (x - 40) = 180$$ $$5x - 40 = 180$$ $$5x = 220$$ $$x = 44$$ The angles measure $x = 44, 3x = 3 \cdot 44 = 132$, and $x - 40 = 44 - 40 = 4.$ The angles measure 44°, 132°, and 4°.
Two complementary angles have measures whose sum is 90°. Two supplementary angles have measures whose sum is 180°. If an angle measures x, its complement measures $90 - x$, and its supplement measures $180 - x$.	An angle measures five times its complement. Find the angle's measure. $$x = \text{angle's measure}$$ $$90 - x = \text{measure of complement}$$ $$x = 5(90 - x)$$ $$x = 450 - 5x$$ $$6x = 450$$ $$x = 75$$ The angle measures 75°.

CHAPTER 3 REVIEW EXERCISES

3.1

1. You invested $10,000 in two accounts paying 8% and 10% annual interest. At the end of the year, the total interest from these investments was $940. How much was invested at each rate?

2. You invested money in two accounts that paid 10% and 12% annual interest. The amount invested at 12% exceeded the amount invested at 10% by $6000. At the end of the year, the total interest from these investments was $2480. How much was invested at each rate?

3. A chemist needs to mix a 75% saltwater solution with a 50% saltwater solution to obtain 10 gallons of a 60% saltwater solution. How many gallons of each of the solutions must be used?

4. Is it possible to mix an alloy that is 30% copper with one that is 50% copper to obtain 50 ounces of an alloy that is 70% copper? If it is possible, how many ounces of each kind of alloy must be used?

5. A university is merging two departments into one. At the north campus, 5% of the students in the department are men. At the south campus, 80% of the students in the department are men. After the merger, 50% of the 150 students in the department are men. How many students were in the department at the north and south campuses before the merger?

6. Two trains start from the same place at the same time. They travel in opposite directions. One averages 60 miles per hour and the other averages 80 miles per hour. After how long will they be 420 miles apart?

7. Two buses leave a station at the same time, traveling in opposite directions. The rate of the faster bus exceeds that of the slower bus by 10 miles per hour. After 3 hours, they are 210 miles apart. What are the rates of the buses?

8. A wallet contains $5 bills and $10 bills. There are 15 bills in the wallet with a total value of $120. Determine the number of $5 bills and the number of $10 bills in the wallet.

*In Exercises 9–11, determine if there is **exactly enough information** or **not enough information** to solve each problem. It is not necessary actually to solve any of these problems.*

9. You invested $10,000 in an account paying 6% annual interest and the remainder of your savings in a bond paying 8% annual interest. At the end of the year, the total interest from these investments was $730. How much was invested at 8%?

10. A chemist needs a 15% benzene solution. The stockroom has only 12% and 20% solutions. How much of each of these solutions must be used?

11. Downtown Books is having a sale on hardcover and soft-cover books. All hardcover books are priced at $5 per book and all softcover books at $2 per book. The store sold 30 more softcover than hardcover books and had total sales of $570. How many books of each type did the store sell?

3.2

12. In a class, there are 15 men and 10 women. Find the ratio of the number of women to the number of students in the class. First express the ratio as a fraction reduced to lowest terms. Then rewrite the ratio using a second method.

The circle graphs show the percentage of men and women in the United States, for two age groups. Use the graphs to find each of the ratios in Exercises 13–14. First express the ratio as a fraction reduced to lowest terms. Then rewrite the ratio using a second method.

Percentage of Men and Women in the United States, by Age Group

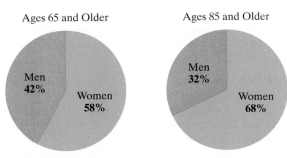

Source: U.S. Census Bureau

13. For Americans 65 and older, the percentage of men to the percentage of women

14. For Americans 85 and older, the percentage of women to the percentage of men

Solve each proportion in Exercises 15–18.

15. $\dfrac{3}{x} = \dfrac{15}{25}$

16. $\dfrac{-7}{5} = \dfrac{91}{x}$

17. $\dfrac{x+2}{3} = \dfrac{4}{5}$

18. $\dfrac{5}{x+7} = \dfrac{3}{x+3}$

Use a proportion to solve each problem in Exercises 19–21.

19. If a school board determines that there should be 3 teachers for every 50 students, how many teachers are needed for an enrollment of 5400 students?

20. To determine the number of trout in a lake, a conservationist catches 112 trout, tags them, and returns them to the lake. Later, 82 trout are caught, and 32 of them are found to be tagged. How many trout are in the lake?

21. The owners of Skaters Now determine that the monthly sales of their skates is proportional to their advertising budget. When $60,000 is spent on advertising, the company sells 12,000 skates in a month. What will be the monthly sales if they spend $96,000 on advertising?

3.3

Use a formula for area to find the area of each figure in Exercises 22–25.

22.

23.

24.

25.

26. Find the circumference and the area of a circle with a diameter of 20 meters. Round answers to the nearest whole number.

27. A sailboat has a triangular sail with an area of 42 square feet and a base that measures 14 feet. Find the height of the sail.

28. A rectangular kitchen floor measures 12 feet by 15 feet. A stove on the floor has a rectangular base measuring 3 feet by 4 feet, and a refrigerator covers a rectangular area of the floor measuring 3 feet by 4 feet. How many square feet of tile will be needed to cover the kitchen floor not counting the area used by the stove and the refrigerator?

29. A yard that is to be covered with mats of grass is shaped like a trapezoid. The bases are 80 feet and 100 feet, and the height is 60 feet. What is the cost of putting the grass mats on the yard if the landscaper charges $0.35 per square foot?

30. Which one of the following is a better buy: a medium pizza with a 14-inch diameter for $6.00 or two small pizzas, each with an 8-inch diameter, for $6.00?

Use a formula for volume to find the volume of each figure in Exercises 31–33. Where applicable, express answers in terms of π. Then round to the nearest whole number.

31.

4 cm

3 cm

5 cm

32.

8 yd

4 yd

33.

6 m

34. A train is being loaded with freight containers. Each box is 8 meters long, 4 meters wide, and 3 meters high. If there are 50 freight containers, how much space is needed?

35. A cylindrical fish tank has a diameter of 6 feet and a height of 3 feet. How many tropical fish can be put in the tank if each fish needs 5 cubic feet of water?

36. Find the measure of each angle of the triangle shown in the figure.

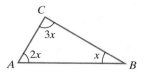

37. In a triangle, the measure of the first angle is 15° more than twice the measure of the second angle. The measure of the third angle exceeds that of the second angle by 25°. What is the measure of each angle?

38. Find the measure of the complement of a 57° angle.

39. Find the measure of the supplement of a 75° angle.

40. How many degrees are there in an angle that measures 25° more than the measure of its complement?

41. The measure of the supplement of an angle is 45° less than four times the measure of the angle. Find the measure of the angle and its supplement.

CHAPTER 3 TEST

Remember to use your Chapter Test Prep Video CD to see the worked-out solutions to the test questions you want to review.

1. You invested $6000 in two accounts paying 9% and 6% annual interest. At the end of the year, the total interest from these investments was $480. How much was invested at each rate?

2. A chemist needs to mix a 50% acid solution with an 80% acid solution to obtain 100 milliliters of a 68% acid solution. How many milliliters of each of the solutions must be used?

3. Two cars that are 400 miles apart are traveling directly toward each other on the same road. One is averaging 45 miles per hour and the other averages 35 miles per hour. After how long will they meet?

4. In a class, there are 20 men and 15 women. Find the ratio of the number of men to the number of students in the class. First express the ratio as a fraction reduced to lowest terms. Then rewrite the ratio using a second method

5. Solve the proportion: $\dfrac{-5}{8} = \dfrac{x}{12}$.

6. Park rangers catch, tag, and release 200 tule elk back into a wildlife refuge. Two weeks later they observe a sample of 150 elk, of which 5 are tagged. Assuming that the ratio of tagged elk in the sample holds for all elk in the refuge, how many elk are there in the park?

7. The pressure of water on an object below the surface is proportional to its distance below the surface. If a submarine experiences a pressure of 25 pounds per square inch 60 feet below the surface, how much pressure will it experience 330 feet below the surface?

In Exercises 8–10, find the area of each figure.

8.

22 m

47 m

9.

30 in.

15 in.

40 in.

10.

3 ft

13 ft

6 ft

3 ft

In Exercises 11–12, find the volume of each figure. Where applicable, express answers in terms of π. Then round to the nearest whole number.

11.

3 in.

2 in.

3 in.

12.

5 cm

7 cm

13. What will it cost to cover a rectangular floor measuring 40 feet by 50 feet with square tiles that measure 2 feet on each side if a package of 10 tiles costs $13 per package?

14. A sailboat has a triangular sail with an area of 56 square feet and a base that measures 8 feet. Find the height of the sail.

15. In a triangle, the measure of the first angle is three times that of the second angle. The measure of the third angle is 30° less than the measure of the second angle. What is the measure of each angle?

16. How many degrees are there in an angle that measures 16° more than the measure of its complement?

CUMULATIVE REVIEW EXERCISES (CHAPTERS 1–3)

1. Perform the indicated operations:
$$\frac{-9(3-6)}{(-12)(3)+(-3-5)(8-4)}.$$

2. Simplify:
$$8 - 3[2(x-1)+5].$$

3. List all the integers in this set:
$$\left\{-3, -\frac{1}{2}, 0, \sqrt{3}, \sqrt{9}\right\}.$$

4. Evaluate $x^2 - 10x$ for $x = -3$.

5. Insert either $<$ or $>$ in the shaded area to make a true statement:
$$-2000 \quad \rule{1cm}{0.4pt} \quad -3.$$

6. On February 8, the temperature in Fairbanks, Alaska at 10 P.M. was $-4°$F. By 3 A.M. the next day, the temperature had fallen 11°, but by noon the temperature increased by 21°. What was the temperature at noon?

In Exercises 7–8, solve each equation.

7. $10(2x - 1) = 8(2x + 1) + 14$

8. $\dfrac{x}{5} + \dfrac{2x}{3} = x + \dfrac{1}{15}$

9. Solve for m: $A = \dfrac{m+n}{2}.$

Your credit score is based on your financial behavior, such as debt, credit usage, and whether you pay bills on time. The average credit score for people in the United States is 677 (out of a possible 830). The mathematical model

$$C = 2.2a + 589$$

describes the average credit score, C, of Americans who are a years old. Use this information to solve Exercises 10–11.

10. What is the average credit score for Americans who are 20 years old?

11. What age corresponds to an average credit score of 699?

12. 144 is 60% of what number?

13. The length of a rectangular football field is 14 meters more than twice the width. If the perimeter is 346 meters, find the field's dimensions.

14. After a 10% weight loss, a person weighed 180 pounds. What was the weight before the loss?

In Exercises 15–16, solve each inequality and graph the solution set on a number line.

15. $5x - 5 \leq -5$

16. $-5x + 9 > -2x + 6$

17. Two runners start at the same point and run in opposite directions. One runs at 6 miles per hour and the other runs at 8 miles per hour. In how many hours will they be 21 miles apart?

18. Solve the proportion: $\dfrac{-5}{20} = \dfrac{x}{21}$.

19. At the time they took office, Ronald Reagan and James Buchanan were among the oldest U.S. presidents. Reagan was 4 years older than Buchanan. The sum of their ages was 134. Determine Reagan's age and Buchanan's age at the time each man took office.

20. John Tesh, while he was still coanchoring *Entertainment Tonight*, reported that the PBS series *The Civil War* had an audience of 13% versus the usual 4% PBS audience, "an increase of more than 300%." Did Tesh report the percent increase correctly?

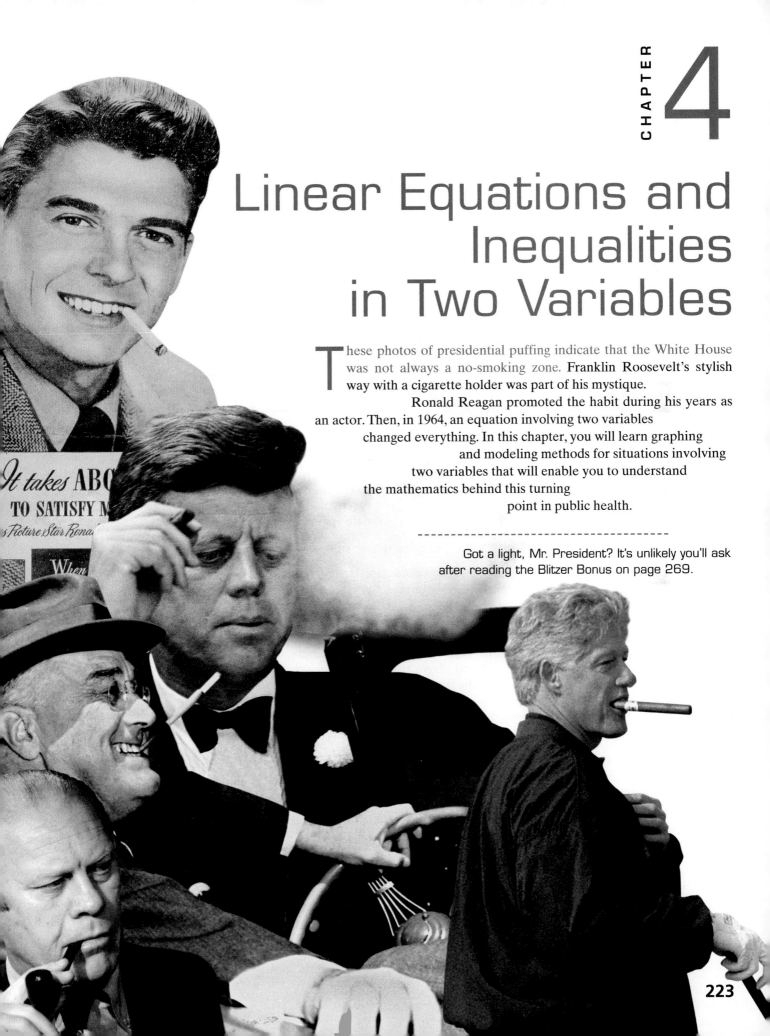

Linear Equations and Inequalities in Two Variables

These photos of presidential puffing indicate that the White House was not always a no-smoking zone. Franklin Roosevelt's stylish way with a cigarette holder was part of his mystique.

Ronald Reagan promoted the habit during his years as an actor. Then, in 1964, an equation involving two variables changed everything. In this chapter, you will learn graphing and modeling methods for situations involving two variables that will enable you to understand the mathematics behind this turning point in public health.

Got a light, Mr. President? It's unlikely you'll ask after reading the Blitzer Bonus on page 269.

4.1

Graphing Linear Equations in Two Variables

Objectives

1 Plot ordered pairs in the rectangular coordinate system.

2 Find coordinates of points in the rectangular coordinate system.

3 Determine whether an ordered pair is a solution of an equation.

4 Find solutions of an equation in two variables.

5 Use point plotting to graph linear equations.

6 Use graphs of linear equations to solve problems.

Despite high interest rates, fees, and penalties, American consumers are paying more with credit cards, and less with checks and cash. In this section, we will use geometric figures to visualize mathematical models describing our primary spending methods. The idea of visualizing equations as geometric figures was developed by the French philosopher and mathematician René Descartes (1596–1650). We begin by looking at Descartes's idea that brought together algebra and geometry, called the **rectangular coordinate system** or (in his honor) the **Cartesian coordinate system**.

1 Plot ordered pairs in the rectangular coordinate system.

Points and Ordered Pairs

The rectangular coordinate system consists of two number lines that intersect at right angles at their zero points, as shown in **Figure 4.1**. The horizontal number line is the **x-axis**. The vertical number line is the **y-axis**. The point of intersection of these axes is their zero points, called the **origin**. Positive numbers are shown to the right and above the origin. Negative numbers are shown to the left and below the origin. The axes divide the plane into four regions, called **quadrants**. The points located on the axes are not in any quadrant.

Each point in the rectangular coordinate system corresponds to an **ordered pair** of real numbers, (x, y). Examples of such pairs are $(4, 2)$ and $(-5, -3)$. The first number in each pair, called the **x-coordinate**, denotes the distance and direction from the origin along the x-axis. The second number, called the **y-coordinate**, denotes vertical distance and direction along a line parallel to the y-axis or along the y-axis itself.

Figure 4.2 shows how we **plot**, or locate, the points corresponding to the ordered pairs $(4, 2)$ and $(-5, -3)$. We plot $(4, 2)$ by going 4 units from 0 to the right along the x-axis. Then we go 2 units up parallel to the y-axis. We plot $(-5, -3)$ by going 5 units from 0 to the left along the x-axis and 3 units down parallel to the y-axis. The phrase "the point corresponding to the ordered pair $(-5, -3)$" is often abbreviated as "the point $(-5, -3)$."

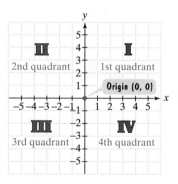

FIGURE 4.1 The rectangular coordinate system

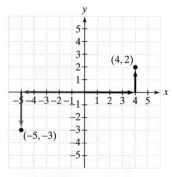

FIGURE 4.2 Plotting $(4, 2)$ and $(-5, -3)$

EXAMPLE 1 Plotting Points in the Rectangular Coordinate System

Plot the points: $A(-3, 5), B(2, -4), C(5, 0), D(-5, -3), E(0, 4)$ and $F(0, 0)$.

Solution See **Figure 4.3**. We move from the origin and plot the points in the following way:

$A(-3, 5)$:	3 units left, 5 units up
$B(2, -4)$:	2 units right, 4 units down
$C(5, 0)$:	5 units right, 0 units up or down
$D(-5, -2)$:	5 units left, 2 units down
$E(0, 4)$:	0 units right or left, 4 units up
$F(0, 0)$:	0 units right or left, 0 units up or down

Notice that the origin is represented by $(0, 0)$.

FIGURE 4.3 Plotting points

Study Tip

Any point on the x-axis, such as $(5, 0)$, has a y-coordinate of 0.
Any point on the y-axis, such as $(0, 4)$, has an x-coordinate of 0.

The phrase *ordered pair* is used because **order is important**. For example, the points $(2, 5)$ and $(5, 2)$ are not the same. To plot $(2, 5)$, move 2 units right and 5 units up. To plot $(5, 2)$, move 5 units right and 2 units up. The points $(2, 5)$ and $(5, 2)$ are in different locations. **The order in which coordinates appear makes a difference in a point's location.**

☑ **CHECK POINT 1** Plot the points: $A(-2, 4), B(4, -2), C(-3, 0)$, and $D(0, -3)$.

---

2 Find coordinates of points in the rectangular coordinate system.

In the rectangular coordinate system, each ordered pair corresponds to exactly one point. Example 2 illustrates that each point in the rectangular coordinate system corresponds to exactly one ordered pair.

EXAMPLE 2 Finding Coordinates of Points

Determine the coordinates of points A, B, C, and D shown in **Figure 4.4**.

Solution

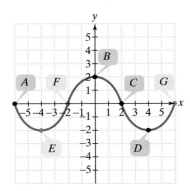

FIGURE 4.4 Finding coordinates of points

Point	Position from the Origin	Coordinates
A	6 units left, 0 units up or down	$(-6, 0)$
B	0 units right or left, 2 units up	$(0, 2)$
C	2 units right, 0 units up or down	$(2, 0)$
D	4 units right, 2 units down	$(4, -2)$

☑ **CHECK POINT 2** Determine the coordinates of points E, F, and G shown in **Figure 4.4**.

③ Determine whether an ordered pair is a solution of an equation.

Solutions of Equations in Two Variables

The rectangular coordinate system allows us to visualize relationships between two variables by connecting any equation in two variables with a geometric figure. Consider, for example, the following equation in two variables:

$$x + y = 10.$$

The sum of two numbers, x and y, is 10.

Many pairs of numbers fit the description in the voice balloon, such as $x = 1$ and $y = 9$, or $x = 3$ and $y = 7$. The phrase "$x = 1$ and $y = 9$" is abbreviated using the ordered pair $(1, 9)$. Similarly, the phrase "$x = 3$ and $y = 7$" is abbreviated using the ordered pair $(3, 7)$.

A **solution of an equation in two variables**, x and y, is an ordered pair of real numbers with the following property: When the x-coordinate is substituted for x and the y-coordinate is substituted for y in the equation, we obtain a true statement. For example, $(1, 9)$ is a solution of the equation $x + y = 10$. When 1 is substituted for x and 9 is substituted for y, we obtain the true statement $1 + 9 = 10$, or $10 = 10$. Because there are infinitely many pairs of numbers that have a sum of 10, the equation $x + y = 10$ has infinitely many solutions. Each ordered-pair solution is said to **satisfy** the equation. Thus, $(1, 9)$ satisfies the equation $x + y = 10$.

EXAMPLE 3 Deciding Whether an Ordered Pair Satisfies an Equation

Determine whether each ordered pair is a solution of the equation

$$x - 4y = 14:$$

a. $(2, -3)$ **b.** $(12, 1)$.

Solution

a. To determine whether $(2, -3)$ is a solution of the equation, we substitute 2 for x and -3 for y.

$$x - 4y = 14 \quad \text{This is the given equation.}$$
$$2 - 4(-3) \stackrel{?}{=} 14 \quad \text{Substitute 2 for x and } -3 \text{ for y.}$$
$$2 - (-12) \stackrel{?}{=} 14 \quad \text{Multiply: } 4(-3) = -12.$$
$$14 = 14 \quad \text{Subtract: } 2 - (-12) = 2 + 12 = 14.$$

This statement is true.

Because we obtain a true statement, we conclude that $(2, -3)$ is a solution of the equation $x - 4y = 14$. Thus, $(2, -3)$ satisfies the equation.

b. To determine whether $(12, 1)$ is a solution of the equation, we substitute 12 for x and 1 for y.

$$x - 4y = 14 \quad \text{This is the given equation.}$$
$$12 - 4(1) \stackrel{?}{=} 14 \quad \text{Substitute 12 for x and 1 for y.}$$
$$12 - 4 \stackrel{?}{=} 14 \quad \text{Multiply: } 4(1) = 4.$$
$$8 = 14 \quad \text{Subtract: } 12 - 4 = 8.$$

This statement is false.

Because we obtain a false statement, we conclude that $(12, 1)$ is not a solution of $x - 4y = 14$. The ordered pair $(12, 1)$ does not satisfy the equation. ∎

✓ **CHECK POINT 3** Determine whether each ordered pair is a solution of the equation $x - 3y = 9$:

a. $(3, -2)$ b. $(-2, 3)$.

In this chapter, we will use x and y to represent the variables of an equation in two variables. However, any two letters can be used. Solutions are still ordered pairs. Although it depends on the role variables play in equations, the first number in an ordered pair often replaces the variable that occurs first alphabetically. The second number in an ordered pair often replaces the variable that occurs last alphabetically.

4 Find solutions of an equation in two variables.

How do we find ordered pairs that are solutions of an equation in two variables, x and y?

- Select a value for one of the variables.
- Substitute that value into the equation and find the corresponding value of the other variable.
- Use the values of the two variables to form an ordered pair (x, y). This pair is a solution of the equation.

EXAMPLE 4 **Finding Solutions of an Equation**

Find five solutions of

$$y = 2x - 1.$$

Select integers for x, starting with -2 and ending with 2.

Solution We organize the process of finding solutions in the following table of values.

	Start with these values of x.	Substitute x into $y = 2x - 1$ and compute y.	Use values for x and y to form an ordered-pair solution.
	x	$y = 2x - 1$	(x, y)
Any numbers can be selected for x. There is nothing special about integers from -2 to 2, inclusive. We chose these values to include two negative numbers, 0, and two positive numbers. We also wanted to keep the resulting computations for y relatively simple.	-2	$y = 2(-2) - 1 = -4 - 1 = -5$	$(-2, -5)$
	-1	$y = 2(-1) - 1 = -2 - 1 = -3$	$(-1, -3)$
	0	$y = 2 \cdot 0 - 1 = 0 - 1 = -1$	$(0, -1)$
	1	$y = 2 \cdot 1 - 1 = 2 - 1 = 1$	$(1, 1)$
	2	$y = 2 \cdot 2 - 1 = 4 - 1 = 3$	$(2, 3)$

Look at the ordered pairs in the last column. Five solutions of $y = 2x - 1$ are $(-2, -5), (-1, -3), (0, -1), (1, 1),$ and $(2, 3)$. ■

✓ **CHECK POINT 4** Find five solutions of $y = 3x + 2$. Select integers for x, starting with -2 and ending with 2.

---

5 Use point plotting to graph linear equations.

Graphing Linear Equations in the Form $y = mx + b$

In Example 4, we found five solutions of $y = 2x - 1$. We can generate as many ordered-pair solutions as desired to $y = 2x - 1$ by substituting numbers for x and then finding the corresponding values for y. The **graph of the equation** is the set of all points whose coordinates satisfy the equation.

One method for graphing an equation such as $y = 2x - 1$ is the **point-plotting method**.

> ### The Point-Plotting Method for Graphing an Equation in Two Variables
>
> **1.** Find several ordered pairs that are solutions of the equation.
>
> **2.** Plot these ordered pairs as points in the rectangular coordinate system.
>
> **3.** Connect the points with a smooth curve or line.

EXAMPLE 5 Graphing an Equation Using the Point-Plotting Method

Graph the equation: $y = 3x$.

Solution

Step 1. Find several ordered pairs that are solutions of the equation. Because there are infinitely many solutions, we cannot list them all. To find a few solutions of the equation, we select integers for x, starting with -2 and ending with 2.

Start with these values of x.	Substitute x into $y = 3x$ and compute y.	These are solutions of $y = 3x$.
x	$y = 3x$	(x, y)
-2	$y = 3(-2) = -6$	$(-2, -6)$
-1	$y = 3(-1) = -3$	$(-1, -3)$
0	$y = 3 \cdot 0 = 0$	$(0, 0)$
1	$y = 3 \cdot 1 = 3$	$(1, 3)$
2	$y = 3 \cdot 2 = 6$	$(2, 6)$

Step 2. Plot these ordered pairs as points in the rectangular coordinate system. The five ordered pairs in the table of values are plotted in **Figure 4.5(a)**.

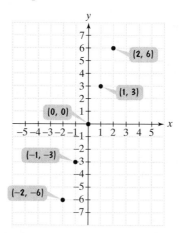

FIGURE 4.5(a) A few solutions of $y = 3x$ plotted as points

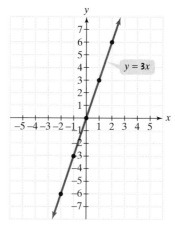

FIGURE 4.5(b) The graph of $y = 3x$

Step 3. Connect the points with a smooth curve or line. The points lie along a straight line. The graph of $y = 3x$ is shown in **Figure 4.5(b)**. The arrows on both ends of the line indicate that it extends indefinitely in both directions. ▬

✓ **CHECK POINT 5** Graph the equation: $y = 2x$.

Equations like $y = 3x$ and $y = 2x$ are called **linear equations in two variables** because the graph of each equation is a line. Any equation that can be written in the form $y = mx + b$, where m and b are constants, is a linear equation in two variables. Here are examples of linear equations in two variables:

$$y = 3x \qquad\qquad y = 3x - 2$$
$$\text{or} \quad y = 3x + 0 \qquad \text{or} \quad y = 3x + (-2).$$

This is in the form of $y = mx + b$, with $m = 3$ and $b = 0$.

This is in the form of $y = mx + b$, with $m = 3$ and $b = -2$.

Study Tip

For the linear equation $y = mx + b$, any real numbers can be used for m and b. In Section 4.4, you will see how these numbers affect the equation's graph in special ways. The number m determines the line's steepness. The number b determines the point where the line crosses the y-axis.

Can you guess how the graph of the linear equation $y = 3x - 2$ compares with the graph of $y = 3x$? In Example 5, we graphed $y = 3x$. Now, let's graph the equation $y = 3x - 2$.

EXAMPLE 6 Graphing a Linear Equation in Two Variables

Graph the equation: $y = 3x - 2$.

Solution

Step 1. Find several ordered pairs that are solutions of the equation. To find a few solutions, we select integers for x, starting with -2 and ending with 2.

Start with x. Compute y. Form the ordered pair (x, y).

x	$y = 3x - 2$	(x, y)
-2	$y = 3(-2) - 2 = -6 - 2 = -8$	$(-2, -8)$
-1	$y = 3(-1) - 2 = -3 - 2 = -5$	$(-1, -5)$
0	$y = 3 \cdot 0 - 2 = 0 - 2 = -2$	$(0, -2)$
1	$y = 3 \cdot 1 - 2 = 3 - 2 = 1$	$(1, 1)$
2	$y = 3 \cdot 2 - 2 = 6 - 2 = 4$	$(2, 4)$

Step 2. Plot these ordered pairs as points in the rectangular coordinate system. The five ordered pairs in the table of values are plotted in **Figure 4.6(a)**.

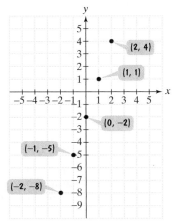

FIGURE 4.6(a) A few solutions of $y = 3x - 2$ plotted as points

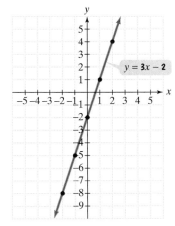

FIGURE 4.6(b) The graph of $y = 3x - 2$

Step 3. Connect the points with a smooth curve or line. The points lie along a straight line. The graph of $y = 3x - 2$ is shown in **Figure 4.6(b)**.

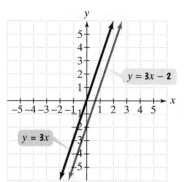

FIGURE 4.7

Now we are ready to compare the graphs of $y = 3x - 2$ and $y = 3x$. The graphs of both linear equations are shown in the same rectangular coordinate system in **Figure 4.7**. Can you see that the blue graph of $y = 3x - 2$ is parallel to the red graph of $y = 3x$ and 2 units lower than the graph of $y = 3x$? Instead of crossing the y-axis at $(0, 0)$, the graph of $y = 3x - 2$ crosses the y-axis at $(0, -2)$.

Comparing Graphs of Linear Equations

If the value of m does not change,

- The graph of $y = mx + b$ is the graph of $y = mx$ shifted b units up when b is a positive number.
- The graph of $y = mx + b$ is the graph of $y = mx$ shifted b units down when b is a negative number.

✓ **CHECK POINT 6**   Graph the equation: $y = 2x - 2$.

EXAMPLE 7   Graphing a Linear Equation in Two Variables

Graph the equation: $y = \dfrac{2}{3}x + 1$.

Solution

Step 1. Find several ordered pairs that are solutions of the equation.   Notice that m, the coefficient of x, is $\frac{2}{3}$. When m is a fraction, we will select values of x that are multiples of the denominator. In this way, we can avoid values of y that are fractions. Because the denominator of $\frac{2}{3}$ is 3, we select multiples of 3 for x. Let's use $-6, -3, 0, 3$, and 6.

Start with multiples of 3 for x. x	Compute y. $y = \dfrac{2}{3}x + 1$	Form the ordered pair (x, y). (x, y)
-6	$y = \dfrac{2}{3}(-6) + 1 = -4 + 1 = -3$	$(-6, -3)$
-3	$y = \dfrac{2}{3}(-3) + 1 = -2 + 1 = -1$	$(-3, -1)$
0	$y = \dfrac{2}{3} \cdot 0 + 1 = 0 + 1 = 1$	$(0, 1)$
3	$y = \dfrac{2}{3} \cdot 3 + 1 = 2 + 1 = 3$	$(3, 3)$
6	$y = \dfrac{2}{3} \cdot 6 + 1 = 4 + 1 = 5$	$(6, 5)$

Step 2. Plot these ordered pairs as points in the rectangular coordinate system. The five ordered pairs in the table of values are plotted in **Figure 4.8**.

Step 3. Connect the points with a smooth curve or line. The points lie along a straight line. The graph of $y = \frac{2}{3}x + 1$ is shown in **Figure 4.8**.

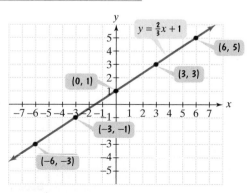

FIGURE 4.8   The graph of $y = \frac{2}{3}x + 1$

✓ **CHECK POINT 7**   Graph the equation: $y = \dfrac{1}{2}x + 2$.

6 Use graphs of linear equations to solve problems.

Applications

Part of the beauty of the rectangular coordinate system is that it allows us to "see" mathematical models and visualize the solution to a problem. This idea is demonstrated in Example 8.

EXAMPLE 8   Graphing a Mathematical Model

The graph in **Figure 4.9** shows that over time fewer consumers are paying with checks and cash, and more are paying with plastic.

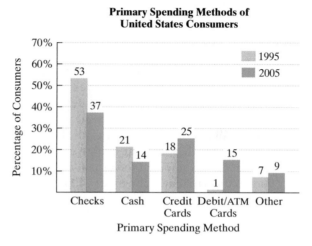

FIGURE 4.9

Source: Mintel/Visa USA Research Services

The mathematical model

$$C = -1.6n + 53$$

describes the percentage of consumers, C, who paid primarily with checks n years after 1995.

a. Let $n = 0, 5, 10, 15,$ and 20. Make a table of values showing five solutions of the equation.

b. Graph the formula in a rectangular coordinate system.

c. Use the graph to estimate the percentage of consumers who will pay primarily by check in 2025.

d. Use the formula to project the percentage of consumers who will pay primarily by check in 2025.

Solution

a. The table of values shows five solutions of $C = -1.6n + 53$.

| Start with these values of n. | Substitute n into $C = -1.6n + 53$ and compute C. | Form the ordered pair (n, C). |

n	$C = -1.6n + 53$	(n, C)
0	$C = -1.6(0) + 53 = 0 + 53 = 53$	$(0, 53)$
5	$C = -1.6(5) + 53 = -8 + 53 = 45$	$(5, 45)$
10	$C = -1.6(10) + 53 = -16 + 53 = 37$	$(10, 37)$
15	$C = -1.6(15) + 53 = -24 + 53 = 29$	$(15, 29)$
20	$C = -1.6(20) + 53 = -32 + 53 = 21$	$(20, 21)$

b. Now we are ready to graph $C = -1.6n + 53$. Because of the letters used to represent the variables, we label the horizontal axis n and the vertical axis C. We plot the five ordered pairs from the table of values and connect them with a line. The graph of $C = -1.6n + 53$ is shown in **Figure 4.10**.

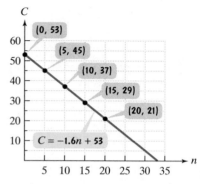

As you examine **Figure 4.10**, are you wondering why we did not extend the graph into the other quadrants of the rectangular coordinate system? Keep in mind that n represents the number of years after 1995 and C represents the percentage of consumers paying by check. Only nonnegative values of each variable are meaningful. Thus, when drawing the graph, we started on the positive portion of the vertical axis and ended on the positive horizontal axis.

FIGURE 4.10

c. Now we can use the graph to project the percentage of consumers who will pay primarily by check in 2025. The year 2025 is 30 years after 1995. We need to determine what second coordinate is paired with 30. **Figure 4.11** shows how to do this. Locate the point on the line that is above 30 and then find the value on the vertical axis that corresponds to that point. **Figure 4.11** shows that this value is 5.

We see that 30 years after 1995, or in 2025, only 5% of consumers will pay primarily by check.

d. Now let's use the mathematical model to verify our graphical observation in part (c).

FIGURE 4.11

$$C = -1.6n + 53 \qquad \text{This is the given mathematical model.}$$
$$C = -1.6(30) + 53 \qquad \text{Substitute 30 for } n.$$
$$C = -48 + 53 \qquad \text{Multiply: } -1.6(30) = -48.$$
$$C = 5 \qquad \text{Add.}$$

This verifies that 30 years after 1995, in 2025, 5% of consumers will pay primarily by check. ∎

✓ **CHECK POINT 8** The mathematical model

$$D = 1.4n + 1$$

describes the percentage of consumers, D, who paid primarily with debit cards n years after 1995.

a. Let $n = 0, 5, 10,$ and 15. Make a table of values showing four solutions of the equation.

b. Graph the formula in a rectangular coordinate system.

c. Use the graph to estimate the percentage of consumers who will pay primarily with debit cards in 2015.

d. Use the formula to project the percentage of consumers who will pay primarily by debit cards in 2015.

Using Technology

Graphing calculators or graphing software packages for computers are referred to as **graphing utilities** or graphers. A graphing utility is a powerful tool that quickly generates the graph of an equation in two variables. **Figure 4.12** shows two such graphs for the equations in Examples 5 and 6.

FIGURE 4.12(a)
The graph of $y = 3x$

FIGURE 4.12(b)
The graph of $y = 3x - 2$

What differences do you notice between these graphs and the graphs that we drew by hand? They do seem a bit "jittery." Arrows do not appear on both ends of the graphs. Furthermore, numbers are not given along the axes. For both graphs in **Figure 4.12**, the x-axis extends from -10 to 10 and the y-axis also extends from -10 to 10. The distance represented by each consecutive tick mark is one unit. We say that the **viewing rectangle**, or the **viewing window**, is $[-10, 10, 1]$ by $[-10, 10, 1]$.

$$[-10, \quad 10, \quad 1] \quad \text{by} \quad [-10, \quad 10, \quad 1]$$

| The minimum x-value along the x-axis is -10. | The maximum x-value along the x-axis is 10. | The scale on the x-axis is 1 unit per tick mark. | The minimum y-value along the y-axis is -10. | The maximum y-value along the y-axis is 10. | The scale on the y-axis is 1 unit per tick mark. |

To graph an equation in x and y using a graphing utility, enter the equation and specify the size of the viewing rectangle. The size of the viewing rectangle sets minimum and maximum values for both the x- and y-axes. Enter these values, as well as the values for the distances between consecutive tick marks on the respective axes. The $[-10, 10, 1]$ by $[-10, 10, 1]$ viewing rectangle used in **Figure 4.12** is called the **standard viewing rectangle**.

On most graphing utilities, the display screen is two-thirds as high as it is wide. By using a square setting, you can equally space the x and y tick marks. (This does not occur in the standard viewing rectangle.) Graphing utilities can also *zoom in* and *zoom out*. When you zoom in, you see a smaller portion of the graph, but you do so in greater detail. When you zoom out, you see a larger portion of the graph. Thus, zooming out may help you to develop a better understanding of the overall character of the graph. With practice, you will become more comfortable with graphing equations in two variables using your graphing utility. You will also develop a better sense of the size of the viewing rectangle that will reveal needed information about a particular graph.

4.1 EXERCISE SET *MyMathLab*

PRACTICE   WATCH   DOWNLOAD   READ   REVIEW

Practice Exercises

In Exercises 1–8, plot the given point in a rectangular coordinate system. Indicate in which quadrant each point lies.

1. $(3, 5)$ **2.** $(5, 3)$ **3.** $(-5, 1)$

4. $(1, -5)$ **5.** $(-3, -1)$ **6.** $(-1, -3)$

7. $(6, -3.5)$ **8.** $(-3.5, 6)$

In Exercises 9–24, plot the given point in a rectangular coordinate system.

9. $(-3, -3)$ **10.** $(-5, -5)$ **11.** $(-2, 0)$

12. $(-5, 0)$ **13.** $(0, 2)$ **14.** $(0, 5)$

15. $(0, -3)$ **16.** $(0, -5)$ **17.** $\left(\dfrac{5}{2}, \dfrac{7}{2}\right)$

18. $\left(\dfrac{7}{2}, \dfrac{5}{2}\right)$ **19.** $\left(-5, \dfrac{3}{2}\right)$ **20.** $\left(-\dfrac{9}{2}, -4\right)$

21. $(0, 0)$ **22.** $\left(-\dfrac{5}{2}, 0\right)$ **23.** $\left(0, -\dfrac{5}{2}\right)$

24. $\left(0, \dfrac{7}{2}\right)$

In Exercises 25–32, give the ordered pairs that correspond to the points labeled in the figure.

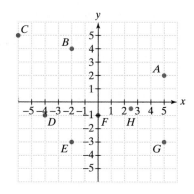

25. A **26.** B **27.** C

28. D **29.** E **30.** F

31. G **32.** H

33. In which quadrants are the y-coordinates positive?

34. In which quadrants are the x-coordinates negative?

35. In which quadrants do the x-coordinates and the y-coordinates have the same sign?

36. In which quadrants do the x-coordinates and the y-coordinates have opposite signs?

In Exercises 37–48, determine whether each ordered pair is a solution of the given equation.

37. $y = 3x$ $(2, 3), (3, 2), (-4, -12)$

38. $y = 4x$ $(3, 12), (12, 3), (-5, -20)$

39. $y = -4x$ $(-5, -20), (0, 0), (9, -36)$

40. $y = -3x$ $(-5, 15), (0, 0), (7, -21)$

41. $y = 2x + 6$ $(0, 6), (-3, 0), (2, -2)$

42. $y = 8 - 4x$ $(8, 0), (16, -2), (3, -4)$

43. $3x + 5y = 15$ $(-5, 6), (0, 5), (10, -3)$

44. $2x - 5y = 0$ $(-2, 0), (-10, 6), (5, 0)$

45. $x + 3y = 0$ $(0, 0), \left(1, \dfrac{1}{3}\right), \left(2, -\dfrac{2}{3}\right)$

46. $x + 5y = 0$ $(0, 0), \left(1, \dfrac{1}{5}\right), \left(2, -\dfrac{2}{5}\right)$

47. $x - 4 = 0$ $(4, 7), (3, 4), (0, -4)$

48. $y + 2 = 0$ $(0, 2), (2, 0), (0, -2)$

In Exercises 49–56, find five solutions of each equation. Select integers for x, starting with -2 and ending with 2. Organize your work in a table of values.

49. $y = 12x$ **50.** $y = 14x$

51. $y = -10x$ **52.** $y = -20x$

53. $y = 8x - 5$ **54.** $y = 6x - 4$

55. $y = -3x + 7$ **56.** $y = -5x + 9$

In Exercises 57–80, graph each linear equation in two variables. Find at least five solutions in your table of values for each equation.

57. $y = x$ **58.** $y = x + 1$

59. $y = x - 1$ **60.** $y = x - 2$

61. $y = 2x + 1$ **62.** $y = 2x - 1$

63. $y = -x + 2$ **64.** $y = -x + 3$

65. $y = -3x - 1$ **66.** $y = -3x - 2$

67. $y = \dfrac{1}{2}x$ **68.** $y = -\dfrac{1}{2}x$

69. $y = -\dfrac{1}{4}x$ **70.** $y = \dfrac{1}{4}x$

71. $y = \dfrac{1}{3}x + 1$ **72.** $y = \dfrac{1}{3}x - 1$

73. $y = -\dfrac{3}{2}x + 1$ **74.** $y = -\dfrac{3}{2}x + 2$

75. $y = -\dfrac{5}{2}x - 1$ **76.** $y = -\dfrac{5}{2}x + 1$

77. $y = x + \dfrac{1}{2}$ **78.** $y = x - \dfrac{1}{2}$

79. $y = 4$, or $y = 0x + 4$ **80.** $y = 3$, or $y = 0x + 3$

Practice PLUS

In Exercises 81–84, write each sentence as a linear equation in two variables. Then graph the equation.

81. The y-variable is 3 more than the x-variable.

82. The y-variable exceeds the x-variable by 4.

83. The *y*-variable exceeds twice the *x*-variable by 5.

84. The *y*-variable is 2 less than 3 times the *x*-variable.

85. At the beginning of a semester, a student purchased eight pens and six pads for a total cost of $14.50.

    a. If *x* represents the cost of one pen and *y* represents the cost of one pad, write an equation in two variables that reflects the given conditions.

    b. If pads cost $0.75 each, find the cost of one pen.

86. A nursery offers a package of three small orange trees and four small grapefruit trees for $22.

    a. If *x* represents the cost of one orange tree and *y* represents the cost of one grapefruit tree, write an equation in two variables that reflects the given conditions.

    b. If a grapefruit tree costs $2.50, find the cost of an orange tree.

Application Exercises

A football is thrown by a quarterback to a receiver. The points in the figure show the height of the football, in feet, above the ground in terms of its distance, in yards, from the quarterback. Use this information to solve Exercises 87–92.

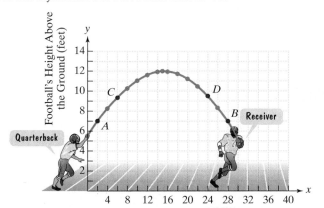

87. Find the coordinates of point *A*. Then interpret the coordinates in terms of the information given.

88. Find the coordinates of point *B*. Then interpret the coordinates in terms of the information given.

89. Estimate the coordinates of point *C*.

90. Estimate the coordinates of point *D*.

91. What is the football's maximum height? What is its distance from the quarterback when it reaches its maximum height?

92. What is the football's height when it is caught by the receiver? What is the receiver's distance from the quarterback when he catches the football?

The graph shows the average number of pounds of chicken and fish that each American consumed in 1960 and 2005. Exercises 93–94 involve models, graphs of models, and projections based on the data. When solving these problems, you will need to estimate the value on the vertical axis that corresponds to a particular point on a graph.

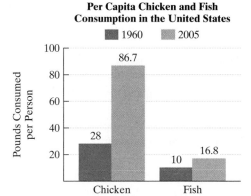

Sources: Department of Agriculture and National Marine Fisheries Service

93. The graph shows that in 1960, per capita chicken consumption was 28 pounds. This increased by approximately 1.3 pounds per year from 1960 through 2005. These conditions can be described by the mathematical model

$$C = 1.3n + 28,$$

where *C* is per capita chicken consumption *n* years after 1960.

    a. Let *n* = 0, 10, 20, 30, and 40. Make a table of values showing five solutions of the equation.

    b. Graph the formula in a rectangular coordinate system. Suggestion: Let each tick mark on the horizontal axis, labeled *n*, represent 10 units. Extend the axis to include *n* = 60. Let each tick mark on the vertical axis, labeled *C*, represent 10 units and extend the axis to include *C* = 110.

    c. Use the graph to estimate per capita chicken consumption in 2015.

    d. Use the formula to project per capita chicken consumption in 2015.

94. The graph shows that in 1960, per capita fish consumption was 10 pounds. This increased by approximately 0.15 pound per year from 1960 through 2005. These conditions can be described by the mathematical model

$$F = 0.15n + 10,$$

where *F* is per capita fish consumption *n* years after 1960.

    a. Let *n* = 0, 10, 20, 30, and 40. Make a table of values showing five solutions of the equation.

    b. Graph the formula in a rectangular coordinate system. Suggestion: Let each tick mark on the horizontal axis, labeled *n*, represent 10 units. Extend the axis to include *n* = 80. Let each tick mark on the vertical axis, labeled *F*, represent 1 unit and extend the axis to include *F* = 25.

(*This exercise continues on the next page.*)

c. Use the graph to estimate per capita fish consumption in 2020.

d. Use the formula to project per capita fish consumption in 2020.

Writing in Mathematics

95. What is the rectangular coordinate system?

96. Explain how to plot a point in the rectangular coordinate system. Give an example with your explanation.

97. Explain why $(5, -2)$ and $(-2, 5)$ do not represent the same point.

98. Explain how to find the coordinates of a point in the rectangular coordinate system.

99. How do you determine whether an ordered pair is a solution of an equation in two variables, x and y?

100. Explain how to find ordered pairs that are solutions of an equation in two variables, x and y.

101. What is the graph of an equation?

102. Explain how to graph an equation in two variables in the rectangular coordinate system.

Critical Thinking Exercises

Make Sense? *In Exercises 103–106, determine whether each statement "makes sense" or "does not make sense" and explain your reasoning.*

103. When I know that an equation's graph is a straight line, I don't need to plot more than two points, although I sometimes plot three just to check that the points line up.

104. The graph that I'm looking at is U-shaped, so its equation cannot be of the form $y = mx + b$.

105. I'm working with a linear equation in two variables and found that $(-2, 2)$, $(0, 0)$, and $(2, 2)$ are solutions.

106. When a real-world situation is modeled with a linear equation in two variables, I can use its graph to predict specific information about the situation.

In Exercises 107–110, determine whether each statement is true or false. If the statement is false, make the necessary change(s) to produce a true statement.

107. The graph of $y = 3x + 1$ is the graph of $y = 2x$ shifted up 1 unit.

108. The graph of any equation in the form $y = mx + b$ passes through the point $(0, b)$.

109. The ordered pair $(3, 4)$ satisfies the equation
$$2y - 3x = -6.$$

110. If $(2, 5)$ satisfies an equation, then $(5, 2)$ also satisfies the equation.

111. a. Graph each of the following points:
$$\left(1, \frac{1}{2}\right), \quad (2, 1), \quad \left(3, \frac{3}{2}\right), \quad (4, 2).$$

Parts (b)–(d) can be answered by changing the sign of one or both coordinates of the points in part (a).

b. What must be done to the coordinates so that the resulting graph is a mirror-image reflection about the y-axis of your graph in part (a)?

c. What must be done to the coordinates so that the resulting graph is a mirror-image reflection about the x-axis of your graph in part (a)?

d. What must be done to the coordinates so that the resulting graph is a straight-line extension of your graph in part (a)?

Technology Exercises

Use a graphing utility to graph each equation in Exercises 112–115 in a standard viewing rectangle, $[-10, 10, 1]$ by $[-10, 10, 1]$. Then use the TRACE *feature to trace along the line and find the coordinates of two points.*

112. $y = 2x - 1$

113. $y = -3x + 2$

114. $y = \frac{1}{2}x$

115. $y = \frac{3}{4}x - 2$

116. Use a graphing utility to verify any five of your hand-drawn graphs in Exercises 57–80. Use an appropriate viewing rectangle and the ZOOM SQUARE feature to make the graph look like the one you drew by hand.

Review Exercises

117. Solve: $3x + 5 = 4(2x - 3) + 7$. (Section 2.3, Example 3)

118. Simplify: $3(1 - 2 \cdot 5) - (-28)$. (Section 1.8, Example 7)

119. Solve for h: $V = \frac{1}{3}Ah$. (Section 2.4, Example 4)

Preview Exercises

Exercises 120–122 will help you prepare for the material covered in the next section. Remember that a solution of an equation in two variables is an ordered pair.

120. Let $y = 0$ and find a solution of $3x - 4y = 24$.

121. Let $x = 0$ and find a solution of $3x - 4y = 24$.

122. Let $x = 0$ and find a solution of $x + 2y = 0$.

SECTION 4.2

Graphing Linear Equations Using Intercepts

It's hard to believe that this gas-guzzler, with its huge fins and overstated design, was available in 1957 for approximately $1800. Sadly, its elegance quickly faded, depreciating linearly by $300 per year, often sold for scrap just six years after its glorious emergence from the dealer's showroom.

From these casual observations, we can obtain a mathematical model and its graph. The model is

$$y = -300x + 1800.$$

The car is depreciating by $300 per year for x years.

The new car is worth $1800.

In this model, y is the car's value after x years. **Figure 4.13** shows the equation's graph.

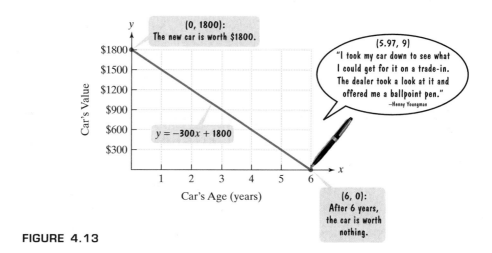

FIGURE 4.13

Here are some important observations about the model and its graph:

- The points in **Figure 4.13** show where the graph intersects the y-axis and where it intersects the x-axis. These important points, called the *intercepts*, are the topic of this section.

- We can rewrite the model, $y = -300x + 1800$, by adding $300x$ to both sides:

$$300x + y = 1800.$$

The form of this equation is $Ax + By = C$.

$$300x + y = 1800$$

| A, the coefficient of x, is 300. | B, the coefficient of y, is 1. | C, the constant on the right, is 1800. |

All equations of the form $Ax + By = C$ are straight lines when graphed as long as A and B are not both zero. To graph linear equations of this form, we will use the intercepts.

1 Use a graph to identify intercepts.

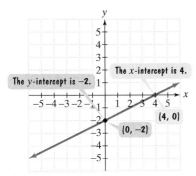

FIGURE 4.14
The graph of $2x - 4y = 8$

Intercepts

An *x*-intercept of a graph is the *x*-coordinate of a point where the graph intersects the *x*-axis. For example, look at the graph of $2x - 4y = 8$ in **Figure 4.14**. The graph crosses the *x*-axis at $(4, 0)$. Thus, the *x*-intercept is 4. **The *y*-coordinate corresponding to an *x*-intercept is always zero.**

A *y*-intercept of a graph is the *y*-coordinate of a point where the graph intersects the *y*-axis. The graph of $2x - 4y = 8$ in **Figure 4.14** shows that the graph crosses the *y*-axis at $(0, -2)$. Thus, the *y*-intercept is -2. **The *x*-coordinate corresponding to a *y*-intercept is always zero.**

Study Tip

Mathematicians tend to use two ways to describe intercepts. Did you notice that we are using single numbers? If a graph's *x*-intercept is a, it passes through the point $(a, 0)$. If a graph's *y*-intercept is b, it passes through the point $(0, b)$.

Some books state that the *x*-intercept is the *point* $(a, 0)$ and the *x*-intercept is *at a* on the *x*-axis. Similarly, the *y*-intercept is the *point* $(0, b)$ and the *y*-intercept is *at b* on the *y*-axis. In these descriptions, the intercepts are the actual points where a graph intersects the axes.

Although we'll describe intercepts as single numbers, we'll immediately state the point on the *x*- or *y*-axis that the graph passes through. Here's the important thing to keep in mind:

x-intercept: The corresponding *y* is 0.

y-intercept: The corresponding *x* is 0.

EXAMPLE 1 Identifying Intercepts

Identify the *x*- and *y*-intercepts.

a.

b.

c.

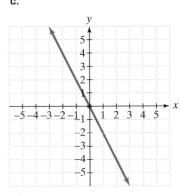

Solution

a. The graph crosses the x-axis at $(-1, 0)$. Thus, the x-intercept is -1. The graph crosses the y-axis at $(0, 2)$. Thus, the y-intercept is 2.

b. The graph crosses the x-axis at $(3, 0)$, so the x-intercept is 3. This vertical line does not cross the y-axis. Thus, there is no y-intercept.

c. This graph crosses the x- and y-axes at the same point, the origin. Because the graph crosses both axes at $(0, 0)$, the x-intercept is 0 and the y-intercept is 0. ■

☑ **CHECK POINT 1**   Identify the x- and y-intercepts.

a.

b.

c.

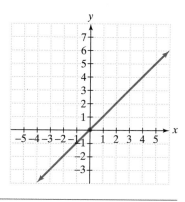

---

② Graph a linear equation in two variables using intercepts.

Graphing Using Intercepts

An equation of the form $Ax + By = C$, where A, B, and C are integers, is called the **standard form** of the equation of a line. The equation can be graphed by finding the x- and y-intercepts, plotting the intercepts, and drawing a straight line through these points. How do we find the intercepts of a line, given its equation? Because the y-coordinate of the x-intercept is 0, to find the x-intercept,

- Substitute 0 for y in the equation.
- Solve for x.

EXAMPLE 2   Finding the x-Intercept

Find the x-intercept of the graph of $3x - 4y = 24$.

Solution   To find the x-intercept, let $y = 0$ and solve for x.

$$3x - 4y = 24 \quad \text{This is the given equation.}$$
$$3x - 4 \cdot 0 = 24 \quad \text{Let } y = 0.$$
$$3x = 24 \quad \text{Simplify: } 4 \cdot 0 = 0 \text{ and } 3x - 0 = 3x.$$
$$x = 8 \quad \text{Divide both sides by 3.}$$

The x-intercept is 8. The graph of $3x - 4y = 24$ passes through the point $(8, 0)$. ■

☑ **CHECK POINT 2**   Find the x-intercept of the graph of $4x - 3y = 12$.

Because the x-coordinate of the y-intercept is 0, to find the y-intercept,

- Substitute 0 for x in the equation.
- Solve for y.

EXAMPLE 3 Finding the *y*-Intercept

Find the *y*-intercept of the graph of $3x - 4y = 24$.

Solution To find the *y*-intercept, let $x = 0$ and solve for *y*.

$$3x - 4y = 24 \quad \text{This is the given equation.}$$
$$3 \cdot 0 - 4y = 24 \quad \text{Let } x = 0.$$
$$-4y = 24 \quad \text{Simplify: } 3 \cdot 0 = 0 \text{ and } 0 - 4y = -4y.$$
$$y = -6 \quad \text{Divide both sides by } -4.$$

The *y*-intercept is -6. The graph of $3x - 4y = 24$ passes through the point $(0, -6)$. ■

☑ **CHECK POINT 3** Find the *y*-intercept of the graph of $4x - 3y = 12$.

When graphing using intercepts, it is a good idea to use a third point, a check-point, before drawing the line. A checkpoint can be obtained by selecting a value for either variable, other than 0, and finding the corresponding value for the other variable. The checkpoint should lie on the same line as the *x*- and *y*-intercepts. If it does not, recheck your work and find the error.

> **Using Intercepts to Graph $Ax + By = C$**
>
> **1.** Find the *x*-intercept. Let $y = 0$ and solve for *x*.
> **2.** Find the *y*-intercept. Let $x = 0$ and solve for *y*.
> **3.** Find a checkpoint, a third ordered-pair solution.
> **4.** Graph the equation by drawing a line through the three points.

EXAMPLE 4 Using Intercepts to Graph a Linear Equation

Graph: $3x + 2y = 6$.

Solution

Step 1. Find the *x*-intercept. Let $y = 0$ and solve for *x*.

$$3x + 2 \cdot 0 = 6 \quad \text{Replace } y \text{ with 0 in } 3x + 2y = 6.$$
$$3x = 6 \quad \text{Simplify.}$$
$$x = 2 \quad \text{Divide both sides by 3.}$$

The *x*-intercept is 2, so the line passes through $(2, 0)$.

Step 2. Find the *y*-intercept. Let $x = 0$ and solve for *y*.

$$3 \cdot 0 + 2y = 6 \quad \text{Replace } x \text{ with 0 in } 3x + 2y = 6.$$
$$2y = 6 \quad \text{Simplify.}$$
$$y = 3 \quad \text{Divide both sides by 2.}$$

The *y*-intercept is 3, so the line passes through $(0, 3)$.

Step 3. Find a checkpoint, a third ordered-pair solution. For our checkpoint, we will let $x = 1$ and find the corresponding value for *y*.

$$3x + 2y = 6 \quad \text{This is the given equation.}$$
$$3 \cdot 1 + 2y = 6 \quad \text{Substitute 1 for } x.$$
$$3 + 2y = 6 \quad \text{Simplify.}$$
$$2y = 3 \quad \text{Subtract 3 from both sides.}$$
$$y = \frac{3}{2} \quad \text{Divide both sides by 2.}$$

The checkpoint is the ordered pair $\left(1, \dfrac{3}{2}\right)$, or $(1, 1.5)$.

Step 4. Graph the equation by drawing a line through the three points. The three points in **Figure 4.15** lie along the same line. Drawing a line through the three points results in the graph of $3x + 2y = 6$. ■

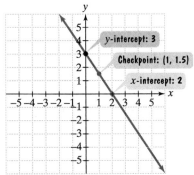

FIGURE 4.15
The graph of $3x + 2y = 6$

Using Technology

You can use a graphing utility to graph equations of the form $Ax + By = C$. Begin by solving the equation for y. For example, to graph $3x + 2y = 6$, solve the equation for y.

$$3x + 2y = 6 \qquad \text{This is the equation to be graphed.}$$

$$3x - 3x + 2y = -3x + 6 \qquad \text{Subtract 3x from both sides.}$$

$$2y = -3x + 6 \qquad \text{Simplify.}$$

$$\frac{2y}{2} = \frac{-3x + 6}{2} \qquad \text{Divide both sides by 2.}$$

$$y = -\frac{3}{2}x + 3 \qquad \text{Divide each term of } -3x + 6 \text{ by 2.}$$

This is the equation to enter into your graphing utility. The graph of $y = -\frac{3}{2}x + 3$ or, equivalently, $3x + 2y = 6$, is shown below in a $[-6, 6, 1]$ by $[-6, 6, 1]$ viewing rectangle.

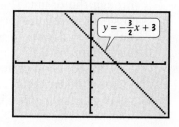

☑ **CHECK POINT 4**   Graph: $2x + 3y = 6$.

EXAMPLE 5   Using Intercepts to Graph a Linear Equation

Graph: $2x - y = 4$.

Solution

Step 1. Find the x-intercept. Let $y = 0$ and solve for x.

$$2x - 0 = 4 \qquad \text{Replace y with 0 in 2x - y = 4.}$$

$$2x = 4 \qquad \text{Simplify.}$$

$$x = 2 \qquad \text{Divide both sides by 2.}$$

The x-intercept is 2, so the line passes through $(2, 0)$.

Step 2. Find the y-intercept. Let $x = 0$ and solve for y.

$$2 \cdot 0 - y = 4 \qquad \text{Replace x with 0 in 2x - y = 4.}$$

$$-y = 4 \qquad \text{Simplify.}$$

$$y = -4 \qquad \text{Divide (or multiply) both sides by -1.}$$

The y-intercept is -4, so the line passes through $(0, -4)$.

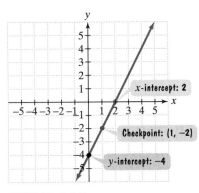

FIGURE 4.16
The graph of $2x - y = 4$

Step 3. Find a checkpoint, a third ordered-pair solution. For our checkpoint, we will let $x = 1$ and find the corresponding value for y.

$2x - y = 4$	This is the given equation.
$2 \cdot 1 - y = 4$	Substitute 1 for x.
$2 - y = 4$	Simplify.
$-y = 2$	Subtract 2 from both sides.
$y = -2$	Divide (or multiply) both sides by -1.

The checkpoint is $(1, -2)$.

Step 4. Graph the equation by drawing a line through the three points. The three points in **Figure 4.16** lie along the same line. Drawing a line through the three points results in the graph of $2x - y = 4$. ■

✓ **CHECK POINT 5** Graph: $x - 2y = 4$.

We have seen that not all lines have two different intercepts. Some lines pass through the origin. Thus, they have an x-intercept of 0 and a y-intercept of 0. Is it possible to recognize these lines by their equations? Yes. **The graph of the linear equation $Ax + By = 0$ passes through the origin.** Notice that the constant on the right side of this equation is 0.

An equation of the form $Ax + By = 0$ can be graphed by using the origin as one point on the line. Find two other points by finding two other solutions of the equation. Select values for either variable, other than 0, and find the corresponding values for the other variable.

EXAMPLE 6 **Graphing a Linear Equation of the Form $Ax + By = 0$**

Graph: $x + 2y = 0$.

Solution Because the constant on the right is 0, the graph passes through the origin. The x- and y-intercepts are both 0. Remember that we are using two points and a checkpoint to graph a line. Thus, we still want to find two other points. Let $y = -1$ to find a second ordered-pair solution. Let $y = 1$ to find a third ordered-pair (checkpoint) solution.

$$x + 2y = 0 \qquad\qquad x + 2y = 0$$

Let $y = -1$. Let $y = 1$.

$$x + 2(-1) = 0 \qquad\qquad x + 2 \cdot 1 = 0$$
$$x + (-2) = 0 \qquad\qquad x + 2 = 0$$
$$x = 2 \qquad\qquad\qquad x = -2$$

The solutions are $(2, -1)$ and $(-2, 1)$. Plot these two points, as well as the origin—that is, $(0, 0)$. The three points in **Figure 4.17** lie along the same line. Drawing a line through the three points results in the graph of $x + 2y = 0$. ■

✓ **CHECK POINT 6** Graph: $x + 3y = 0$.

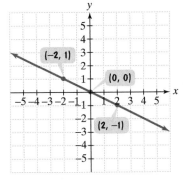

FIGURE 4.17
The graph of $x + 2y = 0$

3 Graph horizontal or vertical lines.

Equations of Horizontal and Vertical Lines

We know that the graph of any equation of the form $Ax + By = C$ is a line as long as A and B are not both zero. What happens if A or B, but not both, is zero? Example 7 shows that if $A = 0$, the equation $Ax + By = C$ has no x-term and the graph is a horizontal line.

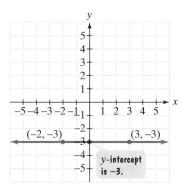

FIGURE 4.18
The graph of $y = -3$

EXAMPLE 7 Graphing a Horizontal Line

Graph: $y = -3$.

Solution All ordered pairs that are solutions of $y = -3$ have a value of y that is -3. Any value can be used for x. In the table on the right, we have selected three of the possible values for x: $-2, 0$, and 3. The table shows that three ordered pairs that are solutions of $y = -3$ are $(-2, -3), (0, -3)$, and $(3, -3)$. Drawing a line that passes through the three points gives the horizontal line shown in **Figure 4.18**.

x	$y = -3$	(x, y)
-2	-3	$(-2, -3)$
0	-3	$(0, -3)$
3	-3	$(3, -3)$

For all choices of x, y is a constant -3.

☑ **CHECK POINT 7** Graph: $y = 3$.

Example 8 shows that if $B = 0$, the equation $Ax + By = C$ has no y-term and the graph is a vertical line.

EXAMPLE 8 Graphing a Vertical Line

Graph: $x = 5$.

Solution All ordered pairs that are solutions of $x = 5$ have a value of x that is 5. Any value can be used for y. In the table on the right, we have selected three of the possible values for y: $-2, 0$, and 3. The table shows that three ordered pairs that are solutions of $x = 5$ are $(5, -2), (5, 0)$, and $(5, 3)$. Drawing a line that passes through the three points gives the vertical line shown in **Figure 4.19**.

For all choices of y,

$x = 5$	y	(x, y)
5	-2	$(5, -2)$
5	0	$(5, 0)$
5	3	$(5, 3)$

x is always 5.

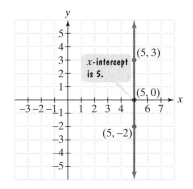

FIGURE 4.19
The graph of $x = 5$

Study Tip

Do not confuse two-dimensional graphing and one-dimensional graphing of $x = 5$. The graph of $x = 5$ in a two-dimensional rectangular coordinate system is the vertical line in **Figure 4.19**. By contrast, the graph of $x = 5$ on a one-dimensional number line representing values of x is a single point at 5:

☑ **CHECK POINT 8** Graph: $x = -2$.

Horizontal and Vertical Lines

The graph of $y = b$ is a horizontal line. The y-intercept is b.

The graph of $x = a$ is a vertical line. The x-intercept is a.

4.2 EXERCISE SET **MyMathLab**

Practice Exercises

In Exercises 1–8, use the graph to identify the

  a. *x-intercept, or state that there is no x-intercept;*

  b. *y-intercept, or state that there is no y-intercept.*

1.

2.

3.

4.

5.

6.

7.

8.

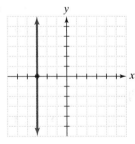

In Exercises 9–18, find the x-intercept and the y-intercept of the graph of each equation. Do not graph the equation.

  9. $2x + 5y = 20$

10. $2x + 6y = 30$

11. $2x - 3y = 15$

12. $4x - 5y = 10$

13. $-x + 3y = -8$

14. $-x + 3y = -10$

15. $7x - 9y = 0$

16. $8x - 11y = 0$

17. $2x = 3y - 11$

18. $2x = 4y - 13$

In Exercises 19–40, use intercepts and a checkpoint to graph each equation.

19. $x + y = 5$ **20.** $x + y = 6$

21. $x + 3y = 6$ **22.** $2x + y = 4$

23. $6x - 9y = 18$ **24.** $6x - 2y = 12$

25. $-x + 4y = 6$ **26.** $-x + 3y = 10$

27. $2x - y = 7$ **28.** $2x - y = 5$

29. $3x = 5y - 15$ **30.** $2x = 3y + 6$

31. $25y = 100 - 50x$ **32.** $10y = 60 - 40x$

33. $2x - 8y = 12$ **34.** $3x - 6y = 15$

35. $x + 2y = 0$ **36.** $2x + y = 0$

37. $y - 3x = 0$ **38.** $y - 4x = 0$

39. $2x - 3y = -11$ **40.** $3x - 2y = -7$

In Exercises 41–46, write an equation for each graph.

41.

42.

43.

44.

45.

46.

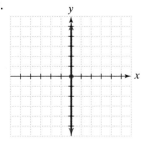

70. The perimeter of the shaded trapezoid is 84 feet.

In Exercises 47–62, graph each equation.

47. $y = 4$

48. $y = 2$

49. $y = -2$

50. $y = -3$

51. $x = 2$

52. $x = 4$

53. $x + 1 = 0$

54. $x + 5 = 0$

55. $y - 3.5 = 0$

56. $y - 2.5 = 0$

57. $x = 0$

58. $y = 0$

59. $3y = 9$

60. $5y = 20$

61. $12 - 3x = 0$

62. $12 - 4x = 0$

Practice PLUS

In Exercises 63–68, match each equation with one of the graphs shown in Exercises 1–8.

63. $3x + 2y = -6$

64. $x + 2y = -4$

65. $y = -2$

66. $x = -3$

67. $4x + 3y = 12$

68. $2x + 5y = 10$

In Exercises 69–70,

  a. *Write a linear equation in standard form satisfying the given condition. Assume that all measurements shown in each figure are in feet.*

  b. *Graph the equation in part (a). Because x and y must be non-negative (why?), limit your final graph to quadrant I and its boundaries.*

69. The perimeter of the larger rectangle is 58 feet.

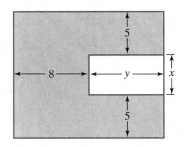

Application Exercises

The flight of an eagle is observed for 30 seconds. The graph shows the eagle's height, in meters, during this period of time. Use the graph to solve Exercises 71–75.

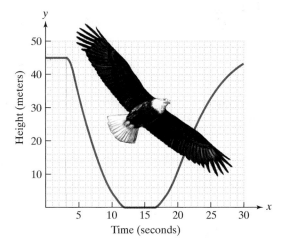

71. During which period of time is the eagle's height decreasing?

72. During which period of time is the eagle's height increasing?

73. What is the *y*-intercept? What does this mean about the eagle's height at the beginning of the observation?

74. During the first three seconds of observation, the eagle's flight is graphed as a horizontal line. Write the equation of the line. What does this mean about the eagle's flight pattern during this time?

75. Use integers to write five *x*-intercepts of the graph. What is the eagle doing during these times?

76. A new car worth $24,000 is depreciating in value by $3000 per year. The mathematical model

$$y = -3000x + 24,000$$

describes the car's value, *y*, in dollars, after *x* years.

  a. Find the *x*-intercept. Describe what this means in terms of the car's value.

(This exercise continues on the next page.)

b. Find the y-intercept. Describe what this means in terms of the car's value.

c. Use the intercepts to graph the linear equation. Because x and y must be nonnegative (why?), limit your graph to quadrant I and its boundaries.

d. Use your graph to estimate the car's value after five years.

77. A new car worth $45,000 is depreciating in value by $5000 per year. The mathematical model

$$y = -5000x + 45,000$$

describes the car's value, y, in dollars, after x years.

a. Find the x-intercept. Describe what this means in terms of the car's value.

b. Find the y-intercept. Describe what this means in terms of the car's value.

c. Use the intercepts to graph the linear equation. Because x and y must be nonnegative (why?), limit your graph to quadrant I and its boundaries.

d. Use your graph to estimate the car's value after five years.

Writing in Mathematics

78. What is an x-intercept of a graph?

79. What is a y-intercept of a graph?

80. If you are given an equation of the form $Ax + By = C$, explain how to find the x-intercept.

81. If you are given an equation of the form $Ax + By = C$, explain how to find the y-intercept.

82. Explain how to graph $Ax + By = C$ if C is not equal to zero.

83. Explain how to graph a linear equation of the form $Ax + By = 0$.

84. How many points are needed to graph a line? How many should actually be used? Explain.

85. Describe the graph of $y = 200$.

86. Describe the graph of $x = -100$.

Critical Thinking Exercises

Make Sense?   *In Exercises 87–90, determine whether each statement "makes sense" or "does not make sense" and explain your reasoning.*

87. If I could be absolutely certain that I have not made an algebraic error in obtaining intercepts, I would not need to use checkpoints.

88. I like to select a point represented by one of the intercepts as my checkpoint.

89. The graphs of $2x - 3y = -18$ and $-2x + 3y = 18$ must have the same intercepts because I can see that the equations are equivalent.

90. From 1997 through 2007, the federal minimum wage remained constant at $5.15 per hour, so I modeled the situation with $y = 5.15$ and the graph of a vertical line.

In Exercises 91–92, find the coefficients that must be placed in each shaded area so that the equation's graph will be a line with the specified intercepts.

91. ▢$x +$ ▢$y = 10$; x-intercept $= 5$; y-intercept $= 2$

92. ▢$x +$ ▢$y = 12$; x-intercept $= -2$; y-intercept $= 4$

Technology Exercises

93. Use a graphing utility to verify any five of your hand-drawn graphs in Exercises 19–40. Solve the equation for y before entering it.

In Exercises 94–97, use a graphing utility to graph each equation. You will need to solve the equation for y before entering it. Use the graph displayed on the screen to identify the x-intercept and the y-intercept.

94. $2x + y = 4$

95. $3x - y = 9$

96. $2x + 3y = 30$

97. $4x - 2y = -40$

Review Exercises

98. Find the absolute value: $|-13.4|$. (Section 1.3, Example 8)

99. Simplify: $7x - (3x - 5)$. (Section 1.7, Example 7)

100. Solve: $8(x - 2) - 2(x - 3) \le 8x$. (Section 2.6, Example 8)

Preview Exercises

Exercises 101–103 will help you prepare for the material covered in the next section. In each exercise, evaluate

$$\frac{y_2 - y_1}{x_2 - x_1}$$

for the given ordered pairs (x_1, y_1) *and* (x_2, y_2).

101. $(x_1, y_1) = (1, 3)$; $(x_2, y_2) = (6, 13)$

102. $(x_1, y_1) = (4, -2)$; $(x_2, y_2) = (6, -4)$

103. $(x_1, y_1) = (3, 4)$; $(x_2, y_2) = (5, 4)$

SECTION
4.3

Slope

Objectives

1 Compute a line's slope.

2 Use slope to show that lines are parallel.

3 Use slope to show that lines are perpendicular.

4 Calculate rate of change in applied situations.

A best guess at the future of our nation indicates that the numbers of men and women living alone will increase each year. **Figure 4.20** shows that in 2005, 12.7 million men and 17.2 million women lived alone, an increase over the numbers displayed in the graph for 1990.

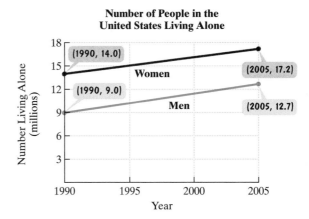

Number of People in the United States Living Alone

(1990, 14.0)

Women

(2005, 17.2)

(1990, 9.0)

Men

(2005, 12.7)

FIGURE 4.20

Source: U.S. Census Bureau

By looking at **Figure 4.20**, can you tell that the green graph representing men is steeper than the red graph representing women? This indicates a greater rate of increase in the millions of men living alone than in the millions of women living alone over the period from 1990 through 2005. In this section, we will study the idea of a line's steepness and see what this has to do with how its variables are changing.

1 Compute a line's slope.

The Slope of a Line

Mathematicians have developed a useful measure of the steepness of a line, called the **slope** of the line. Slope compares the vertical change (the **rise**) to the horizontal change (the **run**) when moving from one fixed point to another along the line. To calculate the slope of a line, we use a ratio that compares the change in y (the rise) to the change in x (the run).

Definition of Slope

The **slope** of the line through the distinct points (x_1, y_1) and (x_2, y_2) is

$$\frac{\text{Change in } y}{\text{Change in } x} = \frac{\text{Rise}}{\text{Run}}$$

$$= \frac{y_2 - y_1}{x_2 - x_1}$$

where $x_2 - x_1 \neq 0$.

It is common notation to let the letter m represent the slope of a line. The letter m is used because it is the first letter of the French verb *monter*, meaning to rise, or to ascend.

EXAMPLE 1   Using the Definition of Slope

Find the slope of the line passing through each pair of points:

   a. $(-3, -1)$ and $(-2, 4)$

   b. $(-3, 4)$ and $(2, -2)$.

Solution

   a. Let $(x_1, y_1) = (-3, -1)$ and $(x_2, y_2) = (-2, 4)$. We obtain the slope as follows:

$$m = \frac{\text{Change in } y}{\text{Change in } x} = \frac{y_2 - y_1}{x_2 - x_1} = \frac{4 - (-1)}{-2 - (-3)} = \frac{5}{1} = 5.$$

The situation is illustrated at the top of the next page in **Figure 4.21**. The slope of the line is 5, indicating that there is a vertical change, a rise, of 5 units for each horizontal change, a run, of 1 unit. The slope is positive and the line rises from left to right.

Study Tip

When computing slope, it makes no difference which point you call (x_1, y_1) and which point you call (x_2, y_2). If we let $(x_1, y_1) = (-2, 4)$ and $(x_2, y_2) = (-3, -1)$, the slope is still 5:

$$m = \frac{\text{Change in } y}{\text{Change in } x} = \frac{y_2 - y_1}{x_2 - x_1} = \frac{-1 - 4}{-3 - (-2)} = \frac{-5}{-1} = 5.$$

However, you should not subtract in one order in the numerator $(y_2 - y_1)$ and then in the opposite order in the denominator $(x_1 - x_2)$.

$$\frac{-1 - 4}{-2 - (-3)} = \frac{-5}{1} = -5 \quad \text{Incorrect! The slope is not } -5.$$

   b. We can let $(x_1, y_1) = (-3, 4)$ and $(x_2, y_2) = (2, -2)$. The slope is computed as follows:

$$m = \frac{\text{Change in } y}{\text{Change in } x} = \frac{y_2 - y_1}{x_2 - x_1} = \frac{-2 - 4}{2 - (-3)} = \frac{-6}{5} = -\frac{6}{5}.$$

The situation is illustrated at the top of the next page in **Figure 4.22**. The slope of the line is $-\frac{6}{5}$. For every vertical change of -6 units (6 units down), there is a corresponding horizontal change of 5 units. The slope is negative and the line falls from left to right.

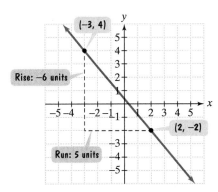

FIGURE 4.21 Visualizing a slope of 5

FIGURE 4.22 Visualizing a slope of $-\frac{6}{5}$

✓ **CHECK POINT 1**   Find the slope of the line passing through each pair of points:

a. $(-3, 4)$ and $(-4, -2)$

b. $(4, -2)$ and $(-1, 5)$.

EXAMPLE 2   **Using the Definition of Slope for Horizontal and Vertical Lines**

Find the slope of the line passing through each pair of points:

a. $(5, 4)$ and $(3, 4)$ b. $(2, 5)$ and $(2, 1)$.

Solution

a. Let $(x_1, y_1) = (5, 4)$ and $(x_2, y_2) = (3, 4)$. We obtain the slope as follows:

$$m = \frac{\text{Change in } y}{\text{Change in } x} = \frac{y_2 - y_1}{x_2 - x_1} = \frac{4 - 4}{3 - 5} = \frac{0}{-2} = 0.$$

The situation is illustrated in **Figure 4.23**. Can you see that the line is horizontal? Because any two points on a horizontal line have the same y-coordinate, these lines neither rise nor fall from left to right. The change in y, $y_2 - y_1$, is always zero. Thus, **the slope of any horizontal line is zero.**

b. We can let $(x_1, y_1) = (2, 5)$ and $(x_2, y_2) = (2, 1)$. **Figure 4.24** shows that these points are on a vertical line. We attempt to compute the slope as follows:

$$m = \frac{\text{Change in } y}{\text{Change in } x} = \frac{1 - 5}{2 - 2} = \frac{-4}{0}. \quad \text{Division by zero is undefined.}$$

Because division by zero is undefined, the slope of the vertical line in **Figure 4.24** is undefined. In general, **the slope of any vertical line is undefined.**

FIGURE 4.23 Horizontal lines have no vertical change.

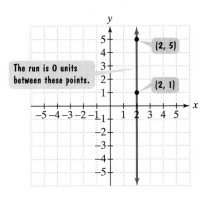

FIGURE 4.24 Vertical lines have no horizontal change.

Table 4.1 summarizes four possibilities for the slope of a line.

Table 4.1	Possibilities for a Line's Slope			
	Positive Slope	**Negative Slope**	**Zero Slope**	**Undefined Slope**
	$m > 0$	$m < 0$	$m = 0$	m is undefined.
	Line rises from left to right.	Line falls from left to right.	Line is horizontal.	Line is vertical.

✓ **CHECK POINT 2** Find the slope of the line passing through each pair of points or state that the slope is undefined:

    a. $(6, 5)$ and $(2, 5)$ **b.** $(1, 6)$ and $(1, 4)$.

2 Use slope to show that lines are parallel.

Slope and Parallel Lines

Two nonintersecting lines that lie in the same plane are **parallel**. If two lines do not intersect, the ratio of the vertical change to the horizontal change is the same for each line. Because two parallel lines have the same "steepness," they must have the same slope.

> **Slope and Parallel Lines**
>
> **1.** If two nonvertical lines are parallel, then they have the same slope.
> **2.** If two distinct nonvertical lines have the same slope, then they are parallel.
> **3.** Two distinct vertical lines, each with undefined slope, are parallel.

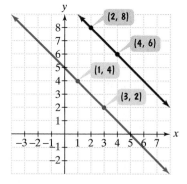

FIGURE 4.25 Using slope to show that lines are parallel

EXAMPLE 3 Using Slope to Show That Lines Are Parallel

Show that the line passing through $(1, 4)$ and $(3, 2)$ is parallel to the line passing through $(2, 8)$ and $(4, 6)$.

Solution The situation is illustrated in **Figure 4.25**. The lines certainly look like they are parallel. Let's use equal slopes to confirm this fact. For each line, we compute the ratio of the difference in y-coordinates to the difference in x-coordinates. Be sure to subtract the coordinates in the same order.

We begin with the slope of the blue line through $(1, 4)$ and $(3, 2)$.

Change in y is $2 - 4$.

$$(1, 4) \qquad (3, 2)$$

Change in x is $3 - 1$.

$$m = \frac{\text{Change in } y}{\text{Change in } x} = \frac{2 - 4}{3 - 1} = \frac{-2}{2} = -1$$

Now we find the slope of the red line through $(2, 8)$ and $(4, 6)$.

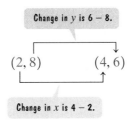

Change in y is $6 - 8$.

$(2, 8)$ $(4, 6)$

Change in x is $4 - 2$.

$$m = \frac{\text{Change in } y}{\text{Change in } x} = \frac{6 - 8}{4 - 2} = \frac{-2}{2} = -1$$

Because the slopes are equal, the lines are parallel.

✓ **CHECK POINT 3** Show that the line passing through $(4, 2)$ and $(6, 6)$ is parallel to the line passing through $(0, -2)$ and $(1, 0)$.

---

3 Use slope to show that lines are perpendicular.

Slope and Perpendicular Lines

Two lines that intersect at a right angle (90°) are said to be **perpendicular**, as shown in **Figure 4.26**. There is a relationship between the slopes of perpendicular lines.

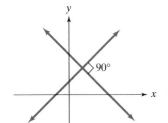

FIGURE 4.26
Perpendicular lines

> ### Slope and Perpendicular Lines
> 1. If two nonvertical lines are perpendicular, then the product of their slopes is -1.
> 2. If the product of the slopes of two lines is -1, then the lines are perpendicular.
> 3. A horizontal line having zero slope is perpendicular to a vertical line having undefined slope.

EXAMPLE 4 **Using Slope to Show That Lines Are Perpendicular**

Show that the line passing through $(-6, -9)$ and $(3, 6)$ is perpendicular to the line passing through $(10, -8)$ and $(-5, 1)$.

Solution The situation is illustrated in **Figure 4.27**. The lines certainly look like they intersect at a right angle. Let's show that the product of their slopes is -1 to confirm this fact.

We begin with the slope of the red line through $(-6, -9)$ and $(3, 6)$.

Change in y is $6 - (-9)$.

$(-6, -9)$ $(3, 6)$

Change in x is $3 - (-6)$.

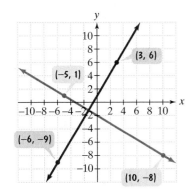

FIGURE 4.27 Using slope to show that lines are perpendicular

$$m = \frac{\text{Change in } y}{\text{Change in } x} = \frac{6 - (-9)}{3 - (-6)} = \frac{6 + 9}{3 + 6} = \frac{15}{9} = \frac{5}{3}$$

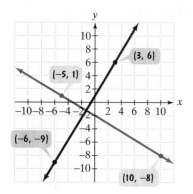

FIGURE 4.27   (repeated)

Now we find the slope of the blue line through $(10, -8)$ and $(-5, 1)$.

Change in y is $1 - (-8)$.

$(10, -8)$ $(-5, 1)$

Change in x is $-5 - 10$.

$$m = \frac{\text{Change in } y}{\text{Change in } x} = \frac{1 - (-8)}{-5 - 10} = \frac{1 + 8}{-5 - 10} = \frac{9}{-15} = -\frac{3}{5}$$

We have determined that the slope of the red line in **Figure 4.27** is $\frac{5}{3}$ and the slope of the blue line is $-\frac{3}{5}$. Now we can find the product of the slopes:

$$\frac{5}{3}\left(-\frac{3}{5}\right) = -1.$$

Because the product of the slopes is -1, the lines are perpendicular.   ◼

In this example, the slopes of the perpendicular lines, $\frac{5}{3}$ and $-\frac{3}{5}$, are reciprocals with opposite signs, called **negative reciprocals**. This gives us an equivalent way of stating the relationship between slope and perpendicular lines:

Two nonvertical lines are perpendicular if the slope of one is the negative reciprocal of the slope of the other.

☑   **CHECK POINT 4**   Show that the line passing through $(-1, 4)$ and $(3, 2)$ is perpendicular to the line passing through $(-2, -1)$ and $(2, 7)$.

---

4 Calculate rate of change in applied situations.

Slope as Rate of Change

Slope is defined as the ratio of a change in y to a corresponding change in x. It tells how fast y is changing with respect to x. Thus, the slope of a line represents its rate of change.

Our next example shows how slope can be interpreted as a rate of change in an applied situation. When calculating slope in applied problems, keep track of the units in the numerator and the denominator.

EXAMPLE 5   **Slope as a Rate of Change**

The line graphs for the numbers of women and men living alone are shown again in **Figure 4.28**. Find the slope of the line segment for the women. Describe what this slope represents.

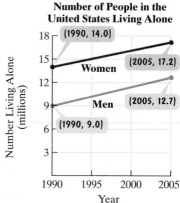

FIGURE 4.28

Source: U.S. Census Bureau

Solution   We let *x* represent a year and *y* the number of women living alone in that year. The two points shown on the red line segment for women have the following coordinates:

(1990, 14.0)   (2005, 17.2)

In 1990, 14 million U.S. women lived alone.

In 2005, 17.2 million U.S. women lived alone.

Now we compute the slope:

$$m = \frac{\text{Change in } y}{\text{Change in } x} = \frac{17.2 - 14.0}{2005 - 1990}$$

The unit in the numerator is *million women*.

The unit in the denominator is *year*.

$$= \frac{3.2}{15} \approx \frac{0.21 \text{ million women}}{\text{year}}.$$

The slope indicates that the number of American women living alone increased at a rate of approximately 0.21 million each year for the period from 1990 through 2005. The rate of change is 0.21 million women per year.

✓ **CHECK POINT 5**   Use the ordered pairs in **Figure 4.28** to find the slope of the green line segment for the men. Express the slope correct to two decimal places and describe what it represents.

In Check Point 5, did you find that the slope of the line segment for the men is different from that of the women? The rate of change for men living alone is greater than the rate of change for women living alone. The line segment representing men in **Figure 4.28** is steeper than the line segment representing women. If you extend the line segments far enough, the resulting lines will intersect. They are not parallel.

Blitzer Bonus

Railroads and Highways

The steepest part of Mt. Washington Cog Railway in New Hampshire has a 37% grade. This is equivalent to a slope of $\frac{37}{100}$. For every horizontal change of 100 feet, the railroad ascends 37 feet vertically. Engineers denote slope by grade, expressing slope as a percentage.

4 feet
5 feet
100 feet
4% grade
100 feet
−5% grade

Railroad grades are usually less than 2%, although in the mountains they may go as high as 4%. The grade of the Mt. Washington Cog Railway is phenomenal, making it necessary for locomotives to *push* single cars up its steepest part.

A Mount Washington Cog Railway locomotive pushing a single car up the steepest part of the railroad. The locomotive is about 120 years old.

4.3 EXERCISE SET

MyMathLab

PRACTICE   WATCH   DOWNLOAD   READ   REVIEW

Practice Exercises

In Exercises 1–10, find the slope of the line passing through each pair of points or state that the slope is undefined. Then indicate whether the line through the points rises, falls, is horizontal, or is vertical.

1. $(4, 7)$ and $(8, 10)$

2. $(2, 1)$ and $(3, 4)$

3. $(-2, 1)$ and $(2, 2)$

4. $(-1, 3)$ and $(2, 4)$

5. $(4, -2)$ and $(3, -2)$

6. $(4, -1)$ and $(3, -1)$

7. $(-2, 4)$ and $(-1, -1)$

8. $(6, -4)$ and $(4, -2)$

9. $(5, 3)$ and $(5, -2)$

10. $(3, -4)$ and $(3, 5)$

In Exercises 11–22, find the slope of each line, or state that the slope is undefined.

11.

12.

13.

14.

15.

16.

17.

18.

19.

20.

21.

22.

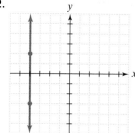

In Exercises 23–26, determine whether the distinct lines through each pair of points are parallel.

23. $(-2, 0)$ and $(0, 6); (1, 8)$ and $(0, 5)$

24. $(2, 4)$ and $(6, 1); (-3, 1)$ and $(1, -2)$

25. $(0, 3)$ and $(1, 5); (-1, 7)$ and $(1, 10)$

26. $(-7, 6)$ and $(0, 4); (-9, -3)$ and $(1, 5)$

In Exercises 27–30, determine whether the lines through each pair of points are perpendicular.

27. $(1, 5)$ and $(0, 3); (-2, 8)$ and $(2, 6)$

28. $(3, 2)$ and $(-2, -2); (3, -2)$ and $(-1, 3)$

29. $(-1, -6)$ and $(2, 9); (-15, -1)$ and $(5, 3)$

30. $(-1, -6)$ and $(2, 6); (-8, -1)$ and $(4, 2)$

In Exercises 31–36, determine whether the lines through each pair of points are parallel, perpendicular, or neither.

31. $(-2, -5)$ and $(3, 10)$; $(-1, -9)$ and $(4, 6)$

32. $(-2, -7)$ and $(3, 13)$; $(-1, -9)$ and $(5, 15)$

33. $(-4, -12)$ and $(0, -4)$; $(0, -5)$ and $(2, -4)$

34. $(-1, -11)$ and $(0, -5)$; $(0, -8)$ and $(12, -6)$

35. $(-5, -1)$ and $(0, 2)$; $(-6, 9)$ and $(3, -6)$

36. $(-2, -15)$ and $(0, -3)$; $(-12, 6)$ and $(6, 3)$

Practice PLUS

37. On the same set of axes, draw lines passing through the origin with slopes $-1, -\frac{1}{2}, 0, \frac{1}{3},$ and 2.

38. On the same set of axes, draw lines with y-intercept 4 and slopes $-1, -\frac{1}{2}, 0, \frac{1}{3},$ and 2.

Use slopes to solve Exercises 39–40.

39. Show that the points whose coordinates are $(-3, -3)$, $(2, -5)$, $(5, -1)$, and $(0, 1)$ are the vertices of a four-sided figure whose opposite sides are parallel. (Such a figure is called a *parallelogram*.)

40. Show that the points whose coordinates are $(-3, 6)$, $(2, -3)$, $(11, 2)$, and $(6, 11)$ are the vertices of a four-sided figure whose opposite sides are parallel.

41. The line passing through $(5, y)$ and $(1, 0)$ is parallel to the line joining $(2, 3)$ and $(-2, 1)$. Find y.

42. The line passing through $(1, y)$ and $(7, 12)$ is parallel to the line joining $(-3, 4)$ and $(-5, -2)$. Find y.

43. The line passing through $(-1, y)$ and $(1, 0)$ is perpendicular to the line joining $(2, 3)$ and $(-2, 1)$. Find y.

44. The line passing through $(-2, y)$ and $(-4, 4)$ is perpendicular to the line passing through $(-1, -2)$ and $(4, -1)$. Find y.

Application Exercises

45. Exercise is useful not only in preventing depression, but also as a treatment. The graphs show the percentage of patients with depression in remission when exercise (brisk walking) was used as a treatment. (The control group that engaged in no exercise had 11% of the patients in remission.)

Exercise and Percentage of Patients with Depression in Remission

Source: Newsweek, March 26, 2007

a. Find the slope of the line passing through the two points shown by the voice balloons. Express the slope as a decimal.

b. Use your answer from part (a) to complete this statement:

For each minute of brisk walking, the percentage of patients with depression in remission increased by ____ %. The rate of change is ____ % per _____ _____.

46. With offers for everybody, including 5% cash back on bookstore purchases for college students, credit cards and their rewards can leave you dizzy. (Check out the interest rates and annual fees before you grab the next glitzy offer.) The graphs show the number of credit cards offering rewards in 2003 and 2006.

Number of Credit Cards Offering Rewards

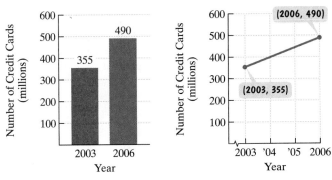

Source: CardTrak.com

a. Find the slope of the line passing through the two points shown by the voice balloons.

b. Use your answer from part (a) to complete this statement:

For each year from 2003 through 2006, the number of credit cards offering rewards increased by ___ million. The rate of change was _____ rewards cards per ____.

The pitch of a roof refers to the absolute value of its slope. In Exercises 47–48, find the pitch of each roof shown.

47.

48.

The grade of a road or ramp refers to its slope expressed as a percent. Use this information to solve Exercises 49–50.

49. Construction laws are very specific when it comes to access ramps for the disabled. Every vertical rise of 1 foot requires a horizontal run of 12 feet. What is the grade of such a ramp? Round to the nearest tenth of a percent.

50. A college campus goes beyond the standards described in Exercise 49. All wheelchair ramps on campus are designed so that every vertical rise of 1 foot is accompanied by a horizontal run of 14 feet. What is the grade of such a ramp? Round to the nearest tenth of a percent.

Writing in Mathematics

51. What is the slope of a line?

52. Describe how to calculate the slope of a line passing through two points.

53. What does it mean if the slope of a line is zero?

54. What does it mean if the slope of a line is undefined?

55. If two lines are parallel, describe the relationship between their slopes.

56. If two lines are perpendicular, describe the relationship between their slopes.

Critical Thinking Exercises

Make Sense? *In Exercises 57–60, determine whether each statement "makes sense" or "does not make sense" and explain your reasoning.*

57. When finding the slope of the line passing through $(-1, 5)$ and $(2, -3)$, I must let (x_1, y_1) be $(-1, 5)$ and (x_2, y_2) be $(2, -3)$.

58. When applying the slope formula, it is important to subtract corresponding coordinates in the same order.

59. I visualize slope as walking along a line from left to right. If I'm walking uphill, the slope is positive, and if I'm walking downhill, the slope is negative.

60. I computed the slope of one line to be $-\frac{3}{5}$ and the slope of a second line to be $-\frac{5}{3}$, so the lines must be perpendicular.

In Exercises 61–64, determine whether each statement is true or false. If the statement is false, make the necessary change(s) to produce a true statement.

61. Slope is run divided by rise.

62. The line through $(2, 2)$ and the origin has slope 1.

63. A line with slope 3 can be parallel to a line with slope -3.

64. The line through $(3, 1)$ and $(3, -5)$ has zero slope.

In Exercises 65–66, use the figure shown to make the indicated list.

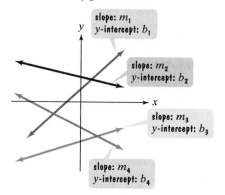

65. List the slopes $m_1, m_2, m_3,$ and m_4 in order of decreasing size.

66. List the y-intercepts $b_1, b_2, b_3,$ and b_4 in order of decreasing size.

Technology Exercises

Use a graphing utility to graph each equation in Exercises 67–70. Then use the $\boxed{\text{TRACE}}$ *feature to trace along the line and find the coordinates of two points. Use these points to compute the line's slope.*

67. $y = 2x + 4$

68. $y = -3x + 6$

69. $y = -\frac{1}{2}x - 5$

70. $y = \frac{3}{4}x - 2$

71. In Exercises 67–70, compare the slope that you found with the line's equation. What relationship do you observe between the line's slope and one of the constants in the equation?

Review Exercises

72. A 36-inch board is cut into two pieces. One piece is twice as long as the other. How long are the pieces? (Section 2.5, Example 2)

73. Simplify: $-10 + 16 \div 2(-4)$. (Section 1.8, Example 4)

74. Solve and graph the solution set on a number line: $2x - 3 \le 5$. (Section 2.6, Example 6)

Preview Exercises

Exercises 75–77 will help you prepare for the material covered in the next section.

75. From $(0, -3)$, move 4 units up and 1 unit to the right. What point do you obtain?

76. From $(0, 1)$, move 2 units down and 3 units to the right. What point do you obtain?

77. Solve for y: $2x + 5y = 0$.

<table>
<tr><td>SECTION 4.4</td></tr>
</table>

The Slope-Intercept Form of the Equation of a Line

Objectives

1 Find a line's slope and y-intercept from its equation.

2 Graph lines in slope-intercept form.

3 Use slope and y-intercept to graph $Ax + By = C$.

4 Use slope and y-intercept to model data.

"For a while we pondered whether to take a vacation or get a divorce. We decided that a trip to Bermuda is over in two weeks, but a divorce is something you always have."

—*Woody Allen*

The *Comedy Thesaurus* (Quirk Books, 2005) contains 23 jokes on the subject of divorce. This section contains two mathematical models on the subject, intended less for hilarious observations and more for serious projections. To develop these models, we turn to a new form for a line's equation.

The Slope-Intercept Form of the Equation of a Line

1 Find a line's slope and y-intercept from its equation.

Let's begin with an example that shows how easy it is to find a line's slope and y-intercept from its equation.

Figure 4.29 shows the graph of $y = 2x + 4$. Verify that the x-intercept is -2 by setting y equal to 0 and solving for x. Similarly, verify that the y-intercept is 4 by setting x equal to 0 and solving for y.

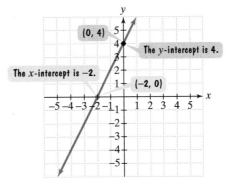

The x-intercept is −2.   (0, 4)   The y-intercept is 4.   (−2, 0)

FIGURE 4.29
The graph of $y = 2x + 4$

Now that we have two points on the line, $(-2, 0)$ and $(0, 4)$, we can calculate the slope of the graph of $y = 2x + 4$.

$$\text{Slope} = \frac{\text{Change in } y}{\text{Change in } x}$$
$$= \frac{4 - 0}{0 - (-2)} = \frac{4}{2} = 2$$

We see that the slope of the line is 2, the same as the coefficient of x in the equation $y = 2x + 4$. The y-intercept is 4, the same as the constant in the equation $y = 2x + 4$.

$$y = 2x + 4$$

The slope is 2.   The y-intercept is 4.

$$y = 2x + 4$$

The slope is 2. The y-intercept is 4.

A linear equation like $y = 2x + 4$ that is solved for y is said to be in *slope-intercept form*. This is because the slope and the y-intercept can be immediately determined from the equation. The x-coefficient is the line's slope and the constant term is the y-intercept.

> ### Slope-Intercept Form of the Equation of a Line
>
> The **slope-intercept form of the equation** of a nonvertical line with slope m and y-intercept b is
> $$y = mx + b.$$

EXAMPLE 1 Finding a Line's Slope and y-Intercept from Its Equation

Find the slope and the y-intercept of the line with the given equation:

a. $y = 2x - 4$ **b.** $y = \dfrac{1}{2}x + 2$ **c.** $5x + y = 4$.

Study Tip

The variables in $y = mx + b$ vary in different ways. The variables for slope, m, and y-intercept, b, vary from one line's equation to another. However, they remain constant in the equation of a single line. By contrast, the variables x and y represent the infinitely many points, (x, y), on a single line. Thus, these variables vary in both the equation of a single line, as well as from one equation to another.

Solution

a. We write $y = 2x - 4$ as $y = 2x + (-4)$. The slope is the x-coefficient and the y-intercept is the constant term.

$$y = 2x + (-4)$$

The slope is 2. The y-intercept is –4.

b. The equation $y = \frac{1}{2}x + 2$ is in the form $y = mx + b$. We can find the slope, m, by identifying the coefficient of x. We can find the y-intercept, b, by identifying the constant term.

$$y = \frac{1}{2}x + 2$$

The slope is $\frac{1}{2}$. The y-intercept is 2.

c. The equation $5x + y = 4$ is not in the form $y = mx + b$. We can obtain this form by isolating y on one side. We isolate y on the left side by subtracting $5x$ from both sides.

$$5x + y = 4 \qquad \text{This is the given equation.}$$
$$5x - 5x + y = -5x + 4 \qquad \text{Subtract 5x from both sides.}$$
$$y = -5x + 4 \qquad \text{Simplify.}$$

Now, the equation is in the form $y = mx + b$. The slope is the coefficient of x and the y-intercept is the constant term.

$$y = -5x + 4$$

The slope is –5. The y-intercept is 4.

✓ **CHECK POINT 1** Find the slope and the y-intercept of the line with the given equation:

a. $y = 5x - 3$ **b.** $y = \dfrac{2}{3}x + 4$

c. $7x + y = 6$.

② Graph lines in slope-intercept form.

Graphing *y* = *mx* + *b* by Using the Slope and *y*-Intercept

If a line's equation is written with *y* isolated on one side, we can use the *y*-intercept and the slope to obtain its graph.

> ### Graphing *y* = *mx* + *b* by Using the Slope and *y*-Intercept
> 1. Plot the point containing the *y*-intercept on the *y*-axis. This is the point $(0, b)$.
> 2. Obtain a second point using the slope, *m*. Write *m* as a fraction, and use rise over run, starting at the point containing the *y*-intercept, to plot this point.
> 3. Use a straightedge to draw a line through the two points. Draw arrowheads at the ends of the line to show that the line continues indefinitely in both directions.

EXAMPLE 2   Graphing by Using the Slope and *y*-Intercept

Graph the line whose equation is $y = 4x - 3$.

Solution   We write $y = 4x - 3$ in the form $y = mx + b$.

$$y = 4x + (-3)$$

The slope is 4.   The *y*-intercept is −3.

Now that we have identified the slope and the *y*-intercept, we use the three steps in the box to graph the equation.

Step 1. Plot the point containing the *y*-intercept on the *y*-axis. The *y*-intercept is −3. We plot the point $(0, -3)$, shown in **Figure 4.30(a)**.

Step 2. Obtain a second point using the slope, *m*. Write *m* as a fraction, and use rise over run, starting at the point containing the *y*-intercept, to plot this point. The slope, 4, written as a fraction is $\frac{4}{1}$.

$$m = \frac{4}{1} = \frac{\text{Rise}}{\text{Run}}$$

We plot the second point on the line by starting at $(0, -3)$, the first point. Based on the slope, we move 4 units *up* (the rise) and 1 unit to the *right* (the run). This puts us at a second point on the line, $(1, 1)$, shown in **Figure 4.30(b)**.

Step 3. Use a straightedge to draw a line through the two points. The graph of $y = 4x - 3$ is shown in **Figure 4.30(c)**.

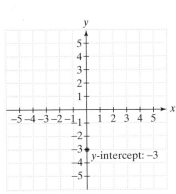

(a) The *y*-intercept is –3, so (0, –3) is a point on the line.

(b) The slope is 4.

(c) The graph of $y = 4x - 3$

FIGURE 4.30 Graphing $y = 4x - 3$ using the *y*-intercept and slope

☑ **CHECK POINT 2**   Graph the line whose equation is $y = 3x - 2$.

EXAMPLE 3 Graphing by Using the Slope and *y*-Intercept

Graph the line whose equation is $y = \frac{2}{3}x + 2$.

Solution The equation of the line is in the form $y = mx + b$. We can find the slope, *m*, by identifying the coefficient of *x*. We can find the *y*-intercept, *b*, by identifying the constant term.

$$y = \frac{2}{3}x + 2$$

The slope is $\frac{2}{3}$.   The *y*-intercept is 2.

Now that we have identified the slope and the *y*-intercept, we use the three-step procedure to graph the equation.

Step 1. Plot the point containing the *y*-intercept on the *y*-axis. The *y*-intercept is 2. We plot $(0, 2)$, shown in **Figure 4.31**.

Step 2. Obtain a second point using the slope, *m*. Write *m* as a fraction, and use rise over run, starting at the point containing the *y*-intercept, to plot this point. The slope, $\frac{2}{3}$, is already written as a fraction.

$$m = \frac{2}{3} = \frac{\text{Rise}}{\text{Run}}$$

We plot the second point on the line by starting at $(0, 2)$, the first point. Based on the slope, we move 2 units *up* (the rise) and 3 units to the *right* (the run). This puts us at a second point on the line, $(3, 4)$, shown in **Figure 4.31**.

Step 3. Use a straightedge to draw a line through the two points. The graph of $y = \frac{2}{3}x + 2$ is shown in **Figure 4.31**. ∎

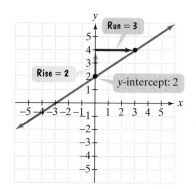

FIGURE 4.31
The graph of $y = \frac{2}{3}x + 2$

✓ **CHECK POINT 3** Graph the line whose equation is $y = \frac{3}{5}x + 1$.

---

3 Use slope and *y*-intercept to graph $Ax + By = C$.

Graphing $Ax + By = C$ by Using the Slope and *y*-Intercept

Earlier in this chapter, we considered linear equations of the form $Ax + By = C$. We used *x*- and *y*-intercepts, as well as checkpoints, to graph these equations. It is also possible to obtain the graphs by using the slope and *y*-intercept. To do this, begin by solving $Ax + By = C$ for *y*. This will put the equation in slope-intercept form. Then use the three-step procedure to graph the equation. This is illustrated in Example 4.

EXAMPLE 4 Graphing by Using the Slope and *y*-Intercept

Graph the linear equation $2x + 5y = 0$ by using the slope and *y*-intercept.

Solution We put the equation in slope-intercept form by solving for *y*.

$$2x + 5y = 0 \qquad \text{This is the given equation.}$$
$$2x - 2x + 5y = -2x + 0 \qquad \text{Subtract 2x from both sides.}$$
$$5y = -2x + 0 \qquad \text{Simplify.}$$
$$\frac{5y}{5} = \frac{-2x + 0}{5} \qquad \text{Divide both sides by 5.}$$
$$y = \frac{-2x}{5} + \frac{0}{5} \qquad \text{Divide each term in the numerator by 5.}$$
$$y = -\frac{2}{5}x + 0 \qquad \text{Simplify.}$$

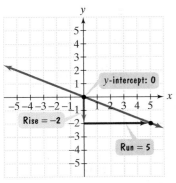

FIGURE 4.32 The graph of $2x + 5y = 0$, or $y = -\frac{2}{5}x + 0$

Now that the equation is in slope-intercept form, we can use the slope and y-intercept to obtain its graph. Examine the slope-intercept form:

$$y = -\frac{2}{5}x + 0.$$

slope: $-\frac{2}{5}$ y-intercept: 0

Note that the slope is $-\frac{2}{5}$ and the y-intercept is 0. Use the y-intercept to plot $(0, 0)$ on the y-axis. Then locate a second point by using the slope.

$$m = -\frac{2}{5} = \frac{-2}{5} = \frac{\text{Rise}}{\text{Run}}$$

Because the rise is -2 and the run is 5, move *down* 2 units and to the *right* 5 units, starting at the point $(0, 0)$. This puts us at a second point on the line, $(5, -2)$. The graph of $2x + 5y = 0$ is the line drawn through these points, shown in **Figure 4.32**. ■

Discover for Yourself

You can obtain a second point in Example 4 by writing the slope as follows:

$$m = \frac{2}{-5} = \frac{\text{Rise}}{\text{Run}}.$$

$-\frac{2}{5}$ can be expressed as $\frac{-2}{5}$ or $\frac{2}{-5}$.

Now obtain this second point in **Figure 4.32** by moving *up* 2 units and to the *left* 5 units, starting at $(0, 0)$. What do you observe once you graph the line?

☑ **CHECK POINT 4**   Graph the linear equation $3x + 4y = 0$ by using the slope and y-intercept.

---

4 Use slope and y-intercept to model data.

Modeling with the Slope-Intercept Form of the Equation of a Line

The slope-intercept form for equations of lines is useful for obtaining mathematical models for data that fall on or near a line. For example, the bar graph in **Figure 4.33(a)** shows the percentage of divorced Americans, by race. The data for African Americans and Hispanics are displayed as points in a rectangular coordinate system in **Figure 4.33(b)**.

Divorced Americans, by Race

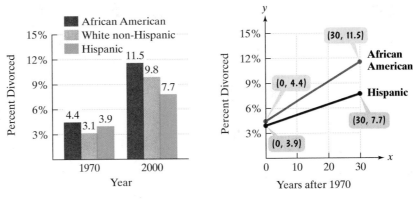

FIGURE 4.33(a)

FIGURE 4.33(b)

Source: U.S. Census Bureau

Example 5 illustrates how we can use the equation $y = mx + b$ to obtain a model for the data and make predictions about what might occur in the future.

Divorced Americans, by Race

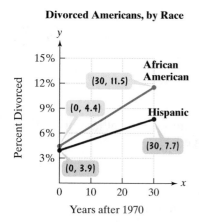

FIGURE 4.33(b) (repeated)

EXAMPLE 5 **Modeling with the Slope-Intercept Form of the Equation**

a. Use the two points for African Americans in **Figure 4.33(b)** to find an equation in the form $y = mx + b$ that models the percentage of divorced African Americans, y, x years after 1970.

b. Use the model to predict the percentage of divorced African Americans in 2010.

Solution

a. We will use the line segment for African Americans using the points $(0, 4.4)$ and $(30, 11.5)$ to obtain a model. We need values for m, the slope, and b, the y-intercept.

$$y = mx + b$$

$$m = \frac{\text{Change in } y}{\text{Change in } x}$$

$$= \frac{11.5 - 4.4}{30 - 0} \approx 0.24$$

The point $(0, 4.4)$ lies on the line segment, so the y-intercept is 4.4: $b = 4.4$.

The percentage of divorced African Americans, y, x years after 1970 can be modeled by the linear equation

$$y = 0.24x + 4.4.$$

The slope, approximately 0.24, indicates an increase in the percentage of divorced African Americans of about 0.24% per year from 1970 through 2000.

b. Now let's use this model to predict the percentage of divorced African Americans in 2010. Because 2010 is 40 years after 1970, substitute 40 for x in $y = 0.24x + 4.4$ and evaluate the formula.

$$y = 0.24(40) + 4.4 = 9.6 + 4.4 = 14$$

Our model projects that 14% of African Americans will be divorced in 2010. ∎

☑ **CHECK POINT 5**

a. Use the two points for Hispanics in **Figure 4.33(b)** to find an equation in the form $y = mx + b$ that models the percentage of divorced Hispanics, y, x years after 1970. Round m to two decimal places.

b. Use the model to predict the percentage of divorced Hispanics in 2010.

4.4 EXERCISE SET **MyMathLab**

PRACTICE **WATCH** **DOWNLOAD** **READ** **REVIEW**

Practice Exercises

In Exercises 1–12, find the slope and the y-intercept of the line with the given equation.

1. $y = 3x + 2$

2. $y = 9x + 4$

3. $y = 3x - 5$

4. $y = 4x - 2$

5. $y = -\frac{1}{2}x + 5$

6. $y = -\frac{3}{4}x + 6$

7. $y = 7x$

8. $y = 10x$

9. $y = 10$

10. $y = 7$

11. $y = 4 - x$

12. $y = 5 - x$

In Exercises 13–26, begin by solving the linear equation for y. This will put the equation in slope-intercept form. Then find the slope and the y-intercept of the line with this equation.

13. $-5x + y = 7$

14. $-9x + y = 5$

15. $x + y = 6$

16. $x + y = 8$

17. $6x + y = 0$

18. $8x + y = 0$

19. $3y = 6x$

20. $3y = -9x$

21. $2x + 7y = 0$

22. $2x + 9y = 0$

23. $3x + 2y = 3$

24. $4x + 3y = 4$

25. $3x - 4y = 12$

26. $5x - 2y = 10$

In Exercises 27–38, graph each linear equation using the slope and y-intercept

27. $y = 2x + 4$

28. $y = 3x + 1$

29. $y = -3x + 5$

30. $y = -2x + 4$

31. $y = \frac{1}{2}x + 1$

32. $y = \frac{1}{3}x + 2$

33. $y = \frac{2}{3}x - 5$

34. $y = \frac{3}{4}x - 4$

35. $y = -\frac{3}{4}x + 2$

36. $y = -\frac{2}{3}x + 4$

37. $y = -\frac{5}{3}x$

38. $y = -\frac{4}{3}x$

In Exercises 39–46,

  a. *Put the equation in slope-intercept form by solving for y.*

  b. *Identify the slope and the y-intercept.*

  c. *Use the slope and y-intercept to graph the equation.*

39. $3x + y = 0$

40. $2x + y = 0$

41. $3y = 4x$

42. $4y = 5x$

43. $2x + y = 3$

44. $3x + y = 4$

45. $7x + 2y = 14$

46. $5x + 3y = 15$

In Exercises 47–56, graph both linear equations in the same rectangular coordinate system. If the lines are parallel or perpendicular, explain why.

47. $y = 3x + 1$
$y = 3x - 3$

48. $y = 2x + 4$
$y = 2x - 3$

49. $y = -3x + 2$
$y = 3x + 2$

50. $y = -2x + 1$
$y = 2x + 1$

51. $y = x + 3$
$y = -x + 1$

52. $y = x + 2$
$y = -x - 1$

53. $x - 2y = 2$
$2x - 4y = 3$

54. $x - 3y = 9$
$3x - 9y = 18$

55. $2x - y = -1$
$x + 2y = -6$

56. $3x - y = -2$
$x + 3y = -9$

Practice PLUS

In Exercises 57–64, write an equation in the form $y = mx + b$ of the line that is described.

57. The y-intercept is 5 and the line is parallel to the line whose equation is $3x + y = 6$.

58. The y-intercept is -4 and the line is parallel to the line whose equation is $2x + y = 8$.

59. The y-intercept is 6 and the line is perpendicular to the line whose equation is $y = 5x - 1$.

60. The y-intercept is 7 and the line is perpendicular to the line whose equation is $y = 8x - 3$.

61. The line has the same y-intercept as the line whose equation is $16y = 8x + 32$ and is parallel to the line whose equation is $3x + 3y = 9$.

62. The line has the same y-intercept as the line whose equation is $2y = 6x + 8$ and is parallel to the line whose equation is $4x + 4y = 20$.

63. The line rises from left to right. It passes through the origin and a second point with equal x- and y-coordinates.

64. The line falls from left to right. It passes through the origin and a second point with opposite x- and y-coordinates.

Application Exercises

The bar graph shows the racial and ethnic composition of the United States population in 2003, with projections for 2050. Exercises 65–66 involve the graphs shown in the rectangular coordinate system for two of the racial/ethnic groups.

Racial and Ethnic Composition of the United States Population

Source: Urban Institute

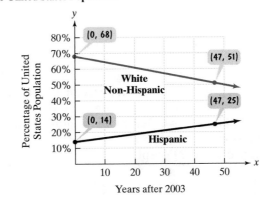

65. a. Use the two points for white non-Hispanics to find an equation in the form $y = mx + b$ that models the percentage of whites, y, in the United States population x years after 2003. Round m to two decimal places.

    b. Use the model from part (a) to project the percentage of whites in the United States in 3003.

66. a. Use the two points for Hispanics to find an equation in the form $y = mx + b$ that models the percentage of Hispanics, y, in the United States population x years after 2003. Round m to two decimal places.

    b. Use the model from part (a) to project the percentage of Hispanics in the United States in 3003.

Writing in Mathematics

67. Describe how to find the slope and the *y*-intercept of a line whose equation is given.

68. Describe how to graph a line using the slope and *y*-intercept. Provide an original example with your description.

69. A formula in the form $y = mx + b$ models the cost, *y*, of a four-year college *x* years after 2007. Would you expect *m* to be positive, negative, or zero? Explain your answer.

Critical Thinking Exercises

Make Sense?   *In Exercises 70–73, determine whether each statement "makes sense" or "does not make sense" and explain your reasoning.*

70. The slope-intercept form of a line's equation makes it possible for me to determine immediately the slope and the *y*-intercept.

71. Because the variable *m* does not appear in $Ax + By = C$, equations in this form make it impossible to determine the line's slope.

72. If I drive *m* miles in a year, the formula $c = 0.25m + 3500$ models the annual cost, *c*, in dollars, of operating my car, so the equation shows that with no driving at all, the cost is $3500, and the rate of increase in this cost is $0.25 for each mile that I drive.

73. Some lines are impossible to graph because when obtaining a second point using the slope, this point lands outside the edge of my graph paper.

In Exercises 74–77, determine whether each statement is true or false. If the statement is false, make the necessary change(s) to produce a true statement.

74. The equation $y = mx + b$ shows that no line can have a *y*-intercept that is numerically equal to its slope.

75. Every line in the rectangular coordinate system has an equation that can be expressed in slope-intercept form.

76. The line $3x + 2y = 5$ has slope $-\frac{3}{2}$.

77. The line $2y = 3x + 7$ has a *y*-intercept of 7.

78. The relationship between Celsius temperature, *C*, and Fahrenheit temperature, *F*, can be described by a linear equation in the form $F = mC + b$. The graph of this equation contains the point $(0, 32)$: Water freezes at 0°C or at 32°F. The line also contains the point $(100, 212)$: Water boils at 100°C or at 212°F. Write the linear equation expressing Fahrenheit temperature in terms of Celsius temperature.

Review Exercises

79. Solve: $\frac{x}{2} + 7 = 13 - \frac{x}{4}$. (Section 2.3, Example 4)

80. Simplify: $3(12 \div 2^2 - 3)^2$. (Section 1.8, Example 6)

81. 14 is 25% of what number? (Section 2.4, Example 8)

Preview Exercises

Exercises 82–84 will help you prepare for the material covered in the next section. In each exercise, solve for y and put the equation in slope-intercept form.

82. $y - 3 = 4(x + 1)$

83. $y + 3 = -\frac{3}{2}(x - 4)$

84. $y - 30.0 = 0.265(x - 10)$

MID-CHAPTER CHECK POINT   Section 4.1–Section 4.4

✓ **What You Know:** We learned to graph equations in two variables using point plotting, as well as a variety of other techniques. We used intercepts and a checkpoint to graph linear equations in the form $Ax + By = C$. We saw that $y = b$ graphs as a horizontal line and $x = a$ graphs as a vertical line. We determined a line's steepness, or rate of change, by computing its slope. We saw that lines with the same slope are parallel and lines with slopes that have a product of -1 are perpendicular. Finally, we learned to graph linear equations in slope-intercept form, $y = mx + b$, using the slope, *m*, and the *y*-intercept, *b*.

In Exercises 1–3, use each graph to determine

   a. *the x-intercept, or state that there is no x-intercept.*

   b. *the y-intercept, or state that there is no y-intercept.*

   c. *the line's slope, or state that the slope is undefined.*

1.

2.

3.

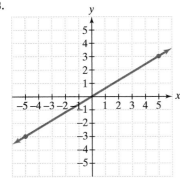

In Exercises 4–15, graph each equation in a rectangular coordinate system.

4. $y = -2x$ **5.** $y = -2$ **6.** $x + y = -2$

7. $y = \frac{1}{3}x - 2$ **8.** $x = 3.5$ **9.** $4x - 2y = 8$

10. $y = 3x + 2$ **11.** $3x + y = 0$ **12.** $y = -x + 4$

13. $y = x - 4$ **14.** $5y = -3x$ **15.** $5y = 20$

16. Find the slope and the y-intercept of the line whose equation is $5x - 2y = 10$.

In Exercises 17–19, determine whether the lines through each pair of points are parallel, perpendicular, or neither.

17. $(-5, -3)$ and $(0, -4)$; $(-2, -8)$ and $(1, 7)$

18. $(-4, 1)$ and $(2, 7)$; $(-5, 13)$ and $(4, -5)$

19. $(2, -4)$ and $(7, 0)$; $(-4, 2)$ and $(1, 6)$

20. Higher prices at the gas pump are eating up hefty portions of American paychecks. The graphs indicate that the gas pump gulped more family pay in 2006 than in 1997.

Average Annual Household Gasoline Bill in the United States

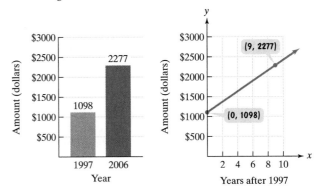

Source: Consumer Federation of America, Consumers Union

a. Use the two points labeled by the voice balloons to find an equation in the form $y = mx + b$ that models the average annual household gasoline bill, y, in dollars, x years after 1997.

b. Describe the meaning of the slope in the model in part (a) in terms of a rate of change.

c. If present trends continue, use the model in part (a) to project the average annual household gasoline bill in 2013.

SECTION 4.5

The Point-Slope Form of the Equation of a Line

Objectives

1 Use the point-slope form to write equations of a line.

2 Write linear equations that model data and make predictions.

Surprised by the number of people smoking cigarettes in movies and television shows made in the 1940s and 1950s? At that time, there was little awareness of the relationship between tobacco use and numerous diseases. Cigarette smoking was seen as a healthy way to relax and help digest a hearty meal. Then, in 1964, a linear equation changed everything. To understand the mathematics behind this turning point in public health, we explore another form of a line's equation.

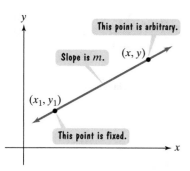

FIGURE 4.34 A line passing through (x_1, y_1) with slope m

Point-Slope Form

We can use the slope of a line to obtain another useful form of the line's equation. Consider a nonvertical line that has slope m and contains the point (x_1, y_1). Now let (x, y) represent any other point on the line, shown in **Figure 4.34**. Keep in mind that the point (x, y) is arbitrary and is not in one fixed position. By contrast, the point (x_1, y_1) is fixed.

Regardless of where the point (x, y) is located on the line, the steepness of the line in **Figure 4.34** remains the same. Thus, the ratio for slope stays a constant m. This means that for all points (x, y) along the line,

$$m = \frac{\text{Change in } y}{\text{Change in } x} = \frac{y - y_1}{x - x_1}.$$

We can clear the fraction by multiplying both sides by $x - x_1$, the least common denominator, where $x - x_1 \neq 0$.

$$m = \frac{y - y_1}{x - x_1} \qquad \text{This is the slope of the line in Figure 4.34.}$$

$$m(x - x_1) = \frac{y - y_1}{x - x_1} \cdot (x - x_1) \qquad \text{Multiply both sides by } x - x_1.$$

$$m(x - x_1) = y - y_1 \qquad \text{Simplify: } \frac{y - y_1}{x - x_1} \cdot (x - x_1) = y - y_1.$$

Now, if we reverse the two sides, we obtain the **point-slope form** of the equation of a line.

> ### Point-Slope Form of the Equation of a Line
>
> The **point-slope form of the equation** of a nonvertical line with slope m that passes through the point (x_1, y_1) is
> $$y - y_1 = m(x - x_1).$$

For example, the point-slope form of the equation of the line passing through $(1, 4)$ with slope 2 ($m = 2$) is

$$y - 4 = 2(x - 1).$$

1 Use the point-slope form to write equations of a line.

Using the Point-Slope Form to Write a Line's Equation

If we know the slope of a line and a point not containing the y-intercept through which the line passes, the point-slope form is the equation that we should use. Once we have obtained this equation, it is customary to solve for y and write the equation in slope-intercept form. Examples 1 and 2 illustrate these ideas.

EXAMPLE 1 Writing the Point-Slope Form and the Slope-Intercept Form

Write the point-slope form and the slope-intercept form of the equation of the line with slope 4 that passes through the point $(-1, 3)$.

Solution We begin with the point-slope form of the equation of a line with $m = 4$, $x_1 = -1$, and $y_1 = 3$.

$$y - y_1 = m(x - x_1) \qquad \text{This is the point-slope form of the equation.}$$

$$y - 3 = 4[x - (-1)] \qquad \text{Substitute: } (x_1, y_1) = (-1, 3) \text{ and } m = 4.$$

$$y - 3 = 4(x + 1) \qquad \text{We now have the point-slope form of the equation of the given line.}$$

Now we solve this equation for y and write an equivalent equation in slope-intercept form ($y = mx + b$).

| We need to isolate y. | $y - 3 = 4(x + 1)$ | This is the point-slope form of the equation. |

$$y - 3 = 4x + 4 \quad \text{Use the distributive property.}$$
$$y = 4x + 7 \quad \text{Add 3 to both sides.}$$

The slope-intercept form of the line's equation is $y = 4x + 7$.

✓ **CHECK POINT 1** Write the point-slope form and the slope-intercept form of the equation of the line with slope 6 that passes through the point $(2, -5)$.

EXAMPLE 2 Writing the Point-Slope Form and the Slope-Intercept Form

A line passes through the points $(4, -3)$ and $(-2, 6)$. (See **Figure 4.35**.) Find the equation of the line

   a. in point-slope form.   **b.** in slope-intercept form.

Solution

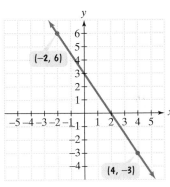

FIGURE 4.35

   a. To use the point-slope form, we need to find the slope. The slope is the change in the y-coordinates divided by the corresponding change in the x-coordinates.

$$m = \frac{6 - (-3)}{-2 - 4} = \frac{9}{-6} = -\frac{3}{2} \quad \begin{array}{l} \text{This is the definition of slope} \\ \text{using } (4, -3) \text{ and } (-2, 6). \end{array}$$

We can take either point on the line to be (x_1, y_1). Let's use $(x_1, y_1) = (4, -3)$. Now, we are ready to write the point-slope form of the equation.

$$y - y_1 = m(x - x_1) \quad \text{This is the point-slope form of the equation.}$$
$$y - (-3) = -\frac{3}{2}(x - 4) \quad \text{Substitute: } (x_1, y_1) = (4, -3) \text{ and } m = -\frac{3}{2}.$$
$$y + 3 = -\frac{3}{2}(x - 4) \quad \text{Simplify.}$$

This equation is one point-slope form of the equation of the line shown in **Figure 4.35**.

   b. Now, we solve this equation for y and write an equivalent equation in slope-intercept form ($y = mx + b$).

| We need to isolate y. | $y + 3 = -\frac{3}{2}(x - 4)$ | This is the point-slope form of the equation. |

$$y + 3 = -\frac{3}{2}x + 6 \quad \text{Use the distributive property.}$$
$$y = -\frac{3}{2}x + 3 \quad \text{Subtract 3 from both sides.}$$

This equation is the slope-intercept form of the equation of the line shown in **Figure 4.35**.

✓ **CHECK POINT 2** A line passes through the points $(-2, -1)$ and $(-1, -6)$. Find the equation of the line

   a. in point-slope form.

   b. in slope-intercept form.

Discover for Yourself

If you are given two points on a line, you can use either point for (x_1, y_1) when you write its point-slope equation. Rework Example 2 using $(-2, 6)$ for (x_1, y_1). Once you solve for y, you should obtain the same slope-intercept equation as the one shown in the last line of the solution to Example 2.

The many forms of a line's equation can be a bit overwhelming. **Table 4.2** summarizes the various forms and contains the most important things you should remember about each form.

Table 4.2 **Equations of Lines**

Form	What You Should Know
Standard Form $Ax + By = C$	• Graph equations in this form using intercepts (x-intercept: set $y = 0$; y-intercept: set $x = 0$) and a checkpoint.
$y = b$	• Graph equations in this form as horizontal lines with b as the y-intercept.
$x = a$	• Graph equations in this form as vertical lines with a as the x-intercept.
Slope-Intercept Form $y = mx + b$	• Graph equations in this form using the y-intercept, b, and the slope, m. • Start with this form when writing a linear equation if you know a line's slope and y-intercept.
Point-Slope Form $y - y_1 = m(x - x_1)$	• Start with this form when writing a linear equation if you know the slope of the line and a point on the line not containing the y-intercept or two points on the line, neither of which contains the y-intercept. Calculate the slope using $$m = \frac{\text{Change in } y}{\text{Change in } x} = \frac{y_2 - y_1}{x_2 - x_1}.$$ Although you begin with point-slope form, you usually solve for y and convert to slope-intercept form.

2 Write linear equations that model data and make predictions.

Applications

Linear equations are useful for modeling data that fall on or near a line. For example, the bar graph in **Figure 4.36(a)** gives the median age of the U.S. population in the indicated year. (The median age is the age in the middle when all the ages of the U.S. population are arranged from youngest to oldest.) The data are displayed as a set of five points in a rectangular coordinate system in **Figure 4.36(b)**.

The Graying of America: Median Age of the United States Population

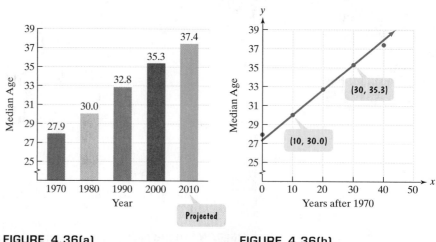

FIGURE 4.36(a)

Source: U.S. Census Bureau

FIGURE 4.36(b)

A set of points representing data is called a **scatter plot**. Also shown on the scatter plot in **Figure 4.36(b)** is a line that passes through or near the five points. By writing the equation of this line, we can obtain a model of the data and make predictions about the median age of the U.S. population in the future.

Using Technology

You can use a graphing utility to obtain a model for a scatter plot in which the data points fall on or near a straight line. The line that best fits the data is called the **regression line**. After entering the data in **Figure 4.36(b)**, a graphing utility displays a scatter plot of the data and the regression line.

[−10, 40, 5] by [25, 39, 2]

Also displayed is the regression line's equation.

LinReg
y=ax+b
a=.243
b=27.82

EXAMPLE 3   Modeling the Graying of America

Write the slope-intercept form of the equation of the line shown in **Figure 4.36(b)**. Use the equation to predict the median age of the U.S. population in 2020.

Solution   The line in **Figure 4.36(b)** passes through $(10, 30.0)$ and $(30, 35.3)$. We start by finding its slope.

$$m = \frac{\text{Change in } y}{\text{Change in } x} = \frac{35.3 - 30.0}{30 - 10} = \frac{5.3}{20} = 0.265$$

The slope indicates that each year the median age of the U.S. population is increasing by 0.265 year.

Now, we write the line's slope-intercept form of the equation.

$y - y_1 = m(x - x_1)$ Begin with the point-slope form.

$y - 30.0 = 0.265(x - 10)$ Either ordered pair can be (x_1, y_1). Let $(x_1, y_1) = (10, 30.0)$. From above, $m = 0.265$.

$y - 30.0 = 0.265x - 2.65$ Apply the distributive property.

$y = 0.265x + 27.35$ Add 30 to both sides and solve for y.

A linear equation that models the median age of the U.S. population, y, x years after 1970 is

$$y = 0.265x + 27.35.$$

Now, let's use this equation to predict the median age in 2020. Because 2020 is 50 years after 1970, substitute 50 for x and compute y.

$$y = 0.265(50) + 27.35 = 40.6$$

Our model predicts that the median age of the U.S. population in 2020 will be 40.6.   ■

☑ **CHECK POINT 3**   Use the data points $(10, 30.0)$ and $(20, 32.8)$ from **Figure 4.36(b)** to write the slope-intercept form of an equation that models the median age of the U.S. population x years after 1970. Use this model to predict the median age in 2020.

Blitzer Bonus

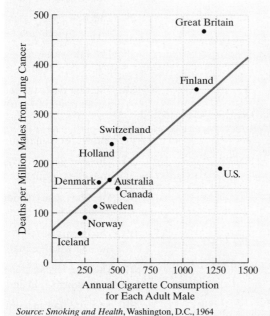

Source: Smoking and Health, Washington, D.C., 1964

Cigarettes and Lung Cancer

This scatter plot shows a relationship between cigarette consumption among males and deaths due to lung cancer per million males. The data are from 11 countries and date back to a 1964 report by the U.S. Surgeon General. The scatter plot can be modeled by a line whose slope indicates an increasing death rate from lung cancer with increased cigarette consumption. At that time, the tobacco industry argued that in spite of this regression line, tobacco use is not the cause of cancer. Recent data do, indeed, show a causal effect between tobacco use and numerous diseases.

4.5 EXERCISE SET   *MyMathLab*

Practice Exercises

Write the point-slope form of the equation of the line satisfying each of the conditions in Exercises 1–28. Then use the point-slope form of the equation to write the slope-intercept form of the equation.

1. Slope = 3, passing through $(2, 5)$

2. Slope = 6, passing through $(3, 1)$

3. Slope = 5, passing through $(-2, 6)$

4. Slope = 7, passing through $(-4, 9)$

5. Slope = -8, passing through $(-3, -2)$

6. Slope = -4, passing through $(-5, -2)$

7. Slope = -12, passing through $(-8, 0)$

8. Slope = -11, passing through $(0, -3)$

9. Slope = -1, passing through $\left(-\frac{1}{2}, -2\right)$

10. Slope = -1, passing through $\left(-4, -\frac{1}{4}\right)$

11. Slope = $\frac{1}{2}$, passing through the origin

12. Slope = $\frac{1}{3}$, passing through the origin

13. Slope = $-\frac{2}{3}$, passing through $(6, -2)$

14. Slope = $-\frac{3}{5}$, passing through $(10, -4)$

15. Passing through $(1, 2)$ and $(5, 10)$

16. Passing through $(3, 5)$ and $(8, 15)$

17. Passing through $(-3, 0)$ and $(0, 3)$

18. Passing through $(-2, 0)$ and $(0, 2)$

19. Passing through $(-3, -1)$ and $(2, 4)$

20. Passing through $(-2, -4)$ and $(1, -1)$

21. Passing through $(-4, -1)$ and $(3, 4)$

22. Passing through $(-6, 1)$ and $(2, -5)$

23. Passing through $(-3, -1)$ and $(4, -1)$

24. Passing through $(-2, -5)$ and $(6, -5)$

25. Passing through $(2, 4)$ with x-intercept $= -2$

26. Passing through $(1, -3)$ with x-intercept $= -1$

27. x-intercept $= -\frac{1}{2}$ and y-intercept $= 4$

28. x-intercept $= 4$ and y-intercept $= -2$

Practice PLUS

In Exercises 29–38, write an equation in slope-intercept form of the line satisfying the given conditions.

29. The line passes through $(-3, 2)$ and is parallel to the line whose equation is $y = 4x + 1$.

30. The line passes through $(5, -3)$ and is parallel to the line whose equation is $y = 2x + 1$.

31. The line passes through $(-1, -5)$ and is parallel to the line whose equation is $3x + y = 6$.

32. The line passes through $(-4, -7)$ and is parallel to the line whose equation is $6x + y = 8$.

33. The line passes through $(4, -7)$ and is perpendicular to the line whose equation is $x - 2y = 3$.

34. The line passes through $(5, -9)$ and is perpendicular to the line whose equation is $x + 7y = 12$.

35. The line passes through $(2, 4)$ and has the same y-intercept as the line whose equation is $x - 4y = 8$.

36. The line passes through $(2, 6)$ and has the same y-intercept as the line whose equation is $x - 3y = 18$.

37. The line has an x-intercept at -4 and is parallel to the line containing $(3, 1)$ and $(2, 6)$.

38. The line has an x-intercept at -6 and is parallel to the line containing $(4, -3)$ and $(2, 2)$.

Application Exercises

The wage gap is used to compare the status of women's earnings relative to men's. The wage gap is expressed as a percent and is calculated by dividing the median, or middlemost, annual earnings for women by the median annual earnings for men. The bar graph shows the wage gap for selected years from 1980 through 2004. In Exercises 39–40, you will use a scatter plot to develop models for the data.

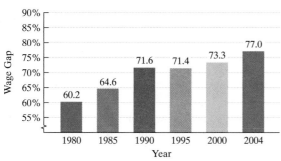

Median Women's Earnings as a Percentage of Median Men's Earnings in the United States

Source: U. S. Women's Bureau

39. The data for the wage gap are displayed as six points in the following scatter plot.

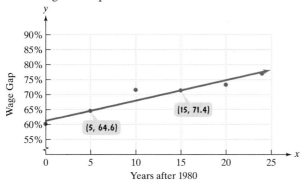

a. Use the two points whose coordinates are shown by the voice balloons to write the slope-intercept form of an equation that models the wage gap, *y*, *x* years after 1980.

b. Use the model from part (a) to predict the wage gap (that is, women's earnings as a percentage of men's) in 2015.

40. The data for the wage gap are displayed as six points in the following scatter plot.

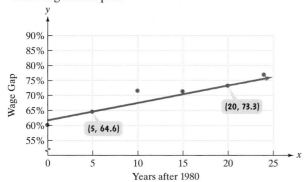

a. Use the two points whose coordinates are shown by the voice balloons to write the slope-intercept form of an equation that models the wage gap, *y*, *x* years after 1980.

b. Use the model from part (a) to predict the wage gap (that is, women's earnings as a percentage of men's) in 2015.

41. It appears that the young see flashing lights the most. The bar graph shows the percentage of drivers at various ages who were stopped by police in 2005.

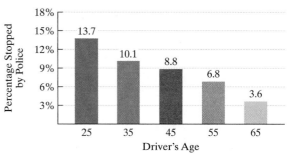

Percentage of United States Drivers of Various Ages Stopped by Police

Source: Bureau of Justice Statistics

a. Let *x* represent a driver's age and let *y* represent the percentage of drivers of that age stopped by police. Create a scatter plot that displays the data as a set of five points in a rectangular coordinate system.

b. Draw a line through the two points that show the data for 45-year-old and 55-year-old drivers. Use the coordinates of these points to write the slope-intercept form of an equation that models the percentage of drivers, *y*, at age *x* who are stopped by police.

c. Does your model overestimate or underestimate the percentage of 35-year-old drivers who are stopped by police? By how much?

42. Is there a relationship between education and prejudice? With increased education, does a person's level of prejudice tend to decrease? The scatter plot shows ten data points, each representing the number of years of school completed and the score on a test measuring prejudice for each subject. Higher scores on this 1-to-10 test indicate greater prejudice. Also shown is the regression line, the line that best fits the data. Use two points on this line to write its equation both in point-slope form and slope-intercept form. Then use the slope-intercept form of the equation to predict the score on the prejudice test for a person with seven years of education.

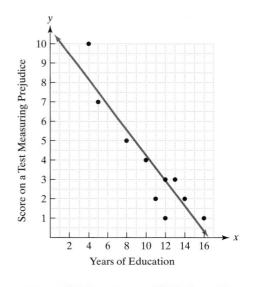

Writing in Mathematics

43. Describe how to write the equation of a line if its slope and a point on the line are known.

44. Describe how to write the equation of a line if two points on the line are known.

45. Take a second look at the scatter plot in Exercise 42. Although there is a relationship between education and prejudice, we cannot necessarily conclude that increased education causes a person's level of prejudice to decrease. Offer two or more possible explanations for the data in the scatter plot.

Critical Thinking Exercises

Make Sense?   *In Exercises 46–49, determine whether each statement "makes sense" or "does not make sense" and explain your reasoning.*

46. I use $y = mx + b$ to write equations of lines passing through two points when neither contains the y-intercept.

47. In many examples, I use the slope-intercept form of a line's equation to obtain an equivalent equation in point-slope form.

48. When I read the derivation of the point-slope form of a line's equation, I cleared the fraction in $m = \dfrac{y - y_1}{x - x_1}$ by rewriting the equation as $\dfrac{m}{1} = \dfrac{y - y_1}{x - x_1}$ and then using the cross-products principle for proportions.

49. The shape of this scatter plot showing per capita U.S. adult wine consumption suggests that I should not use a line or any of its equations to obtain a model.

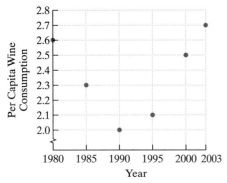

Wine Consumption per United States Adult

Source: Adams Business Media

In Exercises 50–53, determine whether each statement is true or false. If the statement is false, make the necessary change(s) to produce a true statement.

50. If a line has undefined slope, then it has no equation.

51. The line whose equation is $y - 3 = 7(x + 2)$ passes through $(-3, 2)$.

52. The point-slope form can be applied to obtain the equation of the line through the points $(2, -5)$ and $(2, 6)$.

53. The slope of the line whose equation is $3x + y = 7$ is 3.

54. Excited about the success of celebrity stamps, post office officials were rumored to have put forth a plan to institute two new types of thermometers. On these new scales, $°E$ represents degrees Elvis and $°M$ represents degrees Madonna. If it is known that $40°E = 25°M$, $280°E = 125°M$, and degrees Elvis is linearly related to degrees Madonna, write an equation expressing E in terms of M.

Technology Exercises

55. Use a graphing utility to graph $y = 1.75x - 2$. Select the best viewing rectangle possible by experimenting with the range settings to show that the line's slope is $\frac{7}{4}$.

56. a. Use the statistical menu of a graphing utility to enter the six data points shown in the scatter plot in Exercise 39. Refer to the bar graph to obtain coordinates of points that are not given in the scatter plot.

  b. Use the DRAW menu and the scatter plot capability to draw a scatter plot of the data points like the one shown in Exercise 39.

  c. Select the linear regression option. Use your utility to obtain values for a and b for the equation of the regression line, $y = ax + b$. You may also be given a *correlation coefficient, r*. Values of r close to 1 indicate that the points can be described by a linear relationship and the regression line has a positive slope. Values of r close to -1 indicate that the points can be described by a linear relationship and the regression line has a negative slope. Values of r close to 0 indicate no linear relationship between the variables.

  d. Use the appropriate sequence (consult your manual) to graph the regression equation on top of the points in the scatter plot.

Review Exercises

57. How many sheets of paper, weighing 2 grams each, can be put in an envelope weighing 4 grams if the total weight must not exceed 29 grams? (Section 2.6, Example 11)

58. List all the natural numbers in this set:

$$\left\{ -2, 0, \frac{1}{2}, 1, \sqrt{3}, \sqrt{4} \right\}.$$

(Section 1.3, Example 5)

59. Use intercepts to graph $3x - 5y = 15$. (Section 4.2, Example 4)

Preview Exercises

Exercises 60–62 will help you prepare for the material covered in the next section.

60. Is $2x - 3y \geq 6$ a true statement for $x = 3$ and $y = -1$?

61. Is $2x - 3y \geq 6$ a true statement for $x = 0$ and $y = 0$?

62. Is $y \leq \frac{2}{3}x$ a true statement for $x = 1$ and $y = 1$?

SECTION
4.6

Linear Inequalities In Two Variables

Objectives

1. Determine whether an ordered pair is a solution of an inequality.

2. Graph a linear inequality in two variables.

3. Solve applied problems involving linear inequalities in two variables.

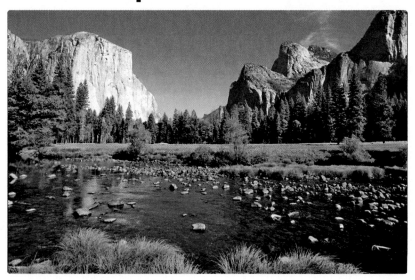

Temperature and precipitation are two variables that affect whether regions are forests, grasslands, or deserts. In this section, you will see how linear inequalities in two variables describe some of the most magnificent places in our nation's landscape.

Linear Inequalities in Two Variables and Their Solutions

1 Determine whether an ordered pair is a solution of an inequality.

We have seen that equations in the form $Ax + By = C$ are straight lines when graphed. If we change the $=$ sign to $>$, $<$, $\geq$, or $\leq$, we obtain a **linear inequality in two variables**. Some examples of linear inequalities in two variables are $x + y > 2$, $3x - 5y \leq 15$, and $2x - y < 4$.

A **solution of an inequality in two variables**, x and y, is an ordered pair of real numbers with the following property: When the x-coordinate is substituted for x and the y-coordinate is substituted for y in the inequality, we obtain a true statement. For example, $(3, 2)$ is a solution of the inequality $x + y > 1$. When 3 is substituted for x and 2 is substituted for y, we obtain the true statement $3 + 2 > 1$, or $5 > 1$. Because there are infinitely many pairs of numbers that have a sum greater than 1, the inequality $x + y > 1$ has infinitely many solutions. Each ordered-pair solution is said to **satisfy** the inequality. Thus, $(3, 2)$ satisfies the inequality $x + y > 1$.

EXAMPLE 1 Deciding Whether an Ordered Pair Satisfies an Inequality

Determine whether each ordered pair is a solution of the inequality
$$2x - 3y \geq 6:$$
a. $(0, 0)$ **b.** $(3, -1)$.

Solution

a. To determine whether $(0, 0)$ is a solution of the inequality, we substitute 0 for x and 0 for y.

$$2x - 3y \geq 6 \quad \text{This is the given inequality.}$$
$$2 \cdot 0 - 3 \cdot 0 \overset{?}{\geq} 6 \quad \text{Substitute 0 for x and 0 for y.}$$
$$0 - 0 \overset{?}{\geq} 6 \quad \text{Multiply: } 2 \cdot 0 = 0 \text{ and } 3 \cdot 0 = 0.$$

This statement is false. $\quad 0 \geq 6 \quad \text{Subtract: } 0 - 0 = 0.$

Because we obtain a false statement, we conclude that $(0, 0)$ is not a solution of $2x - 3y \geq 6$. The ordered pair $(0, 0)$ does not satisfy the inequality.

b. To determine whether $(3, -1)$ is a solution of the inequality, we substitute 3 for x and -1 for y.

$$2x - 3y \geq 6 \quad \text{This is the given inequality.}$$

$$2 \cdot 3 - 3(-1) \stackrel{?}{\geq} 6 \quad \text{Substitute 3 for } x \text{ and } -1 \text{ for } y.$$

$$6 - (-3) \stackrel{?}{\geq} 6 \quad \text{Multiply: } 2 \cdot 3 = 6 \text{ and } 3(-1) = -3.$$

> This statement is true.
$$9 \geq 6 \quad \text{Subtract: } 6 - (-3) = 6 + 3 = 9.$$

Because we obtain a true statement, we conclude that $(3, -1)$ is a solution of $2x - 3y \geq 6$. The ordered pair $(3, -1)$ satisfies the inequality. ▬

☑ **CHECK POINT 1** Determine whether each ordered pair is a solution of the inequality $5x + 4y \leq 20$:

a. $(0, 0)$ **b.** $(6, 2)$.

---

2 Graph a linear inequality in two variables.

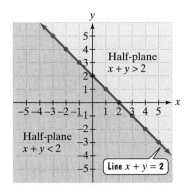

FIGURE 4.37

The Graph of a Linear Inequality in Two Variables

We know that the graph of an equation in two variables is the set of all points whose coordinates satisfy the equation. Similarly, the **graph of an inequality in two variables** is the set of all points whose coordinates satisfy the inequality.

Let's use **Figure 4.37** to get an idea of what the graph of a linear inequality in two variables looks like. Part of the figure shows the graph of the linear equation $x + y = 2$. The line divides the points in the rectangular coordinate system into three sets. First, there is the set of points along the line, satisfying $x + y = 2$. Next, there is the set of points in the green region above the line. Points in the green region satisfy the linear inequality $x + y > 2$. Finally, there is the set of points in the purple region below the line. Points in the purple region satisfy the linear inequality $x + y < 2$.

A **half-plane** is the set of all the points on one side of a line. In **Figure 4.37**, the green region is a half-plane. The purple region is also a half-plane. A half-plane is the graph of a linear inequality that involves $>$ or $<$. The graph of an inequality that involves $\geq$ or $\leq$ is a half-plane and a line. A solid line is used to show that the line is part of the graph. A dashed line is used to show that a line is not part of a graph.

Graphing a Linear Inequality in Two Variables

1. Replace the inequality symbol with an equal sign and graph the corresponding linear equation. Draw a solid line if the original inequality contains a $\leq$ or $\geq$ symbol. Draw a dashed line if the original inequality contains a $<$ or $>$ symbol.
2. Choose a test point from one of the half-planes. (Do not choose a point that is on the line.) Substitute the coordinates of the test point into the inequality.
3. If a true statement results, shade the half-plane containing this test point. If a false statement results, shade the half-plane not containing this test point.

EXAMPLE 2 Graphing a Linear Inequality in Two Variables

Graph: $3x - 5y < 15$.

Solution

Step 1. Replace the inequality symbol with = and graph the linear equation. We need to graph $3x - 5y = 15$. We can use intercepts to graph this line.

We set $y = 0$ to find the x-intercept:	We set $x = 0$ to find the y-intercept:
$3x - 5y = 15$	$3x - 5y = 15$
$3x - 5 \cdot 0 = 15$	$3 \cdot 0 - 5y = 15$
$3x = 15$	$-5y = 15$
$x = 5.$	$y = -3.$

The x-intercept is 5, so the line passes through $(5, 0)$. The y-intercept is -3, so the line passes through $(0, -3)$. The graph of the equation is indicated by a dashed line because the inequality $3x - 5y < 15$ contains a $<$ symbol, rather than $\le$. The graph of the line is shown in **Figure 4.38**.

Step 2. Choose a test point from one of the half-planes and not from the line. Substitute its coordinates into the inequality. The line $3x - 5y = 15$ divides the plane into three parts—the line itself and two half-planes. The points in one half-plane satisfy $3x - 5y > 15$. The points in the other half-plane satisfy $3x - 5y < 15$. We need to find which half-plane is the solution. To do so, we test a point from either half-plane. The origin, $(0, 0)$, is the easiest point to test.

$$3x - 5y < 15 \quad \text{\small This is the given inequality.}$$
$$3 \cdot 0 - 5 \cdot 0 \overset{?}{<} 15 \quad \text{\small Test (0, 0) by substituting 0 for x and 0 for y.}$$
$$0 - 0 \overset{?}{<} 15 \quad \text{\small Multiply.}$$
$$0 < 15 \quad \text{\small This statement is true.}$$

Step 3. If a true statement results, shade the half-plane containing the test point. Because 0 is less than 15, the test point $(0, 0)$ is part of the solution set. All the points on the same side of the line $3x - 5y = 15$ as the point $(0, 0)$ are members of the solution set. The solution set is the half-plane that contains the point $(0, 0)$, indicated by shading this half-plane. The graph is shown using green shading and a dashed blue line in **Figure 4.39**.

FIGURE 4.38 Preparing to graph $3x - 5y < 15$

FIGURE 4.39
The graph of $3x - 5y < 15$

✓ **CHECK POINT 2** Graph: $2x - 4y < 8$.

When graphing a linear inequality, test a point that lies in one of the half-planes and *not on the line dividing the half-planes*. The test point $(0, 0)$ is convenient because it is easy to calculate when 0 is substituted for each variable. However, if $(0, 0)$ lies on the dividing line and not in a half-plane, a different test point must be selected.

EXAMPLE 3 Graphing a Linear Inequality

Graph: $y \le \dfrac{2}{3}x$.

Solution

Step 1. Replace the inequality symbol with = and graph the linear equation. Because we are interested in graphing $y \le \frac{2}{3}x$, we begin by graphing $y = \frac{2}{3}x$. We can use the slope and the y-intercept to graph this line.

$$y = \frac{2}{3}x + 0$$

slope $= \frac{2}{3} = \frac{\text{rise}}{\text{run}}$ y-intercept is 0.

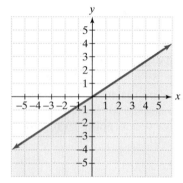

FIGURE 4.40
Preparing to graph $y \le \frac{2}{3}x$

The y-intercept of $y = \frac{2}{3}x$ is 0, so the line passes through (0, 0). Using the y-intercept and the slope, $\frac{2}{3}$, the line is shown in **Figure 4.40** as a solid line. This is because the inequality $y \le \frac{2}{3}x$ contains a $\le$ symbol, in which equality is included.

Step 2. Choose a test point from one of the half-planes and not from the line. Substitute its coordinates into the inequality.   We cannot use (0, 0) as a test point because it lies on the line and not in a half-plane. Let's use (1, 1), which lies in the half-plane above the line.

$$y \le \frac{2}{3}x \qquad \text{This is the given inequality.}$$

$$1 \overset{?}{\le} \frac{2}{3} \cdot 1 \qquad \text{Test (1, 1) by substituting 1 for } x \text{ and 1 for } y.$$

$$1 \le \frac{2}{3} \qquad \text{This statement is false.}$$

Step 3. If a false statement results, shade the half-plane not containing the test point.   Because 1 is not less than or equal to $\frac{2}{3}$, the test point (1, 1) is not part of the solution set. Thus, the half-plane below the solid line $y = \frac{2}{3}x$ is part of the solution set. The solution set is the line and the half-plane that does not contain the point (1, 1), indicated by shading this half-plane. The graph is shown using green shading and a blue line in **Figure 4.41**. ∎

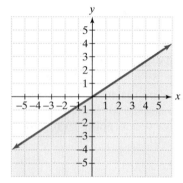

FIGURE 4.41
The graph of $y \le \frac{2}{3}x$

☑ **CHECK POINT 3**   Graph: $y \ge \frac{1}{2}x$.

Graphing Linear Inequalities without Using Test Points

You can graph inequalities in the form $y > mx + b$ or $y < mx + b$ without using test points. The inequality symbol indicates which half-plane to shade.

- If $y > mx + b$, shade the half-plane above the line $y = mx + b$.
- If $y < mx + b$, shade the half-plane below the line $y = mx + b$.

Observe how this is illustrated in **Figure 4.41**. The graph of $y \le \frac{2}{3}x$ contains the half-plane below the line $y = \frac{2}{3}x$.

It is also not necessary to use test points when graphing inequalities involving half-planes on one side of a vertical or a horizontal line.

Study Tip

Continue using test points to graph inequalities in the form $Ax + By > C$ or $Ax + By < C$. The graph of $Ax + By > C$ can lie above or below the line of $Ax + By = C$, depending on the value of B. The same comment applies to the graph of $Ax + By < C$.

For the Vertical Line $x = a$:

- If $x > a$, shade the half-plane to the right of $x = a$.

- If $x < a$, shade the half-plane to the left of $x = a$.

For the Horizontal Line $y = b$:

- If $y > b$, shade the half-plane above $y = b$.

- If $y < b$, shade the half-plane below $y = b$.

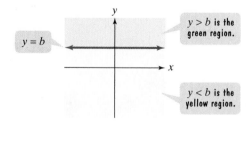

EXAMPLE 4   Graphing Inequalities without Using Test Points

Graph each inequality in a rectangular coordinate system:

   a. $y \leq -3$ **b.** $x > 2$.

Solution

   a. $y \leq -3$ **b.** $x > 2$

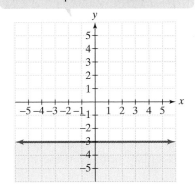

Graph $y = -3$, a horizontal line with y-intercept -3. The line is solid because equality is included in $y \leq -3$. Because of the less than part of $\leq$, shade the half-plane below the horizontal line.

Graph $x = 2$, a vertical line with x-intercept **2**. The line is dashed because equality is not included in $x > 2$. Because of $>$, the greater than symbol, shade the half-plane to the right of the vertical line.

✓ **CHECK POINT 4**   Graph each inequality in a rectangular coordinate system:

   a. $y > 1$ **b.** $x \leq -2$.

---

3 Solve applied problems involving linear inequalities in two variables.

Applications

Temperature and precipitation affect whether or not trees and forests can grow. At certain levels of precipitation and temperature, only grasslands and deserts will exist. **Figure 4.42** shows three kinds of regions—deserts, grasslands, and forests—that result from various ranges of temperature and precipitation. Notice that the horizontal axis is labeled T, for temperature, rather than x. The vertical axis is labeled P, for precipitation, rather than y. We can use inequalities in two variables, T and P, to describe the regions in the figure.

Regions Resulting from Ranges of Temperature and Precipitation

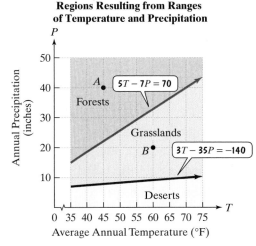

FIGURE 4.42

Source: A. Miller and J. Thompson, Elements of Meteorology

EXAMPLE 5   Forests and Inequalities

   a. Use **Figure 4.42** to find the coordinates of point A. What does this mean in terms of the kind of region that occurs?

   b. For average annual temperatures that exceed 35°F, the inequality

$$5T - 7P < 70$$

models where forests occur. Show that the coordinates of point A satisfy this inequality.

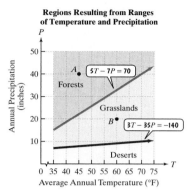

FIGURE 4.42   (repeated)

Solution

a. Point A has coordinates $(45, 40)$. This means that if a region has an average annual temperature of $45°F$ and an average annual precipitation of 40 inches, then a forest occurs.

b. We can show that $(45, 40)$ satisfies the inequality for forests by substituting 45 for T and 40 for P.

$$5T - 7P < 70 \quad \text{This is the given inequality.}$$
$$5 \cdot 45 - 7 \cdot 40 \overset{?}{<} 70 \quad \text{Substitute 45 for } T \text{ and 40 for } P.$$
$$225 - 280 \overset{?}{<} 70 \quad \text{Multiply: } 5 \cdot 45 = 225 \text{ and } 7 \cdot 40 = 280.$$

This statement is true.
$$-55 < 70 \quad \text{Subtract: } 225 - 280 = -55.$$

The coordinates $(45, 40)$ make the inequality true. Thus, $(45, 40)$ satisfies the inequality.

✓ CHECK POINT 5

a. Use **Figure 4.42** to find the coordinates of point B. What does this mean in terms of the kind of region that occurs?

b. For average annual temperatures that exceed $35°F$, the inequalities $5T - 7P \geq 70$ and $3T - 35P \leq -140$ model where grasslands occur. Show that the coordinates of point B satisfy both of these inequalities.

4.6 EXERCISE SET   MyMathLab

MathXL PRACTICE   WATCH   DOWNLOAD   READ   REVIEW

Practice Exercises

In Exercises 1–8, determine whether each ordered pair is a solution of the given inequality.

1. $x + y > 4$: $(2, 2), (3, 2), (-3, 8)$
2. $2x - y < 3$: $(0, 0), (3, 0), (-4, -15)$
3. $2x + y \geq 5$: $(4, 0), (1, 3), (0, 0)$
4. $3x - 5y \geq -12$: $(2, -3), (2, 8), (0, 0)$
5. $y \geq -2x + 4$: $(4, 0), (1, 3), (-2, -4)$
6. $y \leq -x + 5$: $(5, 0), (0, 5), (8, -4)$
7. $y > -2x + 1$: $(2, 3), (0, 0), (0, 5)$
8. $x < -y - 2$: $(-1, -1), (0, 0), (4, -5)$

In Exercises 9–36, graph each inequality.

9. $x + y \geq 3$
10. $x + y \geq 4$
11. $x - y < 5$
12. $x - y < 2$
13. $x + 2y > 4$
14. $2x + y > 6$
15. $3x - y \leq 6$
16. $x - 3y \leq -6$
17. $3x - 2y \leq 8$
18. $2x - 3y \geq 8$
19. $4x + 3y > 15$
20. $5x + 10y > 15$
21. $5x - y < -7$
22. $x - 5y < -7$
23. $y \leq \frac{1}{3}x$
24. $y \leq \frac{1}{4}x$
25. $y > 2x$
26. $y > 4x$
27. $y > 3x + 2$
28. $y > 2x - 1$
29. $y < \frac{3}{4}x - 3$
30. $y < \frac{2}{3}x - 1$
31. $x \leq 1$
32. $x \leq -3$
33. $y > 1$
34. $y > -3$
35. $x \geq 0$
36. $y \leq 0$

Practice PLUS

In Exercises 37–44, write each sentence as a linear inequality in two variables. Then graph the inequality.

37. The sum of the x-variable and the y-variable is at least 2.
38. The difference between the x-variable and the y-variable is at least 3.
39. The difference between 5 times the x-variable and 2 times the y-variable is at most 10.
40. The sum of 4 times the x-variable and 2 times the y-variable is at most 8.
41. The y-variable is no less than $\frac{1}{2}$ of the x-variable.
42. The y-variable is no less than $\frac{1}{4}$ of the x-variable.
43. The y-variable is no more than -1.
44. The y-variable is no more than -2.

Application Exercises

45. Bottled water and medical supplies are to be shipped to survivors of a hurricane by plane. Each plane can carry no more than 80,000 pounds. The bottled water weights 20 pounds per container and each medical kit weighs 10 pounds. Let x represent the number of bottles of water to be shipped. Let y represent the number of medical kits. The plane's weight limitations can be described by the following inequality:

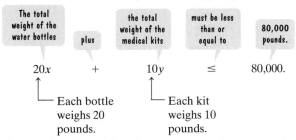

a. Graph the inequality. Because x and y must be nonnegative, limit the graph to quadrant I and its boundary only.

b. Select an ordered pair satisfying the inequality. What are its coordinates and what do they represent in this situation?

46. Bottled water and medical supplies are to be shipped to survivors of a hurricane by plane. Each plane can carry a total volume that does not exceed 6000 cubic feet. Each water bottle is 2 cubic feet and each medical kit has a volume of 1 cubic foot. Let x represent the number of bottles of water to be shipped. Let y represent the number of medical kits. The plane's volume limitations can be described by the following inequality:

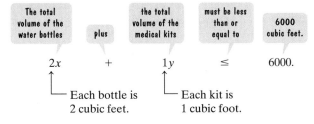

a. Graph the inequality. Because x and y must be nonnegative, limit the graph to quadrant I and its boundary only.

b. Select an ordered pair satisfying the inequality. What are its coordinates and what do they represent in this situation?

47. Many elevators have a capacity of 2000 pounds.

a. If a child averages 50 pounds and an adult 150 pounds, write an inequality that describes when x children and y adults will cause the elevator to be overloaded.

b. Graph the inequality. Because x and y must be nonnegative, limit the graph to quadrant I and its boundary only.

c. Select an ordered pair satisfying the inequality. What are its coordinates and what do they represent in this situation?

48. A patient is not allowed to have more than 330 milligrams of cholesterol per day from a diet of eggs and meat. Each egg provides 165 milligrams of cholesterol. Each ounce of meat provides 110 milligrams of cholesterol.

a. Write an inequality that describes the patient's dietary restrictions for x eggs and y ounces of meat.

b. Graph the inequality. Because x and y must be nonnegative, limit the graph to quadrant I and its boundary only.

c. Select an ordered pair satisfying the inequality. What are its coordinates and what do they represent in this situation?

Not all regions in rectangular coordinates are bounded by straight lines. Although these regions cannot be described using linear inequalities, they often have numerous applications. For example, the regions in the following two graphs indicate whether a person is obese, overweight, borderline overweight, normal weight, or underweight.

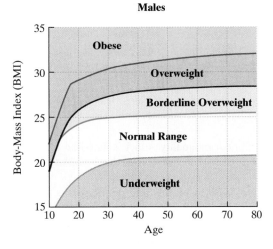

Source: Centers for Disease Control and Prevention

The horizontal axis of each graph on the previous page shows a person's age. The vertical axis shows that person's body-mass index (BMI), computed using the following formula:

$$\text{BMI} = \frac{703W}{H^2}.$$

The variable W represents weight, in pounds. The variable H represents height, in inches. Use this information to solve Exercises 49–50.

49. A man is 20 years old, 72 inches (6 feet) tall, and weighs 200 pounds.

    a. Compute the man's BMI. Round to the nearest tenth.

    b. Use the man's age and his BMI to locate this information as a point in the coordinate system for males shown on the previous page. Is this man obese, overweight, borderline overweight, normal weight, or underweight?

50. A woman is 25 years old, 66 inches (5 feet, 6 inches) tall, and weighs 105 pounds.

    a. Compute the woman's BMI. Round to the nearest tenth.

    b. Use the woman's age and her BMI to locate this information as a point in the coordinate system for females shown on the previous page. Is this woman obese, overweight, borderline overweight, normal weight, or underweight?

Writing in Mathematics

51. What is a linear inequality in two variables? Provide an example with your description.

52. How do you determine whether an ordered pair is a solution of an inequality in two variables, x and y?

53. What is a half-plane?

54. What does a solid line mean in the graph of an inequality?

55. What does a dashed line mean in the graph of an inequality?

56. Explain how to graph $2x - 3y < 6$.

57. Compare the graphs of $3x - 2y > 6$ and $3x - 2y \le 6$. Discuss similarities and differences between the graphs.

Critical Thinking Exercises

Make Sense?   In Exercises 58–61, determine whether each statement "makes sense" or "does not make sense" and explain your reasoning.

58. By looking at a linear inequality in two variables, I can immediately determine whether the boundary line of its graph should be solid or dashed.

59. The inequality $2x - 3y < 6$ contains a "less than" symbol, so its graph lies below the boundary line.

60. When I write a linear inequality with y isolated on the left, $<$ indicates a region that lies below the boundary line.

61. I have less than $5.00 in nickels and dimes, so the linear inequality

$$0.05n + 0.10d < 5.00$$

models how many nickels, n, and how many dimes, d, that I might have.

In Exercises 62–65, determine whether each statement is true or false. If the statement is false, make the necessary change(s) to produce a true statement.

62. The ordered pair $(0, -3)$ satisfies $y > 2x - 3$.

63. The graph of $x < y + 1$ is the half-plane below the line $x = y + 1$.

64. In graphing $y \ge 4x$, a dashed line is used.

65. The graph of $x < 4$ is the half-plane to the left of the vertical line described by $x = 4$.

In Exercises 66–67, write an inequality that represents each graph.

66.   **67.**

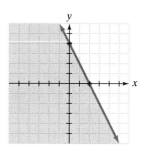

Technology Exercises

Graphing utilities can be used to shade regions in the rectangular coordinate system, thereby graphing an inequality in two variables. Read the section of the user's manual for your graphing utility that describes how to shade a region. Then use your graphing utility to graph the inequalities in Exercises 68–71.

68. $y \le 4x + 4$ **69.** $y \ge x - 2$

70. $y \ge \dfrac{1}{2}x + 4$ **71.** $y \le -\dfrac{1}{2}x + 4$

Review Exercises

72. Solve for h: $V = lwh$. (Section 2.4, Example 1)

73. Find the quotient: $\dfrac{2}{3} \div \left(-\dfrac{5}{4}\right)$. (Section 1.7, Example 5)

74. Evaluate $x^2 - 4$ for $x = -3$. (Section 1.8, Example 10)

Preview Exercises

Exercises 75–77 will help you prepare for the material covered in the first section of the next chapter.

75. Is $(4, -1)$ a solution of both $x + 2y = 2$ and $x - 2y = 6$?

76. Is $(-4, 3)$ a solution of both $x + 2y = 2$ and $x - 2y = 6$?

77. Determine the point of intersection of the graphs of $2x + 3y = 6$ and $2x + y = -2$ by graphing both equations in the same rectangular coordinate system.

GROUP PROJECT

CHAPTER 4

In Example 3 on page 269, we used the data in **Figure 4.36** on page 268 to develop a linear equation that modeled the graying of America. For this group exercise, you might find it helpful to pattern your work after **Figure 4.36** and the solution to Example 3. Group members should begin by consulting an almanac, newspaper, magazine, or the Internet to find data that lie approximately on or near a straight line. Working by hand or using a graphing utility, construct a scatter plot for the data. If working by hand, draw a line that passes through or near the data points and then write its equation. If using a graphing utility, obtain the equation of the regression line. Then use the equation of the line to make a prediction about what might happen in the future. Are there circumstances that might affect the accuracy of this prediction? List some of these circumstances.

Chapter 4 Summary

Definitions and Concepts	Examples

Section 4.1 Graphing Linear Equations in Two Variables

The rectangular coordinate system consists of a horizontal number line, the x-axis, and a vertical number line, the y-axis, intersecting at their zero points, the origin. Each point in the system corresponds to an ordered pair of real numbers (x, y). The first number in the pair is the x-coordinate; the second number is the y-coordinate.

Plot: $(2, 3), (-5, 4), (-4, -3)$, and $(5, -2)$.

An ordered pair is a solution of an equation in two variables if replacing the variables by the coordinates of the ordered pair results in a true statement.

Is $(-1, 4)$ a solution of $2x + 5y = 18$?
$$2(-1) + 5 \cdot 4 \stackrel{?}{=} 18$$
$$-2 + 20 \stackrel{?}{=} 18$$
$$18 = 18, \text{ true}$$

Thus, $(-1, 4)$ is a solution.

One method for graphing an equation in two variables is point plotting. Find several ordered-pair solutions, plot them as points, and connect the points with a smooth curve or line.

Graph: $y = 2x + 1$.

x	$y = 2x + 1$	(x, y)
-2	$y = 2(-2) + 1 = -3$	$(-2, -3)$
-1	$y = 2(-1) + 1 = -1$	$(-1, -1)$
0	$y = 2 \cdot 0 + 1 = 1$	$(0, 1)$
1	$y = 2 \cdot 1 + 1 = 3$	$(1, 3)$
2	$y = 2 \cdot 2 + 1 = 5$	$(2, 5)$

The graph of $y = 2x + 1$

Definitions and Concepts	**Examples**

Section 4.2 Graphing Linear Equations Using Intercepts

If a graph intersects the x-axis at $(a, 0)$, then a is an x-intercept.
If a graph intersects the y-axis at $(0, b)$, then b is a y-intercept.

An equation of the form $Ax + By = C$, where A, B, and C are integers, is called the standard form of the equation of a line. The graph of $Ax + By = C$ is a line that can be obtained using intercepts. To find the x-intercept, let $y = 0$ and solve for x. To find the y-intercept, let $x = 0$ and solve for y. Find a checkpoint, a third ordered-pair solution. Graph the equation by drawing a line through the three points.

Graph using intercepts: $4x + 3y = 12$.

x-intercept:
$$4x = 12$$
$$x = 3$$

y-intercept:
$$3y = 12$$
$$y = 4$$

Checkpoint: Let $x = 2$.
$$8 + 3y = 12$$
$$3y = 4$$
$$y = \tfrac{4}{3}$$

The graph of $Ax + By = 0$ is a line that passes through the origin. Find two other points by finding two other solutions of the equation. Graph the equation by drawing a line through the origin and these two points.

Graph: $x + 2y = 0$.
$$x = 2: \quad 2 + 2y = 0$$
$$2y = -2$$
$$y = -1$$
$$y = 1: \quad x + 2(1) = 0$$
$$x = -2$$

Horizontal and Vertical Lines

The graph of $y = b$ is a horizontal line. The y-intercept is b.
The graph of $x = a$ is a vertical line. The x-intercept is a.

Definitions and Concepts	**Examples**

Section 4.3 Slope

The slope, m, of the line through the points (x_1, y_1) and (x_2, y_2) is

$$m = \frac{y_2 - y_1}{x_2 - x_1}, \quad x_2 - x_1 \neq 0.$$

If the slope is positive, the line rises from left to right. If the slope is negative, the line falls from left to right.
The slope of a horizontal line is 0. The slope of a vertical line is undefined.
If two distinct nonvertical lines have the same slope, then the lines are parallel.
If the product of the slopes of two lines is -1, then the lines are perpendicular. Equivalently, if the slopes are negative reciprocals, then the lines are perpendicular.
The slope of a line represents its rate of change.

Find the slope of the line passing through the points shown.

Let $(x_1, y_1) = (-1, 2)$ and $(x_2, y_2) = (2, -2)$.

$$m = \frac{y_2 - y_1}{x_2 - x_1} = \frac{-2 - 2}{2 - (-1)} = \frac{-4}{3} = -\frac{4}{3}$$

Section 4.4 The Slope-Intercept Form of the Equation of a Line

The slope-intercept form of the equation of a nonvertical line with slope m and y-intercept b is

$$y = mx + b.$$

Find the slope and the y-intercept of the line with the given

- $y = -2x + 5$

  Slope is −2. y-intercept is 5.

- $2x + 3y = 9$ (Solve for y.)

  $3y = -2x + 9$ Subtract 2x.

  $y = -\frac{2}{3}x + 3$ Divide by 3.

  Slope is $-\frac{2}{3}$. y-intercept is 3.

Graphing $y = mx + b$ Using the Slope and y-Intercept
1. Plot the point containing the y-intercept on the y-axis. This is the point $(0, b)$.
2. Use the slope, m, to obtain a second point. Write m as a fraction, and use rise over run, starting at the point containing the y-intercept, to plot this point.
3. Graph the equation by drawing a line through the two points.

Graph: $y = -\dfrac{3}{4}x + 1$.

Slope is $-\frac{3}{4}$. y-intercept is 1.

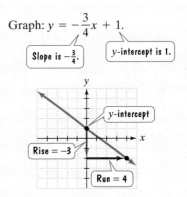

Section 4.5 The Point-Slope Form of the Equation of a Line

The point-slope form of the equation of a nonvertical line with slope m that passes through the point (x_1, y_1) is
$$y - y_1 = m(x - x_1).$$

Slope = -3, passing through $(-1, 5)$

$m = -3$ $x_1 = -1$ $y_1 = 5$

The point-slope form of the line's equation is

$$y - 5 = -3[x - (-1)].$$

Simplify:

$$y - 5 = -3(x + 1).$$

Definitions and Concepts	**Examples**

Section 4.5 The Point-Slope Form of the Equation of a Line (continued)

To write the point-slope form of the equation of a line passing through two points, begin by using the points to compute the slope, m. Use either given point as (x_1, y_1) and write the point-slope form of the equation:

$$y - y_1 = m(x - x_1).$$

Solving this equation for y gives the slope-intercept form of the line's equation.

Write an equation in point-slope form and slope-intercept form of the line passing through $(-1, -3)$ and $(4, 2)$.

$$m = \frac{2 - (-3)}{4 - (-1)} = \frac{2 + 3}{4 + 1} = \frac{5}{5} = 1$$

Using $(4, 2)$ as (x_1, y_1), the point-slope form of the equation is

$$y - 2 = 1(x - 4).$$

Solve for y to obtain the slope-intercept form.

$$y = x - 2 \quad \text{Add 2 to both sides.}$$

Section 4.6 Linear Inequalities in Two Variables

A linear inequality in two variables can be written in one of the following forms:

$$Ax + By < C \quad Ax + By \le C$$
$$Ax + By > C \quad Ax + By \ge C.$$

An ordered pair is a solution if replacing the variables by the coordinates of the ordered pair results in a true statement.

Is $(2, -6)$ a solution of $3x - 4y \le 7$?

$$3 \cdot 2 - 4(-6) \overset{?}{\le} 7$$

$$6 - (-24) \overset{?}{\le} 7$$

$$6 + 24 \overset{?}{\le} 7$$

$$30 \le 7, \text{false}$$

Thus, $(2, -6)$ is not a solution.

Graphing a Linear Inequality in Two Variables

1. Replace the inequality symbol with an equal sign and graph the boundary line. Use a solid line for $\le$ or $\ge$ and a dashed line for $<$ or $>$.
2. Choose a test point not on the line and substitute its coordinates into the inequality.
3. If a true statement results, shade the half-plane containing the test point. If a false statement results, shade the half-plane not containing the test point.

Graph: $x - 2y \le 4$.

1. Graph $x - 2y = 4$. Use a solid line because the inequality symbol is $\le$.
2. Test $(0, 0)$.

$$0 - 2 \cdot 0 \overset{?}{\le} 4$$

$$0 \le 4, \text{true}$$

3. The inequality is true. Shade the half-plane containing $(0, 0)$.

CHAPTER 4 REVIEW EXERCISES

4.1 *In Exercises 1–4, plot the given point in a rectangular coordinate system. Indicate in which quadrant each point lies.*

1. $(1, -5)$

2. $(4, -3)$

3. $\left(\dfrac{7}{2}, \dfrac{3}{2}\right)$

4. $(-5, 2)$

5. Give the ordered pairs that correspond to the points labeled in the figure.

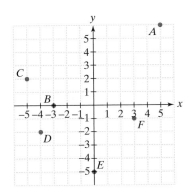

In Exercises 6–7, determine whether each ordered pair is a solution of the given equation.

6. $y = 3x + 6$ $(-3, 3)$, $(0, 6)$, $(1, 9)$

7. $3x - y = 12$ $(0, 4), (4, 0), (-1, 15)$

In Exercises 8–9,

   a. *Find five solutions of each equation. Organize your work in a table of values.*

   b. *Use the five solutions in the table to graph each equation.*

8. $y = 2x - 3$

9. $y = \frac{1}{2}x + 1$

4.2 *In Exercises 10–12, use the graph to identify the*

   a. *x-intercept, or state that there is no x-intercept.*

   b. *y-intercept, or state that there is no y-intercept.*

10.

11.

12.

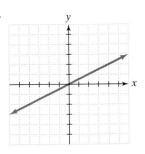

In Exercises 13–16, use intercepts to graph each equation.

13. $2x + y = 4$

14. $3x - 2y = 12$

15. $3x = 6 - 2y$

16. $3x - y = 0$

In Exercises 17–20, graph each equation.

17. $x = 3$ **18.** $y = -5$ **19.** $y + 3 = 5$ **20.** $2x = -8$

21. The graph shows the Fahrenheit temperature, y, x hours after noon.

   a. At what time did the minimum temperature occur? What is the minimum temperature?

   b. At what time did the maximum temperature occur? What is the maximum temperature?

   c. What are the x-intercepts? In terms of time and temperature, interpret the meaning of these intercepts.

   d. What is the y-intercept? What does this mean in terms of time and temperature?

   e. From 9 P.M. until midnight, the graph is shown as a horizontal line. What does this mean about the temperature over this period of time?

4.3 *In Exercises 22–25, calculate the slope of the line passing through the given points. If the slope is undefined, so state. Then indicate whether the line rises, falls, is horizontal, or is vertical.*

22. $(3, 2)$ and $(5, 1)$

23. $(-1, 2)$ and $(-3, -4)$

24. $(-3, 4)$ and $(6, 4)$

25. $(5, 3)$ and $(5, -3)$

In Exercises 26–29, find the slope of each line, or state that the slope is undefined.

26.

27.

28.

29.

In Exercises 30–32, determine whether the lines through each pair of points are parallel, perpendicular, or neither.

30. $(-1, -3)$ and $(2, -8)$; $(8, -7)$ and $(9, 10)$

31. $(0, -4)$ and $(5, -1)$; $(-6, 8)$ and $(3, -7)$

32. $(5, 4)$ and $(9, 7)$; $(-6, 0)$ and $(-2, 3)$

33. The graphs show the percentage of female medical school graduates in the United States in 1965 and 2005.

Percentage of Female Medical School Graduates in the United States

Source: Association of American Medical Colleges

a. Find the slope of the line passing through the two points shown by the voice balloons.

b. Use your answer from part (a) to complete this statement.

For each year from 1965 through 2005, the percentage of female medical school graduates increased by ___. The rate of change was ___ per ____.

4.4 *In Exercises 34–37, find the slope and the y-intercept of the line with the given equation.*

34. $y = 5x - 7$

35. $y = 6 - 4x$

36. $y = 3$

37. $2x + 3y = 6$

In Exercises 38–40, graph each linear equation using the slope and y-intercept.

38. $y = 2x - 4$

39. $y = \frac{1}{2}x - 1$

40. $y = -\frac{2}{3}x + 5$

In Exercises 41–42, write each equation in slope-intercept form. Then use the slope and y-intercept to graph the equation.

41. $y - 2x = 0$

42. $\frac{1}{3}x + y = 2$

43. Graph $y = -\frac{1}{2}x + 4$ and $y = -\frac{1}{2}x - 1$ in the same rectangular coordinate system. Are the lines parallel? If so, explain why.

44. The graphs show the percentage of federal taxes that were filed electronically in 1995 and 2005.

Percentage of Tax Returns Filed Electronically

Source: IRS

a. Use the two points shown by the voice balloons to find an equation in the form $y = mx + b$ that models the percentage of tax returns, y, that, were e-filed x years after 1995.

b. Use the model from part (a) to project the percentage of tax returns that will be e-filed in 2015.

4.5 *Write the point-slope form of the equation of the line satisfying the conditions in Exercises 45–46. Then use the point-slope form of the equation to write the slope-intercept form.*

45. Slope = 6, passing through $(-4, 7)$

46. Passing through $(3, 4)$ and $(2, 1)$

47. The bar graph shows world population, in billions, for seven selected years from 1950 through 2006. Also shown is a scatter plot with a line passing through two of the data points.

World Population, 1950–2006

Source: U.S. Census Bureau, International Database

a. Use the two points whose coordinates are shown by the voice balloons to write the slope-intercept form of an equation that models world population, *y*, in billions, *x* years after 1950.

b. Use the model from part (a) to project world population in 2025.

4.6

48. Determine whether each ordered pair is a solution of $3x - 4y > 7$:

$$(0, 0), \quad (3, -6), \quad (-2, -5), \quad (-3, 4).$$

In Exercises 49–56, graph each inequality.

49. $x - 2y > 6$

50. $4x - 6y \leq 12$

51. $y > 3x + 2$

52. $y \leq \dfrac{1}{3}x - 1$

53. $y < -\dfrac{1}{2}x$

54. $x < 4$

55. $y \geq -2$

56. $x + 2y \leq 0$

CHAPTER 4 TEST

Remember to use your Chapter Test Prep Video CD to see the worked-out solutions to the test questions you want to review.

1. Determine whether each ordered pair is a solution of $4x - 2y = 10$:

$$(0, -5), \quad (-2, 1), \quad (4, 3).$$

2. Find five solutions of $y = 3x + 1$. Organize your work in a table of values. Then use the five solutions in the table to graph the equation.

3. Use the graph to identify the

   a. *x*-intercept, or state that there is no *x*-intercept.

   b. *y*-intercept, or state that there is no *y*-intercept.

4. Use intercepts to graph $4x - 2y = -8$.

5. Graph $y = 4$ in a rectangular coordinate system.

In Exercises 6–7, calculate the slope of the line passing through the given points. If the slope is undefined, so state. Then indicate whether the line rises, falls, is horizontal, or is vertical.

6. $(-3, 4)$ and $(-5, -2)$

7. $(6, -1)$ and $(6, 3)$

8. Find the slope of the line in the figure shown or state that the slope is undefined.

In Exercises 9–10 determine whether the lines through each pair of points are parallel, perpendicular, or neither.

   9. $(-2, 10)$ and $(0, 2)$; $(-8, -7)$ and $(24, 1)$

   10. $(2, 4)$ and $(6, 1)$; $(-3, 1)$ and $(1, -2)$

In Exercises 11–12, find the slope and the y-intercept of the line with the given equation.

11. $y = -x + 10$

12. $2x + y = 6$

In Exercises 13–14, graph each linear equation using the slope and y-intercept.

13. $y = \dfrac{2}{3}x - 1$

14. $y = -2x + 3$

In Exercises 15–16, use the given conditions to write an equation for each line in point-slope form and slope-intercept form.

15. Slope $= -2$, passing through $(-1, 4)$

16. Passing through $(2, 1)$ and $(-1, -8)$

In Exercises 17–19, graph each inequality.

17. $3x - 2y < 6$ **18.** $y \geq 2x - 2$ **19.** $x > -1$

20. The bar graph shows the growth in corporate profits, in billions of dollars, from 2002 through 2006.

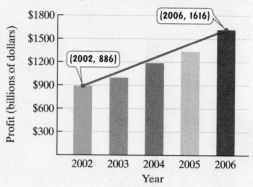

Corporate Profits, in Billions of Dollars

Source: Bureau of Economic Analysis

a. Find the slope of the line passing through the two points shown by the voice balloons.

b. Use your answer from part (a) to complete this statement.

For the period shown, corporate profits increased each year by approximately _____. The rate of change was _____ per ____.

CUMULATIVE REVIEW EXERCISES (CHAPTERS 1–4)

1. Perform the indicated operations:
$$\frac{10 - (-6)}{3^2 - (4 - 3)}.$$

2. Simplify: $6 - 2[3(x - 1) + 4]$.

3. List all the irrational numbers in this set:
$\left\{-3, 0, 1, \sqrt{4}, \sqrt{5}, \frac{11}{2}\right\}$.

In Exercises 4–5, solve each equation.

4. $6(2x - 1) - 6 = 11x + 7$

5. $x - \dfrac{3}{4} = \dfrac{1}{2}$

6. Solve for x: $y = mx + b$.

7. 120 is 15% of what number?

8. The formula $y = 4.5x - 46.7$ models the stopping distance, y, in feet, for a car traveling x miles per hour. If the stopping distance is 133.3 feet, how fast was the car traveling?

In Exercises 9–10, solve each inequality and graph the solution set on a number line.

9. $2 - 6x \geq 2(5 - x)$

10. $6(2 - x) > 12$

11. A plumber charged a customer $228, listing $18 for parts and the remainder for labor. If the cost of the labor is $35 per hour, how many hours did the plumber work?

12. A chemist needs to mix a 40% acid solution with a 70% acid solution to obtain 12 liters of a 50% acid solution. How many liters of each of the acid solutions must be used?

13. Two people leave by car from the same point of departure and travel in opposite directions. One car averages 40 miles per hour and the other averages 60 miles per hour. After how many hours will the cars be 350 miles apart?

14. In a triangle, the measure of the second angle is 20° greater than the measure of the first angle. The measure of the third angle is twice that of the first. What is the measure of each angle?

15. Solve the proportion: $\dfrac{45}{2} = \dfrac{360}{x}$.

In Exercises 16–20, graph each equation or inequality in the rectangular coordinate system.

16. $2x - y = 4$ **17.** $x = -5$ **18.** $y = -4x + 3$

19. $3x - 2y < -6$ **20.** $y \geq -1$

Systems of Linear Equations and Inequalities

Television, movies, and magazines place great emphasis on physical beauty. **Our culture emphasizes physical appearance to such an** extent that it is a central factor in the perception and judgment of others. The modern emphasis on thinness as the ideal body shape has been suggested as a major cause of eating disorders among adolescent women.

What is the relationship between our changing cultural values of physical attractiveness and undernutrition? Given the importance of culture in setting standards of attractiveness, how can you establish a healthy weight range for your age and height? In this chapter, we will use systems of linear equations and inequalities to explore these skin-deep issues.

The Blitzer Bonus on page 298 contains a graphical look at the shifting shape of that icon of American beauty, Miss America. You'll find a weight that fits you using the models (mathematical, not fashion) in Example 1 of Section 5.5 and Exercises 45–48 in Exercise Set 5.5.

Solving Systems of Linear Equations by Graphing

Objectives

1 Decide whether an ordered pair is a solution of a linear system.

2 Solve systems of linear equations by graphing.

3 Use graphing to identify systems with no solution or infinitely many solutions.

4 Use graphs of linear systems to solve problems.

Visitors to the world's great bridges are frequently inspired by their beauty. This feeling is not always shared by daily commuters dealing with escalating toll costs and peak-hour congestion. For these frequent users, most bridge authorities provide the option of a fixed cost that reduces the toll. In this section, you will see how toll options in these situations can be modeled by two linear equations in two variables and their graphs.

1 Decide whether an ordered pair is a solution of a linear system.

Systems of Linear Equations and Their Solutions

We have seen that all equations in the form $Ax + By = C$ are straight lines when graphed. Two such equations are called a **system of linear equations** or a **linear system**. A **solution to a system of two linear equations in two variables** is an ordered pair that satisfies both equations in the system. For example, $(3, 4)$ satisfies the system

$$x + y = 7 \quad \text{(3 + 4 is, indeed, 7.)}$$
$$x - y = -1. \quad \text{(3 − 4 is, indeed, −1.)}$$

Thus, $(3, 4)$ satisfies both equations and is a solution of the system. The solution can be described by saying that $x = 3$ and $y = 4$. The solution can also be described using the ordered pair $(3, 4)$. The solution set of the system is $\{(3, 4)\}$—that is, the set consisting of the ordered pair $(3, 4)$.

A system of linear equations can have exactly one solution, no solution, or infinitely many solutions. We begin with systems having exactly one solution.

EXAMPLE 1 Determining Whether Ordered Pairs Are Solutions of a System

Consider the system:

$$x + 2y = 2$$
$$x - 2y = 6.$$

Determine if each ordered pair is a solution of the system:

a. $(4, -1)$ **b.** $(-4, 3)$.

Solution

a. We begin by determining whether $(4, -1)$ is a solution. Because 4 is the x-coordinate and -1 is the y-coordinate of $(4, -1)$, we replace x with 4 and y with -1.

$$x + 2y = 2$$
$$4 + 2(-1) \stackrel{?}{=} 2$$
$$4 + (-2) \stackrel{?}{=} 2$$
$$2 = 2, \quad \text{true}$$

$$x - 2y = 6$$
$$4 - 2(-1) \stackrel{?}{=} 6$$
$$4 - (-2) \stackrel{?}{=} 6$$
$$4 + 2 \stackrel{?}{=} 6$$
$$6 = 6, \quad \text{true}$$

The pair $(4, -1)$ satisfies both equations: It makes each equation true. Thus, the ordered pair is a solution of the system.

b. To determine whether $(-4, 3)$ is a solution, we replace x with -4 and y with 3.

$$x + 2y = 2$$
$$-4 + 2 \cdot 3 \stackrel{?}{=} 2$$
$$-4 + 6 \stackrel{?}{=} 2$$
$$2 = 2, \quad \text{true}$$

$$x - 2y = 6$$
$$-4 - 2 \cdot 3 \stackrel{?}{=} 6$$
$$-4 - 6 \stackrel{?}{=} 6$$
$$-10 = 6, \quad \text{false}$$

The pair $(-4, 3)$ fails to satisfy *both* equations: It does not make both equations true. Thus, the ordered pair is not a solution of the system. ■

✓ **CHECK POINT 1**   Consider the system:

$$2x - 3y = -4$$
$$2x + y = 4.$$

Determine if each ordered pair is a solution of the system:

a. $(1, 2)$ b. $(7, 6)$.

---

2 Solve systems of linear equations by graphing.

Solving Linear Systems by Graphing

The solution of a system of two linear equations in two variables can be found by graphing both of the equations in the same rectangular coordinate system. For a system with one solution, **the coordinates of the point of intersection give the system's solution**. For example, the system in Example 1,

$$x + 2y = 2$$
$$x - 2y = 6$$

is graphed in **Figure 5.1**. The solution of the system, $(4, -1)$, corresponds to the point of intersection of the lines.

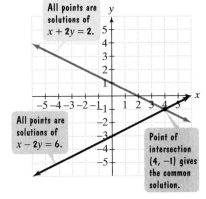

FIGURE 5.1
Visualizing a system's solution

Solving Systems of Two Linear Equations in Two Variables, *x* and *y*, by Graphing

1. Graph the first equation.

2. Graph the second equation on the same set of axes.

3. If the lines representing the two graphs intersect at a point, determine the coordinates of this point of intersection. The ordered pair is the solution of the system.

4. Check the solution in both equations.

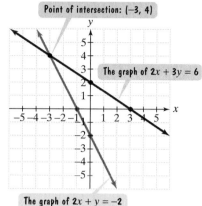

Point of intersection: (−3, 4)

The graph of 2x + 3y = 6

The graph of 2x + y = −2

FIGURE 5.2

EXAMPLE 2 Solving a Linear System by Graphing

Solve by graphing:

$$2x + 3y = 6$$
$$2x + y = -2.$$

Solution

Step 1. Graph the first equation. We use intercepts to graph $2x + 3y = 6$.

x-intercept (Set $y = 0$.)	*y*-intercept (Set $x = 0$.)
$2x + 3 \cdot 0 = 6$	$2 \cdot 0 + 3y = 6$
$2x = 6$	$3y = 6$
$x = 3$	$y = 2$

The *x*-intercept is 3, so the line passes through $(3, 0)$. The *y*-intercept is 2, so the line passes through $(0, 2)$. The graph of $2x + 3y = 6$ is shown as the red line in **Figure 5.2**.

Step 2. Graph the second equation on the same axes. We use intercepts to graph $2x + y = -2$.

x-intercept (Set $y = 0$.)	*y*-intercept (Set $x = 0$.)
$2x + 0 = -2$	$2 \cdot 0 + y = -2$
$2x = -2$	$y = -2$
$x = -1$	

The *x*-intercept is -1, so the line passes through $(-1, 0)$. The *y*-intercept is -2, so the line passes through $(0, -2)$. The graph of $2x + y = -2$ is shown as the blue line in **Figure 5.2**.

Step 3. Determine the coordinates of the intersection point. This ordered pair is the system's solution. Using **Figure 5.2**, it appears that the lines intersect at $(-3, 4)$. The "apparent" solution of the system is $(-3, 4)$.

Step 4. Check the solution in both equations.

Check $(-3, 4)$ in	Check $(-3, 4)$ in
2x + 3y = 6:	**2x + y = −2:**
$2(-3) + 3 \cdot 4 \stackrel{?}{=} 6$	$2(-3) + 4 \stackrel{?}{=} -2$
$-6 + 12 \stackrel{?}{=} 6$	$-6 + 4 \stackrel{?}{=} -2$
$6 = 6,$ true	$-2 = -2,$ true

Because both equations are satisfied, $(-3, 4)$ is the solution of the system and $\{(-3, 4)\}$ is the solution set. ▬

✓ **CHECK POINT 2** Solve by graphing:

$$2x + y = 6$$
$$2x - y = -2.$$

Discover for Yourself

Must two lines intersect at exactly one point? Sketch two lines that have less than one intersection point. Now sketch two lines that have more than one intersection point. What does this say about each of these systems?

Study Tip

When solving linear systems by graphing, neatly drawn graphs are essential for determining points of intersection.

- Use rectangular coordinate graph paper.
- Use a ruler or straightedge.
- Use a pencil with a sharp point.

EXAMPLE 3 **Solving a Linear System by Graphing**

Solve by graphing:

$$y = -3x + 2$$
$$y = 5x - 6.$$

Solution Each equation is in the form $y = mx + b$. Thus, we use the y-intercept, b, and the slope, m, to graph each line.

Step 1. Graph the first equation.

$$y = -3x + 2$$

> The slope is -3.   The y-intercept is 2.

The y-intercept is 2, so the line passes through $(0, 2)$. The slope is $-\frac{3}{1}$. Start at the y-intercept and move 3 units down (the rise) and 1 unit to the right (the run). The graph of $y = -3x + 2$ is shown as the red line in **Figure 5.3**.

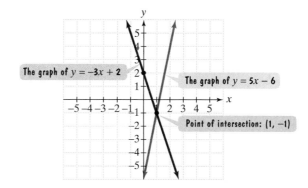

The graph of $y = -3x + 2$

The graph of $y = 5x - 6$

Point of intersection: $(1, -1)$

FIGURE 5.3

Step 2. Graph the second equation on the same axes.

$$y = 5x - 6$$

> The slope is 5.   The y-intercept is -6.

The y-intercept is -6, so the line passes through $(0, -6)$. The slope is $\frac{5}{1}$. Start at the y-intercept and move 5 units up (the rise) and 1 unit to the right (the run). The graph of $y = 5x - 6$ is shown as the blue line in **Figure 5.3**.

Step 3. Determine the coordinates of the intersection point. This ordered pair is the system's solution. Using **Figure 5.3**, it appears that the lines intersect at $(1, -1)$. The "apparent" solution of the system is $(1, -1)$.

Step 4. Check the solution in both equations.

Check $(1, -1)$ in	**Check $(1, -1)$ in**
$y = -3x + 2$:	$y = 5x - 6$:
$-1 \overset{?}{=} -3 \cdot 1 + 2$	$-1 \overset{?}{=} 5 \cdot 1 - 6$
$-1 \overset{?}{=} -3 + 2$	$-1 \overset{?}{=} 5 - 6$
$-1 = -1$, true	$-1 = -1$, true

Because both equations are satisfied, $(1, -1)$ is the solution and the solution set is $\{(1, -1)\}$. ▬

☑ **CHECK POINT 3** Solve by graphing:

$$y = -x + 6$$
$$y = 3x - 6.$$

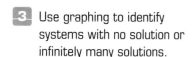

Use graphing to identify systems with no solution or infinitely many solutions.

Linear Systems Having No Solution or Infinitely Many Solutions

We have seen that a system of linear equations in two variables represents a pair of lines. The lines either intersect at one point, are parallel, or are identical. Thus, there are three possibilities for the number of solutions to a system of two linear equations.

The Number of Solutions to a System of Two Linear Equations

The number of solutions to a system of two linear equations in two variables is given by one of the following. (See **Figure 5.4**.)

Number of Solutions	What This Means Graphically
Exactly one ordered-pair solution	The two lines intersect at one point.
No solution	The two lines are parallel.
Infinitely many solutions	The two lines are identical.

Exactly one solution   No solution (parallel lines)   Infinitely many solutions (Lines are identical, or coincide.)

FIGURE 5.4 Possible graphs for a system of two linear equations in two variables

A linear system with no solution is called an **inconsistent system**. If you attempt to solve such a system by graphing, you will obtain two parallel lines. The solution set is the empty set, $\varnothing$.

EXAMPLE 4   A System with No Solution

Solve by graphing:

$$y = 2x - 1$$
$$y = 2x + 3.$$

Solution   Compare the slopes and y-intercepts in the two equations.

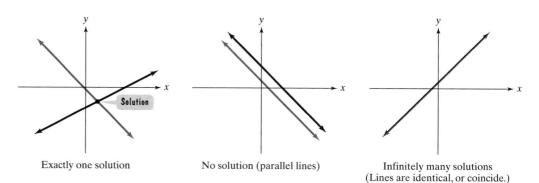

The lines have the same slope, 2.

$$y = 2x - 1 \qquad y = 2x + 3$$

The lines have different y-intercepts, −1 and 3.

Figure 5.5 shows the graphs of the two equations. Because both equations have the same slope, 2, but different y-intercepts, the lines are parallel. The system is inconsistent and has no solution. The solution set is the empty set, $\varnothing$. ∎

FIGURE 5.5 The graph of an inconsistent system

☑ **CHECK POINT 4**   Solve by graphing:

$$y = 3x - 2$$
$$y = 3x + 1.$$

EXAMPLE 5   A System with Infinitely Many Solutions

Solve by graphing:

$$2x + y = 3$$
$$4x + 2y = 6.$$

Solution   We use intercepts to graph each equation.

• $2x + y = 3$

x-intercept

$$2x + 0 = 3$$
$$2x = 3$$
$$x = \frac{3}{2}$$

Graph $(\frac{3}{2}, 0)$ and $(0, 3)$.

y-intercept

$$2 \cdot 0 + y = 3$$
$$y = 3$$

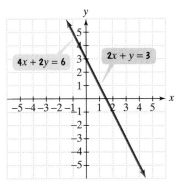

FIGURE 5.6   The graph of a system with infinitely many solutions

• $4x + 2y = 6$

x-intercept

$$4x + 2 \cdot 0 = 6$$
$$4x = 6$$
$$x = \frac{6}{4} = \frac{3}{2}$$

y-intercept

$$4 \cdot 0 + 2y = 6$$
$$2y = 6$$
$$y = 3$$

Graph $(\frac{3}{2}, 0)$ and $(0, 3)$.

Both lines have the same x-intercept, $\frac{3}{2}$, or 1.5, and the same y-intercept, 3. Thus, the graphs of the two equations in the system are the same line, shown in **Figure 5.6**. The two equations have the same solutions. Any ordered pair that is a solution of one equation is a solution of the other, and, consequently, a solution of the system. The system has an infinite number of solutions, namely all points that are solutions of either line.

We express the solution set for the system in one of two equivalent ways:

$$\{(x, y) \,|\, 2x + y = 3\} \quad \text{or} \quad \{(x, y) \,|\, 4x + 2y = 6\}.$$

The set of all ordered pairs (x, y) such that $2x + y = 3$

The set of all ordered pairs (x, y) such that $4x + 2y = 6$

Study Tip

Although the system in Example 5 has infinitely many solutions, this does not mean that any ordered pair of numbers you can form will be a solution. The ordered pair (x, y) must satisfy one of the system's equations, $2x + y = 3$ or $4x + 2y = 6$, and there are infinitely many such ordered pairs. Because the graphs are coinciding lines, the ordered pairs that are solutions of one of the equations are also solutions of the other equation.

Take a second look at the two equations, $2x + y = 3$ and $4x + 2y = 6$, in Example 5. If you multiply both sides of the first equation, $2x + y = 3$, by 2, you will obtain the second equation, $4x + 2y = 6$.

$$2x + y = 3 \qquad \text{This is the first equation in the system.}$$
$$2(2x + y) = 2 \cdot 3 \qquad \text{Multiply both sides by 2.}$$
$$2 \cdot 2x + 2y = 2 \cdot 3 \qquad \text{Use the distributive property.}$$

This is the second equation in the system. $4x + 2y = 6$ Simplify.

Because $2x + y = 3$ and $4x + 2y = 6$ are different forms of the same equation, these equations are called *dependent equations*. In general, the equations in a linear system with infinitely many solutions are called **dependent equations**.

☑ **CHECK POINT 5** Solve by graphing:

$$x + y = 3$$
$$2x + 2y = 6.$$

---

4 Use graphs of linear systems to solve problems.

Applications

One advantage of solving linear systems by graphing is that it allows us to "see" the equations and visualize the solution. This idea is demonstrated in Example 6.

EXAMPLE 6 Modeling Options for a Toll

The toll to a bridge costs $2.50. Commuters who use the bridge frequently have the option of purchasing a monthly coupon book for $21.00. With the coupon book, the toll is reduced to $1.00. The monthly cost, y, of using the bridge x times can be modeled by the following linear equations:

Without the coupon book:

$$y = 2.50x$$ The monthly cost, y, is $2.50 times the number of times, x, that the bridge is used.

With the coupon book:

$$y = 21 + 1 \cdot x$$ The monthly cost, y, is $21 for the book plus $1 times the number of times, x, that the bridge is used.

$$y = 21 + x$$ Simplify: $1 \cdot x = x$.

Expressing 2.50 as $\frac{5}{2}$, the options for the toll can be modeled by the linear system

$$y = \frac{5}{2}x$$ Without coupon book
$$y = x + 21.$$ With coupon book

a. Use graphing to solve the system.

b. Interpret the coordinates of the solution in practical terms.

Solution

a. Notice that x, the number of times the bridge is used in a month, and y, the monthly cost, are nonnegative. Thus, we will use only the first quadrant and its boundary to solve the system. We begin by identifying slopes and y-intercepts.

$$y = \frac{5}{2}x + 0 \qquad\qquad y = x + 21$$

The slope is $\frac{5}{2}$.   The y-intercept is 0.   The slope is 1.   The y-intercept is 21.

Using slopes and *y*-intercepts, we graph the two lines, as shown in **Figure 5.7**.

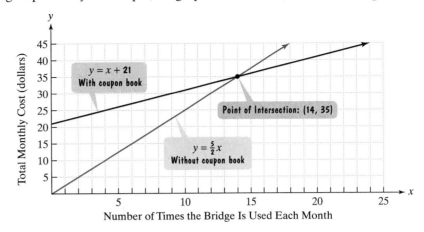

FIGURE 5.7   Options for a Toll

Using **Figure 5.7**, it appears that the lines intersect at $(14, 35)$. Let's check this solution in both equations.

Check (14, 35) in

$$y = \frac{5}{2}x:$$

$$35 \stackrel{?}{=} \frac{5}{2} \cdot 14$$

$$35 \stackrel{?}{=} \frac{5}{\underset{1}{2}} \cdot \frac{\overset{7}{\cancel{14}}}{1}$$

$$35 = 35, \quad \text{true}$$

Check (14, 35), in

$$y = x + 21:$$

$$35 \stackrel{?}{=} 14 + 21$$

$$35 = 35, \quad \text{true}$$

Because both equations are satisfied, $(14, 35)$ is the solution and the solution set is $\{(14, 35)\}$.

b. The graphs intersect at $(14, 35)$. This means that if the bridge is used 14 times in a month, the total monthly cost without the coupon book is the same as the total monthly cost with the coupon book, namely \$35.   ■

In **Figure 5.7**, look at the two graphs to the right of the intersection point $(14, 35)$. The red graph of $y = x + 21$ lies below the blue graph of $y = \frac{5}{2}x$. This means that if the bridge is used more than 14 times in a month ($x > 14$), the (red) monthly cost, y, with the coupon book is cheaper than the (blue) monthly cost, y, without the coupon book.

✓ **CHECK POINT 6**   The toll to a bridge costs \$2.00. If you use the bridge x times in a month, the monthly cost, y, is $y = 2x$. With a \$10 coupon book, the toll is reduced to \$1.00. The monthly cost, y, of using the bridge x times in a month with the coupon book is $y = x + 10$.

a. Solve the system by graphing:

$$y = 2x$$
$$y = x + 10.$$

Suggestion: Let the *x*-axis extend from 0 to 20 and let the *y*-axis extend from 0 to 40.

b. Interpret the coordinates of the solution in practical terms.

Blitzer Bonus

Missing America

Mary Katherine Campbell,
Miss America 1922

Here she is, Miss America, the icon of American beauty. Always thin, she is becoming more so. The scatter plot in the figure shows Miss America's body-mass index, a ratio comparing weight divided by the square of height. Two lines are also shown: a line that passes near the data points and a horizontal line representing the World Health Organization's cutoff point for undernutrition. The intersection point indicates that in approximately 1978, Miss America reached this cutoff. There she goes: If the trend continues, Miss America's body-mass index could reach zero in about 320 years.

Body-Mass Index of Miss America

Source: Johns Hopkins School of Public Health

5.1 EXERCISE SET   MyMathLab

PRACTICE   WATCH   DOWNLOAD   READ   REVIEW

Practice Exercises

In Exercises 1–10, determine whether the given ordered pair is a solution of the system.

1. $(2, -3)$
$2x + 3y = -5$
$7x - 3y = 23$

2. $(-2, -5)$
$6x - 2y = -2$
$3x + y = -11$

3. $\left(\dfrac{2}{3}, \dfrac{1}{9}\right)$
$x + 3y = 1$
$4x + 3y = 3$

4. $\left(\dfrac{7}{25}, -\dfrac{1}{25}\right)$
$4x + 3y = 1$
$3x - 4y = 1$

5. $(-5, 9)$
$5x + 3y = 2$
$x + 4y = 14$

6. $(10, 7)$
$6x - 5y = 25$
$4x + 15y = 13$

7. $(1400, 450)$
$x - 2y = 500$
$0.03x + 0.02y = 51$

8. $(200, 700)$
$-4x + y = -100$
$0.05x - 0.06y = -32$

9. $(8, 5)$
$5x - 4y = 20$
$3y = 2x + 1$

10. $(5, -2)$
$4x - 3y = 26$
$x = 15 - 5y$

In Exercises 11–42, solve each system by graphing. If there is no solution or an infinite number of solutions, so state. Use set notation to express solution sets.

11. $x + y = 6$
$x - y = 2$

12. $x + y = 2$
$x - y = 4$

13. $x + y = 1$
$y - x = 3$

14. $x + y = 4$
$y - x = 4$

15. $2x - 3y = 6$
$4x + 3y = 12$

16. $x + 2y = 2$
$x - y = 2$

17. $4x + y = 4$
$3x - y = 3$

18. $5x - y = 10$
$2x + y = 4$

19. $y = x + 5$
$y = -x + 3$

20. $y = x + 1$
$y = 3x - 1$

21. $y = 2x$
$y = -x + 6$

22. $y = 2x + 1$
$y = -2x - 3$

23. $y = -2x + 3$
$y = -x + 1$

24. $y = 3x - 4$
$y = -2x + 1$

25. $y = 2x - 1$
$y = 2x + 1$

26. $y = 3x - 1$
$y = 3x + 2$

27. $x + y = 4$
$x = -2$

28. $x + y = 6$
$y = -3$

29. $x - 2y = 4$
$2x - 4y = 8$

30. $2x + 3y = 6$
$4x + 6y = 12$

31. $y = 2x - 1$
$x - 2y = -4$

32. $y = -2x - 4$
$4x - 2y = 8$

33. $x + y = 5$
$2x + 2y = 12$

34. $x - y = 2$
$3x - 3y = -6$

35. $x - y = 0$
$y = x$

36. $2x - y = 0$
$y = 2x$

37. $x = 2$
$y = 4$

38. $x = 3$
$y = 5$

39. $x = 2$
$x = -1$

40. $x = 3$
$x = -2$

41. $y = 0$
$y = 4$

42. $y = 0$
$y = 5$

Practice PLUS

In Exercises 43–50, find the slope and the y-intercept for the graph of each equation in the given system. Use this information (and not the equations' graphs) to determine if the system has no solution, one solution, or an infinite number of solutions.

43. $y = \dfrac{1}{2}x - 3$

$y = \dfrac{1}{2}x - 5$

44. $y = \dfrac{3}{4}x - 2$

$y = \dfrac{3}{4}x + 1$

45. $y = -\dfrac{1}{2}x + 4$

$3x - y = -4$

46. $y = -\dfrac{1}{4}x + 3$

$4x - y = -3$

47. $3x - y = 6$

$x = \dfrac{y}{3} + 2$

48. $2x - y = 4$

$x = \dfrac{y}{2} + 2$

49. $3x + y = 0$
$y = -3x + 1$

50. $2x + y = 0$
$y = -2x + 1$

Application Exercises

51. A rental company charges $40.00 a day plus $0.35 per mile to rent a moving truck. The total cost, y, for a day's rental if x miles are driven is described by $y = 0.35x + 40$. A second company charges $36.00 a day plus $0.45 per mile, so the daily cost, y, if x miles are driven is described by $y = 0.45x + 36$. The graphs of the two equations are shown in the same rectangular coordinate system.

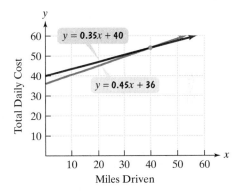

a. What is the x-coordinate of the intersection point of the graphs? Describe what this x-coordinate means in practical terms.

b. What is a reasonable estimate for the y-coordinate of the intersection point?

c. Substitute the x-coordinate of the intersection point into each of the equations and find the corresponding value for y. Describe what this value represents in practical terms. How close is this value to your estimate from part (b)?

52. A band plans to record a demo. Studio A rents for $100 plus $50 per hour. Studio B rents for $50 plus $75 per hour. The total cost, y, in dollars, of renting the studios for x hours can be modeled by the linear system

$$y = 50x + 100 \quad \text{Studio A}$$
$$y = 75x + 50. \quad \text{Studio B}$$

a. Use graphing to solve the system. Extend the x-axis from 0 to 4 and let each tick mark represent 1 unit (one hour in a recording studio). Extend the y-axis from 0 to 400 and let each tick mark represent 100 units (a rental cost of $100).

b. Interpret the coordinates of the solution in practical terms.

53. You plan to start taking an aerobics class. Nonmembers pay $4 per class. Members pay a $10 monthly fee plus an additional $2 per class. The monthly cost, y, of taking x classes can be modeled by the linear system

$$y = 4x \qquad \text{Nonmembers}$$
$$y = 2x + 10. \qquad \text{Members}$$

a. Use graphing to solve the system.

b. Interpret the coordinates of the solution in practical terms.

Writing in Mathematics

54. What is a system of linear equations? Provide an example with your description.

55. What is a solution of a system of linear equations?

56. Explain how to determine if an ordered pair is a solution of a system of linear equations.

57. Explain how to solve a system of linear equations by graphing.

58. What is an inconsistent system? What happens if you attempt to solve such a system by graphing?

59. Explain how a linear system can have infinitely many solutions.

60. What are dependent equations? Provide an example with your description.

61. The following system models the winning times, y, in seconds, in the Olympic 500-meter speed skating event x years after 1970:

$$y = -0.19x + 43.7 \qquad \text{Women}$$
$$y = -0.16x + 39.9. \qquad \text{Men}$$

Use the slope of each model to explain why the system has a solution. What does this solution represent?

Critical Thinking Exercises

Make Sense? *In Exercises 62–65, determine whether each statement "makes sense" or "does not make sense" and explain your reasoning.*

62. Each equation in a system of linear equations has infinitely many ordered-pair solutions.

63. Every linear system has infinitely many ordered-pair solutions.

64. In dependent systems, the two equations represent the same line.

65. When I use graphing to solve an inconsistent system, the lines should look parallel, and I can always use slope to confirm that they really are.

In Exercises 66–69, determine whether each statement is true or false. If the statement is false, make the necessary change(s) to produce a true statement.

66. If a linear system has graphs with equal slopes, the system must be inconsistent.

67. If a linear system has graphs with equal y-intercepts, the system must have infinitely many solutions.

68. If a linear system has two distinct points that are solutions, then the graphs of the system's equations have equal slopes and equal y-intercepts.

69. It is possible for a linear system with one solution to have graphs with equal slopes.

70. Write a system of linear equations whose solution is $(5, 1)$. How many different systems are possible? Explain.

71. Write a system of equations with one solution, a system of equations with no solution, and a system of equations with infinitely many solutions. Explain how you were able to think of these systems.

Technology Exercises

72. Verify your solutions to any five exercises from Exercises 11 through 36 by using a graphing utility to graph the two equations in the system in the same viewing rectangle. After entering the two equations, one as y_1 and the other as y_2, and graphing them, use the INTERSECTION feature to find the system's solution. (It may first be necessary to solve the equations for y before entering them.)

Read Exercise 72. Then use a graphing utility to solve the systems in Exercises 73–80.

73.
$$y = 2x + 2$$
$$y = -2x + 6$$

74.
$$y = -x + 5$$
$$y = x - 7$$

75.
$$x + 2y = 2$$
$$x - y = 2$$

76.
$$2x - 3y = 6$$
$$4x + 3y = 12$$

77.
$$3x - y = 5$$
$$-5x + 2y = -10$$

78.
$$2x - 3y = 7$$
$$3x + 5y = 1$$

79.
$$y = \frac{1}{3}x + \frac{2}{3}$$
$$y = \frac{5}{7}x - 2$$

80.
$$y = -\frac{1}{2}x + 2$$
$$y = \frac{3}{4}x + 7$$

Review Exercises

In Exercises 81–83, perform the indicated operation.

81. $-3 + (-9)$ (Section 1.7, **Table 1.7**)

82. $-3 - (-9)$ (Section 1.7, **Table 1.7**)

83. $-3(-9)$ (Section 1.7, **Table 1.7**)

Preview Exercises

Exercises 84–86 will help you prepare for the material covered in the next section. In each exercise, solve the given equation.

84. $4x - 3(-x - 1) = 24$

85. $5(2y - 3) - 4y = 9$

86. $(5x - 1) + 1 = 5x + 5$

Solving Systems of Linear Equations by the Substitution Method

Objectives

1 Solve linear systems by the substitution method.

2 Use the substitution method to identify systems with no solution or infinitely many solutions.

3 Solve problems using the substitution method.

Other than outrage, what is going on at the gas pumps? Is surging demand creating the increasing oil prices? Like all things in a free market economy, the price of a commodity is based on supply and demand. In this section, we use a second method for solving linear systems, the *substitution method*, to understand this economic phenomenon.

1 Solve linear systems by the substitution method.

Eliminating a Variable Using the Substitution Method

Finding the solution of a linear system by graphing equations may not be easy to do. For example, a solution of $\left(-\frac{2}{3}, \frac{157}{29}\right)$ would be difficult to "see" as an intersection point on a graph.

Let's consider a method that does not depend on finding a system's solution visually: the substitution method. This method involves converting the system to one equation in one variable by an appropriate substitution.

Study Tip

In step 1, you can choose which variable to isolate in which equation. If possible, solve for a variable whose coefficient is 1 or −1 to avoid working with fractions.

Solving Linear Systems by Substitution

1. Solve either of the equations for one variable in terms of the other. (If one of the equations is already in this form, you can skip this step.)
2. Substitute the expression found in step 1 into the *other* equation. This will result in an equation in one variable.
3. Solve the equation containing one variable.
4. Back-substitute the value found in step 3 into the equation from step 1. Simplify and find the value of the remaining variable.
5. Check the proposed solution in both of the system's given equations.

EXAMPLE 1 Solving a System by Substitution

Solve by the substitution method:

$$y = -x - 1$$
$$4x - 3y = 24.$$

Solution

Step 1. Solve either of the equations for one variable in terms of the other. This step has already been done for us. The first equation, $y = -x - 1$, is solved for y in terms of x.

Step 2. Substitute the expression from step 1 into the other equation. We substitute the expression $-x - 1$ for y into the second equation, $4x - 3y = 24$:

$$y = \boxed{-x - 1} \qquad 4x - 3\boxed{y} = 24. \quad \text{Substitute } -x - 1 \text{ for } y.$$

This gives us an equation in one variable, namely

$$4x - 3(-x - 1) = 24.$$

The variable y has been eliminated.

Step 3. Solve the resulting equation containing one variable.

$$
\begin{aligned}
4x - 3(-x - 1) &= 24 & &\text{This is the equation containing one variable.} \\
4x + 3x + 3 &= 24 & &\text{Apply the distributive property.} \\
7x + 3 &= 24 & &\text{Combine like terms.} \\
7x &= 21 & &\text{Subtract 3 from both sides.} \\
x &= 3 & &\text{Divide both sides by 7.}
\end{aligned}
$$

Using Technology

A graphing utility can be used to solve the system in Example 1. Graph each equation and use the intersection feature. The utility displays the solution $(3, -4)$ as $x = 3$, $y = -4$.

Intersection
X=3   Y=-4

[–10, 10, 1] by [–10, 10, 1]

Step 4. Back-substitute the obtained value into the equation from step 1. We now know that the x-coordinate of the solution is 3. To find the y-coordinate, we back-substitute the x-value in the equation from step 1,

$$y = -x - 1. \qquad \text{This is the equation from step 1.}$$

Substitute 3 for x.

$$
\begin{aligned}
y &= -3 - 1 \\
y &= -4 \qquad \text{Simplify.}
\end{aligned}
$$

With $x = 3$ and $y = -4$, the proposed solution is $(3, -4)$.

Step 5. Check the proposed solution in both of the system's given equations. Replace x with 3 and y with -4.

$$
\begin{array}{ll}
y = -x - 1 & 4x - 3y = 24 \\
-4 \overset{?}{=} -3 - 1 & 4(3) - 3(-4) \overset{?}{=} 24 \\
-4 = -4, \quad \text{true} & 12 + 12 \overset{?}{=} 24 \\
& 24 = 24, \quad \text{true}
\end{array}
$$

The pair $(3, -4)$ satisfies both equations. The system's solution is $(3, -4)$ and the solution set is $\{(3, -4)\}$. ∎

✓ **CHECK POINT 1** Solve by the substitution method:

$$
\begin{aligned}
y &= 5x - 13 \\
2x + 3y &= 12.
\end{aligned}
$$

EXAMPLE 2 Solving a System by Substitution

Solve by the substitution method:

$$
\begin{aligned}
5x - 4y &= 9 \\
x - 2y &= -3.
\end{aligned}
$$

Solution

Step 1. Solve either of the equations for one variable in terms of the other. We begin by isolating one of the variables in either of the equations. By solving for x in the second equation, which has a coefficient of 1, we can avoid fractions.

$$
\begin{aligned}
x - 2y &= -3 & &\text{This is the second equation in the given system.} \\
x &= 2y - 3 & &\text{Solve for } x \text{ by adding } 2y \text{ to both sides.}
\end{aligned}
$$

Step 2. Substitute the expression from step 1 into the other equation. We substitute $2y - 3$ for x in the first equation, $5x - 4y = 9$.

$$x = \boxed{2y - 3} \quad 5\boxed{x} - 4y = 9$$

This gives us an equation in one variable, namely

$$5(2y - 3) - 4y = 9.$$

The variable x has been eliminated.

Step 3. Solve the resulting equation containing one variable.

$$
\begin{array}{ll}
5(2y - 3) - 4y = 9 & \text{This is the equation containing one variable.} \\
10y - 15 - 4y = 9 & \text{Apply the distributive property.} \\
6y - 15 = 9 & \text{Combine like terms.} \\
6y = 24 & \text{Add 15 to both sides.} \\
y = 4 & \text{Divide both sides by 6.}
\end{array}
$$

Step 4. Back-substitute the obtained value into the equation from step 1. Now that we have the y-coordinate of the solution, we back-substitute 4 for y in the equation $x = 2y - 3$.

$$
\begin{array}{ll}
x = 2y - 3 & \text{Use the equation obtained in step 1.} \\
x = 2(4) - 3 & \text{Substitute 4 for y.} \\
x = 8 - 3 & \text{Multiply.} \\
x = 5 & \text{Subtract.}
\end{array}
$$

Study Tip

Get into the habit of checking ordered-pair solutions in *both* equations of the system.

With $x = 5$ and $y = 4$, the proposed solution is $(5, 4)$.

Step 5. Check. Take a moment to show that $(5, 4)$ satisfies both given equations. The solution is $(5, 4)$ and the solution set is $\{(5, 4)\}$. ∎

☑ **CHECK POINT 2**   Solve by the substitution method:

$$
\begin{aligned}
3x + 2y &= -1 \\
x - y &= 3.
\end{aligned}
$$

2 Use the substitution method to identify systems with no solution or infinitely many solutions.

The Substitution Method with Linear Systems Having No Solution or Infinitely Many Solutions

Recall that a linear system with no solution is called an **inconsistent system**. If you attempt to solve such a system by substitution, you will eliminate both variables. A false statement such as $0 = 5$ will be the result.

EXAMPLE 3   **Using the Substitution Method on an Inconsistent System**

Solve the system:

$$
\begin{aligned}
y + 1 &= 5(x + 1) \\
y &= 5x - 1.
\end{aligned}
$$

Using Technology

A graphing utility was used to graph the equations in Example 3. The lines are parallel and have no point of intersection. This verifies that the system is inconsistent.

$[-5, 5, 1]$ by $[-5, 5, 1]$

Solution   The variable y is isolated in the second equation, $y = 5x - 1$. We use the substitution method and substitute the expression for y in the first equation, $y + 1 = 5(x + 1)$.

$$\boxed{y} + 1 = 5(x + 1) \qquad y = \boxed{5x - 1}$$ Substitute $5x - 1$ for y.

$$(5x - 1) + 1 = 5(x + 1)$$ This substitution gives an equation in one variable.

$$5x = 5x + 5$$ Simplify on the left side. Use the distributive property on the right side.

There are no values of x and y for which $0 = 5$. $\qquad 0 = 5,$ false Subtract 5x from both sides.

The false statement $0 = 5$ indicates that the system is inconsistent and has no solution. The solution set is the empty set, $\varnothing$.   ■

✓ **CHECK POINT 3**   Solve the system:

$$3x + y = -5$$
$$y = -3x + 3.$$

Do you remember that the equations in a linear system with infinitely many solutions are called **dependent**? If you attempt to solve such a system by substitution, you will eliminate both variables. However, a true statement such as $6 = 6$ will be the result.

EXAMPLE 4   Using the Substitution Method on a System with Infinitely Many Solutions

Solve the system:

$$y = 3 - 2x$$
$$4x + 2y = 6.$$

Solution   The variable y is isolated in the first equation. We use the substitution method and substitute the expression for y in the second equation.

$$y = \boxed{3 - 2x} \qquad 4x + 2\boxed{y} = 6$$ Substitute $3 - 2x$ for y.

$$4x + 2(3 - 2x) = 6$$ This substitution gives an equation in one variable.

$$4x + 6 - 4x = 6$$ Apply the distributive property:
$$2(3 - 2x) = 2 \cdot 3 - 2 \cdot 2x = 6 - 4x.$$

$$6 = 6, \quad \text{true}$$ Simplify: $4x - 4x = 0$.

This true statement indicates that the system contains dependent equations and has infinitely many solutions. We express the solution set for the system in one of two equivalent ways:

$$\{(x, y) \,|\, y = 3 - 2x\}$$ The set of all ordered pairs (x, y) such that $y = 3 - 2x$

or $\{(x, y) \,|\, 4x + 2y = 6\}$. The set of all ordered pairs (x, y) such that $4x + 2y = 6$ ■

✓ **CHECK POINT 4**   Solve the system:

$$y = 3x - 4$$
$$9x - 3y = 12.$$

3 Solve problems using the substitution method.

Applications

An important application of systems of equations arises in connection with supply and demand. As the price of a product increases, the demand for that product decreases. However, at higher prices suppliers are willing to produce greater quantities of the product.

EXAMPLE 5   Supply and Demand Models

Table 5.1 shows the price of a slice of pizza. For each price, the table lists the number of slices that consumers are willing to buy and the number of slices that pizzerias are willing to supply.

Table 5.1   Supply and Demand for Pizza Slices

Price of a slice of pizza	Quantity demanded per day	Quantity supplied per day	Result
$0.50	300	100	There is a shortage from excess demand.
$1.00	250	150	
$2.00	150	250	There is a surplus from excess supply.
$3.00	50	350	

Source: O'Sullivan and Sheffrin, *Economics*, Prentice Hall, 2007

The data in **Table 5.1** can be described by the following demand and supply models:

Demand Model	**Supply Model**
$p = -0.01x + 3.5$	$p = 0.01x - 0.5.$

Price per slice   Number of slices demanded per day   Price per slice   Number of slices supplied per day

The price at which supply and demand are equal is called the **equilibrium price**. The quantity supplied and demanded at that price is called the **equilibrium quantity**. Find the equilibrium quantity and the equilibrium price for pizza slices.

Solution   We can find the equilibrium quantity and the equilibrium price by solving the demand-supply linear system. We will use substitution, substituting $-0.01x + 3.5$ for p in the second equation.

$$p = \boxed{-0.01x + 3.5} \quad \boxed{p} = 0.01x - 0.5 \qquad \text{Substitute } -0.01x + 3.5 \text{ for } p.$$

$$-0.01x + 3.5 = 0.01x - 0.5 \qquad \text{The resulting equation contains only one variable.}$$

$$-0.02x + 3.5 = -0.5 \qquad \text{Subtract 0.01x from both sides.}$$

$$-0.02x = -4.0 \qquad \text{Subtract 3.5 from both sides:}$$
$$-0.5 - 3.5 = -4.0.$$

$$x = 200 \qquad \text{Divide both sides by } -0.02: \dfrac{-4.0}{-0.02} = 200.$$

Because $x = 200$, the equilibrium quantity is 200 slices per day. To find the equilibrium price, we back-substitute 200 for x into either the demand or the supply model. We'll use both models to make sure we get the same number in each case.

Demand Model	**Supply Model**
$p = -0.01x + 3.5$	$p = 0.01x - 0.5$
Substitute 200 for x.	Substitute 200 for x.
$p = -0.01(200) + 3.5$	$p = 0.01(200) - 0.5$
$= -2 + 3.5 = 1.5$	$= 2 - 0.5 = 1.5$

The equilibrium price is $1.50 per slice. At that price, and only at that price, the quantity demanded and the quantity supplied are equal, at 200 slices per day. **Figure 5.8** shows the equilibrium point, (200, 1.50), as the intersection point for the graphs of the demand and supply models.

FIGURE 5.8

☑ **CHECK POINT 5** The following models describe demand and supply for two-bedroom rental apartments.

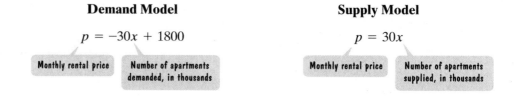

Demand Model	**Supply Model**
$p = -30x + 1800$	$p = 30x$
Monthly rental price	Number of apartments demanded, in thousands	Monthly rental price	Number of apartments supplied, in thousands

a. Solve the system and find the equilibrium quantity and the equilibrium price.

b. Use your answer from part (a) to complete this statement:

When rents are ____ per month, consumers will demand _____ apartments and suppliers will offer _____ apartments for rent.

Practice Exercises

In Exercises 1–32, solve each system by the substitution method. If there is no solution or an infinite number of solutions, so state. Use set notation to express solution sets.

1. $x + y = 4$
$y = 3x$

2. $x + y = 6$
$y = 2x$

3. $x + 3y = 8$
$y = 2x - 9$

4. $2x - 3y = -13$
$y = 2x + 7$

5. $x + 3y = 5$
$4x + 5y = 13$

6. $x + 2y = 5$
$2x - y = -15$

7. $2x - y = -5$
$x + 5y = 14$

8. $2x + 3y = 11$
$x - 4y = 0$

9. $2x - y = 3$
$5x - 2y = 10$

10. $-x + 3y = 10$
$2x + 8y = -6$

11. $-3x + y = -1$
$x - 2y = 4$

12. $-4x + y = -11$
$2x - 3y = 5$

13. $x = 9 - 2y$
$x + 2y = 13$

14. $6x + 2y = 7$
$y = 2 - 3x$

15. $y = 3x - 5$
$21x - 35 = 7y$

16. $9x - 3y = 12$
$y = 3x - 4$

17. $5x + 2y = 0$
$x - 3y = 0$

18. $4x + 3y = 0$
$2x - y = 0$

19. $2x - y = 6$
$3x + 2y = 5$

20. $2x - y = 4$
$3x - 5y = 2$

21. $2(x - 1) - y = -3$
$y = 2x + 3$

22. $x + y - 1 = 2(y - x)$
$y = 3x - 1$

23. $x = 2y + 9$
$x = 7y + 10$

24. $x = 5y - 3$
$x = 8y + 4$

25. $4x - y = 100$
$0.05x - 0.06y = -32$

26. $x + 6y = 8000$
$0.3x - 0.6y = 0$

27. $y = \frac{1}{3}x + \frac{2}{3}$
$y = \frac{5}{7}x - 2$

28. $y = -\frac{1}{2}x + 2$
$y = \frac{3}{4}x + 7$

29. $\frac{x}{6} - \frac{y}{2} = \frac{1}{3}$
$x + 2y = -3$

30. $\frac{x}{4} - \frac{y}{4} = -1$
$x + 4y = -9$

31. $2x - 3y = 8 - 2x$
$3x + 4y = x + 3y + 14$

32. $3x - 4y = x - y + 4$
$2x + 6y = 5y - 4$

Practice PLUS

In Exercises 33–38, write a system of equations modeling the given conditions. Then solve the system by the substitution method and find the two numbers.

33. The sum of two numbers is 81. One number is 41 more than the other. Find the numbers.

34. The sum of two numbers is 62. One number is 12 more than the other. Find the numbers.

35. The difference between two numbers is 5. Four times the larger number is 6 times the smaller number. Find the numbers.

36. The difference between two numbers is 25. Two times the larger number is 12 times the smaller number. Find the numbers.

37. The difference between two numbers is 1. The sum of the larger number and twice the smaller number is 7. Find the numbers.

38. The difference between two numbers is 5. The sum of the larger number and twice the smaller number is 14. Find the numbers.

In Exercises 39–40, multiply each equation in the system by an appropriate number so that the coefficients are integers. Then solve the system by the substitution method.

39. $0.7x - 0.1y = 0.6$
$0.8x - 0.3y = -0.8$

40. $1.25x - 0.01y = 4.5$
$0.5x - 0.02y = 1$

Application Exercises

41. The following models describe wages for low-skilled labor.

Demand Model

$$p = -0.325x + 5.8$$

| Price of labor (per hour) | Millions of workers employers will hire |

Supply Model

$$p = 0.375x + 3$$

| Price of labor (per hour) | Millions of available workers |

(*Source:* O'Sullivan and Sheffrin, *Economics*, Prentice Hall, 2007)

(*The exercise continues on the next page.*)

a. Solve the system on the bottom of the previous page, and find the equilibrium number of workers, in millions, and the equilibrium hourly wage.

b. Use your answer from part (a) to complete this statement:

If workers are paid _____ per hour, there will be __ million available workers and __ million workers will be hired.

c. In 2007, the federal minimum wage was set at \$5.15 per hour. Substitute 5.15 for p in the demand model, $p = -0.325x + 5.8$, and determine the millions of workers employers will hire at this price.

d. At a minimum wage of \$5.15 per hour, use the supply model, $p = 0.375x + 3$, to determine the millions of available workers. Round to one decimal place.

e. At a minimum wage of \$5.15 per hour, use your answers from parts (c) and (d) to determine how many more people are looking for work than employers are willing to hire.

42. The music business is evolving into a digital marketplace. The bar graph shows that from 2003 through 2005, CD album sales declined, while digital track sales grew.

Sales of Albums and Digital Tracks in the United States

Source: Nielsen Soundscan

The data can be described by the following mathematical models:

Album sales → $y = -23x + 656$

Sales, in millions / Number of years after 2003

Track sales → $y = 189x + 20$.

a. Solve the system.

b. Based on your solution from part (a), when did digital track sales catch up with album sales? How many digital tracks and how many albums were sold in that year?

Writing in Mathematics

43. Describe a problem that might arise when solving a system of equations using graphing. Assume that both equations in the system have been graphed correctly and the system has exactly one solution.

44. Explain how to solve a system of equations using the substitution method. Use $y = 3 - 3x$ and $3x + 4y = 6$ to illustrate your explanation.

45. When using the substitution method, how can you tell if a system of linear equations has no solution?

46. When using the substitution method, how can you tell if a system of linear equations has infinitely many solutions?

47. The law of supply and demand states that, in a free market economy, a commodity tends to be sold at its equilibrium price. At this price, the amount that the seller will supply is the same amount that the consumer will buy. Explain how systems of equations can be used to determine the equilibrium price.

Critical Thinking Exercises

Make Sense? *In Exercises 48–51, determine whether each statement "makes sense" or "does not make sense" and explain your reasoning.*

48. The substitution method provides me with solutions that might be awkward to determine by graphing.

49. When using substitution to solve

$$5x - 4y = 9$$
$$x - 2y = -3,$$

I find it easiest to solve for x in the first equation.

50. While solving a system using substitution, I eliminated both variables and obtained $6 = 6$, so the system has no solution.

51. I think of equilibrium as the point at which quantity demanded exceeds quantity supplied.

In Exercises 52–55, determine whether each statement is true or false. If the statement is false, make the necessary change(s) to produce a true statement.

52. Solving an inconsistent system by substitution results in a true statement.

53. The line passing through the intersection of the graphs of $x + y = 4$ and $x - y = 0$ with slope = 3 has an equation given by $y - 2 = 3(x - 2)$.

54. Unlike the graphing method, where solutions cannot be seen, the substitution method provides a way to visualize solutions as intersection points.

55. To solve the system

$$2x - y = 5$$
$$3x + 4y = 7$$

by substitution, replace y in the second equation with $5 - 2x$.

56. If $x = 3 - y - z$, $2x + y - z = -6$, and $3x - y + z = 11$, find the values for x, y, and z.

57. Find the value of m that makes

$$y = mx + 3$$
$$5x - 2y = 7$$

an inconsistent system.

Review Exercises

58. Graph: $4x + 6y = 12$. (Section 4.2, Example 4)

59. Solve: $4(x + 1) = 25 + 3(x - 3)$. (Section 2.3, Example 3)

60. List all the integers in this set: $\left\{ -73, -\dfrac{2}{3}, 0, \dfrac{3}{1}, \dfrac{3}{2}, \dfrac{\pi}{1} \right\}$.

(Section 1.3, Example 5)

Preview Exercises

Exercises 61–63 will help you prepare for the material covered in the next section.

61. Use both equations in the system

$$3x + 2y = 48$$
$$9x - 8y = -24$$

to find x for $y = 12$. What do you observe?

62. Solve: $-14y = -168$.

63. Multiply both sides of $x - 5y = 3$ by -4 and simplify.

SECTION 5.3

Solving Systems of Linear Equations by the Addition Method

Objectives

1 Solve linear systems by the addition method.

2 Use the addition method to identify systems with no solution or infinitely many solutions.

3 Determine the most efficient method for solving a linear system.

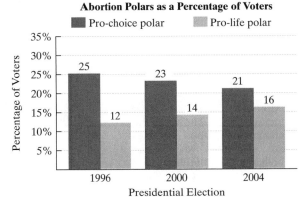

FIGURE 5.9

Source: Newsweek

In each of the 1996, 2000, and 2004 presidential elections, 37% of voters believed abortion should be "always legal" or "always illegal." **Figure 5.9** shows the breakdown of these abortion "polars" as a percentage of voters in each election. If these trends continue, when will the percentage of voters who are pro-choice polars be the same as the percentage of voters who are pro-life polars? The question can be answered by modeling the data with a system of linear equations and solving the system. However, the substitution method is not always the easiest way to solve linear systems. In this section, we consider a third method for solving these systems.

1 Solve linear systems by the addition method.

Eliminating a Variable Using the Addition Method

The substitution method is most useful if one of the given equations has an isolated variable. A third, and frequently the easiest, method for solving a linear system is the addition method. Like the substitution method, the addition method involves eliminating a variable and ultimately solving an equation containing only one variable. However, this time we eliminate a variable by adding the equations.

For example, consider the following system of linear equations:

$$3x - 4y = 11$$
$$-3x + 2y = -7.$$

When we add these two equations, the x-terms are eliminated. This occurs because the coefficients of the x-terms, 3 and -3, are opposites (additive inverses) of each other.

$$\begin{array}{r} 3x - 4y = 11 \\ -3x + 2y = -7 \\ \hline \text{Add:} \quad -2y = 4 \end{array}$$

> The sum is an equation in one variable.

$$y = -2 \quad \text{Divide both sides by } -2 \text{ and solve for } y.$$

Now we can back-substitute -2 for y into one of the original equations to find x. It does not matter which equation you use; you will obtain the same value for x in either case. If we use either equation, we can show that $x = 1$. The solution $(1, -2)$ satisfies both equations in the system.

When we use the addition method, we want to obtain two equations whose sum is an equation containing only one variable. The key step is to **obtain, for one of the variables, coefficients that differ only in sign**. To do this, we may need to multiply one or both equations by some nonzero number so that the coefficients of one of the variables, x or y, become opposites. Then when the two equations are added, this variable is eliminated.

EXAMPLE 1   Solving a System by the Addition Method

Solve by the addition method:

$$x + y = 4$$
$$x - y = 6.$$

Study Tip

Although the addition method is also known as the elimination method, variables are eliminated when using both the substitution and addition methods. The name *addition method* specifically tells us that the elimination of a variable is accomplished by adding two equations.

Solution   The coefficients of y in the two equations, 1 and -1, differ only in sign and are opposites. Therefore, by adding the two left sides and the two right sides, we can eliminate the y-terms.

$$\begin{array}{r} x + y = 4 \\ x - y = 6 \\ \hline \text{Add:} \quad 2x + 0y = 10 \\ 2x = 10 \quad \text{Simplify.} \end{array}$$

Now y is eliminated and we can solve $2x = 10$ for x.

$$2x = 10$$
$$x = 5 \quad \text{Divide both sides by 2 and solve for } x.$$

We back-substitute 5 for x into one of the original equations to find y. We will use both equations to show that we obtain the same value for y in either case.

Use the first equation:	**Use the second equation:**
$x + y = 4$	$x - y = 6$
$5 + y = 4$	$5 - y = 6$ Replace x with 5.
$y = -1.$	$-y = 1$ Solve for y.
	$y = -1.$

Thus, $x = 5$ and $y = -1$. The proposed solution, $(5, -1)$, can be shown to satisfy both equations in the system. Consequently, the solution is $(5, -1)$ and the solution set is $\{(5, -1)\}$. ∎

☑ **CHECK POINT 1**   Solve by the addition method:

$$x + y = 5$$
$$x - y = 9.$$

EXAMPLE 2 Solving a System by the Addition Method

Solve by the addition method:

$$3x - y = 11$$
$$2x + 5y = 13.$$

Solution We must rewrite one or both equations in equivalent forms so that the coefficients of the same variable (either x or y) are opposites. Consider the terms in y in each equation, that is, $-1y$ and $5y$. To eliminate y, we can multiply each term of the first equation by 5 and then add the equations.

$$3x - y = 11 \xrightarrow{\text{Multiply by 5.}} 15x - 5y = 55$$
$$2x + 5y = 13 \xrightarrow{\text{No change}} \underline{2x + 5y = 13}$$
$$\text{Add:} \quad 17x + 0y = 68$$
$$17x = 68 \quad \text{Simplify.}$$
$$x = 4 \quad \text{Divide both sides by 17}$$
$$\text{and solve for } x.$$

Thus, $x = 4$. To find y, we back-substitute 4 for x into either one of the given equations. We'll use the second equation.

$$2x + 5y = 13 \quad \text{This is the second equation in the given system.}$$
$$2 \cdot 4 + 5y = 13 \quad \text{Substitute 4 for x.}$$
$$8 + 5y = 13 \quad \text{Multiply: } 2 \cdot 4 = 8.$$
$$5y = 5 \quad \text{Subtract 8 from both sides.}$$
$$y = 1 \quad \text{Divide both sides by 5.}$$

The solution is $(4, 1)$. Check to see that it satisfies both of the original equations in the system. The solution set is $\{(4, 1)\}$. ■

✓ **CHECK POINT 2** Solve by the addition method:

$$4x - y = 22$$
$$3x + 4y = 26.$$

Before considering additional examples, let's summarize the steps for solving linear systems by the addition method.

Solving Linear Systems by Addition

1. If necessary, rewrite both equations in the form $Ax + By = C$.
2. If necessary, multiply either equation or both equations by appropriate nonzero numbers so that the sum of the x-coefficients or the sum of the y-coefficients is 0.
3. Add the equations in step 2. The sum is an equation in one variable.
4. Solve the equation in one variable.
5. Back-substitute the value obtained in step 4 into either of the given equations and solve for the other variable.
6. Check the solution in both of the original equations.

EXAMPLE 3   Solving a System by the Addition Method

Solve by the addition method:

$$3x + 2y = 48$$
$$9x - 8y = -24.$$

Solution

Step 1. Rewrite both equations in the form $Ax + By = C$. Both equations are already in this form. Variable terms appear on the left and constants appear on the right.

Step 2. If necessary, multiply either equation or both equations by appropriate numbers so that the sum of the x-coefficients or the sum of the y-coefficients is 0. We can eliminate x or y. Let's eliminate x. Consider the terms in x in each equation, that is, $3x$ and $9x$. To eliminate x, we can multiply each term of the first equation by -3 and then add the equations.

$$3x + 2y = 48 \xrightarrow{\text{Multiply by } -3.} -9x - 6y = -144$$
$$9x - 8y = -24 \xrightarrow{\text{No change}} \underline{9x - 8y = -24}$$

Step 3. Add the equations.   Add: $-14y = -168$

Step 4. Solve the equation in one variable. We solve $-14y = -168$ by dividing both sides by -14.

$$\frac{-14y}{-14} = \frac{-168}{-14} \quad \text{Divide both sides by } -14.$$
$$y = 12 \qquad \text{Simplify.}$$

Step 5. Back-substitute and find the value for the other variable. We can back-substitute 12 for y in either one of the given equations. We'll use the first one.

$$3x + 2y = 48 \quad \text{This is the first equation in the given system.}$$
$$3x + 2(12) = 48 \quad \text{Substitute 12 for } y.$$
$$3x + 24 = 48 \quad \text{Multiply.}$$
$$3x = 24 \quad \text{Subtract 24 from both sides.}$$
$$x = 8 \quad \text{Divide both sides by 3.}$$

The solution is $(8, 12)$.

Step 6. Check. Take a minute or so to show that $(8, 12)$ satisfies both of the original equations in the system. The solution set is $\{(8, 12)\}$.   ■

✓ **CHECK POINT 3**   Solve by the addition method:

$$4x + 5y = 3$$
$$2x - 3y = 7.$$

Some linear systems have solutions that are not integers. If the value of one variable turns out to be a "messy" fraction, back-substitution might lead to cumbersome arithmetic. If this happens, you can return to the original system and use addition to find the value of the other variable.

<table><tr><td>**EXAMPLE 4**</td><td>**Solving a System by the Addition Method**</td></tr></table>

Solve by the addition method:

$$2x = 7y - 17$$
$$5y = 17 - 3x.$$

Solution

Step 1. Rewrite both equations in the form $Ax + By = C$. We first arrange the system so that variable terms appear on the left and constants appear on the right. We obtain

$$2x - 7y = -17 \qquad \text{Subtract } 7y \text{ from both sides of the first equation.}$$
$$3x + 5y = 17. \qquad \text{Add } 3x \text{ to both sides of the second equation.}$$

Step 2. If necessary, multiply either equation or both equations by appropriate numbers so that the sum of the x-coefficients or the sum of the y-coefficients is 0. We can eliminate x or y. Let's eliminate x by multiplying the first equation by 3 and the second equation by -2.

$$2x - 7y = -17 \xrightarrow{\text{Multiply by 3.}} 6x - 21y = -51$$
$$3x + 5y = 17 \xrightarrow{\text{Multiply by } -2.} -6x - 10y = -34$$

Step 3. Add the equations. $\qquad$ Add: $\qquad -31y = -85$

Step 4. Solve the equation in one variable. We solve $-31y = -85$ by dividing both sides by -31.

$$\frac{-31y}{-31} = \frac{-85}{-31} \qquad \text{Divide both sides by } -31.$$
$$y = \frac{85}{31} \qquad \text{Simplify.}$$

Step 5. Back-substitute and find the value for the other variable. Back-substitution of $\frac{85}{31}$ for y into either of the given equations results in cumbersome arithmetic. Instead, let's use the addition method on the given system in the form $Ax + By = C$ to find the value for x. Thus, we eliminate y by multiplying the first equation by 5 and the second equation by 7.

$$2x - 7y = -17 \xrightarrow{\text{Multiply by 5.}} 10x - 35y = -85$$
$$3x + 5y = 17 \xrightarrow{\text{Multiply by 7.}} 21x + 35y = 119$$
$$\text{Add: } 31x = 34$$
$$x = \frac{34}{31} \qquad \text{Divide both sides by 31.}$$

The solution is $\left(\dfrac{34}{31}, \dfrac{85}{31}\right)$.

Step 6. Check. For this system, a calculator is helpful in showing that $\left(\frac{34}{31}, \frac{85}{31}\right)$ satisfies both of the original equations in the system. The solution set is $\left\{\left(\frac{34}{31}, \frac{85}{31}\right)\right\}$. ▬

☑ **CHECK POINT 4**   Solve by the addition method:

$$2x = 9 + 3y$$
$$4y = 8 - 3x.$$

2 Use the addition method to identify systems with no solution or infinitely many solutions.

The Addition Method with Linear Systems Having No Solution or Infinitely Many Solutions

As with the substitution method, if the addition method results in a false statement, the linear system is inconsistent and has no solution.

EXAMPLE 5   **Using the Addition Method on an Inconsistent System**

Solve the system:

$$4x + 6y = 12$$
$$6x + 9y = 12.$$

Solution   We can eliminate x or y. Let's eliminate x by multiplying the first equation by 3 and the second equation by -2.

$$
\begin{array}{ll}
4x + 6y = 12 & \xrightarrow{\text{Multiply by 3.}} \quad 12x + 18y = 36 \\
6x + 9y = 12 & \xrightarrow{\text{Multiply by -2.}} \quad \underline{-12x - 18y = -24} \\
& \qquad\qquad\text{Add:} \qquad\qquad 0 = 12
\end{array}
$$

There are no values of x and y for which $0 = 12$.

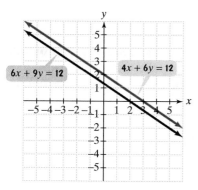

FIGURE 5.10 Visualizing the inconsistent system in Example 5

The false statement $0 = 12$ indicates that the system is inconsistent and has no solution. The solution set is the empty set, $\varnothing$. The graphs of the system's equations are shown in **Figure 5.10**. The lines are parallel and have no point of intersection.   ■

✓ **CHECK POINT 5**   Solve the system:

$$x + 2y = 4$$
$$3x + 6y = 13.$$

If you use the addition method, how can you tell if a system has infinitely many solutions? As with the substitution method, you will eliminate both variables and obtain a true statement.

EXAMPLE 6   **Using the Addition Method on a System with Infinitely Many Solutions**

Solve by the addition method:

$$2x - y = 3$$
$$-4x + 2y = -6.$$

Solution   We can eliminate y by multiplying the first equation by 2.

$$
\begin{array}{ll}
2x - y = 3 & \xrightarrow{\text{Multiply by 2.}} \quad 4x - 2y = 6 \\
-4x + 2y = -6 & \xrightarrow{\text{No change}} \quad \underline{-4x + 2y = -6} \\
& \qquad\qquad\text{Add:} \qquad\qquad 0 = 0
\end{array}
$$

The true statement $0 = 0$ indicates that the system contains dependent equations and has infinitely many solutions. Any ordered pair that satisfies the first equation also satisfies the second equation. We express the solution set for the system in one of two equivalent ways:

$$\{(x, y)\,|\,2x - y = 3\} \text{ or } \{(x, y)\,|\,-4x + 2y = -6\}.$$ ■

☑ **CHECK POINT 6** Solve by the addition method:

$$x - 5y = 7$$
$$3x - 15y = 21.$$

---

③ Determine the most efficient method for solving a linear system.

Comparing the Three Solution Methods

Table 5.2 compares the graphing, substitution, and addition methods for solving systems of linear equations. With increased practice, it will become easier for you to select the best method for solving a particular linear system.

Table 5.2	Comparing Solution Methods	
Method	**Advantages**	**Disadvantages**
Graphing	You can see the solutions.	If the solutions do not involve integers or are too large to be seen on the graph, it's impossible to tell exactly what the solutions are.
Substitution	Gives exact solutions. Easy to use if a variable is on one side by itself.	Solutions cannot be seen. Can introduce extensive work with fractions when no variable has a coefficient of 1 or −1.
Addition	Gives exact solutions. Easy to use even if no variable has a coefficient of 1 or −1.	Solutions cannot be seen.

5.3 EXERCISE SET

MyMathLab | **Math XL** PRACTICE | WATCH | DOWNLOAD | READ | REVIEW

Practice Exercises

In Exercises 1–44, solve each system by the addition method. If there is no solution or an infinite number of solutions, so state. Use set notation to express solution sets.

1. $x + y = -3$
$x - y = 11$

2. $x + y = 6$
$x - y = -2$

3. $2x + 3y = 6$
$2x - 3y = 6$

4. $3x + 2y = 14$
$3x - 2y = 10$

5. $x + 2y = 7$
$-x + 3y = 18$

6. $2x + y = -2$
$-2x - 3y = -6$

7. $5x - y = 14$
$-5x + 2y = -13$

8. $7x - 4y = 13$
$-7x + 6y = -11$

9. $3x + y = 7$
$2x - 5y = -1$

10. $3x - y = 11$
$2x + 5y = 13$

11. $x + 3y = 4$
$4x + 5y = 2$

12. $x + 2y = -1$
$4x - 5y = 22$

13. $-3x + 7y = 14$
$2x - y = -13$

14. $2x - 5y = -1$
$3x + y = 7$

15. $3x - 14y = 6$
$5x + 7y = 10$

16. $5x - 4y = 19$
$3x + 2y = 7$

17. $3x - 4y = 11$
$2x + 3y = -4$

18. $2x + 3y = -16$
$5x - 10y = 30$

19. $3x + 2y = -1$
$-2x + 7y = 9$

20. $5x + 3y = 27$
$7x - 2y = 13$

21. $3x = 2y + 7$
$5x = 2y + 13$

22. $9x = 25 + y$
$2y = 4 - 9x$

23. $2x = 3y - 4$
$-6x + 12y = 6$

24. $5x = 4y - 8$
$3x + 7y = 14$

25. $2x - y = 3$
$4x + 4y = -1$

26. $3x - y = 22$
$4x + 5y = -21$

27. $4x = 5 + 2y$
$2x + 3y = 4$

28. $3x = 4y + 1$
$4x + 3y = 1$

29. $3x - y = 1$
$3x - y = 2$

30. $4x - 9y = -2$
$-4x + 9y = -2$

31. $x + 3y = 2$
$3x + 9y = 6$

32. $4x - 2y = 2$
$2x - y = 1$

33. $7x - 3y = 4$
$-14x + 6y = -7$

34. $2x + 4y = 5$
$3x + 6y = 6$

35. $5x + y = 2$
$3x + y = 1$

36. $2x - 5y = -1$
$2x - y = 1$

37. $x = 5 - 3y$
$2x + 6y = 10$

38. $4x = 36 + 8y$
$3x - 6y = 27$

39. $4(3x - y) = 0$
$3(x + 3) = 10y$

40. $2(2x + 3y) = 0$
$7x = 3(2y + 3) + 2$

41. $x + y = 11$
$\dfrac{x}{5} + \dfrac{y}{7} = 1$

42. $x - y = -3$
$\dfrac{x}{9} - \dfrac{y}{7} = -1$

43. $\dfrac{4}{5}x - y = -1$
$\dfrac{2}{5}x + y = 1$

44. $\dfrac{x}{3} + y = 3$
$\dfrac{x}{2} - \dfrac{y}{4} = 1$

In Exercises 45–56, solve each system by the method of your choice. If there is no solution or an infinite number of solutions, so state. Use set notation to express solution sets. Explain why you selected one method over the other two.

45. $3x - 2y = 8$
$x = -2y$

46. $2x - y = 10$
$y = 3x$

47. $3x + 2y = -3$
$2x - 5y = 17$

48. $2x - 7y = 17$
$4x - 5y = 25$

49. $3x - 2y = 6$
$y = 3$

50. $2x + 3y = 7$
$x = 2$

51. $y = 2x + 1$
$y = 2x - 3$

52. $y = 2x + 4$
$y = 2x - 1$

53. $2(x + 2y) = 6$
$3(x + 2y - 3) = 0$

54. $2(x + y) = 4x + 1$
$3(x - y) = x + y - 3$

55. $3y = 2x$
$2x + 9y = 24$

56. $4y = -5x$
$5x + 8y = 20$

Practice PLUS

In Exercises 57–64, solve each system or state that the system is inconsistent or dependent.

57. $\dfrac{3x}{5} + \dfrac{4y}{5} = 1$
$\dfrac{x}{4} - \dfrac{3y}{8} = -1$

58. $\dfrac{x}{3} - \dfrac{y}{2} = \dfrac{2}{3}$
$\dfrac{2x}{3} + y = \dfrac{4}{3}$

59. $5(x + 1) = 7(y + 1) - 7$
$6(x + 1) + 5 = 5(y + 1)$

60. $6x = 5(x + y + 3) - x$
$3(x - y) + 4y = 5(y + 1)$

61. $0.4x + y = 2.2$
$0.5x - 1.2y = 0.3$

62. $1.25x - 1.5y = 2$
$3.5x - 1.75y = 10.5$

63. $\dfrac{x}{2} = \dfrac{y + 8}{3}$
$\dfrac{x + 2}{2} = \dfrac{y + 11}{3}$

64. $\dfrac{x}{2} = \dfrac{y + 8}{4}$
$\dfrac{x + 3}{2} = \dfrac{y + 5}{4}$

Application Exercises

65. We opened this section with data from the 1996, 2000, and 2004 presidential elections that showed abortion polars as a percentage of voters. (See **Figure 5.9** on page 309.) The data can be modeled by the following system of equations:

$$x + 2y = 50$$ The percentage of pro-choice polars, y, x years after 1996

$$-x + 2y = 24.$$ The percentage of pro-life polars, y, x years after 1996

Use the addition method to determine in which year the percentage of voters who are pro-choice polars will be the same as the percentage of voters who are pro-life polars. For that year, what percent will be pro-choice and what percent will be pro-life?

66. The bar graph shows the percentage of Americans for and against the death penalty for a person convicted of murder.

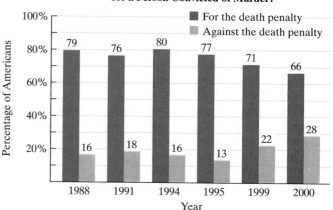

Are You in Favor of the Death Penalty for a Person Convicted of Murder?

Source: Newsweek poll

The data can be modeled by the following system of equations:

$$13x + 12y = 992$$ The percent, y, in favor of the death penalty x years after 1988

$$-x + y = 16.$$ The percent, y, against the death penalty x years after 1988

Use the addition method to determine in which year the percentage of Americans in favor of the death penalty will be the same as the percentage of Americans who oppose it. For that year, what percent will be for the death penalty and what percent will be against it?

Writing in Mathematics

67. Explain how to solve a system of equations using the addition method. Use $3x + 5y = -2$ and $2x + 3y = 0$ to illustrate your explanation.

68. When using the addition method, how can you tell if a system of linear equations has no solution?

69. When using the addition method, how can you tell if a system of linear equations has infinitely many solutions?

70. Take a second look at the data about the death penalty shown in Exercise 66. Do you think that these trends will continue? Explain your answer.

71. The formula $3239x + 96y = 134{,}014$ models the number of daily evening newspapers, y, x years after 1980. The formula $-665x + 36y = 13{,}800$ models the number of daily morning newspapers, y, x years after 1980. What is the most efficient method for solving this system? Explain why. What does the solution mean in terms of the variables in the formulas? (It is not necessary to actually solve the system.)

Critical Thinking Exercises

Make Sense? *In Exercises 72–75, determine whether each statement "makes sense" or "does not make sense" and explain your reasoning.*

72. When I use the addition method, I add the equations only if the x-coefficients or the y-coefficients are opposites.

73. Unlike substitution, the addition method lets me see solutions as intersection points of graphs.

74. When I use the addition method, I sometimes need to multiply more than one equation by a nonzero number before adding the equations.

75. I find it easiest to use the addition method when one of the equations has a variable on one side by itself.

In Exercises 76–79, determine whether each statement is true or false. If the statement is false, make the necessary change(s) to produce a true statement.

76. If x can be eliminated by the addition method, y cannot be eliminated by using the original equations of the system.

77. If $Ax + 2y = 2$ and $2x + By = 10$ have graphs that intersect at $(2, -2)$, then $A = -3$ and $B = 3$.

78. The equations $y = x - 1$ and $x = y + 1$ are dependent.

79. If the two equations in a linear system are $5x - 3y = 7$ and $4x + 9y = 11$, multiplying the first equation by 4, the second by 5, and then adding equations will eliminate x.

80. Solve by expressing x and y in terms of a and b:

$$x - y = a$$
$$y = 2x + b.$$

81. The point of intersection of the graphs of the equations $Ax - 3y = 16$ and $3x + By = 7$ is $(5, -2)$. Find A and B.

Review Exercises

82. For which number is 5 times the number equal to the number increased by 40? (Section 2.5, Example 1)

83. In which quadrant is $\left(-\frac{3}{2}, 15\right)$ located? (Section 4.1, Example 1)

84. Solve: $29{,}700 + 150x = 5000 + 1100x$. (Section 2.2, Example 7)

Preview Exercises

Exercises 85–87 will help you prepare for the material covered in the next section.

85. The sum of two numbers, x and y, is 28. The difference between the numbers is 6.

   a. Write a system of linear equations that models these conditions.

   b. Solve the system and find the numbers.

86. If a slice of cheese contains x calories and a glass of wine contains y calories, write an algebraic expression for the number of calories in 3 slices of cheese and 2 glasses of wine.

87. A long-distance telephone plan has a monthly fee of $20 with a charge of $0.05 per minute.

   a. What is the total monthly cost for the plan if there are 200 minutes of long-distance calls?

   b. Write a formula that describes the total monthly cost of the plan, y, for x minutes of long-distance calls.

MID-CHAPTER CHECK POINT | Section 5.1–Section 5.3

 What You Know: We learned how to solve systems of linear equations by graphing, by the substitution method, and by the addition method. We saw that some systems, called inconsistent systems, have no solution, whereas other systems, called dependent systems, have infinitely many solutions.

In Exercises 1–3, solve each system by graphing.

1. $3x + 2y = 6$
    $2x - y = 4$

2. $y = 2x - 1$
    $y = 3x - 2$

3. $y = 2x - 1$
    $6x - 3y = 12$

In Exercises 4–15, solve each system by the method of your choice.

4. $5x - 3y = 1$
$\quad y = 3x - 7$

5. $6x + 5y = 7$
$\quad 3x - 7y = 13$

6. $x = \dfrac{y}{3} - 1$
$\quad 6x + y = 21$

7. $3x - 4y = 6$
$\quad 5x - 6y = 8$

8. $3x - 2y = 32$
$\quad \dfrac{x}{5} + 3y = -1$

9. $x - y = 3$
$\quad 2x = 4 + 2y$

10. $x = 2(y - 5)$
$\quad 4x + 40 = y - 7$

11. $y = 3x - 2$
$\quad y = 2x - 9$

12. $2x - 3y = 4$
$\quad 3x + 4y = 0$

13. $y - 2x = 7$
$\quad 4x = 2y - 14$

14. $4(x + 3) = 3y + 7$
$\quad 2(y - 5) = x + 5$

15. $\dfrac{x}{2} - \dfrac{y}{5} = 1$
$\quad y - \dfrac{x}{3} = 8$

SECTION 5.4

Problem Solving Using Systems of Equations

Objective

1 Solve problems using linear systems.

"Bad television is three things: a bullet train to a morally bankrupt youth, a slow spiral into an intellectual void, and of course, a complete blast to watch."

—*Dennis Miller*

A blast, indeed. Americans spend more of their free time watching TV than any other activity. We open this section with a system of equations that models how we spend free time, including how many of those 1440 minutes in a day we spend glued to the tube.

1 Solve problems using linear systems.

A Strategy for Solving Word Problems Using Systems of Equations

When we solved problems in Chapters 2 and 3, we let *x* represent a quantity that was unknown. Problems in this section involve two unknown quantities. We will let *x* and *y* represent these quantities. We then translate from the verbal conditions of the problem to a *system* of linear equations.

> **EXAMPLE 1** Free Time

Figure 5.11 on the next page shows the average time per day on weekends that Americans spend on various leisure and sports activities.

**Average Time per Day on Weekends
Americans Devote to Leisure Activities**

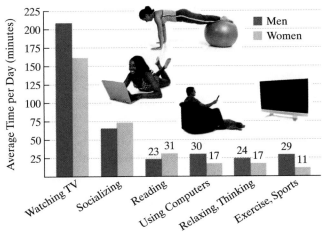

FIGURE 5.11

Source: Bureau of Labor Statistics American Time Use Survey

It appears that TV is our preferred way of goofing off. Each weekend day, the sum of the average times spent watching TV for men and women is 369 minutes. (That's more than 6 hours.) The difference between the average times spent watching TV for men and women is 47 minutes. How many minutes per day on weekends do men and women devote to watching TV?

Solution

Step 1. Use variables to represent unknown quantities.

Let x = average time per day men watch TV.

Let y = average time per day women watch TV.

Step 2. Write a system of equations that models the problem's conditions.

The sum of average times for men and women	is	369 minutes.

$$x + y = 369$$

The difference between average times for men and women	is	47 minutes.

$$x - y = 47$$

Step 3. Solve the system and answer the problem's question. The system

$$x + y = 369$$
$$x - y = 47$$

can easily be solved by addition because the y-coefficients are opposites. Adding equations will eliminate y.

$$
\begin{aligned}
x + y &= 369 \\
\underline{x - y} &= \underline{47} \\
\text{Add:} \quad 2x \quad &= 416 \\
x &= 208 \quad \text{Divide both sides by 2.}
\end{aligned}
$$

Because x represents average time per day men watch TV, we see that men devote 208 minutes per day on weekends (or 3 hours 28 minutes) to this leisure activity. Now we can find y, the time for women. We do so by back-substituting 208 for x in either of the system's equations.

$$x + y = 369 \quad \text{We'll use the first equation.}$$
$$208 + y = 369 \quad \text{Back-substitute 208 for x.}$$
$$y = 161 \quad \text{Subtract 208 from both sides.}$$

Because y represents average time per day women watch TV, we see that women devote 161 minutes per day on weekends (or 2 hours 41 minutes) to this leisure activity.

Step 4. Check the proposed answers in the original wording of the problem. We check the times, 208 minutes for men and 161 minutes for women, in each of the problem's conditions.

The sum of the average times spent watching TV for men and women is 369 minutes:

$$208 \text{ minutes} + 161 \text{ minutes} = 369 \text{ minutes}.$$

The difference between the average times spent watching TV for men and women is 47 minutes:

$$208 \text{ minutes} - 161 \text{ minutes} = 47 \text{ minutes}.$$

This verifies that men spend 208 minutes per day on weekends watching TV and women spend 161 minutes per day. ▬

☑ **CHECK POINT 1** **Figure 5.11** on page 319 indicates that socializing is our second-favorite leisure activity. Each weekend day, the sum of the average times spent socializing for men and women is 138 minutes. The difference between the average times spent socializing for women and men is 8 minutes. How many minutes per day on weekends do men and women devote to socializing?

EXAMPLE 2 Cholesterol and Heart Disease

The verdict is in. After years of research, the nation's health experts agree that high cholesterol in the blood is a major contributor to heart disease. Thus, cholesterol intake should be limited to 300 milligrams or less each day. Fast foods provide a cholesterol carnival. All together, two McDonald's Quarter Pounders and three Burger King Whoppers with cheese contain 520 milligrams of cholesterol. Three Quarter Pounders and one Whopper with cheese exceed the suggested daily cholesterol intake by 53 milligrams. Determine the cholesterol content in each item.

Solution

Step 1. Use variables to represent the unknown quantities.

Let x = the cholesterol content, in milligrams, of a Quarter Pounder.

Let y = the cholesterol content, in milligrams, of a Whopper with cheese.

Step 2. Write a system of equations that models the problem's conditions.

The amount of cholesterol in 2 Quarter Pounders	plus	the amount of cholesterol in 3 Whoppers with cheese	is	520 mg.
$2x$	$+$	$3y$	$=$	520

The amount of cholesterol in 3 Quarter Pounders	plus	the amount of cholesterol in 1 Whopper with cheese	is	the suggested daily limit plus 53 mg.
$3x$	$+$	y	$=$	$300 + 53$

Step 3. Solve the system and answer the problem's question. The system

$$2x + 3y = 520$$
$$3x + y = 353$$

can be solved by substitution or addition. We will use addition; if we multiply the second equation by −3, adding equations will eliminate y.

$$2x + 3y = 520 \xrightarrow{\text{No change}} 2x + 3y = 520$$

$$3x + y = 353 \xrightarrow{\text{Multiply by } -3.} -9x - 3y = -1059$$

$$\text{Add: } -7x = -539$$

$$x = \frac{-539}{-7} = 77$$

Because x represents the cholesterol content of a Quarter Pounder, we see that a Quarter Pounder contains 77 milligrams of cholesterol. Now we can find y, the cholesterol content of a Whopper with cheese. We do so by back-substituting 77 for x in either of the system's equations.

$$3x + y = 353 \quad \text{We'll use the second equation.}$$

$$3(77) + y = 353 \quad \text{Back-substitute 77 for x.}$$

$$231 + y = 353 \quad \text{Multiply.}$$

$$y = 122 \quad \text{Subtract 231 from both sides.}$$

Because $x = 77$ and $y = 122$, a Quarter Pounder contains 77 milligrams of cholesterol and a Whopper with cheese contains 122 milligrams of cholesterol.

Step 4. Check the proposed answers in the original wording of the problem. Two Quarter Pounders and three Whoppers with cheese contain

$$2(77 \text{ mg}) + 3(122 \text{ mg}) = 520 \text{ mg},$$

which checks with the given conditions. Furthermore, three Quarter Pounders and one Whopper with cheese contain

$$3(77 \text{ mg}) + 1(122 \text{ mg}) = 353 \text{ mg},$$

which does exceed the daily limit of 300 milligrams by 53 milligrams. ■

✓ **CHECK POINT 2**   How do the Quarter Pounder and Whopper with cheese measure up in the calorie department? Actually, not too well. Two Quarter Pounders and three Whoppers with cheese provide 2607 calories. Even combining one of each provides enough calories to bring tears to Jenny Craig's eyes—9 calories in excess of what is allowed on a 1000-calorie-a-day diet. Find the caloric content of each item.

EXAMPLE 3   Fencing a Waterfront Lot

You just purchased a rectangular waterfront lot along a river's edge. The perimeter of the lot is 1000 feet. To create a sense of privacy, you decide to fence along three sides, excluding the side that fronts the river (see **Figure 5.12**). An expensive fencing along the lot's front length costs $25 per foot. An inexpensive fencing along the two side widths costs only $5 per foot. The total cost of the fencing along three sides comes to $9500.

a. What are the lot's dimensions?

b. You are considering using the expensive fencing on all three sides of the lot. However, you are limited by a $16,000 budget. Can this be done within your budget constraints?

Width : y

Width : y

Length : x

FIGURE 5.12

Width : y

Width : y

Length : x

FIGURE 5.12   (repeated)

Solution

a. We begin by finding the lot's dimensions.

Step 1. Use variables to represent unknown quantities.

Let x = the lot's length, in feet.

Let y = the lot's width, in feet.

(If you prefer, you can use the variable l for length and w for width.)

Step 2. Write a system of equations that models the problem's conditions.

- The lot's perimeter is 1000 feet.

Twice the length	plus	twice the width	is	the perimeter.
$2x$	$+$	$2y$	$=$	1000

- The cost of fencing three sides of the lot is \$9500.

Fencing along the front length	plus	fencing along the two side widths	costs	\$9500.

Cost per foot	·	number of feet	+	cost per foot	·	number of feet	is	\$9500.
25	·	x	$+$	5	·	$2y$	$=$	9500

Step 3. Solve the system and answer the problem's question.   The system

$$2x + 2y = 1000$$
$$25x + 10y = 9500$$

can be solved most easily by addition. If we multiply the first equation by -5, adding equations will eliminate y.

$$2x + 2y = 1000 \xrightarrow{\text{Multiply by } -5.} -10x - 10y = -5000$$
$$25x + 10y = 9500 \xrightarrow{\text{No change}} \underline{25x + 10y = 9500}$$
$$\text{Add:} \quad 15x = 4500$$
$$x = \frac{4500}{15} = 300$$

Because x represents length, we see that the lot is 300 feet long. Now we can find y, the lot's width. We do so by back-substituting 300 for x in either of the system's equations.

$2x + 2y = 1000$	**We'll use the first equation.**
$2 \cdot 300 + 2y = 1000$	**Back-substitute 300 for x.**
$600 + 2y = 1000$	**Multiply: 2 · 300 = 600.**
$2y = 400$	**Subtract 600 from both sides.**
$y = 200$	**Divide both sides by 2.**

Because $x = 300$ and $y = 200$, the lot is 300 feet long and 200 feet wide. Its dimensions are 300 feet by 200 feet.

Width : y

Width : y

Length : x

proposed: $y = 200$

proposed: $x = 300$

FIGURE 5.12 (repeated)
Showing proposed values for x and y

Step 4. Check the proposed answers in the original wording of the problem.
The perimeter should be 1000 feet:

$$2(300 \text{ ft}) + 2(200 \text{ ft}) = 600 \text{ ft} + 400 \text{ ft} = 1000 \text{ ft}.$$

The cost of fencing three sides of the lot should be $9500:

Fencing cost:
front length

Fencing cost:
two side widths

$$300 \cdot \$25 + 400 \cdot \$5 = \$7500 + \$2000 = \$9500.$$

b. Can you use the expensive fencing along all three sides of the lot and stay within a $16,000 budget? The fencing is to be placed along the length, 300 feet, and the two side widths, $2 \cdot 200$ feet, or 400 feet. Thus, you will fence 700 feet. The expensive fencing costs $25 per foot. The cost is

$$700 \text{ feet} \cdot \frac{\$25}{\text{feet}} = 700 \cdot \$25 = \$17,500.$$

Limited by a $16,000 budget, you cannot use the expensive fencing on three sides of the lot. ∎

✓ **CHECK POINT 3** A rectangular lot whose perimeter is 360 feet is fenced along three sides. An expensive fencing along the lot's length costs $20 per foot. An inexpensive fencing along the two side widths costs only $8 per foot. The total cost of the fencing along the three sides comes to $3280. What are the lot's dimensions?

EXAMPLE 4 Solar and Electric Heating Systems

The costs for two different kinds of heating systems for a two-bedroom home are given in the following table.

System	Cost to Install	Operating Cost/Year
Solar	$29,700	$150
Electric	$5000	$1100

After how many years will total costs for solar heating and electric heating be the same? What will be the cost at that time?

Solution

Step 1. Use variables to represent unknown quantities.
Let $x =$ the number of years the heating system is used.
Let $y =$ the total cost for the heating system.

Step 2. Write a system of equations that models the problem's conditions.

Total cost for the solar system	equals	installation cost	plus	yearly operating cost	times	the number of years the system is used.
y	$=$	29,700	$+$	150	$\cdot$	x

Total cost for the electric system	equals	installation cost	plus	yearly operating cost	times	the number of years the system is used.
y	$=$	5000	$+$	1100	$\cdot$	x

Step 3. Solve the system and answer the problem's question. We want to know after how many years the total costs for the two systems will be the same. We must solve the system

$$y = 29,700 + 150x$$
$$y = 5000 + 1100x.$$

Substitution works well because y is isolated in each equation.

$y = \boxed{29,700 + 150x}\;\boxed{y} = 5000 + 1100x$ Substitute 29,700 + 150x for y.

$29,700 + 150x = 5000 + 1100x$ This substitution gives an equation in one variable.

$29,700 = 5000 + 950x$ Subtract 150x from both sides.

$24,700 = 950x$ Subtract 5000 from both sides.

$26 = x$ Divide both sides by 950: $\dfrac{24,700}{950} = 26$.

Because x represents the number of years the heating system is used, we see that after 26 years, the total costs for the two systems will be the same. Now we can find y, the total cost. Back-substitute 26 for x in either of the system's equations. We will use the second equation, $y = 5000 + 1100x$.

$$y = 5000 + 1100 \cdot 26 = 5000 + 28,600 = 33,600$$

Because $x = 26$ and $y = 33,600$, after 26 years, the total costs for the two systems will be the same. The cost for each system at that time will be $33,600.

Step 4. Check the proposed answers in the original wording of the problem. Let's verify that after 26 years, the two systems will cost the same amount. The installation cost for the solar system is $29,700 and the yearly operating cost is $150. Thus, the total cost after 26 years is

$$\$29,700 + \$150(26) = \$29,700 + \$3900 = \$33,600.$$

The installation cost for the electric system is $5000 and the yearly operating cost is $1100. Thus, the total cost after 26 years is

$$\$5000 + \$1100(26) = \$5000 + \$28,600 = \$33,600.$$

This verifies that after 26 years the two systems will cost the same amount, $33,600.

The graphs in **Figure 5.13** give us a way of visualizing the solution to Example 4. The total cost of solar heating over 40 years is represented by the blue line. The total cost of electric heating over 40 years is represented by the red line. The lines intersect at (26, 33,600): After 26 years, the cost for each system is the same, $33,600.

Can you see that to the right of the intersection point, (26, 33,600), the blue graph representing solar costs lies below the red graph representing electric costs? Thus, after 26 years, or when $x > 26$, the cost for solar heating is less than the cost for electric heating.

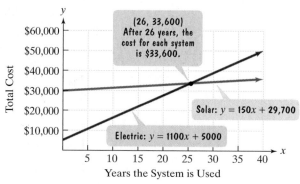

FIGURE 5.13 Visualizing total costs of solar and electric heating

✓ **CHECK POINT 4** Costs for two different kinds of heating systems for a home are given in the following table.

System	Cost to Install	Operating Cost/Year
Electric	$5000	$1100
Gas	$12,000	$700

After how long will total costs for electric heating and gas heating be the same? What will be the cost at that time?

5.4 EXERCISE SET

MyMathLab

Practice Exercises

In Exercises 1–4, let x represent one number and let y represent the other number. Use the given conditions to write a system of equations. Solve the system and find the numbers.

1. The sum of two numbers is 17. If one number is subtracted from the other, their difference is −3. Find the numbers.

2. The sum of two numbers is 5. If one number is subtracted from the other, their difference is 13. Find the numbers.

3. Three times a first number decreased by a second number is −1. The first number increased by twice the second number is 23. Find the numbers.

4. The sum of three times a first number and twice a second number is 43. If the second number is subtracted from twice the first number, the result is −4. Find the numbers.

Application Exercises

The bar graph shows the average time per day that Americans devote to sprucing up. Exercises 5–6 are based on the graph.

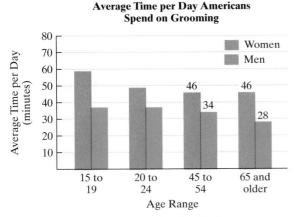

Source: Bureau of Labor Statistics American Time Use Survey

5. Each day, the sum of the average times spent on grooming for 20- to 24-year-old women and men is 86 minutes. The difference between grooming times for 20- to 24-year-old women and men is 12 minutes. How many minutes per day do 20- to 24-year-old women and men spend on grooming?

6. Each day, the sum of the average times spent on grooming for 15- to 19-year-old women and men is 96 minutes. The difference between grooming times for 15- to 19-year-old women and men is 22 minutes. How many minutes per day do 15- to 19-year-old women and men spend on grooming?

Looking for Mr. Goodbar? It's probably not a good idea if you want to look like Mr. Universe or Julia Roberts. The graph shows the four candy bars with the highest fat content, representing grams of fat and calories in each bar. Exercises 7–10 are based on the graph.

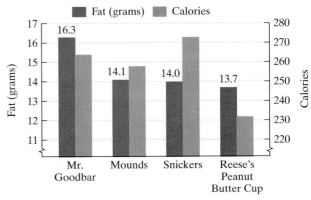

Source: Krantz and Sveum, *The World's Worsts*, HarperCollins, 2005

7. One Mr. Goodbar and two Mounds bars contain 780 calories. Two Mr. Goodbars and one Mounds bar contain 786 calories. Find the caloric content of each candy bar.

(In Exercises 8–10, refer to the bar graph on the previous page to visualize that your solution is reasonable.)

8. One Snickers bar and two Reese's Peanut Butter Cups contain 737 calories. Two Snickers bars and one Reese's Peanut Butter Cup contain 778 calories. Find the caloric content of each candy bar.

9. A collection of Halloween candy contains a total of five Mr. Goodbars and Mounds bars. Chew on this: The grams of fat in these candy bars exceed the daily maximum desirable fat intake of 70 grams by 7.1 grams. How many bars of each kind of candy are contained in the Halloween collection?

10. A collection of Halloween candy contains a total of 12 Snickers bars and Reese's Peanut Butter Cups. Chew on this: The grams of fat in these candy bars exceeds twice the daily maximum desirable fat intake of 70 grams by 26.5 grams. How many bars of each kind of candy are contained in the Halloween collection?

11. In a discount clothing store, all sweaters are sold at one fixed price and all shirts are sold at another fixed price. If one sweater and three shirts cost $42, while three sweaters and two shirts cost $56, find the price of one sweater and the price of one shirt.

12. A restaurant purchased eight tablecloths and five napkins for $106. A week later, a tablecloth and six napkins were bought for $24. Find the cost of one tablecloth and the cost of one napkin, assuming the same prices for both purchases.

13. The perimeter of a badminton court is 128 feet. After a game of badminton, a player's coach estimates that the athlete has run a total of 444 feet, which is equivalent to six times the court's length plus nine times its width. What are the dimensions of a standard badminton court?

14. The perimeter of a tennis court is 228 feet. After a round of tennis, a player's coach estimates that the athlete has run a total of 690 feet, which is equivalent to 7 times the court's length plus four times its width. What are the dimensions of a standard tennis court?

15. A rectangular lot whose perimeter is 320 feet is fenced along three sides. An expensive fencing along the lot's length costs $16 per foot. An inexpensive fencing along the two side widths costs only $5 per foot. The total cost of the fencing along the three sides comes to $2140. What are the lot's dimensions?

16. A rectangular lot whose perimeter is 1600 feet is fenced along three sides. An expensive fencing along the lot's length costs $20 per foot. An inexpensive fencing along the two side widths costs only $5 per foot. The total cost of the fencing along the three sides comes to $13,000. What are the lot's dimensions?

17. You are choosing between two long-distance telephone plans. Plan A has a monthly fee of $20 with a charge of $0.05 per minute for all long-distance calls. Plan B has a monthly fee of $5 with a charge of $0.10 per minute for all long-distance calls.

    a. For how many minutes of long-distance calls will the costs for the two plans be the same? What will be the cost for each plan?

    b. If you make approximately 10 long-distance calls per month, each averaging 20 minutes, which plan should you select? Explain your answer.

18. You are choosing between two long-distance telephone plans. Plan A has a monthly fee of $15 with a charge of $0.08 per minute for all long-distance calls. Plan B has a monthly fee of $3 with a charge of $0.12 per minute for all long-distance calls.

    a. For how many minutes of long-distance calls will the costs for the two plans be the same? What will be the cost for each plan?

    b. If you make approximately 15 long-distance calls per month, each averaging 30 minutes, which plan should you select? Explain your answer.

19. You are choosing between two plans at a discount warehouse. Plan A offers an annual membership fee of $100 and you pay 80% of the manufacturer's recommended list price. Plan B offers an annual membership fee of $40 and you pay 90% of the manufacturer's recommended list price. How many dollars of merchandise would you have to purchase in a year to pay the same amount under both plans? What will be the cost for each plan?

20. You are choosing between two plans at a discount warehouse. Plan A offers an annual membership fee of $300 and you pay 70% of the manufacturer's recommended list price. Plan B offers an annual membership fee of $40 and you pay 90% of the manufacturer's recommended list price. How many dollars of merchandise would you have to purchase in a year to pay the same amount under both plans? What will be the cost for each plan?

21. A community center sells a total of 301 tickets for a basketball game. An adult ticket costs $3. A student ticket costs $1. The sponsors collect $487 in ticket sales. Find the number of each type of ticket sold.

22. On a special day, tickets for a minor league baseball game cost $5 for adults and $1 for students. The attendance that day was 1281 and $3425 was collected. Find the number of each type of ticket sold.

23. At the 1987 annual meeting of the National Council of Teachers of Mathematics, the system below was found on a menu card at a restaurant in California. Find the cost of each item in column A and the cost of each item in column B.

24. The perimeter of the rectangle is 34 inches. The perimeter of the triangle is 30 inches.

   a. Find the values of x and y.

   b. Find the area of the rectangle and the area of the triangle.

 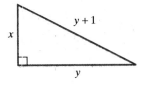

25. Nutritional information for macaroni and broccoli is given in the table. How many servings of each would it take to get exactly 14 grams of protein and 48 grams of carbohydrates?

	Macaroni	Broccoli
Protein (grams/serving)	3	2
Carbohydrates (grams/serving)	16	4

26. The calorie-nutrient information for an apple and an avocado is given in the table. How many of each should be eaten to get exactly 1000 calories and 100 grams of carbohydrates?

	One Apple	One Avocado
Calories	100	350
Carbohydrates (grams)	24	14

In Exercises 27–28, an isosceles triangle in which angles B and C have the same measure is shown. Find the measure of each angle whose degree measure is represented with variables.

27.

28.

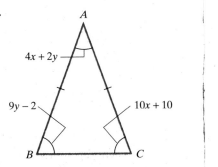

29. When a crew rows with the current, it travels 16 miles in 2 hours. Against the current, the crew rows 8 miles in 2 hours. Let x = the crew's rowing rate in still water and let y = the rate of the current. The following chart summarizes this information.

	Rate	×	Time	=	Distance
Rowing with current	$x + y$		2		16
Rowing against current	$x - y$		2		8

Find the rate of rowing in still water and the rate of the current.

30. When an airplane flies with the wind, it travels 800 miles in 4 hours. Against the wind, it takes 5 hours to cover the same distance. Let x = the plane's rate in still air and let y = the rate of the wind. The following chart summarizes this information.

	Rate	×	Time	=	Distance
Flying with the wind	$x + y$		4		800
Flying against the wind	$x - y$		5		800

Find the plane's rate in still air and the rate of the wind.

Writing in Mathematics

31. Describe the conditions in a problem that enable it to be solved using a system of linear equations.

32. Use **Figure 5.11** on page 319 to write a word problem similar to Example 1 or Check Point 1 that involves the leisure activity of reading. Then solve the problem.

33. Exercises 17–20 involve using systems of linear equations to compare costs of long-distance telephone plans and plans at a discount warehouse. Describe another situation that involves choosing between two options that can be modeled and solved with a linear system.

Critical Thinking Exercises

Make Sense? *In Exercises 34–37, determine whether each statement "makes sense" or "does not make sense" and explain your reasoning.*

34. The Sea Drift Hotel charges $650 for 7 nights and 10 meals, so I modeled these conditions using $7n + 10m = 650$, where n is the cost of one night and m is the cost of one meal.

35. A long-distance calling plan charges $4 per month plus 7¢ per minute, so I modeled these conditions using $y = 0.07 + 4x$, where y is the monthly charge when x minutes of long-distance calls are made.

36. The perimeter of the large rectangle shown in the figure is 58 meters, so I modeled this condition using $2(x + 8) + 2(y + 10) = 58$.

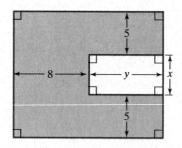

37. Angles A and B are supplementary, so I modeled this condition using

$$4x - 2y + 4 = 12x + 6y + 12 + 180.$$

38. A set of identical twins can only be recognized by the characteristic that one always tells the truth and the other always lies. One twin tells you of a lucky number pair: "When I multiply my first lucky number by 3 and my second lucky number by 6, the addition of the resulting numbers produces a sum of 12. When I add my first lucky number and twice my second lucky number, the sum is 5." Which twin is talking?

39. Tourist: "How many birds and lions do you have in your zoo?" Zookeeper: "There are 30 heads and 100 feet." Tourist: "I can't tell from that." Zookeeper: "Oh, yes, you can!" Can you? Find the number of each.

40. Find the measure of each angle whose degree measure is represented with a variable.

41. One apartment is directly above a second apartment. The resident living downstairs calls his neighbor living above him and states, "If one of you is willing to come downstairs, we'll have the same number of people in both apartments." The upstairs resident responds, "We're all too tired to move. Why don't one of you come up here? Then we'll have twice as many people up here as you've got down there." How many people are in each apartment?

42. In Lewis Carroll's *Through the Looking Glass*, the following dialogue takes place:

> *Tweedledum (to Tweedledee):* The sum of your weight and twice mine is 361 pounds.

> *Tweedledee (to Tweedledum):* Contrawise, the sum of your weight and twice mine is 362 pounds.

Find the weight of each of the two characters.

Technology Exercise

43. Select any two problems that you solved from Exercises 5–26. Use a graphing utility to graph the system of equations that you wrote for that problem. Then use the ⎡TRACE⎤ or ⎡INTERSECTION⎤ feature to show the point on the graphs that corresponds to the problem's solution.

Review Exercises

44. Solve: $2(x + 3) = 24 - 2(x + 4)$. (Section 2.3, Example 3)

45. Simplify: $5 + 6(x + 1)$. (Section 1.7, Example 7)

46. Write the slope-intercept form of the equation of the line passing through $(-5, 6)$ and $(3, -10)$. (Section 4.5, Example 2)

Preview Exercises

Exercises 47–49 will help you prepare for the material covered in the next section. In each exercise, graph the given inequality in a rectangular coordinate system.

47. $2x - y < 4$

48. $y \geq x + 1$

49. $x \geq 2$

Systems of Linear Inequalities

Objectives

1 Use mathematical models involving systems of linear inequalities.

2 Graph the solution sets of systems of linear inequalities.

We opened the chapter noting that the modern emphasis on thinness as the ideal body shape has been suggested as a major cause of eating disorders. In this section, as well as in the exercise set, we use systems of linear inequalities in two variables that will enable you to establish a healthy weight range for your height and age.

1 Use mathematical models involving systems of linear inequalities.

Systems of Linear Inequalities

In Section 4.6, we graphed the solutions of a linear inequality in two variables, such as $4.9x - y \geq 165$. Just as two linear equations make up a system of linear equations, two (or more) linear inequalities make up a **system of linear inequalities**. Here is an example of a system of linear inequalities:

$$4.9x - y \geq 165$$

$$3.7x - y \leq 125.$$

A **solution of a system of linear inequalities** in two variables is an ordered pair that satisfies each inequality in the system.

EXAMPLE 1 Does Your Weight Fit You?

The latest guidelines, which apply to both men and women, give healthy weight ranges, rather than specific weights, for your height. **Figure 5.14** shows the healthy weight region for various heights for people between the ages of 19 and 34, inclusive.

FIGURE 5.14

Source: U.S. Department of Health and Human Services

If x represents height, in inches, and y represents weight, in pounds, the healthy weight region in **Figure 5.14** can be modeled by the following system of linear inequalities:

$$4.9x - y \geq 165$$
$$3.7x - y \leq 125.$$

Show that point A in **Figure 5.14** is a solution of the system of inequalities that describes healthy weight.

Solution Point A has coordinates $(70, 170)$. This means that if a person is 70 inches tall, or 5 feet 10 inches, and weighs 170 pounds, then that person's weight is within the healthy weight region. We can show that $(70, 170)$ satisfies the system of inequalities by substituting 70 for x and 170 for y in each inequality in the system.

$$4.9x - y \geq 165 \qquad\qquad 3.7x - y \leq 125$$
$$4.9(70) - 170 \overset{?}{\geq} 165 \qquad\qquad 3.7(70) - 170 \overset{?}{\leq} 125$$
$$343 - 170 \overset{?}{\geq} 165 \qquad\qquad 259 - 170 \overset{?}{\leq} 125$$
$$173 \geq 165, \quad \text{true} \qquad\qquad 89 \leq 125, \quad \text{true}$$

The coordinates $(70, 170)$ make each inequality true. Thus, $(70, 170)$ satisfies the system for the healthy weight region and is a solution of the system. ▬

☑ **CHECK POINT 1** Show that point B in **Figure 5.14** is a solution of the system of inequalities that describes healthy weight.

2 Graph the solution sets of systems of linear inequalities.

Graphing Systems of Linear Inequalities

The **solution set of a system of linear inequalities in two variables** is the set of all ordered pairs that satisfy each inequality in the system. Thus, to graph a system of inequalities in two variables, begin by graphing each individual inequality in the same rectangular coordinate system. Then find the region, if there is one, that is common to every graph in the system. This region of intersection gives a picture of the system's solution set.

EXAMPLE 2   Graphing a System of Linear Inequalities

Graph the solution set of the system:

$$2x - y < 4$$
$$x + y \geq -1.$$

Solution   Replacing each inequality symbol with an equal sign indicates that we need to graph $2x - y = 4$ and $x + y = -1$. We can use intercepts to graph these lines.

$2x - y = 4$		**$x + y = -1$**
x-intercept: $2x - 0 = 4$	Set $y = 0$ in each equation.	x-intercept: $x + 0 = -1$
$2x = 4$		$x = -1$
$x = 2$		

The line passes through $(2, 0)$. The line passes through $(-1, 0)$.

y-intercept: $2 \cdot 0 - y = 4$	Set $x = 0$ in each equation.	y-intercept: $0 + y = -1$
$-y = 4$		$y = -1$
$y = -4$		

The line passes through $(0, -4)$. The line passes through $(0, -1)$.

Now we are ready to graph the solution set of the system of linear inequalities.

Graph $2x - y < 4$. The blue line, $2x - y = 4$, is dashed: Equality is not included in $2x - y < 4$. Because $(0, 0)$ makes the inequality true $(2 \cdot 0 - 0 < 4$, or $0 < 4$, is true), shade the half-plane containing $(0, 0)$ in yellow.

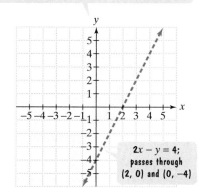

$2x - y = 4$; passes through $(2, 0)$ and $(0, -4)$

The graph of $2x - y < 4$

Add the graph of $x + y \geq -1$. The red line, $x + y = -1$, is solid: Equality is included in $x + y \geq -1$. Because $(0, 0)$ makes the inequality true $(0 + 0 \geq -1$, or $0 \geq -1$, is true), shade the half-plane containing $(0, 0)$ using green vertical shading.

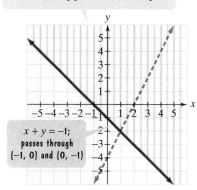

$x + y = -1$; passes through $(-1, 0)$ and $(0, -1)$

Adding the graph of $x + y \geq -1$

The solution set of the system is graphed as the intersection (the overlap) of the two half-planes. The is the region in which the yellow shading and the green vertical shading overlap.

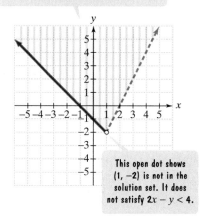

This open dot shows $(1, -2)$ is not in the solution set. It does not satisfy $2x - y < 4$.

The graph of the system $2x - y < 4, x + y \geq -1$

✓ **CHECK POINT 2**   Graph the solution set of the system:

$$x + 2y > 4$$
$$2x - 3y \leq -6.$$

EXAMPLE 3   Graphing a System of Linear Inequalities

Graph the solution set of the system:

$$y \geq x + 1$$
$$x \geq 2.$$

Solution   Replacing each inequality symbol with an equal sign indicates that we need to graph $y = x + 1$ and $x = 2$.

$$y = 1x + 1 \qquad\qquad x = 2$$

Slope is $\frac{1}{1}$. y-intercept is 1.

The graph is a vertical line with an x-intercept of **2**.

Now we are ready to graph the solution set of the system of inequalities.

Graph $y \geq x + 1$. The blue line, $y = x + 1$, is solid because equality is included in $y \geq x + 1$. Because of $\geq$, shade the half-plane above the blue line in yellow.

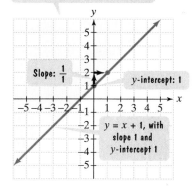

The graph of $y \geq x + 1$

Add the graph of $x \geq 2$. The red line, $x = 2$, is solid: Equality is included in $x \geq 2$. Because of $\geq$, shade the half-plane to the right of $x = 2$ using green vertical shading.

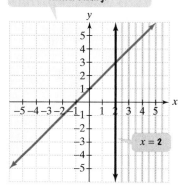

Adding the graph of $x \geq 2$

The solution set of the system is graphed as the intersection (the overlap) of the two half-planes. This is the region in which the yellow shading and the green vertical shading overlap.

This closed dot shows (2, 3) is in the solution set, satisfying $y \geq x + 1$ and $x \geq 2$.

The graph of the system
$y \geq x + 1, x \geq 2$

✓ **CHECK POINT 3**   Graph the solution set of the system:

$$y \geq x + 2$$
$$x \geq 1.$$

5.5 EXERCISE SET   **MyMathLab**   Math XL PRACTICE   WATCH   DOWNLOAD   READ   REVIEW

Practice Exercises

In Exercises 1–36, graph the solution set of each system of linear inequalities.

1. $x + y \leq 4$
  $x - y \leq 2$

2. $x + y \geq 4$
  $x - y \leq 2$

3. $2x - 4y \leq 8$
  $x + y \geq -1$

4. $4x + 3y \leq 12$
  $x - 2y \leq 4$

5. $x + 3y \leq 6$
  $x - 2y \leq 4$

6. $2x + y \leq 4$
  $2x - y \leq 6$

7. $x - 2y > 4$
  $2x + y \geq 6$

8. $3x + y < 6$
  $x + 2y \geq 2$

9. $x + y > 1$
  $x + y < 4$

10. $x - y > 1$
  $x - y < 3$

11. $y \geq 2x + 1$
  $y \leq 4$

12. $y \geq \frac{1}{2}x + 2$
  $y \leq 2$

13. $y > x - 1$
  $x > 5$

14. $y > x - 2$
  $x > 3$

15. $y \geq 2x - 3$
  $y \leq 2x + 1$

16. $y \geq 3x - 2$
  $y \leq 3x + 1$

17. $y > 2x - 3$
  $y \leq -x + 6$

18. $y < -2x + 4$
  $y \leq x - 4$

19. $x + 2y \leq 4$
  $y \geq x - 3$

20. $x + y \leq 4$
  $y \geq 2x - 4$

21. $x \leq 3$
  $y \geq -2$

22. $x \leq 4$
  $y \leq -3$

23. $x \geq 3$
$y < 2$

24. $x \geq -2$
$y < -1$

25. $x \geq 0$
$y \leq 0$

26. $x \leq 0$
$y \geq 0$

27. $x \geq 0$
$y > 0$

28. $x \leq 0$
$y < 0$

29. $x + y \leq 5$
$x \geq 0$
$y \geq 0$

30. $2x + y \leq 4$
$x \geq 0$
$y \geq 0$

31. $4x - 3y > 12$
$x \geq 0$
$y \leq 0$

32. $2x - 6y > 12$
$x \leq 0$
$y \leq 0$

33. $0 \leq x \leq 3$
$0 \leq y \leq 3$

34. $0 \leq x \leq 5$
$0 \leq y \leq 5$

35. $x - y \leq 4$
$x + 2y \leq 4$

36. $x - y \leq 3$
$2x + y \leq 4$

Practice PLUS

In Exercises 37–44, graph the solution set of each system of linear inequalities. If the system has no solutions, state this and explain why.

37. $x + y \geq 1$
$x - y \geq 1$
$x \geq 4$

38. $x - y \leq 3$
$x + y \leq 3$
$x \geq -2$

39. $x + 2y < 6$
$y > 2x - 2$
$y \geq 2$

40. $2x - 3y < 6$
$2x - 3y > -6$
$-3 \leq x \leq 2$

41. $y \leq -3x + 3$
$y \geq -x - 1$
$y < x + 7$

42. $y \geq -3x + 5$
$y \geq -x + 3$
$y \geq \frac{1}{2}x$
$x \geq 0$
$y \geq 0$

43. $y \geq 2x + 2$
$y < 2x - 3$
$x \geq 2$

44. $y \geq -3x + 2$
$y < -3x$
$x \geq 1$

Application Exercises

The figure shows the healthy weight region for various heights for people ages 35 and older.

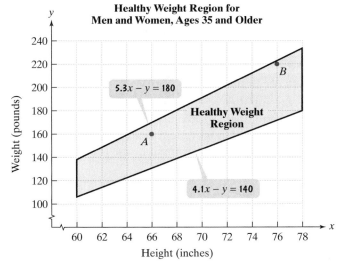

Healthy Weight Region for Men and Women, Ages 35 and Older

Weight (pounds) vs Height (inches)

5.3x − y = 180

Healthy Weight Region

4.1x − y = 140

Source: U.S. Department of Health and Human Services

If x represents height, in inches, and y represents weight, in pounds, the healthy weight region can be modeled by the following system of linear inequalities:

$$5.3x - y \geq 180$$
$$4.1x - y \leq 140.$$

Use this information to solve Exercises 45–48.

45. Show that point A is a solution of the system of inequalities that describes healthy weight for this age group.

46. Show that point B is a solution of the system of inequalities that describes healthy weight for this age group.

47. Is a person in this age group who is 6 feet tall weighing 205 pounds within the healthy weight region?

48. Is a person in this age group who is 5 feet 8 inches tall weighing 135 pounds within the healthy weight region?

49. Suppose a patient is not allowed to have more than 330 milligrams of cholesterol per day from a diet of eggs and meat. Each egg provides 165 milligrams of cholesterol and each ounce of meat provides 110 milligrams of cholesterol. Thus, $165x + 110y \leq 330$, where x is the number of eggs and y the number of ounces of meat. Furthermore, the patient must have at least 165 milligrams of cholesterol from the diet. Graph the system of inequalities in the first quadrant. Give the coordinates of any two points in the solution set. Describe what each set of coordinates means in terms of the variables in the problem.

Writing in Mathematics

50. What does the graph of a system of linear inequalities represent?

51. Look at the shaded region showing recommended weight and height combinations in the figure for Exercises 45–48. Describe why a system of inequalities, rather than an equation, is better suited to give the recommended combinations.

Critical Thinking Exercises

Make Sense? *In Exercises 52–55, determine whether each statement "makes sense" or "does not make sense" and explain your reasoning.*

52. The reason that systems of linear inequalities are appropriate for modeling healthy weight is because guidelines give healthy weight ranges, rather than specific weights, for various heights.

53. I graphed the solution set of $y \geq x + 2$ and $x \geq 1$ without using test points.

54. I graphed the solution set of $2x - y < 4$ and $x + y > -1$ without using test points.

55. I use two different colors to graph solution sets of systems of inequalities, selecting the region where the colors overlap.

In Exercises 56–57, write a system of inequalities whose solution set is shown by the region in which the yellow shading and the green vertical shading overlap.

56.

57.

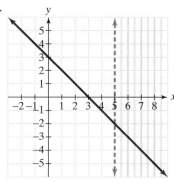

58. Write a system of inequalities that has no solutions.

59. Write a system of inequalities that describes all the points in quadrant III of a rectangular coordinate system.

60. A person plans to invest no more than $15,000, placing the money in two investments. One investment is high risk, high yield; the other is low risk, low yield. At least $2000 is to be placed in the high-risk investment. Furthermore, the amount invested at low risk should be at least three times the amount invested at high risk. Write and graph a system of inequalities that describes all possibilities for placing the money in the high- and low-risk investments.

61. Promoters of a rock concert must sell at least 25,000 tickets priced at $35 and $50 per ticket. Furthermore, the promoters must take in at least $1,025,000 in ticket sales. Write and graph a system of inequalities that describes all possibilities for selling the $35 tickets and the $50 tickets.

Review Exercises

62. Find the slope of the line containing the points $(-6, 1)$ and $(2, -1)$. (Section 4.3, Example 1)

63. Add: $\dfrac{1}{5} + \left(-\dfrac{3}{4}\right)$. (Section 1.5, Example 4)

64. Solve: $7x = 10 + 6(11 - 2x)$. (Section 2.3, Example 3)

Preview Exercises

Exercises 65–67 will help you prepare for the material covered in the first section of the next chapter.

65. Add: $5x^3 + 12x^3$.

66. Add: $-8x^2 + 6x^2$.

67. Subtract: $-9y^4 - (-2y^4)$.

GROUP PROJECT

CHAPTER 5

Group members should go online and obtain a list of telecommunication companies that provide residential long-distance service. The list should contain the monthly fee, the monthly minimum, and the rate per minute for each service provider.

a. For each provider in the list, write an equation that describes the total monthly cost, y, of x minutes of long-distance phone calls.

b. Compare two of the plans. After how many minutes of long-distance calls will the costs for the two plans be the same? Solve a linear system to obtain your answer. What will be the cost for each plan?

c. Repeat part (b) for another two of the plans.

d. Each person should estimate the number of minutes he or she spends talking long distance each month. Group members should assist that person in selecting the plan that will save the most amount of money. Be sure to factor in the monthly minimum, if any, when choosing a plan. Whenever possible, use the equations that you wrote in part (a).

Caution: We've left something out! Your comparisons do not take into account the in-state rates of the plan. Furthermore, if you make international calls, international rates should also be one of your criteria in choosing a plan. Problem solving in real life can get fairly complicated.

Chapter 5 Summary

Definitions and Concepts	**Examples**

Section 5.1 Solving Systems of Linear Equations by Graphing

A system of linear equations in two variables, x and y, consists of two equations of the form $Ax + By = C$. A solution is an ordered pair of numbers that satisfies both equations.

Determine whether $(3, -1)$ is a solution of

$$2x + 5y = 1$$
$$4x + y = 11.$$

Replace x with 3 and y with -1 in both equations.

$$2x + 5y = 1 \qquad\qquad 4x + y = 11$$
$$2 \cdot 3 + 5(-1) \stackrel{?}{=} 1 \qquad\qquad 4 \cdot 3 + (-1) \stackrel{?}{=} 11$$
$$6 + (-5) \stackrel{?}{=} 1 \qquad\qquad 12 + (-1) \stackrel{?}{=} 11$$
$$1 = 1, \ \ \text{true} \qquad\qquad 11 = 11, \ \ \text{true}$$

Thus, $(3, -1)$ is a solution of the system.

To solve a linear system by graphing,
1. Graph the first equation.
2. Graph the second equation on the same set of axes.
3. If the graphs intersect, determine the coordinates of this point. The ordered pair is the solution of the system.
4. Check the solution in both equations. If the graphs are parallel lines, the system has no solution and is called inconsistent. If the graphs are the same line, the system has infinitely many solutions. The equations are called dependent.

Solve by graphing:

$$2x + y = 4$$
$$x + y = 2.$$

The solution is $(2, 0)$ and the solution set is $\{(2, 0)\}$.

Section 5.2 Solving Systems of Linear Equations by the Substitution Method

To solve a linear system by the substitution method,
1. Solve one equation for one variable in terms of the other.
2. Substitute the expression for that variable into the other equation. This will result in an equation in one variable.
3. Solve the equation in one variable.
4. Back-substitute the value of the variable found in step 3 in the equation from step 1. Simply and find the value of the remaining variable.
5. Check the proposed solution in both of the system's given equations.

If both variables are eliminated and a false statement results, the system has no solution. If both variables are eliminated and a true statement results, the system has infinitely many solutions.

Solve by the substitution method:

$$y = 2x + 3$$
$$7x - 5y = -18.$$

Substitute $2x + 3$ for y in the second equation.

$$7x - 5(2x + 3) = -18$$
$$7x - 10x - 15 = -18$$
$$-3x - 15 = -18$$
$$-3x = -3$$
$$x = 1$$

Find y. Substitute 1 for x in $y = 2x + 3$.

$$y = 2 \cdot 1 + 3 = 2 + 3 = 5$$

The solution, $(1, 5)$, checks. The solution set is $\{(1, 5)\}$.

Definitions and Concepts	**Examples**

Section 5.3 Solving Systems of Linear Equations by the Addition Method

To solve a linear system by the addition method,

1. Write equations in $Ax + By = C$ form.
2. Multiply one or both equations by nonzero numbers so that coefficients of one of the variables are opposites.
3. Add equations.
4. Solve the resulting equation for the remaining variable.
5. Back-substitute the value of the variable into either original equation and find the value of the other variable.
6. Check the proposed solution in both the original equations.

If both variables are eliminated and a false statement results, the system has no solution. If both variables are eliminated and a true statement results, the system has infinitely many solutions.

Solve by the addition method:

$$3x + y = -11$$

$$6x - 2y = -2.$$

Eliminate y. Multiply both sides of the first equation by 2.

$$6x + 2y = -22$$

$$\underline{6x - 2y = -2}$$

Add: $12x = -24$

$$x = -2$$

Find y. Back-substitute -2 for x. We'll use the first equation.

$$3(-2) + y = -11$$

$$-6 + y = -11$$

$$y = -5$$

The solution, $(-2, -5)$, checks. The solution set is $\{(-2, -5)\}$.

Section 5.4 Problem Solving Using Systems of Equations

A Problem-Solving Strategy

1. Use variables, usually x and y, to represent unknown quantities.
2. Write a system of equations that models the problem's conditions.
3. Solve the system and answer the problem's question.
4. Check proposed answers in the problem's wording.

A rectangular garden has a perimeter of 50 yards. Fencing along the length costs \$3 per yard and along the width costs \$2 per yard. The total cost of the fencing is \$140. Find the length and width of the rectangle.

Let x = length and y = width.

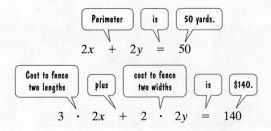

Solve

$$2x + 2y = 50$$

$$6x + 4y = 140$$

by addition. Multiply the first equation by -2 and add. Solving gives $x = 20$ and $y = 5$, so the length is 20 yards and the width is 5 yards.

Definitions and Concepts	**Examples**

Section 5.5 Systems of Linear Inequalities

Two (or more) linear inequalities make up a system of linear inequalities. A solution is an ordered pair satisfying each of the inequalities in the system. The solution set is the set of all such ordered pairs. To graph the solution set of a system of inequalities, graph each inequality in the system in the same rectangular coordinate system. The overlapping region represents the solution set of the system.

Graph the solution set of the system:
$$y \leq -2x$$
$$x - y \geq 3.$$

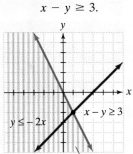

CHAPTER 5 REVIEW EXERCISES

5.1 *In Exercises 1–2, determine whether the given ordered pair is a solution of the system.*

1. $(1, -5)$
$4x - y = 9$
$2x + 3y = -13$

2. $(-5, 2)$
$2x + 3y = -4$
$x - 4y = -10$

3. Does the graphing-utility screen show the solution for the following system? Explain.

$$x + y = 2$$
$$2x + y = -5$$

In Exercises 4–14, solve each system by graphing. If there is no solution or an infinite number of solutions, so state. Use set notation to express solution sets.

4. $x + y = 2$
$x - y = 6$

5. $2x - 3y = 12$
$-2x + y = -8$

6. $3x + 2y = 6$
$3x - 2y = 6$

7. $y = \frac{1}{2}x$
$y = 2x - 3$

8. $x + 2y = 2$
$y = x - 5$

9. $x + 2y = 8$
$3x + 6y = 12$

10. $2x - 4y = 8$
$x - 2y = 4$

11. $y = 3x - 1$
$y = 3x + 2$

12. $x - y = 4$
$x = -2$

13. $x = 2$
$y = 5$

14. $x = 2$
$x = 5$

5.2 *In Exercises 15–23, solve each system by the substitution method. If there is no solution or an infinite number of solutions, so state. Use set notation to express solution sets.*

15. $2x - 3y = 7$
$y = 3x - 7$

16. $2x - y = 6$
$x = 13 - 2y$

17. $2x - 5y = 1$
$3x + y = -7$

18. $3x + 4y = -13$
$5y - x = -21$

19. $y = 39 - 3x$
$y = 2x - 61$

20. $4x + y = 5$
$12x + 3y = 15$

21. $4x - 2y = 10$
$y = 2x + 3$

22. $x - 4 = 0$
$9x - 2y = 0$

23. $8y = 4x$
$7x + 2y = -8$

24. The following models describe demand and supply for three-bedroom rental apartments.

Demand Model	**Supply Model**
$p = -50x + 2000$	$p = 50x$

Monthly rental price (Demand) — Number of apartments demanded, in thousands

Monthly rental price (Supply) — Number of apartments supplied, in thousands

a. Solve the system and find the equilibrium quantity and the equilibrium price.

b. Use your answer from part (a) to complete this statement:

When rents are _____ per month, consumers will demand _____ apartments and suppliers will offer _____ apartments for rent.

5.3 *In Exercises 25–35, solve each system by the addition method. If there is no solution or an infinite number of solutions, so state. Use set notation to express solution sets.*

25. $x + y = 6$
$2x + y = 8$

26. $3x - 4y = 1$
$12x - y = -11$

27. $3x - 7y = 13$
$6x + 5y = 7$

28. $8x - 4y = 16$
$4x + 5y = 22$

29. $5x - 2y = 8$
$3x - 5y = 1$

30. $2x + 7y = 0$
$7x + 2y = 0$

31. $x + 3y = -4$
$3x + 2y = 3$

32. $2x + y = 5$
$2x + y = 7$

33. $3x - 4y = -1$
$-6x + 8y = 2$

34. $2x = 8y + 24$
$3x + 5y = 2$

35. $5x - 7y = 2$
$3x = 4y$

In Exercises 36–41, solve each system by the method of your choice. If there is no solution or an infinite number of solutions, so state. Use set notation to express solution sets.

36. $3x + 4y = -8$
$2x + 3y = -5$

37. $6x + 8y = 39$
$y = 2x - 2$

38. $x + 2y = 7$
$2x + y = 8$

39. $y = 2x - 3$
$y = -2x - 1$

40. $3x - 6y = 7$
$3x = 6y$

41. $y - 7 = 0$
$7x - 3y = 0$

5.4

42.

Gustav Klimt (1826–1918) "Mrs. Adele Bloch-Bauer, I", 1907. Oil on canvas, 138 × 138 cm. Private Collection. Photo: Erich Lessing/Art Resource, NY.

Pablo Picasso (1881–1973) "Boy with a Pipe", 1905 (oil on canvas). Collection of Mr. and Mrs. John Hay Whitney, New York, USA. © DACS/The Bridgeman Art Library. © 2008 Estate of Pablo Picasso/Artists Rights Society (ARS), New York.

Talk about paintings by numbers: In 2007, this glittering Klimt set the record for the most ever paid for a painting, a title that had been held by Picasso's *Boy with a Pipe*. Combined, the two paintings sold for $239 million. The difference between the selling price for Klimt's work and the selling price for Picasso's work was $31 million. Find the amount paid for each painting.

43. Health experts agree that cholesterol intake should be limited to 300 milligrams or less each day. Three ounces of shrimp and 2 ounces of scallops contain 156 milligrams of cholesterol. Five ounces of shrimp and 3 ounces of scallops contain 45 milligrams of cholesterol less than the suggested maximum daily intake. Determine the cholesterol content in an ounce of each item.

44. The perimeter of a table tennis top is 28 feet. The difference between 4 times the length and 3 times the width is 21 feet. Find the dimensions.

Length: x   Width: y

45. A rectangular garden has a perimeter of 24 yards. Fencing across the length costs $3 per yard and along the width $2 per yard. The total cost of the fencing is $62. Find the length and width of the rectangle.

46. A travel agent offers two package vacation plans. The first plan costs $360 and includes 3 days at a hotel and a rental car for 2 days. The second plan costs $500 and includes 4 days at a hotel and a rental car for 3 days. The daily charge for the room is the same under each plan, as is the daily charge for the car. Find the cost per day for the room and for the car.

47. You are choosing between two long-distance telephone plans. One plan has a monthly fee of $15 with a charge of $0.05 per minute for all long-distance calls. The other plan has a monthly fee of $10 with a charge of $0.075 per minute for all long-distance calls. For how many minutes of long-distance calls will the costs for the two plans be the same? What will be the cost for each plan?

48. Tickets for a touring production of *Sweeney Todd* cost $90 for orchestra seats and $60 for balcony seats. You purchase nine tickets, some in the orchestra and some in the balcony, for a total cost of $720. How many orchestra tickets and how many balcony tickets were purchased?

5.5 *In Exercises 49–55, graph the solution set of each system of linear inequalities.*

49. $3x - y \le 6$
$x + y \ge 2$

50. $x + y < 4$
$x - y < 4$

51. $y < 2x - 2$
$x \ge 3$

52. $4x + 6y \le 24$
$y > 2$

53. $x \le 3$
$y \ge -2$

54. $y \ge \frac{1}{2}x - 2$
$y \le \frac{1}{2}x + 1$

55. $x \le 0$
$y \ge 0$

CHAPTER 5 TEST

Remember to use your Chapter Test Prep Video CD to see the worked-out solutions to the test questions you want to review.

In Exercises 1–2, determine whether the given ordered pair is a solution of the system.

1. $(5, -5)$
$2x + y = 5$
$x + 3y = -10$

2. $(-3, 2)$
$x + 5y = 7$
$3x - 4y = 1$

In Exercises 3–4, solve each system by graphing. If there is no solution or an infinite number of solutions, so state. Use set notation to express solution sets.

3. $x + y = 6$
$4x - y = 4$

4. $2x + y = 8$
$y = 3x - 2$

In Exercises 5–7, solve each system by the substitution method. If there is no solution or an infinite number of solutions, so state. Use set notation to express solution sets.

5. $x = y + 4$
$3x + 7y = -18$

6. $2x - y = 7$
$3x + 2y = 0$

7. $2x - 4y = 3$
$x = 2y + 4$

In Exercises 8–10, solve each system by the addition method. If there is no solution or an infinite number of solutions, so state. Use set notation to express solution sets.

8. $2x + y = 2$
$4x - y = -8$

9. $2x + 3y = 1$
$3x + 2y = -6$

10. $3x - 2y = 2$
$-9x + 6y = -6$

11. According to the U.S. Census Bureau, the two most popular female first names in the United States are Mary and Patricia. Combined, these two names account for 3.7% of all female first names. The difference between the percentage of women named Mary and the percentage of women named Patricia is 1.5%. What percentage of all female first names in the United States are Mary and what percentage are Patricia?

12. A rectangular garden has a perimeter of 34 yards. Fencing across the length costs $2 per yard and fencing along the width costs $1 per yard. The total cost of the fencing is $58. Find the length and width of the rectangle.

13. You are choosing between two long-distance telephone plans. One plan has a monthly fee of $15 with a charge of $0.05 per minute. The other plan has a monthly fee of $5 with a charge of $0.07 per minute. For how many minutes of long-distance calls will the costs for the two plans be the same? What will be the cost for each plan?

In Exercises 14–15, graph the solution set of each system of linear inequalities.

14. $x - 3y > 6$
$2x + 4y \le 8$

15. $y \ge 2x - 4$
$x < 2$

CUMULATIVE REVIEW EXERCISES (CHAPTERS 1–5)

1. Perform the indicated operations:

$$-14 - [18 - (6 - 10)].$$

2. Simplify: $6(3x - 2) - (x - 1)$.

In Exercises 3–4, solve each equation.

3. $17(x + 3) = 13 + 4(x - 10)$

4. $\dfrac{x}{4} - 1 = \dfrac{x}{5}$

5. Solve for t: $A = P + Prt$.

6. Solve and graph the solution set on a number line: $2x - 5 < 5x - 11$.

In Exercises 7–9, graph each equation in the rectangular coordinate system.

7. $x - 3y = 6$

8. $y = 4$

9. $y = -\dfrac{3}{5}x + 2$

In Exercises 10–11, solve each linear system.

10. $3x - 4y = 8$
$4x + 5y = -10$

11. $2x - 3y = 9$
$y = 4x - 8$

12. Find the slope of the line passing through $(5, -6)$ and $(6, -5)$.

13. Write the point-slope form and the slope-intercept form of the equation of the line passing through $(-1, 6)$ with slope $= -4$.

14. The area of a triangle is 80 square feet. Find the height if the base is 16 feet.

15. If 10 pens and 15 pads of paper cost $26, and 5 of the same pens and 10 of the same pads cost $16, find the cost of a pen and a pad.

16. List all the integers in this set:

$$\left\{ -93, -\frac{7}{3}, 0, \sqrt{3}, \frac{7}{1}, \sqrt{100} \right\}.$$

Americans are souring on whole milk. The graphs show per capita consumption of whole milk and low-fat skim milk, in gallons, over time. Exercises 17–20 are based on these graphs.

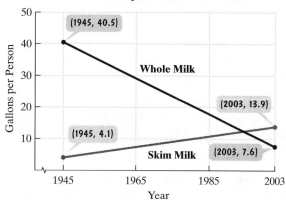

Milk Consumption in the United States

Source: United States Dairy Association

17. Which line has a positive slope? What does this mean in terms of the variables in this situation?

18. Which line has a negative slope? What does this mean in terms of the variables in this situation?

19. In 2003, Americans drank 7.6 gallons of whole milk per person. If the trend shown by the graph continues and this decreases by approximately 0.57 gallon per year, in how many years after 2003 will no whole milk be consumed? Round to the nearest whole number. In which year will that occur?

20. The mathematical model $y = 0.17x + 13.9$ describes skim-milk consumption, y, in gallons per person, x years after 2003. Use the formula to project in which year Americans will drink 15.6 gallons of skim milk per person.

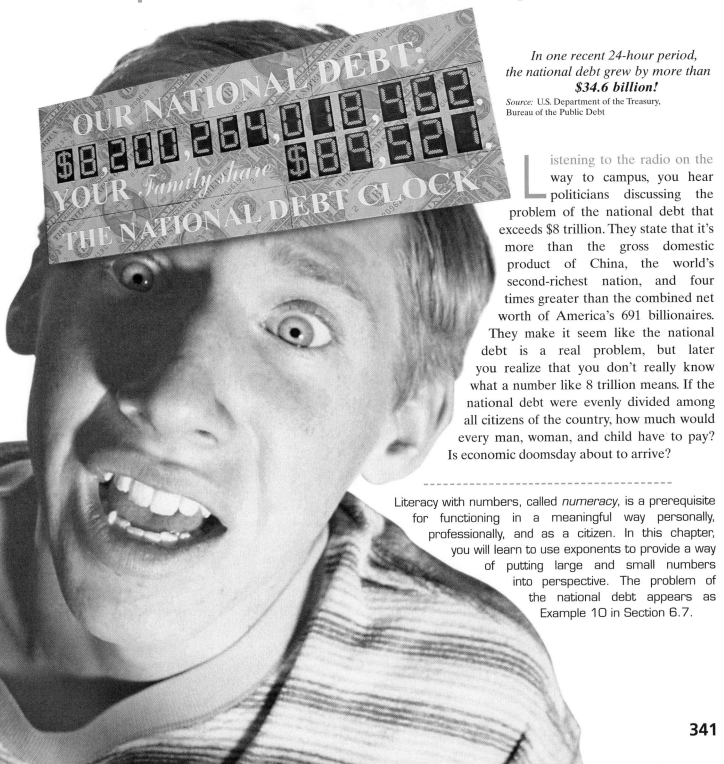

Exponents and Polynomials

In one recent 24-hour period, the national debt grew by more than $34.6 billion!
Source: U.S. Department of the Treasury, Bureau of the Public Debt

Listening to the radio on the way to campus, you hear politicians discussing the problem of the national debt that exceeds $8 trillion. They state that it's more than the gross domestic product of China, the world's second-richest nation, and four times greater than the combined net worth of America's 691 billionaires. They make it seem like the national debt is a real problem, but later you realize that you don't really know what a number like 8 trillion means. If the national debt were evenly divided among all citizens of the country, how much would every man, woman, and child have to pay? Is economic doomsday about to arrive?

Literacy with numbers, called *numeracy*, is a prerequisite for functioning in a meaningful way personally, professionally, and as a citizen. In this chapter, you will learn to use exponents to provide a way of putting large and small numbers into perspective. The problem of the national debt appears as Example 10 in Section 6.7.

341

Adding and Subtracting Polynomials

Objectives

1. Understand the vocabulary used to describe polynomials.
2. Add polynomials.
3. Subtract polynomials.
4. Graph equations defined by polynomials of degree 2.

More education results in a higher income. The mathematical models

$$M = -18x^3 + 923x^2 - 9603x + 48,446$$
$$\text{and} \quad W = 17x^3 - 450x^2 + 6392x - 14,764$$

describe the median, or middlemost, annual income for men, M, and women, W, who have completed x years of education. We'll be working with these models and the data upon which they are based in the exercise set.

The algebraic expressions that appear on the right side of the models are examples of *polynomials*. A **polynomial** is a single term or the sum of two or more terms containing variables with whole-number exponents. These particular polynomials each contain four terms. Equations containing polynomials are used in such diverse areas as science, business, medicine, psychology, and sociology. In this section, we present basic ideas about polynomials. We then use our knowledge of combining like terms to find sums and differences of polynomials.

Describing Polynomials

1. Understand the vocabulary used to describe polynomials.

Consider the polynomial

$$7x^3 - 9x^2 + 13x - 6.$$

We can express this polynomial as

$$7x^3 + (-9x^2) + 13x + (-6).$$

The polynomial contains four terms. It is customary to write the terms in the order of descending powers of the variables. This is the **standard form** of a polynomial.

We begin this chapter by limiting our discussion to polynomials containing only one variable. Each term of such a polynomial in x is of the form ax^n. The **degree** of ax^n is n. For example, the degree of the term $7x^3$ is 3.

Study Tip

We can express 0 in many ways, including $0x$, $0x^2$, and $0x^3$. It is impossible to assign a unique exponent to the variable. This is why 0 has no defined degree.

The Degree of ax^n

If $a \neq 0$ and n is a whole number, the degree of ax^n is n. The degree of a nonzero constant term is 0. The constant 0 has no defined degree.

Here is an example of a polynomial and the degree of each of its four terms:

$$6x^4 - 3x^3 - 2x - 5.$$

| degree 4 | degree 3 | degree 1 | degree of nonzero constant: 0 |

Notice that the exponent on x for the term $-2x$, meaning $-2x^1$, is understood to be 1. For this reason, the degree of $-2x$ is 1.

A polynomial is simplified when it contains no grouping symbols and no like terms. A simplified polynomial that has exactly one term is called a **monomial**. A **binomial** is a simplified polynomial that has two terms. A **trinomial** is a simplified polynomial with three terms. Simplified polynomials with four or more terms have no special names.

The **degree of a polynomial** is the greatest degree of all the terms of the polynomial. For example, $4x^2 + 3x$ is a binomial of degree 2 because the degree of the first term is 2, and the degree of the other term is less than 2. Also, $7x^5 - 2x^2 + 4$ is a trinomial of degree 5 because the degree of the first term is 5, and the degrees of the other terms are less than 5.

Up to now, we have used x to represent the variable in a polynomial. However, any letter can be used. For example,

- $7x^5 - 3x^3 + 8$ is a polynomial (in x) of degree 5. Because there are three terms, the polynomial is a trinomial.

- $6y^3 + 4y^2 - y + 3$ is a polynomial (in y) of degree 3. Because there are four terms, the polynomial has no special name.

- $z^7 + \sqrt{2}$ is a polynomial (in z) of degree 7. Because there are two terms, the polynomial is a binomial.

2   Add polynomials.

Adding Polynomials

Recall that *like terms* are terms containing exactly the same variables to the same powers. Polynomials are added by combining like terms. For example, we can add the monomials $-9x^3$ and $13x^3$ as follows:

$$-9x^3 + 13x^3 = (-9 + 13)x^3 = 4x^3.$$

These like terms both contain x to the third power.	Add coefficients and keep the same variable factor, x^3.

EXAMPLE 1   Adding Polynomials

Add: $(-9x^3 + 7x^2 - 5x + 3) + (13x^3 + 2x^2 - 8x - 6)$.

Solution   The like terms are $-9x^3$ and $13x^3$, containing the same variable to the same power (x^3), as well as $7x^2$ and $2x^2$ (both containing x^2), $-5x$ and $-8x$ (both containing x), and the constant terms 3 and -6. We begin by grouping these pairs of like terms.

$$(-9x^3 + 7x^2 - 5x + 3) + (13x^3 + 2x^2 - 8x - 6)$$

$$= (-9x^3 + 13x^3) + (7x^2 + 2x^2) + (-5x - 8x) + (3 - 6) \quad \text{Group like terms.}$$

$$= 4x^3 + 9x^2 + (-13x) + (-3) \quad\quad\quad\quad\quad\quad\quad \text{Combine like terms.}$$

$$= 4x^3 + 9x^2 - 13x - 3 \quad\quad\quad\quad\quad\quad\quad\quad\quad \text{Express addition of opposites as subtraction.} \quad \blacksquare$$

☑ **CHECK POINT 1**   Add: $(-11x^3 + 7x^2 - 11x - 5) + (16x^3 - 3x^2 + 3x - 15)$.

Polynomials can also be added by arranging like terms in columns. Then combine like terms, column by column.

| EXAMPLE 2 | **Adding Polynomials Vertically** |

Add: $(-9x^3 + 7x^2 - 5x + 3) + (13x^3 + 2x^2 - 8x - 6)$.

Solution

$$\begin{array}{r}
-9x^3 + 7x^2 - 5x + 3 \\
\underline{13x^3 + 2x^2 - 8x - 6} \\
4x^3 + 9x^2 - 13x - 3
\end{array}$$

Line up like terms vertically.

Add the like terms in each column.

This is the same answer that we found in Example 1.

☑ **CHECK POINT 2** Add the polynomials in Check Point 1 using a vertical format. Begin by arranging like terms in columns.

3 Subtract polynomials.

Study Tip

A vertical format often makes it easier to see the like terms.

Subtracting Polynomials

We subtract real numbers by adding the opposite, or additive inverse, of the number being subtracted. For example,

$$8 - 3 = 8 + (-3) = 5.$$

Subtraction of polynomials also involves opposites. If the sum of two polynomials is 0, the polynomials are **opposites**, or **additive inverses**, of each other. Here is an example:

$$(4x^2 - 6x - 7) + (-4x^2 + 6x + 7) = 0.$$

The opposite of $4x^2 - 6x - 7$ is $-4x^2 + 6x + 7$, and vice-versa.

Observe that the opposite of $4x^2 - 6x - 7$ can be obtained by changing the sign of each of its coefficients:

| **Polynomial** | Change 4 to −4, change −6 to 6, and change −7 to 7. | **Opposite** |
| $4x^2 - 6x - 7$ | | $-4x^2 + 6x + 7$ |

In general, **the opposite of a polynomial is that polynomial with the sign of every coefficient changed.** Just as we did with real numbers, we subtract one polynomial from another by adding the opposite of the polynomial being subtracted.

Subtracting Polynomials

To subtract two polynomials, add the first polynomial and the opposite of the polynomial being subtracted.

| EXAMPLE 3 | **Subtracting Polynomials** |

Subtract: $(7x^2 + 3x - 4) - (4x^2 - 6x - 7)$.

Solution

$$(7x^2 + 3x - 4) - (4x^2 - 6x - 7)$$

Change the sign of each coefficient.

$$= (7x^2 + 3x - 4) + (-4x^2 + 6x + 7)$$

Add the opposite of the polynomial being subtracted.

$$= (7x^2 - 4x^2) + (3x + 6x) + (-4 + 7)$$ Group like terms.

$$= 3x^2 + 9x + 3$$ Combine like terms.

☑ **CHECK POINT 3** Subtract: $(9x^2 + 7x - 2) - (2x^2 - 4x - 6)$.

EXAMPLE 4 Subtracting Polynomials

Subtract $2x^3 - 6x^2 - 3x + 9$ from $7x^3 - 8x^2 + 9x - 6$.

Solution

$$(7x^3 - 8x^2 + 9x - 6) - (2x^3 - 6x^2 - 3x + 9)$$

Change the sign of each coefficient.

$$= (7x^3 - 8x^2 + 9x - 6) + (-2x^3 + 6x^2 + 3x - 9)$$ Add the opposite of the polynomial being subtracted.

$$= (7x^3 - 2x^3) + (-8x^2 + 6x^2)$$
$$+ (9x + 3x) + (-6 - 9)$$ Group like terms.

$$= 5x^3 + (-2x^2) + 12x + (-15)$$ Combine like terms.

$$= 5x^3 - 2x^2 + 12x - 15$$ Express addition of opposites as subtraction. ■

☑ **CHECK POINT 4** Subtract $3x^3 - 8x^2 - 5x + 6$ from $10x^3 - 5x^2 + 7x - 2$.

Subtraction can also be performed in columns.

EXAMPLE 5 Subtracting Polynomials Vertically

Use the method of subtracting in columns to find

$$(12y^3 - 9y^2 - 11y - 3) - (4y^3 - 5y + 8).$$

Solution Arrange like terms in columns.

$$\begin{array}{r} 12y^3 - 9y^2 - 11y - 3 \\ -(4y^3 \qquad - 5y + 8) \\ \hline \end{array}$$

Add the opposite of the polynomial being subtracted.

Leave space for the missing term.

$$\begin{array}{r} 12y^3 - 9y^2 - 11y - 3 \\ + -4y^3 \qquad + 5y - 8 \\ \hline 8y^3 - 9y^2 - 6y - 11 \end{array}$$

Change the sign of each coefficient of $4y^3 - 5y + 8$.

Combine like terms.

■

☑ **CHECK POINT 5** Use the method of subtracting in columns to find

$$(8y^3 - 10y^2 - 14y - 2) - (5y^3 - 3y + 6).$$

Graphing Equations Defined by Polynomials

Look at the picture of this gymnast. He has created a perfect balance in which the two halves of his body are mirror images of each other. Graphs of equations defined by polynomials of degree 2, such as $y = x^2 - 4$, have this mirrorlike quality. We can obtain their graphs, shaped like bowls or inverted bowls, using the point-plotting method for graphing an equation in two variables.

4 Graph equations defined by polynomials of degree 2.

Graphing an Equation Defined by a Polynomial of Degree 2

Graph the equation: $y = x^2 - 4$.

Solution The given equation involves two variables, x and y. However, because the variable x is squared, it is not a linear equation in two variables.

$$y = x^2 - 4$$

> This is not in the form $y = mx + b$ because x is squared.

Although the graph is not a line, it is still a picture of all the ordered-pair solutions of $y = x^2 - 4$. Thus, we can use the point-plotting method to obtain the graph.

Step 1. Find several ordered pairs that are solutions of the equation. To find some solutions of $y = x^2 - 4$, we select integers for x, starting with -3 and ending with 3.

Study Tip

If the graph of an equation is not a straight line, use more solutions than when graphing lines. These extra solutions are needed to get a better general idea of the graph's shape.

Start with x.	Compute y.	Form the ordered pair (x, y).
x	$y = x^2 - 4$	(x, y)
-3	$y = (-3)^2 - 4 = 9 - 4 = 5$	$(-3, 5)$
-2	$y = (-2)^2 - 4 = 4 - 4 = 0$	$(-2, 0)$
-1	$y = (-1)^2 - 4 = 1 - 4 = -3$	$(-1, -3)$
0	$y = 0^2 - 4 = 0 - 4 = -4$	$(0, -4)$
1	$y = 1^2 - 4 = 1 - 4 = -3$	$(1, -3)$
2	$y = 2^2 - 4 = 4 - 4 = 0$	$(2, 0)$
3	$y = 3^2 - 4 = 9 - 4 = 5$	$(3, 5)$

Step 2. Plot these ordered pairs as points in the rectangular coordinate system. The seven ordered pairs in the table of values are plotted in **Figure 6.1(a)**.

Step 3. Connect the points with a smooth curve. The seven points are joined with a smooth curve in **Figure 6.1(b)**. The graph of $y = x^2 - 4$ is a curve where the part of the graph to the right of the y-axis is a reflection of the part to the left of it, and vice versa. The arrows on both ends of the curve indicate that it extends indefinitely in both directions.

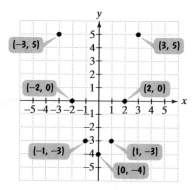

FIGURE 6.1(a) Some solutions of $y = x^2 - 4$ plotted as points

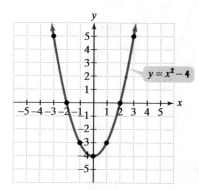

FIGURE 6.1(b) The graph of $y = x^2 - 4$

☑ **CHECK POINT 6** Graph the equation: $y = x^2 - 1$. Select integers for x, starting with -3 and ending with 3.

6.1 EXERCISE SET   **MyMathLab**

PRACTICE   WATCH   DOWNLOAD   READ   REVIEW

Practice Exercises

In Exercises 1–16, identify each polynomial as a monomial, a binomial, or a trinomial. Give the degree of the polynomial.

1. $3x + 7$

2. $5x - 2$

3. $x^3 - 2x$

4. $x^5 - 7x$

5. $8x^2$

6. $10x^2$

7. 5

8. 9

9. $x^2 - 3x + 4$

10. $x^2 - 9x + 2$

11. $7y^2 - 9y^4 + 5$

12. $3y^2 - 14y^5 + 6$

13. $15x - 7x^3$

14. $9x - 5x^3$

15. $-9y^{23}$

16. $-11y^{26}$

In Exercises 17–38, add the polynomials.

17. $(9x + 8) + (-17x + 5)$

18. $(8x - 5) + (-13x + 9)$

19. $(4x^2 + 6x - 7) + (8x^2 + 9x - 2)$

20. $(11x^2 + 7x - 4) + (27x^2 + 10x - 20)$

21. $(7x^2 - 11x) + (3x^2 - x)$

22. $(-3x^2 + x) + (4x^2 + 8x)$

23. $(4x^2 - 6x + 12) + (x^2 + 3x + 1)$

24. $(-7x^2 + 8x + 3) + (2x^2 + x + 8)$

25. $(4y^3 + 7y - 5) + (10y^2 - 6y + 3)$

26. $(2y^3 + 3y + 10) + (3y^2 + 5y - 22)$

27. $(2x^2 - 6x + 7) + (3x^3 - 3x)$

28. $(4x^3 + 5x + 13) + (-4x^2 + 22)$

29. $(4y^2 + 8y + 11) + (-2y^3 + 5y + 2)$

30. $(7y^3 + 5y - 1) + (2y^2 - 6y + 3)$

31. $(-2y^6 + 3y^4 - y^2) + (-y^6 + 5y^4 + 2y^2)$

32. $(7r^4 + 5r^2 + 2r) + (-18r^4 - 5r^2 - r)$

33. $\left(9x^3 - x^2 - x - \dfrac{1}{3}\right) + \left(x^3 + x^2 + x + \dfrac{4}{3}\right)$

34. $\left(12x^3 - x^2 - x + \dfrac{4}{3}\right) + \left(x^3 + x^2 + x - \dfrac{1}{3}\right)$

35. $\left(\dfrac{1}{5}x^4 + \dfrac{1}{3}x^3 + \dfrac{3}{8}x^2 + 6\right) +$
$\left(-\dfrac{3}{5}x^4 + \dfrac{2}{3}x^3 - \dfrac{1}{2}x^2 - 6\right)$

36. $\left(\dfrac{2}{5}x^4 + \dfrac{2}{3}x^3 + \dfrac{5}{8}x^2 + 7\right) +$
$\left(-\dfrac{4}{5}x^4 + \dfrac{1}{3}x^3 - \dfrac{1}{4}x^2 - 7\right)$

37. $(0.03x^5 - 0.1x^3 + x + 0.03) +$
$(-0.02x^5 + x^4 - 0.7x + 0.3)$

38. $(0.06x^5 - 0.2x^3 + x + 0.05) +$
$(-0.04x^5 + 2x^4 - 0.8x + 0.5)$

In Exercises 39–54, use a vertical format to add the polynomials.

39. $5y^3 - 7y^2$
$\underline{6y^3 + 4y^2}$

40. $13x^4 - x^2$
$\underline{7x^4 + 2x^2}$

41. $3x^2 - 7x + 4$
$\underline{-5x^2 + 6x - 3}$

42. $7x^2 - 5x - 6$
$\underline{-9x^2 + 4x + 6}$

43. $\dfrac{1}{4}x^4 - \dfrac{2}{3}x^3 - 5$
$\underline{-\dfrac{1}{2}x^4 + \dfrac{1}{5}x^3 + 4.7}$

44. $\dfrac{1}{3}x^9 - \dfrac{1}{5}x^5 - 2.7$
$\underline{-\dfrac{3}{4}x^9 + \dfrac{2}{3}x^5 + 1}$

45. $y^3 + 5y^2 - 7y - 3$
$\underline{-2y^3 + 3y^2 + 4y - 11}$

46. $y^3 + y^2 - 7y + 9$
$\underline{-y^3 - 6y^2 - 8y + 11}$

47. $4x^3 - 6x^2 + 5x - 7$
$\underline{-9x^3 \qquad - 4x + 3}$

48. $-4y^3 + 6y^2 - 8y + 11$
$\underline{2y^3 \qquad + 9y - 3}$

49. $7x^4 - 3x^3 + x^2$
$\underline{\qquad x^3 - x^2 + 4x - 2}$

50. $7y^5 - 3y^3 + y^2$
$\underline{\qquad 2y^3 - y^2 - 4y - 3}$

51. $7x^2 - 9x + 3$
$4x^2 + 11x - 2$
$\underline{-3x^2 + 5x - 6}$

52. $7y^2 - 11y - 6$
$8y^2 + 3y + 4$
$\underline{-9y^2 - 5y + 2}$

53. $1.2x^3 - 3x^2 + 9.1$
$7.8x^3 - 3.1x^2 + 8$
$\underline{\qquad 1.2x^2 - 6}$

54. $7.9x^3 - 6.8x^2 + 3.3$
$6.1x^3 - 2.2x^2 + 7$
$\underline{\qquad 4.3x^2 - 5}$

In Exercises 55–74, subtract the polynomials.

55. $(x - 8) - (3x + 2)$

56. $(x - 2) - (7x + 9)$

57. $(x^2 - 5x - 3) - (6x^2 + 4x + 9)$

58. $(3x^2 - 8x - 2) - (11x^2 + 5x + 4)$

59. $(x^2 - 5x) - (6x^2 - 4x)$

60. $(3x^2 - 2x) - (5x^2 - 6x)$

61. $(x^2 - 8x - 9) - (5x^2 - 4x - 3)$

62. $(x^2 - 5x + 3) - (x^2 - 6x - 8)$

63. $(y - 8) - (3y - 2)$

64. $(y - 2) - (7y - 9)$

65. $(6y^3 + 2y^2 - y - 11) - (y^2 - 8y + 9)$

66. $(5y^3 + y^2 - 3y - 8) - (y^2 - 8y + 11)$

67. $(7n^3 - n^7 - 8) - (6n^3 - n^7 - 10)$

68. $(2n^2 - n^7 - 6) - (2n^3 - n^7 - 8)$

69. $(y^6 - y^3) - (y^2 - y)$

70. $(y^5 - y^3) - (y^4 - y^2)$

71. $(7x^4 + 4x^2 + 5x) - (-19x^4 - 5x^2 - x)$

72. $(-3x^6 + 3x^4 - x^2) - (-x^6 + 2x^4 + 2x^2)$

73. $\left(\frac{3}{7}x^3 - \frac{1}{5}x - \frac{1}{3}\right) - \left(-\frac{2}{7}x^3 + \frac{1}{4}x - \frac{1}{3}\right)$

74. $\left(\frac{3}{8}x^2 - \frac{1}{3}x - \frac{1}{4}\right) - \left(-\frac{1}{8}x^2 + \frac{1}{2}x - \frac{1}{4}\right)$

In Exercises 75–88, use a vertical format to subtract the polynomials.

75. $7x + 1$
  $-(3x - 5)$

76. $4x + 2$
  $-(3x - 5)$

77. $7x^2 - 3$
  $-(3x^2 + 4)$

78. $9y^2 - 6$
  $-(-5y^2 + 2)$

79. $7y^2 - 5y + 2$
  $-(11y^2 + 2y - 3)$

80. $3x^5 - 5x^3 + 6$
  $-(7x^5 + 4x^3 - 2)$

81. $7x^3 + 5x^2 - 3$
  $-(-2x^3 - 6x^2 + 5)$

82. $3y^4 - 4y^2 + 7$
  $-(-5y^4 - 6y^2 - 13)$

83. $5y^3 + 6y^2 - 3y + 10$
  $-(6y^3 - 2y^2 - 4y - 4)$

84. $4y^3 + 5y^2 + 7y + 11$
  $-(-5y^3 + 6y^2 - 9y - 3)$

85. $7x^4 - 3x^3 + 2x^2$
  $-(\ \ - x^3 - x^2 + x - 2)$

86. $5y^6 - 3y^3 + 2y^2$
  $-(\ \ - y^3 - y^2 - y - 1)$

87. $0.07x^3 - 0.01x^2 + 0.02x$
  $-(0.02x^3 - 0.03x^2 - x)$

88. $0.04x^3 - 0.03x^2 + 0.05x$
  $-(0.02x^3 - 0.06x^2 - x)$

Graph each equation in Exercises 89–94. Find seven solutions in your table of values for each equation by using integers for x, starting with −3 and ending with 3.

89. $y = x^2$

90. $y = x^2 - 2$

91. $y = x^2 + 1$

92. $y = x^2 + 2$

93. $y = 4 - x^2$

94. $y = 9 - x^2$

Practice PLUS

In Exercises 95–98, perform the indicated operations.

95. $[(4x^2 + 7x - 5) - (2x^2 - 10x + 3)] - (x^2 + 5x - 8)$

96. $[(10x^3 - 5x^2 + 4x + 3) - (-3x^3 - 4x^2 + x)] - (7x^3 - 5x + 4)$

97. $[(4y^2 - 3y + 8) - (5y^2 + 7y - 4)] - [(8y^2 + 5y - 7) + (-10y^2 + 4y + 3)]$

98. $[(7y^2 - 4y + 2) - (12y^2 + 3y - 5)] - [(5y^2 - 2y - 8) + (-7y^2 + 10y - 13)]$

99. Subtract $x^3 - 2x^2 + 2$ from the sum of $4x^3 + x^2$ and $-x^3 + 7x - 3$.

100. Subtract $-3x^3 - 7x + 5$ from the sum of $2x^2 + 4x - 7$ and $-5x^3 - 2x - 3$.

101. Subtract $-y^2 + 7y^3$ from the difference between $-5 + y^2 + 4y^3$ and $-8 - y + 7y^3$. Express the answer in standard form.

102. Subtract $-2y^2 + 8y^3$ from the difference between $-6 + y^2 + 5y^3$ and $-12 - y + 13y^3$. Express the answer in standard form.

Application Exercises

As you complete more years of education, you can count on a greater income. The bar graph shows the median, or middle-most, annual income for Americans, by level of education, in 2004.

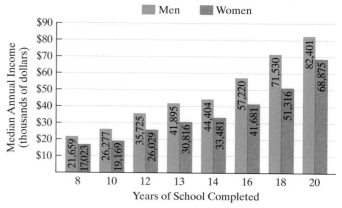

Median Annual Income, by Level of Education, 2004

Source: Bureau of the Census

Here are polynomial models that describe the median annual income for men, M, and for women, W, who have completed x years of education:

$$M = 177x^2 + 288x + 7075$$
$$W = 255x^2 - 2956x + 24{,}336$$

$$M = -18x^3 + 923x^2 - 9603x + 48{,}446$$
$$W = 17x^3 - 450x^2 + 6392x - 14{,}764.$$

Exercises 103–106 are based on these models and the data displayed by the graph.

103. a. Use the equations defined by polynomials of degree 3 to find a mathematical model for $M - W$.

   b. According to the model in part (a), what is the difference in the median annual income between men and women with 14 years of education?

   c. According to the data displayed by the graph on the previous page, what is the actual difference in the median annual income between men and women with 14 years of education? Did the model in part (b) underestimate or overestimate this difference? By how much?

104. a. Use the equations defined by polynomials of degree 3 to find a mathematical model for $M - W$.

   b. According to the model in part (a), what is the difference in the median annual income between men and women with 16 years of education?

   c. According to the data displayed by the graph on the previous page, what is the actual difference in the median annual income between men and women with 16 years of education? Did the model in part (b) underestimate or overestimate this difference? By how much?

105. a. Use the equation defined by a polynomial of degree 2 to find the median annual income for a man with 16 years of education. Does this underestimate or overestimate the median income shown by the bar graph on the previous page? By how much?

   b. Shown in a rectangular coordinate system are the graphs of the polynomial models of degree 2 that describe median annual income, by level of education. Identify your solution from part (a) as a point on the appropriate graph.

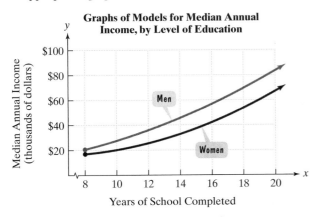

Graphs of Models for Median Annual Income, by Level of Education

   c. Use the appropriate graph in part (b) to estimate, to the nearest thousand dollars, the median annual income for a woman with 16 years of education.

106. a. Use the equation defined by a polynomial of degree 2 to find the median annual income for a woman with 18 years of education. Does this underestimate or overestimate the median income shown by the bar graph on the previous page? By how much?

   b. Shown in Exercise 105(b) are rectangular coordinate graphs of the polynomial models of degree 2 that describe median annual income, by level of education. Identify your solution from part (a) as a point on the appropriate graph.

   c. Use the appropriate graph in Exercise 105(b) to estimate, to the nearest thousand dollars, the median annual income for a man with 18 years of education.

Writing in Mathematics

107. What is a polynomial?

108. What is a monomial? Give an example with your explanation.

109. What is a binomial? Give an example with your explanation.

110. What is a trinomial? Give an example with your explanation.

111. What is the degree of a polynomial? Provide an example with your explanation.

112. Explain how to add polynomials.

113. Explain how to subtract polynomials.

Critical Thinking Exercises

Make Sense? *In Exercises 114–117, determine whether each statement "makes sense" or "does not make sense" and explain your reasoning.*

114. I add like monomials by adding both their coefficients and the exponents that appear on their common variable factor.

115. By looking at the first terms of a polynomial, I can determine its degree.

116. As long as I understand how to add and subtract polynomials, I can select the format, horizontal or vertical, that works best for me.

117. I used two points and a checkpoint to graph $y = x^2 - 4$.

In Exercises 118–121, determine whether each statement is true or false. If the statement is false, make the necessary change(s) to produce a true statement.

118. It is not possible to write a binomial with degree 0.

119. $\dfrac{1}{5x^2} + \dfrac{1}{3x}$ is a binomial.

120. $(2x^2 - 8x + 6) - (x^2 - 3x + 5) = x^2 - 5x + 1$ for any value of x.

121. In the polynomial $3x^2 - 5x + 13$, the coefficient of x is 5.

122. What polynomial must be subtracted from $5x^2 - 2x + 1$ so that the difference is $8x^2 - x + 3$?

123. The number of people who catch a cold t weeks after January 1 is $5t - 3t^2 + t^3$. The number of people who recover t weeks after January 1 is $t - t^2 + \frac{1}{3}t^3$. Write a polynomial in standard form for the number of people who are still ill with a cold t weeks after January 1.

124. Explain why it is not possible to add two polynomials of degree 3 and get a polynomial of degree 4.

Review Exercises

125. Simplify: $(-10)(-7) \div (1 - 8)$. (Section 1.8, Example 8)

126. Subtract: $-4.6 - (-10.2)$. (Section 1.6, Example 2)

127. Solve: $3(x - 2) = 9(x + 2)$. (Section 2.3, Example 3)

Preview Exercises

Exercises 128–130 will help you prepare for the material covered in the next section.

128. Find the missing exponent, designated by the question mark, in the final step.

$$x^3 \cdot x^4 = (x \cdot x \cdot x) \cdot (x \cdot x \cdot x \cdot x) = x^?$$

129. Use the distributive property to multiply: $3x(x + 5)$.

130. Simplify: $x(x + 2) + 3(x + 2)$.

6.2

Objectives

1 Use the product rule for exponents.

2 Use the power rule for exponents.

3 Use the products-to-powers rule.

4 Multiply monomials.

5 Multiply a monomial and a polynomial.

6 Multiply polynomials when neither is a monomial.

Multiplying Polynomials

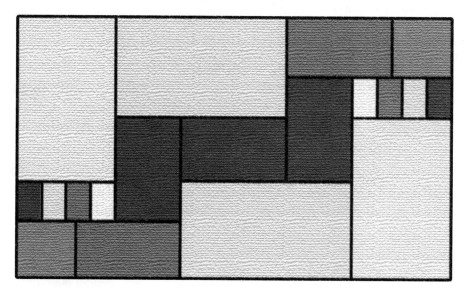

The ancient Greeks believed that the most visually pleasing rectangles have a ratio of length to width of approximately 1.618 to 1. With the exception of the squares on the lower left and the upper right, the interior of this geometric figure is filled entirely with these *golden rectangles*. Furthermore, the large rectangle is also a golden rectangle.

The total area of the large rectangle shown above can be found in many ways. This is because the area of any large rectangular region is related to the areas of the smaller rectangles that make up that region. In this section, we apply areas of rectangles as a way to picture the multiplication of polynomials. Before studying how polynomials are multiplied, we must develop some rules for working with exponents.

1 Use the product rule for exponents.

The Product Rule for Exponents

We have seen that exponents are used to indicate repeated, multiplication. For example, 2^4, where 2 is the base and 4 is the exponent, indicates that 2 occurs as a factor four times:

$$2^4 = 2 \cdot 2 \cdot 2 \cdot 2.$$

Now consider the multiplication of two exponential expressions, such as $2^4 \cdot 2^3$. We are multiplying 4 factors of 2 and 3 factors of 2. We have a total of 7 factors of 2:

4 factors of 2	3 factors of 2

$$2^4 \cdot 2^3 = (2 \cdot 2 \cdot 2 \cdot 2) \cdot (2 \cdot 2 \cdot 2)$$

Total: 7 factors of 2

Thus,

$$2^4 \cdot 2^3 = 2^7.$$

Caution: $2^4 \cdot 2^3$ is not equal to $2^{4\cdot3}$, or 2^{12}, as might be expected.

We can quickly find the exponent, 7, of the product by adding 4 and 3, the original exponents:

$$2^4 \cdot 2^3 = 2^{4+3} = 2^7.$$

This suggests the following rule:

The Product Rule

$$b^m \cdot b^n = b^{m+n}$$

When multiplying exponential expressions with the same base, add the exponents. Use this sum as the exponent of the common base.

EXAMPLE 1   Using the Product Rule

Multiply each expression using the product rule:

a. $2^2 \cdot 2^3$ **b.** $x^7 \cdot x^9$ **c.** $y \cdot y^5$ **d.** $y^3 \cdot y^2 \cdot y^5$.

Solution

a. $2^2 \cdot 2^3 = 2^{2+3} = 2^5$ or 32
b. $x^7 \cdot x^9 = x^{7+9} = x^{16}$
c. $y \cdot y^5 = y^1 \cdot y^5 = y^{1+5} = y^6$
d. $y^3 \cdot y^2 \cdot y^5 = y^{3+2+5} = y^{10}$

✓ **CHECK POINT 1**   Multiply each expression using the product rule:

a. $2^2 \cdot 2^4$ **b.** $x^6 \cdot x^4$ **c.** $y \cdot y^7$ **d.** $y^4 \cdot y^3 \cdot y^2$.

Study Tip

The product rule does not apply to exponential expressions with different bases:

• $x^7 \cdot y^9$, or $x^7 y^9$, cannot be simplified.

2 Use the power rule for exponents.

The Power Rule for Exponents

The next property of exponents applies when an exponential expression is raised to a power. Here is an example:

$$(3^2)^4.$$

The exponential expression 3^2 is raised to the fourth power.

There are 4 factors 3^2. Thus,

$$(3^2)^4 = 3^2 \cdot 3^2 \cdot 3^2 \cdot 3^2 = 3^{2+2+2+2} = 3^8.$$

Add exponents when multiplying with the same base.

We can obtain the answer, 3^8, by multiplying the exponents:

$$(3^2)^4 = 3^{2\cdot4} = 3^8.$$

By generalizing $(3^2)^4 = 3^{2 \cdot 4} = 3^8$, we obtain the following rule:

> ### The Power Rule (Powers to Powers)
>
> $$(b^m)^n = b^{mn}$$
>
> When an exponential expression is raised to a power, multiply the exponents. Place the product of the exponents on the base and remove the parentheses.

EXAMPLE 2 Using the Power Rule

Simplify each expression using the power rule:

    a. $(2^3)^5$ **b.** $(x^6)^4$ **c.** $[(-3)^7]^5$.

Solution

    a. $(2^3)^5 = 2^{3 \cdot 5} = 2^{15}$

    b. $(x^6)^4 = x^{6 \cdot 4} = x^{24}$

    c. $[(-3)^7]^5 = (-3)^{7 \cdot 5} = (-3)^{35}$

✓ **CHECK POINT 2** Simplify each expression using the power rule:

    a. $(3^4)^5$ **b.** $(x^9)^{10}$ **c.** $[(-5)^7]^3$.

Study Tip

Do not confuse the product and power rules. Note the following differences:

- $x^4 \cdot x^7 = x^{4+7} = x^{11}$
- $(x^4)^7 = x^{4 \cdot 7} = x^{28}$.

3 Use the products-to-powers rule.

The Products-to-Powers Rule for Exponents

The next property of exponents applies when we are raising a product to a power. Here is an example:

$$(2x)^4.$$

> The product $2x$ is raised to the fourth power.

There are four factors of $2x$. Thus,

$$(2x)^4 = 2x \cdot 2x \cdot 2x \cdot 2x = 2 \cdot 2 \cdot 2 \cdot 2 \cdot x \cdot x \cdot x \cdot x = 2^4 x^4.$$

We can obtain the answer, $2^4 x^4$, by raising each factor within the parentheses to the fourth power:

$$(2x)^4 = 2^4 x^4.$$

This suggests the following rule:

> ### Products to Powers
>
> $$(ab)^n = a^n b^n$$
>
> When a product is raised to a power, raise each factor to the power.

EXAMPLE 3 Using the Products-to-Powers Rule

Simplify each expression using the products-to-powers rule:

    a. $(5x)^3$ **b.** $(-2y^4)^5$.

Solution

a. $(5x)^3 = 5^3 x^3$ Raise each factor to the third power.

$\qquad = 125x^3$ $5^3 = 5 \cdot 5 \cdot 5 = 125$

b. $(-2y^4)^5 = (-2)^5 (y^4)^5$ Raise each factor to the fifth power.

$\qquad = (-2)^5 y^{4 \cdot 5}$ To raise an exponential expression to a power, multiply exponents: $(b^m)^n = b^{mn}$.

$\qquad = -32y^{20}$ $(-2)^5 = (-2)(-2)(-2)(-2)(-2) = -32$ ∎

☑ **CHECK POINT 3**   Simplify each expression using the products-to-powers rule:

a. $(2x)^4$ **b.** $(-4y^2)^3$.

Study Tip

Try to avoid the following common errors that can occur when simplifying exponential expressions.

Correct	Incorrect	Description of Error
$b^3 \cdot b^4 = b^{3+4} = b^7$	$b^3 \cdot b^4 = b^{12}$	Exponents should be added, not multiplied.
$3^2 \cdot 3^4 = 3^{2+4} = 3^6$	$3^2 \cdot 3^4 = 9^{2+4} = 9^6$	The common base should be retained, not multiplied.
$(x^5)^3 = x^{5 \cdot 3} = x^{15}$	$(x^5)^3 = x^{5+3} = x^8$	Exponents should be multiplied, not added, when raising a power to a power.
$(4x)^3 = 4^3 x^3 = 64x^3$	$(4x)^3 = 4x^3$	Both factors should be cubed.

4   Multiply monomials.

Multiplying Monomials

Now that we have developed three properties of exponents, we are ready to turn to polynomial multiplication. We begin with the product of two monomials, such as $-8x^6$ and $5x^3$. This product is obtained by multiplying the coefficients, -8 and 5, and then multiplying the variables using the product rule for exponents.

$$(-8x^6)(5x^3) = -8 \cdot 5 \cdot x^6 \cdot x^3 = -8 \cdot 5x^{6+3} = -40x^9$$

Multiply coefficients and add exponents.

Multiplying Monomials

To multiply monomials with the same variable base, multiply the coefficients and then multiply the variables. Use the product rule for exponents to multiply the variables: Keep the variable and add the exponents.

EXAMPLE 4   Multiplying Monomials

Multiply: **a.** $(2x)(4x^2)$ **b.** $(-10x^6)(6x^{10})$.

Study Tip

Don't confuse adding and multiplying monomials.

Addition:

$$5x^4 + 6x^4 = 11x^4$$

Multiplication:

$$(5x^4)(6x^4) = (5 \cdot 6)(x^4 \cdot x^4)$$
$$= 30x^{4+4}$$
$$= 30x^8$$

Only like terms can be added or subtracted, but unlike terms may be multiplied.

Addition:

$5x^4 + 3x^2$ cannot be simplified.

Multiplication:

$$(5x^4)(3x^2) = (5 \cdot 3)(x^4 \cdot x^2)$$
$$= 15x^{4+2}$$
$$= 15x^6$$

Solution

a. $(2x)(4x^2) = (2 \cdot 4)(x \cdot x^2)$ *Multiply the coefficients and multiply the variables.*

$\qquad\qquad\quad = 8x^{1+2}$ *Add exponents: $b^m \cdot b^n = b^{m+n}$.*

$\qquad\qquad\quad = 8x^3$ *Simplify.*

b. $(-10x^6)(6x^{10}) = (-10 \cdot 6)(x^6 \cdot x^{10})$ *Multiply the coefficients and multiply the variables.*

$\qquad\qquad\qquad\quad = -60x^{6+10}$ *Add exponents: $b^m \cdot b^n = b^{m+n}$.*

$\qquad\qquad\qquad\quad = -60x^{16}$ *Simplify.*   ■

☑ **CHECK POINT 4**   Multiply: **a.** $(7x^2)(10x)$ **b.** $(-5x^4)(4x^5)$.

Multiplying a Monomial and a Polynomial That Is Not a Monomial

We use the distributive property to multiply a monomial and a polynomial that is not a monomial. For example,

$$3x^2(2x^3 + 5x) = 3x^2 \cdot 2x^3 + 3x^2 \cdot 5x = 3 \cdot 2x^{2+3} + 3 \cdot 5x^{2+1} = 6x^5 + 15x^3.$$

Monomial   Binomial   Multiply coefficients and add exponents.

5 Multiply a monomial and a polynomial.

Multiplying a Monomial and a Polynomial That Is Not a Monomial

To multiply a monomial and a polynomial, use the distributive property to multiply each term of the polynomial by the monomial.

EXAMPLE 5   Multiplying a Monomial and a Polynomial

Multiply: **a.** $2x(x + 4)$ **b.** $3x^2(4x^3 - 5x + 2)$.

Solution

a. $2x(x + 4) = 2x \cdot x + 2x \cdot 4$ *Use the distributive property.*

$\qquad\qquad\quad = 2 \cdot 1x^{1+1} + 2 \cdot 4x$ *To multiply the monomials, multiply coefficients and add exponents.*

$\qquad\qquad\quad = 2x^2 + 8x$ *Simplify.*

b. $3x^2(4x^3 - 5x + 2)$

$\qquad = 3x^2 \cdot 4x^3 - 3x^2 \cdot 5x + 3x^2 \cdot 2$ *Use the distributive property.*

$\qquad = 3 \cdot 4x^{2+3} - 3 \cdot 5x^{2+1} + 3 \cdot 2x^2$ *To multiply the monomials, multiply coefficients and add exponents.*

$\qquad = 12x^5 - 15x^3 + 6x^2$ *Simplify.*   ■

Rectangles often make it possible to visualize polynomial multiplication. For example, **Figure 6.2** shows a rectangle with length $2x$ and width $x + 4$. The area of the large rectangle is

$$2x(x + 4).$$

The sum of the areas of the two smaller rectangles is

$$2x^2 + 8x.$$

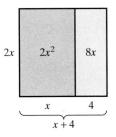

FIGURE 6.2

Conclusion:

$$2x(x + 4) = 2x^2 + 8x$$

☑ **CHECK POINT 5**   Multiply:

a. $3x(x + 5)$ **b.** $6x^2(5x^3 - 2x + 3)$.

6 Multiply polynomials when neither is a monomial.

Multiplying Polynomials When Neither Is a Monomial

How do we multiply two polynomials if neither is a monomial? For example, consider

$$(2x + 3)(x^2 + 4x + 5).$$

Binomial   Trinomial

One way to perform this multiplication is to distribute $2x$ throughout the trinomial

$$2x(x^2 + 4x + 5)$$

and 3 throughout the trinomial

$$3(x^2 + 4x + 5).$$

Then combine the like terms that result. In general, the product of two polynomials is the polynomial obtained by multiplying each term of one polynomial by each term of the other polynomial and then combining like terms.

Multiplying Polynomials When Neither Is a Monomial

Multiply each term of one polynomial by each term of the other polynomial. Then combine like terms.

Using Technology

Graphic Connections

The graphs of

$$y_1 = (x + 3)(x + 2)$$
and $$y_2 = x^2 + 5x + 6$$

are the same.

$[-6, 2, 1]$ by $[-1, 10, 1]$

This verifies that

$$(x + 3)(x + 2) = x^2 + 5x + 6.$$

EXAMPLE 6   **Multiplying Binomials**

Multiply: **a.** $(x + 3)(x + 2)$ **b.** $(3x + 7)(2x - 4)$.

Solution   We begin by multiplying each term of the second binomial by each term of the first binomial.

a. $(x + 3)(x + 2)$

$\quad = x(x + 2) + 3(x + 2)$ Multiply the second binomial by each term of the first binomial.

$\quad = x \cdot x + x \cdot 2 + 3 \cdot x + 3 \cdot 2$ Use the distributive property.

$\quad = x^2 + 2x + 3x + 6$ Multiply. Note that $x \cdot x = x^1 \cdot x^1 = x^{1+1} = x^2$.

$\quad = x^2 + 5x + 6$ Combine like terms.

b. $(3x + 7)(2x - 4)$

$= 3x(2x - 4) + 7(2x - 4)$ Multiply the second binomial by each term of the first binomial.

$= 3x \cdot 2x - 3x \cdot 4 + 7 \cdot 2x - 7 \cdot 4$ Use the distributive property.

$= 6x^2 - 12x + 14x - 28$ Multiply.

$= 6x^2 + 2x - 28$ Combine like terms.

☑ **CHECK POINT 6**   Multiply:

a. $(x + 4)(x + 5)$ **b.** $(5x + 3)(2x - 7)$.

You can visualize the polynomial multiplication in Example 6(a), $(x + 3)(x + 2) = x^2 + 5x + 6$, by analyzing the areas in **Figure 6.3**.

Area of the large rectangle $(x + 3)(x + 2)$

Sum of the areas of the four smaller rectangles inside the large rectangle $x^2 + 3x + 2x + 6$
$= x^2 + 5x + 6$

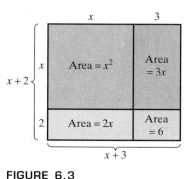

FIGURE 6.3

Conclusion:

$$(x + 3)(x + 2) = x^2 + 5x + 6$$

EXAMPLE 7   **Multiplying a Binomial and a Trinomial**

Multiply: $(2x + 3)(x^2 + 4x + 5)$.

Solution

$(2x + 3)(x^2 + 4x + 5)$

$= 2x(x^2 + 4x + 5) + 3(x^2 + 4x + 5)$ Multiply the trinomial by each term of the binomial.

$= 2x \cdot x^2 + 2x \cdot 4x + 2x \cdot 5 + 3x^2 + 3 \cdot 4x + 3 \cdot 5$ Use the distributive property.

$= 2x^3 + 8x^2 + 10x + 3x^2 + 12x + 15$ Multiply monomials: Multiply coefficients and add exponents.

$= 2x^3 + 11x^2 + 22x + 15$ Combine like terms:
$8x^2 + 3x^2 = 11x^2$ and
$10x + 12x = 22x$.

☑ **CHECK POINT 7**   Multiply: $(5x + 2)(x^2 - 4x + 3)$.

Another method for solving Example 7 is to use a vertical format similar to that used for multiplying whole numbers.

$$
\begin{array}{r}
x^2 + 4x + 5 \\
2x + 3 \\
\hline
3x^2 + 12x + 15 \\
2x^3 + 8x^2 + 10x \\
\hline
2x^3 + 11x^2 + 22x + 15
\end{array}
$$

Write like terms in the same column.

$3(x^2 + 4x + 5)$

$2x(x^2 + 4x + 5)$

Combine like terms.

EXAMPLE 8   **Multiplying Polynomials Using a Vertical Format**

Multiply: $(2x^2 - 3x)(5x^3 - 4x^2 + 7x)$.

Solution To use the vertical format, it is most convenient to write the polynomial with the greater number of terms in the top row.

$$5x^3 - 4x^2 + 7x$$
$$2x^2 - 3x$$

We now multiply each term in the top polynomial by the last term in the bottom polynomial.

$$
\begin{array}{r}
5x^3 - 4x^2 + 7x \\
2x^2 - 3x \\
\hline
-15x^4 + 12x^3 - 21x^2
\end{array}
$$
 $-3x(5x^3 - 4x^2 + 7x)$

Then we multiply each term in the top polynomial by $2x^2$, the first term in the bottom polynomial. Like terms are placed in columns because the final step involves combining them.

Write like terms in the same column.
$$
\begin{array}{r}
5x^3 - 4x^2 + 7x \\
2x^2 - 3x \\
\hline
-15x^4 + 12x^3 - 21x^2 \\
10x^5 - 8x^4 + 14x^3 \\
\hline
10x^5 - 23x^4 + 26x^3 - 21x^2
\end{array}
$$
 $-3x(5x^3 - 4x^2 + 7x)$

 $2x^2(5x^3 - 4x^2 + 7x)$

 Combine like terms, which are lined up in columns.

☑ **CHECK POINT 8** Multiply using a vertical format: $(3x^2 - 2x)(2x^3 - 5x^2 + 4x)$.

6.2 EXERCISE SET **MyMathLab** | Math XL PRACTICE | WATCH | DOWNLOAD | READ | REVIEW

Practice Exercises

In Exercises 1–8, multiply each expression using the product rule.

1. $x^{15} \cdot x^3$ **2.** $x^{12} \cdot x^4$

3. $y \cdot y^{11}$ **4.** $y \cdot y^{19}$

5. $x^2 \cdot x^6 \cdot x^3$ **6.** $x^4 \cdot x^3 \cdot x^5$

7. $7^9 \cdot 7^{10}$ **8.** $8^7 \cdot 8^{10}$

In Exercises 9–14, simplify each expression using the power rule.

9. $(6^9)^{10}$ **10.** $(6^7)^{10}$

11. $(x^{15})^3$ **12.** $(x^{12})^4$

13. $[(-20)^3]^3$ **14.** $[(-50)^4]^4$

In Exercises 15–24, simplify each expression using the products-to-powers rule.

15. $(2x)^3$ **16.** $(4x)^3$

17. $(-5x)^2$ **18.** $(-6x)^2$

19. $(4x^3)^2$ **20.** $(6x^3)^2$

21. $(-2y^6)^4$ **22.** $(-2y^5)^4$

23. $(-2x^7)^5$ **24.** $(-2x^{11})^5$

In Exercises 25–34, multiply the monomials.

25. $(7x)(2x)$ **26.** $(8x)(3x)$

27. $(6x)(4x^2)$ **28.** $(10x)(3x^2)$

29. $(-5y^4)(3y^3)$ **30.** $(-6y^4)(2y^3)$

31. $\left(-\dfrac{1}{2}a^3\right)\left(-\dfrac{1}{4}a^2\right)$ **32.** $\left(-\dfrac{1}{3}a^4\right)\left(-\dfrac{1}{2}a^2\right)$

33. $(2x^2)(-3x)(8x^4)$ **34.** $(3x^3)(-2x)(5x^6)$

In Exercises 35–54, find each product of the monomial and the polynomial.

35. $4x(x + 3)$ **36.** $6x(x + 5)$

37. $x(x - 3)$ **38.** $x(x - 7)$

39. $2x(x - 6)$ **40.** $3x(x - 5)$

41. $-4y(3y + 5)$

42. $-5y(6y + 7)$

43. $4x^2(x + 2)$

44. $5x^2(x + 6)$

45. $2y^2(y^2 + 3y)$

46. $4y^2(y^2 + 2y)$

47. $2y^2(3y^2 - 4y + 7)$

48. $4y^2(5y^2 - 6y + 3)$

49. $(3x^3 + 4x^2)(2x)$

50. $(4x^3 + 5x^2)(2x)$

51. $(x^2 + 5x - 3)(-2x)$

52. $(x^3 - 2x + 2)(-4x)$

53. $-3x^2(-4x^2 + x - 5)$

54. $-6x^2(3x^2 - 2x - 7)$

In Exercises 55–78, find each product. In each case, neither factor is a monomial.

55. $(x + 3)(x + 5)$

56. $(x + 4)(x + 6)$

57. $(2x + 1)(x + 4)$

58. $(2x + 5)(x + 3)$

59. $(x + 3)(x - 5)$

60. $(x + 4)(x - 6)$

61. $(x - 11)(x + 9)$

62. $(x - 12)(x + 8)$

63. $(2x - 5)(x + 4)$

64. $(3x - 4)(x + 5)$

65. $\left(\dfrac{1}{4}x + 4\right)\left(\dfrac{3}{4}x - 1\right)$

66. $\left(\dfrac{1}{5}x + 5\right)\left(\dfrac{3}{5}x - 1\right)$

67. $(x + 1)(x^2 + 2x + 3)$

68. $(x + 2)(x^2 + x + 5)$

69. $(y - 3)(y^2 - 3y + 4)$

70. $(y - 2)(y^2 - 4y + 3)$

71. $(2a - 3)(a^2 - 3a + 5)$

72. $(2a - 1)(a^2 - 4a + 3)$

73. $(x + 1)(x^3 + 2x^2 + 3x + 4)$

74. $(x + 1)(x^3 + 4x^2 + 7x + 3)$

75. $\left(x - \dfrac{1}{2}\right)(4x^3 - 2x^2 + 5x - 6)$

76. $\left(x - \dfrac{1}{3}\right)(3x^3 - 6x^2 + 5x - 9)$

77. $(x^2 + 2x + 1)(x^2 - x + 2)$

78. $(x^2 + 3x + 1)(x^2 - 2x - 1)$

In Exercises 79–92, use a vertical format to find each product.

79.
$$\begin{array}{r} x^2 - 5x + 3 \\ \underline{x + 8} \end{array}$$

80.
$$\begin{array}{r} x^2 - 7x + 9 \\ \underline{x + 4} \end{array}$$

81.
$$\begin{array}{r} x^2 - 3x + 9 \\ \underline{2x - 3} \end{array}$$

82.
$$\begin{array}{r} y^2 - 5y + 3 \\ \underline{4y - 5} \end{array}$$

83.
$$\begin{array}{r} 2x^3 + x^2 + 2x + 3 \\ \underline{x + 4} \end{array}$$

84.
$$\begin{array}{r} 3y^3 + 2y^2 + y + 4 \\ \underline{y + 3} \end{array}$$

85.
$$\begin{array}{r} 4z^3 - 2z^2 + 5z - 4 \\ \underline{3z - 2} \end{array}$$

86.
$$\begin{array}{r} 5z^3 - 3z^2 + 4z - 3 \\ \underline{2z - 4} \end{array}$$

87.
$$\begin{array}{r} 7x^3 - 5x^2 + 6x \\ \underline{3x^2 - 4x} \end{array}$$

88.
$$\begin{array}{r} 9y^3 - 7y^2 + 5y \\ \underline{3y^2 + 5y} \end{array}$$

89.
$$\begin{array}{r} 2y^5 - 3y^3 + y^2 - 2y + 3 \\ \underline{2y - 1} \end{array}$$

90.
$$\begin{array}{r} n^4 - n^3 + n^2 - n + 1 \\ \underline{2n + 3} \end{array}$$

91.
$$\begin{array}{r} x^2 + 7x - 3 \\ \underline{x^2 - x - 1} \end{array}$$

92.
$$\begin{array}{r} x^2 + 6x - 4 \\ \underline{x^2 - x - 2} \end{array}$$

Practice PLUS

In Exercises 93–100, perform the indicated operations.

93. $(x + 4)(x - 5) - (x + 3)(x - 6)$

94. $(x + 5)(x - 6) - (x + 2)(x - 9)$

95. $4x^2(5x^3 + 3x - 2) - 5x^3(x^2 - 6)$

96. $3x^2(6x^3 + 2x - 3) - 4x^3(x^2 - 5)$

97. $(y + 1)(y^2 - y + 1) + (y - 1)(y^2 + y + 1)$

98. $(y + 1)(y^2 - y + 1) - (y - 1)(y^2 + y + 1)$

99. $(y + 6)^2 - (y - 2)^2$

100. $(y + 5)^2 - (y - 4)^2$

Application Exercises

101. Find a trinomial for the area of the rectangular rug shown below whose sides are $x + 5$ feet and $2x - 3$ feet.

102. The base of a triangular sail is $4x$ feet and its height is $3x + 10$ feet. Write a binomial in terms of x for the area of the sail.

In Exercises 103–104,

  a. *Express the area of the large rectangle as the product of two binomials.*

  b. *Find the sum of the areas of the four smaller rectangles.*

  c. *Use polynomial multiplication to show that your expressions for area in parts (a) and (b) are equal.*

103.

104.

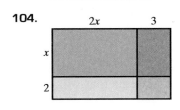

Writing in Mathematics

105. Explain the product rule for exponents. Use $2^3 \cdot 2^5$ in your explanation.

106. Explain the power rule for exponents. Use $(3^2)^4$ in your explanation.

107. Explain how to simplify an expression that involves a product raised to a power. Provide an example with your explanation.

108. Explain how to multiply monomials. Give an example.

109. Explain how to multiply a monomial and a polynomial that is not a monomial. Give an example.

110. Explain how to multiply polynomials when neither is a monomial. Give an example.

111. Explain the difference between performing these two operations:

$$2x^2 + 3x^2 \quad \text{and} \quad (2x^2)(3x^2).$$

112. Discuss situations in which a vertical format, rather than a horizontal format, is useful for multiplying polynomials.

Critical Thinking Exercises

Make Sense? *In Exercises 113–116, determine whether each statement "makes sense" or "does not make sense" and explain your reasoning.*

113. I'm working with two monomials that I cannot add, although I can multiply them.

114. I'm working with two monomials that I can add, although I cannot multiply them.

115. Other than multiplying monomials, the distributive property is used to multiply other kinds of polynomials.

116. I used the product rule for exponents to multiply x^7 and y^9.

In Exercises 117–120, determine whether each statement is true or false. If the statement is false, make the necessary change(s) to produce a true statement.

117. $4x^3 \cdot 3x^4 = 12x^{12}$

118. $5x^2 \cdot 4x^6 = 9x^8$

119. $(y - 1)(y^2 + y + 1) = y^3 - 1$

120. Some polynomial multiplications can only be performed by using a vertical format.

121. Find a polynomial in descending powers of x representing the area of the shaded region.

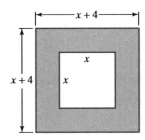

122. Find each of the products in parts (a)–(c).

  a. $(x - 1)(x + 1)$

  b. $(x - 1)(x^2 + x + 1)$

  c. $(x - 1)(x^3 + x^2 + x + 1)$

  d. Using the pattern found in parts (a)–(c), find $(x - 1)(x^4 + x^3 + x^2 + x + 1)$ without actually multiplying.

123. Find the missing factor.

$$(\underline{\quad})\left(-\frac{1}{4}xy^3\right) = 2x^5y^3$$

Review Exercises

124. Solve: $4x - 7 > 9x - 2$. (Section 2.6, Example 7)

125. Graph $3x - 2y = 6$ using intercepts. (Section 4.2, Example 4)

126. Find the slope of the line passing through the points $(-2, 8)$ and $(1, 6)$. (Section 4.3, Example 1)

Preview Exercises

Exercises 127–129 will help you prepare for the material covered in the next section. In each exercise, find the indicated products. Then, if possible, state a fast method for finding these products. (You may already be familiar with some of these methods from a high school algebra course.)

127. a. $(x + 3)(x + 4)$
    b. $(x + 5)(x + 20)$

128. a. $(x + 3)(x - 3)$
    b. $(x + 5)(x - 5)$

129. a. $(x + 3)^2$
    b. $(x + 5)^2$

6.3

Special Products

Objectives

1. Use FOIL in polynomial multiplication.
2. Multiply the sum and difference of two terms.
3. Find the square of a binomial sum.
4. Find the square of a binomial difference.

Let's cut to the chase. Are there fast methods for finding products of polynomials? Yes. In this section, we use the distributive property to develop patterns that will let you multiply certain binomials quite rapidly.

1️⃣ **Use FOIL in polynomial multiplication.**

The Product of Two Binomials: FOIL

Frequently, we need to find the product of two binomials. One way to perform this multiplication is to distribute each term in the first binomial through the second binomial. For example, we can find the product of the binomials $3x + 2$ and $4x + 5$ as follows:

$$(3x + 2)(4x + 5) = 3x(4x + 5) + 2(4x + 5)$$
$$= 3x(4x) + 3x(5) + 2(4x) + 2(5)$$
$$= 12x^2 + 15x + 8x + 10.$$

> Distribute $3x$ over $4x + 5$.   Distribute 2 over $4x + 5$.

> We'll combine these like terms later. For now, our interest is in how to obtain *each* of these four terms.

We can also find the product of $3x + 2$ and $4x + 5$ using a method called FOIL, which is based on our work shown above. Any two binomials can be quickly multiplied by using the FOIL method, in which **F** represents the product of the **first** terms in each binomial, **O** represents the product of the **outside** terms, **I** represents the product of the **inside** terms, and **L** represents the product of the **last**, or second, terms in each binomial. For example, we can use the FOIL method to find the product of the binomials $3x + 2$ and $4x + 5$ as follows:

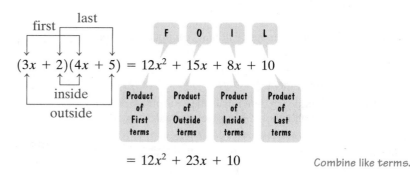

$$(3x + 2)(4x + 5) = 12x^2 + 15x + 8x + 10$$

| F | O | I | L |
| Product of First terms | Product of Outside terms | Product of Inside terms | Product of Last terms |

$$= 12x^2 + 23x + 10$$ *Combine like terms.*

In general, here's how to use the FOIL method to find the product of $ax + b$ and $cx + d$:

Using the FOIL Method to Multiply Binomials

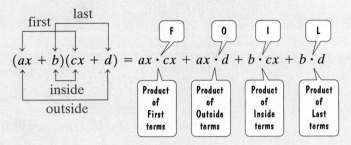

EXAMPLE 1 Using the FOIL Method

Multiply: $(x + 3)(x + 4)$.

Solution

F: First terms $= x \cdot x = x^2$ $(x + 3)(x + 4)$

O: Outside terms $= x \cdot 4 = 4x$ $(x + 3)(x + 4)$

I: Inside terms $= 3 \cdot x = 3x$ $(x + 3)(x + 4)$

L: Last terms $= 3 \cdot 4 = 12$ $(x + 3)(x + 4)$

$$
\begin{aligned}
(x + 3)(x + 4) &= x \cdot x + x \cdot 4 + 3 \cdot x + 3 \cdot 4 \\
&= x^2 + 4x + 3x + 12 \\
&= x^2 + 7x + 12 \quad \text{Combine like terms.}
\end{aligned}
$$

☑ **CHECK POINT 1** Multiply: $(x + 5)(x + 6)$.

EXAMPLE 2 Using the FOIL Method

Multiply: $(3x + 4)(5x - 3)$.

Solution

$$
\begin{aligned}
(3x + 4)(5x - 3) &= 3x \cdot 5x + 3x(-3) + 4 \cdot 5x + 4(-3) \\
&= 15x^2 - 9x + 20x - 12 \\
&= 15x^2 + 11x - 12 \quad \text{Combine like terms.}
\end{aligned}
$$

☑ **CHECK POINT 2** Multiply: $(7x + 5)(4x - 3)$.

EXAMPLE 3   Using the FOIL Method

Multiply: $(2 - 5x)(3 - 4x)$.

Solution

$$(2 - 5x)(3 - 4x) = 2 \cdot 3 + 2(-4x) + (-5x)(3) + (-5x)(-4x)$$
$$= 6 - 8x - 15x + 20x^2$$
$$= 6 - 23x + 20x^2 \qquad \text{Combine like terms.}$$

The product can also be expressed in standard form as $20x^2 - 23x + 6$.

☑ **CHECK POINT 3**   Multiply: $(4 - 2x)(5 - 3x)$.

---

2 Multiply the sum and difference of two terms.

Multiplying the Sum and Difference of Two Terms

We can use the FOIL method to multiply $A + B$ and $A - B$ as follows:

$$(A + B)(A - B) = A^2 - AB + AB - B^2 = A^2 - B^2.$$

Notice that the outside and inside products have a sum of 0 and the terms cancel. The FOIL multiplication provides us with a quick rule for multiplying the sum and difference of two terms, referred to as a *special-product formula*.

The Product of the Sum and Difference of Two Terms

$$(A + B)(A - B) = A^2 - B^2$$

| The product of the sum and the difference of the same two terms | is | the square of the first term minus the square of the second term. |

EXAMPLE 4   Finding the Product of the Sum and Difference of Two Terms

Multiply: **a.** $(4y + 3)(4y - 3)$ **b.** $(3x - 7)(3x + 7)$ **c.** $(5a^4 + 6)(5a^4 - 6)$.

Solution   Use the special-product formula shown.

$$(A + B)(A - B) = A^2 - B^2$$

First term squared − Second term squared = Product

a. $(4y + 3)(4y - 3) = (4y)^2 - 3^2 = 16y^2 - 9$
b. $(3x - 7)(3x + 7) = (3x)^2 - 7^2 = 9x^2 - 49$
c. $(5a^4 + 6)(5a^4 - 6) = (5a^4)^2 - 6^2 = 25a^8 - 36$

☑ **CHECK POINT 4**   Multiply: **a.** $(7y + 8)(7y - 8)$
b. $(4x - 5)(4x + 5)$ **c.** $(2a^3 + 3)(2a^3 - 3)$.

3 Find the square of a binomial sum.

The Square of a Binomial

Let's now find $(A + B)^2$, the square of a binomial sum. To do so, we begin with the FOIL method and look for a general rule.

$$(A + B)^2 = (A + B)(A + B) = A \cdot A + A \cdot B + A \cdot B + B \cdot B$$
$$= A^2 + 2AB + B^2$$

This result implies the following rule, which is another example of a special-product formula:

The Square of a Binomial Sum

$$(A + B)^2 = A^2 + 2AB + B^2$$

| The square of a binomial sum | is | first term squared | plus | 2 times the product of the terms | plus | last term squared. |

EXAMPLE 5 **Finding the Square of a Binomial Sum**

Multiply:

a. $(x + 3)^2$ **b.** $(3x + 7)^2$.

Solution Use the special-product formula shown.

$(A + B)^2 =$	A^2	$+$	$2AB$	$+$	B^2	
	(First Term)^2	$+$	2 · Product of the Terms	$+$	(Last Term)^2	= Product
a. $(x + 3)^2 =$	x^2	$+$	$2 \cdot x \cdot 3$	$+$	3^2	$= x^2 + 6x + 9$
b. $(3x + 7)^2 =$	$(3x)^2$	$+$	$2(3x)(7)$	$+$	7^2	$= 9x^2 + 42x + 49$

✓ CHECK POINT 5 Multiply:

a. $(x + 10)^2$ **b.** $(5x + 4)^2$.

Study Tip

Caution! The square of a sum is *not* the sum of the squares.

Show that $(x + 3)^2$ and $x^2 + 9$ are not equal by substituting 5 for x in each expression and simplifying.

The formula for the square of a binomial sum can be interpreted geometrically by analyzing the areas in **Figure 6.4**.

Area of the large square $(A + B)^2$

Sum of the areas of the four smaller rectangles inside the large square
$$A^2 + AB + AB + B^2$$
$$= A^2 + 2AB + B^2$$

Conclusion:

$$(A + B)^2 = A^2 + 2AB + B^2$$

FIGURE 6.4

4 Find the square of a binomial difference.

A similar pattern occurs for $(A - B)^2$, the square of a binomial difference. Using the FOIL method on $(A - B)^2$, we obtain the following rule:

The Square of a Binomial Difference

$$(A - B)^2 = A^2 - 2AB + B^2$$

| The square of a binomial difference | is | first term squared | minus | 2 times the product of the terms | plus | last term squared. |

EXAMPLE 6   Finding the Square of a Binomial Difference

Multiply:

a. $(x - 4)^2$ b. $(5y - 6)^2$.

Solution   Use the special-product formula shown.

$$(A - B)^2 = A^2 - 2AB + B^2$$

	(First Term)^2	−	2 · Product of the Terms	+	(Last Term)^2	= Product
a. $(x - 4)^2 =$	x^2	−	$2 \cdot x \cdot 4$	+	4^2	$= x^2 - 8x + 16$
b. $(5y - 6)^2 =$	$(5y)^2$	−	$2(5y)(6)$	+	6^2	$= 25y^2 - 60y + 36$

✓ **CHECK POINT 6**   Multiply:

a. $(x - 9)^2$ b. $(7x - 3)^2$.

The following table summarizes the FOIL method and the three special products. The special products occur so frequently in algebra that it is convenient to memorize the form or pattern of these formulas.

FOIL and Special Products
Let *A*, *B*, *C*, and *D* be real numbers, variables, or algebraic expressions.

FOIL	*Example*
F O I L	F O I L
$(A + B)(C + D) = AC + AD + BC + BD$	$(2x + 3)(4x + 5) = (2x)(4x) + (2x)(5) + (3)(4x) + (3)(5)$
	$= 8x^2 + 10x + 12x + 15$
	$= 8x^2 + 22x + 15$
Sum and Difference of Two Terms	*Example*
$(A + B)(A - B) = A^2 - B^2$	$(2x + 3)(2x - 3) = (2x)^2 - 3^2$
	$= 4x^2 - 9$
Square of a Binomial	*Example*
$(A + B)^2 = A^2 + 2AB + B^2$	$(2x + 3)^2 = (2x)^2 + 2(2x)(3) + 3^2$
	$= 4x^2 + 12x + 9$
$(A - B)^2 = A^2 - 2AB + B^2$	$(2x - 3)^2 = (2x)^2 - 2(2x)(3) + 3^2$
	$= 4x^2 - 12x + 9$

6.3 EXERCISE SET *MyMathLab*

Practice Exercises

In Exercises 1–24, use the FOIL method to find each product. Express the product in descending powers of the variable.

1. $(x + 4)(x + 6)$

2. $(x + 8)(x + 2)$

3. $(y - 7)(y + 3)$

4. $(y - 3)(y + 4)$

5. $(2x - 3)(x + 5)$

6. $(3x - 5)(x + 7)$

7. $(4y + 3)(y - 1)$

8. $(5y + 4)(y - 2)$

9. $(2x - 3)(5x + 3)$

10. $(2x - 5)(7x + 2)$

11. $(3y - 7)(4y - 5)$

12. $(4y - 5)(7y - 4)$

13. $(7 + 3x)(1 - 5x)$

14. $(2 + 5x)(1 - 4x)$

15. $(5 - 3y)(6 - 2y)$

16. $(7 - 2y)(10 - 3y)$

17. $(5x^2 - 4)(3x^2 - 7)$

18. $(7x^2 - 2)(3x^2 - 5)$

19. $(6x - 5)(2 - x)$

20. $(4x - 3)(2 - x)$

21. $(x + 5)(x^2 + 3)$

22. $(x + 4)(x^2 + 5)$

23. $(8x^3 + 3)(x^2 + 5)$

24. $(7x^3 + 5)(x^2 + 2)$

In Exercises 25–44, multiply using the rule for finding the product of the sum and difference of two terms.

25. $(x + 3)(x - 3)$

26. $(y + 5)(y - 5)$

27. $(3x + 2)(3x - 2)$

28. $(2x + 5)(2x - 5)$

29. $(3r - 4)(3r + 4)$

30. $(5z - 2)(5z + 2)$

31. $(3 + r)(3 - r)$

32. $(4 + s)(4 - s)$

33. $(5 - 7x)(5 + 7x)$

34. $(4 - 3y)(4 + 3y)$

35. $\left(2x + \dfrac{1}{2}\right)\left(2x - \dfrac{1}{2}\right)$

36. $\left(3y + \dfrac{1}{3}\right)\left(3y - \dfrac{1}{3}\right)$

37. $(y^2 + 1)(y^2 - 1)$

38. $(y^2 + 2)(y^2 - 2)$

39. $(r^3 + 2)(r^3 - 2)$

40. $(m^3 + 4)(m^3 - 4)$

41. $(1 - y^4)(1 + y^4)$

42. $(2 - s^5)(2 + s^5)$

43. $(x^{10} + 5)(x^{10} - 5)$

44. $(x^{12} + 3)(x^{12} - 3)$

In Exercises 45–62, multiply using the rules for the square of a binomial.

45. $(x + 2)^2$

46. $(x + 5)^2$

47. $(2x + 5)^2$

48. $(5x + 2)^2$

49. $(x - 3)^2$

50. $(x - 6)^2$

51. $(3y - 4)^2$

52. $(4y - 3)^2$

53. $(4x^2 - 1)^2$

54. $(5x^2 - 3)^2$

55. $(7 - 2x)^2$

56. $(9 - 5x)^2$

57. $\left(2x + \dfrac{1}{2}\right)^2$

58. $\left(3x + \dfrac{1}{3}\right)^2$

59. $\left(4y - \dfrac{1}{4}\right)^2$

60. $\left(2y - \dfrac{1}{2}\right)^2$

61. $(x^8 + 3)^2$

62. $(x^8 + 5)^2$

In Exercises 63–82, multiply using the method of your choice.

63. $(x - 1)(x^2 + x + 1)$

64. $(x + 1)(x^2 - x + 1)$

65. $(x - 1)^2$

66. $(x + 1)^2$

67. $(3y + 7)(3y - 7)$

68. $(4y + 9)(4y - 9)$

69. $3x^2(4x^2 + x + 9)$

70. $5x^2(7x^2 + x + 6)$

71. $(7y + 3)(10y - 4)$

72. $(8y + 3)(10y - 5)$

73. $(x^2 + 1)^2$

74. $(x^2 + 2)^2$

75. $(x^2 + 1)(x^2 + 2)$

76. $(x^2 + 2)(x^2 + 3)$

77. $(x^2 + 4)(x^2 - 4)$

78. $(x^2 + 5)(x^2 - 5)$

79. $(2 - 3x^5)^2$

80. $(2 - 3x^6)^2$

81. $\left(\frac{1}{4}x^2 + 12\right)\left(\frac{3}{4}x^2 - 8\right)$

82. $\left(\frac{1}{4}x^2 + 16\right)\left(\frac{3}{4}x^2 - 4\right)$

In Exercises 83–88, find the area of each shaded region. Write the answer as a polynomial in descending powers of x.

83.

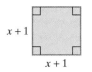

$x + 1$
$x + 1$

84.

$x + 3$
$x + 3$

85.

$2x + 3$
$2x - 3$

86.

$4x + 3$
$4x - 3$

87.

$x + 9$
$x + 5$
$x + 3$ $x + 1$

88.

$x + 4$
$x + 2$
$x + 3$ $x + 1$

Practice PLUS

In Exercises 89–96, multiply by the method of your choice.

89. $[(2x + 3)(2x - 3)]^2$

90. $[(3x + 2)(3x - 2)]^2$

91. $(4x^2 + 1)[(2x + 1)(2x - 1)]$

92. $(9x^2 + 1)[(3x + 1)(3x - 1)]$

93. $(x + 2)^3$

94. $(x + 4)^3$

95. $[(x + 3) - y][(x + 3) + y]$

96. $[(x + 5) - y][(x + 5) + y]$

Application Exercises

The square garden shown in the figure measures x yards on each side. The garden is to be expanded so that one side is increased by 2 yards and an adjacent side is increased by 1 yard. The graph shows the area of the expanded garden, y, in terms of the length of one of its original sides, x. Use this information to solve Exercises 97–100.

97. Write a product of two binomials that expresses the area of the larger garden.

98. Write a polynomial in descending powers of x that expresses the area of the larger garden.

99. If the original garden measures 6 yards on a side, use your expression from Exercise 97 to find the area of the larger garden. Then identify your solution as a point on the graph shown.

100. If the original garden measures 8 yards on a side, use your polynomial from Exercise 98 to find the area of the larger garden. Then identify your solution as a point on the graph shown.

The square painting in the figure measures x inches on each side. The painting is uniformly surrounded by a frame that measures 1 inch wide. Use this information to solve Exercises 101–102.

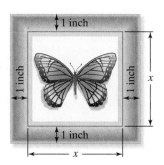

101. Write a polynomial in descending powers of x that expresses the area of the square that includes the painting and the frame.

102. Write an algebraic expression that describes the area of the frame. (*Hint*: The area of the frame is the area of the square that includes the painting and the frame minus the area of the painting.)

Writing in Mathematics

103. Explain how to multiply two binomials using the FOIL method. Give an example with your explanation.

104. Explain how to find the product of the sum and difference of two terms. Give an example with your explanation.

105. Explain how to square a binomial sum. Give an example with your explanation.

106. Explain how to square a binomial difference. Give an example with your explanation.

107. Explain why the graph for Exercises 97–100 is shown only in quadrant I.

Critical Thinking Exercises

Make Sense? *In Exercises 108–111, determine whether each statement "makes sense" or "does not make sense" and explain your reasoning.*

108. Squaring a binomial sum is as simple as squaring each of the two terms and then writing their sum.

109. I can distribute the exponent 2 on each factor of $(5x)^2$, but I cannot do the same thing on each term of $(x + 5)^2$.

110. Instead of using the formula for the square of a binomial sum, I prefer to write the binomial sum twice and then apply the FOIL method.

111. Special-product formulas for $(A + B)(A - B)$, $(A + B)^2$, and $(A - B)^2$ have patterns that make their multiplications quicker than using the FOIL method.

In Exercises 112–115, determine whether each statement is true or false. If the statement is false, make the necessary change(s) to produce a true statement.

112. $(3 + 4)^2 = 3^2 + 4^2$

113. $(2y + 7)^2 = 4y^2 + 28y + 49$

114. $(3x^2 + 2)(3x^2 - 2) = 9x^2 - 4$

115. $(x - 5)^2 = x^2 - 5x + 25$

116. What two binomials must be multiplied using the FOIL method to give a product of $x^2 - 8x - 20$?

117. Express the volume of the box as a polynomial in standard form.

118. Express the area of the plane figure shown as a polynomial in standard form.

Technology Exercises

In Exercises 119–122, use a graphing utility to graph each side of the equation in the same viewing rectangle. (Call the left side y_1 and the right side y_2.) If the graphs coincide, verify that the multiplication has been performed correctly. If the graphs do not appear to coincide, this indicates that the multiplication is incorrect. In these exercises, correct the right side of the equation. Then graph the left side and the corrected right side to verify that the graphs coincide.

119. $(x + 1)^2 = x^2 + 1$; Use a $[-5, 5, 1]$ by $[0, 20, 1]$ viewing rectangle.

120. $(x + 2)^2 = x^2 + 2x + 4$; Use a $[-6, 5, 1]$ by $[0, 20, 1]$ viewing rectangle.

121. $(x + 1)(x - 1) = x^2 - 1$; Use a $[-6, 5, 1]$ by $[-2, 18, 1]$ viewing rectangle.

122. $(x - 2)(x + 2) + 4 = x^2$; Use a $[-6, 5, 1]$ by $[-2, 18, 1]$ viewing rectangle.

Review Exercises

In Exercises 123–124, solve each system by the method of your choice.

123. $2x + 3y = 1$
$y = 3x - 7$
(Section 5.2, Example 1)

124. $3x + 4y = 7$
$2x + 7y = 9$
(Section 5.3, Example 3)

125. Graph: $y \le \frac{1}{3}x$.
(Section 4.6, Example 3)

Preview Exercises

Exercises 126–128 will help you prepare for the material covered in the next section.

126. Use the order of operations to evaluate
$$x^3y + 2xy^2 + 5x - 2$$
for $x = -2$ and $y = 3$.

127. Use the second step to combine the like terms.
$$5xy + 6xy = (5 + 6)xy = ?$$

128. Multiply using FOIL: $(x + 2y)(3x + 5y)$.

6.4

Objectives

1 Evaluate polynomials in several variables.

2 Understand the vocabulary of polynomials in two variables.

3 Add and subtract polynomials in several variables.

4 Multiply polynomials in several variables.

Polynomials in Several Variables

The next time you visit a lumberyard and go rummaging through piles of wood, think *polynomials*, although polynomials a bit different from those we have encountered so far. The construction industry uses a polynomial in two variables to determine the number of board feet that can be manufactured from a tree with a diameter of x inches and a length of y feet. This polynomial is

$$\frac{1}{4}x^2y - 2xy + 4y.$$

We call a polynomial containing two or more variables a **polynomial in several variables**. These polynomials can be evaluated, added, subtracted, and multiplied just like polynomials that contain only one variable.

1 Evaluate polynomials in several variables.

Evaluating a Polynomial in Several Variables

Two steps can be used to evaluate a polynomial in several variables.

Evaluating a Polynomial in Several Variables

1. Substitute the given value for each variable.
2. Perform the resulting computation using the order of operations.

EXAMPLE 1 **Evaluating a Polynomial in Two Variables**

Evaluate $2x^3y + xy^2 + 7x - 3$ for $x = -2$ and $y = 3$.

Solution We begin by substituting -2 for x and 3 for y in the polynomial.

$$2x^3y + xy^2 + 7x - 3 \qquad \text{This is the given polynomial.}$$
$$= 2(-2)^3 \cdot 3 + (-2) \cdot 3^2 + 7(-2) - 3 \qquad \text{Replace } x \text{ with } -2 \text{ and } y \text{ with } 3.$$
$$= 2(-8) \cdot 3 + (-2) \cdot 9 + 7(-2) - 3 \qquad \text{Evaluate exponential expressions:}$$
$$(-2)^3 = (-2)(-2)(-2) = -8 \text{ and}$$
$$3^2 = 3 \cdot 3 = 9.$$
$$= -48 + (-18) + (-14) - 3 \qquad \text{Perform the indicated multiplications.}$$
$$= -83 \qquad \text{Add from left to right.} \quad \blacksquare$$

☑ **CHECK POINT 1** Evaluate $3x^3y + xy^2 + 5y + 6$ for $x = -1$ and $y = 5$.

2 Understand the vocabulary of polynomials in two variables.

Describing Polynomials in Two Variables

In this section, we will limit our discussion of polynomials in several variables to two variables.

In general, a **polynomial in two variables**, x and y, contains the sum of one or more monomials in the form $ax^n y^m$. The constant, a, is the **coefficient**. The exponents, n and m, represent whole numbers. The **degree** of the monomial $ax^n y^m$ is $n + m$. We'll use the polynomial from the construction industry to illustrate these ideas.

The coefficients are $\frac{1}{4}$, -2, and 4.

$$\frac{1}{4}x^2 y \quad - 2xy \quad + 4y$$

Degree of monomial: $2 + 1 = 3$

Degree of monomial: $1 + 1 = 2$

Degree of monomial: 1

The **degree of a polynomial in two variables** is the highest degree of all its terms. For the preceding polynomial, the degree is 3.

EXAMPLE 2 Using the Vocabulary of Polynomials

Determine the coefficient of each term, the degree of each term, and the degree of the polynomial:

$$7x^2 y^3 - 17x^4 y^2 + xy - 6y^2 + 9.$$

Solution

Term	Coefficient	Degree (Sum of Exponents on the Variables)
$7x^2 y^3$	7	$2 + 3 = 5$
$-17x^4 y^2$	-17	$4 + 2 = 6$
xy	1	$1 + 1 = 2$
$-6y^2$	-6	2
9	9	0

Think of xy as $1x^1 y^1$.

The degree of the polynomial is the highest degree of all its terms, which is 6. ■

☑ **CHECK POINT 2** Determine the coefficient of each term, the degree of each term, and the degree of the polynomial:

$$8x^4 y^5 - 7x^3 y^2 - x^2 y - 5x + 11.$$

3 Add and subtract polynomials in several variables.

Adding and Subtracting Polynomials in Several Variables

Polynomials in several variables are added by combining like terms. For example, we can add the monomials $-7xy^2$ and $13xy^2$ as follows:

$$-7xy^2 + 13xy^2 = (-7 + 13)xy^2 = 6xy^2.$$

These like terms both contain the variable factors x and y^2.

Add coefficients and keep the same variable factors, xy^2.

> **EXAMPLE 3** Adding Polynomials in Two Variables

Add: $(6xy^2 - 5xy + 7) + (9xy^2 + 2xy - 6)$.

Solution

$$(6xy^2 - 5xy + 7) + (9xy^2 + 2xy - 6)$$
$$= (6xy^2 + 9xy^2) + (-5xy + 2xy) + (7 - 6) \quad \text{Group like terms.}$$
$$= 15xy^2 - 3xy + 1 \quad \text{Combine like terms by adding coefficients and keeping the same variable factors.}$$

☑ **CHECK POINT 3** Add: $(-8x^2y - 3xy + 6) + (10x^2y + 5xy - 10)$.

We subtract polynomials in two variables just as we did when subtracting polynomials in one variable. Add the first polynomial and the opposite of the polynomial being subtracted.

> **EXAMPLE 4** Subtracting Polynomials in Two Variables

Subtract:

$$(5x^3 - 9x^2y + 3xy^2 - 4) - (3x^3 - 6x^2y - 2xy^2 + 3).$$

Solution

$$(5x^3 - 9x^2y + 3xy^2 - 4) - (3x^3 - 6x^2y - 2xy^2 + 3)$$

Change the sign of each coefficient.

$$= (5x^3 - 9x^2y + 3xy^2 - 4) + (-3x^3 + 6x^2y + 2xy^2 - 3) \quad \text{Add the opposite of the polynomial being subtracted.}$$

$$= (5x^3 - 3x^3) + (-9x^2y + 6x^2y) + (3xy^2 + 2xy^2) + (-4 - 3) \quad \text{Group like terms.}$$
$$= 2x^3 - 3x^2y + 5xy^2 - 7 \quad \text{Combine like terms by adding coefficients and keeping the same variable factors.}$$

☑ **CHECK POINT 4** Subtract:

$$(7x^3 - 10x^2y + 2xy^2 - 5) - (4x^3 - 12x^2y - 3xy^2 + 5).$$

---

4 Multiply polynomials in several variables.

Multiplying Polynomials in Several Variables

The product of monomials forms the basis of polynomial multiplication. As with monomials in one variable, multiplication can be done mentally by multiplying coefficients and adding exponents on variables with the same base.

> **EXAMPLE 5** Multiplying Monomials

Multiply: $(7x^2y)(5x^3y^2)$.

Solution

$$(7x^2y)(5x^3y^2)$$
$$= (7 \cdot 5)(x^2 \cdot x^3)(y \cdot y^2) \quad \text{This regrouping can be worked mentally.}$$
$$= 35x^{2+3}y^{1+2} \quad \text{Multiply coefficients and add exponents on variables with same base.}$$

$$= 35x^5y^3 \quad \text{Simplify.}$$

✓ **CHECK POINT 5** Multiply: $(6xy^3)(10x^4y^2)$.

How do we multiply a monomial and a polynomial that is not a monomial? As we did with polynomials in one variable, multiply each term of the polynomial by the monomial.

EXAMPLE 6 Multiplying a Monomial and a Polynomial

Multiply: $3x^2y(4x^3y^2 - 6x^2y + 2)$.

Solution

$$3x^2y(4x^3y^2 - 6x^2y + 2)$$

$$= 3x^2y \cdot 4x^3y^2 - 3x^2y \cdot 6x^2y + 3x^2y \cdot 2 \quad \text{Use the distributive property.}$$

$$= 12x^{2+3}y^{1+2} - 18x^{2+2}y^{1+1} + 6x^2y \quad \text{Multiply coefficients and add exponents on variables with the same base.}$$

$$= 12x^5y^3 - 18x^4y^2 + 6x^2y \quad \text{Simplify.}$$ ▬

✓ **CHECK POINT 6** Multiply: $6xy^2(10x^4y^5 - 2x^2y + 3)$.

FOIL and the special-products formulas can be used to multiply polynomials in several variables.

EXAMPLE 7 Multiplying Polynomials in Two Variables

Multiply: **a.** $(x + 4y)(3x - 5y)$ **b.** $(5x + 3y)^2$.

Solution We will perform the multiplication in part (a) using the FOIL method. We will multiply in part (b) using the formula for the square of a binomial, $(A + B)^2$.

a. $(x + 4y)(3x - 5y)$ Multiply these binomials using the FOIL method.

F   O   I   L

$$= (x)(3x) + (x)(-5y) + (4y)(3x) + (4y)(-5y)$$

$$= 3x^2 - 5xy + 12xy - 20y^2$$

$$= 3x^2 + 7xy - 20y^2 \quad \text{Combine like terms.}$$

$$(A + B)^2 = A^2 + 2 \cdot A \cdot B + B^2$$

b. $(5x + 3y)^2 = (5x)^2 + 2(5x)(3y) + (3y)^2$

$$= 25x^2 + 30xy + 9y^2$$ ▬

✓ **CHECK POINT 7** Multiply:

a. $(7x - 6y)(3x - y)$ **b.** $(2x + 4y)^2$.

EXAMPLE 8 Multiplying Polynomials in Two Variables

Multiply: **a.** $(4x^2y + 3y)(4x^2y - 3y)$ **b.** $(x + y)(x^2 - xy + y^2)$.

Solution We perform the multiplication in part (a) using the formula for the product of the sum and difference of two terms. We perform the multiplication in part (b) by multiplying each term of the trinomial, $x^2 - xy + y^2$, by x and y, respectively, and then adding like terms.

$$(A + B) \cdot (A - B) = A^2 - B^2$$

a. $(4x^2y + 3y)(4x^2y - 3y) = (4x^2y)^2 - (3y)^2$
$$= 16x^4y^2 - 9y^2$$

b. $(x + y)(x^2 - xy + y^2)$

$= x(x^2 - xy + y^2) + y(x^2 - xy + y^2)$ Multiply the trinomial by each term of the binomial.

$= x \cdot x^2 - x \cdot xy + x \cdot y^2 + y \cdot x^2 - y \cdot xy + y \cdot y^2$ Use the distributive property.

$= x^3 - x^2y + xy^2 + x^2y - xy^2 + y^3$ Add exponents on variables with the same base.

$= x^3 + y^3$ Combine like terms: $-x^2y + x^2y = 0$ and $xy^2 - xy^2 = 0$.

☑ **CHECK POINT 8**   Multiply:

   a. $(6xy^2 + 5x)(6xy^2 - 5x)$

   b. $(x - y)(x^2 + xy + y^2)$.

6.4 EXERCISE SET   *MyMathLab*   PRACTICE   WATCH   DOWNLOAD   READ   REVIEW

Practice Exercises

In Exercises 1–6, evaluate each polynomial for x = 2 and y = −3.

1. $x^2 + 2xy + y^2$
2. $x^2 + 3xy + y^2$
3. $xy^3 - xy + 1$
4. $x^3y - xy + 2$
5. $2x^2y - 5y + 3$
6. $3x^2y - 4y + 5$

In Exercises 7–8, determine the coefficient of each term, the degree of each term, and the degree of the polynomial.

7. $x^3y^2 - 5x^2y^7 + 6y^2 - 3$
8. $12x^4y - 5x^3y^7 - x^2 + 4$

In Exercises 9–20, add or subtract as indicated.

9. $(5x^2y - 3xy) + (2x^2y - xy)$
10. $(-2x^2y + xy) + (4x^2y + 7xy)$
11. $(4x^2y + 8xy + 11) + (-2x^2y + 5xy + 2)$
12. $(7x^2y + 5xy + 13) + (-3x^2y + 6xy + 4)$
13. $(7x^4y^2 - 5x^2y^2 + 3xy) + (-18x^4y^2 - 6x^2y^2 - xy)$
14. $(6x^4y^2 - 10x^2y^2 + 7xy) + (-12x^4y^2 - 3x^2y^2 - xy)$
15. $(x^3 + 7xy - 5y^2) - (6x^3 - xy + 4y^2)$

16. $(x^4 - 7xy - 5y^3) - (6x^4 - 3xy + 4y^3)$

17. $(3x^4y^2 + 5x^3y - 3y) - (2x^4y^2 - 3x^3y - 4y + 6x)$

18. $(5x^4y^2 + 6x^3y - 7y) - (3x^4y^2 - 5x^3y - 6y + 8x)$

19. $(x^3 - y^3) - (-4x^3 - x^2y + xy^2 + 3y^3)$

20. $(x^3 - y^3) - (-6x^3 + x^2y - xy^2 + 2y^3)$

21. Add: $5x^2y^2 - 4xy^2 + 6y^2$
               $-8x^2y^2 + 5xy^2 - y^2$

22. Add: $7a^2b^2 - 5ab^2 + 6b^2$
               $-10a^2b^2 + 6ab^2 + 6b^2$

23. Subtract: $3a^2b^4 - 5ab^2 + 7ab$
                    $-(-5a^2b^4 - 8ab^2 - ab)$

24. Subtract: $13x^2y^4 - 17xy^2 + xy$
                    $-(-7x^2y^4 - 8xy^2 - xy)$

25. Subtract $11x - 5y$ from the sum of $7x + 13y$ and $-26x + 19y$.

26. Subtract $23x - 5y$ from the sum of $6x + 15y$ and $x - 19y$.

In Exercises 27–76, find each product.

27. $(5x^2y)(8xy)$
28. $(10x^2y)(5xy)$

29. $(-8x^3y^4)(3x^2y^5)$
30. $(7x^4y^5)(-10x^7y^{11})$
31. $9xy(5x + 2y)$
32. $7xy(8x + 3y)$
33. $5xy^2(10x^2 - 3y)$
34. $6x^2y(5x^2 - 9y)$
35. $4ab^2(7a^2b^3 + 2ab)$
36. $2ab^2(20a^2b^3 + 11ab)$
37. $-b(a^2 - ab + b^2)$
38. $-b(a^3 - ab + b^3)$
39. $(x + 5y)(7x + 3y)$
40. $(x + 9y)(6x + 7y)$
41. $(x - 3y)(2x + 7y)$
42. $(3x - y)(2x + 5y)$
43. $(3xy - 1)(5xy + 2)$
44. $(7xy + 1)(2xy - 3)$
45. $(2x + 3y)^2$
46. $(2x + 5y)^2$
47. $(xy - 3)^2$
48. $(xy - 5)^2$
49. $(x^2 + y^2)^2$
50. $(2x^2 + y^2)^2$
51. $(x^2 - 2y^2)^2$
52. $(x^2 - y^2)^2$
53. $(3x + y)(3x - y)$
54. $(x + 5y)(x - 5y)$
55. $(ab + 1)(ab - 1)$
56. $(ab + 2)(ab - 2)$
57. $(x + y^2)(x - y^2)$
58. $(x^2 + y)(x^2 - y)$
59. $(3a^2b + a)(3a^2b - a)$
60. $(5a^2b + a)(5a^2b - a)$
61. $(3xy^2 - 4y)(3xy^2 + 4y)$
62. $(7xy^2 - 10y)(7xy^2 + 10y)$
63. $(a + b)(a^2 - b^2)$
64. $(a - b)(a^2 + b^2)$
65. $(x + y)(x^2 + 3xy + y^2)$
66. $(x + y)(x^2 + 5xy + y^2)$
67. $(x - y)(x^2 - 3xy + y^2)$
68. $(x - y)(x^2 - 4xy + y^2)$
69. $(xy + ab)(xy - ab)$
70. $(xy + ab^2)(xy - ab^2)$
71. $(x^2 + 1)(x^4y + x^2 + 1)$
72. $(x^2 + 1)(xy^4 + y^2 + 1)$
73. $(x^2y^2 - 3)^2$
74. $(x^2y^2 - 5)^2$
75. $(x + y + 1)(x + y - 1)$
76. $(x + y + 1)(x - y + 1)$

In Exercises 77–80, write a polynomial in two variables that describes the total area of each shaded region. Express each polynomial as the sum or difference of terms.

77.
78.
79.
80.

Practice PLUS

In Exercises 81–86, find each product. As we said in the Section 6.3 opener, cut to the chase in each part of the polynomial multiplication: Use only the special-product formula for the sum and difference of two terms or the formulas for the square of a binomial.

81. $[(x^3y^3 + 1)(x^3y^3 - 1)]^2$
82. $[(1 - a^3b^3)(1 + a^3b^3)]^2$
83. $(xy - 3)^2(xy + 3)^2$ (Do not begin by squaring a binomial.)
84. $(ab - 4)^2(ab + 4)^2$ (Do not begin by squaring a binomial.)
85. $[x + y + z][x - (y + z)]$
86. $(a - b - c)(a + b + c)$

Application Exercises

87. The number of board feet, N, that can be manufactured from a tree with a diameter of x inches and a length of y feet is modeled by the formula

$$N = \frac{1}{4}x^2y - 2xy + 4y.$$

A building contractor estimates that 3000 board feet of lumber is needed for a job. The lumber company has just milled a fresh load of timber from 20 trees that averaged 10 inches in diameter and 16 feet in length. Is this enough to complete the job? If not, how many additional board feet of lumber is needed?

88. The storage shed shown in the figure has a volume given by the polynomial

$$2x^2y + \frac{1}{2}\pi x^2 y.$$

a. A small business is considering having a shed installed like the one shown in the figure. The shed's height, $2x$, is 26 feet and its length, y, is 27 feet. Using $x = 13$ and $y = 27$, find the volume of the storage shed.

b. The business requires at least 18,000 cubic feet of storage space. Should they construct the storage shed described in part (a)?

An object that is falling or vertically projected into the air has its height, in feet, above the ground given by

$$s = -16t^2 + v_0 t + s_0,$$

where s is the height, in feet, v_0 is the original velocity of the object, in feet per second, t is the time the object is in motion, in seconds, and s_0 is the height, in feet, from which the object is dropped or projected. The figure shows that a ball is thrown straight up from a rooftop at an original velocity of 80 feet per second from a height of 96 feet. The ball misses the rooftop on its way down and eventually strikes the ground. Use the formula and this information to solve Exercises 89–91.

89. How high above the ground will the ball be 2 seconds after being thrown?

90. How high above the ground will the ball be 4 seconds after being thrown?

91. How high above the ground will the ball be 6 seconds after being thrown? Describe what this means in practical terms.

The graph visually displays the information about the thrown ball described in Exercises 89–91. The horizontal axis represents the ball's time in motion, in seconds. The vertical axis represents the ball's height above the ground, in feet. Use the graph to solve Exercises 92–97.

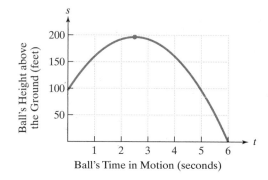

92. During which time period is the ball rising?

93. During which time period is the ball falling?

94. Identify your answer from Exercise 90 as a point on the graph.

95. Identify your answer from Exercise 89 as a point on the graph.

96. After how many seconds does the ball strike the ground?

97. After how many seconds does the ball reach its maximum height above the ground? What is a reasonable estimate of this maximum height?

Writing in Mathematics

98. What is a polynomial in two variables? Provide an example with your description.

99. Explain how to find the degree of a polynomial in two variables.

100. Suppose that you take up sky diving. Explain how to use the formula for Exercises 89–91 to determine your height above the ground at every instant of your fall.

Critical Thinking Exercises

Make Sense? *In Exercises 101–104, determine whether each statement "makes sense" or "does not make sense" and explain your reasoning.*

101. I use the same procedures for operations with polynomials in two variables as I did when performing these operations with polynomials in one variable.

102. Adding polynomials in several variables is the same as adding like terms.

103. I used FOIL to find the product of $x + y$ and $x^2 - xy + y^2$.

104. I used FOIL to multiply $5xy$ and $3xy + 4$.

In Exercises 105–108, determine whether each statement is true or false. If the statement is false, make the necessary change(s) to produce a true statement.

105. The degree of $5x^{24} - 3x^{16}y^9 - 7xy^2 + 6$ is 24.

106. In the polynomial $4x^2y + x^3y^2 + 3x^2y^3 + 7y$, the term x^3y^2 has degree 5 and no numerical coefficient.

107. $(2x + 3 - 5y)(2x + 3 + 5y) = 4x^2 + 12x + 9 - 25y^2$

108. $(6x^2y - 7xy - 4) - (6x^2y + 7xy - 4) = 0$

In Exercises 109–110, find a polynomial in two variables that describes the area of the shaded region of each figure. Write the polynomial as the sum or difference of terms.

109.

110.

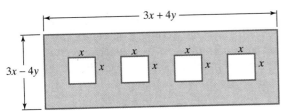

111. Use the formulas for the volume of a rectangular solid and a cylinder to derive the polynomial in Exercise 88 that describes the volume of the storage building.

Review Exercises

112. Solve for W: $\quad R = \dfrac{L + 3W}{2}$. (Section 2.4, Example 4)

113. Subtract: $-6.4 - (-10.2)$. (Section 1.6, Example 2)

114. Solve the proportion: $\dfrac{63}{x} = \dfrac{3}{5}$. (Section 3.2, Example 2)

Preview Exercises

Exercises 115–117 will help you prepare for the material covered in the next section.

115. Find the missing exponent, designated by the question mark, in the final step.

$$\frac{x^7}{x^3} = \frac{x \cdot x \cdot x \cdot x \cdot x \cdot x \cdot x}{x \cdot x \cdot x} = x^?$$

116. Simplify: $\dfrac{(x^2)^3}{5^3}$.

117. Simplify: $\dfrac{(2a^3)^5}{(b^4)^5}$.

| MID-CHAPTER CHECK POINT | Section 6.1–Section 6.4 |

✓ **What You Know:** We learned to add, subtract, and multiply polynomials. We used a number of fast methods for finding products of polynomials, including the FOIL method for multiplying binomials, a special-product formula for the product of the sum and difference of two terms $[(A + B)(A - B) = A^2 - B^2]$, and special-product formulas for squaring binomials $[(A + B)^2 = A^2 + 2AB + B^2; (A - B)^2 = A^2 - 2AB + B^2]$. Finally, we applied all of these operations to polynomials in several variables.

In Exercises 1–21, perform the indicated operations.

1. $(11x^2y^3)(-5x^2y^3)$

2. $11x^2y^3 - 5x^2y^3$

3. $(3x + 5)(4x - 7)$

4. $(3x + 5) - (4x - 7)$

5. $(2x - 5)(x^2 - 3x + 1)$

6. $(2x - 5) + (x^2 - 3x + 1)$

7. $(8x - 3)^2$

8. $(-10x^4)(-7x^5)$

9. $(x^2 + 2)(x^2 - 2)$

10. $(x^2 + 2)^2$

11. $(9a - 10b)(2a + b)$

12. $7x^2(10x^3 - 2x + 3)$

13. $(3a^2b^3 - ab + 4b^2) - (-2a^2b^3 - 3ab + 5b^2)$

14. $2(3y - 5)(3y + 5)$

15. $(-9x^3 + 5x^2 - 2x + 7) + (11x^3 - 6x^2 + 3x - 7)$

16. $10x^2 - 8xy - 3(y^2 - xy)$

17. $(-2x^5 + x^4 - 3x + 10) - (2x^5 - 6x^4 + 7x - 13)$

18. $(x + 3y)(x^2 - 3xy + 9y^2)$

19. $(5x^4 + 4)(2x^3 - 1)$

20. $(y - 6z)^2$

21. $(2x + 3)(2x - 3) - (5x + 4)(5x - 4)$

22. Graph: $y = 1 - x^2$.

6.5

Dividing Polynomials

Objectives

1. Use the quotient rule for exponents.
2. Use the zero-exponent rule.
3. Use the quotients-to-powers rule.
4. Divide monomials.
5. Check polynomial division.
6. Divide a polynomial by a monomial.

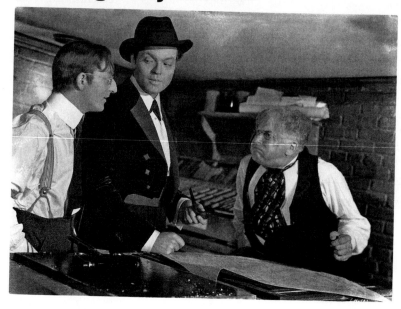

Orson Welles (center), star and director of the highly innovative Citizen Kane

In the dramatic arts, ours is the era of the movies. As individuals and as a nation, we've grown up with them. Our images of love, war, family, country—even of things that terrify us—owe much to what we've seen on screen. In this section's exercise set, we'll model our love for movies with polynomials and polynomial division. Before discussing polynomial division, we must develop some additional rules for working with exponents.

1 Use the quotient rule for exponents.

The Quotient Rule for Exponents

Consider the quotient of two exponential expressions, such as the quotient of 2^7 and 2^3. We are dividing 7 factors of 2 by 3 factors of 2. We are left with 4 factors of 2:

$$\frac{2^7}{2^3} = \frac{\overbrace{2 \cdot 2 \cdot 2 \cdot 2 \cdot 2 \cdot 2 \cdot 2}^{7 \text{ factors of } 2}}{\underbrace{2 \cdot 2 \cdot 2}_{3 \text{ factors of } 2}} = \frac{\cancel{2} \cdot \cancel{2} \cdot \cancel{2} \cdot 2 \cdot 2 \cdot 2 \cdot 2}{\cancel{2} \cdot \cancel{2} \cdot \cancel{2}} = \underbrace{2 \cdot 2 \cdot 2 \cdot 2}_{4 \text{ factors of } 2}$$

Divide out pairs of factors: $\frac{2}{2} = 1$.

Thus,

$$\frac{2^7}{2^3} = 2^4.$$

We can quickly find the exponent, 4, on the quotient by subtracting the original exponents:

$$\frac{2^7}{2^3} = 2^{7-3}.$$

This suggests the following rule:

The Quotient Rule

$$\frac{b^m}{b^n} = b^{m-n}, \quad b \neq 0$$

When dividing exponential expressions with the same nonzero base, subtract the exponent in the denominator from the exponent in the numerator. Use this difference as the exponent of the common base.

EXAMPLE 1 Using the Quotient Rule

Divide each expression using the quotient rule:

a. $\dfrac{2^8}{2^4}$ b. $\dfrac{x^{13}}{x^3}$ c. $\dfrac{y^{15}}{y}$.

Solution

a. $\dfrac{2^8}{2^4} = 2^{8-4} = 2^4$ or 16

b. $\dfrac{x^{13}}{x^3} = x^{13-3} = x^{10}$

c. $\dfrac{y^{15}}{y} = \dfrac{y^{15}}{y^1} = y^{15-1} = y^{14}$

✓ **CHECK POINT 1** Divide each expression using the quotient rule:

a. $\dfrac{5^{12}}{5^4}$ b. $\dfrac{x^9}{x^2}$ c. $\dfrac{y^{20}}{y}$.

---

2 Use the zero-exponent rule.

Zero as an Exponent

A nonzero base can be raised to the 0 power. The quotient rule can be used to help determine what zero as an exponent should mean. Consider the quotient of b^4 and b^4, where b is not zero. We can determine this quotient in two ways.

$$\dfrac{b^4}{b^4} = 1 \qquad\qquad \dfrac{b^4}{b^4} = b^{4-4} = b^0$$

Any nonzero expression divided by itself is 1.   Use the quotient rule and subtract exponents.

This means that b^0 must equal 1.

> **The Zero-Exponent Rule**
>
> If b is any real number other than 0,
>
> $$b^0 = 1.$$

EXAMPLE 2 Using the Zero-Exponent Rule

Use the zero-exponent rule to simplify each expression:

a. 7^0 b. $(-5)^0$ c. -5^0 d. $10x^0$ e. $(10x)^0$.

Solution

a. $7^0 = 1$ Any nonzero number raised to the 0 power is 1.

b. $(-5)^0 = 1$ Any nonzero number raised to the 0 power is 1.

c. $-5^0 = -1$ $-5^0 = -(5^0) = -1$

Only 5 is raised to the 0 power.

d. $10x^0 = 10 \cdot 1 = 10$

Only x is raised to the 0 power.

e. $(10x)^0 = 1$

The entire expression, $10x$, is raised to the 0 power.

☑ **CHECK POINT 2**   Use the zero-exponent rule to simplify each expression:

a. 14^0 **b.** $(-10)^0$ **c.** -10^0 **d.** $20x^0$ **e.** $(20x)^0$.

3 Use the quotients-to-powers rule.

The Quotients-to-Powers Rule for Exponents

We have seen that when a product is raised to a power, we raise every factor in the product to the power:

$$(ab)^n = a^n b^n.$$

There is a similar property for raising a quotient to a power.

Quotients to Powers

If a and b are real numbers and b is nonzero, then

$$\left(\frac{a}{b}\right)^n = \frac{a^n}{b^n}.$$

When a quotient is raised to a power, raise the numerator to the power and divide by the denominator raised to the power.

EXAMPLE 3   Using the Quotients-to-Powers Rule

Simplify each expression using the quotients-to-powers rule:

a. $\left(\dfrac{x}{4}\right)^2$ **b.** $\left(\dfrac{x^2}{5}\right)^3$ **c.** $\left(\dfrac{2a^3}{b^4}\right)^5$.

Solution

a. $\left(\dfrac{x}{4}\right)^2 = \dfrac{x^2}{4^2} = \dfrac{x^2}{16}$ Square the numerator and the denominator.

b. $\left(\dfrac{x^2}{5}\right)^3 = \dfrac{(x^2)^3}{5^3} = \dfrac{x^{2\cdot3}}{5\cdot5\cdot5} = \dfrac{x^6}{125}$ Cube the numerator and the denominator.

c. $\left(\dfrac{2a^3}{b^4}\right)^5 = \dfrac{(2a^3)^5}{(b^4)^5}$ Raise the numerator and the denominator to the fifth power.

$= \dfrac{2^5(a^3)^5}{(b^4)^5}$ Raise each factor in the numerator to the fifth power.

$= \dfrac{2^5 a^{3\cdot5}}{b^{4\cdot5}}$ To raise exponential expressions to powers, multiply exponents: $(b^m)^n = b^{mn}$.

$= \dfrac{32a^{15}}{b^{20}}$ Simplify.

☑ **CHECK POINT 3**   Simplify each expression using the quotients-to-powers rule:

a. $\left(\dfrac{x}{5}\right)^2$ b. $\left(\dfrac{x^4}{2}\right)^3$ c. $\left(\dfrac{2a^{10}}{b^3}\right)^4$.

Study Tip

Try to avoid the following common errors that can occur when simplifying exponential expressions.

Correct	Incorrect	Description of Error
$\dfrac{2^{20}}{2^4} = 2^{20-4} = 2^{16}$	$\dfrac{2^{20}}{2^4} = 2^5$	Exponents should be subtracted, not divided.
$-8^0 = -1$	$-8^0 = 1$	Only 8 is raised to the 0 power.
$\left(\dfrac{x}{5}\right)^2 = \dfrac{x^2}{5^2} = \dfrac{x^2}{25}$	$\left(\dfrac{x}{5}\right)^2 = \dfrac{x^2}{5}$	The numerator and denominator must both be squared.

4 Divide monomials.

Dividing Monomials

Now that we have developed three additional properties of exponents, we are ready to turn to polynomial division. We begin with the quotient of two monomials, such as $16x^{14}$ and $8x^2$. This quotient is obtained by dividing the coefficients, 16 and 8, and then dividing the variables using the quotient rule for exponents.

$$\frac{16x^{14}}{8x^2} = \frac{16}{8}x^{14-2} = 2x^{12}$$

Divide coefficients and subtract exponents.

Dividing Monomials

To divide monomials, divide the coefficients and then divide the variables. Use the quotient rule for exponents to divide the variables: Keep the variable and subtract the exponents.

EXAMPLE 4   Dividing Monomials

Divide: a. $\dfrac{-12x^8}{4x^2}$ b. $\dfrac{2x^3}{8x^3}$ c. $\dfrac{15x^5y^4}{3x^2y}$.

Study Tip

Look at the solution to Example 4(b). Rather than subtracting exponents for division that results in a 0 exponent, you might prefer to divide out x^3.

$$\frac{2x^3}{8x^3} = \frac{2}{8} = \frac{1}{4}$$

Solution

a. $\dfrac{-12x^8}{4x^2} = \dfrac{-12}{4}x^{8-2} = -3x^6$

b. $\dfrac{2x^3}{8x^3} = \dfrac{2}{8}x^{3-3} = \dfrac{1}{4}x^0 = \dfrac{1}{4} \cdot 1 = \dfrac{1}{4}$

c. $\dfrac{15x^5y^4}{3x^2y} = \dfrac{15}{3}x^{5-2}y^{4-1} = 5x^3y^3$

☑ **CHECK POINT 4**   Divide:

a. $\dfrac{-20x^{12}}{10x^4}$ b. $\dfrac{3x^4}{15x^4}$ c. $\dfrac{9x^6y^5}{3xy^2}$.

5 Check polynomial division.

Checking Division of Polynomial Problems

The answer to a division problem can be checked. For example, consider the following problem:

Dividend: the polynomial you are dividing into

$$\frac{15x^5y^4}{3x^2y} = 5x^3y^3.$$

Quotient: the answer to your division problem

Divisor: the polynomial you are dividing by

The quotient is correct if the product of the divisor and the quotient is the dividend. Is the quotient shown in the preceding equation correct?

$$(3x^2y)(5x^3y^3) = 3 \cdot 5x^{2+3}y^{1+3} = 15x^5y^4$$

Divisor   Quotient   This is the dividend.

Because the product of the divisor and the quotient is the dividend, the answer to the division problem is correct.

Checking Division of Polynomials

To check a quotient in a division problem, multiply the divisor and the quotient. If this product is the dividend, the quotient is correct.

6 Divide a polynomial by a monomial.

Dividing a Polynomial That Is Not a Monomial by a Monomial

To divide a polynomial by a monomial, we divide each term of the polynomial by the monomial. For example,

polynomial dividend

$$\frac{10x^8 + 15x^6}{5x^3} = \frac{10x^8}{5x^3} + \frac{15x^6}{5x^3} = \frac{10}{5}x^{8-3} + \frac{15}{5}x^{6-3} = 2x^5 + 3x^3.$$

monomial divisor   Divide the first term by $5x^3$.   Divide the second term by $5x^3$.

Is the quotient correct? Multiply the divisor and the quotient.

$$5x^3(2x^5 + 3x^3) = 5x^3 \cdot 2x^5 + 5x^3 \cdot 3x^3$$
$$= 5 \cdot 2x^{3+5} + 5 \cdot 3x^{3+3} = 10x^8 + 15x^6$$

Because this product gives the dividend, the quotient is correct.

Dividing a Polynomial That Is Not a Monomial by a Monomial

To divide a polynomial by a monomial, divide each term of the polynomial by the monomial.

Study Tip

Try to avoid this common error:

Incorrect:

$$\frac{x^4 - x}{x} = \frac{x^4 - 1}{1} = x^4 - 1$$

Correct:

$$\frac{x^4 - x}{x} = \frac{x^4}{x} - \frac{x}{x}$$

$$= x^{4-1} - x^{1-1}$$

Don't leave out the 1.

$$= x^3 - x^0$$

$$= x^3 - 1$$

EXAMPLE 5 Dividing a Polynomial by a Monomial

Find the quotient: $(-12x^8 + 4x^6 - 8x^3) \div 4x^2$.

Solution

$$\frac{-12x^8 + 4x^6 - 8x^3}{4x^2} \qquad \text{Rewrite the division in a vertical format.}$$

$$= \frac{-12x^8}{4x^2} + \frac{4x^6}{4x^2} - \frac{8x^3}{4x^2} \qquad \text{Divide each term of the polynomial by the monomial.}$$

$$= \frac{-12}{4}x^{8-2} + \frac{4}{4}x^{6-2} - \frac{8}{4}x^{3-2} \qquad \text{Divide coefficients and subtract exponents.}$$

$$= -3x^6 + x^4 - 2x \qquad \text{Simplify.}$$

To check the answer, multiply the divisor and the quotient.

$$4x^2(-3x^6 + x^4 - 2x) = 4x^2(-3x^6) + 4x^2 \cdot x^4 - 4x^2(2x)$$

Divisor · Quotient

$$= 4(-3)x^{2+6} + 4x^{2+4} - 4 \cdot 2x^{2+1}$$

$$= -12x^8 + 4x^6 - 8x^3$$

This is the dividend.

Because the product of the divisor and the quotient is the dividend, the answer—that is, the quotient—is correct.

✓ **CHECK POINT 5** Find the quotient: $(-15x^9 + 6x^5 - 9x^3) \div 3x^2$.

EXAMPLE 6 Dividing a Polynomial by a Monomial

Divide: $\dfrac{16x^5 - 9x^4 + 8x^3}{2x^3}$.

Solution

$$\frac{16x^5 - 9x^4 + 8x^3}{2x^3} \qquad \text{This is the given polynomial division.}$$

$$= \frac{16x^5}{2x^3} - \frac{9x^4}{2x^3} + \frac{8x^3}{2x^3} \qquad \text{Divide each term by } 2x^3.$$

$$= \frac{16}{2}x^{5-3} - \frac{9}{2}x^{4-3} + \frac{8}{2}x^{3-3} \qquad \begin{array}{l}\text{Divide coefficients and subtract exponents.} \\ \text{Did you immediately write the last term as 4?}\end{array}$$

$$= 8x^2 - \frac{9}{2}x + 4x^0 \qquad \text{Simplify.}$$

$$= 8x^2 - \frac{9}{2}x + 4 \qquad x^0 = 1, \text{ so } 4x^0 = 4 \cdot 1 = 4.$$

Check the answer by showing that the product of the divisor and the quotient is the dividend.

✓ **CHECK POINT 6** Divide: $\dfrac{25x^9 - 7x^4 + 10x^3}{5x^3}$.

EXAMPLE 7 Dividing Polynomials in Two Variables

Divide: $(15x^5y^4 - 3x^3y^2 + 9x^2y) \div 3x^2y$.

Solution

$$\frac{15x^5y^4 - 3x^3y^2 + 9x^2y}{3x^2y}$$ Rewrite the division in a vertical format.

$$= \frac{15x^5y^4}{3x^2y} - \frac{3x^3y^2}{3x^2y} + \frac{9x^2y}{3x^2y}$$ Divide each term of the polynomial by the monomial.

$$= \frac{15}{3}x^{5-2}y^{4-1} - \frac{3}{3}x^{3-2}y^{2-1} + \frac{9}{3}x^{2-2}y^{1-1}$$ Divide coefficients and subtract exponents.

$$= 5x^3y^3 - xy + 3$$ Simplify.

Check the answer by showing that the product of the divisor and the quotient is the dividend.

☑ **CHECK POINT 7**   Divide: $(18x^7y^6 - 6x^2y^3 + 60xy^2) \div 6xy^2$.

6.5 EXERCISE SET   MyMathLab

 Math XL PRACTICE   WATCH   DOWNLOAD   READ   REVIEW

Practice Exercises

In Exercises 1–10, divide each expression using the quotient rule. Express any numerical answers in exponential form.

1. $\dfrac{3^{20}}{3^5}$ 2. $\dfrac{3^{30}}{3^{10}}$ 3. $\dfrac{x^6}{x^2}$

4. $\dfrac{x^8}{x^4}$ 5. $\dfrac{y^{13}}{y^5}$ 6. $\dfrac{y^{19}}{y^6}$

7. $\dfrac{5^6 \cdot 2^8}{5^3 \cdot 2^4}$ 8. $\dfrac{3^6 \cdot 2^8}{3^3 \cdot 2^4}$ 9. $\dfrac{x^{100}y^{50}}{x^{25}y^{10}}$

10. $\dfrac{x^{200}y^{40}}{x^{25}y^{10}}$

In Exercises 11–24, use the zero-exponent rule to simplify each expression.

11. 2^0 12. 4^0 13. $(-2)^0$

14. $(-4)^0$ 15. -2^0 16. -4^0

17. $100y^0$ 18. $200y^0$ 19. $(100y)^0$

20. $(200y)^0$ 21. $-5^0 + (-5)^0$ 22. $-6^0 + (-6)^0$

23. $-\pi^0 - (-\pi)^0$ 24. $-\sqrt{3}^0 - (-\sqrt{3})^0$

In Exercises 25–36, simplify each expression using the quotients-to-powers rule. If possible, evaluate exponential expressions.

25. $\left(\dfrac{x}{3}\right)^2$ 26. $\left(\dfrac{x}{5}\right)^2$

27. $\left(\dfrac{x^2}{4}\right)^3$ 28. $\left(\dfrac{x^2}{3}\right)^3$

29. $\left(\dfrac{2x^3}{5}\right)^2$ 30. $\left(\dfrac{3x^4}{7}\right)^2$

31. $\left(\dfrac{-4}{3a^3}\right)^3$ 32. $\left(\dfrac{-5}{2a^3}\right)^3$

33. $\left(\dfrac{-2a^7}{b^4}\right)^5$ 34. $\left(\dfrac{-2a^8}{b^3}\right)^5$

35. $\left(\dfrac{x^2y^3}{2z}\right)^4$ 36. $\left(\dfrac{x^3y^2}{2z}\right)^4$

In Exercises 37–52, divide the monomials. Check each answer by showing that the product of the divisor and the quotient is the dividend.

37. $\dfrac{30x^{10}}{10x^5}$ 38. $\dfrac{45x^{12}}{15x^4}$

39. $\dfrac{-8x^{22}}{4x^2}$ 40. $\dfrac{-15x^{40}}{3x^4}$

41. $\dfrac{-9y^8}{18y^5}$ 42. $\dfrac{-15y^{13}}{45y^9}$

43. $\dfrac{7y^{17}}{5y^5}$ 44. $\dfrac{9y^{19}}{7y^{11}}$

45. $\dfrac{30x^7y^5}{5x^2y}$ 46. $\dfrac{40x^9y^5}{2x^2y}$

47. $\dfrac{-18x^{14}y^2}{36x^2y^2}$ 48. $\dfrac{-15x^{16}y^2}{45x^2y^2}$

49. $\dfrac{9x^{20}y^{20}}{7x^{20}y^{20}}$ 50. $\dfrac{7x^{30}y^{30}}{15x^{30}y^{30}}$

51. $\dfrac{-5x^{10}y^{12}z^6}{50x^2y^3z^2}$ 52. $\dfrac{-8x^{12}y^{10}z^4}{40x^2y^3z^2}$

In Exercises 53–78, divide the polynomial by the monomial. Check each answer by showing that the product of the divisor and the quotient is the dividend.

53. $\dfrac{10x^4 + 2x^3}{2}$ 54. $\dfrac{20x^4 + 5x^3}{5}$

55. $\dfrac{14x^4 - 7x^3}{7x}$

56. $\dfrac{24x^4 - 8x^3}{8x}$

57. $\dfrac{y^7 - 9y^2 + y}{y}$

58. $\dfrac{y^8 - 11y^3 + y}{y}$

59. $\dfrac{24x^3 - 15x^2}{-3x}$

60. $\dfrac{10x^3 - 20x^2}{-5x}$

61. $\dfrac{18x^5 + 6x^4 + 9x^3}{3x^2}$

62. $\dfrac{18x^5 + 24x^4 + 12x^3}{6x^2}$

63. $\dfrac{12x^4 - 8x^3 + 40x^2}{4x}$

64. $\dfrac{49x^4 - 14x^3 + 70x^2}{-7x}$

65. $(4x^2 - 6x) \div x$

66. $(16y^2 - 8y) \div y$

67. $\dfrac{30z^3 + 10z^2}{-5z}$

68. $\dfrac{12y^4 - 42y^2}{-4y}$

69. $\dfrac{8x^3 + 6x^2 - 2x}{2x}$

70. $\dfrac{9x^3 + 12x^2 - 3x}{3x}$

71. $\dfrac{25x^7 - 15x^5 - 5x^4}{5x^3}$

72. $\dfrac{49x^7 - 28x^5 - 7x^4}{7x^3}$

73. $\dfrac{18x^7 - 9x^6 + 20x^5 - 10x^4}{-2x^4}$

74. $\dfrac{25x^8 - 50x^7 + 3x^6 - 40x^5}{-5x^5}$

75. $\dfrac{12x^2y^2 + 6x^2y - 15xy^2}{3xy}$

76. $\dfrac{18a^3b^2 - 9a^2b - 27ab^2}{9ab}$

77. $\dfrac{20x^7y^4 - 15x^3y^2 - 10x^2y}{-5x^2y}$

78. $\dfrac{8x^6y^3 - 12x^8y^2 - 4x^{14}y^6}{-4x^6y^2}$

Practice PLUS

In Exercises 79–82, simplify each expression.

79. $\dfrac{2x^3(4x + 2) - 3x^2(2x - 4)}{2x^2}$

80. $\dfrac{6x^3(3x - 1) + 5x^2(6x - 3)}{3x^2}$

81. $\left(\dfrac{18x^2y^4}{9xy^2}\right) - \left(\dfrac{15x^5y^6}{5x^4y^4}\right)$

82. $\left(\dfrac{9x^3 + 6x^2}{3x}\right) - \left(\dfrac{12x^2y^2 - 4xy^2}{2xy^2}\right)$

83. Divide the sum of $(y + 5)^2$ and $(y + 5)(y - 5)$ by $2y$.

84. Divide the sum of $(y + 4)^2$ and $(y + 4)(y - 4)$ by $2y$.

In Exercises 85–86, the variable n in each exponent represents a natural number. Divide the polynomial by the monomial. Then use polynomial multiplication to check the quotient.

85. $\dfrac{12x^{15n} - 24x^{12n} + 8x^{3n}}{4x^{3n}}$

86. $\dfrac{35x^{10n} - 15x^{8n} + 25x^{2n}}{5x^{2n}}$

Application Exercises

The bar graphs show U.S. film box-office receipts, in millions of dollars, and box-office admissions, in millions of tickets sold, for five selected years.

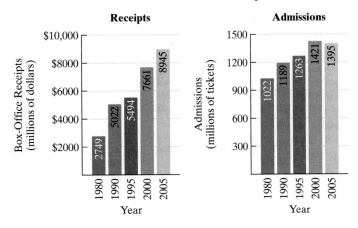

United States Film Box-Office Receipts and Admissions

Sources: U.S. Department of Commerce, Motion Picture Association of America, National Association of Theatre Owners

The following polynomial models of degree 2 can be used to describe the data in the bar graphs on the previous page:

Receipts, in millions of dollars	Admissions, in millions of tickets sold

$$R = 3.6x^2 + 158x + 2790 \qquad A = -0.2x^2 + 21x + 1015.$$

The variable x represents the number of years after 1980.

Use this information to solve Exercises 87–88.

87. a. Use the data displayed by the bar graphs on the previous page to find the average admission charge for a film ticket in 2000. Round to two decimal places, or to the nearest cent.

   b. Use the models to write an algebraic expression that describes the average admission charge for a film ticket x years after 1980.

   c. Use the model from part (b) to find the average admission charge for a film ticket in 2000. Round to the nearest cent. Does the model underestimate or overestimate the actual average charge that you found in part (a)? By how much?

   d. Can the polynomial division for the model in part (b) be performed using the methods that you learned in this section? Explain your answer.

88. a. Use the data displayed by the bar graphs on the previous page to find the average admission charge for a film ticket in 2005. Round to two decimal places, or to the nearest cent.

   b. Use the models to write an algebraic expression that describes the average admission charge for a film ticket x years after 1980.

   c. Use the model from part (b) to find the average admission charge for a film ticket in 2005. Round to the nearest cent. Does the model underestimate or overestimate the actual average charge that you found in part (a)? By how much?

   d. Can the polynomial division for the model in part (b) be performed using the methods that you learned in this section? Explain your answer.

Writing in Mathematics

89. Explain the quotient rule for exponents. Use $\dfrac{3^6}{3^2}$ in your explanation.

90. Explain how to find any nonzero number to the 0 power.

91. Explain the difference between $(-7)^0$ and -7^0.

92. Explain how to simplify an expression that involves a quotient raised to a power. Provide an example with your explanation.

93. Explain how to divide monomials. Give an example.

94. Explain how to divide a polynomial that is not a monomial by a monomial. Give an example.

95. Are the expressions

$$\frac{12x^2 + 6x}{3x} \quad \text{and} \quad 4x + 2$$

equal for every value of x? Explain.

Critical Thinking Exercises

Make Sense? *In Exercises 96–99, determine whether each statement "makes sense" or "does not make sense" and explain your reasoning.*

96. Because division by 0 is undefined, numbers to 0 powers should not be written in denominators.

97. The quotient rule is applied by dividing the exponent in the numerator by the exponent in the denominator.

98. I divide monomials by dividing coefficients and subtracting exponents.

99. I divide a polynomial by a monomial by dividing each term of the monomial by the polynomial.

In Exercises 100–103, determine whether each statement is true or false. If the statement is false, make the necessary change(s) to produce a true statement.

100. $x^{10} \div x^2 = x^5$ for all nonzero real numbers x.

101. $\dfrac{12x^3 - 6x}{2x} = 6x^2 - 6x$

102. $\dfrac{x^2 + x}{x} = x$

103. If a polynomial in x of degree 6 is divided by a monomial in x of degree 2, the degree of the quotient is 4.

104. What polynomial, when divided by $3x^2$, yields the trinomial $6x^6 - 9x^4 + 12x^2$ as a quotient?

In Exercises 105–106, find the missing coefficients and exponents designated by question marks.

105. $\dfrac{?x^8 - ?x^6}{3x^?} = 3x^5 - 4x^3$

106. $\dfrac{3x^{14} - 6x^{12} - ?x^7}{?x^?} = -x^7 + 2x^5 + 3$

Review Exercises

107. Find the absolute value: $|-20.3|$. (Section 1.3, Example 8)

108. Express $\frac{7}{8}$ as a decimal. (Section 1.3, Example 4)

109. Graph: $y = \dfrac{1}{3}x + 2$. (Section 4.4, Example 3)

Preview Exercises

Exercises 110–112 will help you prepare for the material covered in the next section. In each exercise, perform the long division without using a calculator, and then state the quotient and the remainder.

110. $19\overline{)494}$

111. $24\overline{)2958}$

112. $98\overline{)25{,}187}$

Dividing Polynomials by Binomials

Objective

1 Divide polynomials by binomials.

Marisol Escobar (American, b. 1930) "The Family" 1963, wood, metal, graphite, textiles, paint, plaster, and other accessories, $79\frac{1}{2}$ in. × 63 in. × 73 in. (201.93 cm × 160.02 cm × 185.42 cm). Currier Museum of Art, Manchester, New Hampshire. Museum Purchase: The Henry Melville Fuller Acquisition Fund, 2005.12. © 2007 Marisol Escobar/Licensed by VAGA, New York, NY.

"He was an arithmetician rather than a mathematician. None of the humor, the music, or the mysticism of higher mathematics ever entered his head."
—*John Steinbeck*

"Arithmetic has a very great and elevating effect. He who can properly divide is to be considered a god."
—*Plato*

"You cannot ask us to take sides against arithmetic."
—*Winston Churchill*

So, what's the deal? Will performing the repetitive procedure of long division (don't reach for that calculator!) have an elevating effect? Or will it confine you to a computational box that allows neither humor nor music to enter? Forget the box: Mathematician Wilhelm Leibniz believed that music is nothing but unconscious arithmetic. But do think elevation, if not to the level of an ancient Greek god, then to new, algebraic highs. The bottom line: Understanding long division of whole numbers lays the foundation for performing the division of a polynomial by a binomial, such as

$$x + 3 \overline{)x^2 + 10x + 21}.$$

Divisor has two terms and is a binomial.   **The polynomial dividend has three terms and is a trinomial.**

In this section, you will learn how to perform such divisions.

The Steps in Dividing a Polynomial by a Binomial

Dividing a polynomial by a binomial may remind you of long division. Let's review long division of whole numbers by dividing 3983 by 26.

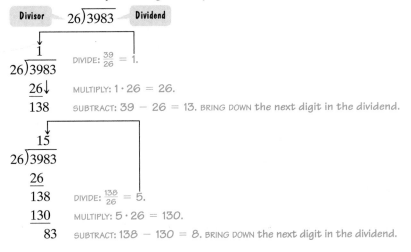

Divisor $26 \overline{)3983}$ **Dividend**

$$26 \overline{)3983}^{\;1}$$

DIVIDE: $\frac{39}{26} = 1$.

$$\underline{26}\downarrow$$
$$138$$

MULTIPLY: $1 \cdot 26 = 26$.

SUBTRACT: $39 - 26 = 13$. BRING DOWN the next digit in the dividend.

$$26 \overline{)3983}^{\;15}$$
$$\underline{26}$$
$$138$$

DIVIDE: $\frac{138}{26} = 5$.

$$\underline{130}$$
$$83$$

MULTIPLY: $5 \cdot 26 = 130$.

SUBTRACT: $138 - 130 = 8$. BRING DOWN the next digit in the dividend.

$$\begin{array}{r} 15\overset{\downarrow}{3} \\ 26\overline{)3983} \\ \underline{26} \\ 138 \\ \underline{130} \\ 83 \\ \underline{78} \\ 5 \end{array}$$

DIVIDE: $\frac{83}{26} = 3$.

MULTIPLY: $3 \cdot 26 = 78$.

SUBTRACT: $83 - 78 = 5$. There are no more digits to bring down, so the remainder is 5.

The quotient is 153 and the remainder is 5. This can be written as

Quotient $\quad 153\dfrac{5}{26}.$ — Remainder / Divisor

This answer can be checked. Multiply the divisor and the quotient. Then add the remainder. If the result is the dividend, the answer is correct. In this case, we have

$$26(153) \ + \ 5 \ = \ 3978 + 5 \ = \ 3983.$$

Divisor   Quotient   Remainder   This is the dividend

Because we obtained the dividend, the answer to the division problem, $153\frac{5}{26}$, is correct.

When a divisor is a binomial, the four steps used to divide whole numbers—**divide, multiply, subtract, bring down the next term**—form the repetitive procedure for dividing a polynomial by a binomial.

1 Divide polynomials by binomials.

EXAMPLE 1 Dividing a Polynomial by a Binomial

Divide $x^2 + 10x + 21$ by $x + 3$.

Solution The following steps illustrate how polynomial division is very similar to numerical division.

$x + 3\overline{)x^2 + 10x + 21}$

Arrange the terms of the dividend ($x^2 + 10x + 21$) and the divisor ($x + 3$) in descending powers of x.

$\dfrac{x\quad\quad\quad}{x + 3\overline{)x^2 + 10x + 21}}$

DIVIDE x^2 (the first term in the dividend) by x (the first term in the divisor): $\frac{x^2}{x} = x$. Align like terms.

$x(x + 3) = x^2 + 3x$ $\quad \dfrac{x\quad\quad\quad}{x + 3\overline{)x^2 + 10x + 21}}$ $x^2 + 3x$

MULTIPLY each term in the divisor ($x + 3$) by x, aligning terms of the product under like terms in the dividend.

$\dfrac{x\quad\quad\quad}{x + 3\overline{)x^2 + 10x + 21}}$ $\underset{\ominus\quad\ominus}{x^2 + 3x}$ / $7x$

SUBTRACT $x^2 + 3x$ from $x^2 + 10x$ by changing the sign of each term in the lower expression and adding.

Change signs of the polynomial being subtracted.

$\dfrac{x\quad\quad\quad}{x + 3\overline{)x^2 + 10x + 21}}$ $\underline{x^2 + 3x}\quad\downarrow$ / $7x + 21$

BRING DOWN 21 from the original dividend and add algebraically to form a new dividend.

$$
\begin{array}{r}
x + 7 \\
x + 3 \overline{) x^2 + 10x + 21} \\
x^2 + 3x \\
\hline
7x + 21
\end{array}
$$

Find the second term of the quotient. Divide the first term of $7x + 21$ by x, the first term of the divisor: $\dfrac{7x}{x} = 7$.

$7(x + 3) = 7x + 21$

$$
\begin{array}{r}
x + 7 \\
x + 3 \overline{) x^2 + 10x + 21} \\
x^2 + 3x \\
\hline
7x + 21 \\
\ominus 7x \ominus 21 \\
\hline
0
\end{array}
$$

Remainder

MULTIPLY the divisor $(x + 3)$ by 7, aligning under like terms in the new dividend. Then subtract to obtain the remainder of 0.

The quotient is $x + 7$ and the remainder is 0. We will not list a remainder of 0 in the answer. Thus,

$$
\frac{x^2 + 10x + 21}{x + 3} = x + 7.
$$

When dividing polynomials by binomials, the answer can be checked. Find the product of the divisor and the quotient and add the remainder. If the result is the dividend, the answer to the division problem is correct. For example, let's check our work in Example 1.

Dividend

$$
\frac{x^2 + 10x + 21}{x + 3} = x + 7
$$

Divisor

Quotient to be checked

Multiply the divisor and the quotient and add the remainder, 0:

$$
(x + 3)(x + 7) + 0 = x^2 + 7x + 3x + 21 + 0 = x^2 + 10x + 21.
$$

Divisor   Quotient   Remainder   This is the dividend.

Because we obtained the dividend, the quotient is correct.

✓ **CHECK POINT 1** Divide $x^2 + 14x + 45$ by $x + 9$.

Before considering additional examples, let's summarize the general procedure for dividing a polynomial by a binomial.

Dividing A Polynomial by a Binomial

1. **Arrange the terms** of both the dividend and the divisor in descending powers of the variable.
2. **Divide** the first term in the dividend by the first term in the divisor. The result is the first term of the quotient.
3. **Multiply** every term in the divisor by the first term in the quotient. Write the resulting product beneath the dividend with like terms lined up.
4. **Subtract** the product from the dividend.
5. **Bring down** the next term in the original dividend and write it next to the remainder to form a new dividend.
6. Use this new expression as the dividend and repeat this process until the remainder can no longer be divided. This will occur when the degree of the remainder (the highest exponent on a variable in the remainder) is less than the degree of the divisor.

In our next division, we will obtain a nonzero remainder.

EXAMPLE 2 **Dividing a Polynomial by a Binomial**

Divide: $\dfrac{7x - 9 - 4x^2 + 4x^3}{2x - 1}$.

Solution We begin by writing the dividend in descending powers of x.

$$7x - 9 - 4x^2 + 4x^3 = 4x^3 - 4x^2 + 7x - 9$$

> Think of 9 as $9x^0$. The powers descend from 3 to 0.

$2x - 1 \overline{)4x^3 - 4x^2 + 7x - 9}$ This is the problem with the dividend in descending powers of x.

$$\begin{array}{r} 2x^2 \\ 2x - 1 \overline{)4x^3 - 4x^2 + 7x - 9} \end{array}$$ DIVIDE: $\dfrac{4x^3}{2x} = 2x^2$.

$2x^2(2x - 1) = 4x^3 - 2x^2$ $\begin{array}{r} 2x^2 \\ 2x - 1 \overline{)4x^3 - 4x^2 + 7x - 9} \\ 4x^3 - 2x^2 \end{array}$ MULTIPLY: $2x^2(2x - 1) = 4x^3 - 2x^2$.

$$\begin{array}{r} 2x^2 \\ 2x - 1 \overline{)4x^3 - 4x^2 + 7x - 9} \\ \ominus 4x^3 \oplus 2x^2 \\ \hline - 2x^2 \end{array}$$

SUBTRACT: $4x^3 - 4x^2 - (4x^3 - 2x^2)$
$= 4x^3 - 4x^2 - 4x^3 + 2x^2$
$= -2x^2$.

> Change signs of the polynomial being subtracted.

$$\begin{array}{r} 2x^2 \\ 2x - 1 \overline{)4x^3 - 4x^2 + 7x - 9} \\ 4x^3 - 2x^2 \downarrow \\ \hline - 2x^2 + 7x \end{array}$$

BRING DOWN $7x$. The new dividend is $-2x^2 + 7x$.

$$\begin{array}{r} 2x^2 - x \\ 2x - 1 \overline{)4x^3 - 4x^2 + 7x - 9} \\ 4x^3 - 2x^2 \\ \hline - 2x^2 + 7x \end{array}$$

DIVIDE: $\dfrac{-2x^2}{2x} = -x$.

$-x(2x - 1) = -2x^2 + x$ $\begin{array}{r} 2x^2 - x \\ 2x - 1 \overline{)4x^3 - 4x^2 + 7x - 9} \\ 4x^3 - 2x^2 \\ \hline - 2x^2 + 7x \\ - 2x^2 + x \end{array}$ MULTIPLY: $-x(2x - 1) = -2x^2 + x$.

$$\begin{array}{r} 2x^2 - x \\ 2x - 1 \overline{)4x^3 - 4x^2 + 7x - 9} \\ 4x^3 - 2x^2 \\ \hline - 2x^2 + 7x \\ \oplus - 2x^2 \ominus + x \\ \hline 6x \end{array}$$

SUBTRACT: $-2x^2 + 7x - (-2x^2 + x)$
$= -2x^2 + 7x + 2x^2 - x$
$= 6x$.

$$
\begin{array}{r}
2x^2 - x \\
2x - 1 \overline{) 4x^3 - 4x^2 + 7x - 9} \\
\underline{4x^3 - 2x^2} \\
-2x^2 + 7x \\
\underline{-2x^2 + x} \\
6x - 9
\end{array}
$$

BRING DOWN -9. The new dividend is $6x - 9$.

$$
\begin{array}{r}
2x^2 - x + 3 \\
2x - 1 \overline{) 4x^3 - 4x^2 + 7x - 9} \\
\underline{4x^3 - 2x^2} \\
-2x^2 + 7x \\
\underline{-2x^2 + x} \\
6x - 9
\end{array}
$$

DIVIDE: $\dfrac{6x}{2x} = 3.$

$3(2x - 1) = 6x - 3$

$$
\begin{array}{r}
2x^2 - x + 3 \\
2x - 1 \overline{) 4x^3 - 4x^2 + 7x - 9} \\
\underline{4x^3 - 2x^2} \\
-2x^2 + 7x \\
\underline{-2x^2 + x} \\
6x - 9 \\
6x - 3
\end{array}
$$

MULTIPLY: $3(2x - 1) = 6x - 3.$

$$
\begin{array}{r}
2x^2 - x + 3 \\
2x - 1 \overline{) 4x^3 - 4x^2 + 7x - 9} \\
\underline{4x^3 - 2x^2} \\
-2x^2 + 7x \\
\underline{-2x^2 + x} \\
6x - 9 \\
\underline{\ominus 6x \oplus 3} \\
-6
\end{array}
$$

SUBTRACT: $6x - 9 - (6x - 3)$
$= 6x - 9 - 6x + 3$
$= -6.$

Remainder

The quotient is $2x^2 - x + 3$ and the remainder is -6. When there is a nonzero remainder, as in this example, list the quotient, plus the remainder above the divisor. Thus,

$$
\frac{7x - 9 - 4x^2 + 4x^3}{2x - 1} = 2x^2 - x + 3 + \frac{-6}{2x - 1}
$$

Remainder above divisor

Quotient

or

$$
\frac{7x - 9 - 4x^2 + 4x^3}{2x - 1} = 2x^2 - x + 3 - \frac{6}{2x - 1}.
$$

Check this result by showing that the product of the divisor and the quotient,

$$
(2x - 1)(2x^2 - x + 3),
$$

plus the remainder, -6, is the dividend, $7x - 9 - 4x^2 + 4x^3$.

☑ **CHECK POINT 2** Divide: $\dfrac{6x + 8x^2 - 12}{2x + 3}.$

If a power of the variable is missing in a dividend, add that power of the variable with a coefficient of 0 and then divide. In this way, like terms will be aligned as you carry out the division.

EXAMPLE 3 Dividing a Polynomial with Missing Terms

Divide: $\dfrac{8x^3 - 1}{2x - 1}$.

Solution We write the dividend, $8x^3 - 1$, as

$$8x^3 + 0x^2 + 0x - 1.$$

> Use a coefficient of 0 with missing terms.

By doing this, we will keep all like terms aligned.

$$4x^2(2x-1) = 8x^3 - 4x^2$$

$$\begin{array}{r} 4x^2 \\ 2x - 1 \overline{)8x^3 + 0x^2 + 0x - 1} \\ \underline{8x^3 \ominus 4x^2} \\ 4x^2 + 0x \end{array}$$

Divide $\left(\dfrac{8x^3}{2x} = 4x^2\right)$, multiply $[4x^2(2x - 1) = 8x^3 - 4x^2]$, subtract, and bring down the next term.
The new dividend is $4x^2 + 0x$.

$$2x(2x-1) = 4x^2 - 2x$$

$$\begin{array}{r} 4x^2 + 2x \\ 2x - 1 \overline{)8x^3 + 0x^2 + 0x - 1} \\ \underline{8x^3 - 4x^2} \\ 4x^2 + 0x \\ \underline{\ominus 4x^2 \oplus 2x} \\ 2x - 1 \end{array}$$

Divide $\left(\dfrac{4x^2}{2x} = 2x\right)$, multiply $[2x(2x - 1) = 4x^2 - 2x]$, subtract, and bring down the next term.
The new dividend is $2x - 1$.

$$1(2x-1) = 2x - 1$$

$$\begin{array}{r} 4x^2 + 2x + 1 \\ 2x - 1 \overline{)8x^3 + 0x^2 + 0x - 1} \\ \underline{8x^3 - 4x^2} \\ 4x^2 + 0x \\ \underline{4x^2 - 2x} \\ 2x - 1 \\ \underline{\ominus 2x \oplus 1} \\ 0 \end{array}$$

Divide $\left(\dfrac{2x}{2x} = 1\right)$, multiply $[1(2x - 1) = 2x - 1]$, and subtract.
The remainder is 0.

Thus,

$$\frac{8x^3 - 1}{2x - 1} = 4x^2 + 2x + 1.$$

Check this result by showing that the product of the divisor and the quotient

$$(2x - 1)(4x^2 + 2x + 1)$$

plus the remainder, 0, is the dividend, $8x^3 - 1$.

Using Technology

Graphic Connections

The graphs of $y_1 = \dfrac{8x^3 - 1}{2x - 1}$ and $y_2 = 4x^2 + 2x + 1$ are shown below.

$y_1 = \dfrac{8x^3 - 1}{2x - 1}$

$y_2 = 4x^2 + 2x + 1$

$[-3, 3, 1]$ by $[-1, 15, 1]$

The graphs coincide. Thus,

$$\frac{8x^3 - 1}{2x - 1} = 4x^2 + 2x + 1.$$

☑ **CHECK POINT 3** Divide: $\dfrac{x^3 - 1}{x - 1}$.

6.6 EXERCISE SET   *MyMathLab*

PRACTICE   WATCH   DOWNLOAD   READ   REVIEW

Practice Exercises

In Exercises 1–36, divide as indicated. Check each answer by showing that the product of the divisor and the quotient, plus the remainder, is the dividend.

1. $\dfrac{x^2 + 6x + 8}{x + 2}$

2. $\dfrac{x^2 + 7x + 10}{x + 5}$

3. $\dfrac{2x^2 + x - 10}{x - 2}$

4. $\dfrac{2x^2 + 13x + 15}{x + 5}$

5. $\dfrac{x^2 - 5x + 6}{x - 3}$

6. $\dfrac{x^2 - 2x - 24}{x + 4}$

7. $\dfrac{2y^2 + 5y + 2}{y + 2}$

8. $\dfrac{2y^2 - 13y + 21}{y - 3}$

9. $\dfrac{x^2 - 3x + 4}{x + 2}$

10. $\dfrac{x^2 - 7x + 5}{x + 3}$

11. $\dfrac{5y + 10 + y^2}{y + 2}$

12. $\dfrac{-8y + y^2 - 9}{y - 3}$

13. $\dfrac{x^3 - 6x^2 + 7x - 2}{x - 1}$

14. $\dfrac{x^3 + 3x^2 + 5x + 3}{x + 1}$

15. $\dfrac{12y^2 - 20y + 3}{2y - 3}$

16. $\dfrac{4y^2 - 8y - 5}{2y + 1}$

17. $\dfrac{4a^2 + 4a - 3}{2a - 1}$

18. $\dfrac{2b^2 - 9b - 5}{2b + 1}$

19. $\dfrac{3y - y^2 + 2y^3 + 2}{2y + 1}$

20. $\dfrac{9y + 18 - 11y^2 + 12y^3}{4y + 3}$

21. $\dfrac{6x^2 - 5x - 30}{2x - 5}$

22. $\dfrac{4y^2 + 8y + 3}{2y - 1}$

23. $\dfrac{x^3 + 4x - 3}{x - 2}$

24. $\dfrac{x^3 + 2x^2 - 3}{x - 2}$

25. $\dfrac{4y^3 + 8y^2 + 5y + 9}{2y + 3}$

26. $\dfrac{2y^3 - y^2 + 3y + 2}{2y + 1}$

27. $\dfrac{6y^3 - 5y^2 + 5}{3y + 2}$

28. $\dfrac{4y^3 + 3y + 5}{2y - 3}$

29. $\dfrac{27x^3 - 1}{3x - 1}$

30. $\dfrac{8x^3 + 27}{2x + 3}$

31. $\dfrac{81 - 12y^3 + 54y^2 + y^4 - 108y}{y - 3}$

32. $\dfrac{8y^3 + y^4 + 16 + 32y + 24y^2}{y + 2}$

33. $\dfrac{4y^2 + 6y}{2y - 1}$

34. $\dfrac{10x^2 - 3x}{x + 3}$

35. $\dfrac{y^4 - 2y^2 + 5}{y - 1}$

36. $\dfrac{y^4 - 6y^2 + 3}{y - 1}$

Practice PLUS

In Exercises 37–42, divide as indicated.

37. $\dfrac{4x^3 - 3x^2 + x + 1}{x^2 + 2}$

38. $\dfrac{3x^3 + 4x^2 + x + 7}{x^2 + 1}$

39. $\dfrac{x^3 - a^3}{x - a}$

40. $\dfrac{x^4 - a^4}{x - a}$

41. $\dfrac{6x^4 - 5x^3 - 8x^2 + 16x - 8}{3x^2 + 2x - 4}$

42. $\dfrac{2x^4 + 5x^3 - 11x^2 - 20x + 12}{x^2 + x - 6}$

43. Divide the difference between $4x^3 + x^2 - 2x + 7$ and $3x^3 - 2x^2 - 7x + 4$ by $x + 1$.

44. Divide the difference between $4x^3 + 2x^2 - x - 1$ and $2x^3 - x^2 + 2x - 5$ by $x + 2$.

Application Exercises

45. Write a simplified polynomial that represents the length of the rectangle.

The width is $x + 1$ units.

The area is $x^3 + 3x^2 + 5x + 3$ square units.

46. Write a simplified polynomial that represents the measure of the base of the parallelogram.

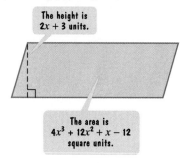

The height is $2x + 3$ units.

The area is $4x^3 + 12x^2 + x - 12$ square units.

You just signed a contract for a new job. The salary for the first year is $30,000 and there is to be a percent increase in your salary each year. The algebraic expression

$$\frac{30{,}000x^n - 30{,}000}{x - 1}$$

describes your total salary over n years, where x is the sum of 1 and the yearly percent increase, expressed as a decimal. Use this information to solve Exercises 47–48.

47. a. Use the given expression and write a quotient of polynomials that describes your total salary over three years.

   b. Simplify the expression in part (a) by performing the division.

   c. Suppose you are to receive an increase of 5% per year. Thus, x is the sum of 1 and 0.05, or 1.05. Substitute 1.05 for x in the expression in part (a) as well as in the simplified form of the expression in part (b). Evaluate each expression. What is your total salary over the three-year period?

48. a. Use the given expression and write a quotient of polynomials that describes your total salary over four years.

   b. Simplify the expression in part (a) by performing the division.

   c. Suppose you are to receive an increase of 8% per year. Thus, x is the sum of 1 and 0.08, or 1.08. Substitute 1.08 for x in the expression in part (a) as well as in the simplified form of the expression in part (b). Evaluate each expression. What is your total salary over the four-year period?

Writing in Mathematics

49. In your own words, explain how to divide a polynomial by a binomial. Use $\dfrac{x^2 + 4}{x + 2}$ in your explanation.

50. When dividing a polynomial by a binomial, explain when to stop dividing.

51. After dividing a polynomial by a binomial, explain how to check the answer.

52. When dividing a binomial into a polynomial with missing terms, explain the advantage of writing the missing terms with zero coefficients.

Critical Thinking Exercises

Make Sense? *In Exercises 53–56, determine whether each statement "makes sense" or "does not make sense" and explain your reasoning. Each statement applies to the division problem*

$$\frac{x^3 + 1}{x + 1}.$$

53. The purpose of writing $x^3 + 1$ as $x^3 + 0x^2 + 0x + 1$ is to keep all like terms aligned.

54. Rewriting $x^3 + 1$ as $x^3 + 0x^2 + 0x + 1$ can change the value of the variable expression for certain values of x.

55. There's no need to apply the long-division process to this problem because I can work the problem in my head and see that the quotient must be $x^2 + 1$.

56. The degree of the quotient must be $3 - 1$.

In Exercises 57–60, determine whether each statement is true or false. If the statement is false, make the necessary change(s) to produce a true statement.

57. If $4x^2 + 25x - 3$ is divided by $4x + 1$, the remainder is 9.

58. If polynomial division results in a remainder of zero, then the product of the divisor and the quotient is the dividend.

59. A nonzero remainder indicates that the answer to a polynomial long-division problem is not a polynomial.

60. When a polynomial is divided by a binomial, the division process stops when the last term of the dividend is brought down.

61. When a certain polynomial is divided by $2x + 4$, the quotient is

$$x - 3 + \frac{17}{2x + 4}.$$

What is the polynomial?

62. Find the number k such that when $16x^2 - 2x + k$ is divided by $2x - 1$, the remainder is 0.

63. Describe the pattern that you observe in the following quotients and remainders.

$$\frac{x^3 - 1}{x + 1} = x^2 - x + 1 - \frac{2}{x + 1}$$

$$\frac{x^5 - 1}{x + 1} = x^4 - x^3 + x^2 - x + 1 - \frac{2}{x + 1}$$

Use this pattern to find $\dfrac{x^7 - 1}{x + 1}$. Verify your result by dividing.

Technology Exercises

In Exercises 64–68, use a graphing utility to determine whether the divisions have been performed correctly. Graph each side of the given equation in the same viewing rectangle. The graphs should coincide. If they do not, correct the expression on the right side by using polynomial division. Then use your graphing utility to show that the division has been performed correctly.

64. $\dfrac{x^2 - 4}{x - 2} = x + 2$

65. $\dfrac{x^2 - 25}{x - 5} = x - 5$

66. $\dfrac{2x^2 + 13x + 15}{x - 5} = 2x + 3$

67. $\dfrac{6x^2 + 16x + 8}{3x + 2} = 2x - 4$

68. $\dfrac{x^3 + 3x^2 + 5x + 3}{x + 1} = x^2 - 2x + 3$

Review Exercises

69. Graph the solution set of the system:

$$2x - y \geq 4$$
$$x + y \leq -1.$$

(Section 5.5, Example 2)

70. What is 6% of 20? (Section 2.4, Example 7)

71. Solve: $\dfrac{x}{3} + \dfrac{2}{5} = \dfrac{x}{5} - \dfrac{2}{5}$ (Section 2.3, Example 4)

Preview Exercises

Exercises 72–74 will help you prepare for the material covered in the next section.

72. a. Find the missing exponent, designated by the question mark, in each final step.

$$\frac{7^3}{7^5} = \frac{7 \cdot 7 \cdot 7}{7 \cdot 7 \cdot 7 \cdot 7 \cdot 7} = \frac{1}{7^?}$$

$$\frac{7^3}{7^5} = 7^{3-5} = 7^?$$

b. Based on your two results for $\dfrac{7^3}{7^5}$, what can you conclude?

73. Simplify: $\dfrac{(2x^3)^4}{x^{10}}$.

74. Simplify: $\left(\dfrac{x^5}{x^2}\right)^3$.

Negative Exponents and Scientific Notation

Objectives

1. Use the negative exponent rule.
2. Simplify exponential expressions.
3. Convert from scientific notation to decimal notation.
4. Convert from decimal notation to scientific notation.
5. Compute with scientific notation.
6. Solve applied problems using scientific notation.

We frequently encounter very large and very small numbers. Governments throughout the world are concerned about the billions of tons of carbon dioxide that the global population of 6.5 billion people release into the atmosphere each year. In the photo shown above, the national debt of the United States was about $8.2 trillion. A typical atom has a diameter of about one-ten-billionth of a meter. Exponents provide a way of putting these large and small numbers in perspective.

1 Use the negative exponent rule.

Negative Integers as Exponents

A nonzero base can be raised to a negative power. The quotient rule can be used to help determine what a negative integer as an exponent should mean. Consider the quotient of b^3 and b^5, where b is not zero. We can determine this quotient in two ways.

$$\frac{b^3}{b^5} = \frac{1 \cdot \cancel{b} \cdot \cancel{b} \cdot \cancel{b}}{\cancel{b} \cdot \cancel{b} \cdot \cancel{b} \cdot b \cdot b} = \frac{1}{b^2}$$

After dividing out pairs of factors, we have two factors of b in the denominator.

$$\frac{b^3}{b^5} = b^{3-5} = b^{-2}$$

Use the quotient rule and subtract exponents.

Notice that $\frac{b^3}{b^5}$ equals both b^{-2} and $\frac{1}{b^2}$. This means that b^{-2} must equal $\frac{1}{b^2}$. This example is a special case of the **negative exponent rule**.

The Negative Exponent Rule

If b is any real number other than 0 and n is a natural number, then

$$b^{-n} = \frac{1}{b^n}.$$

Study Tip

A negative exponent does not make the value of an expression negative. For example,

$$7^{-2} = \frac{1}{7^2} = \frac{1}{49}$$

is positive. Avoid these common errors:

Incorrect!

$$\cancel{7^{-2} = -7^2}$$

$$\cancel{7^{-2} = -\frac{1}{7^2}}$$

EXAMPLE 1   Using the Negative Exponent Rule

Use the negative exponent rule to write each expression with a positive exponent. Then simplify the expression.

a. 7^{-2} **b.** 4^{-3} **c.** $(-2)^{-4}$ **d.** -2^{-4} **e.** 5^{-1}

Solution

a. $7^{-2} = \dfrac{1}{7^2} = \dfrac{1}{7 \cdot 7} = \dfrac{1}{49}$

b. $4^{-3} = \dfrac{1}{4^3} = \dfrac{1}{4 \cdot 4 \cdot 4} = \dfrac{1}{64}$

c. $(-2)^{-4} = \dfrac{1}{(-2)^4} = \dfrac{1}{(-2)(-2)(-2)(-2)} = \dfrac{1}{16}$

d. $-2^{-4} = -\dfrac{1}{2^4} = -\dfrac{1}{2 \cdot 2 \cdot 2 \cdot 2} = -\dfrac{1}{16}$

The negative is not inside parentheses and is not taken to the -4 power.

e. $5^{-1} = \dfrac{1}{5^1} = \dfrac{1}{5}$

✓ **CHECK POINT 1**   Use the negative exponent rule to write each expression with a positive exponent. Then simplify the expression.

a. 6^{-2} **b.** 5^{-3} **c.** $(-3)^{-4}$

d. -3^{-4} **e.** 8^{-1}

Negative exponents can also appear in denominators. For example,

$$\frac{1}{2^{-10}} = \frac{1}{\frac{1}{2^{10}}} = 1 \div \frac{1}{2^{10}} = 1 \cdot \frac{2^{10}}{1} = 2^{10}.$$

In general, if a negative exponent appears in a denominator, an expression can be written with a positive exponent using

$$\frac{1}{b^{-n}} = b^n.$$

For example,

$$\frac{1}{2^{-3}} = 2^3 = 8 \qquad \text{and} \qquad \frac{1}{(-6)^{-2}} = (-6)^2 = 36.$$

> Change only the sign of the exponent and not the sign of the base, −6.

Negative Exponents in Numerators and Denominators

If b is any real number other than 0 and n is a natural number, then

$$b^{-n} = \frac{1}{b^n} \qquad \text{and} \qquad \frac{1}{b^{-n}} = b^n.$$

When a negative number appears as an exponent, switch the position of the base (from numerator to denominator or from denominator to numerator) and make the exponent positive. The sign of the base does not change.

EXAMPLE 2   Using Negative Exponents

Write each expression with positive exponents only. Then simplify, if possible.

a. $\dfrac{4^{-3}}{5^{-2}}$　　**b.** $\left(\dfrac{3}{4}\right)^{-2}$　　**c.** $\dfrac{1}{4x^{-3}}$　　**d.** $\dfrac{x^{-5}}{y^{-1}}$

Solution

a. $\dfrac{4^{-3}}{5^{-2}} = \dfrac{5^2}{4^3} = \dfrac{5 \cdot 5}{4 \cdot 4 \cdot 4} = \dfrac{25}{64}$

> Switch the position of the bases and make the exponents positive.

b. $\left(\dfrac{3}{4}\right)^{-2} = \dfrac{3^{-2}}{4^{-2}} = \dfrac{4^2}{3^2} = \dfrac{4 \cdot 4}{3 \cdot 3} = \dfrac{16}{9}$

> Switch the position of the bases and make the exponents positive.

c. $\dfrac{1}{4x^{-3}} = \dfrac{x^3}{4}$

> Switch the position of the base and make the exponent positive. Note that only x is raised to the −3 power.

d. $\dfrac{x^{-5}}{y^{-1}} = \dfrac{y^1}{x^5} = \dfrac{y}{x^5}$

✓ **CHECK POINT 2**   Write each expression with positive exponents only. Then simplify, if possible.

a. $\dfrac{2^{-3}}{7^{-2}}$　　**b.** $\left(\dfrac{4}{5}\right)^{-2}$　　**c.** $\dfrac{1}{7y^{-2}}$　　**d.** $\dfrac{x^{-1}}{y^{-8}}$

2 Simplify exponential expressions.

Simplifying Exponential Expressions

Properties of exponents are used to simplify exponential expressions. An exponential expression is **simplified** when

- Each base occurs only once.
- No parentheses appear.
- No powers are raised to powers.
- No negative or zero exponents appear.

Simplifying Exponential Expressions

	Example
1. If necessary, be sure that each base appears only once, using $$b^m \cdot b^n = b^{m+n} \quad \text{or} \quad \frac{b^m}{b^n} = b^{m-n}.$$	$x^4 \cdot x^3 = x^{4+3} = x^7$
2. If necessary, remove parentheses using $$(ab)^n = a^n b^n \quad \text{or} \quad \left(\frac{a}{b}\right)^n = \frac{a^n}{b^n}.$$	$(xy)^3 = x^3 y^3$
3. If necessary, simplify powers to powers using $$(b^m)^n = b^{mn}.$$	$(x^4)^3 = x^{4\cdot3} = x^{12}$
4. If necessary, rewrite exponential expressions with zero powers as 1 ($b^0 = 1$). Furthermore, write the answer with positive exponents using $$b^{-n} = \frac{1}{b^n} \quad \text{or} \quad \frac{1}{b^{-n}} = b^n.$$	$\dfrac{x^5}{x^8} = x^{5-8} = x^{-3} = \dfrac{1}{x^3}$

The following examples show how to simplify exponential expressions. In each example, assume that any variable in a denominator is not equal to zero.

Study Tip

There is often more than one way to simplify an exponential expression. For example, you may prefer to simplify Example 3 as follows:

$$x^{-9} \cdot x^4 = \frac{x^4}{x^9} = x^{4-9} = x^{-5} = \frac{1}{x^5}.$$

EXAMPLE 3 Simplifying an Exponential Expression

Simplify: $x^{-9} \cdot x^4$.

Solution

$$x^{-9} \cdot x^4 = x^{-9+4} \qquad b^m \cdot b^n = b^{m+n}$$
$$= x^{-5} \qquad \text{The base, } x, \text{ now appears only once.}$$
$$= \frac{1}{x^5} \qquad b^{-n} = \frac{1}{b^n}$$

✓ **CHECK POINT 3** Simplify: $x^{-12} \cdot x^2$.

EXAMPLE 4 Simplifying Exponential Expressions

Simplify:

a. $\dfrac{x^4}{x^{20}}$ b. $\dfrac{25x^6}{5x^8}$ c. $\dfrac{10y^7}{-2y^{10}}$.

Solution

a. $\dfrac{x^4}{x^{20}} = x^{4-20} = x^{-16} = \dfrac{1}{x^{16}}$

b. $\dfrac{25x^6}{5x^8} = \dfrac{25}{5} \cdot \dfrac{x^6}{x^8} = 5x^{6-8} = 5x^{-2} = \dfrac{5}{x^2}$

c. $\dfrac{10y^7}{-2y^{10}} = \dfrac{10}{-2} \cdot \dfrac{y^7}{y^{10}} = -5y^{7-10} = -5y^{-3} = -\dfrac{5}{y^3}$ ■

☑ **CHECK POINT 4** Simplify:

a. $\dfrac{x^2}{x^{10}}$ b. $\dfrac{75x^3}{5x^9}$ c. $\dfrac{50y^8}{-25y^{14}}$.

EXAMPLE 5 Simplifying an Exponential Expression

Simplify: $\dfrac{(5x^3)^2}{x^{10}}$.

Solution

$$\dfrac{(5x^3)^2}{x^{10}} = \dfrac{5^2(x^3)^2}{x^{10}}$$ Raise each factor in the product to the second power using $(ab)^n = a^n b^n$.

$$= \dfrac{5^2 x^{3 \cdot 2}}{x^{10}}$$ Raise powers to powers using $(b^m)^n = b^{mn}$.

$$= \dfrac{25x^6}{x^{10}}$$ Simplify.

$$= 25x^{6-10}$$ When dividing with the same base, subtract exponents: $\dfrac{b^m}{b^n} = b^{m-n}$.

$$= 25x^{-4}$$ Simplify. The base, x, now appears only once.

$$= \dfrac{25}{x^4}$$ Rewrite with a positive exponent using $b^{-n} = \dfrac{1}{b^n}$. ■

☑ **CHECK POINT 5** Simplify: $\dfrac{(6x^4)^2}{x^{11}}$.

EXAMPLE 6 Simplifying an Exponential Expression

Simplify: $\left(\dfrac{x^5}{x^2}\right)^{-3}$.

Solution

Method 1. First perform the division within the parentheses.

$$\left(\dfrac{x^5}{x^2}\right)^{-3} = (x^{5-2})^{-3}$$ Within parentheses, divide by subtracting exponents: $\dfrac{b^m}{b^n} = b^{m-n}$.

$$= (x^3)^{-3}$$ Simplify. The base, x, now appears only once.

$$= x^{3(-3)}$$ Raise powers to powers: $(b^m)^n = b^{mn}$.

$$= x^{-9}$$ Simplify.

$$= \dfrac{1}{x^9}$$ Rewrite with a positive exponent using $b^{-n} = \dfrac{1}{b^n}$.

Method 2. Remove parentheses first by raising the numerator and the denominator to the -3 power.

$$\left(\frac{x^5}{x^2}\right)^{-3} = \frac{(x^5)^{-3}}{(x^2)^{-3}} \qquad \text{Use } \left(\frac{a}{b}\right)^n = \frac{a^n}{b^n} \text{ and raise the numerator and denominator to the } -3 \text{ power.}$$

$$= \frac{x^{5(-3)}}{x^{2(-3)}} \qquad \text{Raise powers to powers using } (b^m)^n = b^{mn}.$$

$$= \frac{x^{-15}}{x^{-6}} \qquad \text{Simplify.}$$

$$= x^{-15-(-6)} \qquad \text{When dividing with the same base, subtract the exponent in the denominator from the exponent in the numerator: } \frac{b^m}{b^n} = b^{m-n}.$$

$$= x^{-9} \qquad \text{Subtract: } -15 - (-6) = -15 + 6 = -9. \text{ The base, } x, \text{ now appears only once.}$$

$$= \frac{1}{x^9} \qquad \text{Rewrite with a positive exponent using } b^{-n} = \frac{1}{b^n}.$$

Which method do you prefer?

✓ **CHECK POINT 6** Simplify: $\left(\dfrac{x^8}{x^4}\right)^{-5}$.

Table 6.1	Names of Large Numbers
10^2	hundred
10^3	thousand
10^6	million
10^9	billion
10^{12}	trillion
10^{15}	quadrillion
10^{18}	quintillion
10^{21}	sextillion
10^{24}	septillion
10^{27}	octillion
10^{30}	nonillion
10^{100}	googol
10^{googol}	googolplex

Scientific Notation

As of December 2006, the national debt of the United States was about $8.6 trillion. This is the amount of money the government has had to borrow over the years, mostly by selling bonds, because it has spent more than it has collected in taxes. A stack of $1 bills equaling the national debt would rise to twice the distance from Earth to the moon. Because a trillion is 10^{12} (see **Table 6.1**), the national debt can be expressed as

$$8.6 \times 10^{12}.$$

The number 8.6×10^{12} is written in a form called *scientific notation*.

> ### Scientific Notation
>
> A positive number is written in **scientific notation** when it is expressed in the form
>
> $$a \times 10^n,$$
>
> where a is a number greater than or equal to 1 and less than 10 ($1 \leq a < 10$) and n is an integer.

It is customary to use the multiplication symbol, $\times$, rather than a dot, when writing a number in scientific notation.

Here are two examples of numbers in scientific notation:

- Each day, 2.6×10^7 pounds of dust from the atmosphere settle on Earth.
- The length of the AIDS virus is 1.1×10^{-4} millimeter.

3 Convert from scientific notation to decimal notation.

We can use n, the exponent on the 10 in $a \times 10^n$, to change a number in scientific notation to decimal notation. If n is **positive**, move the decimal point in a to the **right** n places. If n is **negative**, move the decimal point in a to the **left** $|n|$ places.

EXAMPLE 7 Converting from Scientific to Decimal Notation

Write each number in decimal notation:

a. 2.6×10^7 **b.** 1.1×10^{-4}.

Solution In each case, we use the exponent on the 10 to move the decimal point. In part (a), the exponent is positive, so we move the decimal point to the right. In part (b), the exponent is negative, so we move the decimal point to the left.

a. $2.6 \times 10^7 = 26,000,000$

$n = 7$

> Move the decimal point 7 places to the right.

b. $1.1 \times 10^{-4} = 0.00011$

$n = -4$

> Move the decimal point $|-4|$ places, or 4 places, to the left.

✓ **CHECK POINT 7** Write each number in decimal notation:

a. 7.4×10^9 **b.** 3.017×10^{-6}.

---

4 Convert from decimal notation to scientific notation.

To convert a positive number from decimal notation to scientific notation, we reverse the procedure of Example 7.

> **Converting from Decimal to Scientific Notation**
>
> Write the number in the form $a \times 10^n$.
>
> - Determine a, the numerical factor. Move the decimal point in the given number to obtain a number greater than or equal to 1 and less than 10.
> - Determine n, the exponent on 10^n. The absolute value of n is the number of places the decimal point was moved. The exponent n is positive if the given number is greater than 10 and negative if the given number is between 0 and 1.

Using Technology

You can use your calculator's EE (enter exponent) or EXP key to convert from decimal to scientific notation. Here is how it's done for 0.000023:

Many Scientific Calculators

Keystrokes	Display
.000023 EE =	2.3 − 05

Many Graphing Calculators

Use the mode setting for scientific notation.

Keystrokes	Display
.000023 ENTER	2.3E-5

EXAMPLE 8 Converting from Decimal Notation to Scientific Notation

Write each number in scientific notation:

a. 4,600,000 **b.** 0.000023.

Solution

a. $4,600,000 = 4.6 \times 10^6$

> This number is greater than 10, so n is positive in $a \times 10^n$.

> Move the decimal point in 4,600,000 to get $1 \le a < 10$.

> The decimal point moved 6 places from 4,600,000 to 4.6.

b. $0.000023 = 2.3 \times 10^{-5}$

> This number is less than 1, so n is negative in $a \times 10^n$.

> Move the decimal point in 0.000023 to get $1 \le a < 10$.

> The decimal point moved 5 places from 0.000023 to 2.3.

☑ **CHECK POINT 8** Write each number in scientific notation:

a. 7,410,000,000 b. 0.000000092.

---

5 Compute with scientific notation.

Computations with Scientific Notation

Properties of exponents are used to perform computations with numbers that are expressed in scientific notation.

> ### Computations with Numbers in Scientific Notation
>
> **Multiplication**
>
> $$(a \times 10^n) \times (b \times 10^m) = (a \times b) \times 10^{n+m}$$
>
> > Add the exponents on 10 and multiply the other parts of the numbers separately.
>
> **Division**
>
> $$\frac{a \times 10^n}{b \times 10^m} = \left(\frac{a}{b}\right) \times 10^{n-m}$$
>
> > Subtract the exponents on 10 and divide the other parts of the numbers separately.
>
> **Exponentiation**
>
> $$(a \times 10^n)^m = a^n \times 10^{nm}$$
>
> > Multiply exponents on 10 and raise the other part of the number to the power.
>
> After the computation is completed, the answer may require an adjustment before it is expressed in scientific notation.

Using Technology

$(4 \times 10^5)(2 \times 10^9)$ On a Calculator:

Many Scientific Calculators

4 [EE] 5 [×] 2 [EE] 9 [=]

Display: 8. 14

Many Graphing Calculators

4 [EE] 5 [×] 2 [EE] 9 [ENTER]

Display: 8E14

EXAMPLE 9 **Computations with Scientific Notation**

Perform the indicated computations, writing the answers in scientific notation:

a. $(4 \times 10^5)(2 \times 10^9)$ b. $\dfrac{1.2 \times 10^6}{4.8 \times 10^{-3}}$ c. $(5 \times 10^{-4})^3$.

Solution

a. $(4 \times 10^5)(2 \times 10^9) = (4 \times 2) \times (10^5 \times 10^9)$ Regroup.

$= 8 \times 10^{5+9}$ Add the exponents on 10 and multiply the other parts.

$= 8 \times 10^{14}$ Simplify.

b. $\dfrac{1.2 \times 10^6}{4.8 \times 10^{-3}} = \left(\dfrac{1.2}{4.8}\right) \times \left(\dfrac{10^6}{10^{-3}}\right)$ Regroup.

$= 0.25 \times 10^{6-(-3)}$ Subtract the exponents on 10 and divide the other parts.

$= 0.25 \times 10^9$ Simplify. Because 0.25 is not between 1 and 10, it must be written in scientific notation.

$= 2.5 \times 10^{-1} \times 10^9$ $0.25 = 2.5 \times 10^{-1}$

$= 2.5 \times 10^{-1+9}$ Add the exponents on 10.

$= 2.5 \times 10^8$ Simplify.

c. $(5 \times 10^{-4})^3 = 5^3 \times (10^{-4})^3$ $(ab)^n = a^n b^n$. Cube each factor in parentheses.
$\qquad\qquad = 125 \times 10^{-4 \cdot 3}$ Multiply the exponents and cube the other part of
$\qquad\qquad\qquad\qquad\qquad\qquad\qquad$ the number.
$\qquad\qquad = 125 \times 10^{-12}$ Simplify. 125 must be written in scientific notation.
$\qquad\qquad = 1.25 \times 10^2 \times 10^{-12}$ $125 = 1.25 \times 10^2$
$\qquad\qquad = 1.25 \times 10^{2+(-12)}$ Add the exponents on 10.
$\qquad\qquad = 1.25 \times 10^{-10}$ Simplify.   ■

✓ **CHECK POINT 9**   Perform the indicated computations, writing the answers in scientific notation:

a. $(3 \times 10^8)(2 \times 10^2)$

b. $\dfrac{8.4 \times 10^7}{4 \times 10^{-4}}$

c. $(4 \times 10^{-2})^3$.

---

6 Solve applied problems using scientific notation.

Applications: Putting Numbers in Perspective

Due to tax cuts and spending increases, the United States began accumulating large deficits in the 1980s. To finance the deficit, the government had borrowed $8.6 trillion as of December 2006. The graph in **Figure 6.5** shows the national debt increasing over time.

The National Debt

FIGURE 6.5

Source: Office of Management and Budget

Example 10 shows how we can use scientific notation to comprehend the meaning of a number such as 8.6 trillion.

EXAMPLE 10   The National Debt

As of December 2006, the national debt was $8.6 trillion, or 8.6×10^{12} dollars. At that time, the U.S. population was approximately 300,000,000 (300 million), or 3×10^8. If the national debt was evenly divided among every individual in the United States, how much would each citizen have to pay?

Solution   The amount each citizen must pay is the total debt, 8.6×10^{12} dollars, divided by the number of citizens, 3×10^8.

$$\frac{8.6 \times 10^{12}}{3 \times 10^8} = \left(\frac{8.6}{3}\right) \times \left(\frac{10^{12}}{10^8}\right)$$
$$\approx 2.87 \times 10^{12-8}$$
$$= 2.87 \times 10^4$$
$$= 28,700$$

Every U.S. citizen would have to pay approximately $28,700 to the federal government to pay off the national debt.   ■

Using Technology

Here is the keystroke sequence for solving Example 10, $\dfrac{8.6 \times 10^{12}}{3 \times 10^{8}}$, using a calculator:

$$8.6 \boxed{\text{EE}}\ 12\ \div\ 3\ \boxed{\text{EE}}\ 8.$$

The quotient is displayed by pressing $\boxed{=}$ on a scientific calculator or $\boxed{\text{ENTER}}$ on a graphing calculator. The answer can be displayed in scientific or decimal notation. Consult your manual.

☑ **CHECK POINT 10** Pell Grants help low-income undergraduate students pay for college. In 2006, the federal cost of this program was $13 billion ($13 \times 10^{9}$) and there were 5.1 million (5.1×10^{6}) grant recipients. How much, to the nearest hundred dollars, was the average grant?

Blitzer Bonus

Earthquakes and Exponents

The earthquake that ripped through northern California on October 17, 1989, measured 7.1 on the Richter scale, killed more than 60 people, and injured more than 2400. Shown here is San Francisco's Marina district, where shock waves tossed houses off their foundations and into the street.

The Richter scale is misleading because it is not actually a 1 to 8, but rather a 1 to 10 million scale. Each level indicates a tenfold increase in magnitude from the previous level, making a 7.0 earthquake a million times greater than a 1.0 quake.

The following is a translation of the Richter scale:

Richter number (R)	Magnitude (10^{R-1})
1	$10^{1-1} = 10^{0} = 1$
2	$10^{2-1} = 10^{1} = 10$
3	$10^{3-1} = 10^{2} = 100$
4	$10^{4-1} = 10^{3} = 1000$
5	$10^{5-1} = 10^{4} = 10{,}000$
6	$10^{6-1} = 10^{5} = 100{,}000$
7	$10^{7-1} = 10^{6} = 1{,}000{,}000$
8	$10^{8-1} = 10^{7} = 10{,}000{,}000$

6.7 EXERCISE SET

MyMathLab PRACTICE WATCH DOWNLOAD READ REVIEW

Practice Exercises

In Exercises 1–28, write each expression with positive exponents only. Then simplify, if possible.

1. 8^{-2}

2. 9^{-2}

3. 5^{-3}

4. 4^{-3}

5. $(-6)^{-2}$

6. $(-7)^{-2}$

7. -6^{-2}

8. -7^{-2}

9. 4^{-1}

10. 6^{-1}

11. $2^{-1} + 3^{-1}$

12. $3^{-1} - 6^{-1}$

13. $\dfrac{1}{3^{-2}}$

14. $\dfrac{1}{4^{-3}}$

15. $\dfrac{1}{(-3)^{-2}}$

16. $\dfrac{1}{(-2)^{-2}}$

17. $\dfrac{2^{-3}}{8^{-2}}$

18. $\dfrac{4^{-3}}{2^{-2}}$

19. $\left(\dfrac{1}{4}\right)^{-2}$

20. $\left(\dfrac{1}{5}\right)^{-2}$

21. $\left(\dfrac{3}{5}\right)^{-3}$

22. $\left(\dfrac{3}{4}\right)^{-3}$

23. $\dfrac{1}{6x^{-5}}$

24. $\dfrac{1}{8x^{-6}}$

25. $\dfrac{x^{-8}}{y^{-1}}$

26. $\dfrac{x^{-12}}{y^{-1}}$

27. $\dfrac{3}{(-5)^{-3}}$

28. $\dfrac{4}{(-3)^{-3}}$

In Exercises 29–78, simplify each exponential expression. Assume that variables represent nonzero real numbers.

29. $x^{-8} \cdot x^3$

30. $x^{-11} \cdot x^5$

31. $(4x^{-5})(2x^2)$

32. $(5x^{-7})(3x^3)$

33. $\dfrac{x^3}{x^9}$

34. $\dfrac{x^5}{x^{12}}$

35. $\dfrac{y}{y^{100}}$

36. $\dfrac{y}{y^{50}}$

37. $\dfrac{30z^5}{10z^{10}}$

38. $\dfrac{45z^4}{15z^{12}}$

39. $\dfrac{-8x^3}{2x^7}$

40. $\dfrac{-15x^4}{3x^9}$

41. $\dfrac{-9a^5}{27a^8}$

42. $\dfrac{-15a^8}{45a^{13}}$

43. $\dfrac{7w^5}{5w^{13}}$

44. $\dfrac{7w^8}{9w^{14}}$

45. $\dfrac{x^3}{(x^4)^2}$

46. $\dfrac{x^5}{(x^3)^2}$

47. $\dfrac{y^{-3}}{(y^4)^2}$

48. $\dfrac{y^{-5}}{(y^3)^2}$

49. $\dfrac{(4x^3)^2}{x^8}$

50. $\dfrac{(5x^3)^2}{x^7}$

51. $\dfrac{(6y^4)^3}{y^{-5}}$

52. $\dfrac{(4y^5)^3}{y^{-4}}$

53. $\left(\dfrac{x^4}{x^2}\right)^{-3}$

54. $\left(\dfrac{x^6}{x^2}\right)^{-3}$

55. $\left(\dfrac{4x^5}{2x^2}\right)^{-4}$

56. $\left(\dfrac{6x^7}{2x^2}\right)^{-4}$

57. $(3x^{-1})^{-2}$

58. $(4x^{-1})^{-2}$

59. $(-2y^{-1})^{-3}$

60. $(-3y^{-1})^{-3}$

61. $\dfrac{2x^5 \cdot 3x^7}{15x^6}$

62. $\dfrac{3x^3 \cdot 5x^{14}}{20x^{14}}$

63. $(x^3)^5 \cdot x^{-7}$

64. $(x^4)^3 \cdot x^{-5}$

65. $(2y^3)^4 y^{-6}$

66. $(3y^4)^3 y^{-7}$

67. $\dfrac{(y^3)^4}{(y^2)^7}$

68. $\dfrac{(y^2)^5}{(y^3)^4}$

69. $(y^{10})^{-5}$

70. $(y^{20})^{-5}$

71. $(a^4 b^5)^{-3}$

72. $(a^5 b^3)^{-4}$

73. $(a^{-2} b^6)^{-4}$

74. $(a^{-7} b^2)^{-5}$

75. $\left(\dfrac{x^2}{2}\right)^{-2}$

76. $\left(\dfrac{x^2}{2}\right)^{-3}$

77. $\left(\dfrac{x^2}{y^3}\right)^{-3}$

78. $\left(\dfrac{x^3}{y^2}\right)^{-4}$

In Exercises 79–90, write each number in decimal notation without the use of exponents.

79. 8.7×10^2

80. 2.75×10^3

81. 9.23×10^5

82. 7.24×10^4

83. 3.4×10^0

84. 9.115×10^0

85. 7.9×10^{-1}

86. 8.6×10^{-1}

87. 2.15×10^{-2}

88. 3.14×10^{-2}

89. 7.86×10^{-4}

90. 4.63×10^{-5}

In Exercises 91–106, write each number in scientific notation.

91. 32,400

92. 327,000

93. 220,000,000

94. 370,000,000,000

95. 713

96. 623

97. 6751

98. 9832

99. 0.0027

100. 0.00083

101. 0.0000202

102. 0.00000103

103. 0.005

104. 0.006

105. 3.14159

106. 2.71828

In Exercises 107–126, perform the indicated computations. Write the answers in scientific notation.

107. $(2 \times 10^3)(3 \times 10^2)$

108. $(3 \times 10^4)(3 \times 10^2)$

109. $(2 \times 10^5)(8 \times 10^3)$

110. $(4 \times 10^3)(5 \times 10^4)$

111. $\dfrac{12 \times 10^6}{4 \times 10^2}$

112. $\dfrac{20 \times 10^{20}}{10 \times 10^{10}}$

113. $\dfrac{15 \times 10^4}{5 \times 10^{-2}}$

114. $\dfrac{18 \times 10^2}{9 \times 10^{-3}}$

115. $\dfrac{15 \times 10^{-4}}{5 \times 10^2}$

116. $\dfrac{18 \times 10^{-2}}{9 \times 10^3}$

117. $\dfrac{180 \times 10^6}{2 \times 10^3}$

118. $\dfrac{180 \times 10^8}{2 \times 10^4}$

119. $\dfrac{3 \times 10^4}{12 \times 10^{-3}}$

120. $\dfrac{5 \times 10^2}{20 \times 10^{-3}}$

121. $(5 \times 10^2)^3$

122. $(4 \times 10^3)^2$

123. $(3 \times 10^{-2})^4$

124. $(2 \times 10^{-3})^5$

125. $(4 \times 10^6)^{-1}$

126. $(5 \times 10^4)^{-1}$

Practice PLUS

In Exercises 127–134, simplify each exponential expression. Assume that variables represent nonzero real numbers.

127. $\dfrac{\left(x^{-2}y\right)^{-3}}{\left(x^2y^{-1}\right)^3}$ **128.** $\dfrac{\left(xy^{-2}\right)^{-2}}{\left(x^{-2}y\right)^{-3}}$

129. $\left(2x^{-3}yz^{-6}\right)\left(2x\right)^{-5}$

130. $\left(3x^{-4}yz^{-7}\right)\left(3x\right)^{-3}$

131. $\left(\dfrac{x^3y^4z^5}{x^{-3}y^{-4}z^{-5}}\right)^{-2}$

132. $\left(\dfrac{x^4y^5z^6}{x^{-4}y^{-5}z^{-6}}\right)^{-4}$

133. $\dfrac{\left(2^{-1}x^{-2}y^{-1}\right)^{-2}\left(2x^{-4}y^3\right)^{-2}\left(16x^{-3}y^3\right)^0}{\left(2x^{-3}y^{-5}\right)^2}$

134. $\dfrac{\left(2^{-1}x^{-3}y^{-1}\right)^{-2}\left(2x^{-6}y^4\right)^{-2}\left(9x^3y^{-3}\right)^0}{\left(2x^{-4}y^{-6}\right)^2}$

In Exercises 135–138, perform the indicated computations. Express answers in scientific notation.

135. $\left(5 \times 10^3\right)\left(1.2 \times 10^{-4}\right) \div \left(2.4 \times 10^2\right)$

136. $\left(2 \times 10^2\right)\left(2.6 \times 10^{-3}\right) \div \left(4 \times 10^3\right)$

137. $\dfrac{\left(1.6 \times 10^4\right)\left(7.2 \times 10^{-3}\right)}{\left(3.6 \times 10^8\right)\left(4 \times 10^{-3}\right)}$

138. $\dfrac{\left(1.2 \times 10^6\right)\left(8.7 \times 10^{-2}\right)}{\left(2.9 \times 10^6\right)\left(3 \times 10^{-3}\right)}$

Application Exercises

In Exercises 139–142, rewrite the number in each statement in scientific notation.

139. King Mongkut of Siam (the king in the musical *The King and I*) had 9200 wives.

140. The top-selling music album of all time is "Their Greatest Hits" by the Eagles, selling 28,000,000 copies.

141. The volume of a bacterium is 0.00000000000000025 cubic meter.

142. Home computers can perform a multiplication in 0.00000000036 second.

In Exercises 143–146, use 10^6 for one million and 10^9 for one billion to rewrite the number in each statement in scientific notation.

Pet Ownership in the United States

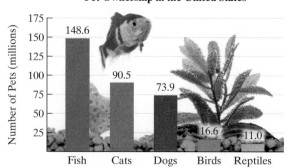

Source: American Pet Product Manufacturers Association

143. Americans are caregivers to 90.5 million cats.

144. Americans are caregivers to 73.9 million dogs.

Countries with the Greatest Defense Spending

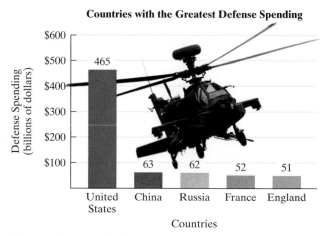

Source: The International Institute for Strategic Studies, "The Military Balance 2005–2006"

145. In 2005, the United States spent $465 billion on defense.

146. In 2005, the United States spent more on defense than the four next top-spending nations combined. These nations spent $228 billion on defense.

The bar graph shows the total amount Americans paid in federal taxes, in trillions of dollars, and the U.S. population, in millions, from 2002 through 2005. Exercises 147–148 are based on the numbers displayed by the graph.

Federal Taxes and the United States Population

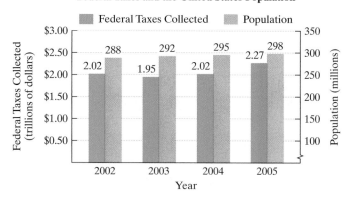

Sources: Internal Revenue Service and U.S. Census Bureau

147. a. In 2005, the United States government collected $2.27 trillion in taxes. Express this number in scientific notation.

    b. In 2005, the population of the United States was approximately 298 million. Express this number in scientific notation.

    c. Use your scientific notation answers from parts (a) and (b) to answer this question: If the total 2005 tax collections were evenly divided among all Americans, how much would each citizen pay? Express the answer in decimal notation, rounded to the nearest dollar.

148. a. In 2004, the United States government collected $2.02 trillion in taxes. Express this number in scientific notation.

b. In 2004, the population of the United States was approximately 295 million. Express this number in scientific notation.

c. Use your scientific notation answers from parts (a) and (b) to answer this question: If the total 2004 tax collections were evenly divided among all Americans, how much would each citizen pay? Express the answer in decimal notation, rounded to the nearest dollar.

149. In 2007, the population of the United States was approximately 3.1×10^8 and each person spent about $120 per year on ice cream. Express the total annual spending on ice cream for that year in scientific notation.

150. A human brain contains 3×10^{10} neurons and a gorilla brain contains 7.5×10^9 neurons. How many times as many neurons are in the brain of a human as in the brain of a gorilla?

Use the motion formula $d = rt$, distance equals rate times time, and the fact that light travels at the rate of 1.86×10^5 miles per second, to solve Exercises 151–152.

151. If the moon is approximately 2.325×10^5 miles from Earth, how many seconds does it take moonlight to reach Earth?

152. If the sun is approximately 9.14×10^7 miles from Earth, how many seconds, to the nearest tenth of a second, does it take sunlight to reach Earth?

Writing in Mathematics

153. Explain the negative exponent rule and give an example.

154. How do you know if an exponential expression is simplified?

155. How do you know if a number is written in scientific notation?

156. Explain how to convert from scientific to decimal notation and give an example.

157. Explain how to convert from decimal to scientific notation and give an example.

158. Describe one advantage of expressing a number in scientific notation over decimal notation.

Critical Thinking Exercises

Make Sense? *In Exercises 159–162, determine whether each statement "makes sense" or "does not make sense" and explain your reasoning.*

159. There are many exponential expressions that are equal to $36x^{12}$, such as $(6x^6)^2$, $(6x^3)(6x^9)$, $36(x^3)^9$, and $6^2(x^2)^6$.

160. If 5^{-2} is raised to the third power, the result is a number between 0 and 1.

161. The population of Colorado is approximately 4.6×10^{12}.

162. I wrote a number where there is no advantage to using scientific notation instead of decimal notation.

In Exercises 163–170, determine whether each statement is true or false. If the statement is false, make the necessary change(s) to produce a true statement.

163. $4^{-2} < 4^{-3}$

164. $5^{-2} > 2^{-5}$

165. $(-2)^4 = 2^{-4}$

166. $5^2 \cdot 5^{-2} > 2^5 \cdot 2^{-5}$

167. $534.7 = 5.347 \times 10^3$

168. $\dfrac{8 \times 10^{30}}{4 \times 10^{-5}} = 2 \times 10^{25}$

169. $(7 \times 10^5) + (2 \times 10^{-3}) = 9 \times 10^2$

170. $(4 \times 10^3) + (3 \times 10^2) = 4.3 \times 10^3$

171. The mad Dr. Frankenstein has gathered enough bits and pieces (so to speak) for $2^{-1} + 2^{-2}$ of his creature-to-be. Write a fraction that represents the amount of his creature that must still be obtained.

Technology Exercises

172. Use a calculator in a fraction mode to check any five of your answers in Exercises 1–22.

173. Use a calculator to check any three of your answers in Exercises 79–90.

174. Use a calculator to check any three of your answers in Exercises 91–106.

175. Use a calculator with an $\boxed{\text{EE}}$ or $\boxed{\text{EXP}}$ key to check any four of your computations in Exercises 107–126. Display the result of the computation in scientific notation.

Review Exercises

176. Park rangers catch, tag, and then release 25 deer back into a state park. Two weeks later, they select a sample of 36 deer, 4 of which are tagged. Assuming the ratio of tagged deer in the sample holds for all deer in the park, approximately how many deer are in the park? (Section 3.2, Example 4)

177. Simplify: $24 \div 8 \cdot 3 + 28 \div (-7)$. (Section 1.8, Example 8)

178. List the whole numbers in this set:

$$\left\{ -4, -\frac{1}{5}, 0, \pi, \sqrt{16}, \sqrt{17} \right\}.$$

(Section 1.3, Example 5)

Preview Exercises

Exercises 179–181 will help you prepare for the material covered in the first section of the next chapter. In each exercise, find the product.

179. $4x^3(4x^2 - 3x + 1)$

180. $9xy(3xy^2 - y + 9)$

181. $(x + 3)(x^2 + 5)$

GROUP PROJECT

CHAPTER 6

▪ PUTTING NUMBERS INTO PERSPECTIVE

A large number can be put into perspective by comparing it with another number. For example, we put the $8.6 trillion national debt into perspective by comparing it to the number of U.S. citizens. We did the same thing with total tax collections in Exercises 147–148 of Exercise Set 6.7.

    For this project, each group member should consult an almanac, a newspaper, or the World Wide Web to find a number greater than one million. Explain to other members of the group the context in which the large number is used. Express the number in scientific notation. Then put the number into perspective by comparing it with another number.

Chapter 6 Summary

Definitions and Concepts	Examples

Section 6.1 Adding and Subtracting Polynomials

A polynomial is a single term or the sum of two or more terms containing variables with whole number exponents. A monomial is a polynomial with exactly one term; a binomial has exactly two terms; a trinomial has exactly three terms. The degree of a polynomial is the highest power of all the terms. The standard form of a polynomial is written in descending powers of the variable.

Polynomials

Monomial: $2x^5$
> Degree is 5.

Binomial: $6x^3 + 5x$
> Degree is 3.

Trinomial: $7x + 4x^2 - 5$
> Degree is 2.

To add polynomials, add like terms.

$(6x^3 + 5x^2 - 7x) + (-9x^3 + x^2 + 6x)$
$= (6x^3 - 9x^3) + (5x^2 + x^2) + (-7x + 6x)$
$= -3x^3 + 6x^2 - x$

The opposite, or additive inverse, of a polynomial is that polynomial with the sign of every coefficient changed. To subtract two polynomials, add the first polynomial and the opposite of the polynomial being subtracted.

$(5y^3 - 9y^2 - 4) - (3y^3 - 12y^2 - 5)$
$= (5y^3 - 9y^2 - 4) + (-3y^3 + 12y^2 + 5)$
$= (5y^3 - 3y^3) + (-9y^2 + 12y^2) + (-4 + 5)$
$= 2y^3 + 3y^2 + 1$

The graphs of equations defined by polynomials of degree 2, shaped like bowls or inverted bowls, can be obtained using the point-plotting method.

Graph: $y = x^2 - 1$.

x	$y = x^2 - 1$
-2	$(-2)^2 - 1 = 3$
-1	$(-1)^2 - 1 = 0$
0	$0^2 - 1 = -1$
1	$1^2 - 1 = 0$
2	$2^2 - 1 = 3$

Definitions and Concepts	**Examples**

Section 6.2 Multiplying Polynomials

Properties of Exponents

Product Rule: $b^m \cdot b^n = b^{m+n}$

Power Rule: $(b^m)^n = b^{mn}$

Products to Powers: $(ab)^n = a^n b^n$

$$x^3 \cdot x^8 = x^{3+8} = x^{11}$$
$$(x^3)^8 = x^{3 \cdot 8} = x^{24}$$
$$(-5x^2)^3 = (-5)^3(x^2)^3 = -125x^6$$

To multiply monomials, multiply coefficients and add exponents.

$$(-6x^4)(3x^{10}) = -6 \cdot 3x^{4+10} = -18x^{14}$$

To multiply a monomial and a polynomial, multiply each term of the polynomial by the monomial.

$$2x^4(3x^2 - 6x + 5)$$
$$= 2x^4 \cdot 3x^2 - 2x^4 \cdot 6x + 2x^4 \cdot 5$$
$$= 6x^6 - 12x^5 + 10x^4$$

To multiply polynomials when neither is a monomial, multiply each term of one polynomial by each term of the other polynomial. Then combine like terms.

$$(2x + 3)(5x^2 - 4x + 2)$$
$$= 2x(5x^2 - 4x + 2) + 3(5x^2 - 4x + 2)$$
$$= 10x^3 - 8x^2 + 4x + 15x^2 - 12x + 6$$
$$= 10x^3 + 7x^2 - 8x + 6$$

Section 6.3 Special Products

The FOIL method may be used when multiplying two binomials: First terms multiplied. Outside terms multiplied. Inside terms multiplied. Last terms multiplied.

F	O	I	L

$$(3x + 7)(2x - 5) = 3x \cdot 2x + 3x(-5) + 7 \cdot 2x + 7(-5)$$
$$= 6x^2 - 15x + 14x - 35$$
$$= 6x^2 - x - 35$$

The Product of the Sum and Difference of Two Terms
$$(A + B)(A - B) = A^2 - B^2$$

$$(4x + 7)(4x - 7) = (4x)^2 - 7^2$$
$$= 16x^2 - 49$$

The Square of a Binomial Sum
$$(A + B)^2 = A^2 + 2AB + B^2$$

$$(x^2 + 6)^2 = (x^2)^2 + 2 \cdot x^2 \cdot 6 + 6^2$$
$$= x^4 + 12x^2 + 36$$

The Square of a Binomial Difference
$$(A - B)^2 = A^2 - 2AB + B^2$$

$$(9x - 3)^2 = (9x)^2 - 2 \cdot 9x \cdot 3 + 3^2$$
$$= 81x^2 - 54x + 9$$

Section 6.4 Polynomials in Several Variables

To evaluate a polynomial in several variables, substitute the given value for each variable and perform the resulting computation.

Evaluate $4x^2y + 3xy - 2x$ for $x = -1$ and $y = -3$.
$$4x^2y + 3xy - 2x$$
$$= 4(-1)^2(-3) + 3(-1)(-3) - 2(-1)$$
$$= 4(1)(-3) + 3(-1)(-3) - 2(-1)$$
$$= -12 + 9 + 2 = -1$$

| **Definitions and Concepts** | **Examples** |

Section 6.4 Polynomials in Several Variables (continued)

For a polynomial in two variables, the degree of a term is the sum of the exponents on its variables. The degree of the polynomial is the highest degree of all its terms.

$$7x^2y + 12x^4y^3 - 17x^5 + 6$$

degree: $2 + 1 = 3$ | degree: $4 + 3 = 7$ | degree: 5 | degree: 0

Degree of polynomial = 7

Polynomials in several variables are added, subtracted, and multiplied using the same rules for polynomials in one variable.

$$(5x^2y^3 - xy + 4y^2) - (8x^2y^3 - 6xy - 2y^2)$$
$$= (5x^2y^3 - xy + 4y^2) + (-8x^2y^3 + 6xy + 2y^2)$$
$$= (5x^2y^3 - 8x^2y^3) + (-xy + 6xy) + (4y^2 + 2y^2)$$
$$= -3x^2y^3 + 5xy + 6y^2$$

F   O   I   L

$$(3x - 2y)(x - y) = 3x \cdot x + 3x(-y) + (-2y)x + (-2y)(-y)$$
$$= 3x^2 - 3xy - 2xy + 2y^2$$
$$= 3x^2 - 5xy + 2y^2$$

Section 6.5 Dividing Polynomials

Additional Properties of Exponents

Quotient Rule: $\dfrac{b^m}{b^n} = b^{m-n}, \quad b \neq 0$

$$\frac{x^{12}}{x^4} = x^{12-4} = x^8$$

Zero-Exponent Rule: $b^0 = 1, \quad b \neq 0$

$$(-3)^0 = 1 \qquad -3^0 = -(3^0) = -1$$

Quotients to Powers: $\left(\dfrac{a}{b}\right)^n = \dfrac{a^n}{b^n}, \quad b \neq 0$

$$\left(\frac{y^2}{4}\right)^3 = \frac{(y^2)^3}{4^3} = \frac{y^{2 \cdot 3}}{4 \cdot 4 \cdot 4} = \frac{y^6}{64}$$

To divide monomials, divide coefficients and subtract exponents.

$$\frac{-40x^{40}}{20x^{20}} = \frac{-40}{20}x^{40-20} = -2x^{20}$$

To divide a polynomial by a monomial, divide each term of the polynomial by the monomial.

$$\frac{8x^6 - 4x^3 + 10x}{2x}$$
$$= \frac{8x^6}{2x} - \frac{4x^3}{2x} + \frac{10x}{2x}$$
$$= 4x^{6-1} - 2x^{3-1} + 5x^{1-1} = 4x^5 - 2x^2 + 5$$

Section 6.6 Dividing Polynomials by Binomials

To divide a polynomial by a binomial, begin by arranging the polynomial in descending powers of the variable. If a power of a variable is missing, add that power with a coefficient of 0. Repeat the four steps—divide, multiply, subtract, bring down the next term—until the degree of the remainder is less than the degree of the divisor.

Divide: $\dfrac{10x^2 + 13x + 8}{2x + 3}$.

$$
\require{enclose}
\begin{array}{r}
5x - 1 + \dfrac{11}{2x+3} \\[4pt]
2x + 3 \enclose{longdiv}{10x^2 + 13x + 8} \\[2pt]
\underline{10x^2 + 15x} \\
-2x + 8 \\
\underline{-2x - 3} \\
11
\end{array}
$$

Definitions and Concepts	Examples

Section 6.7 Negative Exponents and Scientific Notation

Negative Exponents in Numerators and Denominators

If $b \neq 0$, $b^{-n} = \dfrac{1}{b^n}$ and $\dfrac{1}{b^{-n}} = b^n$.

$$6^{-2} = \frac{1}{6^2} = \frac{1}{36}$$

$$\frac{1}{(-2)^{-4}} = (-2)^4 = 16$$

$$\left(\frac{2}{3}\right)^{-3} = \frac{2^{-3}}{3^{-3}} = \frac{3^3}{2^3} = \frac{27}{8}$$

An exponential expression is simplified when
- Each base occurs only once.
- No parentheses appear.
- No powers are raised to powers.
- No negative or zero exponents appear.

Simplify: $\dfrac{(2x^4)^3}{x^{18}}$.

$$\frac{(2x^4)^3}{x^{18}} = \frac{2^3(x^4)^3}{x^{18}} = \frac{8x^{4\cdot3}}{x^{18}} = \frac{8x^{12}}{x^{18}} = 8x^{12-18} = 8x^{-6} = \frac{8}{x^6}$$

A positive number in scientific notation is expressed as $a \times 10^n$, where $1 \le a < 10$ and n is an integer.

Write 2.9×10^{-3} in decimal notation.

$$2.9 \times 10^{-3} = .0029 = 0.0029$$

Write 16,000 in scientific notation.

$$16{,}000 = 1.6 \times 10^4$$

Use properties of exponents with base 10

$$10^m \cdot 10^n = 10^{m+n}, \quad \frac{10^m}{10^n} = 10^{m-n}, \quad \text{and} \quad (10^m)^n = 10^{mn}$$

to perform computations with scientific notation.

$(5 \times 10^3)(4 \times 10^{-8})$
$= 5 \cdot 4 \times 10^{3-8}$
$= 20 \times 10^{-5}$
$= 2 \times 10^1 \times 10^{-5} = 2 \times 10^{-4}$

CHAPTER 6 REVIEW EXERCISES

6.1 *In Exercises 1–3, identify each polynomial as a monomial, binomial, or trinomial. Give the degree of the polynomial.*

1. $7x^4 + 9x$

2. $3x + 5x^2 - 2$

3. $16x$

In Exercises 4–8, add or subtract as indicated.

4. $(-6x^3 + 7x^2 - 9x + 3) + (14x^3 + 3x^2 - 11x - 7)$

5. $(9y^3 - 7y^2 + 5) + (4y^3 - y^2 + 7y - 10)$

6. $(5y^2 - y - 8) - (-6y^2 + 3y - 4)$

7. $(13x^4 - 8x^3 + 2x^2) - (5x^4 - 3x^3 + 2x^2 - 6)$

8. Subtract $x^4 + 7x^2 - 11x$ from $-13x^4 - 6x^2 + 5x$.

In Exercises 9–11, add or subtract as indicated.

9. Add. $\quad 7y^4 - 6y^3 + 4y^2 - 4y$
$\qquad\qquad \underline{\quad\quad y^3 - \ y^2 + 3y - 4}$

10. Subtract. $\quad 7x^2 - 9x + 2$
$\qquad\qquad\quad \underline{-(4x^2 - 2x - 7)}$

11. Subtract. $\quad 5x^3 - 6x^2 - \ 9x + 14$
$\qquad\qquad\quad \underline{-(-5x^3 + 3x^2 - 11x + \ 3)}$

In Exercises 12–13, graph each equation.

12. $y = x^2 + 3$

13. $y = 1 - x^2$

6.2 *In Exercises 14–18, simplify each expression.*

14. $x^{20} \cdot x^3$

15. $y \cdot y^5 \cdot y^8$

16. $\left(x^{20}\right)^5$

17. $(10y)^2$

18. $\left(-4x^{10}\right)^3$

In Exercises 19–27, find each product.

19. $(5x)(10x^3)$

20. $(-12y^7)(3y^4)$

21. $(-2x^5)(-3x^4)(5x^3)$

22. $7x(3x^2 + 9)$

23. $5x^3(4x^2 - 11x)$

24. $3y^2(-7y^2 + 3y - 6)$

25. $2y^5(8y^3 - 10y^2 + 1)$

26. $(x + 3)(x^2 - 5x + 2)$

27. $(3y - 2)(4y^2 + 3y - 5)$

In Exercises 28–29, use a vertical format to find each product.

28. $y^2 - 4y + 7$
$ \underline{3y - 5}$

29. $4x^3 - 2x^2 - 6x - 1$
$ \underline{2x + 3}$

6.3 *In Exercises 30–42, find each product.*

30. $(x + 6)(x + 2)$

31. $(3y - 5)(2y + 1)$

32. $(4x^2 - 2)(x^2 - 3)$

33. $(5x + 4)(5x - 4)$

34. $(7 - 2y)(7 + 2y)$

35. $(y^2 + 1)(y^2 - 1)$

36. $(x + 3)^2$

37. $(3y + 4)^2$

38. $(y - 1)^2$

39. $(5y - 2)^2$

40. $(x^2 + 4)^2$

41. $(x^2 + 4)(x^2 - 4)$

42. $(x^2 + 4)(x^2 - 5)$

43. Write a polynomial in descending powers of x that represents the area of the shaded region.

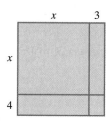

44. The parking garage shown in the figure measures 30 yards by 20 yards. The length and the width are each increased by a fixed amount, x yards. Write a trinomial that describes the area of the expanded garage.

6.4

45. Evaluate $2x^3y - 4xy^2 + 5y + 6$ for $x = -1$ and $y = 2$.

46. Determine the coefficient of each term, the degree of each term, and the degree of the polynomial:

$$4x^2y + 9x^3y^2 - 17x^4 - 12.$$

In Exercises 47–56, perform the indicated operations.

47. $(7x^2 - 8xy + y^2) + (-8x^2 - 9xy + 4y^2)$

48. $(13x^3y^2 - 5x^2y - 9x^2) - (11x^3y^2 - 6x^2y - 3x^2 + 4)$

49. $(-7x^2y^3)(5x^4y^6)$

50. $5ab^2(3a^2b^3 - 4ab)$

51. $(x + 7y)(3x - 5y)$

52. $(4xy - 3)(9xy - 1)$

53. $(3x + 5y)^2$

54. $(xy - 7)^2$

55. $(7x + 4y)(7x - 4y)$

56. $(a - b)(a^2 + ab + b^2)$

6.5 *In Exercises 57–63, simplify each expression.*

57. $\dfrac{6^{40}}{6^{10}}$

58. $\dfrac{x^{18}}{x^3}$

59. $(-10)^0$

60. -10^0

61. $400x^0$

62. $\left(\dfrac{x^4}{2}\right)^3$

63. $\left(\dfrac{-3}{2y^6}\right)^4$

In Exercises 64–68, divide and check each answer.

64. $\dfrac{-15y^8}{3y^2}$

65. $\dfrac{40x^8y^6}{5xy^3}$

66. $\dfrac{18x^4 - 12x^2 + 36x}{6x}$

67. $\dfrac{30x^8 - 25x^7 - 40x^5}{-5x^3}$

68. $\dfrac{27x^3y^2 - 9x^2y - 18xy^2}{3xy}$

6.6 *In Exercises 69–72, divide and check each answer.*

69. $\dfrac{2x^2 + 3x - 14}{x - 2}$

70. $\dfrac{2x^3 - 5x^2 + 7x + 5}{2x + 1}$

71. $\dfrac{x^3 - 2x^2 - 33x - 7}{x - 7}$

72. $\dfrac{y^3 - 27}{y - 3}$

6.7 *In Exercises 73–77, write each expression with positive exponents only and then simplify.*

73. 7^{-2}

74. $(-4)^{-3}$

75. $2^{-1} + 4^{-1}$

76. $\dfrac{1}{5^{-2}}$

77. $\left(\dfrac{2}{5}\right)^{-3}$

In Exercises 78–86, simplify each exponential expression. Assume that variables in denominators do not equal zero.

78. $\dfrac{x^3}{x^9}$

79. $\dfrac{30y^6}{5y^8}$

80. $(5x^{-7})(6x^2)$

81. $\dfrac{x^4 \cdot x^{-2}}{x^{-6}}$

82. $\dfrac{(3y^3)^4}{y^{10}}$

83. $\dfrac{y^{-7}}{(y^4)^3}$

84. $(2x^{-1})^{-3}$

85. $\left(\dfrac{x^7}{x^4}\right)^{-2}$

86. $\dfrac{(y^3)^4}{(y^{-2})^4}$

In Exercises 87–89, write each number in decimal notation without the use of exponents.

87. 2.3×10^4

88. 1.76×10^{-3}

89. 9×10^{-1}

In Exercises 90–93, write each number in scientific notation.

90. 73,900,000

91. 0.00062

92. 0.38

93. 3.8

In Exercises 94–96, perform the indicated computation. Write the answers in scientific notation.

94. $(6 \times 10^{-3})(1.5 \times 10^6)$

95. $\dfrac{2 \times 10^2}{4 \times 10^{-3}}$

96. $(4 \times 10^{-2})^2$

In Exercises 97–98, use 10^6 for one million and 10^9 for one billion to rewrite the number in each statement in scientific notation.

97. According to the Tax Foundation, in 2005, individuals and companies in the United States spent approximately $257 billion on tax preparation costs.

98. According to the Internal Revenue Service, in 2005, approximately 175 million tax returns were filed.

99. Use your scientific notation answers from Exercises 97 and 98 to answer this question: If the total 2005 tax preparation costs were evenly divided among all tax returns, how much would it cost to prepare each return? Express the answer in decimal notation, rounded to the nearest dollar.

CHAPTER 6 TEST Remember to use your Chapter Test Prep Video CD to see the worked-out solutions to the test questions you want to review.

1. Identify $9x + 6x^2 - 4$ as a monomial, binomial, or trinomial. Give the degree of the polynomial.

In Exercises 2–3, add or subtract as indicated.

2. $(7x^3 + 3x^2 - 5x - 11) + (6x^3 - 2x^2 + 4x - 13)$

3. $(9x^3 - 6x^2 - 11x - 4) - (4x^3 - 8x^2 - 13x + 5)$

4. Graph the equation: $y = x^2 - 3$. Select integers for x, starting with -3 and ending with 3.

In Exercises 5–11, find each product.

5. $(-7x^3)(5x^8)$

6. $6x^2(8x^3 - 5x - 2)$

7. $(3x + 2)(x^2 - 4x - 3)$

8. $(3y + 7)(2y - 9)$

9. $(7x + 5)(7x - 5)$

10. $(x^2 + 3)^2$

11. $(5x - 3)^2$

12. Evaluate $4x^2y + 5xy - 6x$ for $x = -2$ and $y = 3$.

In Exercises 13–15, perform the indicated operations.

13. $(8x^2y^3 - xy + 2y^2) - (6x^2y^3 - 4xy - 10y^2)$

14. $(3a - 7b)(4a + 5b)$

15. $(2x + 3y)^2$

In Exercises 16–18, divide and check each answer.

16. $\dfrac{-25x^{16}}{5x^4}$

17. $\dfrac{15x^4 - 10x^3 + 25x^2}{5x}$

18. $\dfrac{2x^3 - 3x^2 + 4x + 4}{2x + 1}$

In Exercises 19–20, write each expression with positive exponents only and then simplify.

19. 10^{-2}

20. $\dfrac{1}{4^{-3}}$

In Exercises 21–26, simplify each expression.

21. $(-3x^2)^3$

22. $\dfrac{20x^3}{5x^8}$

23. $(-7x^{-8})(3x^2)$

24. $\dfrac{(2y^3)^4}{y^8}$

25. $(5x^{-4})^{-2}$

26. $\left(\dfrac{x^{10}}{x^5}\right)^{-3}$

27. Write 3.7×10^{-4} in decimal notation.

28. Write 7,600,000 in scientific notation.

In Exercises 29–30, perform the indicated computation. Write the answers in scientific notation.

29. $(4.1 \times 10^2)(3 \times 10^{-5})$

30. $\dfrac{8.4 \times 10^6}{4 \times 10^{-2}}$

31. Write a polynomial in descending powers of x that represents the area of the figure.

CUMULATIVE REVIEW EXERCISES (CHAPTERS 1–6)

In Exercises 1–2, perform the indicated operation or operations.

1. $(-7)(-5) \div (12 - 9)$

2. $(3 - 7)^2(9 - 11)^3$

3. What is the difference in elevation between a plane flying 14,300 feet above sea level and a submarine traveling 750 feet below sea level?

In Exercises 4–5, solve each equation.

4. $2(x + 3) + 2x = x + 4$

5. $\dfrac{x}{5} - \dfrac{1}{3} = \dfrac{x}{10} - \dfrac{1}{2}$

6. The length of a rectangular sign is 2 feet less than three times its width. If the perimeter of the sign is 28 feet, what are its dimensions?

7. Solve $7 - 8x \le -6x - 5$ and graph the solution set on a number line.

8. You and a friend live 72 miles apart. You decide to meet by riding your bikes directly toward each other on the same road. If you bike at 13 miles per hour and your friend bikes at 11 miles per hour, in how many hours will you meet?

9. If a 20-pound bag of fertilizer covers 5000 square feet, how many pounds are needed to cover an area of 26,000 square feet? How many bags of fertilizer are needed?

10. Graph $y = -\frac{2}{5}x + 2$ using the slope and y-intercept.

11. Graph $x - 2y = 4$ using intercepts.

12. Find the slope of the line passing through the points $(-3, 2)$ and $(2, -4)$. Is the line rising, falling, horizontal, or vertical?

13. The slope of a line is -2 and the line passes through the point $(3, -1)$. Write the line's equation in point-slope form and slope-intercept form.

In Exercises 14–15, solve each system by the method of your choice.

14. $3x + 2y = 10$
$4x - 3y = -15$

15. $2x + 3y = -6$
$y = 3x - 13$

16. You are choosing between two long-distance telephone plans. One has a monthly fee of $15 with a charge of $0.05 per minute for all long-distance calls. The other plan has a monthly fee of $5 with a charge of $0.07 per minute for all long-distance calls. For how many minutes of long-distance calls will the costs for the two plans be the same? What will be the cost for each plan?

17. Graph the solution set for the system of linear inequalities:

$$2x + 5y \le 10$$
$$x - y \ge 4.$$

18. Subtract: $(9x^5 - 3x^3 + 2x - 7) - (6x^5 + 3x^3 - 7x - 9)$.

19. Divide: $\dfrac{x^3 + 3x^2 + 5x + 3}{x + 1}$.

20. Simplify: $\dfrac{(3x^2)^4}{x^{10}}$.

Factoring Polynomials

In Chapter 6, we used areas of large rectangular regions and the smaller rectangles that made up those regions to picture polynomial multiplication. We can also apply areas of rectangles to reverse this process, visualizing polynomials as products. When it comes to polynomials, thinking within boxes is not always a bad idea.

The role of rectangles to visualize reversing polynomial multiplication is developed throughout Section 7.4. These rectangles can be used in landscape design, as illustrated in Exercises 86–87 in Exercise Set 7.6.

The Greatest Common Factor and Factoring by Grouping

Objectives

1. Find the greatest common factor.

2. Factor out the greatest common factor of a polynomial.

3. Factor by grouping.

A two-year-old boy is asked, "Do you have a brother?" He answers, "Yes." "What is your brother's name?" "Tom." Asked if Tom has a brother, the two-year-old replies, "No." The child can go in the direction from self to brother, but he cannot reverse this direction and move from brother back to self.

As our intellects develop, we learn to reverse the direction of our thinking. Reversibility of thought is found throughout algebra. For example, we can multiply polynomials and show that

$$5x(2x + 3) = 10x^2 + 15x.$$

We can also reverse this process and express the resulting polynomial as

$$10x^2 + 15x = 5x(2x + 3).$$

Factoring a polynomial containing the sum of monomials means finding an equivalent expression that is a product.

Factoring $10x^2 + 15x$

Sum of monomials   Equivalent expression that is a product

$$10x^2 + 15x = 5x(2x + 3)$$

The factors of $10x^2 + 15x$ are $5x$ and $2x + 3$.

In this chapter, we will be factoring over the set of integers, meaning that the coefficients in the factors are integers. Polynomials that cannot be factored using integer coefficients are called **prime polynomials** over the set of integers.

Factoring Out the Greatest Common Factor

We use the distributive property to multiply a monomial and a polynomial of two or more terms. When we factor, we reverse this process, expressing the polynomial as a product.

Multiplication	**Factoring**
$a(b + c) = ab + ac$	$ab + ac = a(b + c)$

Here is a specific example:

Multiplication	**Factoring**
$5x(2x + 3)$	$10x^2 + 15x$
$= 5x \cdot 2x + 5x \cdot 3$	$= 5x \cdot 2x + 5x \cdot 3$
$= 10x^2 + 15x$	$= 5x(2x + 3).$

In the process of finding an equivalent expression for $10x^2 + 15x$ that is a product, we used the fact that $5x$ is a factor of the monomials $10x^2$ and $15x$. The factoring on the right shows that $5x$ is a *common factor* for both terms of the binomial $10x^2 + 15x$.

1 Find the greatest common factor.

In any factoring problem, the first step is to look for the *greatest common factor*. The **greatest common factor**, abbreviated GCF, is an expression of the highest degree that divides each term of the polynomial. Can you see that $5x$ is the greatest common factor of $10x^2 + 15x$? 5 is the greatest integer that divides both 10 and 15. Furthermore, x is the greatest expression that divides both x^2 and x.

The variable part of the greatest common factor always contains the smallest power of a variable that appears in all terms of the polynomial. For example, consider the polynomial

$$10x^2 + 15x.$$

> x^1, or x, is the variable raised to the smaller exponent.

We see that x is the variable part of the greatest common factor, $5x$.

EXAMPLE 1 Finding the Greatest Common Factor

Find the greatest common factor of each list of monomials:

a. $6x^3$ and $10x^2$ **b.** $15y^5$, $-9y^4$, and $27y^3$ **c.** x^5y^3, x^4y^4, and x^3y^2.

Solution Use numerical coefficients to determine the coefficient of the GCF. Use variable factors to determine the variable factor of the GCF.

> 2 is the greatest integer that divides 6 and 10.

a. $6x^3$ and $10x^2$

> x^2 is the variable raised to the smaller exponent.

We see that 2 is the coefficient of the GCF and x^2 is the variable factor of the GCF. Thus, the GCF of $6x^3$ and $10x^2$ is $2x^2$.

> 3 is the greatest integer that divides 15, −9, and 27.

b. $15y^5$, $-9y^4$, and $27y^3$

> y^3 is the variable raised to the smallest exponent.

We see that 3 is the coefficient of the GCF and y^3 is the variable factor of the GCF. Thus, the GCF of $15y^5$, $-9y^4$, and $27y^3$ is $3y^3$.

> x^3 is the variable, x, raised to the smallest exponent.

c. x^5y^3, x^4y^4, and x^3y^2

> y^2 is the variable, y, raised to the smallest exponent.

Because all terms have coefficients of 1, 1 is the greatest integer that divides these coefficients. Thus, 1 is the coefficient of the GCF. The voice balloons show that x^3 and y^2 are the variable factors of the GCF. Thus, the GCF of x^5y^3, x^4y^4, and x^3y^2 is x^3y^2. ◼

✓ **CHECK POINT 1** Find the greatest common factor of each list of monomials:

a. $18x^3$ and $15x^2$ **b.** $-20x^2$, $12x^4$, and $40x^3$

c. x^4y, x^3y^2, and x^2y.

2 Factor out the greatest common factor of a polynomial.

When we factor a monomial from a polynomial, we determine the greatest common factor of all terms in the polynomial. Sometimes there may not be a GCF other than 1. When a GCF other than 1 exists, we use the following procedure:

Factoring a Monomial From a Polynomial

1. Determine the greatest common factor of all terms in the polynomial.
2. Express each term as the product of the GCF and its other factor.
3. Use the distributive property to factor out the GCF.

EXAMPLE 2   Factoring Out the Greatest Common Factor

Factor: $5x^2 + 30$.

Solution   The GCF of $5x^2$ and 30 is 5.

$$5x^2 + 30$$
$$= 5 \cdot x^2 + 5 \cdot 6 \qquad \text{Express each term as the product of the GCF and its other factor.}$$
$$= 5(x^2 + 6) \qquad \text{Factor out the GCF.}$$

Because factoring reverses the process of multiplication, all factorizations can be checked by multiplying.

$$5(x^2 + 6) = 5 \cdot x^2 + 5 \cdot 6 = 5x^2 + 30$$

The factorization is correct because multiplication gives us the original polynomial. ■

☑ **CHECK POINT 2**   Factor: $6x^2 + 18$.

EXAMPLE 3   Factoring Out the Greatest Common Factor

Factor: $18x^3 + 27x^2$.

Solution   We begin by determining the greatest common factor.

9 is the greatest integer that divides 18 and 27.

$$18x^3 \qquad \text{and} \qquad 27x^2$$

x^2 is the variable raised to the smaller exponent.

The GCF of the two terms in the polynomial is $9x^2$.

$$18x^3 + 27x^2$$
$$= 9x^2 \cdot 2x + 9x^2 \cdot 3 \qquad \text{Express each term as the product of the GCF and its other factor.}$$
$$= 9x^2(2x + 3) \qquad \text{Factor out the GCF.}$$

We can check this factorization by multiplying $9x^2$ and $2x + 3$, obtaining the original polynomial as the answer. ■

☑ **CHECK POINT 3**   Factor: $25x^2 + 35x^3$.

EXAMPLE 4 Factoring Out the Greatest Common Factor

Factor: $16x^5 - 12x^4 + 4x^3$.

Solution First, determine the greatest common factor.

4 is the greatest integer that divides 16, −12, and 4.

$$16x^5, \quad -12x^4, \quad \text{and} \quad 4x^3$$

x^3 is the variable raised to the smallest exponent.

The GCF of the three terms of the polynomial is $4x^3$.

$$16x^5 - 12x^4 + 4x^3$$

$$= 4x^3 \cdot 4x^2 - 4x^3 \cdot 3x + 4x^3 \cdot 1$$ Express each term as the product of the GCF and its other factor.

You can obtain the factors shown in black by dividing each term of the given polynomial by $4x^3$, the GCF:

$$\frac{16x^5}{4x^3} = 4x^2 \qquad \frac{12x^4}{4x^3} = 3x \qquad \frac{4x^3}{4x^3} = 1.$$

$$= 4x^3(4x^2 - 3x + 1)$$ Factor out the GCF.

Don't leave out the 1.

✓ **CHECK POINT 4** Factor: $15x^5 + 12x^4 - 27x^3$.

EXAMPLE 5 Factoring Out the Greatest Common Factor

Factor: $27x^2y^3 - 9xy^2 + 81xy$.

Solution First, determine the greatest common factor.

9 is the greatest integer that divides 27, −9, and 81.

$$27x^2y^3, \quad -9xy^2, \quad \text{and} \quad 81xy$$

The variables raised to the smallest exponents are x and y.

The GCF of the three terms of the polynomial is $9xy$.

$$27x^2y^3 - 9xy^2 + 81xy$$

$$= 9xy \cdot 3xy^2 - 9xy \cdot y + 9xy \cdot 9$$ Express each term as the product of the GCF and its other factor.

You can obtain the factors shown in black by dividing each term of the given polynomial by $9xy$, the GCF:

$$\frac{27x^2y^3}{9xy} = 3xy^2 \qquad \frac{9xy^2}{9xy} = y \qquad \frac{81xy}{9xy} = 9.$$

$$= 9xy(3xy^2 - y + 9)$$ Factor out the GCF.

✓ **CHECK POINT 5** Factor: $8x^3y^2 - 14x^2y + 2xy$.

3 Factor by grouping.

Factoring by Grouping

Up to now, we have factored a monomial from a polynomial. By contrast, in our next example, the greatest common factor of the polynomial is a binomial.

EXAMPLE 6 Factoring Out the Greatest Common Binomial Factor

Factor:

a. $x^2(x + 3) + 5(x + 3)$ **b.** $x(y + 1) - 2(y + 1)$.

Solution Let's identify the common binomial factor in each part of the problem.

$$x^2(x + 3) \quad \text{and} \quad 5(x + 3) \qquad\qquad x(y + 1) \quad \text{and} \quad -2(y + 1)$$

The GCF, a binomial, is $x + 3$.   The GCF, a binomial, is $y + 1$.

We factor out these common binomial factors as follows.

a. $x^2(x + 3) + 5(x + 3)$

$= (x + 3)x^2 + (x + 3)5$ Express each term as the product of the GCF and its other factor, in that order. Hereafter, we omit this step.

$= (x + 3)(x^2 + 5)$ Factor out the GCF, $x + 3$.

b. $x(y + 1) - 2(y + 1)$ The GCF is $y + 1$.

$= (y + 1)(x - 2)$ Factor out the GCF.   ■

☑ **CHECK POINT 6** Factor:

a. $x^2(x + 1) + 7(x + 1)$ **b.** $x(y + 4) - 7(y + 4)$.

Some polynomials have only a greatest common factor of 1. However, by a suitable grouping of the terms, it still may be possible to factor. This process, called **factoring by grouping**, is illustrated in Example 7.

EXAMPLE 7 Factoring by Grouping

Factor: $x^3 + 4x^2 + 3x + 12$.

Solution There is no factor other than 1 common to all four terms. However, we can group terms that have a common factor:

$$\boxed{x^3 + 4x^2} \quad + \quad \boxed{3x + 12}.$$

Common factor is x^2.   Common factor is 3.

We now factor the given polynomial as follows:

$x^3 + 4x^2 + 3x + 12$

$= (x^3 + 4x^2) + (3x + 12)$ Group terms with common factors.

$= x^2(x + 4) + 3(x + 4)$ Factor out the greatest common factor from the grouped terms. The remaining two terms have $x + 4$ as a common binomial factor.

$= (x + 4)(x^2 + 3).$ Factor out the GCF, $x + 4$.

Thus, $x^3 + 4x^2 + 3x + 12 = (x + 4)(x^2 + 3)$. Check the factorization by multiplying the right side of the equation using the FOIL method. Because the factorization is correct, you should obtain the original polynomial.   ■

Discover for Yourself

In Example 7, group the terms as follows:

$(x^3 + 3x) + (4x^2 + 12)$.

Factor out the greatest common factor from each group and complete the factoring process. Describe what happens. What can you conclude?

☑ **CHECK POINT 7**   Factor: $x^3 + 5x^2 + 2x + 10$.

Factoring by Grouping

1. Group terms that have a common monomial factor. There will usually be two groups. Sometimes the terms must be rearranged.
2. Factor out the common monomial factor from each group.
3. Factor out the remaining common binomial factor (if one exists).

EXAMPLE 8   Factoring by Grouping

Factor: $xy + 5x - 4y - 20$.

Solution   There is no factor other than 1 common to all four terms. However, we can group terms that have a common factor:

$$\boxed{xy + 5x} \quad + \quad \boxed{-4y - 20}.$$

Common factor is x:
$xy + 5x = x(y + 5)$.

Use -4, rather than 4, as the common factor: $-4y - 20 = -4(y + 5)$. In this way, the common binomial factor, $y + 5$, appears.

The voice balloons illustrate that it is sometimes necessary to factor out a negative number from a grouping to obtain a common binomial factor for the two groupings. We now factor the given polynomial as follows:

$$xy + 5x - 4y - 20$$
$$= x(y + 5) - 4(y + 5) \quad \text{Factor } x \text{ and } -4, \text{ respectively, from each grouping.}$$
$$= (y + 5)(x - 4). \quad \text{Factor out the GCF, } y + 5.$$

Thus, $xy + 5x - 4y - 20 = (y + 5)(x - 4)$. Using the commutative property of multiplication, the factorization can also be expressed as $(x - 4)(y + 5)$. Multiply these factors using the FOIL method to verify that, regardless of the order, these are the correct factors.   ▬

☑ **CHECK POINT 8**   Factor: $xy + 3x - 5y - 15$.

7.1 EXERCISE SET   *MyMathLab*   Math XL   PRACTICE   WATCH   DOWNLOAD   READ   REVIEW

Practice Exercises

In Exercises 1–12 find the greatest common factor of each list of monomials.

1. 4 and $8x$

2. 5 and $15x$

3. $12x^2$ and $8x$

4. $20x^2$ and $15x$

5. $-2x^4$ and $6x^3$

6. $-3x^4$ and $6x^3$

7. $9y^5$, $18y^2$, and $-3y$

8. $10y^5$, $20y^2$, and $-5y$

9. xy, xy^2, and xy^3

10. x^2y, $3x^3y$, and $6x^2$

11. $16x^5y^4$, $8x^6y^3$, and $20x^4y^5$

12. $18x^5y^4$, $6x^6y^3$, and $12x^4y^5$

In Exercises 13–48, factor each polynomial using the greatest common factor. If there is no common factor other than 1 and the polynomial cannot be factored, so state.

13. $8x + 8$

14. $9x + 9$

15. $4y - 4$

16. $5y - 5$

17. $5x + 30$

18. $10x + 30$

19. $30x - 12$

20. $32x - 24$

21. $x^2 + 5x$

22. $x^2 + 6x$

23. $18y^2 + 12$

24. $20y^2 + 15$

25. $14x^3 + 21x^2$

26. $6x^3 + 15x^2$

27. $13y^2 - 25y$

28. $11y^2 - 30y$

29. $9y^4 + 27y^6$

30. $10y^4 + 15y^6$

31. $8x^2 - 4x^4$

32. $12x^2 - 4x^4$

33. $12y^2 + 16y - 8$

34. $15y^2 - 3y + 9$

35. $9x^4 + 18x^3 + 6x^2$

36. $32x^4 + 2x^3 + 8x^2$

37. $100y^5 - 50y^3 + 100y^2$

38. $26y^5 - 13y^3 + 39y^2$

39. $10x - 20x^2 + 5x^3$

40. $6x - 4x^2 + 2x^3$

41. $11x^2 - 23$

42. $12x^2 - 25$

43. $6x^3y^2 + 9xy$

44. $4x^2y^3 + 6xy$

45. $30x^2y^3 - 10xy^2 + 20xy$

46. $27x^2y^3 - 18xy^2 + 45x^2y$

47. $32x^3y^2 - 24x^3y - 16x^2y$

48. $18x^3y^2 - 12x^3y - 24x^2y$

In Exercises 49–60, factor each polynomial using the greatest common binomial factor.

49. $x(x + 5) + 3(x + 5)$

50. $x(x + 7) + 10(x + 7)$

51. $x(x + 2) - 4(x + 2)$

52. $x(x + 3) - 8(x + 3)$

53. $x(y + 6) - 7(y + 6)$

54. $x(y + 9) - 11(y + 9)$

55. $3x(x + y) - (x + y)$

56. $7x(x + y) - (x + y)$

57. $4x(3x + 1) + 3x + 1$

58. $5x(2x + 1) + 2x + 1$

59. $7x^2(5x + 4) + 5x + 4$

60. $9x^2(7x + 2) + 7x + 2$

In Exercises 61–78, factor by grouping.

61. $x^2 + 2x + 4x + 8$

62. $x^2 + 3x + 5x + 15$

63. $x^2 + 3x - 5x - 15$

64. $x^2 + 7x - 4x - 28$

65. $x^3 - 2x^2 + 5x - 10$

66. $x^3 - 3x^2 + 4x - 12$

67. $x^3 - x^2 + 2x - 2$

68. $x^3 + 6x^2 - 2x - 12$

69. $xy + 5x + 9y + 45$

70. $xy + 6x + 2y + 12$

71. $xy - x + 5y - 5$

72. $xy - x + 7y - 7$

73. $3x^2 - 6xy + 5xy - 10y^2$

74. $10x^2 - 12xy + 35xy - 42y^2$

75. $3x^3 - 2x^2 - 6x + 4$

76. $4x^3 - x^2 - 12x + 3$

77. $x^2 - ax - bx + ab$

78. $x^2 + ax + bx + ab$

Practice PLUS

In Exercises 79–86, factor each polynomial.

79. $24x^3y^3z^3 + 30x^2y^2z + 18x^2yz^2$

80. $16x^2y^2z^2 + 32x^2yz^2 + 24x^2yz$

81. $x^3 - 4 + 3x^3y - 12y$

82. $x^3 - 5 + 2x^3y - 10y$

83. $4x^5(x + 1) - 6x^3(x + 1) - 8x^2(x + 1)$

84. $8x^5(x + 2) - 10x^3(x + 2) - 2x^2(x + 2)$

85. $3x^5 - 3x^4 + x^3 - x^2 + 5x - 5$

86. $7x^5 - 7x^4 + x^3 - x^2 + 3x - 3$

The figures for Exercises 87–88 show one or more circles drawn inside a square. Write a polynomial that represents the shaded area in each figure. Then factor the polynomial.

87.

88.

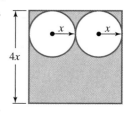

Application Exercises

89. An explosion causes debris to rise vertically with an initial velocity of 64 feet per second. The polynomial $64x - 16x^2$ describes the height of the debris above the ground, in feet, after x seconds.

   a. Find the height of the debris after 3 seconds.

   b. Factor the polynomial.

   c. Use the factored form of the polynomial in part (b) to find the height of the debris after 3 seconds. Do you get the same answer as you did in part (a)? If so, does this prove that your factorization is correct? Explain.

90. An explosion causes debris to rise vertically with an initial velocity of 72 feet per second. The polynomial $72x - 16x^2$ describes the height of the debris above the ground, in feet, after x seconds.

   a. Find the height of the debris after 4 seconds.

b. Factor the polynomial.

c. Use the factored form of the polynomial in part (b) to find the height of the debris after 4 seconds. Do you get the same answer as you did in part (a)? If so, does this prove that your factorization is correct? Explain

In Exercises 91–92, write a polynomial for the length of each rectangle.

91.

Area is $5x^4 - 10x$ square units. — $5x$

?

92.

Area is $7x^4 - 14x$ square units. — $7x$

?

Writing in Mathematics

93. What is factoring?

94. What is a prime polynomial?

95. Explain how to find the greatest common factor of a list of terms. Give an example with your explanation.

96. Use an example and explain how to factor out the greatest common factor of a polynomial.

97. Suppose that a polynomial contains four terms and can be factored by grouping. Explain how to obtain the factorization.

98. Write a sentence that uses the word "factor" as a noun. Then write a sentence that uses the word "factor" as a verb.

Critical Thinking Exercises

Make Sense? *In Exercises 99–102, determine whether each statement "makes sense" or "does not make sense" and explain your reasoning.*

99. After factoring $20x^3 + 8x^2$ and $20x^3 + 10x$, I noticed that I factored the monomial $20x^3$ in two different ways.

100. After I've factored a polynomial, my answer cannot always be checked by multiplication.

101. The word *greatest* in greatest common factor is helpful because it tells me to look for the greatest power of a variable appearing in all terms.

102. You grouped the polynomial's terms using different groupings than I did, yet we both obtained the same factorization.

In Exercises 103–106, determine whether each statement is true or false. If the statement is false, make the necessary change(s) to produce a true statement.

103. Since the GCF of $9x^3 + 6x^2 + 3x$ is $3x$, it is not necessary to write the 1 when $3x$ is factored from the last term.

104. $a(x - 7) + b(7 - x) = a(x - 7) + b(-1)(x - 7)$
$$= a(x - 7) - b(x - 7)$$
$$= (x - 7)(a - b)$$

105. $a^2 + b^2 = a^2 + ab - ab + b^2$
$$= a(a + b) - b(a + b)$$
$$= (a + b)(a - b)$$

106. $-4x^2 + 12x$ can be factored as $-4x(x - 3)$ or $4x(-x + 3)$.

107. Suppose you receive x dollars in January. Each month thereafter, you receive $100 more than you received the month before. Write a factored polynomial that describes the total dollar amount you receive from January through April.

In Exercises 108–109, write a polynomial that fits the given description. Do not use a polynomial that appears in this section or in the exercise set.

108. The polynomial has four terms and can be factored using a greatest common factor that has both a coefficient and a variable.

109. The polynomial has four terms and can be factored by grouping.

Technology Exercises

In Exercises 110–112, use a graphing utility to graph each side of the equation in the same viewing rectangle. Do the graphs coincide? If so, this means that the polynomial on the left side has been factored correctly. If not, factor the polynomial correctly and then use your graphing utility to verify the factorization.

110. $-3x - 6 = -3(x - 2)$

111. $x^2 - 2x + 5x - 10 = (x - 2)(x - 5)$

112. $x^2 + 2x + x + 2 = x(x + 2) + 1$

Review Exercises

113. Multiply: $(x + 7)(x + 10)$. (Section 6.3, Example 1)

114. Solve the system by graphing:
$$2x - y = -4$$
$$x - 3y = 3.$$
(Section 5.1, Example 2)

115. Write the point-slope form of the equation of the line passing through $(-7, 2)$ and $(-4, 5)$. Then use the point-slope form of the equation to write the slope-intercept form of the equation. (Section 4.5, Example 2)

Preview Exercises

Exercises 116–118 will help you prepare for the material covered in the next section.

116. Find two factors of 8 whose sum is 6.

117. Find two factors of 6 whose sum is -5.

118. Find two factors of -35 whose sum is 2.

Factoring Trinomials Whose Leading Coefficient Is 1

Objective

1 Factor trinomials of the form $x^2 + bx + c$.

Not afraid of heights and cutting-edge excitement? How about sky diving? Behind your exhilarating experience is the world of algebra. After you jump from the airplane, your height above the ground at every instant of your fall can be described by a formula involving a variable that is squared. At a height of approximately 2000 feet, you'll need to open your parachute. How can you determine when you must do so?

    The answer to this critical question involves using the factoring technique presented in this section. In Section 7.6, in which applications are discussed, this technique is applied to models involving the height of any free-falling object—in this case, you.

1 Factor trinomials of the form $x^2 + bx + c$.

A Strategy for Factoring $x^2 + bx + c$

In Section 6.3, we used the FOIL method to multiply two binomials. The product was often a trinomial. The following are some examples:

Factored Form	F	O	I	L	Trinomial Form

$$(x + 3)(x + 4) = x^2 + 4x + 3x + 12 = x^2 + 7x + 12$$
$$(x - 3)(x - 4) = x^2 - 4x - 3x + 12 = x^2 - 7x + 12$$
$$(x + 3)(x - 5) = x^2 - 5x + 3x - 15 = x^2 - 2x - 15.$$

Observe that each trinomial is of the form $x^2 + bx + c$, where the coefficient of the squared term, called the **leading coefficient**, is 1. Our goal in this section is to start with the trinomial form and, assuming that it is factorable, return to the factored form.

    The first FOIL multiplication shown above indicates that $(x + 3)(x + 4) = x^2 + 7x + 12$. Let's reverse the sides of this equation:

$$x^2 + 7x + 12 = (x + 3)(x + 4).$$

We can make several important observations about the factors on the right side.

$$x^2 + 7x + 12 = (x + 3)(x + 4)$$

The first term of each factor is x. The product of the First terms is $x \cdot x = x^2$.

$$x^2 + 7x + 12 = (x + 3)(x + 4)$$

3 and 4 are factors of 12. The product of the Last terms is $3 \cdot 4 = 12$.

$$x^2 + 7x + 12 = (x + 3)(x + 4)$$

I: $3x$
O: $4x$

The sum of the Outside and Inside products is $4x + 3x = 7x$.

These observations provide us with a procedure for factoring $x^2 + bx + c$.

A Strategy for Factoring $x^2 + bx + c$

1. Enter x as the first term of each factor.

$$(x \quad)(x \quad) = x^2 + bx + c$$

2. List pairs of factors of the constant c.

3. Try various combinations of these factors as the second term in each set of parentheses. Select the combination in which the sum of the Outside and Inside products is equal to bx.

4. Check your work by multiplying the factors using the FOIL method. You should obtain the original trinomial.

If none of the possible combinations yield an Outside product and an Inside product whose sum is equal to bx, the trinomial cannot be factored using integers and is called **prime** over the set of integers.

EXAMPLE 1 Factoring a Trinomial in $x^2 + bx + c$ Form

Factor: $x^2 + 6x + 8$.

Solution

Step 1. Enter x as the first term of each factor.

$$x^2 + 6x + 8 = (x \quad)(x \quad)$$

Step 2. List pairs of factors of the constant, 8.

Factors of 8	8, 1	4, 2	$-8, -1$	$-4, -2$

Step 3. Try various combinations of these factors. The correct factorization of $x^2 + 6x + 8$ is the one in which the sum of the Outside and Inside products is equal to $6x$. Here is a list of the possible factorizations:

Possible Factorizations of $x^2 + 6x + 8$	Sum of Outside and Inside Products (Should Equal 6x)	
$(x + 8)(x + 1)$	$x + 8x = 9x$	
$(x + 4)(x + 2)$	$2x + 4x = 6x$	This is the required middle term.
$(x - 8)(x - 1)$	$-x - 8x = -9x$	
$(x - 4)(x - 2)$	$-2x - 4x = -6x$	

Thus, $x^2 + 6x + 8 = (x + 4)(x + 2)$.

Step 4. Check this result by multiplying the right side using the FOIL method. You should obtain the original trinomial. Because of the commutative property, the factorization can also be expressed as

$$x^2 + 6x + 8 = (x + 2)(x + 4).$$

Using Technology

Graphic and Numeric Connections

If a polynomial contains one variable, a graphing utility can be used to check its factorization. For example, the factorization in Example 1

$$x^2 + 6x + 8 = (x + 4)(x + 2)$$

can be checked graphically or numerically.

Graphic Check

Use the GRAPH feature. Graph $y_1 = x^2 + 6x + 8$ and $y_2 = (x + 4)(x + 2)$ on the same screen. Because the graphs are identical, the factorization appears to be correct.

$y_1 = x^2 + 6x + 8$

$y_2 = (x + 4)(x + 2)$

$[-7, 1, 1]$ by $[-2, 12, 1]$

Numeric Check

Use the TABLE feature. Enter $y_1 = x^2 + 6x + 8$ and $y_2 = (x + 4)(x + 2)$ and press TABLE. Two columns of values are shown, one for y_1 and one for y_2. Because the corresponding values are equal regardless of how far up or down we scroll, the factorization is correct.

X	Y₁	Y₂
-3	-1	-1
-2	0	0
-1	3	3
0	8	8
1	15	15
2	24	24
3	35	35

X = -3

☑ **CHECK POINT 1**   Factor: $x^2 + 5x + 6$.

EXAMPLE 2   Factoring a Trinomial in $x^2 + bx + c$ Form

Factor: $x^2 - 5x + 6$.

Solution

Step 1. Enter x as the first term of each factor.

$$x^2 - 5x + 6 = (x \quad)(x \quad)$$

Step 2. List pairs of factors of the constant, 6.

Factors of 6	6, 1	3, 2	-6, -1	-3, -2

Step 3. Try various combinations of these factors. The correct factorization of $x^2 - 5x + 6$ is the one in which the sum of the Outside and Inside products is equal to $-5x$. Here is a list of the possible factorizations:

Possible Factorizations of $x^2 - 5x + 6$	Sum of Outside and Inside Products (Should Equal $-5x$)
$(x + 6)(x + 1)$	$x + 6x = 7x$
$(x + 3)(x + 2)$	$2x + 3x = 5x$
$(x - 6)(x - 1)$	$-x - 6x = -7x$
$(x - 3)(x - 2)$	$-2x - 3x = -5x$

This is the required middle term.

Study Tip

To factor $x^2 + bx + c$ when c is positive, find two numbers with the same sign as the middle term.

$x^2 + 6x + 8 = (x + 2)(x + 4)$

Same signs

$x^2 - 5x + 6 = (x - 3)(x - 2)$

Same signs

Thus, $x^2 - 5x + 6 = (x - 3)(x - 2)$. Verify this result using the FOIL method. ■

In factoring a trinomial of the form $x^2 + bx + c$, you can speed things up by listing the factors of c and then finding their sums. We are interested in a sum of b. For example, in factoring $x^2 - 5x + 6$, we are interested in the factors of 6 whose sum is -5.

Factors of 6	6, 1	3, 2	−6, −1	−3, −2
Sum of Factors	7	5	−7	−5

This is the desired sum.

Thus, $x^2 - 5x + 6 = (x - 3)(x - 2)$.

✓ **CHECK POINT 2** Factor: $x^2 - 6x + 8$.

EXAMPLE 3 **Factoring a Trinomial in $x^2 + bx + c$ Form**

Factor: $x^2 + 2x - 35$.

Solution

Step 1. Enter x as the first term of each factor.

$$x^2 + 2x - 35 = (x \quad)(x \quad)$$

To find the second term of each factor, we must find two integers whose product is -35 and whose sum is 2.

Step 2. List pairs of factors of the constant, -35.

Factors of −35	−35, 1	−7, 5	35, −1	7, −5

Step 3. Try various combinations of these factors. We are looking for the pair of factors whose sum is 2.

Factors of −35	−35, 1	−7, 5	35, −1	7, −5
Sum of Factors	−34	−2	34	2

This is the desired sum.

Thus, $x^2 + 2x - 35 = (x + 7)(x - 5)$.

Step 4. Verify the factorization using the FOIL method.

$$(x + 7)(x - 5) = x^2 - 5x + 7x - 35 = x^2 + 2x - 35$$

Because the product of the factors is the original polynomial, the factorization is correct. ∎

✓ **CHECK POINT 3** Factor: $x^2 + 3x - 10$.

EXAMPLE 4 **Factoring a Trinomial Whose Leading Coefficient Is 1**

Factor: $y^2 - 2y - 99$.

Solution

Step 1. Enter y as the first term of each factor.

$$y^2 - 2y - 99 = (y \quad)(y \quad)$$

To find the second term of each factor, we must find two integers whose product is -99 and whose sum is -2.

Study Tip

To factor $x^2 + bx + c$ when c is negative, find two numbers with opposite signs whose sum is the coefficient of the middle term.

$$x^2 + 2x - 35 = (x + 7)(x - 5)$$

Negative   Opposite signs

$$y^2 - 2y - 99 = (y - 11)(y + 9)$$

Negative   Opposite signs

Step 2. List pairs of factors of the constant, -99.

Factors of -99	$-99, 1$	$-11, 9$	$-33, 3$	$99, -1$	$11, -9$	$33, -3$

Step 3. Try various combinations of these factors. In order to factor $y^2 - 2y - 99$, we are interested in the pair of factors of -99 whose sum is -2.

Factors of -99	$-99, 1$	$-11, 9$	$-33, 3$	$99, -1$	$11, -9$	$33, -3$
Sum of Factors	-98	-2	-30	98	2	30

This is the desired sum.

Thus, $y^2 - 2y - 99 = (y - 11)(y + 9)$. Verify this result using the FOIL method. ■

☑ **CHECK POINT 4**   Factor: $y^2 - 6y - 27$.

EXAMPLE 5   Trying to Factor a Trinomial in $x^2 + bx + c$ Form

Factor: $x^2 + x - 5$.

Solution

Step 1. Enter x as the first term of each factor.

$$x^2 + x - 5 = (x \quad)(x \quad)$$

To find the second term of each factor, we must find two integers whose product is -5 and whose sum is 1.

Steps 2 and 3. List pairs of factors of the constant, -5, and try various combinations of these factors. We are interested in a pair of factors whose sum is 1.

Factors of -5	$-5, 1$	$5, -1$
Sum of Factors	-4	4

No pair gives the desired sum, 1.

Because neither pair has a sum of 1, $x^2 + x - 5$ cannot be factored using integers. This trinomial is prime. ■

☑ **CHECK POINT 5**   Factor: $x^2 + x - 7$.

EXAMPLE 6   Factoring a Trinomial in Two Variables

Factor: $x^2 - 5xy + 6y^2$.

Solution

Step 1. Enter x as the first term of each factor. Because the last term of the trinomial contains y^2, the second term of each factor must contain y.

$$x^2 - 5xy + 6y^2 = (x \quad ?y)(x \quad ?y)$$

The question marks indicate that we are looking for the coefficients of y in each factor. To find these coefficients, we must find two integers whose product is 6 and whose sum is -5.

Steps 2 and 3. List pairs of factors of the coefficient of the last term, 6, and try various combinations of these factors. We are interested in the pair of factors whose sum is -5.

Factors of 6	6, 1	3, 2	$-6, -1$	$-3, -2$
Sum of Factors	7	5	-7	-5

This is the desired sum.

Thus, $x^2 - 5xy + 6y^2 = (x - 3y)(x - 2y)$.

Step 4. Verify the factorization using the FOIL method.

$$(x - 3y)(x - 2y) = x^2 - 2xy - 3xy + 6y^2 = x^2 - 5xy + 6y^2$$

Because the product of the factors is the original polynomial, the factorization is correct.

■

☑ **CHECK POINT 6** Factor: $x^2 - 4xy + 3y^2$.

Some polynomials can be factored using more than one technique. **Always begin by looking for a greatest common factor** and, if there is one, factor it out. A polynomial is **factored completely** when it is written as the product of prime polynomials.

EXAMPLE 7 Factoring Completely

Factor: $3x^3 - 15x^2 - 42x$.

Solution The GCF of the three terms of the polynomial is $3x$. We begin by factoring out $3x$. Then we factor the remaining trinomial by the methods of this section.

$$3x^3 - 15x^2 - 42x$$
$$= 3x(x^2 - 5x - 14) \quad \text{Factor out the GCF.}$$
$$= 3x(x \quad)(x \quad) \quad \text{Begin factoring } x^2 - 5x - 14. \text{ Find two integers whose product is } -14 \text{ and whose sum is } -5.$$
$$= 3x(x - 7)(x + 2) \quad \text{The integers are } -7 \text{ and 2.}$$

Thus,

$$3x^3 - 15x^2 - 42x = 3x(x - 7)(x + 2).$$

Be sure to include the GCF in the factorization.

How can we check this factorization? We will multiply the binomials using the FOIL method. Then use the distributive property and multiply each term of this product by $3x$. If the factorization is correct, we should obtain the original polynomial.

$$3x(x - 7)(x + 2) = 3x(x^2 + 2x - 7x - 14) = 3x(x^2 - 5x - 14) = 3x^3 - 15x^2 - 42x$$

Use the FOIL method on $(x - 7)(x + 2)$.

This is the original polynomial.

The factorization is correct.

■

☑ **CHECK POINT 7** Factor: $2x^3 + 6x^2 - 56x$.

7.2 EXERCISE SET *MyMathLab*

PRACTICE WATCH DOWNLOAD READ REVIEW

Practice Exercises

In Exercises 1–42, factor each trinomial, or state that the trinomial is prime. Check each factorization using FOIL multiplication.

1. $x^2 + 7x + 6$

2. $x^2 + 9x + 1$

3. $x^2 + 7x + 10$

4. $x^2 + 9x + 14$

5. $x^2 + 11x + 10$

6. $x^2 + 13x + 12$

7. $x^2 - 7x + 12$

8. $x^2 - 13x + 40$

9. $x^2 - 12x + 36$

10. $x^2 - 8x + 16$

11. $y^2 - 8y + 15$

12. $y^2 - 8y + 7$

13. $x^2 + 3x - 10$

14. $x^2 + 3x - 28$

15. $y^2 + 10y - 39$

16. $y^2 + 5y - 24$

17. $x^2 - 2x - 15$

18. $x^2 - 4x - 5$

19. $x^2 - 2x - 8$

20. $x^2 - 5x - 6$

21. $x^2 + 4x + 12$

22. $x^2 + 4x + 5$

23. $y^2 - 16y + 48$

24. $y^2 - 10y + 21$

25. $x^2 - 3x + 6$

26. $x^2 + 4x - 10$

27. $w^2 - 30w - 64$

28. $w^2 + 12w - 64$

29. $y^2 - 18y + 65$

30. $y^2 - 22y + 72$

31. $r^2 + 12r + 27$

32. $r^2 - 15r - 16$

33. $y^2 - 7y + 5$

34. $y^2 - 15y + 5$

35. $x^2 + 7xy + 6y^2$

36. $x^2 + 6xy + 8y^2$

37. $x^2 - 8xy + 15y^2$

38. $x^2 - 9xy + 14y^2$

39. $x^2 - 3xy - 18y^2$

40. $x^2 - xy - 30y^2$

41. $a^2 - 18ab + 45b^2$

42. $a^2 - 18ab + 80b^2$

In Exercises 43–66, factor completely.

43. $3x^2 + 15x + 18$

44. $3x^2 + 21x + 36$

45. $4y^2 - 4y - 8$

46. $3y^2 + 3y - 18$

47. $10x^2 - 40x - 600$

48. $2x^2 + 10x - 48$

49. $3x^2 - 33x + 54$

50. $2x^2 - 14x + 24$

51. $2r^3 + 6r^2 + 4r$

52. $2r^3 + 8r^2 + 6r$

53. $4x^3 + 12x^2 - 72x$

54. $3x^3 - 15x^2 + 18x$

55. $2r^3 + 8r^2 - 64r$

56. $3r^3 - 9r^2 - 54r$

57. $y^4 + 2y^3 - 80y^2$

58. $y^4 - 12y^3 + 35y^2$

59. $x^4 - 3x^3 - 10x^2$

60. $x^4 - 22x^3 + 120x^2$

61. $2w^4 - 26w^3 - 96w^2$

62. $3w^4 + 54w^3 + 135w^2$

63. $15xy^2 + 45xy - 60x$

64. $20x^2y - 100xy + 120y$

65. $x^5 + 3x^4y - 4x^3y^2$

66. $x^3y - 2x^2y^2 - 3xy^3$

Practice PLUS

In Exercises 67–76, factor completely.

67. $2x^2y^2 - 32x^2yz + 30x^2z^2$

68. $2x^2y^2 - 30x^2yz + 28x^2z^2$

69. $(a + b)x^2 + (a + b)x - 20(a + b)$

70. $(a + b)x^2 - 13(a + b)x + 36(a + b)$

(Hint on Exercises 71–74: Factors contain rational numbers.)

71. $x^2 + 0.5x + 0.06$

72. $x^2 - 0.5x - 0.06$

73. $x^2 - \dfrac{2}{5}x + \dfrac{1}{25}$

74. $x^2 + \dfrac{2}{3}x + \dfrac{1}{9}$

75. $-x^2 - 3x + 40$

76. $-x^2 - 4x + 45$

Application Exercises

77. You dive directly upward from a board that is 32 feet high. After t seconds, your height above the water is described by the polynomial

$$-16t^2 + 16t + 32.$$

a. Factor the polynomial completely. Begin by factoring -16 from each term.

b. Evaluate both the original polynomial and its factored form for $t = 2$. Do you get the same answer for each evaluation? Describe what this answer means.

78. You dive directly upward from a board that is 48 feet high. After t seconds, your height above the water is described by the polynomial

$$-16t^2 + 32t + 48.$$

   a. Factor the polynomial completely. Begin by factoring -16 from each term.

   b. Evaluate both the original polynomial and its factored form for $t = 3$. Do you get the same answer for each evaluation? Describe what this answer means.

Writing in Mathematics

79. Explain how to factor $x^2 + 8x + 15$.

80. Give two helpful suggestions for factoring $x^2 - 5x + 6$.

81. In factoring $x^2 + bx + c$, describe how the last terms in each factor are related to b and c.

82. Without actually factoring and without multiplying the given factors, explain why the following factorization is not correct:

$$x^2 + 46x + 513 = (x - 27)(x - 19).$$

Critical Thinking Exercises

Make Sense?   *In Exercises 83–86, determine whether each statement "makes sense" or "does not make sense" and explain your reasoning.*

83. If I have a correct factorization, switching the order of the factors can give me an incorrect factorization.

84. I began factoring $x^2 - 17x + 72$ by finding all number pairs with a sum of -17.

85. It's easy to factor $x^2 + x + 1$ because of the relatively small numbers for the constant term and the coefficient of x.

86. I factor $x^2 + bx + c$ by finding two numbers that have a product of c and a sum of b.

In Exercises 87–90, determine whether each statement is true or false. If the statement is false, make the necessary change(s) to produce a true statement.

87. One factor of $x^2 + x + 20$ is $x + 5$.

88. A trinomial can never have two identical factors.

89. One factor of $y^2 + 5y - 24$ is $y - 3$.

90. $x^2 + 4 = (x + 2)(x + 2)$

In Exercises 91–92, find all positive integers b so that the trinomial can be factored.

91. $x^2 + bx + 15$

92. $x^2 + 4x + b$

93. Write a trinomial of the form $x^2 + bx + c$ that is prime.

94. Factor: $x^{2n} + 20x^n + 99$.

95. Factor $x^3 + 3x^2 + 2x$. If x represents an integer, use the factorization to describe what the trinomial represents.

96. A box with no top is to be made from an 8-inch by 6-inch piece of metal by cutting identical squares from each corner and turning up the sides (see the figure). The volume of the box is modeled by the polynomial $4x^3 - 28x^2 + 48x$. Factor the polynomial completely. Then use the dimensions given on the box and show that its volume is equivalent to the factorization that you obtain.

Technology Exercises

In Exercises 97–100, use the GRAPH *or* TABLE *feature of a graphing utility to determine if the polynomial on the left side of each equation has been correctly factored. If the graphs of y_1 and y_2 coincide, or if their corresponding table values are equal, this means that the polynomial on the left side has been correctly factored. If not, factor the trinomial correctly and then use your graphing utility to verify the factorization.*

97. $x^2 - 5x + 6 = (x - 2)(x - 3)$

98. $2x^2 + 2x - 12 = 2(x - 3)(x + 2)$

99. $x^2 - 2x + 1 = (x + 1)(x - 1)$

100. $2x^2 + 8x + 6 = (x + 3)(x + 1)$

Review Exercises

101. Solve: $4(x - 2) = 3x + 5$. (Section 2.3, Example 2)

102. Graph: $6x - 5y \le 30$. (Section 4.6, Example 2)

103. Graph: $y = -\frac{1}{2}x + 2$. (Section 4.4, Example 3)

Preview Exercises

Exercises 104–106 will help you prepare for the material covered in the next section.

104. Multiply: $(2x + 3)(x - 2)$.

105. Multiply: $(3x + 4)(3x + 1)$.

106. Factor by grouping: $8x^2 - 2x - 20x + 5$.

Factoring Trinomials Whose Leading Coefficient Is Not 1

Objectives

1 Factor trinomials by trial and error.

2 Factor trinomials by grouping.

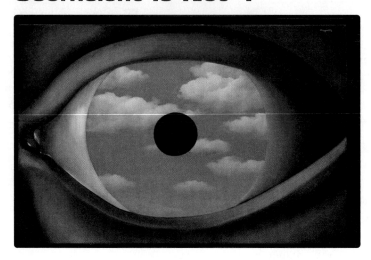

Rene Magritte "The False Mirror" (Le Faux Miroir) 1928, oil on canvas, $21\frac{1}{4} \times 31\frac{7}{8}$ in. (54 × 80.9 cm). The Museum of Modern Art/Licensed by Scala-Art Resource, NY. Purchase. Photograph © 1998 The Museum of Modern Art, New York. © 1998 C. Herscovici, Brusssels/ Artists Rights Society (ARS), New York.

The special significance of the number 1 is reflected in our language. "One," "an," and "a" mean the same thing. The words "unit," "unity," "union," "unique," and "universal" are derived from the Latin word for "one." For the ancient Greeks, 1 was the indivisible unit from which all other numbers arose.

The Greeks' philosophy of 1 applies to our work in this section. Factoring trinomials whose leading coefficient is 1 is the basic technique from which other methods of factoring $ax^2 + bx + c$, where the leading coefficient a is not equal to 1, follow.

1 Factor trinomials by trial and error.

Factoring by the Trial-and-Error Method

How do we factor a trinomial such as $3x^2 - 20x + 28$? Notice that the leading coefficient is 3. We must find two binomials whose product is $3x^2 - 20x + 28$. The product of the First terms must be $3x^2$:

$$(3x \quad)(x \quad).$$

From this point on, the factoring strategy is exactly the same as the one we use to factor trinomials whose leading coefficient is 1.

A Strategy for Factoring $ax^2 + bx + c$

Assume, for the moment, that there is no greatest common factor other than 1.

1. Find two First terms whose product is ax^2:

$$(\Box x + \quad)(\Box x + \quad) = ax^2 + bx + c.$$

2. Find two Last terms whose product is c:

$$(\Box x + \Box)(\Box x + \Box) = ax^2 + bx + c.$$

3. By trial and error, perform steps 1 and 2 until the sum of the Outside product and the Inside product is bx:

$$(\Box x + \Box)(\Box x + \Box) = ax^2 + bx + c.$$

$$\underset{\text{Sum of O + I}}{\underbrace{\quad\quad\quad\quad\quad}}$$

If no such combinations exist, the polynomial is prime.

EXAMPLE 1   Factoring a Trinomial Whose Leading Coefficient Is Not 1

Factor: $3x^2 - 20x + 28$.

Solution

Step 1. Find two First terms whose product is $3x^2$.

$$3x^2 - 20x + 28 = (3x\quad)(x\quad)$$

Step 2. Find two Last terms whose product is 28. The number 28 has pairs of factors that are either both positive or both negative. Because the middle term, $-20x$, is negative, both factors must be negative. The negative factorizations of 28 are $-1(-28)$, $-2(-14)$, and $-4(-7)$.

Step 3. Try various combinations of these factors. The correct factorization of $3x^2 - 20x + 28$ is the one in which the sum of the Outside and Inside products is equal to $-20x$. Here is a list of the possible factorizations:

Possible Factorizations of $3x^2 - 20x + 28$	Sum of Outside and Inside Products (Should Equal $-20x$)
$(3x - 1)(x - 28)$	$-84x - x = -85x$
$(3x - 28)(x - 1)$	$-3x - 28x = -31x$
$(3x - 2)(x - 14)$	$-42x - 2x = -44x$
$(3x - 14)(x - 2)$	$-6x - 14x = -20x$
$(3x - 4)(x - 7)$	$-21x - 4x = -25x$
$(3x - 7)(x - 4)$	$-12x - 7x = -19x$

This is the required middle term.

Thus,

$$3x^2 - 20x + 28 = (3x - 14)(x - 2) \quad \text{or} \quad (x - 2)(3x - 14).$$

Show that this factorization is correct by multiplying the factors using the FOIL method. You should obtain the original trinomial.   ■

✓ **CHECK POINT 1**   Factor: $5x^2 - 14x + 8$.

EXAMPLE 2   Factoring a Trinomial Whose Leading Coefficient Is Not 1

Factor: $8x^2 - 10x - 3$.

Solution

Step 1. Find two First terms whose product is $8x^2$.

$$8x^2 - 10x - 3 \overset{?}{=} (8x\quad)(x\quad)$$
$$8x^2 - 10x - 3 \overset{?}{=} (4x\quad)(2x\quad)$$

Step 2. Find two Last terms whose product is -3. The possible factorizations are $1(-3)$ and $-1(3)$.

Step 3. Try various combinations of these factors. The correct factorization of $8x^2 - 10x - 3$ is the one in which the sum of the Outside and Inside products is equal to $-10x$. Here is a list of the possible factorizations:

Possible Factorizations of $8x^2 - 10x - 3$	Sum of Outside and Inside Products (Should Equal $-10x$)
$(8x + 1)(x - 3)$	$-24x + x = -23x$
$(8x - 3)(x + 1)$	$8x - 3x = 5x$
$(8x - 1)(x + 3)$	$24x - x = 23x$
$(8x + 3)(x - 1)$	$-8x + 3x = -5x$
$(4x + 1)(2x - 3)$	$-12x + 2x = -10x$
$(4x - 3)(2x + 1)$	$4x - 6x = -2x$
$(4x - 1)(2x + 3)$	$12x - 2x = 10x$
$(4x + 3)(2x - 1)$	$-4x + 6x = 2x$

These four factorizations use $(8x\quad)(x\quad)$ with $1(-3)$ and $-1(3)$ as factorizations of -3.

These four factorizations use $(4x\quad)(2x\quad)$ with $1(-3)$ and $-1(3)$ as factorizations of -3.

This is the required middle term.

Thus,

$$8x^2 - 10x - 3 = (4x + 1)(2x - 3) \quad \text{or} \quad (2x - 3)(4x + 1).$$

Use FOIL multiplication to check either of these factorizations.

✓ **CHECK POINT 2** Factor: $6x^2 + 19x - 7$.

Study Tip

Here are some suggestions for reducing the list of possible factorizations for $ax^2 + bx + c$:

1. If b is relatively small, avoid the larger factors of a.

2. If c is positive, the signs in both binomial factors must match the sign of b.

3. If the trinomial has no common factor, no binomial factor can have a common factor.

4. Reversing the signs in the binomial factors changes the sign of bx, the middle term.

EXAMPLE 3 Factoring a Trinomial in Two Variables

Factor: $2x^2 - 7xy + 3y^2$.

Solution

Step 1. Find two First terms whose product is $2x^2$.

$$2x^2 - 7xy + 3y^2 = (2x\quad)(x\quad)$$

Step 2. Find two Last terms whose product is $3y^2$. The possible factorizations are $(y)(3y)$ and $(-y)(-3y)$.

Step 3. Try various combinations of these factors. The correct factorization of $2x^2 - 7xy + 3y^2$ is the one in which the sum of the Outside and Inside products is equal to $-7xy$. Here is a list of possible factorizations:

Possible Factorizations of $2x^2 - 7xy + 3y^2$	Sum of Outside and Inside Products (Should Equal $-7xy$)
$(2x + 3y)(x + y)$	$2xy + 3xy = 5xy$
$(2x + y)(x + 3y)$	$6xy + xy = 7xy$
$(2x - 3y)(x - y)$	$-2xy - 3xy = -5xy$
$(2x - y)(x - 3y)$	$-6xy - xy = -7xy$

This is the required middle term.

Thus,

$$2x^2 - 7xy + 3y^2 = (2x - y)(x - 3y) \quad \text{or} \quad (x - 3y)(2x - y).$$

Use FOIL multiplication to check either of these factorizations.

✓ **CHECK POINT 3** Factor: $3x^2 - 13xy + 4y^2$.

2 Factor trinomials by grouping.

Factoring by the Grouping Method

A second method for factoring $ax^2 + bx + c, a \neq 0$, is called the **grouping method**. The method involves both trial and error, as well as grouping. The trial and error in factoring $ax^2 + bx + c$ depends on finding two numbers, p and q, for which $p + q = b$. Then we factor $ax^2 + px + qx + c$ using grouping.

Let's see how this works by looking at our factorization in Example 2:

$$8x^2 - 10x - 3 = (2x - 3)(4x + 1).$$

If we multiply using FOIL on the right, we obtain:

$$(2x - 3)(4x + 1) = 8x^2 + 2x - 12x - 3.$$

In this case, the desired numbers, p and q, are $p = 2$ and $q = -12$. Compare these numbers to ac and b in the given polynomial:

$$\boxed{a = 8} \quad \boxed{b = -10} \quad \boxed{c = -3}$$
$$8x^2 - 10x - 3.$$
$$\boxed{ac = 8(-3) = -24}$$

Can you see that p and q, 2 and -12, are factors of ac, or -24? Furthermore, p and q have a sum of b, namely -10. By expressing the middle term, $-10x$, in terms of p and q, we can factor by grouping as follows:

$$
\begin{aligned}
8x^2 &- 10x - 3 \\
&= 8x^2 + (2x - 12x) - 3 && \text{Rewrite } -10x \text{ as } 2x - 12x. \\
&= (8x^2 + 2x) + (-12x - 3) && \text{Group terms.} \\
&= 2x(4x + 1) - 3(4x + 1) && \text{Factor from each group.} \\
&= (4x + 1)(2x - 3) && \text{Factor out the common binomial factor.}
\end{aligned}
$$

As we obtained in Example 2,

$$8x^2 - 10x - 3 = (4x + 1)(2x - 3).$$

Generalizing from this example, here's how to factor a trinomial by grouping:

Factoring $ax^2 + bx + c$ Using Grouping ($a \neq 1$)

1. Multiply the leading coefficient, a, and the constant, c.
2. Find the factors of ac whose sum is b.
3. Rewrite the middle term, bx, as a sum or difference using the factors from step 2.
4. Factor by grouping.

EXAMPLE 4   Factoring by Grouping

Factor by grouping: $2x^2 - x - 6$.

Solution   The trinomial is of the form $ax^2 + bx + c$.

$$2x^2 - x - 6$$
$$\boxed{a = 2} \quad \boxed{b = -1} \quad \boxed{c = -6}$$

Step 1. Multiply the leading coefficient, a, and the constant, c. Using $a = 2$ and $c = -6$,
$$ac = 2(-6) = -12.$$

Step 2. Find the factors of ac whose sum is b. We want the factors of -12 whose sum is b, or -1. The factors of -12 whose sum is -1 are -4 and 3.

Discover for Yourself

In step 2, we discovered that the desired numbers were -4 and 3, and we wrote $-x$ as $-4x + 3x$. What happens if we write $-x$ as $3x - 4x$? Use factoring by grouping on

$$2x^2 - x - 6$$
$$= 2x^2 + 3x - 4x - 6.$$

Describe what happens.

Step 3. Rewrite the middle term, $-x$, as a sum or difference using the factors from step 2, -4 and 3.

$$2x^2 - x - 6 = 2x^2 - 4x + 3x - 6$$

Step 4. Factor by grouping.

$$
\begin{aligned}
&= (2x^2 - 4x) + (3x - 6) && \text{Group terms.} \\
&= 2x(x - 2) + 3(x - 2) && \text{Factor from each group.} \\
&= (x - 2)(2x + 3) && \text{Factor out the common binomial factor.}
\end{aligned}
$$

Thus,

$$2x^2 - x - 6 = (x - 2)(2x + 3) \quad \text{or} \quad (2x + 3)(x - 2). \qquad \blacksquare$$

✓ **CHECK POINT 4** Factor by grouping: $3x^2 - x - 10$.

EXAMPLE 5 **Factoring by Grouping**

Factor by grouping: $8x^2 - 22x + 5$.

Solution The trinomial is of the form $ax^2 + bx + c$.

$$8x^2 - 22x + 5$$

$a = 8 \qquad b = -22 \qquad c = 5$

Step 1. Multiply the leading coefficient, a, and the constant, c. Using $a = 8$ and $c = 5$, $ac = 8 \cdot 5 = 40$.

Step 2. Find the factors of ac whose sum is b. We want the factors of 40 whose sum is b, or -22. The factors of 40 whose sum is -22 are -2 and -20.

Step 3. Rewrite the middle term, $-22x$, as a sum or difference using the factors from step 2, -2 and -20.

$$8x^2 - 22x + 5 = 8x^2 - 2x - 20x + 5$$

Step 4. Factor by grouping.

$$
\begin{aligned}
&= (8x^2 - 2x) + (-20x + 5) && \text{Group terms.} \\
&= 2x(4x - 1) - 5(4x - 1) && \text{Factor from each group.} \\
&= (4x - 1)(2x - 5) && \text{Factor out the common binomial factor.}
\end{aligned}
$$

Thus,

$$8x^2 - 22x + 5 = (4x - 1)(2x - 5) \quad \text{or} \quad (2x - 5)(4x - 1). \qquad \blacksquare$$

✓ **CHECK POINT 5** Factor by grouping: $8x^2 - 10x + 3$.

Factoring Completely

Always begin the process of factoring a polynomial by looking for a greatest common factor other than 1. If there is one, **factor out the GCF first**. After doing this, you should attempt to factor the remaining trinomial by one of the methods presented in this section.

EXAMPLE 6 **Factoring Completely**

Factor completely: $15y^4 + 26y^3 + 7y^2$.

Solution We will first factor out a common monomial factor from the polynomial and then factor the resulting trinomial by the methods of this section. The GCF of the three terms is y^2.

$$
\begin{aligned}
15y^4 + 26y^3 + 7y^2 &= y^2(15y^2 + 26y + 7) && \text{Factor out the GCF.} \\
&= y^2(5y + 7)(3y + 1) && \text{Factor } 15y^2 + 26y + 7 \text{ using trial} \\
& && \text{and error or grouping.}
\end{aligned}
$$

Thus,

$$15y^4 + 26y^3 + 7y^2 = y^2(5y + 7)(3y + 1) \ \text{ or } \ y^2(3y + 1)(5y + 7).$$

> Be sure to include the GCF, y^2, in the factorization.

✓ **CHECK POINT 6** Factor completely: $5y^4 + 13y^3 + 6y^2$.

Study Tip

Here is an observation that sometimes helps narrow down the list of possible factorizations. If a polynomial does not have a GCF other than 1 or if you have factored out the GCF, there will be no common factor within any of its binomial factors. Here is an example:

$$6x^2 - 17x + 12.$$ ← There is no GCF other than 1.

$$(2x - 4)(3x - 3)$$ ← This is not a possible factorization.

> This binomial has a common factor of 2.

> This binomial has a common factor of 3.

7.3 EXERCISE SET *MyMathLab*
PRACTICE   WATCH   DOWNLOAD   READ   REVIEW

Practice Exercises

In Exercises 1–58, use the method of your choice to factor each trinomial, or state that the trinomial is prime. Check each factorization using FOIL multiplication.

1. $2x^2 + 5x + 3$

2. $3x^2 + 5x + 2$

3. $3x^2 + 13x + 4$

4. $2x^2 + 7x + 3$

5. $2x^2 + 11x + 12$

6. $2x^2 + 19x + 35$

7. $5y^2 - 16y + 3$

8. $5y^2 - 17y + 6$

9. $3y^2 + y - 4$

10. $3y^2 - y - 4$

11. $3x^2 + 13x - 10$

12. $3x^2 + 14x - 5$

13. $3x^2 - 22x + 7$

14. $3x^2 - 10x + 7$

15. $5y^2 - 16y + 3$

16. $5y^2 - 8y + 3$

17. $3x^2 - 17x + 10$

18. $3x^2 - 25x - 28$

19. $6w^2 - 11w + 4$

20. $6w^2 - 17w + 12$

21. $8x^2 + 33x + 4$

22. $7x^2 + 43x + 6$

23. $5x^2 + 33x - 14$

24. $3x^2 + 22x - 16$

25. $14y^2 + 15y - 9$

26. $6y^2 + 7y - 24$

27. $6x^2 - 7x + 3$

28. $9x^2 + 3x + 2$

29. $25z^2 - 30z + 9$

30. $9z^2 + 12z + 4$

31. $15y^2 - y - 2$

32. $15y^2 + 13y - 2$

33. $5x^2 + 2x + 9$

34. $3x^2 - 5x + 1$

35. $10y^2 + 43y - 9$

36. $16y^2 - 46y + 15$

37. $8x^2 - 2x - 1$

38. $8x^2 - 22x + 5$

39. $9y^2 - 9y + 2$

40. $9y^2 + 5y - 4$

41. $20x^2 + 27x - 8$

42. $15x^2 - 19x + 6$

43. $2x^2 + 3xy + y^2$

44. $3x^2 + 4xy + y^2$

45. $3x^2 + 5xy + 2y^2$

46. $3x^2 + 11xy + 6y^2$

47. $2x^2 - 9xy + 9y^2$

48. $3x^2 + 5xy - 2y^2$

49. $6x^2 - 5xy - 6y^2$ **50.** $6x^2 - 7xy - 5y^2$

51. $15x^2 + 11xy - 14y^2$ **52.** $15x^2 - 31xy + 10y^2$

53. $2a^2 + 7ab + 5b^2$ **54.** $2a^2 + 5ab + 2b^2$

55. $15a^2 - ab - 6b^2$ **56.** $3a^2 - ab - 14b^2$

57. $12x^2 - 25xy + 12y^2$ **58.** $12x^2 + 7xy - 12y^2$

In Exercises 59–86, factor completely.

59. $4x^2 + 26x + 30$ **60.** $4x^2 - 18x - 10$

61. $9x^2 - 6x - 24$ **62.** $12x^2 - 33x + 21$

63. $4y^2 + 2y - 30$ **64.** $36y^2 + 6y - 12$

65. $9y^2 + 33y - 60$ **66.** $16y^2 - 16y - 12$

67. $3x^3 + 4x^2 + x$ **68.** $3x^3 + 14x^2 + 8x$

69. $2x^3 - 3x^2 - 5x$ **70.** $6x^3 + 4x^2 - 10x$

71. $9y^3 - 39y^2 + 12y$ **72.** $10y^3 + 12y^2 + 2y$

73. $60z^3 + 40z^2 + 5z$ **74.** $80z^3 + 80z^2 - 60z$

75. $15x^4 - 39x^3 + 18x^2$ **76.** $24x^4 + 10x^3 - 4x^2$

77. $10x^5 - 17x^4 + 3x^3$ **78.** $15x^5 - 2x^4 - x^3$

79. $6x^2 - 3xy - 18y^2$ **80.** $4x^2 + 14xy + 10y^2$

81. $12x^2 + 10xy - 8y^2$ **82.** $24x^2 + 3xy - 27y^2$

83. $8x^2y + 34xy - 84y$ **84.** $6x^2y - 2xy - 60y$

85. $12a^2b - 46ab^2 + 14b^3$ **86.** $12a^2b - 34ab^2 + 14b^3$

Practice PLUS

In Exercises 87–90, factor completely.

87. $30(y + 1)x^2 + 10(y + 1)x - 20(y + 1)$

88. $6(y + 1)x^2 + 33(y + 1)x + 15(y + 1)$

89. $-32x^2y^4 + 20xy^4 + 12y^4$

90. $-10x^2y^4 + 14xy^4 + 12y^4$

91. **a.** Factor $2x^2 - 5x - 3$.

b. Use the factorization in part (a) to factor

$$2(y + 1)^2 - 5(y + 1) - 3.$$

Then simplify each factor.

92. **a.** Factor $3x^2 + 5x - 2$.

b. Use the factorization in part (a) to factor

$$3(y + 1)^2 + 5(y + 1) - 2.$$

Then simplify each factor.

93. Divide $3x^3 - 11x^2 + 12x - 4$ by $x - 2$. Use the quotient to factor $3x^3 - 11x^2 + 12x - 4$ completely.

94. Divide $2x^3 + x^2 - 13x + 6$ by $x - 2$. Use the quotient to factor $2x^3 + x^2 - 13x + 6$ completely.

Application Exercises

It is possible to construct geometric models for factorizations so that you can see the factoring. This idea is developed in Exercises 95–96.

95. Consider the following figure.

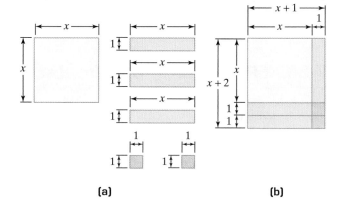

(a)   **(b)**

a. Write a trinomial that expresses the sum of the areas of the six rectangular pieces shown in figure (a).

b. Express the area of the large rectangle in figure (b) as the product of two binomials.

c. Are the pieces in figures (a) and (b) the same? Set the expressions that you wrote in parts (a) and (b) equal to each other. What factorization is illustrated?

96. Copy the figure and cut out the six pieces. Use the pieces to create a geometric model for the factorization

$$2x^2 + 3x + 1 = (2x + 1)(x + 1)$$

by forming a large rectangle using all the pieces.

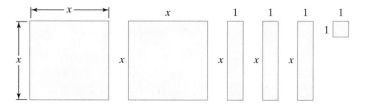

Writing in Mathematics

97. Explain how to factor $2x^2 - x - 1$.

98. Why is it a good idea to factor out the GCF first and then use other methods of factoring? Use $3x^2 - 18x + 15$ as an example. Discuss what happens if one first uses trial and error to factor as two binomials rather than first factoring out the GCF.

99. In factoring $3x^2 - 10x - 8$, a student lists $(3x - 2)(x + 4)$ as a possible factorization. Use FOIL multiplication to determine if this factorization is correct. If it is not correct, describe how the correct factorization can quickly be obtained using these factors.

100. Explain why $2x - 10$ cannot be one of the factors in the correct factorization of $6x^2 - 19x + 10$.

Critical Thinking Exercises

Make Sense? *In Exercises 101–104, determine whether each statement "makes sense" or "does not make sense" and explain your reasoning.*

101. I'm often able to use an incorrect factorization to lead me to the correct factorization.

102. First factoring out the GCF makes it easier for me to determine how to factor the remaining factor, assuming it is not prime.

103. I'm working with a polynomial that has a GCF other than 1, but then it doesn't factor further, so the polynomial that I'm working with is prime.

104. My graphing calculator showed the same graphs for $y_1 = 4x^2 - 20x + 24$ and $y_2 = 4(x^2 - 5x + 6)$, so I can conclude that the complete factorization of $4x^2 - 20x + 24$ is $4(x^2 - 5x + 6)$.

In Exercises 105–108, determine whether each statement is true or false. If the statement is false, make the necessary change(s) to produce a true statement.

105. Once the GCF is factored from $18y^2 - 6y + 6$, the remaining trinomial factor is prime.

106. One factor of $12x^2 - 13x + 3$ is $4x + 3$.

107. One factor of $4y^2 - 11y - 3$ is $y + 3$.

108. The trinomial $3x^2 + 2x + 1$ has relatively small coefficients and therefore can be factored.

In Exercises 109–110, find all integers b so that the trinomial can be factored.

109. $3x^2 + bx + 2$

110. $2x^2 + bx + 3$

111. Factor: $3x^{10} - 4x^5 - 15$.

112. Factor: $2x^{2n} - 7x^n - 4$.

Review Exercises

113. Solve the system:
$$4x - y = 105$$
$$x + 7y = -10. \text{ (Section 5.3, Example 3)}$$

114. Write 0.00086 in scientific notation. (Section 6.7, Example 8)

115. Solve: $8x - \dfrac{x}{6} = \dfrac{1}{6} - 8$. (Section 2.3, Example 4)

Preview Exercises

Exercises 116–118 will help you prepare for the material covered in the next section. In each exercise, perform the indicated operation.

116. $(9x + 10)(9x - 10)$

117. $(4x + 5y)^2$

118. $(x + 2)(x^2 - 2x + 4)$

MID-CHAPTER CHECK POINT Section 7.1–Section 7.3

✓ **What You Know:** We learned to factor out a polynomial's greatest common factor and to use grouping to factor polynomials with four terms. We factored polynomials with three terms, beginning with trinomials with leading coefficient 1 and moving on to $ax^2 + bx + c$, with $a \neq 1$. We saw that the factoring process should begin by looking for a GCF other than 1 and, if there is one, factoring it out first.

In Exercises 1–12, factor completely, or state that the polynomial is prime.

1. $x^5 + x^4$

2. $x^2 + 7x - 18$

3. $x^2y^3 - x^2y^2 + x^2y$

4. $x^2 - 2x + 4$

5. $7x^2 - 22x + 3$

6. $x^3 + 5x^2 + 3x + 15$

7. $2x^3 - 11x^2 + 5x$

8. $xy - 7x - 4y + 28$

9. $x^2 - 17xy + 30y^2$

10. $25x^2 - 25x - 14$

11. $16x^2 - 70x + 24$

12. $3x^2 + 10xy + 7y^2$

7.4

Factoring Special Forms

Objectives

1 Factor the difference of two squares.

2 Factor perfect square trinomials.

3 Factor the sum or difference of two cubes.

Do you enjoy solving puzzles? The process is a natural way to develop problem-solving skills that are important to every area of our lives. Engaging in problem solving for sheer pleasure releases chemicals in the brain that enhance our feeling of well-being. Perhaps this is why puzzles date back 12,000 years.

In this section, we develop factoring techniques by reversing the formulas for special products discussed in Chapter 6. These factorizations can be visualized by fitting pieces of a puzzle together to form rectangles.

1 Factor the difference of two squares.

Factoring the Difference of Two Squares

A method for factoring the difference of two squares is obtained by reversing the special product for the sum and difference of two terms.

> ### The Difference of Two Squares
> If A and B are real numbers, variables, or algebraic expressions, then
> $$A^2 - B^2 = (A + B)(A - B).$$
> In words: The difference of the squares of two terms factors as the product of a sum and a difference of those terms.

EXAMPLE 1   Factoring the Difference of Two Squares

Factor:

a. $x^2 - 4$ b. $81x^2 - 49$.

Solution   We must express each term as the square of some monomial. Then we use the formula for factoring $A^2 - B^2$.

a. $x^2 - 4 = x^2 - 2^2 = (x + 2)(x - 2)$

$A^2 - B^2 = (A + B)(A - B)$

b. $81x^2 - 49 = (9x)^2 - 7^2 = (9x + 7)(9x - 7)$

In order to apply the factoring formula for $A^2 - B^2$, each term must be the square of an integer or a polynomial.

* A number that is the square of an integer is called a **perfect square**. For example, 100 is a perfect square because $100 = 10^2$.
* Any monomial involving a perfect-square coefficient and variables to even powers is a perfect square. For example, $16x^{10}$ is a perfect square because $16x^{10} = (4x^5)^2$.

Study Tip

It's helpful to recognize perfect squares. Here are 16 perfect squares, printed in boldface.

1 $= 1^2$	**25** $= 5^2$	**81** $= 9^2$	**169** $= 13^2$
4 $= 2^2$	**36** $= 6^2$	**100** $= 10^2$	**196** $= 14^2$
9 $= 3^2$	**49** $= 7^2$	**121** $= 11^2$	**225** $= 15^2$
16 $= 4^2$	**64** $= 8^2$	**144** $= 12^2$	**256** $= 16^2$

✓ **CHECK POINT 1**   Factor:

a. $x^2 - 81$ **b.** $36x^2 - 25$.

___

Be careful when determining whether or not to apply the factoring formula for the difference of two squares.

Prime Over the Integers   **Factorable**

* $x^2 - 5$ * $x^7 - 25$ * $9 - 16x^{10}$

> 5 is not a perfect square.

> 7 is an odd power. x^7 is not the square of any integer power of x.

> Perfect square: $9 = 3^2$

> Perfect square: $16x^{10} = (4x^5)^2$

EXAMPLE 2   **Factoring the Difference of Two Squares**

Factor:

a. $9 - 16x^{10}$ **b.** $25x^2 - 4y^2$.

Solution   Begin by expressing each term as the square of some monomial. Then use the formula for factoring $A^2 - B^2$.

a. $9 - 16x^{10} = 3^2 - (4x^5)^2 = (3 + 4x^5)(3 - 4x^5)$

$$A^2 - B^2 = (A + B)(A - B)$$

b. $25x^2 - 4y^2 = (5x)^2 - (2y)^2 = (5x + 2y)(5x - 2y)$

✓ **CHECK POINT 2**   Factor:

a. $25 - 4x^{10}$ **b.** $100x^2 - 9y^2$.

___

When factoring, always check first for common factors. If there are common factors other than 1, factor out the GCF and then factor the resulting polynomial.

EXAMPLE 3 Factoring Out the GCF and Then Factoring the Difference of Two Squares

Factor:

a. $12x^3 - 3x$ **b.** $80 - 125x^2$.

Solution

a. $12x^3 - 3x = 3x(4x^2 - 1) = 3x[(2x)^2 - 1^2] = 3x(2x + 1)(2x - 1)$

                 Factor out the GCF. $A^2 - B^2 = (A + B)(A - B)$

b. $80 - 125x^2 = 5(16 - 25x^2) = 5[4^2 - (5x)^2] = 5(4 + 5x)(4 - 5x)$ ■

✓ **CHECK POINT 3** Factor:

a. $18x^3 - 2x$ **b.** $72 - 18x^2$.

We have seen that a polynomial is factored completely when it is written as the product of prime polynomials. To be sure that you have factored completely, check to see whether any factors with more than one term in the factored polynomial can be factored further. If so, continue factoring.

EXAMPLE 4 A Repeated Factorization

Factor completely: $x^4 - 81$.

Solution

$$x^4 - 81 = (x^2)^2 - 9^2$$ *Express as the difference of two squares.*

$$= (x^2 + 9)(x^2 - 9)$$ *The factors are the sum and the difference of the expressions being squared.*

$$= (x^2 + 9)(x^2 - 3^2)$$ *The factor $x^2 - 9$ is the difference of two squares and can be factored.*

$$= (x^2 + 9)(x + 3)(x - 3)$$ *The factors of $x^2 - 9$ are the sum and the difference of the expressions being squared.* ■

Study Tip

Factoring $x^4 - 81$ as

$$(x^2 + 9)(x^2 - 9)$$

is not a complete factorization. The second factor, $x^2 - 9$, is itself a difference of two squares and can be factored.

Are you tempted to further factor $x^2 + 9$, the sum of two squares, in Example 4? Resist the temptation! **The sum of two squares, $A^2 + B^2$, with no common factor other than 1 is a prime polynomial over the integers.**

✓ **CHECK POINT 4** Factor completely: $81x^4 - 16$.

In Examples 1–4, we used the formula for factoring the difference of two squares. Although we obtained the formula by reversing the special product for the sum and difference of two terms, it can also be obtained geometrically.

The combined area of the three yellow rectangles is $A^2 - B^2$.

We've rearranged the three yellow rectangles. Their combined area is $(A + B)(A - B)$.

Because the three yellow rectangles make up the same combined area in both figures,

$$A^2 - B^2 = (A + B)(A - B).$$

2 Factor perfect square trinomials.

Factoring Perfect Square Trinomials

Our next factoring technique is obtained by reversing the special products for squaring binomials. The trinomials that are factored using this technique are called **perfect square trinomials**.

Factoring Perfect Square Trinomials

Let A and B be real numbers, variables, or algebraic expressions.

1. $A^2 + 2AB + B^2 = (A + B)^2$

Same sign

2. $A^2 - 2AB + B^2 = (A - B)^2$

Same sign

The two formulas in the box show that perfect square trinomials, $A^2 + 2AB + B^2$ and $A^2 - 2AB + B^2$, come in two forms: one in which the coefficient of the middle term is positive and one in which the coefficient of the middle term is negative. Here's how to recognize a perfect square trinomial:

1. The first and last terms are squares of monomials or integers.
2. The middle term is twice the product of the expressions being squared in the first and last terms.

EXAMPLE 5 Factoring Perfect Square Trinomials

Factor:

a. $x^2 + 6x + 9$ **b.** $x^2 - 16x + 64$ **c.** $25x^2 - 60x + 36.$

Solution

a. $x^2 + 6x + 9 = x^2 + 2 \cdot x \cdot 3 + 3^2 = (x + 3)^2$ The middle term has a positive sign.

$A^2 + 2AB + B^2 = (A + B)^2$

b. $x^2 - 16x + 64 = x^2 - 2 \cdot x \cdot 8 + 8^2 = (x - 8)^2$ The middle term has a negative sign.

$A^2 - 2AB + B^2 = (A - B)^2$

c. We suspect that $25x^2 - 60x + 36$ is a perfect square trinomial because $25x^2 = (5x)^2$ and $36 = 6^2$. The middle term can be expressed as twice the product of $5x$ and 6.

$$25x^2 - 60x + 36 = (5x)^2 - 2 \cdot 5x \cdot 6 + 6^2 = (5x - 6)^2$$

$A^2 - 2AB + B^2 = (A - B)^2$

✓ **CHECK POINT 5** Factor:

a. $x^2 + 14x + 49$ **b.** $x^2 - 6x + 9$
c. $16x^2 - 56x + 49.$

Factoring a Perfect Square Trinomial in Two Variables

Factor: $16x^2 + 40xy + 25y^2$.

Solution Observe that $16x^2 = (4x)^2$, $25y^2 = (5y)^2$, and $40xy$ is twice the product of $4x$ and $5y$. Thus, we have a perfect square trinomial.

$$16x^2 + 40xy + 25y^2 = (4x)^2 + 2 \cdot 4x \cdot 5y + (5y)^2 = (4x + 5y)^2$$

$$A^2 + 2AB + B^2 = (A + B)^2$$

✓ **CHECK POINT 6** Factor: $4x^2 + 12xy + 9y^2$.

In Examples 5 and 6, we factored perfect square trinomials using $A^2 + 2AB + B^2 = (A + B)^2$. Although we obtained the formula by reversing the special product for $(A + B)^2$, it can also be obtained geometrically.

The combined area of the four rectangles is $A^2 + 2AB + B^2$.

We've rearranged the four rectangles to form a large square. The area of the square is $(A + B)^2$.

Because the four rectangles make up the same combined area in both figures,

$$A^2 + 2AB + B^2 = (A + B)^2.$$

3 Factor the sum or difference of two cubes.

Factoring the Sum or Difference of Two Cubes

We can use the following formulas to factor the sum or the difference of two cubes:

Factoring the Sum or Difference of Two Cubes

1. Factoring the Sum of Two Cubes

$$A^3 + B^3 = (A + B)(A^2 - AB + B^2)$$

Same signs   Opposite signs

2. Factoring the Difference of Two Cubes

$$A^3 - B^3 = (A - B)(A^2 + AB + B^2)$$

Same signs   Opposite signs

Study Tip

When factoring the sum or difference of cubes, it is helpful to recognize the following cubes:

$$1 = 1^3$$
$$8 = 2^3$$
$$27 = 3^3$$
$$64 = 4^3$$
$$125 = 5^3$$
$$216 = 6^3$$
$$1000 = 10^3.$$

EXAMPLE 7 Factoring the Sum of Two Cubes

Factor: $x^3 + 8$.

Solution We must express each term as the cube of some monomial. Then we use the formula for factoring $A^3 + B^3$.

$$x^3 + 8 = x^3 + 2^3 = (x + 2)(x^2 - x \cdot 2 + 2^2) = (x + 2)(x^2 - 2x + 4)$$

$$A^3 + B^3 = (A + B)(A^2 - AB + B^2)$$

✓ **CHECK POINT 7** Factor: $x^3 + 27$.

EXAMPLE 8 Factoring the Difference of Two Cubes

Factor: $27 - y^3$.

Solution Express each term as the cube of some monomial. Then use the formula for factoring $A^3 - B^3$.

$$27 - y^3 = 3^3 - y^3 = (3 - y)(3^2 + 3y + y^2) = (3 - y)(9 + 3y + y^2)$$

$$A^3 - B^3 = (A - B)(A^2 + AB + B^2)$$

✓ **CHECK POINT 8** Factor: $1 - y^3$.

EXAMPLE 9 Factoring the Sum of Two Cubes

Factor: $64x^3 + 125$.

Solution Express each term as the cube of some monomial. Then use the formula for factoring $A^3 + B^3$.

$$64x^3 + 125 = (4x)^3 + 5^3 = (4x + 5)[(4x)^2 - (4x)(5) + 5^2]$$

$$A^3 + B^3 = (A + B)(A^2 - AB + B^2)$$

$$= (4x + 5)(16x^2 - 20x + 25)$$

✓ **CHECK POINT 9** Factor: $125x^3 + 8$.

Study Tip

A Cube of SOAP

When factoring sums or differences of cubes, observe the sign patterns.

$$A^3 + B^3 = (A + B)(A^2 - AB + B^2)$$

Same signs | Opposite signs | Always positive

$$A^3 - B^3 = (A - B)(A^2 + AB + B^2)$$

Same signs | Opposite signs | Always positive

The word *SOAP* is a way to remember these patterns:

S   O   A   P.

Same signs | Opposite signs | Always Positive

7.4 EXERCISE SET   *MyMathLab*

PRACTICE   WATCH   DOWNLOAD   READ   REVIEW

Practice Exercises

In Exercises 1–26, factor each difference of two squares.

1. $x^2 - 25$ **2.** $x^2 - 16$

3. $y^2 - 1$ **4.** $y^2 - 9$

5. $4x^2 - 9$ **6.** $9x^2 - 25$

7. $25 - x^2$ **8.** $16 - x^2$

9. $1 - 49x^2$ **10.** $1 - 64x^2$

11. $9 - 25y^2$ **12.** $16 - 49y^2$

13. $x^4 - 9$ **14.** $x^4 - 25$

15. $49y^4 - 16$ **16.** $49y^4 - 25$

17. $x^{10} - 9$ **18.** $x^{10} - 1$

19. $25x^2 - 16y^2$ **20.** $9x^2 - 25y^2$

21. $x^4 - y^{10}$ **22.** $x^{14} - y^4$

23. $x^4 - 16$ **24.** $x^4 - 1$

25. $16x^4 - 81$ **26.** $81x^4 - 1$

In Exercises 27–40, factor completely, or state that the polynomial is prime.

27. $2x^2 - 18$ **28.** $5x^2 - 45$

29. $2x^3 - 72x$ **30.** $2x^3 - 8x$

31. $x^2 + 36$ **32.** $x^2 + 4$

33. $3x^3 + 27x$ **34.** $3x^3 + 15x$

35. $18 - 2y^2$ **36.** $32 - 2y^2$

37. $3y^3 - 48y$ **38.** $3y^3 - 75y$

39. $18x^3 - 2x$ **40.** $20x^3 - 5x$

In Exercises 41–62, factor any perfect square trinomials, or state that the polynomial is prime.

41. $x^2 + 2x + 1$ **42.** $x^2 + 4x + 4$

43. $x^2 - 14x + 49$ **44.** $x^2 - 10x + 25$

45. $x^2 - 2x + 1$ **46.** $x^2 - 4x + 4$

47. $x^2 + 22x + 121$ **48.** $x^2 + 24x + 144$

49. $4x^2 + 4x + 1$ **50.** $9x^2 + 6x + 1$

51. $25y^2 - 10y + 1$ **52.** $64y^2 - 16y + 1$

53. $x^2 - 10x + 100$ **54.** $x^2 - 7x + 49$

55. $x^2 + 14xy + 49y^2$ **56.** $x^2 + 16xy + 64y^2$

57. $x^2 - 12xy + 36y^2$ **58.** $x^2 - 18xy + 81y^2$

59. $x^2 - 8xy + 64y^2$ **60.** $x^2 + 9xy + 16y^2$

61. $16x^2 - 40xy + 25y^2$ **62.** $9x^2 + 48xy + 64y^2$

In Exercises 63–70, factor completely.

63. $12x^2 - 12x + 3$ **64.** $18x^2 + 24x + 8$

65. $9x^3 + 6x^2 + x$ **66.** $25x^3 - 10x^2 + x$

67. $2y^2 - 4y + 2$ **68.** $2y^2 - 40y + 200$

69. $2y^3 + 28y^2 + 98y$ **70.** $50y^3 + 20y^2 + 2y$

In Exercises 71–88, factor using the formula for the sum or difference of two cubes.

71. $x^3 + 1$ **72.** $x^3 + 64$

73. $x^3 - 27$ **74.** $x^3 - 64$

75. $8y^3 - 1$ **76.** $27y^3 - 1$

77. $27x^3 + 8$ **78.** $125x^3 + 8$

79. $x^3y^3 - 64$ **80.** $x^3y^3 - 27$

81. $27y^4 + 8y$ **82.** $64y - y^4$

83. $54 - 16y^3$ **84.** $128 - 250y^3$

85. $64x^3 + 27y^3$ **86.** $8x^3 + 27y^3$

87. $125x^3 - 64y^3$ **88.** $125x^3 - y^3$

Practice PLUS

In Exercises 89–96, factor completely
(Hint on Exercises 89–94: Factors contain rational numbers.)

89. $25x^2 - \dfrac{4}{49}$

90. $16x^2 - \dfrac{9}{25}$

91. $y^4 - \dfrac{y}{1000}$

92. $y^4 - \dfrac{y}{8}$

93. $0.25x - x^3$

94. $0.64x - x^3$

95. $(x + 1)^2 - 25$

96. $(x + 2)^2 - 49$

97. Divide $x^3 - x^2 - 5x - 3$ by $x - 3$. Use the quotient to factor $x^3 - x^2 - 5x - 3$ completely.

98. Divide $x^3 + 4x^2 - 3x - 18$ by $x - 2$. Use the quotient to factor $x^3 + 4x^2 - 3x - 18$ completely.

Application Exercises

In Exercises 99–102, find the formula for the area of the shaded region and express it in factored form.

99.

100.

101.

102.

Writing in Mathematics

103. Explain how to factor the difference of two squares. Provide an example with your explanation.

104. What is a perfect square trinomial and how is it factored?

105. Explain why $x^2 - 1$ is factorable, but $x^2 + 1$ is not.

106. Explain how to factor $x^3 + 1$.

Critical Thinking Exercises

Make Sense? *In Exercises 107–110, determine whether each statement "makes sense" or "does not make sense" and explain your reasoning.*

107. I factored $9x^2 - 36$ completely and obtained

$$(3x + 6)(3x - 6).$$

108. Although I can factor the difference of squares and perfect square trinomials using trial-and-error, recognizing these special forms shortens the process.

109. I factored $9 - 25x^2$ as $(3 + 5x)(3 - 5x)$ and then applied the commutative property to rewrite the factorization as $(5x + 3)(5x - 3)$.

110. I compared the factorization for the sum of cubes with the factorization for the difference of cubes and noticed that the only difference between them is the positive and negative signs.

In Exercises 111–114, determine whether each statement is true or false. If the statement is false, make the necessary change(s) to produce a true statement.

111. Because $x^2 - 25 = (x + 5)(x - 5)$, then $x^2 + 25 = (x - 5)(x + 5)$.

112. All perfect square trinomials are squares of binomials.

113. Any polynomial that is the sum of two squares is prime.

114. The polynomial $16x^2 + 20x + 25$ is a perfect square trinomial.

115. Where is the error in this "proof" that $2 = 0$?

$a = b$	Suppose that a and b are any equal real numbers.
$a^2 = b^2$	Square both sides of the equation.
$a^2 - b^2 = 0$	Subtract b^2 from both sides.
$2(a^2 - b^2) = 2 \cdot 0$	Multiply both sides by 2.
$2(a^2 - b^2) = 0$	On the right side, $2 \cdot 0 = 0$.
$2(a + b)(a - b) = 0$	Factor $a^2 - b^2$.
$2(a + b) = 0$	Divide both sides by $a - b$.
$2 = 0$	Divide both sides by $a + b$.

In Exercises 116–119, factor each polynomial.

116. $x^2 - y^2 + 3x + 3y$

117. $x^{2n} - 25y^{2n}$

118. $4x^{2n} + 12x^n + 9$

119. $(x + 3)^2 - 2(x + 3) + 1$

In Exercises 120–121, find all integers k so that the trinomial is a perfect square trinomial.

120. $9x^2 + kx + 1$

121. $64x^2 - 16x + k$

Technology Exercises

In Exercises 122–125, use the ⎢GRAPH⎢ or ⎢TABLE⎢ feature of a graphing utility to determine if the polynomial on the left side of each equation has been correctly factored. If the graphs of y_1 and y_2 coincide, or if their corresponding table values are equal, this means that the polynomial on the left side has been correctly factored. If not, factor the polynomial correctly and then use your graphing utility to verify the factorization.

122. $4x^2 - 9 = (4x + 3)(4x - 3)$

123. $x^2 - 6x + 9 = (x - 3)^2$

124. $4x^2 - 4x + 1 = (4x - 1)^2$

125. $x^3 - 1 = (x - 1)(x^2 - x + 1)$

Review Exercises

126. Simplify: $(2x^2y^3)^4(5xy^2)$. (Section 6.7, Example 5)

127. Subtract: $(10x^2 - 5x + 2) - (14x^2 - 5x - 1)$. (Section 6.1, Example 3)

128. Divide: $\dfrac{6x^2 + 11x - 10}{3x - 2}$. (Section 6.6, Example 1)

Preview Exercises

Exercises 129–131 will help you prepare for the material covered in the next section. In each exercise, factor completely.

129. $3x^3 - 75x$

130. $2x^2 - 20x + 50$

131. $x^3 - 2x^2 - x + 2$

SECTION 7.5

A General Factoring Strategy

Objectives

1 Recognize the appropriate method for factoring a polynomial.

2 Use a general strategy for factoring polynomials.

Jiří Kolář "Untitled" 1971. Collection of Meda Mladek, Courtesy of Museum Kampa

Yogi Berra, catcher and renowned hitter for the New York Yankees (1946–1963), said it best: "If you don't know where you're going, you'll probably end up someplace else." When it comes to factoring, it's easy to know where you're going. Why? In this section, you will learn a step-by-step strategy that provides a plan and direction for solving factoring problems.

A Strategy for Factoring Polynomials

It is important to practice factoring a wide variety of polynomials so that you can quickly select the appropriate technique. The polynomial is factored completely when all its polynomial factors, except possibly the monomial factor, are prime. Because of the commutative property, the order of the factors does not matter.

Here is a general strategy for factoring polynomials:

1 Recognize the appropriate method for factoring a polynomial.

A Strategy for Factoring a Polynomial

1. If there is a common factor other than 1, factor out the GCF.

2. Determine the number of terms in the polynomial and try factoring as follows:

  a. If there are two terms, can the binomial be factored by one of the following special forms?

    Difference of two squares: $A^2 - B^2 = (A + B)(A - B)$

    Sum of two cubes: $A^3 + B^3 = (A + B)(A^2 - AB + B^2)$

    Difference of two cubes: $A^3 - B^3 = (A - B)(A^2 + AB + B^2)$

  b. If there are three terms, is the trinomial a perfect square trinomial? If so, factor by one of the following special forms:

$$A^2 + 2AB + B^2 = (A + B)^2$$
$$A^2 - 2AB + B^2 = (A - B)^2.$$

  If the trinomial is not a perfect square trinomial, try factoring by trial and error or grouping.

  c. If there are four or more terms, try factoring by grouping.

3. Check to see if any factors with more than one term in the factored polynomial can be factored further. If so, factor completely.

4. Check by multiplying.

2 Use a general strategy for factoring polynomials.

The following examples and those in the exercise set are similar to the previous factoring problems. One difference is that although these polynomials may be factored using the techniques we have studied in this chapter, most must be factored using at least two techniques. Also different is that these factorizations are not all of the same type. They are intentionally mixed to promote the development of a general factoring strategy.

EXAMPLE 1   Factoring a Polynomial

Factor: $4x^4 - 16x^2$.

Solution

Step 1. If there is a common factor, factor out the GCF. Because $4x^2$ is common to both terms, we factor it out.

$$4x^4 - 16x^2 = 4x^2(x^2 - 4) \quad \text{Factor out the GCF.}$$

Step 2. Determine the number of terms and factor accordingly. The factor $x^2 - 4$ has two terms. It is the difference of two squares: $x^2 - 2^2$. We factor using the special form for the difference of two squares and rewrite the GCF.

$$4x^4 - 16x^2 = 4x^2(x + 2)(x - 2) \quad \text{Use } A^2 - B^2 = (A + B)(A - B) \text{ on}$$
$$x^2 - 4: A = x \text{ and } B = 2.$$

Step 3. Check to see if any factors with more than one term can be factored further. No factor with more than one term can be factored further, so we have factored completely.

Step 4. Check by multiplying.

$$4x^2(x + 2)(x - 2) = 4x^2(x^2 - 4) = 4x^4 - 16x^2$$

This is the original polynomial, so the factorization is correct.

✓ **CHECK POINT 1**   Factor: $5x^4 - 45x^2$.

EXAMPLE 2 Factoring a Polynomial

Factor: $3x^2 - 6x - 45$.

Solution

Step 1. If there is a common factor, factor out the GCF. Because 3 is common to all terms, we factor it out.

$$3x^2 - 6x - 45 = 3(x^2 - 2x - 15) \qquad \text{Factor out the GCF.}$$

Step 2. Determine the number of terms and factor accordingly. The factor $x^2 - 2x - 15$ has three terms, but it is not a perfect square trinomial. We factor it using trial and error.

$$3x^2 - 6x - 45 = 3(x^2 - 2x - 15) = 3(x - 5)(x + 3)$$

Step 3. Check to see if factors can be factored further. In this case, they cannot, so we have factored completely.

Step 4. Check by multiplying.

$$3(x - 5)(x + 3) = 3(x^2 - 2x - 15) = 3x^2 - 6x - 45$$

FOIL

This is the original polynomial, so the factorization is correct.

✓ **CHECK POINT 2** Factor: $4x^2 - 16x - 48$.

EXAMPLE 3 Factoring a Polynomial

Factor: $7x^5 - 7x$.

Solution

Step 1. If there is a common factor, factor out the GCF. Because $7x$ is common to both terms, we factor it out.

$$7x^5 - 7x = 7x(x^4 - 1) \qquad \text{Factor out the GCF.}$$

Step 2. Determine the number of terms and factor accordingly. The factor $x^4 - 1$ has two terms. This binomial can be expressed as $(x^2)^2 - 1^2$, so it can be factored as the difference of two squares.

$$7x^5 - 7x = 7x(x^4 - 1) = 7x(x^2 + 1)(x^2 - 1) \qquad \text{Use } A^2 - B^2 = (A + B)(A - B)$$
$$\text{on } x^4 - 1\text{: } A = x^2 \text{ and } B = 1.$$

Step 3. Check to see if factors can be factored further. We note that $(x^2 - 1)$ is also the difference of two squares, $x^2 - 1^2$, so we continue factoring.

$$7x^5 - 7x = 7x(x^2 + 1)(x + 1)(x - 1) \qquad \text{Factor } x^2 - 1 \text{ as the difference of}$$
$$\text{two squares.}$$

Step 4. Check by multiplying.

$$7x(x^2 + 1)(x + 1)(x - 1) = 7x(x^2 + 1)(x^2 - 1) = 7x(x^4 - 1) = 7x^5 - 7x$$

We obtain the original polynomial, so the factorization is correct.

✓ **CHECK POINT 3** Factor: $4x^5 - 64x$.

EXAMPLE 4   Factoring a Polynomial

Factor: $x^3 - 5x^2 - 4x + 20$.

Solution

Step 1. If there is a common factor, factor out the GCF. Other than 1, there is no common factor.

Step 2. Determine the number of terms and factor accordingly. There are four terms. We try factoring by grouping.

$$x^3 - 5x^2 - 4x + 20$$
$$= (x^3 - 5x^2) + (-4x + 20) \qquad \text{Group terms with common factors.}$$
$$= x^2(x - 5) - 4(x - 5) \qquad \text{Factor from each group.}$$
$$= (x - 5)(x^2 - 4) \qquad \text{Factor out the common binomial factor, } x - 5.$$

Step 3. Check to see if factors can be factored further. We note that $(x^2 - 4)$ is the difference of two squares, $x^2 - 2^2$, so we continue factoring.

$$x^3 - 5x^2 - 4x + 20 = (x - 5)(x + 2)(x - 2) \qquad \text{Factor } x^2 - 4 \text{ as the}$$
$$\text{difference of two squares.}$$

We have factored completely because no factor with more than one term can be factored further.

Step 4. Check by multiplying.

$$\overset{\text{F \quad O \quad I \quad L}}{(x - 5)(x + 2)(x - 2)} = (x - 5)(x^2 - 4) = x^3 - 4x - 5x^2 + 20$$
$$= x^3 - 5x^2 - 4x + 20$$

We obtain the original polynomial, so the factorization is correct.

✓ **CHECK POINT 4**   Factor: $x^3 - 4x^2 - 9x + 36$.

EXAMPLE 5   Factoring a Polynomial

Factor: $2x^3 - 24x^2 + 72x$.

Solution

Step 1. If there is a common factor, factor out the GCF. Because $2x$ is common to all terms, we factor it out.

$$2x^3 - 24x^2 + 72x = 2x(x^2 - 12x + 36) \qquad \text{Factor out the GCF.}$$

Step 2. Determine the number of terms and factor accordingly. The factor $x^2 - 12x + 36$ has three terms. Is it a perfect square trinomial? Yes. The first term, x^2, is the square of a monomial. The last term, 36 or 6^2, is the square of an integer. The middle term involves twice the product of x and 6. We factor using $A^2 - 2AB + B^2 = (A - B)^2$.

$$2x^3 - 24x^2 + 72x = 2x(x^2 - 12x + 36)$$
$$= 2x(\underbrace{x^2 - 2 \cdot x \cdot 6 + 6^2}_{A^2 - 2 \ A \ B + B^2}) \qquad \text{The second factor is a perfect square trinomial.}$$
$$= 2x(x - 6)^2 \qquad A^2 - 2AB + B^2 = (A - B)^2$$

Step 3. Check to see if factors can be factored further. In this problem, they cannot, so we have factored completely.

Step 4. Check by multiplying. Let's verify that $2x^3 - 24x^2 + 72x = 2x(x - 6)^2$.

$$2x(x - 6)^2 = 2x(x^2 - 12x + 36) = 2x^3 - 24x^2 + 72x$$

We obtain the original polynomial, so the factorization is correct.

☑ **CHECK POINT 5** Factor: $3x^3 - 30x^2 + 75x$.

EXAMPLE 6 **Factoring a Polynomial**

Factor: $3x^5 + 24x^2$.

Solution

Step 1. If there is a common factor, factor out the GCF. Because $3x^2$ is common to both terms, we factor it out.

$$3x^5 + 24x^2 = 3x^2(x^3 + 8) \qquad \text{Factor out the GCF.}$$

Step 2. Determine the number of terms and factor accordingly. The factor $x^3 + 8$ has two terms. This binomial can be expressed as $x^3 + 2^3$, so it can be factored as the sum of two cubes.

$$3x^5 + 24x^2 = 3x^2(x^3 + 2^3) \qquad \text{Express } x^3 + 8 \text{ as the sum of two cubes.}$$

$$A^3 + B^3$$

$$= 3x^2(x + 2)(x^2 - 2x + 4) \qquad \text{Factor the sum of two cubes.}$$

$$(A + B)(A^2 - AB + B^2)$$

Step 3. Check to see if factors can be factored further. In this problem, they cannot, so we have factored completely.

Step 4. Check by multiplying.

$$3x^2(x + 2)(x^2 - 2x + 4) = 3x^2[x(x^2 - 2x + 4) + 2(x^2 - 2x + 4)]$$
$$= 3x^2(x^3 - 2x^2 + 4x + 2x^2 - 4x + 8)$$
$$= 3x^2(x^3 + 8) = 3x^5 + 24x^2$$

We obtain the original polynomial, so the factorization is correct.

☑ **CHECK POINT 6** Factor: $2x^5 + 54x^2$.

Discover for Yourself

In Examples 1–6, substitute 1 for the variable in both the given polynomial and in its factored form. Evaluate each expression. What do you observe? Do this for a second value of the variable. Is this a complete check or only a partial check of the factorization? Explain.

EXAMPLE 7 **Factoring a Polynomial in Two Variables**

Factor: $32x^4y - 2y^5$.

Solution

Step 1. If there is a common factor, factor out the GCF. Because $2y$ is common to both terms, we factor it out.

$$32x^4y - 2y^5 = 2y(16x^4 - y^4) \qquad \text{Factor out the GCF.}$$

Step 2. Determine the number of terms and factor accordingly. The factor $16x^4 - y^4$ has two terms. It is the difference of two squares: $(4x^2)^2 - (y^2)^2$. We factor using the special form for the difference of two squares.

$$32x^4y - 2y^5 = 2y[(4x^2)^2 - (y^2)^2]$$

Express $16x^4 - y^4$ as the difference of two squares.

$$= 2y(4x^2 + y^2)(4x^2 - y^2) \quad A^2 - B^2 = (A + B)(A - B)$$

$(A + B) \quad (A - B)$

Step 3. Check to see if factors can be factored further. We note that the last factor, $4x^2 - y^2$, is also the difference of two squares, $(2x)^2 - y^2$, so we continue factoring.

$$32x^4y - 2y^5 = 2y(4x^2 + y^2)(2x + y)(2x - y)$$

Step 4. Check by multiplying. Multiply the factors in the factorization and verify that you obtain the original polynomial. ■

✓ **CHECK POINT 7** Factor: $3x^4y - 48y^5$.

EXAMPLE 8 Factoring a Polynomial in Two Variables

Factor: $18x^3 + 48x^2y + 32xy^2$.

Solution

Step 1. If there is a common factor, factor out the GCF. Because $2x$ is common to all terms, we factor it out.

$$18x^3 + 48x^2y + 32xy^2 = 2x(9x^2 + 24xy + 16y^2)$$

Step 2. Determine the number of terms and factor accordingly. The factor $9x^2 + 24xy + 16y^2$ has three terms. Is it a perfect square trinomial? Yes. The first term, $9x^2$ or $(3x)^2$, and the last term, $16y^2$ or $(4y)^2$, are squares of monomials. The middle term, $24xy$, is twice the product of $3x$ and $4y$. We factor using $A^2 + 2AB + B^2 = (A + B)^2$.

$$18x^3 + 48x^2y + 32xy^2 = 2x(9x^2 + 24xy + 16y^2)$$
$$= 2x[(3x)^2 + 2 \cdot 3x \cdot 4y + (4y)^2] \quad \text{The second factor is a perfect square trinomial.}$$

$A^2 + 2 \cdot A \cdot B + B^2$

$$= 2x(3x + 4y)^2 \quad A^2 + 2AB + B^2 = (A + B)^2$$

Step 3. Check to see if factors can be factored further. In this problem, they cannot, so we have factored completely.

Step 4. Check by multiplication. Multiply the factors in the factorization and verify that you obtain the original polynomial. ■

✓ **CHECK POINT 8** Factor: $12x^3 + 36x^2y + 27xy^2$.

7.5 EXERCISE SET *MyMathLab*

Practice Exercises

In Exercises 1–64, factor completely, or state that the polynomial is prime. Check factorizations using multiplication or a graphing utility.

1. $5x^3 - 20x$

2. $4x^3 - 100x$

3. $7x^3 + 7x$

4. $6x^3 + 24x$

5. $5x^2 - 5x - 30$

6. $5x^2 - 15x - 50$

7. $2x^4 - 162$

8. $7x^4 - 7$

9. $x^3 + 2x^2 - 9x - 18$

10. $x^3 + 3x^2 - 25x - 75$

11. $3x^3 - 24x^2 + 48x$

12. $5x^3 - 20x^2 + 20x$

13. $2x^5 + 2x^2$

14. $2x^5 + 128x^2$

15. $6x^2 + 8x$

16. $21x^2 - 35x$

17. $2y^2 - 2y - 112$

18. $6x^2 - 6x - 12$

19. $7y^4 + 14y^3 + 7y^2$

20. $2y^4 + 28y^3 + 98y^2$

21. $y^2 + 8y - 16$

22. $y^2 - 18y - 81$

23. $16y^2 - 4y - 2$

24. $32y^2 + 4y - 6$

25. $r^2 - 25r$

26. $3r^2 - 27r$

27. $4w^2 + 8w - 5$

28. $35w^2 - 2w - 1$

29. $x^3 - 4x$

30. $9x^3 - 9x$

31. $x^2 + 64$

32. $y^2 + 36$

33. $9y^2 + 13y + 4$

34. $20y^2 + 12y + 1$

35. $y^3 + 2y^2 - 4y - 8$

36. $y^3 + 2y^2 - y - 2$

37. $16y^2 + 24y + 9$

38. $25y^2 + 20y + 4$

39. $4y^3 - 28y^2 + 40y$

40. $7y^3 - 21y^2 + 14y$

41. $y^5 - 81y$

42. $y^5 - 16y$

43. $20a^4 - 45a^2$

44. $48a^4 - 3a^2$

45. $9x^4 + 18x^3 + 6x^2$

46. $10x^4 + 20x^3 + 15x^2$

47. $12y^2 - 11y + 2$

48. $21x^2 - 25x - 4$

49. $9y^2 - 64$

50. $100y^2 - 49$

51. $9y^2 + 64$

52. $100y^2 + 49$

53. $2y^3 + 3y^2 - 50y - 75$

54. $12y^3 + 16y^2 - 3y - 4$

55. $2r^3 + 30r^2 - 68r$

56. $3r^3 - 27r^2 - 210r$

57. $8x^5 - 2x^3$

58. $y^9 - y^5$

59. $3x^2 + 243$

60. $27x^2 + 75$

61. $x^4 + 8x$

62. $x^4 + 27x$

63. $2y^5 - 2y^2$

64. $2y^5 - 128y^2$

Exercises 65–96 contain polynomials in several variables. Factor each polynomial completely and check using multiplication.

65. $6x^2 + 8xy$

66. $21x^2 - 35xy$

67. $xy - 7x + 3y - 21$

68. $xy - 5x + 2y - 10$

69. $x^2 - 3xy - 4y^2$

70. $x^2 - 4xy - 12y^2$

71. $72a^3b^2 + 12a^2 - 24a^4b^2$

72. $24a^4b + 60a^3b^2 + 150a^2b^3$

73. $3a^2 + 27ab + 54b^2$

74. $3a^2 + 15ab + 18b^2$

75. $48x^4y - 3x^2y$

76. $16a^3b^2 - 4ab^2$

77. $6a^2b + ab - 2b$

78. $16a^2 - 32ab + 12b^2$

79. $7x^5y - 7xy^5$

80. $3x^4y^2 - 3x^2y^2$

81. $10x^3y - 14x^2y^2 + 4xy^3$

82. $18x^3y + 57x^2y^2 + 30xy^3$

83. $2bx^2 + 44bx + 242b$

84. $3xz^2 - 72xz + 432x$

85. $15a^2 + 11ab - 14b^2$

86. $25a^2 + 25ab + 6b^2$

87. $36x^3y - 62x^2y^2 + 12xy^3$

88. $10a^4b^2 - 15a^3b^3 - 25a^2b^4$

89. $a^2y - b^2y - a^2x + b^2x$

90. $bx^2 - 4b + ax^2 - 4a$

91. $9ax^3 + 15ax^2 - 14ax$

92. $4ay^3 - 12ay^2 + 9ay$

93. $2x^4 + 6x^3y + 2x^2y^2$

94. $3x^4 - 9x^3y + 3x^2y^2$

95. $81x^4y - y^5$

96. $16x^4y - y^5$

Practice PLUS

In Exercises 97–106, factor completely.

97. $10x^2(x + 1) - 7x(x + 1) - 6(x + 1)$

98. $12x^2(x - 1) - 4x(x - 1) - 5(x - 1)$

99. $6x^4 + 35x^2 - 6$

100. $7x^4 + 34x^2 - 5$

101. $(x - 7)^2 - 4a^2$

102. $(x - 6)^2 - 9a^2$

103. $x^2 + 8x + 16 - 25a^2$

104. $x^2 + 14x + 49 - 16a^2$

105. $y^7 + y$

106. $(y + 1)^3 + 1$

Application Exercises

107. A rock is dropped from the top of a 256-foot cliff. The height, in feet, of the rock above the water after t seconds is modeled by the polynomial $256 - 16t^2$. Factor this expression completely.

256 feet

108. The building shown in the figure has a height represented by x feet. The building's base is a square and its volume is $x^3 - 60x^2 + 900x$ cubic feet. Express the building's dimensions in terms of x.

x

109. Express the area of the shaded ring shown in the figure in terms of π. Then factor this expression completely.

a

b

Writing in Mathematics

110. Describe a strategy that can be used to factor polynomials.

111. Describe some of the difficulties in factoring polynomials. What suggestions can you offer to overcome these difficulties?

112. You are about to take a great picture of fog rolling into San Francisco from the middle of the Golden Gate Bridge, 400 feet above the water. Whoops! You accidently lean too far over the safety rail and drop your camera. The height, in feet, of the camera after t seconds is modeled by the polynomial $400 - 16t^2$. The factored form of the polynomial is $16(5 + t)(5 - t)$. Describe something about your falling camera that is easier to see from the factored form, $16(5 + t)(5 - t)$, than from the form $400 - 16t^2$.

Critical Thinking Exercises

Make Sense? *In Exercises 113–116, determine whether each statement "makes sense" or "does not make sense" and explain your reasoning.*

113. It takes a great deal of practice to get good at factoring a wide variety of polynomials.

114. Multiplying polynomials is relatively mechanical, but factoring often requires a great deal of thought.

115. The factorable trinomial $4x^2 + 8x + 3$ and the prime trinomial $4x^2 + 8x + 1$ are in the form $ax^2 + bx + c$, but $b^2 - 4ac$ is a perfect square only in the case of the factorable trinomial.

116. When a factorization requires two factoring techniques, I'm less likely to make errors if I show one technique at a time rather than combining the two factorizations into one step.

In Exercises 117–120, determine whether each statement is true or false. If the statement is false, make the necessary change(s) to produce a true statement.

117. $x^2 - 9 = (x - 3)^2$ for any real number x.

118. The polynomial $4x^2 + 100$ is the sum of two squares and therefore cannot be factored.

119. If the general factoring strategy is used to factor a polynomial, at least two factorizations are necessary before the given polynomial is factored completely.

120. Once a common monomial factor is removed from $3xy^3 + 9xy^2 + 21xy$, the remaining trinomial factor cannot be factored further.

In Exercises 121–125, factor completely.

121. $3x^5 - 21x^3 - 54x$

122. $5y^5 - 5y^4 - 20y^3 + 20y^2$

123. $4x^4 - 9x^2 + 5$

124. $(x + 5)^2 - 20(x + 5) + 100$

125. $3x^{2n} - 27y^{2n}$

Technology Exercises

In Exercises 126–130, use the GRAPH *or* TABLE *feature of a graphing utility to determine if the polynomial on the left side of each equation has been correctly factored. If not, factor the polynomial correctly and then use your graphing utility to verify the factorization.*

126. $4x^2 - 12x + 9 = (4x - 3)^2$; $[-5, 5, 1]$ by $[0, 20, 1]$

127. $3x^3 - 12x^2 - 15x = 3x(x + 5)(x - 1)$; $[-5, 7, 1]$ by $[-80, 80, 10]$

128. $6x^2 + 10x - 4 = 2(3x - 1)(x + 2)$; $[-5, 5, 1]$ by $[-20, 20, 2]$

129. $x^4 - 16 = (x^2 + 4)(x + 2)(x - 2)$; $[-5, 5, 1]$ by $[-20, 20, 2]$

130. $2x^3 + 10x^2 - 2x - 10 = 2(x + 5)(x^2 + 1)$; $[-8, 4, 1]$ by $[-100, 100, 10]$

Review Exercises

131. Factor: $9x^2 - 16$. (Section 7.4, Example 1)

132. Graph using intercepts: $5x - 2y = 10$. (Section 4.2, Example 4)

133. The second angle of a triangle measures three times that of the first angle's measure. The third angle measures $80°$ more than the first. Find the measure of each angle. (Section 3.3, Example 6)

Preview Exercises

Exercises 134–136 will help you prepare for the material covered in the next section.

134. Evaluate $(3x - 1)(x + 2)$ for $x = \dfrac{1}{3}$.

135. Evaluate $2x^2 + 7x - 4$ for $x = \dfrac{1}{2}$.

136. Factor: $(x - 2)(x + 3) - 6$.

SECTION **7.6**

Solving Quadratic Equations by Factoring

Objectives

1 Use the zero-product principle.

2 Solve quadratic equations by factoring.

3 Solve problems using quadratic equations.

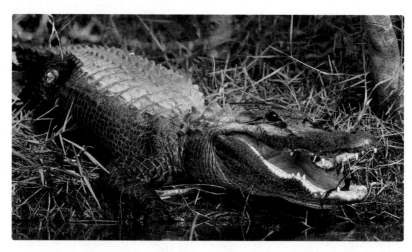

The alligator, at one time an endangered species, was the subject of a protection program at Florida's Everglades National Park. Park rangers used the formula

$$P = -10x^2 + 475x + 3500$$

to estimate the alligator population, P, after x years of the protection program. Their goal was to bring the population up to 7250. To find out how long the program had to be continued for this to happen, we substitute 7250 for P in the formula and solve for x:

$$7250 = -10x^2 + 475x + 3500.$$

Do you see how this equation differs from a linear equation? The highest exponent on x is 2. Solving such an equation involves finding the numbers that will make the equation a true statement. In this section, we use factoring to solve equations in the form $ax^2 + bx + c = 0$. We also look at applications of these equations.

The Standard Form of a Quadratic Equation

We begin by defining a quadratic equation.

> ### Definition of a Quadratic Equation
>
> A **quadratic equation** in x is an equation that can be written in the **standard form**
>
> $$ax^2 + bx + c = 0,$$
>
> where a, b, and c are real numbers, with $a \neq 0$. A quadratic equation in x is also called a **second-degree polynomial equation** in x.

Here is an example of a quadratic equation in standard form:

$$x^2 - 7x + 10 = 0.$$

$a = 1$ $b = -7$ $c = 10$

1 Use the zero-product principle.

Solving Quadratic Equations by Factoring

We can factor the left side of the quadratic equation $x^2 - 7x + 10 = 0$. We obtain $(x - 5)(x - 2) = 0$. If a quadratic equation has zero on one side and a factored expression on the other side, it can be solved using the **zero-product principle**.

> ### The Zero-Product Principle
>
> If the product of two algebraic expressions is zero, then at least one of the factors is equal to zero.
>
> $$\text{If } AB = 0, \text{ then } A = 0 \text{ or } B = 0.$$

For example, consider the equation $(x - 5)(x - 2) = 0$. According to the zero-product principle, this product can be zero only if at least one of the factors is zero. We set each individual factor equal to zero and solve each resulting equation for x.

$$(x - 5)(x - 2) = 0$$
$$x - 5 = 0 \quad \text{or} \quad x - 2 = 0$$
$$x = 5 \qquad\qquad x = 2$$

We can check each of the proposed solutions, 5 and 2, in the original quadratic equation, $x^2 - 7x + 10 = 0$. Substitute each one separately for x in the equation.

Check 5:	Check 2:
$x^2 - 7x + 10 = 0$	$x^2 - 7x + 10 = 0$
$5^2 - 7 \cdot 5 + 10 \stackrel{?}{=} 0$	$2^2 - 7 \cdot 2 + 10 \stackrel{?}{=} 0$
$25 - 35 + 10 \stackrel{?}{=} 0$	$4 - 14 + 10 \stackrel{?}{=} 0$
$0 = 0, \ \text{true}$	$0 = 0, \ \text{true}$

The resulting true statements indicate that the solutions are 5 and 2. Note that with a quadratic equation, we can have two solutions, compared to the linear equation that usually had one.

EXAMPLE 1 Using the Zero-Product Principle

Solve the equation: $(3x - 1)(x + 2) = 0$.

Solution The product $(3x - 1)(x + 2)$ is equal to zero. By the zero-product principle, the only way that this product can be zero is if at least one of the factors is zero. Thus,

$$3x - 1 = 0 \quad \text{or} \quad x + 2 = 0.$$
$$3x = 1 \qquad\qquad x = -2 \qquad \text{\small Solve each equation for x.}$$
$$x = \frac{1}{3}$$

Because each linear equation has a solution, the original equation, $(3x - 1)(x + 2) = 0$, has two solutions, $\frac{1}{3}$ and -2. Check these solutions by substituting each one separately into the given equation. The equation's solution set is $\left\{-2, \frac{1}{3}\right\}$. ■

☑ **CHECK POINT 1** Solve the equation: $(2x + 1)(x - 4) = 0$.

---

2 Solve quadratic equations by factoring.

In Example 1 and Check Point 1, the given equations were in factored form. Here is a procedure for solving a quadratic equation when we must first do the factoring.

> **Solving a Quadratic Equation by Factoring**
>
> **1.** If necessary, rewrite the equation in the standard form $ax^2 + bx + c = 0$, moving all terms to one side, thereby obtaining zero on the other side.
> **2.** Factor.
> **3.** Apply the zero-product principle, setting each factor equal to zero.
> **4.** Solve the equations formed in step 3.
> **5.** Check the solutions in the original equation.

EXAMPLE 2 Solving a Quadratic Equation by Factoring

Solve: $2x^2 + 7x - 4 = 0$.

Solution

Step 1. Move all terms to one side and obtain zero on the other side. All terms are already on the left and zero is on the other side, so we can skip this step.

Step 2. Factor.

$$2x^2 + 7x - 4 = 0$$
$$(2x - 1)(x + 4) = 0$$

Steps 3 and 4. Set each factor equal to zero and solve each resulting equation.

$$2x - 1 = 0 \quad \text{or} \quad x + 4 = 0$$
$$2x = 1 \qquad\qquad x = -4$$
$$x = \frac{1}{2}$$

Step 5. Check the solutions in the original equation.

Check $\frac{1}{2}$:

$$2x^2 + 7x - 4 = 0$$

$$2\left(\frac{1}{2}\right)^2 + 7\left(\frac{1}{2}\right) - 4 \overset{?}{=} 0$$

$$2\left(\frac{1}{4}\right) + 7\left(\frac{1}{2}\right) - 4 \overset{?}{=} 0$$

$$\frac{1}{2} + \frac{7}{2} - 4 \overset{?}{=} 0$$

$$4 - 4 \overset{?}{=} 0$$

$$0 = 0, \;\; \text{true}$$

Check -4:

$$2x^2 + 7x - 4 = 0$$

$$2(-4)^2 + 7(-4) - 4 \overset{?}{=} 0$$

$$2(16) + 7(-4) - 4 \overset{?}{=} 0$$

$$32 + (-28) - 4 \overset{?}{=} 0$$

$$4 - 4 \overset{?}{=} 0$$

$$0 = 0, \;\; \text{true}$$

The solutions are -4 and $\frac{1}{2}$, and the solution set is $\left\{-4, \frac{1}{2}\right\}$.

☑ **CHECK POINT 2** Solve: $x^2 - 6x + 5 = 0$.

Study Tip

Do not confuse factoring a polynomial with solving a quadratic equation by factoring.

Factoring a Polynomial	**Solving a Quadratic Equation**
Factor: $2x^2 + 7x - 4$.	Solve: $2x^2 + 7x - 4 = 0$.

This is not an equation. There is no equal sign.

This is an equation. There is an equal sign.

Solution $(2x - 1)(x + 4)$

Solution $(2x - 1)(x + 4) = 0$

Stop! Avoid the common error of setting each factor equal to zero.

$$2x - 1 = 0 \quad \text{or} \quad x + 4 = 0$$

$$x = \frac{1}{2} \qquad\qquad x = -4$$

The solution set is $\left\{-4, \frac{1}{2}\right\}$.

Using Technology

Graphic Connections

You can use a graphing utility to check the real number solutions of a quadratic equation. **The solutions of $ax^2 + bx + c = 0$ correspond to the x-intercepts for the graph of $y = ax^2 + bx + c$.** For example, to check the solutions of $2x^2 + 7x - 4 = 0$, graph $y = 2x^2 + 7x - 4$. The U-shaped, bowl-like, graph is shown on the right. The x-intercepts are -4 and $\frac{1}{2}$, verifying -4 and $\frac{1}{2}$ as the solutions.

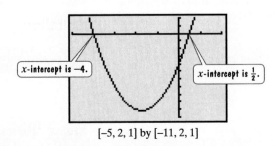

x-intercept is -4.

x-intercept is $\frac{1}{2}$.

$[-5, 2, 1]$ by $[-11, 2, 1]$

EXAMPLE 3 Solving a Quadratic Equation by Factoring

Solve: $3x^2 = 2x$.

Solution

Step 1. Move all terms to one side and obtain zero on the other side. Subtract $2x$ from both sides and write the equation in standard form.

$$3x^2 - 2x = 2x - 2x$$
$$3x^2 - 2x = 0$$

Step 2. Factor. We factor out x from the two terms on the left side.

$$3x^2 - 2x = 0$$
$$x(3x - 2) = 0$$

Steps 3 and 4. Set each factor equal to zero and solve the resulting equations.

$$x = 0 \quad \text{or} \quad 3x - 2 = 0$$
$$3x = 2$$
$$x = \frac{2}{3}$$

Step 5. Check the solutions in the original equation.

Study Tip

Avoid dividing both sides of $3x^2 = 2x$ by x. You will obtain $3x = 2$ and, consequently, $x = \frac{2}{3}$. The other solution, 0, is lost. We can divide both sides of an equation by any *nonzero* real number. If x is zero, we lose the second solution.

Check 0:

$$3x^2 = 2x$$
$$3 \cdot 0^2 \overset{?}{=} 2 \cdot 0$$
$$0 = 0, \quad \text{true}$$

Check $\frac{2}{3}$:

$$3x^2 = 2x$$
$$3\left(\frac{2}{3}\right)^2 \overset{?}{=} 2\left(\frac{2}{3}\right)$$
$$3\left(\frac{4}{9}\right) \overset{?}{=} 2\left(\frac{2}{3}\right)$$
$$\frac{4}{3} = \frac{4}{3}, \quad \text{true}$$

The solutions are 0 and $\frac{2}{3}$, and the solution set is $\left\{0, \frac{2}{3}\right\}$.

✓ **CHECK POINT 3** Solve: $4x^2 = 2x$.

EXAMPLE 4 Solving a Quadratic Equation by Factoring

Solve: $x^2 = 6x - 9$.

Solution

Step 1. Move all terms to one side and obtain zero on the other side. To obtain zero on the right, we subtract $6x$ and add 9 on both sides.

$$x^2 - 6x + 9 = 6x - 6x - 9 + 9$$
$$x^2 - 6x + 9 = 0$$

Step 2. Factor. The trinomial on the left side is a perfect square trinomial: $x^2 - 6x + 9 = x^2 - 2 \cdot x \cdot 3 + 3^2$. We factor using $A^2 - 2AB + B^2 = (A - B)^2$: $A = x$ and $B = 3$.

$$x^2 - 6x + 9 = 0$$
$$(x - 3)^2 = 0$$

Using Technology

Graphic Connections

The graph of $y = x^2 - 6x + 9$ is shown below. Notice that there is only one x-intercept, namely 3, verifying that the solution of

$$x^2 - 6x + 9 = 0$$

is 3.

$[-1, 6, 1]$ by $[-2, 10, 1]$

Steps 3 and 4. Set each factor equal to zero and solve the resulting equations. Because both factors are the same, it is only necessary to set one of them equal to zero.

$$x - 3 = 0$$
$$x = 3$$

Step 5. Check the solution in the original equation.

Check 3:

$$x^2 = 6x - 9$$
$$3^2 \overset{?}{=} 6 \cdot 3 - 9$$
$$9 \overset{?}{=} 18 - 9$$
$$9 = 9, \quad \text{true}$$

The solution is 3 and the solution set is $\{3\}$.

✓ **CHECK POINT 4** Solve: $x^2 = 10x - 25$.

EXAMPLE 5 Solving a Quadratic Equation by Factoring

Solve: $9x^2 = 16$.

Solution

Step 1. Move all terms to one side and obtain zero on the other side. Subtract 16 from both sides and write the equation in standard form.

$$9x^2 - 16 = 16 - 16$$
$$9x^2 - 16 = 0$$

Step 2. Factor. The binomial on the left side is the difference of two squares: $9x^2 - 16 = (3x)^2 - 4^2$. We factor using $A^2 - B^2 = (A + B)(A - B)$: $A = 3x$ and $B = 4$.

$$9x^2 - 16 = 0$$
$$(3x + 4)(3x - 4) = 0$$

Steps 3 and 4. Set each factor equal to zero and solve the resulting equations. We use the zero-product principle to solve $(3x + 4)(3x - 4) = 0$.

$$3x + 4 = 0 \quad \text{or} \quad 3x - 4 = 0$$
$$3x = -4 \qquad\qquad 3x = 4$$
$$x = -\frac{4}{3} \qquad\qquad x = \frac{4}{3}$$

Step 5. Check the solutions in the original equation. Do this now and verify that the solutions of $9x^2 = 16$ are $-\frac{4}{3}$ and $\frac{4}{3}$. The equation's solution set is $\left\{-\frac{4}{3}, \frac{4}{3}\right\}$.

✓ **CHECK POINT 5** Solve: $16x^2 = 25$.

EXAMPLE 6   Solving a Quadratic Equation by Factoring

Solve: $(x - 2)(x + 3) = 6$.

Solution

Step 1. Move all terms to one side and obtain zero on the other side. We write the equation in standard form by multiplying out the product on the left side and then subtracting 6 from both sides.

$(x - 2)(x + 3) = 6$	This is the given equation.
$x^2 + 3x - 2x - 6 = 6$	Use the FOIL method.
$x^2 + x - 6 = 6$	Simplify.
$x^2 + x - 6 - 6 = 6 - 6$	Subtract 6 from both sides.
$x^2 + x - 12 = 0$	Simplify.

Step 2. Factor.

$$x^2 + x - 12 = 0$$
$$(x + 4)(x - 3) = 0$$

Steps 3 and 4. Set each factor equal to zero and solve the resulting equations.

$$x + 4 = 0 \quad \text{or} \quad x - 3 = 0$$
$$x = -4 \qquad\qquad x = 3$$

Step 5. Check the solutions in the original equation. Do this now and verify that the solutions are -4 and 3. The equation's solution set is $\{-4, 3\}$.   ■

☑ **CHECK POINT 6**   Solve: $(x - 5)(x - 2) = 28$.

③ Solve problems using quadratic equations.

Applications of Quadratic Equations

Solving quadratic equations by factoring can be used to answer questions about variables contained in mathematical models.

EXAMPLE 7   Modeling Motion

You throw a ball straight up from a rooftop 160 feet high with an initial speed of 48 feet per second. The formula

$$h = -16t^2 + 48t + 160$$

describes the ball's height above the ground, h, in feet, t seconds after you throw it. The ball misses the rooftop on its way down and eventually strikes the ground. The situation is illustrated in **Figure 7.1**. How long will it take for the ball to hit the ground?

Solution   The ball hits the ground when h, its height above the ground, is 0 feet. Thus, we substitute 0 for h in the given formula and solve for t.

$h = -16t^2 + 48t + 160$	This is the formula that models the ball's height.
$0 = -16t^2 + 48t + 160$	Substitute 0 for h.

It is easier to factor a trinomial with a positive leading coefficient. Thus, if a negative coefficient appears in the second-degree term of a quadratic equation, we make it positive by multiplying both sides of the equation by -1.

$$-1 \cdot 0 = -1(-16t^2 + 48t + 160)$$
$$0 = 16t^2 - 48t - 160$$

Do you see that each term on the right side of the equation changed sign? The left side of the equation remained zero.

FIGURE 7.1

Now we continue to solve the equation.

$$16t^2 - 48t - 160 = 0$$ *Reverse the two sides of the equation. This step is optional.*

$$16(t^2 - 3t - 10) = 0$$ *Factor out the GCF, 16.*

> Do not set the constant, 16, equal to zero: $16 \neq 0$.

$$16(t - 5)(t + 2) = 0$$ *Factor the trinomial.*

$$t - 5 = 0 \quad \text{or} \quad t + 2 = 0$$ *Set each variable factor equal to 0.*

$$t = 5 \qquad\qquad t = -2$$ *Solve for t.*

Because we begin describing the ball's height at $t = 0$, we discard the solution $t = -2$. The ball hits the ground after 5 seconds. ∎

Figure 7.2 shows the graph of the formula $h = -16t^2 + 48t + 160$. The horizontal axis is labeled t, for the ball's time in motion. The vertical axis is labeled h, for the ball's height above the ground. Because time and height are both positive, the model is graphed in quadrant I only.

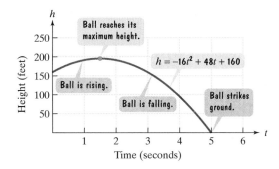

FIGURE 7.2

The graph visually shows what we discovered algebraically: The ball hits the ground after 5 seconds. The graph also reveals that the ball reaches its maximum height, nearly 200 feet, after 1.5 seconds. Then the ball begins to fall.

✓ **CHECK POINT 7**   Use the formula $h = -16t^2 + 48t + 160$ to determine when the ball's height is 192 feet. Identify your solutions as points on the graph in **Figure 7.2**.

In our next example, we use our five-step strategy for solving word problems.

EXAMPLE 8   Solving a Problem About a Rectangle's Area

An architect is allowed no more than 15 square meters to add a small bedroom to a house. Because of the room's design in relationship to the existing structure, the width of its rectangular floor must be 7 meters less than two times the length. Find the precise length and width of the rectangular floor of maximum area that the architect is permitted.

Solution

Step 1.   Let x represent one of the quantities.   We know something about the width: It must be 7 meters less than two times the length. We will let

$$x = \text{the length of the floor.}$$

Step 2.   Represent other unknown quantities in terms of x.   Because the width must be 7 meters less than two times the length, let

$$2x - 7 = \text{the width of the floor.}$$

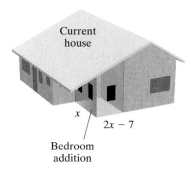

FIGURE 7.3

The problem is illustrated in **Figure 7.3**.

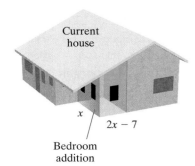

x

$2x - 7$

Bedroom
addition

FIGURE 7.3 (repeated)

Step 3. Write an equation that models the conditions. Because the architect is allowed no more than 15 square meters, an area of 15 square meters is the maximum area permitted. The area of a rectangle is the product of its length and its width.

Length of the floor	times	Width of the floor	is	the area.
x	$\cdot$	$(2x - 7)$	$=$	15

Step 4. Solve the equation and answer the question.

$$x(2x - 7) = 15$$ This is the equation that models the problem's conditions.

$$2x^2 - 7x = 15$$ Use the distributive property.

$$2x^2 - 7x - 15 = 0$$ Subtract 15 from both sides.

$$(2x + 3)(x - 5) = 0$$ Factor.

$$2x + 3 = 0 \quad \text{or} \quad x - 5 = 0$$ Set each factor equal to zero.

$$2x = -3 \qquad\qquad x = 5$$ Solve the resulting equations.

$$x = -\frac{3}{2}$$

A rectangle cannot have a negative length. Thus,

$$\text{Length} = x = 5$$
$$\text{Width} = 2x - 7 = 2 \cdot 5 - 7 = 10 - 7 = 3.$$

The architect is permitted a room of maximum area whose length is 5 meters and whose width is 3 meters.

Step 5. Check the proposed solution in the original wording of the problem. The area of the floor using the dimensions that we found is

$$A = lw = (5 \text{ meters})(3 \text{ meters}) = 15 \text{ square meters}.$$

Because the problem's wording tells us that the maximum area permitted is 15 square meters, our dimensions are correct.

✓ **CHECK POINT 8**   The length of a rectangular sign is 3 feet longer than the width. If the sign's area is 54 square feet, find its length and width.

7.6 EXERCISE SET   **MyMathLab**   Math XL PRACTICE   WATCH   DOWNLOAD   READ   REVIEW

Practice Exercises

In Exercises 1–8, solve each equation using the zero-product principle.

1. $x(x + 7) = 0$

2. $x(x - 3) = 0$

3. $(x - 6)(x + 4) = 0$

4. $(x - 3)(x + 8) = 0$

5. $(x - 9)(5x + 4) = 0$

6. $(x + 7)(3x - 2) = 0$

7. $10(x - 4)(2x + 9) = 0$

8. $8(x - 5)(3x + 11) = 0$

In Exercises 9–56, use factoring to solve each quadratic equation. Check by substitution or by using a graphing utility and identifying x-intercepts.

9. $x^2 + 8x + 15 = 0$

10. $x^2 + 5x + 6 = 0$

11. $x^2 - 2x - 15 = 0$

12. $x^2 + x - 42 = 0$

13. $x^2 - 4x = 21$

14. $x^2 + 7x = 18$

15. $x^2 + 9x = -8$

16. $x^2 - 11x = -10$

17. $x^2 + 4x = 0$

18. $x^2 - 6x = 0$

19. $x^2 - 5x = 0$

20. $x^2 + 3x = 0$

21. $x^2 = 4x$

22. $x^2 = 8x$

23. $2x^2 = 5x$

24. $3x^2 = 5x$

25. $3x^2 = -5x$

26. $2x^2 = -3x$

27. $x^2 + 4x + 4 = 0$

28. $x^2 + 6x + 9 = 0$

29. $x^2 = 12x - 36$

30. $x^2 = 14x - 49$

31. $4x^2 = 12x - 9$

32. $9x^2 = 30x - 25$

33. $2x^2 = 7x + 4$

34. $3x^2 = x + 4$

35. $5x^2 = 18 - x$

36. $3x^2 = 15 + 4x$

37. $x^2 - 49 = 0$

38. $x^2 - 25 = 0$

39. $4x^2 - 25 = 0$

40. $9x^2 - 100 = 0$

41. $81x^2 = 25$

42. $25x^2 = 49$

43. $x(x - 4) = 21$

44. $x(x - 3) = 18$

45. $4x(x + 1) = 15$

46. $x(3x + 8) = -5$

47. $(x - 1)(x + 4) = 14$

48. $(x - 3)(x + 8) = -30$

49. $(x + 1)(2x + 5) = -1$

50. $(x + 3)(3x + 5) = 7$

51. $y(y + 8) = 16(y - 1)$

52. $y(y + 9) = 4(2y + 5)$

53. $4y^2 + 20y + 25 = 0$

54. $4y^2 + 44y + 121 = 0$

55. $64w^2 = 48w - 9$

56. $25w^2 = 80w - 64$

Practice PLUS

In Exercises 57–66, solve each equation and check your solutions.

57. $(x - 4)(x^2 + 5x + 6) = 0$

58. $(x - 5)(x^2 - 3x + 2) = 0$

59. $x^3 - 36x = 0$

60. $x^3 - 4x = 0$

61. $y^3 + 3y^2 + 2y = 0$

62. $y^3 + 2y^2 - 3y = 0$

63. $2(x - 4)^2 + x^2 = x(x + 50) - 46x$

64. $(x - 4)(x - 5) + (2x + 3)(x - 1) = x(2x - 25) - 13$

65. $(x - 2)^2 - 5(x - 2) + 6 = 0$

66. $(x - 3)^2 + 2(x - 3) - 8 = 0$

Application Exercises

A ball is thrown straight up from a rooftop 300 feet high. The formula

$$h = -16t^2 + 20t + 300$$

describes the ball's height above the ground, h, in feet, t seconds after it was thrown. The ball misses the rooftop on its way down and eventually strikes the ground. The graph of the formula is shown, with tick marks omitted along the horizontal axis. Use the formula to solve Exercises 67–69.

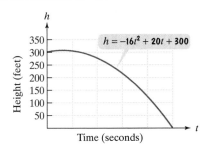

67. How long will it take for the ball to hit the ground? Use this information to provide tick marks with appropriate numbers along the horizontal axis in the figure shown.

68. When will the ball's height be 304 feet? Identify the solution as a point on the graph.

69. When will the ball's height be 276 feet? Identify the solution as a point on the graph.

An explosion causes debris to rise vertically with an initial speed of 72 feet per second. The formula

$$h = -16t^2 + 72t$$

describes the height of the debris above the ground, h, in feet, t seconds after the explosion. Use this information to solve Exercises 70–71.

70. How long will it take for the debris to hit the ground?

71. When will the debris be 32 feet above the ground?

The formula

$$S = 2x^2 - 12x + 82$$

models spending by international travelers to the United States, S, in billions of dollars, x years after 2000. Use this formula to solve Exercises 72–73.

72. In which years did international travelers spend $72 billion?

73. In which years did international travelers spend $66 billion?

The graph of the formula modeling spending by international travelers is shown below. Use the graph to solve Exercises 74–75.

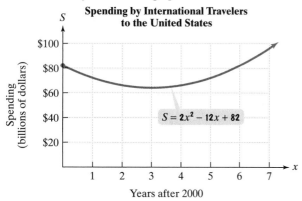

Spending by International Travelers to the United States

$S = 2x^2 - 12x + 82$

Years after 2000

Source: Travel Industry Association of America

74. Identify your solutions from Exercise 72 as points on the graph.

75. Identify your solutions from Exercise 73 as points on the graph.

The alligator, at one time an endangered species, is the subject of a protection program. The formula

$$P = -10x^2 + 475x + 3500$$

models the alligator population, P , after x years of the protection program, where $0 \le x \le 12$. Use the formula to solve Exercises 76–77.

76. After how long is the population up to 5990?

77. After how long is the population up to 7250?

The graph of the alligator population is shown over time. Use the graph to solve Exercises 78–79.

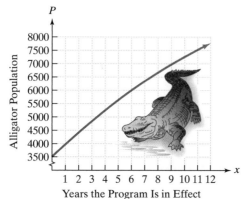

Years the Program Is in Effect

78. Identify your solution from Exercise 76 as a point on the graph.

79. Identify your solution from Exercise 77 as a point on the graph.

The formula

$$N = \frac{t^2 - t}{2}$$

describes the number of football games, N, that must be played in a league with t teams if each team is to play every other team once. Use this information to solve Exercises 80–81.

80. If a league has 36 games scheduled, how many teams belong to the league, assuming that each team plays every other team once?

81. If a league has 45 games scheduled, how many teams belong to the league, assuming that each team plays every other team once?

82. The length of a rectangular garden is 5 feet greater than the width. The area of the rectangle is 300 square feet. Find the length and the width.

$x + 5$

x

83. A rectangular parking lot has a length that is 3 yards greater than the width. The area of the parking lot is 180 square yards. Find the length and the width.

$x + 3$

x

EXIT

84. Each end of a glass prism is a triangle with a height that is 1 inch shorter than twice the base. If the area of the triangle is 60 square inches, how long are the base and height?

$2x - 1$

x

85. Great white sharks have triangular teeth with a height that is 1 centimeter longer than the base. If the area of one tooth is 15 square centimeters, find its base and height.

$x + 1$

x

86. A vacant rectangular lot is being turned into a community vegetable garden measuring 15 meters by 12 meters. A path of uniform width is to surround the garden. If the area of the lot is 378 square meters, find the width of the path surrounding the garden.

87. As part of a landscaping project, you put in a flower bed measuring 10 feet by 12 feet. You plan to surround the bed with a uniform border of low-growing plants.

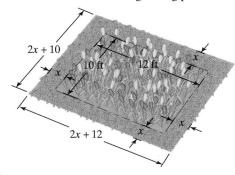

a. Write a polynomial that describes the area of the uniform border that surrounds your flower bed. (*Hint*: The area of the border is the area of the large rectangle shown in the figure minus the area of the flower bed.)

b. The low-growing plants surrounding the flower bed require 1 square foot each when mature. If you have 168 of these plants, how wide a strip around the flower bed should you prepare for the border?

Writing in Mathematics

88. What is a quadratic equation?

89. Explain how to solve $x^2 + 6x + 8 = 0$ using factoring and the zero-product principle.

90. If $(x + 2)(x - 4) = 0$ indicates that $x + 2 = 0$ or $x - 4 = 0$, explain why $(x + 2)(x - 4) = 6$ does not mean $x + 2 = 6$ or $x - 4 = 6$. Could we solve the equation using $x + 2 = 3$ and $x - 4 = 2$ because $3 \cdot 2 = 6$?

Critical Thinking Exercises

Make Sense? *In Exercises 91–94, determine whether each statement "makes sense" or "does not make sense" and explain your reasoning.*

91. When solving $4(x - 3)(x + 2) = 0$ and $4x(x + 2) = 0$, I can ignore the monomial factors.

92. I set the quadratic equation $2x^2 - 5x = 12$ equal to zero and obtained $2x^2 - 5x = 0$.

93. Because some trinomials are prime, some quadratic equations cannot be solved by factoring.

94. I'm looking at a graph with one x-intercept, so it must be the graph of a linear equation.

In Exercises 95–98, determine whether each statement is true or false. If the statement is false, make the necessary change(s) to produce a true statement.

95. If $(x + 3)(x - 4) = 2$, then $x + 3 = 0$ or $x - 4 = 0$.

96. The solutions of the equation $4(x - 5)(x + 3) = 0$ are $4, 5$, and -3.

97. Equations solved by factoring always have two different solutions.

98. Both 0 and $-\pi$ are solutions of the equation $x(x + \pi) = 0$.

99. Write a quadratic equation in standard form whose solutions are -3 and 5.

In Exercises 100–102, solve each equation.

100. $x^3 - x^2 - 16x + 16 = 0$

101. $3^{x^2-9x+20} = 1$

102. $(x^2 - 5x + 5)^3 = 1$

In Exercises 103–106, match each equation with its graph. The graphs are labeled (a) through (d).

103. $y = x^2 - x - 2$ **104.** $y = x^2 + x - 2$

105. $y = x^2 - 4$ **106.** $y = x^2 - 4x$

a.

b.

c.

d.

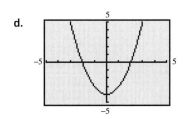

Technology Exercises

In Exercises 107–110, use the x-intercepts for the graph in a $[-10, 10, 1]$ *by* $[-13, 10, 1]$ *viewing rectangle to solve the quadratic equation. Check by substitution.*

107. Use the graph of $y = x^2 + 3x - 4$ to solve

$$x^2 + 3x - 4 = 0.$$

108. Use the graph of $y = x^2 + x - 6$ to solve

$$x^2 + x - 6 = 0.$$

109. Use the graph of $y = (x - 2)(x + 3) - 6$ to solve

$$(x - 2)(x + 3) - 6 = 0.$$

110. Use the graph of $y = x^2 - 2x + 1$ to solve

$$x^2 - 2x + 1 = 0.$$

111. Use the technique of identifying x-intercepts on a graph generated by a graphing utility to check any five equations that you solved in Exercises 9–56.

112. If you have access to a calculator that solves quadratic equations, consult the owner's manual to determine how to use this feature. Then use your calculator to solve any five of the equations in Exercises 9–56.

Review Exercises

113. Graph: $y > -\dfrac{2}{3}x + 1$. (Section 4.6, Example 3)

114. Simplify: $\left(\dfrac{8x^4}{4x^7}\right)^2$. (Section 6.7, Example 6)

115. Solve: $5x + 28 = 6 - 6x$. (Section 2.2, Example 7)

Preview Exercises

Exercises 116–118 will help you prepare for the material covered in the first section of the next chapter.

116. Evaluate $\dfrac{250x}{100 - x}$ for $x = 60$.

117. Why is $\dfrac{6x + 12}{7x - 28}$ undefined for $x = 4$?

118. Factor the numerator and the denominator. Then simplify by dividing out the common factor in the numerator and the denominator.

$$\frac{x^2 + 6x + 5}{x^2 - 25}$$

GROUP PROJECT

CHAPTER 7

Group members are on the board of a condominium association. The condominium has just installed a 35-foot-by-30-foot pool. Your job is to choose a material to surround the pool to create a border of uniform width.

a. Begin by writing an algebraic expression for the area, in square feet, of the border around the pool. (*Hint*: The border's area is the combined area of the pool and border minus the area of the pool.)

b. You must select one of the following options for the border.

Options for the Border	Price
Cement	$6 per square foot
Outdoor carpeting	$5 per square foot plus $10 per foot to install edging around the rectangular border
Brick	$8 per square foot plus a $60 charge for delivering the bricks

Write an algebraic expression for the cost of installing the border for each of these options.

c. You would like the border to be 5 feet wide. Use the algebraic expressions in part (b) to find the cost of the border for each of the three options.

d. You would prefer not to use cement. However, the condominium association is limited by a $5000 budget. Given this limitation, approximately how wide can the border be using outdoor carpeting or brick? Which option should you select and why?

Chapter 7 Summary

| **Definitions and Concepts** | **Examples** |

Section 7.1 The Greatest Common Factor and Factoring by Grouping

Factoring a polynomial containing the sum of monomials means finding an equivalent expression that is a product. The greatest common factor, GCF, is an expression that divides every term of the polynomial. The variable part of the GCF contains the smallest power of a variable that appears in all terms of the polynomial.

Find the GCF of $16x^2y$, $20x^3y^2$, and $8x^2y^3$.

The GCF of 16, 20, and 8 is 4.

The GCF of x^2, x^3, and x^2 is x^2.

The GCF of y, y^2, and y^3 is y.

$$\text{GCF} = 4 \cdot x^2 \cdot y = 4x^2y$$

To factor a monomial from a polynomial, express each term as the product of the GCF and its other factor. Then use the distributive property to factor out the GCF.

$$16x^2y + 20x^3y^2 + 8x^2y^3$$
$$= 4x^2y \cdot 4 + 4x^2y \cdot 5xy + 4x^2y \cdot 2y^2$$
$$= 4x^2y(4 + 5xy + 2y^2)$$

To factor by grouping, factor out the GCF from each group. Then factor out the remaining common factor.

$$xy + 5x - 3y - 15$$
$$= x(y + 5) - 3(y + 5)$$
$$= (y + 5)(x - 3)$$

Section 7.2 Factoring Trinomials Whose Leading Coefficient Is 1

To factor a trinomial of the form $x^2 + bx + c$, find two numbers whose product is c and whose sum is b. The factorization is

$$(x + \text{one number})(x + \text{other number}).$$

Factor: $x^2 + 9x + 20$.

Find two numbers whose product is 20 and whose sum is 9. The numbers are 4 and 5.

$$x^2 + 9x + 20 = (x + 4)(x + 5)$$

Section 7.3 Factoring Trinomials Whose Leading Coefficient Is Not 1

To factor $ax^2 + bx + c$ by trial and error, try various combinations of factors of ax^2 and c until a middle term of bx is obtained for the sum of outside and inside products.

Factor: $3x^2 + 7x - 6$.

Factors of $3x^2$: $3x, x$

Factors of -6: 1 and -6, -1 and 6, 2 and -3, -2 and 3

A possible combination of these factors is

$$(3x - 2)(x + 3).$$

Sum of outside and inside products should equal $7x$.

$$9x - 2x = 7x$$

Thus, $3x^2 + 7x - 6 = (3x - 2)(x + 3)$.

To factor $ax^2 + bx + c$ by grouping, find the factors of ac whose sum is b. Write bx using these factors. Then factor by grouping.

Factor: $3x^2 + 7x - 6$.

Find the factors of $3(-6)$, or -18, whose sum is 7. They are 9 and -2.

$$3x^2 + 7x - 6$$
$$= 3x^2 + 9x - 2x - 6$$
$$= 3x(x + 3) - 2(x + 3) = (x + 3)(3x - 2)$$

Definitions and Concepts	Examples

Section 7.4 Factoring Special Forms

The Difference of Two Squares

$$A^2 - B^2 = (A + B)(A - B)$$

$9x^2 - 25y^2$
$= (3x)^2 - (5y)^2 = (3x + 5y)(3x - 5y)$

Perfect Square Trinomials

$$A^2 + 2AB + B^2 = (A + B)^2$$
$$A^2 - 2AB + B^2 = (A - B)^2$$

$x^2 + 16x + 64 = x^2 + 2 \cdot x \cdot 8 + 8^2 = (x + 8)^2$
$25x^2 - 30x + 9 = (5x)^2 - 2 \cdot 5x \cdot 3 + 3^2 = (5x - 3)^2$

Sum or Difference of Cubes

$$A^3 + B^3 = (A + B)(A^2 - AB + B^2)$$
$$A^3 - B^3 = (A - B)(A^2 + AB + B^2)$$

$8x^3 - 125 = (2x)^3 - 5^3$
$= (2x - 5)[(2x)^2 + 2x \cdot 5 + 5^2]$
$= (2x - 5)(4x^2 + 10x + 25)$

Section 7.5 A General Factoring Strategy

A Factoring Strategy

1. Factor out the GCF.
2. a. If two terms, try
   $$A^2 - B^2 = (A + B)(A - B)$$
   $$A^3 + B^3 = (A + B)(A^2 - AB + B^2)$$
   $$A^3 - B^3 = (A - B)(A^2 + AB + B^2).$$
   b. If three terms, try
   $$A^2 + 2AB + B^2 = (A + B)^2$$
   $$A^2 - 2AB + B^2 = (A - B)^2.$$
   If not a perfect square trinomial, try trial and error or grouping.
   c. If four terms, try factoring by grouping.
3. See if any factors can be factored further.
4. Check by multiplying.

Factor: $2x^4 + 10x^3 - 8x^2 - 40x$.

The GCF is $2x$.

$2x^4 + 10x^3 - 8x^2 - 40x$
$= 2x(x^3 + 5x^2 - 4x - 20)$

> Four terms: Try grouping.

$= 2x[x^2(x + 5) - 4(x + 5)]$
$= 2x(x + 5)(x^2 - 4)$

> This can be factored further.

$= 2x(x + 5)(x + 2)(x - 2)$

Section 7.6 Solving Quadratic Equations by Factoring

The Zero-Product Principle

If $AB = 0$, then $A = 0$ or $B = 0$.

Solve: $(x - 6)(x + 10) = 0$.

$x - 6 = 0$ or $x + 10 = 0$
$x = 6$ $\qquad x = -10$

The solutions are -10 and 6, and the solution set is $\{-10, 6\}$.

A quadratic equation in x is an equation that can be written in the standard form

$$ax^2 + bx + c = 0, \quad a \neq 0.$$

To solve by factoring, write the equation in standard form, factor, set each factor equal to zero, and solve each resulting equation. Check proposed solutions in the original equation.

Solve: $4x^2 + 9x = 9$.

$4x^2 + 9x - 9 = 0$
$(4x - 3)(x + 3) = 0$
$4x - 3 = 0$ or $x + 3 = 0$
$x = \dfrac{3}{4}$ $\qquad x = -3$

The solutions are -3 and $\frac{3}{4}$, and the solution set is $\left\{-3, \frac{3}{4}\right\}$.

CHAPTER 7 REVIEW EXERCISES

7.1 *In Exercises 1–5, factor each polynomial using the greatest common factor. If there is no common factor other than 1 and the polynomial cannot be factored, so state.*

1. $30x - 45$
2. $12x^3 + 16x^2 - 400x$
3. $30x^4y + 15x^3y + 5x^2y$
4. $7(x + 3) - 2(x + 3)$
5. $7x^2(x + y) - (x + y)$

In Exercises 6–9, factor by grouping.

6. $x^3 + 3x^2 + 2x + 6$
7. $xy + y + 4x + 4$
8. $x^3 + 5x + x^2 + 5$
9. $xy + 4x - 2y - 8$

7.2 *In Exercises 10–17, factor completely, or state that the trinomial is prime.*

10. $x^2 - 3x + 2$
11. $x^2 - x - 20$
12. $x^2 + 19x + 48$
13. $x^2 - 6xy + 8y^2$
14. $x^2 + 5x - 9$
15. $x^2 + 16xy - 17y^2$
16. $3x^2 + 6x - 24$
17. $3x^3 - 36x^2 + 33x$

7.3 *In Exercises 18–26, factor completely, or state that the trinomial is prime.*

18. $3x^2 + 17x + 10$
19. $5y^2 - 17y + 6$
20. $4x^2 + 4x - 15$
21. $5y^2 + 11y + 4$
22. $8x^2 + 8x - 6$
23. $2x^3 + 7x^2 - 72x$
24. $12y^3 + 28y^2 + 8y$
25. $2x^2 - 7xy + 3y^2$
26. $5x^2 - 6xy - 8y^2$

7.4 *In Exercises 27–30, factor each difference of two squares completely.*

27. $4x^2 - 1$
28. $81 - 100y^2$
29. $25a^2 - 49b^2$
30. $z^4 - 16$

In Exercises 31–34, factor completely, or state that the polynomial is prime.

31. $2x^2 - 18$
32. $x^2 + 1$
33. $9x^3 - x$
34. $18xy^2 - 8x$

In Exercises 35–41, factor any perfect square trinomials, or state that the polynomial is prime.

35. $x^2 + 22x + 121$
36. $x^2 - 16x + 64$
37. $9y^2 + 48y + 64$
38. $16x^2 - 40x + 25$
39. $25x^2 + 15x + 9$
40. $36x^2 + 60xy + 25y^2$
41. $25x^2 - 40xy + 16y^2$

In Exercises 42–45, factor using the formula for the sum or difference of two cubes.

42. $x^3 - 27$
43. $64x^3 + 1$
44. $54x^3 - 16y^3$
45. $27x^3y + 8y$

In Exercises 46–47, find the formula for the area of the shaded region and express it in factored form.

46.   47.

48. The figure shows a geometric interpretation of a factorization. Use the sum of the areas of the four pieces on the left and the area of the square on the right to write the factorization that is illustrated.

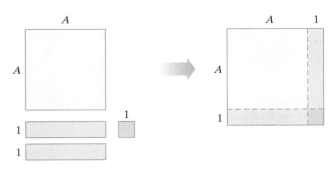

7.5 *In Exercises 49–81, factor completely, or state that the polynomial is prime.*

49. $x^3 - 8x^2 + 7x$
50. $10y^2 + 9y + 2$
51. $128 - 2y^2$
52. $9x^2 + 6x + 1$
53. $20x^7 - 36x^3$
54. $x^3 - 3x^2 - 9x + 27$
55. $y^2 + 16$
56. $2x^3 + 19x^2 + 35x$

57. $3x^3 - 30x^2 + 75x$
58. $3x^5 - 24x^2$
59. $4y^4 - 36y^2$
60. $5x^2 + 20x - 105$
61. $9x^2 + 8x - 3$
62. $10x^5 - 44x^4 + 16x^3$
63. $100y^2 - 49$
64. $9x^5 - 18x^4$
65. $x^4 - 1$
66. $2y^3 - 16$
67. $x^3 + 64$
68. $6x^2 + 11x - 10$
69. $3x^4 - 12x^2$
70. $x^2 - x - 90$
71. $25x^2 + 25xy + 6y^2$
72. $x^4 + 125x$
73. $32y^3 + 32y^2 + 6y$
74. $2y^2 - 16y + 32$
75. $x^2 - 2xy - 35y^2$
76. $x^2 + 7x + xy + 7y$
77. $9x^2 + 24xy + 16y^2$
78. $2x^4y - 2x^2y$
79. $100y^2 - 49z^2$
80. $x^2 + xy + y^2$
81. $3x^4y^2 - 12x^2y^4$

7.6 *In Exercises 82–83, solve each equation using the zero-product principle.*

82. $x(x - 12) = 0$

83. $3(x - 7)(4x + 9) = 0$

In Exercises 84–92, use factoring to solve each quadratic equation.

84. $x^2 + 5x - 14 = 0$
85. $5x^2 + 20x = 0$

86. $2x^2 + 15x = 8$
87. $x(x - 4) = 32$
88. $(x + 3)(x - 2) = 50$
89. $x^2 = 14x - 49$

90. $9x^2 = 100$

91. $3x^2 + 21x + 30 = 0$

92. $3x^2 = 22x - 7$

93. You dive from a board that is 32 feet above the water. The formula
$$h = -16t^2 + 16t + 32$$
describes your height above the water, h, in feet, t seconds after you dive. How long will it take you to hit the water?

94. The length of a rectangular sign is 3 feet longer than the width. If the sign has space for 40 square feet of advertising, find its length and its width.

95. The square lot shown here is being turned into a garden with a 3-meter path at one end. If the area of the garden is 88 square meters, find the dimensions of the square lot.

CHAPTER 7 TEST Remember to use your Chapter Test Prep Video CD to see the worked-out solutions to the test questions you want to review.

In Exercises 1–21, factor completely, or state that the polynomial is prime.

1. $x^2 - 9x + 18$
2. $x^2 - 14x + 49$
3. $15y^4 - 35y^3 + 10y^2$
4. $x^3 + 2x^2 + 3x + 6$
5. $x^2 - 9x$
6. $x^3 + 6x^2 - 7x$
7. $14x^2 + 64x - 30$
8. $25x^2 - 9$
9. $x^3 + 8$
10. $x^2 - 4x - 21$
11. $x^2 + 4$
12. $6y^3 + 9y^2 + 3y$

13. $4y^2 - 36$
14. $16x^2 + 48x + 36$
15. $2x^4 - 32$
16. $36x^2 - 84x + 49$
17. $7x^2 - 50x + 7$
18. $x^3 + 2x^2 - 5x - 10$
19. $12y^3 - 12y^2 - 45y$
20. $y^3 - 125$
21. $5x^2 - 5xy - 30y^2$

In Exercises 22–27, solve each quadratic equation.

22. $x^2 + 2x - 24 = 0$

23. $3x^2 - 5x = 2$

24. $x(x - 6) = 16$

25. $6x^2 = 21x$

26. $16x^2 = 81$

27. $(5x + 4)(x - 1) = 2$

28. Find a formula for the area of the shaded region and express it in factored form.

29. A model rocket is launched from a height of 96 feet. The formula

$$h = -16t^2 + 80t + 96$$

describes the rocket's height, h, in feet, t seconds after it was launched. How long will it take the rocket to reach the ground?

30. The length of a rectangular garden is 6 feet longer than its width. If the area of the garden is 55 square feet, find its length and its width.

CUMULATIVE REVIEW EXERCISES (CHAPTERS 1–7)

1. Simplify: $6[5 + 2(3 - 8) - 3]$.

2. Solve: $4(x - 2) = 2(x - 4) + 3x$.

3. Solve: $\dfrac{x}{2} - 1 = \dfrac{x}{3} + 1$.

4. Solve and graph the solution set on a number line.

$$5 - 5x > 2(5 - x) + 1$$

5. Find the measures of the angles of a triangle whose two base angles have equal measure and whose third angle is 10° less than three times the measure of a base angle.

6. A dinner for six people cost $159, including a 6% tax. What was the dinner's cost before tax?

7. Graph using the slope and y-intercept: $y = -\frac{3}{5}x + 3$.

8. Write the point-slope form of the equation of the line passing through $(2, -4)$ and $(3, 1)$. Then use the point-slope form of the equation to write the slope-intercept form of the equation.

9. Graph: $5x - 6y > 30$.

10. Solve the system:

$$5x + 2y = 13$$
$$y = 2x - 7.$$

11. Solve the system:

$$2x + 3y = 5$$
$$3x - 2y = -4.$$

12. Subtract: $\dfrac{4}{5} - \dfrac{9}{8}$.

In Exercises 13–15, perform the indicated operations.

13. $\dfrac{6x^5 - 3x^4 + 9x^2 + 27x}{3x}$

14. $(3x - 5y)(2x + 9y)$

15. $\dfrac{6x^3 + 5x^2 - 34x + 13}{3x - 5}$

16. Write 0.0071 in scientific notation.

In Exercises 17–19, factor completely.

17. $3x^2 + 11x + 6$

18. $y^5 - 16y$

19. $4x^2 + 12x + 9$

20. The length of a rectangle is 2 feet greater than its width. If the rectangle's area is 24 square feet, find its dimensions.

Rational Expressions

We are plagued by questions about the environment. Will we run out of oil? How hot will it get? Will there be neighborhoods where the air is pristine? Can we make garbage disappear? Will there be any wilderness left? Which wild animals will become extinct? How much will it cost to clean up toxic wastes from our rivers so that they can safely provide food, recreation, and enjoyment of wildlife for the millions who live along and visit their banks?

When making decisions on public policies dealing with the environment, algebraic fractions play an important role in modeling the costs. By learning to work with these fractional expressions, you will gain new insights into phenomena as diverse as the dosage of drugs prescribed for children, inventory costs for a business, the cost of environmental cleanup, and even the shape of our heads.

Here's where you'll find these applications:

- Children and drug dosage: Exercise Set 8.4, Exercises 93–100

- Inventory costs for a business: Exercise Set 8.6, Exercises 61–62

- Costs of environmental cleanup: Section 8.1 opener; Section 8.1, page 480; Exercise Set 8.1, Exercises 85–86; Section 8.6, Example 6; Exercise Set 8.6, Exercises 57–58

- Shape of our heads: Section 8.3 opener; Exercise Set 8.3, Exercise 73.

Rational Expressions and Their Simplification

Objectives

1 Find numbers for which a rational expression is undefined.

2 Simplify rational expressions.

3 Solve applied problems involving rational expressions.

How do we describe the costs of reducing environmental pollution? We often use algebraic expressions involving quotients of polynomials. For example, the algebraic expression

$$\frac{250x}{100 - x}$$

describes the cost, in millions of dollars, to remove x percent of the pollutants that are discharged into a river. Removing a modest percentage of pollutants, say 40%, is far less costly than removing a substantially greater percentage, such as 95%. We see this by evaluating the algebraic expression for $x = 40$ and $x = 95$.

Discover for Yourself

What happens if you try substituting 100 for x in

$$\frac{250x}{100 - x}?$$

What does this tell you about the cost of cleaning up all of the river's pollutants?

Evaluating $\dfrac{250x}{100 - x}$ for

$x = 40$:	$x = 95$:
Cost is $\dfrac{250(40)}{100 - 40} \approx 167.$	Cost is $\dfrac{250(95)}{100 - 95} = 4750.$

The cost increases from approximately \$167 million to a possibly prohibitive \$4750 million, or \$4.75 billion. Costs spiral upward as the percentage of removed pollutants increases.

Many algebraic expressions that describe costs of environmental projects are examples of *rational expressions*. In this section, we introduce rational expressions and their simplification.

Excluding Numbers from Rational Expressions

1 Find numbers for which a rational expression is undefined.

A **rational expression** is the quotient of two polynomials. Some examples are

$$\frac{x - 2}{4}, \quad \frac{4}{x - 2}, \quad \frac{x}{x^2 - 1}, \quad \text{and} \quad \frac{x^2 + 1}{x^2 + 2x - 3}.$$

Rational expressions indicate division and division by zero is undefined. This means that **we must exclude any value or values of the variable that make a denominator zero**. For example, consider the rational expression

$$\frac{4}{x - 2}.$$

When x is replaced with 2, the denominator is 0 and the expression is undefined.

$$\text{If } x = 2: \quad \frac{4}{x - 2} = \frac{4}{2 - 2} = \frac{4}{0}. \quad \boxed{\begin{array}{c}\text{Division by zero}\\\text{is undefined.}\end{array}}$$

Notice that if x is replaced by a number other than 2, such as 1, the expression is defined because the denominator is nonzero.

$$\text{If } x = 1: \quad \frac{4}{x - 2} = \frac{4}{1 - 2} = \frac{4}{-1} = -4.$$

Thus, only 2 must be excluded as a replacement for x in the rational expression $\dfrac{4}{x - 2}$.

Using Technology

We can use the $\boxed{\text{TABLE}}$ feature of a graphing utility to verify our work with $\dfrac{4}{x - 2}$. Enter

$$y_1 = 4 \;\boxed{\div}\; \boxed{(} \;\boxed{x}\; \boxed{-}\; \boxed{2} \;\boxed{)}$$

and press $\boxed{\text{TABLE}}$.

Excluding Values from Rational Expressions

If a variable in a rational expression is replaced by a number that causes the denominator to be 0, that number must be excluded as a replacement for the variable. The rational expression is undefined at any value that produces a denominator of 0.

How do we determine the value or values of the variable for which a rational expression is undefined? Set the denominator equal to 0 and then solve the resulting equation for the variable.

EXAMPLE 1   **Determining Numbers for Which Rational Expressions Are Undefined**

Find all the numbers for which the rational expression is undefined:

a. $\dfrac{6x + 12}{7x - 28}$ b. $\dfrac{2x + 6}{x^2 + 3x - 10}$.

Solution   In each case, we set the denominator equal to 0 and solve.

$$\frac{6x + 12}{7x - 28} \quad \boxed{\begin{array}{c}\text{Exclude values of } x\\\text{that make these}\\\text{denominators 0.}\end{array}} \quad \frac{2x + 6}{x^2 + 3x - 10}$$

a. $7x - 28 = 0$ *Set the denominator of $\dfrac{6x + 12}{7x - 28}$ equal to 0.*

$\qquad 7x = 28$ *Add 28 to both sides.*

$\qquad\;\; x = 4$ *Divide both sides by 7.*

Thus, $\dfrac{6x + 12}{7x - 28}$ is undefined for $x = 4$.

b. $x^2 + 3x - 10 = 0$ *Set the denominator of $\dfrac{2x + 6}{x^2 + 3x - 10}$ equal to 0.*

$\quad (x + 5)(x - 2) = 0$ *Factor.*

$\quad x + 5 = 0 \;\; \text{or} \;\; x - 2 = 0$ *Set each factor equal to 0.*

$\qquad x = -5 \qquad\qquad x = 2$ *Solve the resulting equations.*

Thus, $\dfrac{2x + 6}{x^2 + 3x - 10}$ is undefined for $x = -5$ and $x = 2$. ∎

Using Technology

Graphic Connections

When using a graphing utility to graph an equation containing a rational expression, you might not be pleased with the quality of the display. Compare these two graphs of $y = \dfrac{6x + 12}{7x - 28}$.

The graph on the left was obtained using the $\boxed{\text{DOT}}$ mode in a $[-3, 10, 1]$ by $[-10, 10, 1]$ viewing rectangle. Examine the behavior of the graph near $x = 4$, the number for which the rational expression is undefined. The values of the rational expression are decreasing as the values of x get closer to 4 on the left and increasing as the values of x get closer to 4 on the right. However, there is no point on the graph corresponding to $x = 4$. Would you agree that this behavior is better illustrated in the hand-drawn graph on the right?

☑ **CHECK POINT 1**   Find all the numbers for which the rational expression is undefined:

a. $\dfrac{7x - 28}{8x - 40}$ **b.** $\dfrac{8x - 40}{x^2 + 3x - 28}.$

Is every rational expression undefined for at least one number? No. Consider

$$\frac{x - 2}{x^2 + 1}.$$

Because the denominator, $x^2 + 1$, is not zero for any real-number replacement for x, the rational expression is defined for all real numbers. Thus, it is not necessary to exclude any values for x.

2 Simplify rational expressions.

Simplifying Rational Expressions

A rational expression is **simplified** if its numerator and denominator have no common factors other than 1 or −1. The following principle is used to simplify a rational expression:

Fundamental Principle of Rational Expressions

If P, Q, and R are polynomials, and Q and R are not 0,

$$\frac{PR}{QR} = \frac{P}{Q}.$$

As you read the Fundamental Principle, can you see why $\dfrac{PR}{QR}$ is not simplified?

The numerator and denominator have a common factor, the polynomial R. By dividing the numerator and the denominator by the common factor, R, we obtain the simplified form $\dfrac{P}{Q}$. This is often shown as follows:

$$\frac{P\overset{1}{\cancel{R}}}{Q\underset{1}{\cancel{R}}} = \frac{P}{Q}.$$

> **Observe that**
> $$\frac{PR}{QR} = \frac{P}{Q} \cdot \frac{R}{R} = \frac{P}{Q} \cdot 1 = \frac{P}{Q}.$$

The following procedure can be used to simplify rational expressions:

Simplifying Rational Expressions

1. Factor the numerator and the denominator completely.
2. Divide both the numerator and the denominator by any common factors.

EXAMPLE 2 Simplifying a Rational Expression

Simplify: $\dfrac{5x + 35}{20x}$.

Solution

$$\frac{5x + 35}{20x} = \frac{5(x + 7)}{5 \cdot 4x} \qquad \text{Factor the numerator and denominator.}$$

Because the denominator is 20x, x ≠ 0.

$$= \frac{\overset{1}{\cancel{5}}(x + 7)}{\underset{1}{\cancel{5}} \cdot 4x} \qquad \text{Divide out the common factor of 5.}$$

$$= \frac{x + 7}{4x}$$

✓ **CHECK POINT 2** Simplify: $\dfrac{7x + 28}{21x}$.

EXAMPLE 3 Simplifying a Rational Expression

Simplify: $\dfrac{x^3 + x^2}{x + 1}$.

Solution

$$\dfrac{x^3 + x^2}{x + 1} = \dfrac{x^2(x + 1)}{x + 1}$$ Factor the numerator. Because the denominator is x + 1, x ≠ −1.

$$= \dfrac{x^2 \cancel{(x + 1)}^{1}}{\cancel{x + 1}_{1}}$$ Divide out the common factor of x + 1.

$$= x^2$$

Simplifying a rational expression can change the numbers that make it undefined. For example, we just showed that

$$\dfrac{x^3 + x^2}{x + 1} = x^2.$$

This is undefined for x = −1.

This simplified form is defined for all real numbers.

Thus, to equate the two expressions, we must restrict the values for x in the simplified expression to exclude -1. We can write

$$\dfrac{x^3 + x^2}{x + 1} = x^2, \quad x \neq -1.$$

Hereafter, we will assume that the simplified rational expression is equal to the original rational expression for all real numbers except those for which either denominator is 0.

Using Technology

Graphic and Numeric Connections

A graphing utility can be used to verify that

$$\dfrac{x^3 + x^2}{x + 1} = x^2, \quad x \neq -1.$$

Enter $y_1 = \dfrac{x^3 + x^2}{x + 1}$ and $y_2 = x^2$.

Graphic Check

The graphs of y_1 and y_2 appear to be identical. You can use the ⬚TRACE⬚ feature to trace y_1 and show that it is undefined for $x = -1$.

$[-10, 10, 1]$ by $[-10, 10, 1]$

Numeric Check

No matter how far up or down we scroll, if $x \neq -1$, $y_1 = y_2$. If $x = -1$, y_1 is undefined, although the value of y_2 is 1.

✓ **CHECK POINT 3** Simplify: $\dfrac{x^3 - x^2}{7x - 7}$.

EXAMPLE 4 Simplifying a Rational Expression

Simplify: $\dfrac{x^2 + 6x + 5}{x^2 - 25}$.

Solution

$$\dfrac{x^2 + 6x + 5}{x^2 - 25} = \dfrac{(x + 5)(x + 1)}{(x + 5)(x - 5)}$$

Factor the numerator and denominator. Because the denominator is $(x + 5)(x - 5)$, $x \neq -5$ and $x \neq 5$.

$$= \dfrac{\overset{1}{\cancel{(x + 5)}}(x + 1)}{\underset{1}{\cancel{(x + 5)}}(x - 5)}$$

Divide out the common factor of $x + 5$.

$$= \dfrac{x + 1}{x - 5}$$

✓ **CHECK POINT 4** Simplify: $\dfrac{x^2 - 1}{x^2 + 2x + 1}$.

Study Tip

When simplifying rational expressions, only *factors* that are common to the *entire numerator* and the *entire denominator* can be divided out. **It is incorrect to divide out common terms from the numerator and denominator.**

Incorrect!

The first two expressions, $\dfrac{x + 5}{x + 7}$ and $\dfrac{x^2 - 4}{4}$, have no common factors in their numerators and denominators. Thus, these rational expressions are in simplified form. The rational expression $\dfrac{x^2 - 9}{x - 3}$ can be simplified as follows:

Correct

$$\dfrac{x^2 - 9}{x - 3} = \dfrac{(x + 3)\overset{1}{\cancel{(x - 3)}}}{\underset{1}{\cancel{x - 3}}} = x + 3.$$

Divide out the common factor, $x - 3$.

Factors That Are Opposites

How do we simplify rational expressions that contain factors in the numerator and denominator that are opposites, or additive inverses? Here is an example of such an expression:

$$\dfrac{x - 3}{3 - x}.$$

The numerator and denominator are opposites. They differ only in their signs.

Factor out −1 from either the numerator or the denominator. Then divide out the common factor.

$$\frac{x-3}{3-x} = \frac{-1(-x+3)}{3-x}$$

Factor −1 from the numerator. Notice how the sign of each term in the polynomial $x - 3$ changes.

$$= \frac{-1(3-x)}{3-x}$$

In the numerator, use the commutative property to rewrite $-x + 3$ as $3 - x$.

$$= \frac{-1\overset{1}{(3-x)}}{\underset{1}{3-x}}$$

Divide out the common factor of $3 - x$.

$$= -1$$

Our result, −1, suggests the following useful property:

> ### Simplifying Rational Expressions with Opposite Factors in the Numerator and Denominator
> The quotient of two polynomials that have opposite signs and are additive inverses is −1.

EXAMPLE 5   Simplifying a Rational Expression

Simplify: $\dfrac{4x^2 - 25}{15 - 6x}$.

Solution

$$\frac{4x^2 - 25}{15 - 6x} = \frac{(2x+5)(2x-5)}{3(5-2x)}$$

Factor the numerator and denominator.

$$= \frac{(2x+5)\overset{-1}{(2x-5)}}{3\underset{}{(5-2x)}}$$

The quotient of polynomials with opposite signs is −1.

$$= \frac{-(2x+5)}{3} \quad \text{or} \quad -\frac{2x+5}{3} \quad \text{or} \quad \frac{-2x-5}{3}$$

Each of these forms is an acceptable answer.

☑ **CHECK POINT 5**   Simplify: $\dfrac{9x^2 - 49}{28 - 12x}$.

3 Solve applied problems involving rational expressions.

FIGURE 8.1

Applications

The equation

$$y = \frac{250x}{100 - x}$$

models the cost, in millions of dollars, to remove x percent of the pollutants that are discharged into a river. This equation contains the rational expression that we looked at in the opening to this section. Do you remember how costs were spiraling upward as the percentage of removed pollutants increased?

Is it possible to clean up the river completely? To do this, we must remove 100% of the pollutants. The problem is that the rational expression is undefined for $x = 100$.

$$y = \frac{250x}{100 - x}$$

If $x = 100$, the value of the denominator is 0.

Notice how the graph of $y = \dfrac{250x}{100 - x}$, shown in **Figure 8.1**, approaches but never touches the dashed vertical line $x = 100$, where 100 is the value of x for which $\dfrac{250x}{100 - x}$ is undefined. The graph continues to rise more and more steeply, visually showing the escalating costs. By never touching the dashed vertical line, the graph illustrates that no amount of money will be enough to remove all pollutants from the river.

| 8.1 EXERCISE SET | *MyMathLab* | PRACTICE | WATCH | DOWNLOAD | READ | REVIEW |

Practice Exercises

In Exercises 1–20, find all numbers for which each rational expression is undefined. If the rational expression is defined for all real numbers, so state.

1. $\dfrac{5}{2x}$ **2.** $\dfrac{11}{3x}$

3. $\dfrac{x}{x-8}$ **4.** $\dfrac{x}{x-6}$

5. $\dfrac{13}{5x-20}$ **6.** $\dfrac{17}{6x-30}$

7. $\dfrac{x+3}{(x+9)(x-2)}$ **8.** $\dfrac{x+5}{(x+7)(x-9)}$

9. $\dfrac{4x}{(3x-17)(x+3)}$ **10.** $\dfrac{8x}{(4x-19)(x+2)}$

11. $\dfrac{x+5}{x^2+x-12}$ **12.** $\dfrac{7x-14}{x^2-9x+20}$

13. $\dfrac{x+5}{5}$ **14.** $\dfrac{x+7}{7}$

15. $\dfrac{y+3}{4y^2+y-3}$ **16.** $\dfrac{y+8}{6y^2-y-2}$

17. $\dfrac{y+5}{y^2-25}$ **18.** $\dfrac{y+7}{y^2-49}$

19. $\dfrac{5}{x^2+1}$ **20.** $\dfrac{8}{x^2+4}$

In Exercises 21–76, simplify each rational expression. If the rational expression cannot be simplified, so state.

21. $\dfrac{14x^2}{7x}$ **22.** $\dfrac{9x^2}{6x}$

23. $\dfrac{5x-15}{25}$ **24.** $\dfrac{7x+21}{49}$

25. $\dfrac{2x-8}{4x}$ **26.** $\dfrac{3x-9}{6x}$

27. $\dfrac{3}{3x-9}$ **28.** $\dfrac{12}{6x-18}$

29. $\dfrac{-15}{3x-9}$ **30.** $\dfrac{-21}{7x-14}$

31. $\dfrac{3x+9}{x+3}$ **32.** $\dfrac{5x-10}{x-2}$

33. $\dfrac{x+5}{x^2-25}$ **34.** $\dfrac{x+4}{x^2-16}$

35. $\dfrac{2y-10}{3y-15}$ **36.** $\dfrac{6y+18}{11y+33}$

37. $\dfrac{x+1}{x^2-2x-3}$ **38.** $\dfrac{x+2}{x^2-x-6}$

39. $\dfrac{4x-8}{x^2-4x+4}$ **40.** $\dfrac{x^2-12x+36}{4x-24}$

41. $\dfrac{y^2-3y+2}{y^2+7y-18}$ **42.** $\dfrac{y^2+5y+4}{y^2-4y-5}$

43. $\dfrac{2y^2-7y+3}{2y^2-5y+2}$ **44.** $\dfrac{3y^2+4y-4}{6y^2-y-2}$

45. $\dfrac{2x+3}{2x+5}$ **46.** $\dfrac{3x+7}{3x+10}$

47. $\dfrac{x^2+12x+36}{x^2-36}$ **48.** $\dfrac{x^2-14x+49}{x^2-49}$

49. $\dfrac{x^3-2x^2+x-2}{x-2}$ **50.** $\dfrac{x^3+4x^2-3x-12}{x+4}$

51. $\dfrac{x^3-8}{x-2}$ **52.** $\dfrac{x^3-125}{x^2-25}$

53. $\dfrac{(x-4)^2}{x^2-16}$ **54.** $\dfrac{(x+5)^2}{x^2-25}$

55. $\dfrac{x}{x+1}$ **56.** $\dfrac{x}{x+7}$

57. $\dfrac{x+4}{x^2+16}$ **58.** $\dfrac{x+5}{x^2+25}$

59. $\dfrac{x-5}{5-x}$ **60.** $\dfrac{x-7}{7-x}$

61. $\dfrac{2x-3}{3-2x}$ **62.** $\dfrac{5x-4}{4-5x}$

63. $\dfrac{x-5}{x+5}$ **64.** $\dfrac{x-7}{x+7}$

65. $\dfrac{4x-6}{3-2x}$ **66.** $\dfrac{9x-15}{5-3x}$

67. $\dfrac{4-6x}{3x^2-2x}$ **68.** $\dfrac{9-15x}{5x^2-3x}$

69. $\dfrac{x^2-1}{1-x}$ **70.** $\dfrac{x^2-4}{2-x}$

71. $\dfrac{y^2-y-12}{4-y}$ **72.** $\dfrac{y^2-7y+12}{3-y}$

73. $\dfrac{x^2y - x^2}{x^3 - x^3y}$

74. $\dfrac{xy - 2x}{3y - 6}$

75. $\dfrac{x^2 + 2xy - 3y^2}{2x^2 + 5xy - 3y^2}$

76. $\dfrac{x^2 + 3xy - 10y^2}{3x^2 - 7xy + 2y^2}$

Practice PLUS

In Exercises 77–84, simplify each rational expression.

77. $\dfrac{x^2 - 9x + 18}{x^3 - 27}$

78. $\dfrac{x^3 - 8}{x^2 + 2x - 8}$

79. $\dfrac{9 - y^2}{y^2 - 3(2y - 3)}$

80. $\dfrac{16 - y^2}{y(y - 8) + 16}$

81. $\dfrac{xy + 2y + 3x + 6}{x^2 + 5x + 6}$

82. $\dfrac{xy + 4y - 7x - 28}{x^2 + 11x + 28}$

83. $\dfrac{8x^2 + 4x + 2}{1 - 8x^3}$

84. $\dfrac{x^3 - 3x^2 + 9x}{x^3 + 27}$

Application Exercises

85. The rational expression

$$\frac{130x}{100 - x}$$

describes the cost, in millions of dollars, to inoculate x percent of the population against a particular strain of flu.

  a. Evaluate the expression for $x = 40$, $x = 80$, and $x = 90$. Describe the meaning of each evaluation in terms of percentage inoculated and cost.

  b. For what value of x is the expression undefined?

  c. What happens to the cost as x approaches 100%? How can you interpret this observation?

86. The rational expression

$$\frac{60{,}000x}{100 - x}$$

describes the cost, in dollars, to remove x percent of the air pollutants in the smokestack emissions of a utility company that burns coal to generate electricity.

  a. Evaluate the expression for $x = 20$, $x = 50$, and $x = 80$. Describe the meaning of each evaluation in terms of percentage of pollutants removed and cost.

  b. For what value of x is the expression undefined?

  c. What happens to the cost as x approaches 100%? How can you interpret this observation?

Doctors use the rational expression

$$\frac{DA}{A + 12}$$

to determine the dosage of a drug prescribed for children. In this expression, A = the child's age and D = the adult dosage. Use the expression to solve Exercises 87–88.

87. If the normal adult dosage of medication is 1000 milligrams, what dosage should an 8-year-old child receive?

88. If the normal adult dosage of medication is 1000 milligrams, what dosage should a 4-year-old child receive?

89. A company that manufactures bicycles has costs given by the equation

$$C = \frac{100x + 100{,}000}{x},$$

in which x is the number of bicycles manufactured and C is the cost to manufacture each bicycle.

  a. Find the cost per bicycle when manufacturing 500 bicycles.

  b. Find the cost per bicycle when manufacturing 4000 bicycles.

  c. Does the cost per bicycle increase or decrease as more bicycles are manufactured? Explain why this happens.

90. A company that manufactures small canoes has costs given by the equation

$$C = \frac{20x + 20{,}000}{x},$$

in which x is the number of canoes manufactured and C is the cost to manufacture each canoe.

  a. Find the cost per canoe when manufacturing 100 canoes.

  b. Find the cost per canoe when manufacturing 10,000 canoes.

  c. Does the cost per canoe increase or decrease as more canoes are manufactured? Explain why this happens.

A drug is injected into a patient and the concentration of the drug in the bloodstream is monitored. The drug's concentration, y, in milligrams per liter, after x hours is modeled by

$$y = \frac{5x}{x^2 + 1}.$$

The graph of this equation, obtained with a graphing utility, is shown in the figure in a [0, 10, 1] by [0, 3, 1] viewing rectangle. Use this information to solve Exercises 91–92.

Hours after Injection
[0, 10, 1] by [0, 3, 1]

91. Use the equation to find the drug's concentration after 3 hours. Then identify the point on the equation's graph that conveys this information.

92. Use the graph of the equation to find after how many hours the drug reaches its maximum concentration. Then use the equation to find the drug's concentration at this time.

93. The bar graph shows the estimated number of calories per day needed to maintain energy balance for various gender and age groups for sedentary lifestyles. (Sedentary means a lifestyle that includes only the light physical activity associated with typical day-to-day life.)

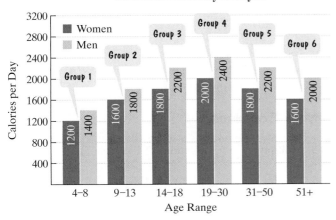

Calories Needed to Maintain Energy Balance for Sedentary Lifestyles

Source: U.S.D.A.

a. The mathematical model

$$W = -82x^2 + 654x + 620$$

describes the number of calories needed per day, W, by women in age group x with sedentary lifestyles. According to the model, how many calories per day are needed by women between the ages of 19 and 30, inclusive, with this lifestyle? Does this underestimate or overestimate the number shown by the graph? By how much?

b. The mathematical model

$$M = -96x^2 + 802x + 660$$

describes the number of calories needed per day, M, by men in age group x with sedentary lifestyles. According to the model, how many calories per day are needed by men between the ages of 19 and 30, inclusive, with this lifestyle? Does this underestimate or overestimate the number shown by the graph? By how much?

c. Write a simplified rational expression that describes the ratio of the number of calories needed per day by women in age group x to the number of calories needed per day by men in age group x for people with sedentary lifestyles.

94. The bar graph shows the estimated number of calories per day needed to maintain energy balance for various gender and age groups for moderately active lifestyles. (Moderately active means a lifestyle that includes physical activity equivalent to walking 1.5 to 3 miles per day at 3 to 4 miles per hour, in addition to the light physical activity associated with typical day-to-day life.)

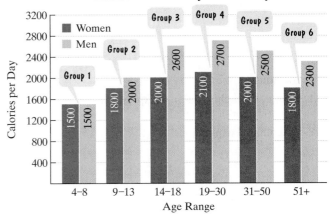

Calories Needed to Maintain Energy Balance for Moderately Active Lifestyles

Source: U.S.D.A.

a. The mathematical model

$$W = -66x^2 + 526x + 1030$$

describes the number of calories needed per day, W, by women in age group x with moderately active lifestyles. According to the model, how many calories per day are needed by women between the ages of 19 and 30, inclusive, with this lifestyle? Does this underestimate or overestimate the number shown by the graph? By how much?

b. The mathematical model

$$M = -120x^2 + 998x + 590$$

describes the number of calories needed per day, M, by men in age group x with moderately active lifestyles. According to the model, how many calories per day are needed by men between the ages of 19 and 30, inclusive, with this lifestyle? Does this underestimate or overestimate the number shown by the graph? By how much?

c. Write a simplified rational expression that describes the ratio of the number of calories needed per day by women in age group x to the number of calories needed per day by men in age group x for people with moderately active lifestyles.

Writing in Mathematics

95. What is a rational expression? Give an example with your explanation.

96. Explain how to find the number or numbers, if any, for which a rational expression is undefined.

97. Explain how to simplify a rational expression.

98. Explain how to simplify a rational expression with opposite factors in the numerator and denominator.

99. Use the graph shown for Exercises 91–92 on page 482 to write a description of the drug's concentration over time. In your description, try to convey as much information as possible that is displayed visually by the graph.

Critical Thinking Exercises

Make Sense? *In Exercises 100–103, determine whether each statement "makes sense" or "does not make sense" and explain your reasoning.*

100. Simplifying rational expressions is similar to reducing fractions.

101. I cannot simplify rational expressions without knowing how to factor polynomials.

102. The rational expressions

$$\frac{7}{14x} \quad \text{and} \quad \frac{7}{14 + x}$$

can both be simplified by dividing each numerator and each denominator by 7.

103. I evaluated $\dfrac{3x - 3}{4x(x - 1)}$ for $x = 1$ and obtained 0.

In Exercises 104–107, determine whether each statement is true or false. If the statement is false, make the necessary change(s) to produce a true statement.

104. $\dfrac{3x + 1}{3} = x + 1$

105. $\dfrac{x^2 + 3}{3} = x^2 + 1$

106. The expression $\dfrac{-3y - 6}{y + 2}$ simplifies to the consecutive integer that follows -4.

107. $\dfrac{3x + 7}{3x + 10} = \dfrac{8}{11}$

108. Write a rational expression that cannot be simplified.

109. Write a rational expression that is undefined for $x = -4$.

110. Write a rational expression with $x^2 - x - 6$ in the numerator that can be simplified to $x - 3$.

Technology Exercises

In Exercises 111–113, use the $\boxed{\text{GRAPH}}$ *or* $\boxed{\text{TABLE}}$ *feature of a graphing utility to determine if the rational expression has been correctly simplified. If the simplification is wrong, correct it and then verify your answer using the graphing utility.*

111. $\dfrac{3x + 15}{x + 5} = 3, \quad x \neq -5$

112. $\dfrac{2x^2 - x - 1}{x - 1} = 2x^2 - 1, \, x \neq 1$

113. $\dfrac{x^2 - x}{x} = x^2 - 1, \quad x \neq 0$

114. Use a graphing utility to verify the graph in **Figure 8.1** on page 480. $\boxed{\text{TRACE}}$ along the graph as x approaches 100. What do you observe?

Review Exercises

115. Multiply: $\dfrac{5}{6} \cdot \dfrac{9}{25}$. (Section 1.2, Example 5)

116. Divide: $\dfrac{2}{3} \div 4$. (Section 1.2, Example 6)

117. Solve by the addition method:

$2x - 5y = -2$
$3x + 4y = 20$. (Section 5.3, Example 3)

Preview Exercises

Exercises 118–120 will help you prepare for the material covered in the next section. In each exercise, perform the indicated operation.

118. $\dfrac{2}{5} \cdot \dfrac{3}{7}$

119. $\dfrac{3}{4} \div \dfrac{1}{2}$

120. $\dfrac{5}{4} \div \dfrac{15}{8}$

SECTION **8.2**

Objectives

1 Multiply rational expressions.

2 Divide rational expressions.

Multiplying and Dividing Rational Expressions

Source: By permission of Johnny Hart FLP and Creators Syndicate, Inc.

Your psychology class is learning various techniques to double what we remember over time. At the beginning of the course, students memorize 40 words in Latin, a language with which they are not familiar. The rational expression

$$\frac{5t + 30}{t}$$

models the class average for the number of words remembered after t days, where $t \geq 1$. If the techniques are successful, what will be the new memory model?

The new model can be found by multiplying the given rational expression by 2. In this section, you will see that we multiply rational expressions in the same way that we multiply rational numbers. Thus, we multiply numerators and multiply denominators. The rational expression for doubling what the class remembers over time is

$$\frac{2}{1} \cdot \frac{5t + 30}{t} = \frac{2(5t + 30)}{1 \cdot t} = \frac{2 \cdot 5t + 2 \cdot 30}{t} = \frac{10t + 60}{t}.$$

1 Multiply rational expressions.

Multiplying Rational Expressions

The product of two rational expressions is the product of their numerators divided by the product of their denominators.

Multiplying Rational Expressions

If $P, Q, R,$ and S are polynomials, where $Q \neq 0$ and $S \neq 0$, then

$$\frac{P}{Q} \cdot \frac{R}{S} = \frac{PR}{QS}.$$

EXAMPLE 1 **Multiplying Rational Expressions**

Multiply: $\dfrac{7}{x + 3} \cdot \dfrac{x - 2}{5}$.

Solution

$$\frac{7}{x + 3} \cdot \frac{x - 2}{5} = \frac{7(x - 2)}{(x + 3)5} \quad \text{Multiply numerators. Multiply denominators. } (x \neq -3)$$

$$= \frac{7x - 14}{5x + 15}$$

☑ **CHECK POINT 1**   Multiply: $\dfrac{9}{x+4} \cdot \dfrac{x-5}{2}$.

Here is a step-by-step procedure for multiplying rational expressions. Before multiplying, divide out any factors common to both a numerator and a denominator.

> **Multiplying Rational Expressions**
> 1. Factor all numerators and denominators completely.
> 2. Divide numerators and denominators by common factors.
> 3. Multiply the remaining factors in the numerators and multiply the remaining factors in the denominators.

EXAMPLE 2   **Multiplying Rational Expressions**

Multiply: $\dfrac{x-3}{x+5} \cdot \dfrac{10x+50}{7x-21}$.

Solution

$\dfrac{x-3}{x+5} \cdot \dfrac{10x+50}{7x-21}$

$= \dfrac{x-3}{x+5} \cdot \dfrac{10(x+5)}{7(x-3)}$ Factor as many numerators and denominators as possible.

$= \dfrac{\overset{1}{\cancel{x-3}}}{\cancel{x+5}} \cdot \dfrac{10\overset{1}{\cancel{(x+5)}}}{7\underset{1}{\cancel{(x-3)}}}$ Divide numerators and denominators by common factors.

$= \dfrac{10}{7}$ Multiply the remaining factors in the numerators and in the denominators. ▬

☑ **CHECK POINT 2**   Multiply: $\dfrac{x+4}{x-7} \cdot \dfrac{3x-21}{8x+32}$.

EXAMPLE 3   **Multiplying Rational Expressions**

Multiply: $\dfrac{x-7}{x-1} \cdot \dfrac{x^2-1}{3x-21}$.

Solution

$\dfrac{x-7}{x-1} \cdot \dfrac{x^2-1}{3x-21}$

$= \dfrac{x-7}{x-1} \cdot \dfrac{(x+1)(x-1)}{3(x-7)}$ Factor as many numerators and denominators as possible.

$= \dfrac{\overset{1}{\cancel{x-7}}}{\cancel{x-1}} \cdot \dfrac{(x+1)\overset{1}{\cancel{(x-1)}}}{3\underset{1}{\cancel{(x-7)}}}$ Divide numerators and denominators by common factors.

$= \dfrac{x+1}{3}$ Multiply the remaining factors in the numerators and in the denominators. ▬

☑ **CHECK POINT 3**   Multiply: $\dfrac{x-5}{x-2} \cdot \dfrac{x^2-4}{9x-45}$.

EXAMPLE 4 Multiplying Rational Expressions

Multiply: $\dfrac{4x + 8}{6x - 3x^2} \cdot \dfrac{3x^2 - 4x - 4}{9x^2 - 4}$.

Solution

$$\dfrac{4x + 8}{6x - 3x^2} \cdot \dfrac{3x^2 - 4x - 4}{9x^2 - 4}$$

$$= \dfrac{4(x + 2)}{3x(2 - x)} \cdot \dfrac{(3x + 2)(x - 2)}{(3x + 2)(3x - 2)}$$

Factor as many numerators and denominators as possible.

$$= \dfrac{4(x + 2)}{3x(2 - x)} \cdot \dfrac{\overset{1}{\cancel{(3x + 2)}}\overset{-1}{\cancel{(x - 2)}}}{\underset{1}{\cancel{(3x + 2)}}(3x - 2)}$$

Divide numerators and denominators by common factors. Because $2 - x$ and $x - 2$ have opposite signs, their quotient is -1.

$$= \dfrac{-4(x + 2)}{3x(3x - 2)} \quad \text{or} \quad -\dfrac{4(x + 2)}{3x(3x - 2)}$$

Multiply the remaining factors in the numerators and in the denominators.

> It is not necessary to carry out these multiplications.

✓ **CHECK POINT 4** Multiply: $\dfrac{5x + 5}{7x - 7x^2} \cdot \dfrac{2x^2 + x - 3}{4x^2 - 9}$.

---

2 Divide rational expressions.

Dividing Rational Expressions

The quotient of two rational expressions is the product of the first expression and the multiplicative inverse, or reciprocal, of the second. The reciprocal is found by interchanging the numerator and the denominator.

> **Dividing Rational Expressions**
>
> If P, Q, R, and S are polynomials, where $Q \neq 0$, $R \neq 0$, and $S \neq 0$, then
>
> $$\dfrac{P}{Q} \div \dfrac{R}{S} = \dfrac{P}{Q} \cdot \dfrac{S}{R} = \dfrac{PS}{QR}.$$
>
> Change division to multiplication.
>
> Replace $\dfrac{R}{S}$ with its reciprocal by interchanging numerator and denominator.

Thus, **we find the quotient of two rational expressions by inverting the divisor and multiplying**. For example,

$$\dfrac{x}{7} \div \dfrac{6}{y} = \dfrac{x}{7} \cdot \dfrac{y}{6} = \dfrac{xy}{42}.$$

Change the division to multiplication.

Replace $\dfrac{6}{y}$ with its reciprocal by interchanging numerator and denominator.

EXAMPLE 5 Dividing Rational Expressions

Divide: $(x + 5) \div \dfrac{x - 2}{x + 9}$.

Study Tip

When performing operations with rational expressions, if a rational expression is written without a denominator, it is helpful to write the expression with a denominator of 1. In Example 5, we wrote $x + 5$ as

$$\frac{x + 5}{1}.$$

Solution

$$(x + 5) \div \frac{x - 2}{x + 9} = \frac{x + 5}{1} \cdot \frac{x + 9}{x - 2} \quad \text{Invert the divisor and multiply.}$$

$$= \frac{(x + 5)(x + 9)}{x - 2} \quad \text{Multiply the factors in the numerators and in the denominators. We need not carry out the multiplication in the numerator.}$$

✓ **CHECK POINT 5** Divide: $(x + 3) \div \dfrac{x - 4}{x + 7}$.

EXAMPLE 6 Dividing Rational Expressions

Divide: $\dfrac{x^2 - 2x - 8}{x^2 - 9} \div \dfrac{x - 4}{x + 3}$.

Solution

$$\frac{x^2 - 2x - 8}{x^2 - 9} \div \frac{x - 4}{x + 3}$$

$$= \frac{x^2 - 2x - 8}{x^2 - 9} \cdot \frac{x + 3}{x - 4} \quad \text{Invert the divisor and multiply.}$$

$$= \frac{(x - 4)(x + 2)}{(x + 3)(x - 3)} \cdot \frac{x + 3}{x - 4} \quad \text{Factor as many numerators and denominators as possible.}$$

$$= \frac{\overset{1}{\cancel{(x - 4)}}(x + 2)}{\cancel{(x + 3)}(x - 3)} \cdot \frac{\overset{1}{\cancel{(x + 3)}}}{\cancel{(x - 4)}} \quad \begin{array}{l}\text{Divide numerators and denominators} \\ \text{by common factors.}\end{array}$$

$$= \frac{x + 2}{x - 3} \quad \begin{array}{l}\text{Multiply the remaining factors in the} \\ \text{numerators and in the denominators.}\end{array}$$

✓ **CHECK POINT 6** Divide: $\dfrac{x^2 + 5x + 6}{x^2 - 25} \div \dfrac{x + 2}{x + 5}$.

EXAMPLE 7 Dividing Rational Expressions

Divide: $\dfrac{y^2 + 7y + 12}{y^2 + 9} \div (7y^2 + 21y)$.

Solution

$$\frac{y^2 + 7y + 12}{y^2 + 9} \div \frac{7y^2 + 21y}{1} \quad \begin{array}{l}\text{It is helpful to write the divisor} \\ \text{with a denominator of 1.}\end{array}$$

$$= \frac{y^2 + 7y + 12}{y^2 + 9} \cdot \frac{1}{7y^2 + 21y} \quad \text{Invert the divisor and multiply.}$$

$$= \frac{(y + 4)(y + 3)}{y^2 + 9} \cdot \frac{1}{7y(y + 3)} \quad \begin{array}{l}\text{Factor as many numerators and} \\ \text{denominators as possible.}\end{array}$$

$$= \frac{(y + 4)\overset{1}{\cancel{(y + 3)}}}{y^2 + 9} \cdot \frac{1}{7y\cancel{(y + 3)}} \quad \begin{array}{l}\text{Divide numerators and denominators} \\ \text{by common factors.}\end{array}$$

$$= \frac{y + 4}{7y(y^2 + 9)} \quad \begin{array}{l}\text{Multiply the remaining factors in the} \\ \text{numerators and in the denominators.}\end{array}$$

✓ **CHECK POINT 7** Divide: $\dfrac{y^2 + 3y + 2}{y^2 + 1} \div (5y^2 + 10y)$.

8.2 EXERCISE SET   *MyMathLab*

Practice Exercises

In Exercises 1–32, multiply as indicated.

1. $\dfrac{4}{x+3} \cdot \dfrac{x-5}{9}$

2. $\dfrac{8}{x-2} \cdot \dfrac{x+5}{3}$

3. $\dfrac{x}{3} \cdot \dfrac{12}{x+5}$

4. $\dfrac{x}{5} \cdot \dfrac{30}{x-4}$

5. $\dfrac{3}{x} \cdot \dfrac{4x}{15}$

6. $\dfrac{7}{x} \cdot \dfrac{5x}{35}$

7. $\dfrac{x-3}{x+5} \cdot \dfrac{4x+20}{9x-27}$

8. $\dfrac{x-2}{x+9} \cdot \dfrac{5x+45}{2x-4}$

9. $\dfrac{x^2+9x+14}{x+7} \cdot \dfrac{1}{x+2}$

10. $\dfrac{x^2+9x+18}{x+6} \cdot \dfrac{1}{x+3}$

11. $\dfrac{x^2-25}{x^2-3x-10} \cdot \dfrac{x+2}{x}$

12. $\dfrac{x^2-49}{x^2-4x-21} \cdot \dfrac{x+3}{x}$

13. $\dfrac{4y+30}{y^2-3y} \cdot \dfrac{y-3}{2y+15}$

14. $\dfrac{9y+21}{y^2-2y} \cdot \dfrac{y-2}{3y+7}$

15. $\dfrac{y^2-7y-30}{y^2-6y-40} \cdot \dfrac{2y^2+5y+2}{2y^2+7y+3}$

16. $\dfrac{3y^2+17y+10}{3y^2-22y-16} \cdot \dfrac{y^2-4y-32}{y^2-8y-48}$

17. $(y^2-9) \cdot \dfrac{4}{y-3}$

18. $(y^2-16) \cdot \dfrac{3}{y-4}$

19. $\dfrac{x^2-5x+6}{x^2-2x-3} \cdot \dfrac{x^2-1}{x^2-4}$

20. $\dfrac{x^2+5x+6}{x^2+x-6} \cdot \dfrac{x^2-9}{x^2-x-6}$

21. $\dfrac{x^3-8}{x^2-4} \cdot \dfrac{x+2}{3x}$

22. $\dfrac{x^2+6x+9}{x^3+27} \cdot \dfrac{1}{x+3}$

23. $\dfrac{(x-2)^3}{(x-1)^3} \cdot \dfrac{x^2-2x+1}{x^2-4x+4}$

24. $\dfrac{(x+4)^3}{(x+2)^3} \cdot \dfrac{x^2+4x+4}{x^2+8x+16}$

25. $\dfrac{6x+2}{x^2-1} \cdot \dfrac{1-x}{3x^2+x}$

26. $\dfrac{8x+2}{x^2-9} \cdot \dfrac{3-x}{4x^2+x}$

27. $\dfrac{25-y^2}{y^2-2y-35} \cdot \dfrac{y^2-8y-20}{y^2-3y-10}$

28. $\dfrac{2y}{3y-y^2} \cdot \dfrac{2y^2-9y+9}{8y-12}$

29. $\dfrac{x^2-y^2}{x} \cdot \dfrac{x^2+xy}{x+y}$

30. $\dfrac{4x-4y}{x} \cdot \dfrac{x^2+xy}{x^2-y^2}$

31. $\dfrac{x^2+2xy+y^2}{x^2-2xy+y^2} \cdot \dfrac{4x-4y}{3x+3y}$

32. $\dfrac{x^2-y^2}{x+y} \cdot \dfrac{x+2y}{2x^2-xy-y^2}$

In Exercises 33–64, divide as indicated.

33. $\dfrac{x}{7} \div \dfrac{5}{3}$

34. $\dfrac{x}{3} \div \dfrac{3}{8}$

35. $\dfrac{3}{x} \div \dfrac{12}{x}$

36. $\dfrac{x}{5} \div \dfrac{20}{x}$

37. $\dfrac{15}{x} \div \dfrac{3}{2x}$

38. $\dfrac{9}{x} \div \dfrac{3}{4x}$

39. $\dfrac{x+1}{3} \div \dfrac{3x+3}{7}$

40. $\dfrac{x+5}{7} \div \dfrac{4x+20}{9}$

41. $\dfrac{7}{x-5} \div \dfrac{28}{3x-15}$

42. $\dfrac{4}{x-6} \div \dfrac{40}{7x-42}$

43. $\dfrac{x^2-4}{x} \div \dfrac{x+2}{x-2}$

44. $\dfrac{x^2-4}{x-2} \div \dfrac{x+2}{4x-8}$

45. $(y^2-16) \div \dfrac{y^2+3y-4}{y^2+4}$

46. $(y^2+4y-5) \div \dfrac{y^2-25}{y+7}$

47. $\dfrac{y^2-y}{15} \div \dfrac{y-1}{5}$

48. $\dfrac{y^2-2y}{15} \div \dfrac{y-2}{5}$

49. $\dfrac{4x^2+10}{x-3} \div \dfrac{6x^2+15}{x^2-9}$

50. $\dfrac{x^2+x}{x^2-4} \div \dfrac{x^2-1}{x^2+5x+6}$

51. $\dfrac{x^2-25}{2x-2} \div \dfrac{x^2+10x+25}{x^2+4x-5}$

52. $\dfrac{x^2-4}{x^2+3x-10} \div \dfrac{x^2+5x+6}{x^2+8x+15}$

53. $\dfrac{y^3+y}{y^2-y} \div \dfrac{y^3-y^2}{y^2-2y+1}$

54. $\dfrac{3y^2 - 12}{y^2 + 4y + 4} \div \dfrac{y^3 - 2y^2}{y^2 + 2y}$

55. $\dfrac{y^2 + 5y + 4}{y^2 + 12y + 32} \div \dfrac{y^2 - 12y + 35}{y^2 + 3y - 40}$

56. $\dfrac{y^2 + 4y - 21}{y^2 + 3y - 28} \div \dfrac{y^2 + 14y + 48}{y^2 + 4y - 32}$

57. $\dfrac{2y^2 - 128}{y^2 + 16y + 64} \div \dfrac{y^2 - 6y - 16}{3y^2 + 30y + 48}$

58. $\dfrac{3y + 12}{y^2 + 3y} \div \dfrac{y^2 + y - 12}{9y - y^3}$

59. $\dfrac{2x + 2y}{3} \div \dfrac{x^2 - y^2}{x - y}$

60. $\dfrac{5x + 5y}{7} \div \dfrac{x^2 - y^2}{x - y}$

61. $\dfrac{x^2 - y^2}{8x^2 - 16xy + 8y^2} \div \dfrac{4x - 4y}{x + y}$

62. $\dfrac{4x^2 - y^2}{x^2 + 4xy + 4y^2} \div \dfrac{4x - 2y}{3x + 6y}$

63. $\dfrac{xy - y^2}{x^2 + 2x + 1} \div \dfrac{2x^2 + xy - 3y^2}{2x^2 + 5xy + 3y^2}$

64. $\dfrac{x^2 - 4y^2}{x^2 + 3xy + 2y^2} \div \dfrac{x^2 - 4xy + 4y^2}{x + y}$

Practice PLUS

In Exercises 65–72, perform the indicated operation or operations.

65. $\left(\dfrac{y - 2}{y^2 - 9y + 18} \cdot \dfrac{y^2 - 4y - 12}{y + 2} \right) \div \dfrac{y^2 - 4}{y^2 + 5y + 6}$

66. $\left(\dfrac{6y^2 + 31y + 18}{3y^2 - 20y + 12} \cdot \dfrac{2y^2 - 15y + 18}{6y^2 + 35y + 36} \right) \div \dfrac{2y^2 - 13y + 15}{9y^2 + 15y + 4}$

67. $\dfrac{3x^2 + 3x - 60}{2x - 8} \div \left(\dfrac{30x^2}{x^2 - 7x + 10} \cdot \dfrac{x^3 + 3x^2 - 10x}{25x^3} \right)$

68. $\dfrac{5x^2 - x}{3x + 2} \div \left(\dfrac{6x^2 + x - 2}{10x^2 + 3x - 1} \cdot \dfrac{2x^2 - x - 1}{2x^2 - x} \right)$

69. $\dfrac{x^2 + xz + xy + yz}{x - y} \div \dfrac{x + z}{x + y}$

70. $\dfrac{x^2 - xz + xy - yz}{x - y} \div \dfrac{x - z}{y - x}$

71. $\dfrac{3xy + ay + 3xb + ab}{9x^2 - a^2} \div \dfrac{y^3 + b^3}{6x - 2a}$

72. $\dfrac{5xy - ay - 5xb + ab}{25x^2 - a^2} \div \dfrac{y^3 - b^3}{15x + 3a}$

Application Exercises

In Exercises 73–76, write a simplified rational expression for the area of each figure. Assume that all measures are in inches.

73.   **74.**

75.   **76.**

Writing in Mathematics

77. Explain how to multiply rational expressions.

78. Explain how to divide rational expressions.

79. In dividing polynomials

$$\dfrac{P}{Q} \div \dfrac{R}{S},$$

why is it necessary to state that polynomial R is not equal to 0?

Critical Thinking Exercises

Make Sense? *In Exercises 80–83, determine whether each statement "makes sense" or "does not make sense" and explain your reasoning.*

80. When dividing rational expressions, I multiply by the reciprocal of the divisor, just as I did when dividing rational numbers.

81. When opposite factors appear in the numerator and the denominator of a multiplication problem, I can put their quotient, −1, in either the numerator or the denominator.

82. When performing the division

$$\dfrac{7x}{x + 3} \div \dfrac{(x + 3)^2}{x - 5},$$

I began by dividing the numerator and the denominator by the common factor, $x + 3$.

83. The quotient

$$\dfrac{x + 2}{x - 5} \div \dfrac{x - 4}{x + 3}$$

is undefined for $x = 5$, $x = -3$, and $x = 4$.

In Exercises 84–87, determine whether each statement is true or false. If the statement is false, make the necessary change(s) to produce a true statement.

84. $5 \div x = \dfrac{1}{5} \cdot x$ for any nonzero number x.

85. $\dfrac{4}{x} \div \dfrac{x - 2}{x} = \dfrac{4}{x - 2}$ if $x \ne 0$ and $x \ne 2$.

86. $\dfrac{x-5}{6} \cdot \dfrac{3}{5-x} = \dfrac{1}{2}$ for any value of x except 5.

87. The quotient of two rational expressions can be found by inverting each expression and multiplying.

88. Find the missing polynomials: $\dfrac{\rule{1cm}{0.3cm}}{\rule{1cm}{0.3cm}} \cdot \dfrac{3x-12}{2x} = \dfrac{3}{2}$.

89. Find the missing polynomials: $-\dfrac{1}{2x-3} \div \dfrac{\rule{1cm}{0.3cm}}{\rule{1cm}{0.3cm}} = \dfrac{1}{3}$.

90. Divide:

$$\dfrac{9x^2 - y^2 + 15x - 5y}{3x^2 + xy + 5x} \div \dfrac{3x+y}{9x^3 + 6x^2y + xy^2}.$$

Technology Exercises

In Exercises 91–94, use the $\boxed{\text{GRAPH}}$ *or* $\boxed{\text{TABLE}}$ *feature of a graphing utility to determine if the multiplication or division has been performed correctly. If the answer is wrong, correct it and then verify your correction using the graphing utility.*

91. $\dfrac{x^2+x}{3x} \cdot \dfrac{6x}{x+1} = 2x$

92. $\dfrac{x^3-25x}{x^2-3x-10} \cdot \dfrac{x+2}{x} = x+5$

93. $\dfrac{x^2-9}{x+4} \div \dfrac{x-3}{x+4} = x-3$

94. $(x-5) \div \dfrac{2x^2-11x+5}{4x^2-1} = 2x-1$

Review Exercises

95. Solve: $2x+3 < 3(x-5)$. (Section 2.6, Example 8)

96. Factor completely: $3x^2 - 15x - 42$. (Section 7.5, Example 2)

97. Solve: $x(2x+9) = 5$. (Section 7.6, Example 6)

Preview Exercises

Exercises 98–100 will help you prepare for the material covered in the next section.

98. Subtract: $\dfrac{7}{9} - \dfrac{1}{9}$.

99. Add: $\dfrac{2x}{3} + \dfrac{x}{3}$.

100. Simplify: $\dfrac{x^2-6x+9}{x^2-9}$.

Adding and Subtracting Rational Expressions with the Same Denominator

Objectives

1 Add rational expressions with the same denominator.

2 Subtract rational expressions with the same denominator.

3 Add and subtract rational expressions with opposite denominators.

Are you long, medium, or round? Your skull, that is? The varying shapes of the human skull create physical diversity in the human species. By learning to add and subtract rational expressions with the same denominator, you will obtain an expression that models this diversity.

1 Add rational expressions with the same denominator.

Addition when Denominators Are the Same

To add rational numbers having the same denominators, such as $\frac{2}{9}$ and $\frac{5}{9}$, we add the numerators and place the sum over the common denominator:

$$\frac{2}{9} + \frac{5}{9} = \frac{2+5}{9} = \frac{7}{9}.$$

We add rational expressions with the same denominator in an identical manner.

Adding Rational Expressions with Common Denominators

If $\dfrac{P}{R}$ and $\dfrac{Q}{R}$ are rational expressions, then

$$\frac{P}{R} + \frac{Q}{R} = \frac{P+Q}{R}.$$

To add rational expressions with the same denominator, add numerators and place the sum over the common denominator. If possible, factor and simplify the result.

EXAMPLE 1 **Adding Rational Expressions when Denominators Are the Same**

Add: $\dfrac{2x-1}{3} + \dfrac{x+4}{3}$.

Solution

$$\frac{2x-1}{3} + \frac{x+4}{3} = \frac{2x-1+x+4}{3} \qquad \text{Add numerators. Place this sum over the common denominator.}$$

$$= \frac{3x+3}{3} \qquad \text{Combine like terms.}$$

$$= \frac{\overset{1}{\cancel{3}}(x+1)}{\underset{1}{\cancel{3}}} \qquad \text{Factor and simplify.}$$

$$= x + 1 \qquad \blacksquare$$

✓ **CHECK POINT 1** Add: $\dfrac{3x-2}{5} + \dfrac{2x+12}{5}$.

EXAMPLE 2 **Adding Rational Expressions when Denominators Are the Same**

Add: $\dfrac{x^2}{x^2-9} + \dfrac{9-6x}{x^2-9}$.

Solution

$$\frac{x^2}{x^2-9} + \frac{9-6x}{x^2-9} = \frac{x^2+9-6x}{x^2-9} \qquad \text{Add numerators. Place this sum over the common denominator.}$$

$$= \frac{x^2-6x+9}{x^2-9} \qquad \text{Write the numerator in descending powers of x.}$$

$$= \frac{(x-3)\overset{1}{\cancel{(x-3)}}}{(x+3)\underset{1}{\cancel{(x-3)}}} \qquad \text{Factor and simplify. What values of x are not permitted?}$$

$$= \frac{x-3}{x+3} \qquad \blacksquare$$

✓ **CHECK POINT 2** Add: $\dfrac{x^2}{x^2 - 25} + \dfrac{25 - 10x}{x^2 - 25}$.

2 Subtract rational expressions with the same denominator.

Subtraction when Denominators Are the Same

The following box shows how to subtract rational expressions with the same denominator:

Subtracting Rational Expressions with Common Denominators

If $\dfrac{P}{R}$ and $\dfrac{Q}{R}$ are rational expressions, then

$$\frac{P}{R} - \frac{Q}{R} = \frac{P - Q}{R}.$$

To subtract rational expressions with the same denominator, subtract numerators and place the difference over the common denominator. If possible, factor and simplify the result.

EXAMPLE 3 **Subtracting Rational Expressions when Denominators Are the Same**

Subtract:

a. $\dfrac{2x + 3}{x + 1} - \dfrac{x}{x + 1}$ **b.** $\dfrac{5x + 1}{x^2 - 9} - \dfrac{4x - 2}{x^2 - 9}$.

Solution

a. $\dfrac{2x + 3}{x + 1} - \dfrac{x}{x + 1} = \dfrac{2x + 3 - x}{x + 1}$ Subtract numerators. Place this difference over the common denominator.

$= \dfrac{x + 3}{x + 1}$ Combine like terms.

b. $\dfrac{5x + 1}{x^2 - 9} - \dfrac{4x - 2}{x^2 - 9} = \dfrac{5x + 1 - (4x - 2)}{x^2 - 9}$ Subtract numerators and include parentheses to indicate that both terms are subtracted. Place this difference over the common denominator.

$= \dfrac{5x + 1 - 4x + 2}{x^2 - 9}$ Remove parentheses and then change the sign of each term.

$= \dfrac{x + 3}{x^2 - 9}$ Combine like terms.

$= \dfrac{\overset{1}{\cancel{x + 3}}}{\underset{1}{\cancel{(x + 3)}}(x - 3)}$ Factor and simplify ($x \neq -3$, and $x \neq 3$).

$= \dfrac{1}{x - 3}$

✓ **CHECK POINT 3** Subtract:

a. $\dfrac{4x + 5}{x + 7} - \dfrac{x}{x + 7}$ **b.** $\dfrac{3x^2 + 4x}{x - 1} - \dfrac{11x - 4}{x - 1}$.

Study Tip

When a numerator is being subtracted, be sure to **subtract every term in that expression**.

| The − sign applies to the entire numerator, $4x - 2$. | Insert parentheses to indicate this. | The sign of every term of $4x - 2$ changes. |

$$\frac{5x + 1}{x^2 - 9} - \frac{4x - 2}{x^2 - 9} = \frac{5x + 1 - (4x - 2)}{x^2 - 9} = \frac{5x + 1 - 4x + 2}{x^2 - 9}$$

The entire numerator of the second rational expression must be subtracted. Avoid the common error of subtracting only the first term.

Incorrect!

$$\frac{5x + 1}{x^2 - 9} - \frac{4x - 2}{x^2 - 9} = \frac{5x + 1 - 4x - 2}{x^2 - 9}$$

−2 must also be subtracted.

EXAMPLE 4   **Subtracting Rational Expressions when Denominators Are the Same**

Subtract: $\dfrac{20y^2 + 5y + 1}{6y^2 + y - 2} - \dfrac{8y^2 - 12y - 5}{6y^2 + y - 2}$.

Solution

$$\frac{20y^2 + 5y + 1}{6y^2 + y - 2} - \frac{8y^2 - 12y - 5}{6y^2 + y - 2}$$

Don't forget the parentheses.

$$= \frac{20y^2 + 5y + 1 - (8y^2 - 12y - 5)}{6y^2 + y - 2}$$

Subtract numerators. Place this difference over the common denominator.

$$= \frac{20y^2 + 5y + 1 - 8y^2 + 12y + 5}{6y^2 + y - 2}$$

Remove parentheses and then change the sign of each term.

$$= \frac{(20y^2 - 8y^2) + (5y + 12y) + (1 + 5)}{6y^2 + y - 2}$$

Group like terms. This step is usually performed mentally.

$$= \frac{12y^2 + 17y + 6}{6y^2 + y - 2}$$

Combine like terms.

$$= \frac{\overset{1}{\cancel{(3y + 2)}}(4y + 3)}{\underset{1}{\cancel{(3y + 2)}}(2y - 1)}$$

Factor and simplify.

$$= \frac{4y + 3}{2y - 1}$$

☑ **CHECK POINT 4**   Subtract: $\dfrac{y^2 + 3y - 6}{y^2 - 5y + 4} - \dfrac{4y - 4 - 2y^2}{y^2 - 5y + 4}$.

3 Add and subtract rational expressions with opposite denominators.

Addition and Subtraction when Denominators Are Opposites

How do we add or subtract rational expressions when denominators are opposites, or additive inverses? Here is an example of this type of addition problem:

$$\frac{x^2}{x-5} + \frac{4x+5}{5-x}.$$

> These denominators are opposites.
> The differ only in their signs.

Multiply the numerator and the denominator of either of the rational expressions by -1. Then they will both have the same denominator.

EXAMPLE 5   Adding Rational Expressions when Denominators Are Opposites

Add: $\dfrac{x^2}{x-5} + \dfrac{4x+5}{5-x}.$

Solution

$$\frac{x^2}{x-5} + \frac{4x+5}{5-x}$$

$$= \frac{x^2}{x-5} + \frac{(-1)}{(-1)} \cdot \frac{4x+5}{5-x} \qquad \text{Multiply the numerator and denominator of the second rational expression by } -1.$$

$$= \frac{x^2}{x-5} + \frac{-4x-5}{-5+x} \qquad \text{Perform the multiplications by } -1 \text{ by changing every term's sign.}$$

$$= \frac{x^2}{x-5} + \frac{-4x-5}{x-5} \qquad \text{Rewrite } -5+x \text{ as } x-5. \text{ Both rational expressions have the same denominator.}$$

$$= \frac{x^2 + (-4x-5)}{x-5} \qquad \text{Add numerators. Place this sum over the common denominator.}$$

$$= \frac{x^2 - 4x - 5}{x-5} \qquad \text{Remove parentheses.}$$

$$= \frac{\overset{1}{\cancel{(x-5)}}(x+1)}{\underset{1}{\cancel{x-5}}} \qquad \text{Factor and simplify.}$$

$$= x + 1 \qquad\qquad\qquad \blacksquare$$

✓ **CHECK POINT 5**   Add: $\dfrac{x^2}{x-7} + \dfrac{4x+21}{7-x}.$

Adding and Subtracting Rational Expressions with Opposite Denominators

When one denominator is the opposite, or additive inverse, of the other, first multiply either rational expression by $\dfrac{-1}{-1}$ to obtain a common denominator.

EXAMPLE 6 **Subtracting Rational Expressions when Denominators Are Opposites**

Subtract: $\dfrac{5x - x^2}{x^2 - 4x - 3} - \dfrac{3x - x^2}{3 + 4x - x^2}$.

Solution We note that $x^2 - 4x - 3$ and $3 + 4x - x^2$ are opposites. We multiply the second rational expression by $\dfrac{-1}{-1}$.

$$\dfrac{(-1)}{(-1)} \cdot \dfrac{3x - x^2}{3 + 4x - x^2} = \dfrac{-3x + x^2}{-3 - 4x + x^2}$$

Multiply the numerator and denominator by −1 by changing every term's sign.

$$= \dfrac{x^2 - 3x}{x^2 - 4x - 3}$$

Write the numerator and the denominator in descending powers of x.

We now return to the original subtraction problem.

$$\dfrac{5x - x^2}{x^2 - 4x - 3} - \dfrac{3x - x^2}{3 + 4x - x^2}$$

This is the given problem.

$$= \dfrac{5x - x^2}{x^2 - 4x - 3} - \dfrac{x^2 - 3x}{x^2 - 4x - 3}$$

Replace the second rational expression by the form obtained through multiplication by $\dfrac{-1}{-1}$.

$$= \dfrac{5x - x^2 - (x^2 - 3x)}{x^2 - 4x - 3}$$

Subtract numerators. Place this difference over the common denominator. Don't forget parentheses!

$$= \dfrac{5x - x^2 - x^2 + 3x}{x^2 - 4x - 3}$$

Remove parentheses and then change the sign of each term.

$$= \dfrac{-2x^2 + 8x}{x^2 - 4x - 3}$$

Combine like terms in the numerator. Although the numerator can be factored, further simplification is not possible. ∎

✓ **CHECK POINT 6** Subtract: $\dfrac{7x - x^2}{x^2 - 2x - 9} - \dfrac{5x - 3x^2}{9 + 2x - x^2}$.

8.3 EXERCISE SET **MyMathLab** | PRACTICE   WATCH   DOWNLOAD READ   REVIEW

Practice Exercises

In Exercises 1–38, add or subtract as indicated. Simplify the result, if possible.

1. $\dfrac{7x}{13} + \dfrac{2x}{13}$

2. $\dfrac{3x}{17} + \dfrac{8x}{17}$

3. $\dfrac{8x}{15} + \dfrac{x}{15}$

4. $\dfrac{9x}{24} + \dfrac{x}{24}$

5. $\dfrac{x - 3}{12} + \dfrac{5x + 21}{12}$

6. $\dfrac{x + 4}{9} + \dfrac{2x - 25}{9}$

7. $\dfrac{4}{x} + \dfrac{2}{x}$

8. $\dfrac{5}{x} + \dfrac{13}{x}$

9. $\dfrac{8}{9x} + \dfrac{13}{9x}$

10. $\dfrac{4}{9x} + \dfrac{11}{9x}$

11. $\dfrac{5}{x + 3} + \dfrac{4}{x + 3}$

12. $\dfrac{8}{x + 6} + \dfrac{10}{x + 6}$

13. $\dfrac{x}{x - 3} + \dfrac{4x + 5}{x - 3}$

14. $\dfrac{x}{x - 4} + \dfrac{9x + 7}{x - 4}$

15. $\dfrac{4x + 1}{6x + 5} + \dfrac{8x + 9}{6x + 5}$

16. $\dfrac{3x + 2}{3x + 4} + \dfrac{3x + 6}{3x + 4}$

17. $\dfrac{y^2 + 7y}{y^2 - 5y} + \dfrac{y^2 - 4y}{y^2 - 5y}$

18. $\dfrac{y^2 - 2y}{y^2 + 3y} + \dfrac{y^2 + y}{y^2 + 3y}$

19. $\dfrac{4y - 1}{5y^2} + \dfrac{3y + 1}{5y^2}$

20. $\dfrac{y + 2}{6y^3} + \dfrac{3y - 2}{6y^3}$

21. $\dfrac{x^2 - 2}{x^2 + x - 2} + \dfrac{2x - x^2}{x^2 + x - 2}$

22. $\dfrac{x^2 + 9x}{4x^2 - 11x - 3} + \dfrac{3x - 5x^2}{4x^2 - 11x - 3}$

23. $\dfrac{x^2 - 4x}{x^2 - x - 6} + \dfrac{4x - 4}{x^2 - x - 6}$

24. $\dfrac{x}{2x + 7} - \dfrac{2}{2x + 7}$

25. $\dfrac{3x}{5x - 4} - \dfrac{4}{5x - 4}$

26. $\dfrac{x}{x - 1} - \dfrac{1}{x - 1}$

27. $\dfrac{4x}{4x - 3} - \dfrac{3}{4x - 3}$

28. $\dfrac{2y + 1}{3y - 7} - \dfrac{y + 8}{3y - 7}$

29. $\dfrac{14y}{7y + 2} - \dfrac{7y - 2}{7y + 2}$

30. $\dfrac{2x + 3}{3x - 6} - \dfrac{3 - x}{3x - 6}$

31. $\dfrac{3x + 1}{4x - 2} - \dfrac{x + 1}{4x - 2}$

32. $\dfrac{x^3 - 3}{2x^4} - \dfrac{7x^3 - 3}{2x^4}$

33. $\dfrac{3y^2 - 1}{3y^3} - \dfrac{6y^2 - 1}{3y^3}$

34. $\dfrac{y^2 + 3y}{y^2 + y - 12} - \dfrac{y^2 - 12}{y^2 + y - 12}$

35. $\dfrac{4y^2 + 5}{9y^2 - 64} - \dfrac{y^2 - y + 29}{9y^2 - 64}$

36. $\dfrac{2y^2 + 6y + 8}{y^2 - 16} - \dfrac{y^2 - 3y - 12}{y^2 - 16}$

37. $\dfrac{6y^2 + y}{2y^2 - 9y + 9} - \dfrac{2y + 9}{2y^2 - 9y + 9} - \dfrac{4y - 3}{2y^2 - 9y + 9}$

38. $\dfrac{3y^2 - 2}{3y^2 + 10y - 8} - \dfrac{y + 10}{3y^2 + 10y - 8} - \dfrac{y^2 - 6y}{3y^2 + 10y - 8}$

In Exercises 39–64, denominators are opposites, or additive inverses. Add or subtract as indicated. Simplify the result, if possible.

39. $\dfrac{4}{x - 3} + \dfrac{2}{3 - x}$

40. $\dfrac{6}{x - 5} + \dfrac{2}{5 - x}$

41. $\dfrac{6x + 7}{x - 6} + \dfrac{3x}{6 - x}$

42. $\dfrac{6x + 5}{x - 2} + \dfrac{4x}{2 - x}$

43. $\dfrac{5x - 2}{3x - 4} + \dfrac{2x - 3}{4 - 3x}$

44. $\dfrac{9x - 1}{7x - 3} + \dfrac{6x - 2}{3 - 7x}$

45. $\dfrac{x^2}{x - 2} + \dfrac{4}{2 - x}$

46. $\dfrac{x^2}{x - 3} + \dfrac{9}{3 - x}$

47. $\dfrac{y - 3}{y^2 - 25} + \dfrac{y - 3}{25 - y^2}$

48. $\dfrac{y - 7}{y^2 - 16} + \dfrac{7 - y}{16 - y^2}$

49. $\dfrac{6}{x - 1} - \dfrac{5}{1 - x}$

50. $\dfrac{10}{x - 2} - \dfrac{6}{2 - x}$

51. $\dfrac{10}{x + 3} - \dfrac{2}{-x - 3}$

52. $\dfrac{11}{x + 7} - \dfrac{5}{-x - 7}$

53. $\dfrac{y}{y - 1} - \dfrac{1}{1 - y}$

54. $\dfrac{y}{y - 4} - \dfrac{4}{4 - y}$

55. $\dfrac{3 - x}{x - 7} - \dfrac{2x - 5}{7 - x}$

56. $\dfrac{4 - x}{x - 9} - \dfrac{3x - 8}{9 - x}$

57. $\dfrac{x - 2}{x^2 - 25} - \dfrac{x - 2}{25 - x^2}$

58. $\dfrac{x - 8}{x^2 - 16} - \dfrac{x - 8}{16 - x^2}$

59. $\dfrac{x}{x - y} + \dfrac{y}{y - x}$

60. $\dfrac{2x - y}{x - y} + \dfrac{x - 2y}{y - x}$

61. $\dfrac{2x}{x^2 - y^2} + \dfrac{2y}{y^2 - x^2}$

62. $\dfrac{2y}{x^2 - y^2} + \dfrac{2x}{y^2 - x^2}$

63. $\dfrac{x^2 - 2}{x^2 + 6x - 7} + \dfrac{19 - 4x}{7 - 6x - x^2}$

64. $\dfrac{2x + 3}{x^2 - x - 30} + \dfrac{x - 2}{30 + x - x^2}$

Practice PLUS

In Exercises 65–72, perform the indicated operation or operations. Simplify the result, if possible.

65. $\dfrac{6b^2 - 10b}{16b^2 - 48b + 27} + \dfrac{7b^2 - 20b}{16b^2 - 48b + 27} - \dfrac{6b - 3b^2}{16b^2 - 48b + 27}$

66. $\dfrac{22b + 15}{12b^2 + 52b - 9} + \dfrac{30b - 20}{12b^2 + 52b - 9} - \dfrac{4 - 2b}{12b^2 + 52b - 9}$

67. $\dfrac{2y}{y - 5} - \left(\dfrac{2}{y - 5} + \dfrac{y - 2}{y - 5} \right)$

68. $\dfrac{3x}{(x + 1)^2} - \left[\dfrac{5x + 1}{(x + 1)^2} - \dfrac{3x + 2}{(x + 1)^2} \right]$

69. $\dfrac{b}{ac + ad - bc - bd} - \dfrac{a}{ac + ad - bc - bd}$

70. $\dfrac{y}{ax + bx - ay - by} - \dfrac{x}{ax + bx - ay - by}$

71. $\dfrac{(y - 3)(y + 2)}{(y + 1)(y - 4)} - \dfrac{(y + 2)(y + 3)}{(y + 1)(4 - y)} - \dfrac{(y + 5)(y - 1)}{(y + 1)(4 - y)}$

72. $\dfrac{(y + 1)(2y - 1)}{(y - 2)(y - 3)} + \dfrac{(y + 2)(y - 1)}{(y - 2)(y - 3)} - \dfrac{(y + 5)(2y + 1)}{(3 - y)(2 - y)}$

Application Exercises

73. Anthropologists and forensic scientists classify skulls using

$$\frac{L + 60W}{L} - \frac{L - 40W}{L},$$

where L is the skull's length and W is its width.

a. Express the classification as a single rational expression.

b. If the value of the rational expression in part (a) is less than 75, a skull is classified as long. A medium skull has a value between 75 and 80, and a round skull has a value over 80. Use your rational expression from part (a) to classify a skull that is 5 inches wide and 6 inches long.

74. The temperature, in degrees Fahrenheit, of a dessert placed in a freezer for t hours is modeled by

$$\frac{t + 30}{t^2 + 4t + 1} - \frac{t - 50}{t^2 + 4t + 1}.$$

a. Express the temperature as a single rational expression.

b. Use your rational expression from part (a) to find the temperature of the dessert, to the nearest hundredth of a degree, after 1 hour and after 2 hours.

In Exercises 75–76, find the perimeter of each rectangle.

75.

$\frac{5}{x+3}$ meters

$\frac{5x+10}{x+3}$ meters

76.

$\frac{7}{x+4}$ inches

$\frac{4x+9}{x+4}$ inches

Writing in Mathematics

77. Explain how to add rational expressions when denominators are the same. Give an example with your explanation.

78. Explain how to subtract rational expressions when denominators are the same. Give an example with your explanation.

79. Describe two similarities between the following problems:

$$\frac{3}{8} + \frac{1}{8} \quad \text{and} \quad \frac{x}{x^2 - 1} + \frac{1}{x^2 - 1}.$$

80. Explain how to add rational expressions when denominators are opposites. Use an example to support your explanation.

Critical Thinking Exercises

Make Sense?   *In Exercises 81–84, determine whether each statement "makes sense" or "does not make sense" and explain your reasoning.*

81. After adding $\dfrac{3x + 1}{4}$ and $\dfrac{x + 2}{4}$, I simplified the sum by dividing the numerator and the denominator by 4.

82. I use similar procedures to find each of the following sums:

$$\frac{3}{8} + \frac{1}{8} \quad \text{and} \quad \frac{x}{x^2 - 1} + \frac{1}{x^2 - 1}.$$

83. I subtracted $\dfrac{3x - 5}{x - 1}$ from $\dfrac{x - 3}{x - 1}$ and obtained a constant.

84. I added $\dfrac{5}{x - 7}$ and $\dfrac{3}{7 - x}$ by first multiplying the second rational expression by -1.

In Exercises 85–88, determine whether each statement is true or false. If the statement is false, make the necessary change(s) to produce a true statement.

85. The sum of two rational expressions with the same denominator can be found by adding numerators, adding denominators, and then simplifying.

86. $\dfrac{4}{b} - \dfrac{2}{-b} = -\dfrac{2}{b}$

87. The difference between two rational expressions with the same denominator can always be simplified.

88. $\dfrac{2x + 1}{x - 7} + \dfrac{3x + 1}{x - 7} - \dfrac{5x + 2}{x - 7} = 0$

In Exercises 89–90, perform the indicated operations. Simplify the result, if possible.

89. $\left(\dfrac{3x - 1}{x^2 + 5x - 6} - \dfrac{2x - 7}{x^2 + 5x - 6} \right) \div \dfrac{x + 2}{x^2 - 1}$

90. $\left(\dfrac{3x^2 - 4x + 4}{3x^2 + 7x + 2} - \dfrac{10x + 9}{3x^2 + 7x + 2} \right) \div \dfrac{x - 5}{x^2 - 4}$

In Exercises 91–95, find the missing expression.

91. $\dfrac{2x}{x + 3} + \dfrac{\rule{0.5cm}{0.4pt}}{x + 3} = \dfrac{4x + 1}{x + 3}$

92. $\dfrac{3x}{x + 2} - \dfrac{\rule{0.5cm}{0.4pt}}{x + 2} = \dfrac{6 - 17x}{x + 2}$

93. $\dfrac{6}{x - 2} + \dfrac{\rule{0.5cm}{0.4pt}}{2 - x} = \dfrac{13}{x - 2}$

94. $\dfrac{a^2}{a - 4} - \dfrac{\rule{0.5cm}{0.4pt}}{a - 4} = a + 3$

95. $\dfrac{3x}{x - 5} + \dfrac{\rule{0.5cm}{0.4pt}}{5 - x} = \dfrac{7x + 1}{x - 5}$

Technology Exercises

In Exercises 96–98, use the $\boxed{\text{GRAPH}}$ *or* $\boxed{\text{TABLE}}$ *feature of a graphing utility to determine if the subtraction has been performed correctly. If the answer is wrong, correct it and then verify your correction using the graphing utility.*

96. $\dfrac{3x + 6}{2} - \dfrac{x}{2} = x + 3$

97. $\dfrac{x^2 + 4x + 3}{x + 2} - \dfrac{5x + 9}{x + 2} = x - 2, x \neq -2$

98. $\dfrac{x^2 - 13}{x + 4} - \dfrac{3}{x + 4} = x + 4, x \neq -4$

Review Exercises

99. Subtract: $\dfrac{13}{15} - \dfrac{8}{45}$. (Section 1.2, Example 9)

100. Factor completely: $81x^4 - 1$. (Section 7.4, Example 4)

101. Divide: $\dfrac{3x^3 + 2x^2 - 26x - 15}{x + 3}$. (Section 6.6, Example 2)

Preview Exercises

Exercises 102–104 will help you prepare for the material covered in the next section.

In Exercises 102–103, perform the indicated operation. Where possible, reduce the answer to its lowest terms.

102. $\dfrac{1}{2} + \dfrac{2}{3}$

103. $\dfrac{1}{8} - \dfrac{5}{6}$

104. Simplify: $\dfrac{(y + 2)y - 2 \cdot 4}{4y(y + 4)}$.

8.4

Adding and Subtracting Rational Expressions with Different Denominators

Objectives

1 Find the least common denominator.

2 Add and subtract rational expressions with different denominators.

When my aunt asked how I liked my five-year-old nephew, I replied "medium rare." Unfortunately, my little joke did not get me out of baby sitting for the Dennis the Menace of our family. Now the little squirt doesn't want to go to bed because his head hurts. Does my aunt have any Tylenol®? What is the proper dosage for a child his age?

In this section's exercise set, you will use two formulas that model drug dosage for children. Before working with these models, we continue drawing on your experience from arithmetic to add and subtract rational expressions that have different denominators.

1 Find the least common denominator.

Finding the Least Common Denominator

We can gain insight into adding rational expressions with different denominators by looking closely at what we do when adding fractions with different denominators. For example, suppose that we want to add $\frac{1}{2}$ and $\frac{2}{3}$. We must first write the fractions with the same denominator. We look for the smallest number that contains both 2 and 3 as factors. This number, 6, is then used as the *least common denominator,* or LCD.

The **least common denominator** of several rational expressions is a polynomial consisting of the product of all prime factors in the denominators, with each factor raised to the greatest power of its occurrence in any denominator.

Finding the Least Common Denominator

1. Factor each denominator completely.
2. List the factors of the first denominator.
3. Add to the list in step 2 any factors of the second denominator that do not appear in the list.
4. Form the product of the factors from the list in step 3. This product is the least common denominator.

EXAMPLE 1 Finding the Least Common Denominator

Find the LCD of $\dfrac{7}{6x^2}$ and $\dfrac{2}{9x}$.

Solution

Step 1. Factor each denominator completely.

$$6x^2 = 3 \cdot 2 \cdot x \cdot x$$
$$9x = 3 \cdot 3 \cdot x$$

$$\frac{7}{6x^2} \qquad\qquad \frac{2}{9x}$$

Factors are 3, 2, x, and x.   Factors are 3, 3, and x.

Step 2. List the factors of the first denominator.

$$3, 2, x, x$$

Step 3. Add any unlisted factors from the second denominator. Two factors from $9x$ are already in our list. These factors include one factor of 3 and x. We add the other factor of 3 to our list. We have

$$3, 3, 2, x, x.$$

Step 4. The least common denominator is the product of all factors in the final list. Thus,

$$3 \cdot 3 \cdot 2 \cdot x \cdot x,$$

or $18x^2$ is the least common denominator.

✓ **CHECK POINT 1** Find the LCD of $\dfrac{3}{10x^2}$ and $\dfrac{7}{15x}$.

EXAMPLE 2 Finding the Least Common Denominator

Find the LCD of $\dfrac{3}{x+1}$ and $\dfrac{5}{x-1}$.

Solution

Step 1. Factor each denominator completely.

$$x + 1 = 1(x + 1)$$
$$x - 1 = 1(x - 1)$$

$$\frac{3}{x+1} \qquad\qquad \frac{5}{x-1}$$

Factors are 1 and $x + 1$.   Factors are 1 and $x - 1$.

Step 2. List the factors of the first denominator.

$$1, x + 1$$

Step 3. Add any unlisted factors from the second denominator. One factor of $x - 1$ is already in our list. That factor is 1. However, the other factor, $x - 1$, is not listed in step 2. We add $x - 1$ to the list. We have

$$1, x + 1, x - 1.$$

Step 4. The least common denominator is the product of all factors in the final list. Thus

$$1(x + 1)(x - 1)$$

or $(x + 1)(x - 1)$ is the least common denominator of $\dfrac{3}{x + 1}$ and $\dfrac{5}{x - 1}$.

✓ **CHECK POINT 2** Find the LCD of $\dfrac{2}{x + 3}$ and $\dfrac{4}{x - 3}$.

EXAMPLE 3   **Finding the Least Common Denominator**

Find the LCD of

$$\frac{7}{5x^2 + 15x} \quad \text{and} \quad \frac{9}{x^2 + 6x + 9}.$$

Solution

Step 1. Factor each denominator completely.

$$5x^2 + 15x = 5x(x + 3)$$
$$x^2 + 6x + 9 = (x + 3)^2 \quad \text{or} \quad (x + 3)(x + 3)$$

$$\frac{7}{5x^2 + 15x} \qquad \frac{9}{x^2 + 6x + 9}$$

Factors are **5**, x, and $x + 3$.   Factors are $x + 3$ and $x + 3$.

Step 2. List the factors of the first denominator.

$$5, x, x + 3$$

Step 3. Add any unlisted factors from the second denominator. One factor of $x^2 + 6x + 9$ is already in our list. That factor is $x + 3$. However, the other factor of $x + 3$ is not listed in step 2. We add $x + 3$ to the list. We have

$$5, x, x + 3, x + 3.$$

Step 4. The least common denominator is the product of all factors in the final list. Thus,

$$5x(x + 3)(x + 3) \quad \text{or} \quad 5x(x + 3)^2$$

is the least common denominator.

✓ **CHECK POINT 3** Find the LCD of $\dfrac{9}{7x^2 + 28x}$ and $\dfrac{11}{x^2 + 8x + 16}$.

---

2 Add and subtract rational expressions with different denominators.

Adding and Subtracting Rational Expressions with Different Denominators

Finding the least common denominator for two (or more) rational expressions is the first step needed to add or subtract the expressions. For example, to add $\frac{1}{2}$ and $\frac{2}{3}$, we first determine that the LCD is 6. Then we write each fraction in terms of the LCD.

$$\frac{1}{2} + \frac{2}{3} = \frac{1}{2} \cdot \frac{3}{3} + \frac{2}{3} \cdot \frac{2}{2}$$

Multiply the numerator and denominator of each fraction by whatever extra factors are required to form 6, the LCD.

> $\frac{3}{3} = 1$ and $\frac{2}{2} = 1$. Multiplying by 1 does not change a fraction's value.

$$= \frac{3}{6} + \frac{4}{6}$$

$$= \frac{3+4}{6} = \frac{7}{6}$$

Add numerators. Place this sum over the LCD.

We follow the same steps in adding or subtracting rational expressions with different denominators.

Adding and Subtracting Rational Expressions That Have Different Denominators

1. Find the LCD of the rational expressions.
2. Rewrite each rational expression as an equivalent expression whose denominator is the LCD. To do so, multiply the numerator and the denominator of each rational expression by any factor(s) needed to convert the denominator into the LCD.
3. Add or subtract numerators, placing the resulting expression over the LCD.
4. If possible, simplify the resulting rational expression.

EXAMPLE 4 **Adding Rational Expressions with Different Denominators**

Add: $\dfrac{7}{6x^2} + \dfrac{2}{9x}$.

Solution

Step 1. Find the least common denominator. In Example 1, we found that the LCD for these rational expressions is $18x^2$.

Step 2. Write equivalent expressions with the LCD as denominators. We must rewrite each rational expression with a denominator of $18x^2$.

$$\frac{7}{6x^2} \cdot \frac{3}{3} = \frac{21}{18x^2} \qquad \frac{2}{9x} \cdot \frac{2x}{2x} = \frac{4x}{18x^2}$$

> Multiply the numerator and denominator by 3 to get $18x^2$, the LCD.

> Multiply the numerator and denominator by $2x$ to get $18x^2$, the LCD.

Because $\dfrac{3}{3} = 1$ and $\dfrac{2x}{2x} = 1$, we are not changing the value of either rational expression, only its appearance.

Now we are ready to perform the indicated addition.

$$\frac{7}{6x^2} + \frac{2}{9x}$$

This is the given problem. The LCD is $18x^2$.

$$= \frac{7}{6x^2} \cdot \frac{3}{3} + \frac{2}{9x} \cdot \frac{2x}{2x}$$

Write equivalent expressions with the LCD.

$$= \frac{21}{18x^2} + \frac{4x}{18x^2}$$

Steps 3 and 4. Add numerators, putting this sum over the LCD. Simplify, if possible.

$$= \frac{21 + 4x}{18x^2} \qquad \text{or} \qquad \frac{4x + 21}{18x^2}$$

The numerator is prime and further simplification is not possible. ∎

Study Tip

It is incorrect to add rational expressions by adding numerators and adding denominators. Avoid this common error.

Incorrect!

☑ **CHECK POINT 4** Add: $\dfrac{3}{10x^2} + \dfrac{7}{15x}$.

EXAMPLE 5 Adding Rational Expressions with Different Denominators

Add: $\dfrac{3}{x+1} + \dfrac{5}{x-1}$.

Solution

Step 1. Find the least common denominator. The factors of the denominators are 1, $x + 1$, and $x - 1$. In Example 2, we found that the LCD is $(x + 1)(x - 1)$.

Step 2. Write equivalent expressions with the LCD as denominators.

$$\dfrac{3}{x+1} + \dfrac{5}{x-1}$$

$$= \dfrac{3(x-1)}{(x+1)(x-1)} + \dfrac{5(x+1)}{(x+1)(x-1)}$$

Multiply each numerator and denominator by the extra factor required to form $(x + 1)(x - 1)$, the LCD.

Steps 3 and 4. Add numerators, putting this sum over the LCD. Simplify, if possible.

$$= \dfrac{3(x-1) + 5(x+1)}{(x+1)(x-1)}$$

$$= \dfrac{3x - 3 + 5x + 5}{(x+1)(x-1)}$$

Use the distributive property to multiply and remove grouping symbols.

$$= \dfrac{8x + 2}{(x+1)(x-1)}$$

Combine like terms: $3x + 5x = 8x$ and $-3 + 5 = 2$. ■

We can factor 2 from the numerator of the answer in Example 5 to obtain

$$\dfrac{2(4x+1)}{(x+1)(x-1)}.$$

Because the numerator and denominator do not have any common factors, further simplification is not possible. In this section, unless there is a common factor in the numerator and denominator, we will leave an answer's numerator in unfactored form and the denominator in factored form.

☑ **CHECK POINT 5** Add: $\dfrac{2}{x+3} + \dfrac{4}{x-3}$.

EXAMPLE 6 Subtracting Rational Expressions with Different Denominators

Subtract: $\dfrac{x}{x+3} - 1$.

Solution

Step 1. Find the least common denominator. We know that 1 means $\frac{1}{1}$. The factors of the first denominator are 1 and $x + 3$:

$$1, x + 3.$$

The factor of the second denominator, 1, is already in this list. Thus, the LCD is $1(x + 3)$, or $x + 3$.

Step 2. Write equivalent expressions with the LCD as denominators.

$$\frac{x}{x+3} - 1$$

$$= \frac{x}{x+3} - \frac{1}{1} \qquad \text{Write 1 as } \frac{1}{1}.$$

$$= \frac{x}{x+3} - \frac{1(x+3)}{1(x+3)} \qquad \begin{array}{l}\text{Multiply the numerator and denominator} \\ \text{of } \frac{1}{1} \text{ by the extra factor required to form} \\ x+3, \text{ the LCD.}\end{array}$$

Steps 3 and 4. Subtract numerators, putting this difference over the LCD. Simplify, if possible.

$$= \frac{x - (x+3)}{x+3}$$

$$= \frac{x - x - 3}{x+3} \qquad \begin{array}{l}\text{Remove parentheses and then} \\ \text{change the sign of each term.}\end{array}$$

$$= \frac{-3}{x+3} \quad \text{or} \quad -\frac{3}{x+3} \qquad \text{Simplify.} \qquad \blacksquare$$

☑ **CHECK POINT 6** Subtract: $\dfrac{x}{x+5} - 1.$

EXAMPLE 7 **Subtracting Rational Expressions with Different Denominators**

Subtract: $\dfrac{y+2}{4y+16} - \dfrac{2}{y^2+4y}.$

Solution

Step 1. Find the least common denominator. Start by factoring the denominators.

$$4y + 16 = 4(y+4)$$
$$y^2 + 4y = y(y+4)$$

The factors of the first denominator are 4 and $y+4$. The only factor from the second denominator that is unlisted is y. Thus, the least common denominator is $4y(y+4)$.

Step 2. Write equivalent expressions with the LCD as denominators.

$$\frac{y+2}{4y+16} - \frac{2}{y^2+4y}$$

$$= \frac{y+2}{4(y+4)} - \frac{2}{y(y+4)} \qquad \begin{array}{l}\text{Factor denominators.} \\ \text{The LCD is } 4y(y+4).\end{array}$$

$$= \frac{(y+2)y}{4y(y+4)} - \frac{2 \cdot 4}{4y(y+4)} \qquad \begin{array}{l}\text{Multiply each numerator and} \\ \text{denominator by the extra factor} \\ \text{required to form } 4y(y+4), \text{ the LCD.}\end{array}$$

Steps 3 and 4. Subtract numerators, putting this difference over the LCD. Simplify, if possible.

$$= \frac{(y+2)y - 2 \cdot 4}{4y(y+4)}$$

$$= \frac{y^2 + 2y - 8}{4y(y+4)} \qquad \text{Use the distributive property.}$$

$$= \frac{\overset{1}{\cancel{(y+4)}}(y-2)}{4y\underset{1}{\cancel{(y+4)}}} \qquad \text{Factor and simplify.}$$

$$= \frac{y-2}{4y} \qquad \blacksquare$$

✓ **CHECK POINT 7** Subtract: $\dfrac{5}{y^2 - 5y} - \dfrac{y}{5y - 25}$.

In some situations, after factoring denominators, a factor in one denominator is the opposite of a factor in the other denominator. When this happens, we can use the following procedure:

> ### Adding and Subtracting Rational Expressions when Denominators Contain Opposite Factors
>
> When one denominator contains the opposite factor of the other, first multiply either rational expression by $\dfrac{-1}{-1}$. Then apply the procedure for adding or subtracting rational expressions that have different denominators to the rewritten problem.

EXAMPLE 8 Adding Rational Expressions with Opposite Factors in the Denominators

Add: $\dfrac{x^2 - 2}{2x^2 - x - 3} + \dfrac{x - 2}{3 - 2x}$.

Solution

Step 1. Find the least common denominator. Start by factoring the denominators.

$$2x^2 - x - 3 = (2x - 3)(x + 1)$$
$$3 - 2x = 1(3 - 2x)$$

Do you see that $2x - 3$ and $3 - 2x$ are opposite factors? Thus, we multiply either rational expression by $\dfrac{-1}{-1}$. We will use the second rational expression, resulting in $2x - 3$ in the denominator.

$$\frac{x^2 - 2}{2x^2 - x - 3} + \frac{x - 2}{3 - 2x}$$

$$= \frac{x^2 - 2}{(2x - 3)(x + 1)} + \frac{(-1)}{(-1)} \cdot \frac{x - 2}{3 - 2x} \qquad \begin{array}{l}\text{Factor the first denominator.} \\ \text{Multiply the second rational} \\ \text{expression by } \frac{-1}{-1}.\end{array}$$

$$= \frac{x^2 - 2}{(2x - 3)(x + 1)} + \frac{-x + 2}{-3 + 2x} \qquad \begin{array}{l}\text{Perform the multiplications by } -1 \text{ by} \\ \text{changing every term's sign.}\end{array}$$

$$= \frac{x^2 - 2}{(2x - 3)(x + 1)} + \frac{2 - x}{2x - 3} \qquad \begin{array}{l}\text{Rewrite } -3 + 2x \text{ as } 2x - 3\end{array}$$

The LCD of our rewritten addition problem is $(2x - 3)(x + 1)$.

Step 2. Write equivalent expressions with the LCD as denominators.

$$= \frac{x^2 - 2}{(2x - 3)(x + 1)} + \frac{(2 - x)(x + 1)}{(2x - 3)(x + 1)} \qquad \begin{array}{l}\text{Multiply the numerator and} \\ \text{denominator of the second rational} \\ \text{expression by the extra factor required} \\ \text{to form } (2x - 3)(x + 1), \text{ the LCD.}\end{array}$$

Discover for Yourself

In Example 8, the denominators can be factored as follows:

$$2x^2 - x - 3 = (2x - 3)(x + 1)$$
$$3 - 2x = -1(2x - 3).$$

Using these factorizations, what is the LCD? Solve Example 8 by obtaining this LCD in each rational expression. Then combine the expressions. How does your solution compare with the one shown on the right?

Here is the original addition problem and our equivalent expressions with the LCD:

$$\frac{x^2 - 2}{2x^2 - x - 3} + \frac{x - 2}{3 - 2x}$$

$$= \frac{x^2 - 2}{(2x - 3)(x + 1)} + \frac{(2 - x)(x + 1)}{(2x - 3)(x + 1)}.$$

Steps 3 and 4. **Add numerators, putting this sum over the LCD. Simplify, if possible.**

$$= \frac{x^2 - 2 + (2 - x)(x + 1)}{(2x - 3)(x + 1)}$$

$$= \frac{x^2 - 2 + 2x + 2 - x^2 - x}{(2x - 3)(x + 1)}$$ Use the FOIL method to multiply $(2 - x)(x + 1)$.

$$= \frac{(x^2 - x^2) + (2x - x) + (-2 + 2)}{(2x - 3)(x + 1)}$$ Group like terms.

$$= \frac{x}{(2x - 3)(x + 1)}$$ Combine like terms.

✓ **CHECK POINT 8** Add: $\dfrac{4x}{x^2 - 25} + \dfrac{3}{5 - x}$.

8.4 EXERCISE SET

MyMathLab Math XL PRACTICE WATCH DOWNLOAD READ REVIEW

Practice Exercises

In Exercises 1–16, find the least common denominator of the rational expressions.

1. $\dfrac{7}{15x^2}$ and $\dfrac{13}{24x}$

2. $\dfrac{11}{25x^2}$ and $\dfrac{17}{35x}$

3. $\dfrac{8}{15x^2}$ and $\dfrac{5}{6x^5}$

4. $\dfrac{7}{15x^2}$ and $\dfrac{11}{24x^5}$

5. $\dfrac{4}{x - 3}$ and $\dfrac{7}{x + 1}$

6. $\dfrac{2}{x - 5}$ and $\dfrac{3}{x + 7}$

7. $\dfrac{5}{7(y + 2)}$ and $\dfrac{10}{y}$

8. $\dfrac{8}{11(y + 5)}$ and $\dfrac{12}{y}$

9. $\dfrac{17}{x + 4}$ and $\dfrac{18}{x^2 - 16}$

10. $\dfrac{3}{x - 6}$ and $\dfrac{4}{x^2 - 36}$

11. $\dfrac{8}{y^2 - 9}$ and $\dfrac{14}{y(y + 3)}$

12. $\dfrac{14}{y^2 - 49}$ and $\dfrac{12}{y(y - 7)}$

13. $\dfrac{7}{y^2 - 1}$ and $\dfrac{y}{y^2 - 2y + 1}$

14. $\dfrac{9}{y^2 - 25}$ and $\dfrac{y}{y^2 - 10y + 25}$

15. $\dfrac{3}{x^2 - x - 20}$ and $\dfrac{x}{2x^2 + 7x - 4}$

16. $\dfrac{7}{x^2 - 5x - 6}$ and $\dfrac{x}{x^2 - 4x - 5}$

In Exercises 17–82, add or subtract as indicated. Simplify the result, if possible.

17. $\dfrac{3}{x} + \dfrac{5}{x^2}$

18. $\dfrac{4}{x} + \dfrac{8}{x^2}$

19. $\dfrac{2}{9x} + \dfrac{11}{6x}$

20. $\dfrac{5}{6x} + \dfrac{7}{8x}$

21. $\dfrac{4}{x} + \dfrac{7}{2x^2}$

22. $\dfrac{10}{x} + \dfrac{3}{5x^2}$

23. $6 + \dfrac{1}{x}$

24. $3 + \dfrac{1}{x}$

25. $\dfrac{2}{x} + 9$

26. $\dfrac{7}{x} + 4$

27. $\dfrac{x - 1}{6} + \dfrac{x + 2}{3}$

28. $\dfrac{x + 3}{2} + \dfrac{x + 5}{4}$

29. $\dfrac{4}{x} + \dfrac{3}{x - 5}$

30. $\dfrac{3}{x} + \dfrac{4}{x - 6}$

31. $\dfrac{2}{x - 1} + \dfrac{3}{x + 2}$

32. $\dfrac{3}{x - 2} + \dfrac{4}{x + 3}$

33. $\dfrac{2}{y + 5} + \dfrac{3}{4y}$

34. $\dfrac{3}{y + 1} + \dfrac{2}{3y}$

35. $\dfrac{x}{x + 7} - 1$

36. $\dfrac{x}{x + 6} - 1$

37. $\dfrac{7}{x + 5} - \dfrac{4}{x - 5}$

38. $\dfrac{8}{x + 6} - \dfrac{2}{x - 6}$

39. $\dfrac{2x}{x^2 - 16} + \dfrac{x}{x - 4}$

40. $\dfrac{4x}{x^2 - 25} + \dfrac{x}{x + 5}$

41. $\dfrac{5y}{y^2 - 9} - \dfrac{4}{y + 3}$

42. $\dfrac{8y}{y^2 - 16} - \dfrac{5}{y + 4}$

43. $\dfrac{7}{x - 1} - \dfrac{3}{(x - 1)^2}$

44. $\dfrac{5}{x + 3} - \dfrac{2}{(x + 3)^2}$

45. $\dfrac{3y}{4y - 20} + \dfrac{9y}{6y - 30}$

46. $\dfrac{4y}{5y - 10} + \dfrac{3y}{10y - 20}$

47. $\dfrac{y + 4}{y} - \dfrac{y}{y + 4}$

48. $\dfrac{y}{y - 5} - \dfrac{y - 5}{y}$

49. $\dfrac{2x + 9}{x^2 - 7x + 12} - \dfrac{2}{x - 3}$

50. $\dfrac{3x + 7}{x^2 - 5x + 6} - \dfrac{3}{x - 3}$

51. $\dfrac{3}{x^2 - 1} + \dfrac{4}{(x + 1)^2}$

52. $\dfrac{6}{x^2 - 4} + \dfrac{2}{(x + 2)^2}$

53. $\dfrac{3x}{x^2 + 3x - 10} - \dfrac{2x}{x^2 + x - 6}$

54. $\dfrac{x}{x^2 - 2x - 24} - \dfrac{x}{x^2 - 7x + 6}$

55. $\dfrac{y}{y^2 + 2y + 1} + \dfrac{4}{y^2 + 5y + 4}$

56. $\dfrac{y}{y^2 + 5y + 6} + \dfrac{4}{y^2 - y - 6}$

57. $\dfrac{x - 5}{x + 3} + \dfrac{x + 3}{x - 5}$

58. $\dfrac{x - 7}{x + 4} + \dfrac{x + 4}{x - 7}$

59. $\dfrac{5}{2y^2 - 2y} - \dfrac{3}{2y - 2}$

60. $\dfrac{7}{5y^2 - 5y} - \dfrac{2}{5y - 5}$

61. $\dfrac{4x + 3}{x^2 - 9} - \dfrac{x + 1}{x - 3}$

62. $\dfrac{2x - 1}{x + 6} - \dfrac{6 - 5x}{x^2 - 36}$

63. $\dfrac{y^2 - 39}{y^2 + 3y - 10} - \dfrac{y - 7}{y - 2}$

64. $\dfrac{y^2 - 6}{y^2 + 9y + 18} - \dfrac{y - 4}{y + 6}$

65. $4 + \dfrac{1}{x - 3}$

66. $7 + \dfrac{1}{x - 5}$

67. $3 - \dfrac{3y}{y + 1}$

68. $7 - \dfrac{4y}{y + 5}$

69. $\dfrac{9x + 3}{x^2 - x - 6} + \dfrac{x}{3 - x}$

70. $\dfrac{x^2 + 9x}{x^2 - 2x - 3} + \dfrac{5}{3 - x}$

71. $\dfrac{x + 3}{x^2 + x - 2} - \dfrac{2}{x^2 - 1}$

72. $\dfrac{x}{x^2 - 10x + 25} - \dfrac{x - 4}{2x - 10}$

73. $\dfrac{y + 3}{5y^2} - \dfrac{y - 5}{15y}$

74. $\dfrac{y - 7}{3y^2} - \dfrac{y - 2}{12y}$

75. $\dfrac{x + 3}{3x + 6} + \dfrac{x}{4 - x^2}$

76. $\dfrac{x + 7}{4x + 12} + \dfrac{x}{9 - x^2}$

77. $\dfrac{y}{y^2 - 1} + \dfrac{2y}{y - y^2}$

78. $\dfrac{y}{y^2 - 1} + \dfrac{5y}{y - y^2}$

79. $\dfrac{x - 1}{x} + \dfrac{y + 1}{y}$

80. $\dfrac{x + 2}{y} + \dfrac{y - 2}{x}$

81. $\dfrac{3x}{x^2 - y^2} - \dfrac{2}{y - x}$

82. $\dfrac{7x}{x^2 - y^2} - \dfrac{3}{y - x}$

Practice PLUS

In Exercises 83–92, perform the indicated operation or operations. Simplify the result, if possible.

83. $\dfrac{x+6}{x^2-4} - \dfrac{x+3}{x+2} + \dfrac{x-3}{x-2}$

84. $\dfrac{x+8}{x^2-9} - \dfrac{x+2}{x+3} + \dfrac{x-2}{x-3}$

85. $\dfrac{5}{x^2-25} + \dfrac{4}{x^2-11x+30} - \dfrac{3}{x^2-x-30}$

86. $\dfrac{3}{x^2-49} + \dfrac{2}{x^2-15x+56} - \dfrac{5}{x^2-x-56}$

87. $\dfrac{x+6}{x^3-27} - \dfrac{x}{x^3+3x^2+9x}$

88. $\dfrac{x+8}{x^3-8} - \dfrac{x}{x^3+2x^2+4x}$

89. $\dfrac{9y+3}{y^2-y-6} + \dfrac{y}{3-y} + \dfrac{y-1}{y+2}$

90. $\dfrac{7y-2}{y^2-y-12} + \dfrac{2y}{4-y} + \dfrac{y+1}{y+3}$

91. $\dfrac{3}{x^2+4xy+3y^2} - \dfrac{5}{x^2-2xy-3y^2} + \dfrac{2}{x^2-9y^2}$

92. $\dfrac{5}{x^2+3xy+2y^2} - \dfrac{7}{x^2-xy-2y^2} + \dfrac{4}{x^2-4y^2}$

Application Exercises

Two formulas that approximate the dosage of a drug prescribed for children are

$$\text{Young's rule:} \quad C = \frac{DA}{A+12}$$

$$\text{and Cowling's rule:} \quad C = \frac{D(A+1)}{24}.$$

In each formula, A = the child's age, in years, D = an adult dosage, and C = the proper child's dosage. The formulas apply for ages 2 through 13, inclusive. Use the formulas to solve Exercises 93–96.

93. Use Young's rule to find the difference in a child's dosage for an 8-year-old child and a 3-year-old child. Express the answer as a single rational expression in terms of D. Then describe what your answer means in terms of the variables in the model.

94. Use Young's rule to find the difference in a child's dosage for a 10-year-old child and a 3-year-old child. Express the answer as a single rational expression in terms of D. Then describe what your answer means in terms of the variables in the model.

95. For a 12-year-old child, what is the difference in the dosage given by Cowling's rule and Young's rule? Express the answer as a single rational expression in terms of D. Then describe what your answer means in terms of the variables in the models.

96. Use Cowling's rule to find the difference in a child's dosage for a 12-year-old child and a 10-year-old child. Express the answer as a single rational expression in terms of D. Then describe what your answer means in terms of the variables in the model.

The graphs illustrate Young's rule and Cowling's rule when the dosage of a drug prescribed for an adult is 1000 milligrams. Use the graphs to solve Exercises 97–100.

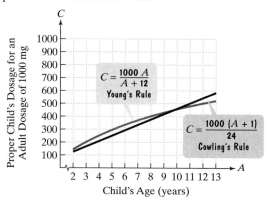

97. Does either formula consistently give a smaller dosage than the other? If so, which one?

98. Is there an age at which the dosage given by one formula becomes greater than the dosage given by the other? If so, what is a reasonable estimate of that age?

99. For what age under 11 is the difference in dosage given by the two formulas the greatest?

100. For what age over 11 is the difference in dosage given by the two formulas the greatest?

In Exercises 101–102, express the perimeter of each rectangle as a single rational expression.

101.

102.

Writing in Mathematics

103. Explain how to find the least common denominator for denominators of $x^2 - 100$ and $x^2 - 20x + 100$.

104. Explain how to add rational expressions that have different denominators. Use $\dfrac{3}{x+5} + \dfrac{7}{x+2}$ in your explanation.

Explain the error in Exercises 105–106. Then rewrite the right side of the equation to correct the error that now exists.

105. $\dfrac{1}{x} + \dfrac{2}{5} = \dfrac{3}{x+5}$

106. $\dfrac{1}{x} + 7 = \dfrac{1}{x+7}$

107. The formulas in Exercises 93–96 relate the dosage of a drug prescribed for children to the child's age. Describe another factor that might be used when determining a child's dosage. Is this factor more or less important than age? Explain why.

Critical Thinking Exercises

Make Sense? *In Exercises 108–111, determine whether each statement "makes sense" or "does not make sense" and explain your reasoning.*

108. It takes me more steps to find $\dfrac{2}{x+5} + \dfrac{7}{x-3}$ than it does to find $\dfrac{2x^3}{x+5} + \dfrac{7x^3}{x+5}$.

109. The reason I can rewrite rational expressions with a common denominator is that 1 is the multiplicative identity.

110. The fastest way for me to find $\dfrac{4}{x-5} + \dfrac{9}{5-x}$ is by using $(x-5)(5-x)$ as the LCD.

111. After adding rational expressions with different denominators, I factored the numerator and found no common factors in the numerator and denominator, so my final answer is incorrect if I leave the numerator in factored form.

In Exercises 112–115, determine whether each statement is true or false. If the statement is false, make the necessary change(s) to produce a true statement.

112. $x - \dfrac{1}{5} = \dfrac{4}{5}x$

113. The LCD of $\dfrac{1}{x}$ and $\dfrac{2x}{x-1}$ is $x^2 - 1$.

114. $\dfrac{1}{x} + \dfrac{x}{1} = \dfrac{1}{\cancel{x}} + \dfrac{\cancel{x}}{1} = 1 + 1 = 2$

115. $\dfrac{2}{x} + 1 = \dfrac{2+x}{x}, \; x \neq 0$

In Exercises 116–117, perform the indicated operations. Simplify the result, if possible.

116. $\dfrac{y^2 + 5y + 4}{y^2 + 2y - 3} \cdot \dfrac{y^2 + y - 6}{y^2 + 2y - 3} - \dfrac{2}{y-1}$

117. $\left(\dfrac{1}{x+h} - \dfrac{1}{x} \right) \div h$

In Exercises 118–119, find the missing rational expression.

118. $\dfrac{2}{x-1} + \rule{1cm}{0.4pt} = \dfrac{2x^2 + 3x - 1}{x^2(x-1)}$

119. $\dfrac{4}{x-2} - \rule{1cm}{0.4pt} = \dfrac{2x+8}{(x-2)(x+1)}$

Review Exercises

120. Multiply: $(3x+5)(2x-7)$. (Section 6.3, Example 2)

121. Graph: $3x - y < 3$. (Section 4.6, Example 2)

122. Write the slope-intercept form of the equation of the line passing through $(-3, -4)$ and $(1, 0)$. (Section 4.5, Example 2)

Preview Exercises

Exercises 123–125 will help you prepare for the material covered in the next section.

123. a. Add: $\dfrac{1}{3} + \dfrac{2}{5}$.

   b. Subtract: $\dfrac{2}{5} - \dfrac{1}{3}$.

   c. Use your answers from parts (a) and (b) to find $\left(\dfrac{1}{3} + \dfrac{2}{5} \right) \div \left(\dfrac{2}{5} - \dfrac{1}{3} \right)$.

124. a. Add: $\dfrac{1}{x} + \dfrac{1}{y}$.

   b. Use your answer from part (a) to find $\dfrac{1}{xy} \div \left(\dfrac{1}{x} + \dfrac{1}{y} \right)$.

125. Multiply and simplify: $xy\left(\dfrac{1}{x} + \dfrac{1}{y} \right)$.

MID-CHAPTER CHECK POINT | Section 8.1–Section 8.4

✓ **What You Know:** We learned that it is necessary to exclude any value or values of a variable that make the denominator of a rational expression zero. We learned to simplify rational expressions by dividing the numerator and the denominator by common factors. We performed a variety of operations with rational expressions, including multiplication, division, addition, and subtraction.

1. Find all numbers for which $\dfrac{x^2 - 4}{x^2 - 2x - 8}$ is undefined.

In Exercises 2–4, simplify each rational expression.

2. $\dfrac{3x^2 - 7x + 2}{6x^2 + x - 1}$

3. $\dfrac{9 - 3y}{y^2 - 5y + 6}$

4. $\dfrac{16w^3 - 24w^2}{8w^4 - 12w^3}$

In Exercises 5–20, perform the indicated operations.
Simplify the result, if possible.

5. $\dfrac{7x - 3}{x^2 + 3x - 4} - \dfrac{3x + 1}{x^2 + 3x - 4}$

6. $\dfrac{x + 2}{2x - 4} \cdot \dfrac{8}{x^2 - 4}$

7. $1 + \dfrac{7}{x - 2}$

8. $\dfrac{2x^2 + x - 1}{2x^2 - 7x + 3} \div \dfrac{x^2 - 3x - 4}{x^2 - x - 6}$

9. $\dfrac{1}{x^2 + 2x - 3} + \dfrac{1}{x^2 + 5x + 6}$

10. $\dfrac{17}{x - 5} + \dfrac{x + 8}{5 - x}$

11. $\dfrac{4y^2 - 1}{9y - 3y^2} \cdot \dfrac{y^2 - 7y + 12}{2y^2 - 7y - 4}$

12. $\dfrac{y}{y + 1} - \dfrac{2y}{y + 2}$

13. $\dfrac{w^2 + 6w + 5}{7w^2 - 63} \div \dfrac{w^2 + 10w + 25}{7w + 21}$

14. $\dfrac{2z}{z^2 - 9} - \dfrac{5}{z^2 + 4z + 3}$

15. $\dfrac{z + 2}{3z - 1} + \dfrac{5}{(3z - 1)^2}$

16. $\dfrac{8}{x^2 + 4x - 21} + \dfrac{3}{x + 7}$

17. $\dfrac{x^4 - 27x}{x^2 - 9} \cdot \dfrac{x + 3}{x^2 + 3x + 9}$

18. $\dfrac{x - 1}{x^2 - x - 2} - \dfrac{x + 2}{x^2 + 4x + 3}$

19. $\dfrac{x^2 - 2xy + y^2}{x + y} \div \dfrac{x^2 - xy}{5x + 5y}$

20. $\dfrac{5}{x + 5} + \dfrac{x}{x - 4} - \dfrac{11x - 8}{x^2 + x - 20}$

SECTION **8.5**

Complex Rational Expressions

Objectives

1 Simplify complex rational expressions by dividing.

2 Simplify complex rational expressions by multiplying by the LCD.

Do you drive to and from campus each day? If the one-way distance of your round-trip commute is d, then your average rate, or speed, is given by the expression

$$\frac{2d}{\dfrac{d}{r_1} + \dfrac{d}{r_2}},$$

in which r_1 and r_2 are your average rates on the outgoing and return trips, respectively. Do you notice anything unusual about this expression? It has two separate rational expressions in its denominator.

Numerator ——

Main fraction bar

$$\frac{2d}{\dfrac{d}{r_1} + \dfrac{d}{r_2}}$$

Denominator ——

Separate rational expressions
occur in the denominator.

Complex rational expressions, also called **complex fractions**, have numerators or denominators containing one or more rational expressions. Here is another example of such an expression:

In this section, we study two methods for simplifying complex rational expressions.

1 Simplify complex rational expressions by dividing.

Simplifying by Rewriting Complex Rational Expressions as a Quotient of Two Rational Expressions

One method for simplifying a complex rational expression is to combine its numerator into a single expression and combine its denominator into a single expression. Then perform the division by inverting the denominator and multiplying.

Simplifying a Complex Rational Expression by Dividing

1. If necessary, add or subtract to get a single rational expression in the numerator.
2. If necessary, add or subtract to get a single rational expression in the denominator.
3. Perform the division indicated by the main fraction bar: Invert the denominator of the complex rational expression and multiply.
4. If possible, simplify.

The following examples illustrate the use of this first method.

EXAMPLE 1 Simplifying a Complex Rational Expression

Simplify:

$$\frac{\frac{1}{3} + \frac{2}{5}}{\frac{2}{5} - \frac{1}{3}}.$$

Solution Let's first identify the parts of this complex rational expression.

Step 1. **Add to get a single rational expression in the numerator.**

$$\frac{1}{3} + \frac{2}{5} = \frac{1 \cdot 5}{3 \cdot 5} + \frac{2 \cdot 3}{5 \cdot 3} = \frac{5}{15} + \frac{6}{15} = \frac{11}{15}$$

The LCD is 3 · 5, or 15.

Numerator

$$\frac{\dfrac{1}{3} + \dfrac{2}{5}}{\dfrac{2}{5} - \dfrac{1}{3}}$$

Main fraction bar

Denominator

The given expression (repeated)

Step 2. **Subtract to get a single rational expression in the denominator.**

$$\frac{2}{5} - \frac{1}{3} = \frac{2 \cdot 3}{5 \cdot 3} - \frac{1 \cdot 5}{3 \cdot 5} = \frac{6}{15} - \frac{5}{15} = \frac{1}{15}$$

The LCD is 15.

Steps 3 and 4. **Perform the division indicated by the main fraction bar: Invert and multiply. If possible, simplify.**

$$\frac{\dfrac{1}{3} + \dfrac{2}{5}}{\dfrac{2}{5} - \dfrac{1}{3}} = \frac{\dfrac{11}{15}}{\dfrac{1}{15}} = \frac{11}{15} \cdot \frac{15}{1} = \frac{11}{\cancel{15}} \cdot \frac{\overset{1}{\cancel{15}}}{1} = 11$$

Invert and multiply.

☑ **CHECK POINT 1** Simplify: $\dfrac{\dfrac{1}{4} + \dfrac{2}{3}}{\dfrac{2}{3} - \dfrac{1}{4}}$.

EXAMPLE 2 **Simplifying a Complex Rational Expression**

Simplify:

$$\frac{1 + \dfrac{1}{x}}{1 - \dfrac{1}{x}}.$$

Solution

Step 1. **Add to get a single rational expression in the numerator.**

$$1 + \frac{1}{x} = \frac{1}{1} + \frac{1}{x} = \frac{1 \cdot x}{1 \cdot x} + \frac{1}{x} = \frac{x}{x} + \frac{1}{x} = \frac{x + 1}{x}$$

The LCD is 1 · x, or x.

Step 2. **Subtract to get a single rational expression in the denominator.**

$$1 - \frac{1}{x} = \frac{1}{1} - \frac{1}{x} = \frac{1 \cdot x}{1 \cdot x} - \frac{1}{x} = \frac{x}{x} - \frac{1}{x} = \frac{x - 1}{x}$$

The LCD is 1 · x, or x.

Steps 3 and 4. **Perform the division indicated by the main fraction bar: Invert and multiply. If possible, simplify.**

$$\frac{1 + \dfrac{1}{x}}{1 - \dfrac{1}{x}} = \frac{\dfrac{x + 1}{x}}{\dfrac{x - 1}{x}} = \frac{x + 1}{x} \cdot \frac{x}{x - 1} = \frac{x + 1}{\cancel{x}} \cdot \frac{\overset{1}{\cancel{x}}}{x - 1} = \frac{x + 1}{x - 1}$$

Invert and multiply.

☑ **CHECK POINT 2** Simplify: $\dfrac{2 - \dfrac{1}{x}}{2 + \dfrac{1}{x}}$.

EXAMPLE 3 Simplifying a Complex Rational Expression

Simplify:

$$\frac{\dfrac{1}{xy}}{\dfrac{1}{x} + \dfrac{1}{y}}.$$

Solution

Step 1. Get a single rational expression in the numerator. The numerator, $\dfrac{1}{xy}$, already contains a single rational expression, so we can skip this step.

Step 2. Add to get a single rational expression in the denominator.

$$\frac{1}{x} + \frac{1}{y} = \frac{1 \cdot y}{x \cdot y} + \frac{1 \cdot x}{y \cdot x} = \frac{y}{xy} + \frac{x}{xy} = \frac{y + x}{xy}$$

The LCD is xy.

Steps 3 and 4. Perform the division indicated by the main fraction bar: Invert and multiply. If possible, simplify.

$$\frac{\dfrac{1}{xy}}{\dfrac{1}{x} + \dfrac{1}{y}} = \frac{\dfrac{1}{xy}}{\dfrac{y + x}{xy}} = \frac{1}{xy} \cdot \frac{xy}{y + x} = \frac{1}{\cancel{xy}} \cdot \frac{\cancel{xy}^{1}}{y + x} = \frac{1}{y + x}$$

Invert and multiply.

☑ **CHECK POINT 3** Simplify: $\dfrac{\dfrac{1}{x} - \dfrac{1}{y}}{\dfrac{1}{xy}}.$

---

② Simplify complex rational expressions by multiplying by the LCD.

Simplifying Complex Rational Expressions by Multiplying by the LCD

A second method for simplifying a complex rational expression is to find the least common denominator of all the rational expressions in its numerator and denominator. Then multiply each term in its numerator and denominator by this least common denominator. Because we are multiplying by a form of 1, we will obtain an equivalent expression that does not contain fractions in its numerator or denominator.

> ### Simplifying a Complex Rational Expression by Multiplying by the LCD
>
> 1. Find the LCD of all rational expressions within the complex rational expression.
> 2. Multiply both the numerator and the denominator of the complex rational expression by this LCD.
> 3. Use the distributive property and multiply each term in the numerator and denominator by this LCD. Simplify each term. No fractional expressions should remain.
> 4. If possible, factor and simplify.

We now rework Examples 1, 2, and 3 using the method of multiplying by the LCD. Compare the two simplification methods to see if there is one method that you prefer.

EXAMPLE 4 Simplifying a Complex Rational Expression by the LCD Method

Simplify:

$$\frac{\frac{1}{3} + \frac{2}{5}}{\frac{2}{5} - \frac{1}{3}}.$$

Solution The denominators in the complex rational expression are 3, 5, 5, and 3. The LCD is $3 \cdot 5$, or 15. Multiply both the numerator and denominator of the complex rational expression by 15.

$$\frac{\frac{1}{3} + \frac{2}{5}}{\frac{2}{5} - \frac{1}{3}} = \frac{15}{15} \cdot \frac{\left(\frac{1}{3} + \frac{2}{5}\right)}{\left(\frac{2}{5} - \frac{1}{3}\right)} = \frac{15 \cdot \frac{1}{3} + 15 \cdot \frac{2}{5}}{15 \cdot \frac{2}{5} - 15 \cdot \frac{1}{3}} = \frac{5 + 6}{6 - 5} = \frac{11}{1} = 11$$

$\frac{15}{15} = 1$, so we are not changing the complex fraction's value.

✓ **CHECK POINT 4** Simplify by the LCD method: $\dfrac{\frac{1}{4} + \frac{2}{3}}{\frac{2}{3} - \frac{1}{4}}.$

EXAMPLE 5 Simplifying a Complex Rational Expression by the LCD Method

Simplify:

$$\frac{1 + \frac{1}{x}}{1 - \frac{1}{x}}.$$

Solution The denominators in the complex rational expression are 1, x, 1, and x.

$$\frac{1 + \frac{1}{x}}{1 - \frac{1}{x}} = \frac{\frac{1}{1} + \frac{1}{x}}{\frac{1}{1} - \frac{1}{x}}$$

Denominators

Denominators

The LCD is $1 \cdot x$, or x. Multiply both the numerator and denominator of the complex rational expression by x.

$$\frac{1 + \frac{1}{x}}{1 - \frac{1}{x}} = \frac{x}{x} \cdot \frac{\left(1 + \frac{1}{x}\right)}{\left(1 - \frac{1}{x}\right)} = \frac{x \cdot 1 + x \cdot \frac{1}{x}}{x \cdot 1 - x \cdot \frac{1}{x}} = \frac{x + 1}{x - 1}$$

✓ **CHECK POINT 5** Simplify by the LCD method: $\dfrac{2 - \frac{1}{x}}{2 + \frac{1}{x}}.$

EXAMPLE 6 **Simplifying a Complex Rational Expression by the LCD Method**

Simplify:

$$\frac{\dfrac{1}{xy}}{\dfrac{1}{x}+\dfrac{1}{y}}.$$

Solution The denominators in the complex rational expression are xy, x, and y. The LCD is xy. Multiply both the numerator and denominator of the complex rational expression by xy.

$$\frac{\dfrac{1}{xy}}{\dfrac{1}{x}+\dfrac{1}{y}}=\frac{xy}{xy}\cdot\frac{\left(\dfrac{1}{xy}\right)}{\left(\dfrac{1}{x}+\dfrac{1}{y}\right)}=\frac{xy\cdot\dfrac{1}{xy}}{xy\cdot\dfrac{1}{x}+xy\cdot\dfrac{1}{y}}=\frac{1}{y+x}.$$

☑ **CHECK POINT 6** Simplify by the LCD method: $\dfrac{\dfrac{1}{x}-\dfrac{1}{y}}{\dfrac{1}{xy}}$.

8.5 EXERCISE SET *MyMathLab*

Practice Exercises

In Exercises 1–40, simplify each complex rational expression by the method of your choice.

1. $\dfrac{\dfrac{1}{2}+\dfrac{1}{4}}{\dfrac{1}{2}+\dfrac{1}{3}}$

2. $\dfrac{\dfrac{1}{3}+\dfrac{1}{4}}{\dfrac{1}{3}+\dfrac{1}{6}}$

3. $\dfrac{5+\dfrac{2}{5}}{7-\dfrac{1}{10}}$

4. $\dfrac{1+\dfrac{3}{5}}{2-\dfrac{1}{4}}$

5. $\dfrac{\dfrac{2}{5}-\dfrac{1}{3}}{\dfrac{2}{3}-\dfrac{3}{4}}$

6. $\dfrac{\dfrac{1}{2}-\dfrac{1}{4}}{\dfrac{3}{8}+\dfrac{1}{16}}$

7. $\dfrac{\dfrac{3}{4}-x}{\dfrac{3}{4}+x}$

8. $\dfrac{\dfrac{2}{3}-x}{\dfrac{2}{3}+x}$

9. $\dfrac{7-\dfrac{2}{x}}{5+\dfrac{1}{x}}$

10. $\dfrac{8+\dfrac{3}{x}}{1-\dfrac{7}{x}}$

11. $\dfrac{2+\dfrac{3}{y}}{1-\dfrac{7}{y}}$

12. $\dfrac{4-\dfrac{7}{y}}{3-\dfrac{2}{y}}$

13. $\dfrac{\dfrac{1}{y}-\dfrac{3}{2}}{\dfrac{1}{y}+\dfrac{3}{4}}$

14. $\dfrac{\dfrac{1}{y}-\dfrac{3}{4}}{\dfrac{1}{y}+\dfrac{2}{3}}$

15. $\dfrac{\dfrac{x}{5}-\dfrac{5}{x}}{\dfrac{1}{5}+\dfrac{1}{x}}$

16. $\dfrac{\dfrac{3}{x}+\dfrac{x}{3}}{\dfrac{x}{3}-\dfrac{3}{x}}$

17. $\dfrac{1+\dfrac{1}{x}}{1-\dfrac{1}{x^2}}$

18. $\dfrac{1+\dfrac{2}{x}}{1-\dfrac{4}{x^2}}$

19. $\dfrac{\dfrac{1}{7}-\dfrac{1}{y}}{7-y}$

20. $\dfrac{\dfrac{1}{9}-\dfrac{1}{y}}{9-y}$

21. $\dfrac{x+\dfrac{2}{y}}{\dfrac{x}{y}}$

22. $\dfrac{x-\dfrac{2}{y}}{\dfrac{x}{y}}$

23. $\dfrac{\dfrac{1}{x}+\dfrac{1}{y}}{xy}$

24. $\dfrac{\dfrac{1}{x}+\dfrac{1}{y}}{x+y}$

25. $\dfrac{\dfrac{x}{y}+\dfrac{1}{x}}{\dfrac{y}{x}+\dfrac{1}{x}}$

26. $\dfrac{\dfrac{1}{x}+\dfrac{1}{y}}{\dfrac{1}{x}-\dfrac{1}{y}}$

27. $\dfrac{\dfrac{1}{y}+\dfrac{2}{y^2}}{\dfrac{2}{y}+1}$

28. $\dfrac{\dfrac{1}{y}+\dfrac{3}{y^2}}{\dfrac{3}{y}+1}$

29. $\dfrac{\dfrac{12}{x^2}-\dfrac{3}{x}}{\dfrac{15}{x}-\dfrac{9}{x^2}}$

30. $\dfrac{\dfrac{8}{x^2}-\dfrac{2}{x}}{\dfrac{10}{x}-\dfrac{6}{x^2}}$

31. $\dfrac{2+\dfrac{6}{y}}{1-\dfrac{9}{y^2}}$

32. $\dfrac{3+\dfrac{12}{y}}{1-\dfrac{16}{y^2}}$

33. $\dfrac{\dfrac{1}{x+2}}{1+\dfrac{1}{x+2}}$

34. $\dfrac{\dfrac{1}{x-2}}{1-\dfrac{1}{x-2}}$

35. $\dfrac{x-5+\dfrac{3}{x}}{x-7+\dfrac{2}{x}}$

36. $\dfrac{x+9-\dfrac{7}{x}}{x-6+\dfrac{4}{x}}$

37. $\dfrac{\dfrac{3}{xy^2}+\dfrac{2}{x^2y}}{\dfrac{1}{x^2y}+\dfrac{2}{xy^3}}$

38. $\dfrac{\dfrac{2}{x^3y}+\dfrac{5}{xy^4}}{\dfrac{5}{x^3y}-\dfrac{3}{xy}}$

39. $\dfrac{\dfrac{3}{x+1}-\dfrac{3}{x-1}}{\dfrac{5}{x^2-1}}$

40. $\dfrac{\dfrac{3}{x+2}-\dfrac{3}{x-2}}{\dfrac{5}{x^2-4}}$

Practice PLUS

In Exercises 41–48, simplify each complex rational expression.

41. $\dfrac{\dfrac{6}{x^2+2x-15}-\dfrac{1}{x-3}}{\dfrac{1}{x+5}+1}$

42. $\dfrac{\dfrac{1}{x-2}-\dfrac{6}{x^2+3x-10}}{1+\dfrac{1}{x-2}}$

43. $\dfrac{y^{-1}-(y+5)^{-1}}{5}$

44. $\dfrac{y^{-1}-(y+2)^{-1}}{2}$

45. $\dfrac{\dfrac{1}{1-\dfrac{1}{x}}-1}$

46. $\dfrac{\dfrac{1}{1-\dfrac{1}{x+1}}-1}$

47. $\dfrac{1}{1+\dfrac{1}{1+\dfrac{1}{x}}}$

48. $\dfrac{1}{1+\dfrac{1}{1+\dfrac{1}{2}}}$

Application Exercises

49. The average rate on a round-trip commute having a one-way distance d is given by the complex rational expression

$$\frac{2d}{\dfrac{d}{r_1}+\dfrac{d}{r_2}},$$

in which r_1 and r_2 are the average rates on the outgoing and return trips, respectively. Simplify the expression. Then find your average rate if you drive to campus averaging 40 miles per hour and return home on the same route averaging 30 miles per hour.

50. If two electrical resistors with resistances R_1 and R_2 are connected in parallel (see the figure), then the total resistance in the circuit is given by the complex rational expression

$$\frac{1}{\dfrac{1}{R_1}+\dfrac{1}{R_2}}.$$

Simplify the expression. Then find the total resistance if $R_1 = 10$ ohms and $R_2 = 20$ ohms.

The bar graph shows the amount, in billions of dollars, that the United States government spent on human resources and total budget outlays for six selected years. (Human resources include education, health, Medicare, Social Security, and veterans benefits and services.)

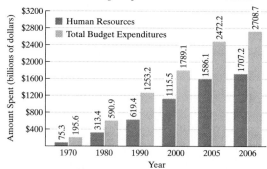

Source: Office of Management and Budget

The fraction of total budget outlays spent on human resources can be modeled by the complex rational expression

$$\frac{\dfrac{11}{10}+\dfrac{4}{x}+\dfrac{104}{x^2}}{\dfrac{6}{5}+\dfrac{23}{x}+\dfrac{219}{x^2}},$$

where x is the number of years after 1970. Use this information to solve Exercises 51–52.

51. a. Simplify the complex rational expression at the bottom of the previous page.

b. Use the data displayed by the bar graph on the previous page to find the percentage of federal expenditures spent on human resources in 2006. Round to the nearest percent.

c. Use the simplified rational expression from part (a) to find the percentage of federal expenditures spent on human resources in 2006. Round to the nearest percent. Does this underestimate or overestimate the actual percent that you found in part (b)? By how much?

52. a. Simplify the complex rational expression at the bottom of the previous page.

b. Use the data displayed by the bar graph on the previous page to find the percentage of federal expenditures spent on human resources in 2005. Round to the nearest percent.

c. Use the simplified rational expression from part (a) to find the percentage of federal expenditures spent on human resources in 2005. Round to the nearest percent. How does this compare with the actual percent that you found in part (b)?

Writing in Mathematics

53. What is a complex rational expression? Give an example with your explanation.

54. Describe two ways to simplify $\dfrac{\dfrac{3}{x} + \dfrac{2}{x^2}}{\dfrac{1}{x^2} + \dfrac{2}{x}}$.

55. Which method do you prefer for simplifying complex rational expressions? Why?

Critical Thinking Exercises

Make Sense? *In Exercises 56–59, determine whether each statement "makes sense" or "does not make sense" and explain your reasoning.*

56. I simplified $\dfrac{\dfrac{1}{2} + \dfrac{x}{3}}{4}$ by multiplying the numerator by 6.

57. I added 1 to $\dfrac{1}{1 + \dfrac{1}{2}}$ and obtained $\dfrac{5}{3}$.

58. I used the LCD method to simplify

$$\frac{3 - \dfrac{6}{x}}{1 + \dfrac{7}{y}}$$

and obtained $\dfrac{3 - 6y}{1 + 7x}$.

59. Because x^{-1} means $\dfrac{1}{x}$ and y^{-1} means $\dfrac{1}{y}$, I simplified $\dfrac{x^{-1} + y^{-1}}{x^{-1} - y^{-1}}$ and obtained $\dfrac{x - y}{x + y}$.

In Exercises 60–63, determine whether each statement is true or false. If the statement is false, make the necessary change(s) to produce a true statement.

60. The fraction $\dfrac{31,729,546}{72,578,112}$ is a complex rational expression.

61. $\dfrac{y - \dfrac{1}{2}}{y + \dfrac{3}{4}} = \dfrac{4y - 2}{4y + 3}$ for any value of y except $-\dfrac{3}{4}$.

62. $\dfrac{\dfrac{1}{4} - \dfrac{1}{3}}{\dfrac{1}{3} + \dfrac{1}{6}} = \dfrac{1}{12} \div \dfrac{3}{6} = \dfrac{1}{6}$

63. Some complex rational expressions cannot be simplified by both methods discussed in this section.

64. In one short sentence, five words or less, explain what

$$\frac{\dfrac{1}{x} + \dfrac{1}{x^2} + \dfrac{1}{x^3}}{\dfrac{1}{x^4} + \dfrac{1}{x^5} + \dfrac{1}{x^6}}$$

does to each number x.

In Exercises 65–66, simplify completely.

65. $\dfrac{2y}{2 + \dfrac{2}{y}} + \dfrac{y}{1 + \dfrac{1}{y}}$

66. $\dfrac{1 + \dfrac{1}{y} - \dfrac{6}{y^2}}{1 - \dfrac{5}{y} + \dfrac{6}{y^2}} - \dfrac{1 - \dfrac{1}{y}}{1 - \dfrac{2}{y} - \dfrac{3}{y^2}}$

Technology Exercises

In Exercises 67–69, use the ⃞GRAPH⃞ or ⃞TABLE⃞ feature of a graphing utility to determine if the simplification is correct. If the answer is wrong, correct it and then verify your corrected simplification using the graphing utility.

67. $\dfrac{x - \dfrac{1}{2x + 1}}{1 - \dfrac{x}{2x + 1}} = 2x - 1$

68. $\dfrac{\dfrac{1}{x} + 1}{\dfrac{1}{x}} = 2$

69. $\dfrac{\dfrac{1}{x} + \dfrac{1}{3}}{\dfrac{1}{3x}} = x + \dfrac{1}{3}$

Review Exercises

70. Factor completely: $2x^3 - 20x^2 + 50x$. (Section 7.5, Example 2)

71. Solve: $2 - 3(x - 2) = 5(x + 5) - 1$. (Section 2.3, Example 3)

72. Multiply: $(x + y)(x^2 - xy + y^2)$. (Section 6.2, Example 7)

Preview Exercises

Exercises 73–75 will help you prepare for the material covered in the next section.

73. Solve: $\dfrac{x}{3} + \dfrac{x}{2} = \dfrac{5}{6}$.

74. Solve: $\dfrac{2x}{3} = \dfrac{14}{3} - \dfrac{x}{2}$.

75. Solve: $2x^2 + 2 = 5x$.

Solving Rational Equations

Objectives

1 Solve rational equations.

2 Solve problems involving formulas with rational expressions.

The time has come to clean up the river. Suppose that the government has committed $375 million for this project. We know that

$$y = \frac{250x}{100 - x}$$

models the cost, y, in millions of dollars, to remove x percent of the river's pollutants. What percentage of pollutants can be removed for $375 million?

In order to determine the percentage, we use the given model. The government has committed $375 million, so substitute 375 for y:

$$375 = \frac{250x}{100 - x} \quad \text{or} \quad \frac{250x}{100 - x} = 375.$$

> The equation contains a rational expression.

Now we need to solve the equation and find the value for x. This variable represents the percentage of pollutants that can be removed for $375 million.

A **rational equation**, also called a **fractional equation**, is an equation containing one or more rational expressions. The preceding equation is an example of a rational equation. Do you see that there is a variable in a denominator? This is a characteristic of many rational equations. In this section, you will learn a procedure for solving such equations.

1 Solve rational equations.

Solving Rational Equations

We have seen that the LCD is used to add and subtract rational expressions. By contrast, when solving rational equations, **the LCD is used as a multiplier that clears an equation of fractions**.

Using Technology

Graphic Connections

We can use a graphing utility to verify the solution to Example 1. Graph each side of the equation:

$$y_1 = \frac{x}{4}$$

$$y_2 = \frac{1}{4} + \frac{x}{6}.$$

Trace along the lines or use the utility's intersection feature. The solution, as shown below, is the first coordinate of the point of intersection. Thus, the solution is 3.

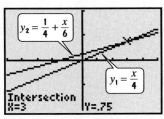

$[-5, 5, 1]$ by $[-2, 2, 1]$

| **EXAMPLE 1** | Solving a Rational Equation |

Solve: $\dfrac{x}{4} = \dfrac{1}{4} + \dfrac{x}{6}$.

Solution   The denominators are 4, 4, and 6. The least common denominator is 12. To clear the equation of fractions, we multiply both sides by 12.

$$\frac{x}{4} = \frac{1}{4} + \frac{x}{6} \qquad \text{This is the given equation.}$$

$$12\left(\frac{x}{4}\right) = 12\left(\frac{1}{4} + \frac{x}{6}\right) \qquad \text{Multiply both sides by 12, the LCD of all the fractions in the equation.}$$

$$12 \cdot \frac{x}{4} = 12 \cdot \frac{1}{4} + 12 \cdot \frac{x}{6} \qquad \text{Use the distributive property on the right side.}$$

$$3x = 3 + 2x \qquad \text{Simplify: } \frac{\overset{3}{\cancel{12}}}{1} \cdot \frac{x}{\cancel{4}} = 3x; \ \frac{\overset{3}{\cancel{12}}}{1} \cdot \frac{1}{\cancel{4}} = 3; \ \frac{\overset{2}{\cancel{12}}}{1} \cdot \frac{x}{\cancel{6}} = 2x.$$

$$x = 3 \qquad \text{Subtract } 2x \text{ from both sides.}$$

Substitute 3 for x in the original equation. You should obtain the true statement $\dfrac{3}{4} = \dfrac{3}{4}$. This verifies that the solution is 3 and the solution set is $\{3\}$.   ■

☑ **CHECK POINT 1**   Solve: $\dfrac{x}{6} = \dfrac{1}{6} + \dfrac{x}{8}$.

In Example 1, we solved a rational equation with constants in the denominators. Now, let's consider an equation such as

$$\frac{1}{x} = \frac{1}{5} + \frac{3}{2x}.$$

Can you see how this equation differs from the rational equation that we solved earlier? The variable, x, appears in two of the denominators. The procedure for solving this equation still involves multiplying each side by the least common denominator. However, we must avoid any values of the variable that make a denominator zero. For example, examine the denominators in the equation:

$$\frac{1}{x} = \frac{1}{5} + \frac{3}{2x}.$$

> This denominator would equal zero if $x = 0$.   This denominator would equal zero if $x = 0$.

We see that x cannot equal zero. With this in mind, let's solve the equation.

| **EXAMPLE 2** | Solving a Rational Equation |

Solve: $\dfrac{1}{x} = \dfrac{1}{5} + \dfrac{3}{2x}$.

Solution   The denominators are x, 5, and $2x$. The least common denominator is $10x$. We begin by multiplying both sides of the equation by $10x$. We will also write the restriction that x cannot equal zero to the right of the equation.

$$\frac{1}{x} = \frac{1}{5} + \frac{3}{2x}, \quad x \neq 0 \qquad \text{This is the given equation.}$$

$$10x \cdot \frac{1}{x} = 10x\left(\frac{1}{5} + \frac{3}{2x}\right) \qquad \text{Multiply both sides by 10x.}$$

$$10x \cdot \frac{1}{x} = 10x \cdot \frac{1}{5} + 10x \cdot \frac{3}{2x} \qquad \text{Use the distributive property. Be sure to multiply all terms by 10x.}$$

$$10\overset{}{x} \cdot \frac{1}{\overset{}{x}} = \overset{2}{10}x \cdot \frac{1}{\underset{1}{\overset{}{5}}} + \overset{5}{10}x \cdot \frac{3}{\underset{1}{2\overset{}{x}}} \qquad \text{Divide out common factors in the multiplications.}$$

$$10 = 2x + 15 \qquad \text{Simplify.}$$

Observe that the resulting equation,

$$10 = 2x + 15$$

is now cleared of fractions. With the variable term, $2x$, already on the right, we will collect constant terms on the left by subtracting 15 from both sides.

$$-5 = 2x \qquad \text{Subtract 15 from both sides.}$$

$$-\frac{5}{2} = x \qquad \text{Divide both sides by 2.}$$

We check our solution by substituting $-\frac{5}{2}$ into the original equation or by using a calculator. With a calculator, evaluate each side of the equation for $x = -\frac{5}{2}$, or for $x = -2.5$. Note that the original restriction that $x \neq 0$ is met. The solution is $-\frac{5}{2}$ and the solution set is $\left\{-\frac{5}{2}\right\}$. ∎

☑ **CHECK POINT 2**   Solve: $\dfrac{5}{2x} = \dfrac{17}{18} - \dfrac{1}{3x}$.

The following steps may be used to solve a rational equation:

Solving Rational Equations

1. List restrictions on the variable. Avoid any values of the variable that make a denominator zero.
2. Clear the equation of fractions by multiplying both sides by the LCD of all rational expressions in the equation.
3. Solve the resulting equation.
4. Reject any proposed solution that is in the list of restrictions on the variable. Check other proposed solutions in the original equation.

EXAMPLE 3   Solving a Rational Equation

Solve: $x + \dfrac{1}{x} = \dfrac{5}{2}$.

Solution

Step 1. List restrictions on the variable.

$$x + \frac{1}{x} = \frac{5}{2}$$

This denominator would
equal 0 if $x = 0$.

The restriction is $x \neq 0$.

Step 2. Multiply both sides by the LCD. The denominators are x and 2. Thus, the LCD is $2x$. We multiply both sides by $2x$.

$$x + \frac{1}{x} = \frac{5}{2}, \quad x \neq 0 \qquad \text{This is the given equation.}$$

$$2x\left(x + \frac{1}{x}\right) = 2x\left(\frac{5}{2}\right) \qquad \text{Multiply both sides by the LCD.}$$

$$2x \cdot x + 2x \cdot \frac{1}{x} = 2x \cdot \frac{5}{2} \qquad \text{Use the distributive property on the left side.}$$

$$2x^2 + 2 = 5x \qquad \text{Simplify.}$$

Step 3. Solve the resulting equation. Can you see that we have a quadratic equation? Write the equation in standard form and solve for x.

$$2x^2 - 5x + 2 = 0 \qquad \text{Subtract 5x from both sides.}$$

$$(2x - 1)(x - 2) = 0 \qquad \text{Factor.}$$

$$2x - 1 = 0 \quad \text{or} \quad x - 2 = 0 \qquad \text{Set each factor equal to 0.}$$

$$2x = 1 \qquad\qquad x = 2 \qquad \text{Solve the resulting equations.}$$

$$x = \frac{1}{2}$$

Step 4. Check proposed solutions in the original equation. The proposed solutions, $\frac{1}{2}$ and 2, are not part of the restriction that $x \neq 0$. Neither makes a denominator in the original equation equal to zero.

Check $\frac{1}{2}$:

$$x + \frac{1}{x} = \frac{5}{2}$$

$$\frac{1}{2} + \frac{1}{\frac{1}{2}} \stackrel{?}{=} \frac{5}{2}$$

$$\frac{1}{2} + 2 \stackrel{?}{=} \frac{5}{2}$$

$$\frac{1}{2} + \frac{4}{2} \stackrel{?}{=} \frac{5}{2}$$

$$\frac{5}{2} = \frac{5}{2}, \quad \text{true}$$

Check 2:

$$x + \frac{1}{x} = \frac{5}{2}$$

$$2 + \frac{1}{2} \stackrel{?}{=} \frac{5}{2}$$

$$\frac{4}{2} + \frac{1}{2} \stackrel{?}{=} \frac{5}{2}$$

$$\frac{5}{2} = \frac{5}{2}, \quad \text{true}$$

The solutions are $\frac{1}{2}$ and 2, and the solution set is $\left\{\frac{1}{2}, 2\right\}$.

☑ **CHECK POINT 3** Solve: $x + \frac{6}{x} = -5$.

EXAMPLE 4   Solving a Rational Equation

Solve: $\dfrac{3x}{x^2 - 9} + \dfrac{1}{x - 3} = \dfrac{3}{x + 3}$.

Solution

Step 1. List restrictions on the variable. By factoring denominators, it makes it easier to see values that make denominators zero.

$$\dfrac{3x}{(x + 3)(x - 3)} + \dfrac{1}{x - 3} = \dfrac{3}{x + 3}$$

| This denominator is zero if $x = -3$ or $x = 3$. | This denominator is zero if $x = 3$. | This denominator is zero if $x = -3$. |

The restrictions are $x \neq -3$ and $x \neq 3$.

Step 2. Multiply both sides by the LCD. The LCD is $(x + 3)(x - 3)$.

$$\dfrac{3x}{(x + 3)(x - 3)} + \dfrac{1}{x - 3} = \dfrac{3}{x + 3}, \quad x \neq -3, x \neq 3$$ This is the given equation with a denominator factored.

$$(x + 3)(x - 3)\left[\dfrac{3x}{(x + 3)(x - 3)} + \dfrac{1}{x - 3}\right] = (x + 3)(x - 3) \cdot \dfrac{3}{x + 3}$$ Multiply both sides by the LCD.

$$\cancel{(x + 3)(x - 3)} \cdot \dfrac{3x}{\cancel{(x + 3)(x - 3)}} + (x + 3)\cancel{(x - 3)} \cdot \dfrac{1}{\cancel{x - 3}}$$

$$= \cancel{(x + 3)}(x - 3) \cdot \dfrac{3}{\cancel{x + 3}}$$ Use the distributive property on the left side.

$$3x + (x + 3) = 3(x - 3)$$ Simplify.

Step 3. Solve the resulting equation.

$$3x + (x + 3) = 3(x - 3)$$ This is the equation cleared of fractions.

$$4x + 3 = 3x - 9$$ Combine like terms on the left side.
Use the distributive property on the right side.

$$x + 3 = -9$$ Subtract 3x from both sides.

$$x = -12$$ Subtract 3 from both sides.

Step 4. Check proposed solutions in the original equation. The proposed solution, -12, is not part of the restriction that $x \neq -3$ and $x \neq 3$. Substitute -12 for x in the given equation and show that -12 is the solution. The equation's solution set is $\{-12\}$. ∎

☑ **CHECK POINT 4**   Solve: $\dfrac{11}{x^2 - 25} + \dfrac{4}{x + 5} = \dfrac{3}{x - 5}$.

EXAMPLE 5   Solving a Rational Equation

Solve: $\dfrac{8x}{x + 1} = 4 - \dfrac{8}{x + 1}$.

Solution

Step 1. List restrictions on the variable.

| This denominator is zero if $x = -1$. | $\dfrac{8x}{x + 1} = 4 - \dfrac{8}{x + 1}$ | This denominator is zero if $x = -1$. |

The restriction is $x \neq -1$.

Step 2. Multiply both sides by the LCD. The LCD is $x + 1$.

$$\frac{8x}{x + 1} = 4 - \frac{8}{x + 1}, \quad x \neq -1 \qquad \text{This is the given equation.}$$

$$(x + 1) \cdot \frac{8x}{x + 1} = (x + 1)\left[4 - \frac{8}{x + 1}\right] \qquad \text{Multiply both sides by the LCD.}$$

$$\cancel{(x + 1)} \cdot \frac{8x}{\cancel{x + 1}} = (x + 1) \cdot 4 - \cancel{(x + 1)} \cdot \frac{8}{\cancel{x + 1}} \qquad \begin{array}{l}\text{Use the distributive property} \\ \text{on the right side.}\end{array}$$

$$8x = 4(x + 1) - 8 \qquad \text{Simplify.}$$

Step 3. Solve the resulting equation.

$$8x = 4(x + 1) - 8 \qquad \text{This is the equation cleared of fractions.}$$
$$8x = 4x + 4 - 8 \qquad \text{Use the distributive property on the right side.}$$
$$8x = 4x - 4 \qquad \text{Simplify.}$$
$$4x = -4 \qquad \text{Subtract 4x from both sides.}$$
$$x = -1 \qquad \text{Divide both sides by 4.}$$

Study Tip

Reject any proposed solution that causes any denominator in a rational equation to equal 0.

Step 4. Check proposed solutions. The proposed solution, -1, is *not* a solution because of the restriction that $x \neq -1$. Notice that -1 makes both of the denominators zero in the original equation. There is *no solution to this equation*. The solution set is $\varnothing$, the empty set. ∎

$\boxed{\checkmark}$ **CHECK POINT 5** Solve: $\dfrac{x}{x - 3} = \dfrac{3}{x - 3} + 9$.

Study Tip

We simplify rational expressions. We solve rational equations. Notice the differences between the procedures.

Adding Rational Expressions

Simplify: $\dfrac{5}{3x} + \dfrac{3}{x}$.

This is not an equation.
There is no equal sign.

Solving a Rational Equation

Solve: $\dfrac{5}{3x} + \dfrac{3}{x} = 1$.

This is an equation.
There is an equal sign.

Solution

The LCD is $3x$. Rewrite each expression with this LCD and retain the LCD.

$$\frac{5}{3x} + \frac{3}{x}$$

$$= \frac{5}{3x} + \frac{3}{x} \cdot \frac{3}{3}$$

$$= \frac{5}{3x} + \frac{9}{3x}$$

$$= \frac{5 + 9}{3x}$$

$$= \frac{14}{3x}$$

Solution

The LCD is $3x$. Multiply both sides by this LCD and clear the fractions.

$$3x\left(\frac{5}{3x} + \frac{3}{x}\right) = 3x \cdot 1$$

$$3x \cdot \frac{5}{3x} + 3x \cdot \frac{3}{x} = 3x$$

$$5 + 9 = 3x$$

$$14 = 3x$$

$$\frac{14}{3} = x$$

The solution set is $\left\{\dfrac{14}{3}\right\}$.

You only eliminate the denominators when solving a rational equation with an equal sign. You should never begin by eliminating the denominators when simplifying a rational expression involving addition or subtraction with no equal sign.

2 Solve problems involving formulas with rational expressions.

Applications of Rational Equations

Rational equations can be solved to answer questions about variables contained in mathematical models.

EXAMPLE 6 **A Government-Funded Cleanup**

The formula

$$y = \frac{250x}{100 - x}$$

models the cost, y, in millions of dollars, to remove x percent of a river's pollutants. If the government commits \$375 million for this project, what percentage of pollutants can be removed?

Solution Substitute 375 for y and solve the resulting rational equation for x.

$$375 = \frac{250x}{100 - x} \qquad \text{The LCD is } 100 - x.$$

$$(100 - x)375 = \cancel{(100 - x)} \cdot \frac{250x}{100 \cancel{- x}} \qquad \text{Multiply both sides by the LCD.}$$

$$375(100 - x) = 250x \qquad \text{Simplify.}$$

$$37{,}500 - 375x = 250x \qquad \text{Use the distributive property on the left side.}$$

$$37{,}500 = 625x \qquad \text{Add 375x to both sides.}$$

$$\frac{37{,}500}{625} = \frac{625x}{625} \qquad \text{Divide both sides by 625.}$$

$$60 = x \qquad \text{Simplify.}$$

If the government spends \$375 million, 60% of the river's pollutants can be removed.

✓ **CHECK POINT 6** Use the model in Example 6 to answer this question: If government funding is increased to \$750 million, what percentage of pollutants can be removed?

8.6 EXERCISE SET *MyMathLab*
PRACTICE · WATCH · DOWNLOAD · READ · REVIEW

Practice Exercises

In Exercises 1–46, solve each rational equation.

1. $\dfrac{x}{3} = \dfrac{x}{2} - 2$

2. $\dfrac{x}{5} = \dfrac{x}{6} + 1$

3. $\dfrac{4x}{3} = \dfrac{x}{18} - \dfrac{x}{6}$

4. $\dfrac{5x}{4} = \dfrac{x}{12} - \dfrac{x}{2}$

5. $2 - \dfrac{8}{x} = 6$

6. $1 - \dfrac{9}{x} = 4$

7. $\dfrac{2}{x} + \dfrac{1}{3} = \dfrac{4}{x}$

8. $\dfrac{5}{x} + \dfrac{1}{3} = \dfrac{6}{x}$

9. $\dfrac{2}{x} + 3 = \dfrac{5}{2x} + \dfrac{13}{4}$

10. $\dfrac{7}{2x} = \dfrac{5}{3x} + \dfrac{22}{3}$

11. $\dfrac{2}{3x} + \dfrac{1}{4} = \dfrac{11}{6x} - \dfrac{1}{3}$

12. $\dfrac{5}{2x} - \dfrac{8}{9} = \dfrac{1}{18} - \dfrac{1}{3x}$

13. $\dfrac{6}{x + 3} = \dfrac{4}{x - 3}$

14. $\dfrac{7}{x + 1} = \dfrac{5}{x - 3}$

15. $\dfrac{x - 2}{2x} + 1 = \dfrac{x + 1}{x}$

16. $\dfrac{7x - 4}{5x} = \dfrac{9}{5} - \dfrac{4}{x}$

17. $x + \dfrac{6}{x} = -7$

18. $x + \dfrac{7}{x} = -8$

19. $\dfrac{x}{5} - \dfrac{5}{x} = 0$

20. $\dfrac{x}{4} - \dfrac{4}{x} = 0$

21. $x + \dfrac{3}{x} = \dfrac{12}{x}$

22. $x + \dfrac{3}{x} = \dfrac{19}{x}$

23. $\dfrac{4}{y} - \dfrac{y}{2} = \dfrac{7}{2}$

24. $\dfrac{4}{3y} - \dfrac{1}{3} = y$

25. $\dfrac{x - 4}{x} = \dfrac{15}{x + 4}$

26. $\dfrac{x - 1}{2x + 3} = \dfrac{6}{x - 2}$

27. $\dfrac{2}{x^2 - 1} = \dfrac{4}{x + 1}$

28. $\dfrac{3}{x + 1} = \dfrac{1}{x^2 - 1}$

29. $\dfrac{1}{x - 1} + 5 = \dfrac{11}{x - 1}$

30. $\dfrac{3}{x + 4} - 7 = \dfrac{-4}{x + 4}$

31. $\dfrac{8y}{y+1} = 4 - \dfrac{8}{y+1}$

32. $\dfrac{2}{y-2} = \dfrac{y}{y-2} - 2$

33. $\dfrac{3}{x-1} + \dfrac{8}{x} = 3$

34. $\dfrac{2}{x-2} + \dfrac{4}{x} = 2$

35. $\dfrac{3y}{y-4} - 5 = \dfrac{12}{y-4}$

36. $\dfrac{10}{y+2} = 3 - \dfrac{5y}{y+2}$

37. $\dfrac{1}{x} + \dfrac{1}{x-3} = \dfrac{x-2}{x-3}$

38. $\dfrac{1}{x-1} + \dfrac{2}{x} = \dfrac{x}{x-1}$

39. $\dfrac{x+1}{3x+9} + \dfrac{x}{2x+6} = \dfrac{2}{4x+12}$

40. $\dfrac{3}{2y-2} + \dfrac{1}{2} = \dfrac{2}{y-1}$

41. $\dfrac{4y}{y^2-25} + \dfrac{2}{y-5} = \dfrac{1}{y+5}$

42. $\dfrac{1}{x+4} + \dfrac{1}{x-4} = \dfrac{22}{x^2-16}$

43. $\dfrac{1}{x-4} - \dfrac{5}{x+2} = \dfrac{6}{x^2-2x-8}$

44. $\dfrac{6}{x+3} - \dfrac{5}{x-2} = \dfrac{-20}{x^2+x-6}$

45. $\dfrac{2}{x+3} - \dfrac{2x+3}{x-1} = \dfrac{6x-5}{x^2+2x-3}$

46. $\dfrac{x-3}{x-2} + \dfrac{x+1}{x+3} = \dfrac{2x^2-15}{x^2+x-6}$

Practice PLUS

In Exercises 47–54, solve or simplify, whichever is appropriate.

47. $\dfrac{x^2-10}{x^2-x-20} = 1 + \dfrac{7}{x-5}$

48. $\dfrac{x^2+4x-2}{x^2-2x-8} = 1 + \dfrac{4}{x-4}$

49. $\dfrac{x^2-10}{x^2-x-20} - 1 - \dfrac{7}{x-5}$

50. $\dfrac{x^2+4x-2}{x^2-2x-8} - 1 - \dfrac{4}{x-4}$

51. $5y^{-2} + 1 = 6y^{-1}$ **52.** $3y^{-2} + 1 = 4y^{-1}$

53. $\dfrac{3}{y+1} - \dfrac{1}{1-y} = \dfrac{10}{y^2-1}$

54. $\dfrac{4}{y-2} - \dfrac{1}{2-y} = \dfrac{25}{y+6}$

Application Exercises

A company that manufactures wheelchairs has fixed costs of $500,000. The average cost per wheelchair, C, for the company to manufacture x wheelchairs per month is modeled by the formula

$$C = \dfrac{400x + 500,000}{x}.$$

Use this mathematical model to solve Exercises 55–56.

55. How many wheelchairs per month can be produced at an average cost of $450 per wheelchair?

56. How many wheelchairs per month can be produced at an average cost of $405 per wheelchair?

In Palo Alto, California, a government agency ordered computer-related companies to contribute to a pool of money to clean up underground water supplies. (The companies had stored toxic chemicals in leaking underground containers.) The formula

$$C = \dfrac{2x}{100-x}$$

models the cost, C, in millions of dollars, for removing x percent of the contaminants. Use this mathematical model to solve Exercises 57–58.

57. What percentage of the contaminants can be removed for $2 million?

58. What percentage of the contaminants can be removed for $8 million?

We have seen that Young's rule

$$C = \dfrac{DA}{A+12}$$

can be used to approximate the dosage of a drug prescribed for children. In this formula, A = the child's age, in years, D = an adult dosage, and C = the proper child's dosage. Use this formula to solve Exercises 59–60.

59. When the adult dosage is 1000 milligrams, a child is given 300 milligrams. What is that child's age? Round to the nearest year.

60. When the adult dosage is 1000 milligrams, a child is given 500 milligrams. What is that child's age?

A grocery store sells 4000 cases of canned soup per year. By averaging costs to purchase soup and to pay storage costs, the owner has determined that if x cases are ordered at a time, the yearly inventory cost, C, can be modeled by

$$C = \dfrac{10,000}{x} + 3x.$$

The graph of this model is shown below. Use this information to solve Exercises 61–62.

61. How many cases should be ordered at a time for yearly inventory costs to be $350? Identify your solutions as points on the graph.

62. How many cases should be ordered at a time for yearly inventory costs to be $790? Identify your solutions as points on the graph.

Writing in Mathematics

63. What is a rational equation?

64. Explain how to solve a rational equation.

65. Explain how to find restrictions on the variable in a rational equation.

66. Why should restrictions on the variable in a rational equation be listed before you begin solving the equation?

67. Describe similarities and differences between the procedures needed to solve the following problems:

$$\text{Add: } \frac{2}{x} + \frac{3}{4}. \qquad \text{Solve: } \frac{2}{x} + \frac{3}{4} = 1.$$

Critical Thinking Exercises

Make Sense? *In Exercises 68–71, determine whether each statement "makes sense" or "does not make sense" and explain your reasoning.*

68. I added two rational expressions and found the solution set.

69. I can solve the equation $\dfrac{6}{x+3} = \dfrac{4}{x-3}$ by multiplying both sides by the LCD or by setting the cross product of 6 and $x - 3$ equal to the cross product of 4 and $x + 3$.

70. I'm solving a rational equation that became a quadratic equation, so my rational equation must have two solutions.

71. I must have made an error if a rational equation produces no solution.

In Exercises 72–75, determine whether each statement is true or false. If the statement is false, make the necessary change(s) to produce a true statement.

72. $\dfrac{1}{x} + \dfrac{1}{6} = 6x\left(\dfrac{1}{x} + \dfrac{1}{6}\right) = 6 + x$

73. If a is any real number, the equation $\dfrac{a}{x} + 1 = \dfrac{a}{x}$ has no solution.

74. All real numbers satisfy the equation $\dfrac{3}{x} - \dfrac{1}{x} = \dfrac{2}{x}$.

75. To solve $\dfrac{5}{3x} + \dfrac{3}{x} = 1$, we must first add the rational expressions on the left side.

76. Solve for f: $\dfrac{1}{p} + \dfrac{1}{q} = \dfrac{1}{f}$.

77. Solve for f_2: $f = \dfrac{f_1 f_2}{f_1 + f_2}$.

In Exercises 78–79, solve each rational equation.

78. $\dfrac{x+1}{2x^2 - 11x + 5} = \dfrac{x-7}{2x^2 + 9x - 5} - \dfrac{2x-6}{x^2 - 25}$

79. $\left(\dfrac{x+1}{x+7}\right)^2 \div \left(\dfrac{x+1}{x+7}\right)^4 = 0.$

80. Find b so that the solution of

$$\frac{7x+4}{b} + 13 = x$$

is -6.

Technology Exercises

In Exercises 81–83, use a graphing utility to solve each rational equation. Graph each side of the equation in the given viewing rectangle. The solution is the first coordinate of the point(s) of intersection. Check by direct substitution.

81. $\dfrac{x}{2} + \dfrac{x}{4} = 6$

$[-5, 10, 1]$ by $[-5, 10, 1]$

82. $\dfrac{50}{x} = 2x$

$[-10, 10, 1]$ by $[-20, 20, 2]$

83. $x + \dfrac{6}{x} = -5$

$[-10, 10, 1]$ by $[-10, 10, 1]$

Review Exercises

84. Factor completely: $x^4 + 2x^3 - 3x - 6$. (Section 7.1, Example 7)

85. Simplify: $(3x^2)(-4x^{-10})$. (Section 6.7, Example 3)

86. Simplify: $-5[4(x - 2) - 3]$. (Section 1.8, Example 11)

Preview Exercises

Exercises 87–89 will help you prepare for the material covered in the next section.

87. Solve: $\dfrac{15}{8+x} = \dfrac{9}{8-x}$.

88. If you can complete a job in 5 hours, what fractional part of the job can you complete in 1 hour? in 3 hours? in x hours?

89. a. If $y = \dfrac{k}{x}$, find the value of k using $x = 8$ and $y = 12$.

b. Substitute the value of k into $y = \dfrac{k}{x}$ and write the resulting equation.

c. Use the equation from part (b) to find y when $x = 3$.

Applications Using Rational Equations and Variation

Objectives

1 Solve problems involving motion.

2 Solve problems involving work.

3 Solve problems involving similar triangles.

4 Solve problems involving variation.

Horror films offer the pleasure of vicarious terror, of being safely scared.

The vampire legend is death as seducer: He/she sucks our blood to take us to a perverse immortality. The vampire resembles us, but appears only at night, hidden among mortals. Can we offer an algebraic application that is particularly seductive to these legendary creatures of the night? Among the examples you'll be reading in this section, there is one that promises to have Dracula panting with anticipatory delight.

1 Solve problems involving motion.

Problems Involving Motion

We have seen that the distance, d, covered by any moving body is the product of its average rate, r, and its time in motion, t: $d = rt$. Rational expressions appear in motion problems when the conditions of the problem involve the time traveled. We can obtain an expression for t, the time traveled, by dividing both sides of $d = rt$ by r.

$$d = rt \qquad \text{Distance equals rate times time.}$$

$$\frac{d}{r} = \frac{rt}{r} \qquad \text{Divide both sides by } r.$$

$$\frac{d}{r} = t \qquad \text{Simplify.}$$

Time in Motion

$$t = \frac{d}{r}$$

$$\text{Time traveled} = \frac{\text{Distance traveled}}{\text{Rate of travel}}$$

Downstream (with the current)

←——15 miles——→

Upstream (against the current)

←—— 9 miles ——→

Traveling 15 miles downstream takes the same time as traveling 9 miles upstream.

| EXAMPLE 1 | A Motion Problem Involving Time |

In still water, your small boat averages 8 miles per hour. It takes you the same amount of time to travel 15 miles downstream, with the current, as 9 miles upstream, against the current. What is the rate of the water's current?

Solution

Step 1. Let x represent one of the quantities. Let

$$x = \text{the rate of the current.}$$

Step 2. Represent other unknown quantities in terms of x. We still need expressions for the rate of your boat with the current and the rate against the current. Traveling with the current, the boat's rate in still water, 8 miles per hour, is increased by the current's rate, x miles per hour. Thus,

$$8 + x = \text{the boat's rate with the current.}$$

Traveling against the current, the boat's rate in still water, 8 miles per hour, is decreased by the current's rate, x miles per hour. Thus,

$$8 - x = \text{the boat's rate against the current.}$$

Step 3. Write an equation that models the conditions. By reading the problem again, we discover that the crucial idea is that the time spent going 15 miles with the current equals the time spent going 9 miles against the current. This information is summarized in the following table.

	Distance	Rate	Time $= \dfrac{\text{Distance}}{\text{Rate}}$
With the current	15	$8 + x$	$\dfrac{15}{8 + x}$
Against the current	9	$8 - x$	$\dfrac{9}{8 - x}$

These two times are equal.

We are now ready to write an equation that models the problem's conditions.

The time spent going 15 miles with the current $\dfrac{15}{8 + x}$ equals $= \dfrac{9}{8 - x}$ the time spent going 9 miles against the current.

Step 4. Solve the equation and answer the question.

$$\frac{15}{8 + x} = \frac{9}{8 - x}$$
This is the equation that models the problem's conditions.

$$(8 + x)(8 - x) \cdot \frac{15}{8 + x} = (8 + x)(8 - x) \cdot \frac{9}{8 - x}$$
Multiply both sides by the LCD, $(8 + x)(8 - x)$.

$$15(8 - x) = 9(8 + x)$$
Simplify.

$$120 - 15x = 72 + 9x$$
Use the distributive property.

$$120 = 72 + 24x$$
Add 15x to both sides.

$$48 = 24x$$
Subtract 72 from both sides.

$$2 = x$$
Divide both sides by 24.

The rate of the water's current is 2 miles per hour.

Step 5. Check the proposed solution in the original wording of the problem. Does it take you the same amount of time to travel 15 miles downstream as 9 miles upstream if the current is 2 miles per hour? Keep in mind that your rate in still water is 8 miles per hour.

$$\text{Time required to travel 15 miles with the current} = \frac{\text{Distance}}{\text{Rate}} = \frac{15}{8+2} = \frac{15}{10} = 1\frac{1}{2} \text{ hours}$$

$$\text{Time required to travel 9 miles against the current} = \frac{\text{Distance}}{\text{Rate}} = \frac{9}{8-2} = \frac{9}{6} = 1\frac{1}{2} \text{ hours}$$

These times are the same, which checks with the original conditions of the problem. ■

☑ **CHECK POINT 1** Forget the small boat! This time we have you canoeing on the Colorado River. In still water, your average canoeing rate is 3 miles per hour. It takes you the same amount of time to travel 10 miles downstream, with the current, as 2 miles upstream, against the current. What is the rate of the water's current?

2 Solve problems involving work.

Problems Involving Work

You are thinking of designing your own Web site. You estimate that it will take 30 hours to do the job. In 1 hour, $\frac{1}{30}$ of the job is completed. In 2 hours, $\frac{2}{30}$, or $\frac{1}{15}$, of the job is completed. In 3 hours, the fractional part of the job done is $\frac{3}{30}$, or $\frac{1}{10}$. In x hours, the fractional part of the job that you can complete is $\frac{x}{30}$.

Your friend, who has experience developing Web sites, took 20 hours working on his own to design an impressive site. You wonder about the possibility of working together. How long would it take both of you to design your Web site?

Problems involving work usually have two people working together to complete a job. The amount of time it takes each person to do the job working alone is frequently known. The question deals with how long it will take both people working together to complete the job.

In work problems, **the number 1 represents one whole job completed.** For example, the completion of your Web site is represented by 1. Equations in work problems are based on the following condition:

| Fractional part of the job done by the first person | + | fractional part of the job done by the second person | = | 1 (one whole job completed). |

EXAMPLE 2 Solving a Problem Involving Work

You can design a Web site in 30 hours. Your friend can design the same site in 20 hours. How long will it take to design the Web site if you both work together?

Solution

Step 1. Let x represent one of the quantities. Let x = the time, in hours, for you and your friend to design the Web site together.

Step 2. Represent other unknown quantities in terms of x. Because there are no other unknown quantities, we can skip this step.

Step 3. Write an equation that models the conditions. We construct a table to help find the fractional part of the task completed by you and your friend in x hours.

	Fractional part of job completed in 1 hour	Time working together	Fractional part of job completed in x hours
You can design the site in 30 hours. **You**	$\dfrac{1}{30}$	x	$\dfrac{x}{30}$
Your friend can design the site in 20 hours. **Your friend**	$\dfrac{1}{20}$	x	$\dfrac{x}{20}$

Fractional part of the job done by you	+	fractional part of the job done by your friend	=	one whole job.
$\dfrac{x}{30}$	+	$\dfrac{x}{20}$	=	1

Step 4. Solve the equation and answer the question.

$$\frac{x}{30} + \frac{x}{20} = 1 \qquad \text{This is the equation that models the problem's conditions.}$$

$$60\left(\frac{x}{30} + \frac{x}{20}\right) = 60 \cdot 1 \qquad \text{Multiply both sides by 60, the LCD.}$$

$$60 \cdot \frac{x}{30} + 60 \cdot \frac{x}{20} = 60 \qquad \text{Use the distributive property on the left side.}$$

$$2x + 3x = 60 \qquad \text{Simplify: } \frac{\overset{2}{\cancel{60}}}{1} \cdot \frac{x}{\underset{1}{\cancel{30}}} = 2x \text{ and } \frac{\overset{3}{\cancel{60}}}{1} \cdot \frac{x}{\underset{1}{\cancel{20}}} = 3x.$$

$$5x = 60 \qquad \text{Combine like terms.}$$

$$x = 12 \qquad \text{Divide both sides by 5.}$$

If you both work together, you can design your Web site in 12 hours.

Step 5. Check the proposed solution in the original wording of the problem. Will you both complete the job in 12 hours? Because you can design the site in 30 hours, in 12 hours, you can complete $\frac{12}{30}$, or $\frac{2}{5}$, of the job. Because your friend can design the site in 20 hours, in 12 hours, he can complete $\frac{12}{20}$, or $\frac{3}{5}$, of the job. Notice that $\frac{2}{5} + \frac{3}{5} = 1$, which represents the completion of the entire job, or one whole job. ∎

Study Tip

Let

a = the time it takes person A to do a job working alone

b = the time it takes person B to do the same job working alone.

If x represents the time it takes for A and B to complete the entire job working together, then the situation can be modeled by the rational equation

$$\frac{x}{a} + \frac{x}{b} = 1.$$

✓ **CHECK POINT 2**   One person can paint the outside of a house in 8 hours. A second person can do it in 4 hours. How long will it take them to do the job if they work together?

---

3 Solve problems involving similar triangles.

Pedestrian Crossing

Similar Triangles

Shown in the margin is an international road sign. This sign is shaped just like the actual sign, although its size is smaller. Figures that have the same shape, but not the same size, are used in **scale drawings**. A scale drawing always pictures the exact shape of the object that the drawing represents. Architects, engineers, landscape gardeners, and interior decorators use scale drawings in planning their work.

Figures that have the same shape, but not necessarily the same size, are called similar figures. In **Figure 8.2**, triangles *ABC* and *DEF* are similar. Angles *A* and *D* measure the same number of degrees and are called **corresponding angles**. Angles *C* and *F* are corresponding angles, as are angles *B* and *E*. Angles with the same number of tick marks in **Figure 8.2** are the corresponding angles.

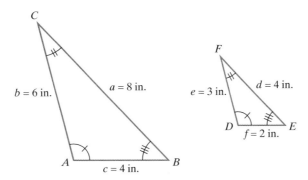

FIGURE 8.2

The sides opposite the corresponding angles are called **corresponding sides**. Although the measures of corresponding angles are equal, corresponding sides may or may not be the same length. For the triangles in **Figure 8.2**, each side in the smaller triangle is half the length of the corresponding side in the larger triangle.

The triangles in **Figure 8.2** illustrate what it means to be **similar triangles**. **Corresponding angles have the same measure and the ratios of the lengths of the corresponding sides are equal.** For the triangles in **Figure 8.2**, each of these ratios is equal to 2:

$$\frac{a}{d} = \frac{8}{4} = 2 \qquad \frac{b}{e} = \frac{6}{3} = 2 \qquad \frac{c}{f} = \frac{4}{2} = 2.$$

In similar triangles, the lengths of the corresponding sides are proportional. Thus,

$$\frac{a}{d} = \frac{b}{e} = \frac{c}{f}.$$

If we know that two triangles are similar, we can set up a proportion to solve for the length of an unknown side.

EXAMPLE 3   **Using Similar Triangles**

The triangles in **Figure 8.3** are similar. Find the missing length, *x*.

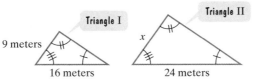

FIGURE 8.3

Solution   Because the triangles in **Figure 8.3** are similar, their corresponding sides are proportional.

FIGURE 8.3 (repeated)

Left side of △ I

$$\frac{9}{x} = \frac{16}{24}$$ Bottom side of △ I

Corresponding side on left of △ II

Corresponding side on bottom of △ II

We solve this rational equation by multiplying both sides by the LCD, 24x. (You can also apply the cross-products principle for solving proportions.)

$$24x \cdot \frac{9}{x} = 24x \cdot \frac{16}{24}$$ Multiply both sides by the LCD, 24x.

$$24 \cdot 9 = 16x$$ Simplify.

$$216 = 16x$$ Multiply: $24 \cdot 9 = 216$.

$$13.5 = x$$ Divide both sides by 16.

The missing length, x, is 13.5 meters.

✓ **CHECK POINT 3**   The similar triangles in the figure are shown with corresponding sides in the same relative position. Find the missing length, x.

3 in. ◺ 8 in. ◺
12 in. x

How can we quickly determine if two triangles are similar? **If the measures of two angles of one triangle are equal to those of two angles of a second triangle, then the two triangles are similar.** If the triangles are similar, then their corresponding sides are proportional.

EXAMPLE 4   **Problem Solving Using Similar Triangles**

A man who is 6 feet tall is standing 10 feet from the base of a lamppost (see **Figure 8.4**). The man's shadow has a length of 4 feet. How tall is the lamppost?

Solution   The drawing in **Figure 8.5** makes the similarity of the triangles easier to see. The large triangle with the lamppost on the left and the small triangle with the man on the left both contain 90° angles. They also share an angle. Thus, two angles of the large triangle are equal in measure to two angles of the small triangle. This means that the triangles are similar and their corresponding sides are proportional. We begin by letting x represent the height of the lamppost, in feet. Because corresponding sides of similar triangles are proportional,

FIGURE 8.4

FIGURE 8.5

Left side of big △

$$\frac{x}{6} = \frac{14}{4}.$$ Bottom side of big △

Corresponding side on left of small △

Corresponding side on bottom of small △

We solve for x by multiplying both sides by the LCD, 12.

$$12 \cdot \frac{x}{6} = 12 \cdot \frac{14}{4}$$ Multiply both sides by the LCD, 12.

$$2x = 42$$ Simplify: $\frac{\overset{2}{\cancel{12}}}{1} \cdot \frac{x}{\cancel{6}} = 2x$ and $\frac{\overset{3}{\cancel{12}}}{1} \cdot \frac{14}{\cancel{4}} = 42$.

$$x = 21$$ Divide both sides by 2.

The lamppost is 21 feet tall.

☑ **CHECK POINT 4**   Find the height of the lookout tower using the figure that lines up the top of the tower with the top of a stick that is 2 yards long.

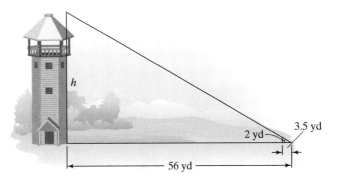

---

4 Solve problems involving variation.

Variation

Certain formulas occur so frequently in applied situations that they are given special names. Variation formulas show how one quantity changes in relation to another quantity. Quantities can vary directly or inversely.

Direct Variation   Because light travels faster than sound, during a thunderstorm we see lightning before we hear thunder. The formula

$$d = 1080t$$

describes your distance, d, in feet, from the storm's center if it takes you t seconds to hear thunder after seeing lightning. Thus,

If $t = 1$,	$d = 1080 \cdot 1 = 1080$:	If it takes 1 second to hear thunder, the storm's center is 1080 feet away.
If $t = 2$,	$d = 1080 \cdot 2 = 2160$:	If it takes 2 seconds to hear thunder, the storm's center is 2160 feet away.
If $t = 3$,	$d = 1080 \cdot 3 = 3240$:	If it takes 3 seconds to hear thunder, the storm's center is 3240 feet away.

The formula $d = 1080t$ illustrates that distance to the storm's center is a constant multiple of how long it takes to hear the thunder. If the time is doubled, the storm's distance is doubled; if the time is tripled, the storm's distance is tripled; and so on. Because of this, the distance is said to **vary directly** as the time. The **equation of variation** is

$$d = 1080t.$$

Generalizing, we obtain the following statement:

Direct Variation

If a situation is described by an equation in the form

$$y = kx,$$

where k is a constant, we say that **y varies directly as x** or **y is directly proportional to x**. The number k is called the **constant of variation**.

Can you see that the direct variation equation, $y = kx$, is a special case of the slope-intercept form of a line's equation, $y = mx + b$? When $m = k$ and $b = 0$, $y = mx + b$ becomes $y = kx$. Thus, the slope of a direct variation equation is k, the constant of variation. Because b, the y-intercept, is 0, the graph of a direct variation equation is a line through the origin. This is illustrated in **Figure 8.6**, which shows the graph of $d = 1080t$: Distance to a storm's center varies directly as the time it takes to hear thunder.

FIGURE 8.6 The graph of a direct variation equation

Problems involving direct variation can be solved using the following procedure. This procedure applies to direct variation problems, as well as to inverse variation problems, the other kind of variation problem that we will discuss.

Solving Variation Problems

1. Write an equation that describes the given English statement.
2. Substitute the given pair of values into the equation in step 1 and find the value of k, the constant of variation.
3. Substitute the value of k into the equation in step 1.
4. Use the equation from step 3 to answer the problem's question.

EXAMPLE 5   Solving a Direct Variation Problem

The volume of blood, B, in a person's body varies directly as body weight, W. A person who weighs 160 pounds has approximately 5 quarts of blood. Estimate the number of quarts of blood in a person who weighs 200 pounds.

Solution

Step 1. Write an equation. We know that *y varies directly as x* is expressed as

$$y = kx.$$

By changing letters, we can write an equation that describes the following English statement: The volume of blood, B, varies directly as body weight, W.

$$B = kW$$

Step 2. Use the given values to find k. A person who weighs 160 pounds has approximately 5 quarts of blood. Substitute 160 for W and 5 for B in the direct variation equation. Then solve for k.

$$B = kW \qquad \text{The volume of blood varies directly as body weight.}$$
$$5 = k \cdot 160 \qquad \text{Substitute 160 for } W \text{ and 5 for } B.$$
$$\frac{5}{160} = \frac{k \cdot 160}{160} \qquad \text{Divide both sides by 160.}$$
$$0.03125 = k \qquad \text{Express } \tfrac{5}{160}, \text{ or } \tfrac{1}{32}, \text{ in decimal form.}$$

Step 3. Substitute the value of k into the equation.

$$B = kW \qquad \text{Use the equation from step 1.}$$
$$B = 0.03125W \qquad \text{Replace } k, \text{ the constant of variation,}$$
$$\text{with } 0.03125.$$

Step 4. Answer the problem's question. We are interested in estimating the number of quarts of blood in a person who weighs 200 pounds. Substitute 200 for W in $B = 0.03125W$ and solve for B.

$$B = 0.03125W \qquad \text{This is the equation from step 3.}$$
$$B = 0.03125(200) \qquad \text{Substitute 200 for } W.$$
$$= 6.25 \qquad \text{Multiply.}$$

A person who weighs 200 pounds has approximately 6.25 quarts of blood. ▬

☑ **CHECK POINT 5** The number of gallons of water, W, used when taking a shower varies directly as the time, t, in minutes, in the shower. A shower lasting 5 minutes uses 30 gallons of water. How much water is used in a shower lasting 8 minutes?

Inverse Variation The distance from Atlanta, Georgia, to Orlando, Florida, is 450 miles. The time that it takes to drive from Atlanta to Orlando depends on the rate at which one drives and is given by

$$\text{Time} = \frac{450}{\text{Rate}}. \qquad \boxed{\text{Time} = \frac{\text{Distance}}{\text{Rate}}}$$

For example, if you average 45 miles per hour, the time for the drive is

$$\text{Time} = \frac{450}{45} = 10$$

or 10 hours. If you average 75 miles per hour, the time for the drive is

$$\text{Time} = \frac{450}{75} = 6$$

FIGURE 8.7

or 6 hours. As your rate (or speed) increases, the time for the trip decreases, and vice versa. This is illustrated by the graph in **Figure 8.7**. Observe the graph's shape. It shows that as the rate increases, the time for the trip decreases quite rapidly.

We can express the time for the Atlanta–Orlando trip using t for time and r for rate:

$$t = \frac{450}{r}.$$

This equation is an example of an **inverse variation** equation. Time, t, **varies inversely** as rate, r. When two quantities vary inversely, one quantity increases as the other decreases, and vice versa.

Generalizing, we obtain the following statement:

Inverse Variation

If a situation is described by an equation in the form

$$y = \frac{k}{x}$$

where k is a constant, we say that **y varies inversely as x** or **y is inversely proportional to x**. The number k is called the **constant of variation**.

We use the same procedure to solve inverse variation problems as we did to solve direct variation problems. Example 6 illustrates this procedure.

EXAMPLE 6 Solving an Inverse Variation Problem

The rate at which heat is lost, L, through a window pane varies inversely as the thickness, T, of the pane. A normal $\frac{1}{8}$-inch-thick pane loses 400 calories per hour. How many calories per hour are lost through a $\frac{3}{8}$-inch-thick window pane?

Solution

Step 1. Write an equation. We know that *y varies inversely as x* is expressed as

$$y = \frac{k}{x}.$$

By changing letters, we can write an equation that describes the following English statement: The rate at which heat is lost, L, varies inversely as the window pane's thickness, T.

$$L = \frac{k}{T}$$

Step 2. Use the given values to find k. A normal $\frac{1}{8}$-inch-thick pane loses 400 calories per hour. Substitute $\frac{1}{8}$ for T and 400 for L in the inverse variation equation. Then solve for k.

$$L = \frac{k}{T} \qquad \text{Heat loss varies inversely as thickness.}$$

$$400 = \frac{k}{\frac{1}{8}} \qquad \begin{array}{l}\text{A } \frac{1}{8}\text{-inch-thick pane loses 400 calories per hour:}\\ \text{If T} = \frac{1}{8}, C = 400.\end{array}$$

$$400 = 8k \qquad \frac{k}{\frac{1}{8}} = \frac{k}{1} \div \frac{1}{8} = \frac{k}{1} \cdot \frac{8}{1} = 8k$$

$$\frac{400}{8} = \frac{8k}{8} \qquad \text{Divide both sides by 8.}$$

$$50 = k \qquad \text{Simplify.}$$

Step 3. Substitute the value of k into the equation.

$$L = \frac{k}{T} \qquad \text{Use the equation from Step 1.}$$

$$L = \frac{50}{T} \qquad \text{Replace k, the constant of variation, with 50.}$$

Step 4. Answer the problem's question. We need to find the number of calories per hour lost through a $\frac{3}{8}$-inch-thick window pane. Substitute $\frac{3}{8}$ for T and solve for L.

$$L = \frac{50}{T} \qquad \text{This is the equation from Step 3.}$$

$$L = \frac{50}{\frac{3}{8}} = 50 \div \frac{3}{8} = \frac{50}{1} \cdot \frac{8}{3} = \frac{400}{3} = 133\frac{1}{3}$$

A $\frac{3}{8}$-inch-thick window pane loses $133\frac{1}{3}$ calories per hour. ■

✓ **CHECK POINT 6** When you use a spray can and press the valve at the top, you decrease the pressure of the gas in the can. This decrease of pressure causes the volume of the gas in the can to increase. Because the gas needs more room than is provided in the can, it expands in spray form through the small hole near the valve. In general, if the temperature is constant, the pressure, P, of a gas in a container varies inversely as the volume, V, of the container. The pressure of a gas sample in a container whose volume is 8 cubic inches is 12 pounds per square inch. If the sample expands to a volume of 20 cubic inches, what is the new pressure of the gas?

Doubling the pressure halves the volume.

8.7 EXERCISE SET *MyMathLab*
PRACTICE   WATCH   DOWNLOAD   READ   REVIEW

Practice and Application Exercises

Use rational equations to solve Exercises 1–10. Each exercise is a problem involving motion.

1. How bad is the heavy traffic? You can walk 10 miles in the same time that it takes to travel 15 miles by car. If the car's rate is 3 miles per hour faster than your walking rate, find the average rate of each.

	Distance	Rate	Time = $\dfrac{\text{Distance}}{\text{Rate}}$
Walking	10	x	$\dfrac{10}{x}$
Car in Heavy Traffic	15	$x + 3$	$\dfrac{15}{x+3}$

2. You can travel 40 miles on motorcycle in the same time that it takes to travel 15 miles on bicycle. If your motorcycle's rate is 20 miles per hour faster than your bicycle's, find the average rate for each.

	Distance	Rate	Time = $\dfrac{\text{Distance}}{\text{Rate}}$
Motorcycle	40	$x + 20$	$\dfrac{40}{x+20}$
Bicycle	15	x	$\dfrac{15}{x}$

3. A jogger runs 4 miles per hour faster downhill than uphill. If the jogger can run 5 miles downhill in the same time that it takes to run 3 miles uphill, find the jogging rate in each direction.

4. A truck can travel 120 miles in the same time that it takes a car to travel 180 miles. If the truck's rate is 20 miles per hour slower than the car's, find the average rate for each.

5. In still water, a boat averages 15 miles per hour. It takes the same amount of time to travel 20 miles downstream, with the current, as 10 miles upstream, against the current. What is the rate of the water's current?

6. In still water, a boat averages 18 miles per hour. It takes the same amount of time to travel 33 miles downstream, with the current, as 21 miles upstream, against the current. What is the rate of the water's current?

7. As part of an exercise regimen, you walk 2 miles on an indoor track. Then you jog at twice your walking speed for another 2 miles. If the total time spent walking and jogging is 1 hour, find the walking and jogging rates.

8. The joys of the Pacific Coast! You drive 90 miles along the Pacific Coast Highway and then take a 5-mile run along a hiking trail in Point Reyes National Seashore. Your driving rate is nine times that of your running rate. If the total time for driving and running is 3 hours, find the average rate driving and the average rate running.

9. The water's current is 2 miles per hour. A boat can travel 6 miles downstream, with the current, in the same amount of time it travels 4 miles upstream, against the current. What is the boat's average rate in still water?

10. The water's current is 2 miles per hour. A canoe can travel 6 miles downstream, with the current, in the same amount of time it travels 2 miles upstream, against the current. What is the canoe's average rate in still water?

Use a rational equation to solve Exercises 11–16. Each exercise is a problem involving work.

11. You must leave for campus in 10 minutes or you will be late for class. Unfortunately, you are snowed in. You can shovel the driveway in 20 minutes and your brother claims he can do it in 15 minutes. If you shovel together, how long will it take to clear the driveway? Will this give you enough time before you have to leave?

12. You promised your parents that you would wash the family car. You have not started the job and they are due home in 16 minutes. You can wash the car in 40 minutes and your sister claims she can do it in 30 minutes. If you work together, how long will it take to do the job? Will this give you enough time before your parents return?

13. The MTV crew will arrive in one week and begin filming the city for *The Real World Kalamazoo*. The mayor is desperate to clean the city streets before filming begins. Two teams are available, one that requires 400 hours and one that requires 300 hours. If the teams work together, how long will it take to clean all of Kalamazoo's streets? Is this enough time before the cameras begin rolling?

14. A hurricane strikes and a rural area is without food or water. Three crews arrive. One can dispense needed supplies in 10 hours, a second in 15 hours, and a third in 20 hours. How long will it take all three crews working together to dispense food and water?

15. A pool can be filled by one pipe in 4 hours and by a second pipe in 6 hours. How long will it take using both pipes to fill the pool?

16. A pool can be filled by one pipe in 3 hours and by a second pipe in 6 hours. How long will it take using both pipes to fill the pool?

In Exercises 17–22, use similar triangles and the fact that corresponding sides are proportional to find the length of the side marked with an x.

17.

18.

19.

20.

21.

22.

Use similar triangles to solve Exercises 23–24.

23. A tree casts a shadow 12 feet long. At the same time, a vertical rod 8 feet high casts a shadow 6 feet long. How tall is the tree?

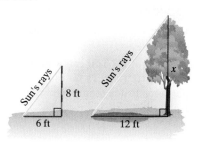

24. A person who is 5 feet tall is standing 80 feet from the base of a tree. The tree casts an 86-foot shadow. The person's shadow is 6 feet in length. What is the tree's height?

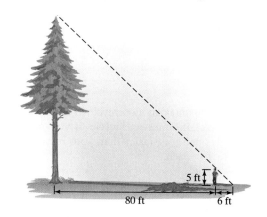

In Exercises 25–28, write an equation that expresses each relationship. Use k as the constant of variation.

25. g varies directly as h.

26. v varies directly as r.

27. w varies inversely as v.

28. a varies inversely as b.

In Exercises 29–32, determine the constant of variation for each stated condition.

29. y varies directly as x, and $y = 80$ when $x = 4$.

30. y varies directly as x, and $y = 108$ when $x = 12$.

31. W varies inversely as r, and $W = 600$ when $r = 10$.

32. T varies inversely as n, and $T = 4$ when $n = 24$.

Use the four-step procedure for solving variation problems given on page 534 to solve Exercises 33–36.

33. y varies directly as x. $y = 35$ when $x = 5$. Find y when $x = 12$.

34. y varies directly as x. $y = 55$ when $x = 5$. Find y when $x = 13$.

35. y varies inversely as x. $y = 10$ when $x = 5$. Find y when $x = 2$.

36. y varies inversely as x. $y = 5$ when $x = 3$. Find y when $x = 9$.

37. A person's fingernail growth, G, in inches, varies directly as the number of weeks it has been growing, W.

   a. Write an equation that expresses this relationship.

   b. Fingernails grow at a rate of about 0.02 inch per week. Substitute 0.02 for k, the constant of variation, in the equation in part (a) and write the equation for fingernail growth.

   c. Substitute 52 for W to determine your fingernail growth at the end of one year if for some bizarre reason you decided not to cut them and they did not break.

38. A person's salary, S, varies directly as the number of hours worked, h.

   a. Write an equation that expresses this relationship.

   b. For a 40-hour work week, Gloria earned $1400. Substitute 1400 for S and 40 for h in the equation from part (a) and find k, the constant of variation.

   c. Substitute the value of k into your equation in part (a) and write the equation that describes Gloria's salary in terms of the number of hours she works.

   d. Use the equation from part (c) to find Gloria's salary for 25 hours of work.

Use the four-step procedure for solving variation problems given on page 534 to solve Exercises 39–48.

39. The cost, C, of an airplane ticket varies directly as the number of miles, M, in the trip. A 3000-mile trip costs $400. What is the cost of a 450-mile trip?

40. An object's weight on the moon, M, varies directly as its weight on Earth, E. A person who weighs 55 kilograms on Earth weighs 8.8 kilograms on the moon. What is the moon weight of a person who weighs 90 kilograms on Earth?

41. The Mach number is a measurement of speed named after the man who suggested it, Ernst Mach (1838-1916). The speed of an aircraft varies directly as its Mach number. Shown here are two aircraft. Use the figures for the Concorde to determine the Blackbird's speed.

Concorde
Mach 2.03
Speed = 1502.2 miles per hour

SR-71 Blackbird
Mach 3.3
Speed = ?

42. A golf ball's bounce height, B, in inches, varies directly as its drop height, d, in inches. A golf ball bounces 36 inches when dropped from a height of 40 inches. What is the ball's bounce height if the drop height is increased to 50 inches?

43. The time that it takes to get to campus varies inversely as your driving rate. Averaging 20 miles per hour in terrible traffic, it takes you 1.5 hours to get to campus. How long would the trip take averaging 60 miles per hour?

44. The weight that can be supported by a 2-inch by 4-inch piece of pine (called a 2-by-4) varies inversely as its length. A 10-foot 2-by-4 can support 500 pounds. What weight can be supported by a 5-foot 2-by-4?

45. The volume of a gas in a container at a constant temperature varies inversely as the pressure. If the volume is 32 cubic centimeters at a pressure of 8 pounds per square centimeter, find the pressure when the volume is 40 cubic centimeters.

46. The current in a circuit varies inversely as the resistance. The current is 20 amperes when the resistance is 5 ohms. Find the current for a resistance of 16 ohms.

47. The number of pens sold varies inversely as the price per pen. If 4000 pens are sold at a price of $1.50 each, find the number of pens sold at a price of $1.20 each.

48. The time required to accomplish a task varies inversely as the number of people working on the task. It takes 6 hours for 20 people to put a new roof on a porch. How long would it take 30 people to do the job?

Writing in Mathematics

49. What is the relationship among time traveled, distance traveled, and rate of travel?

50. If you know how many hours it takes for you to do a job, explain how to find the fractional part of the job you can complete in x hours.

51. If you can do a job in 6 hours and your friend can do the same job in 3 hours, explain how to find how long it takes to complete the job working together. It is not necessary to solve the problem.

52. When two people work together to complete a job, describe one factor that can result in more or less time than the time given by the rational equations we have been using.

53. What are similar triangles?

54. If the ratio of the corresponding sides of two similar triangles is 1 to 1 $\left(\frac{1}{1}\right)$, what must be true about the triangles?

55. What are corresponding angles in similar triangles?

56. Describe how to identify the corresponding sides in similar triangles.

57. What does it mean if two quantities vary directly?

58. In your own words, explain how to solve a variation problem.

59. What does it mean if two quantities vary inversely?

60. Explain the meaning of this statement: A company's monthly sales vary directly as its advertising budget.

61. Explain the meaning of this statement: A company's monthly sales vary inversely as the price of its product.

Critical Thinking Exercises

Make Sense? *In Exercises 62–65, determine whether each statement "makes sense" or "does not make sense" and explain your reasoning.*

62. The time it takes me to drive to campus varies directly as my rate of travel.

63. It took me the same amount of time to travel 10 miles with the current as it did to travel 15 miles against the current.

64. I used $\dfrac{a}{d} = \dfrac{b}{e}$ to show that corresponding sides of similar triangles are proportional, but I could also use $\dfrac{a}{b} = \dfrac{d}{e}$ or $\dfrac{d}{a} = \dfrac{e}{b}$.

65. I can clean my house in 3 hours and my sloppy friend can completely mess it up in 6 hours, so if we both "work" together, the time, x, it takes to clean the house can be modeled by $\dfrac{x}{3} - \dfrac{x}{6} = 1$.

66. Two skiers begin skiing along a trail at the same time. The faster skier averages 9 miles per hour and the slower skier averages 6 miles per hour. The faster skier completes the trail $\frac{1}{4}$ hour before the slower skier. How long is the trail?

67. A snowstorm causes a bus driver to decrease the usual average rate along a 60-mile route by 15 miles per hour. As a result, the bus takes two hours longer than usual to complete the route. At what average rate does the bus usually cover the 60-mile route?

68. One pipe can fill a swimming pool in 2 hours, a second can fill the pool in 3 hours, and a third pipe can fill the pool in 4 hours. How many minutes, to the nearest minute, would it take to fill the pool with all three pipes operating?

69. Ben can prepare a company report in 3 hours. Shane can prepare a report in 4.2 hours. How long will it take them, working together, to prepare *four* company reports?

70. An experienced carpenter can panel a room 3 times faster than an apprentice can. Working together, they can panel the room in 6 hours. How long would it take each person working alone to do the job?

71. It normally takes 2 hours to fill a swimming pool. The pool has developed a slow leak. If the pool were full, it would take 10 hours for all the water to leak out. If the pool is empty, how long will it take to fill it?

72. The intensity of radiation from a machine used to treat tumors varies inversely as the square of the distance from the machine. If the intensity is 140.5 milliroentgens per hour at 2 meters, what is the intensity at a distance of 3 meters?

73. Two investments have interest rates that differ by 1%. An investment for 1 year at the lower rate earns $175. The same principal invested for a year at the higher rate earns $200. What are the two interest rates?

Review Exercises

74. Factor: $25x^2 - 81$. (Section 7.4, Example 1)

75. Solve: $x^2 - 12x + 36 = 0$. (Section 7.6, Example 4)

76. Graph: $y = -\dfrac{2}{3}x + 4$. (Section 4.4, Example 3)

Preview Exercises

Exercises 77–79 will help you prepare for the material covered in the first section of the next chapter.

77. Evaluate $\sqrt{x - 1}$ for $x = 17$.

78. Evaluate $4\sqrt{x} + 30$ for $x = 25$.

79. Simplify: $(-2)^5 - (-1)^3$.

GROUP PROJECT

CHAPTER 8

Group members make up the sales team for a company that makes computer video games. It has been determined that the formula

$$y = \frac{200x}{x^2 + 100}$$

models the monthly sales, y, in thousands of games, of a new video game x months after the game is introduced. The figure shows the graph of the formula. What are the team's recommendations to the company in terms of how long the video game should be on the market before another new video game is introduced? What other factors might members want to take into account in terms of the recommendations? What will eventually happen to sales, and how is this indicated by the graph? What could the company do to change the behavior of this model and continue generating sales? Would this be cost effective?

Monthly Sales of a New Video Game

$y = \dfrac{200x}{x^2 + 100}$

Monthly Sales (thousands)

Months after the Game Is Released

Chapter 8 Summary

| **Definitions and Concepts** | **Examples** |

Section 8.1 Rational Expressions and their Simplification

A rational expression is the quotient of two polynomials. To find values for which a rational expression is undefined, set the denominator equal to 0 and solve.

Find all numbers for which

$$\frac{7x}{x^2 - 3x - 4}$$

is undefined.

$$x^2 - 3x - 4 = 0$$
$$(x - 4)(x + 1) = 0$$
$$x - 4 = 0 \quad \text{or} \quad x + 1 = 0$$
$$x = 4 \qquad\qquad x = -1$$

The expression is undefined at 4 and -1.

To simplify a rational expression:

1. Factor the numerator and the denominator completely.
2. Divide the numerator and the denominator by any common factors.

If factors in the numerator and denominator are opposites, their quotient is -1.

Simplify: $\dfrac{3x + 18}{x^2 - 36}$.

$$\frac{3x + 18}{x^2 - 36} = \frac{3\overset{1}{\cancel{(x + 6)}}}{\underset{1}{\cancel{(x + 6)}}(x - 6)} = \frac{3}{x - 6}$$

Section 8.2 Multiplying and Dividing Rational Expressions

Multiplying Rational Expressions

1. Factor completely.
2. Divide numerators and denominators by common factors.
3. Multiply remaining factors in the numerators and multiply the remaining factors in the denominators.

$$\frac{x^2 + 3x - 10}{x^2 - 2x} \cdot \frac{x^2}{x^2 - 25}$$

$$= \frac{\overset{1}{\cancel{(x + 5)}}\overset{1}{\cancel{(x - 2)}}}{\underset{1}{\cancel{x}}\underset{1}{\cancel{(x - 2)}}} \cdot \frac{\overset{1}{\cancel{x}} \cdot x}{\underset{1}{\cancel{(x + 5)}}(x - 5)}$$

$$= \frac{x}{x - 5}$$

Dividing Rational Expressions

Invert the divisor and multiply.

$$\frac{3y + 3}{(y + 2)^2} \div \frac{y^2 - 1}{y + 2}$$

$$= \frac{3y + 3}{(y + 2)^2} \cdot \frac{y + 2}{y^2 - 1}$$

$$= \frac{3\overset{1}{\cancel{(y + 1)}}}{(y + 2)\underset{1}{\cancel{(y + 2)}}} \cdot \frac{\overset{1}{\cancel{(y + 2)}}}{\underset{1}{\cancel{(y + 1)}}(y - 1)}$$

$$= \frac{3}{(y + 2)(y - 1)}$$

Definitions and Concepts	Examples

Section 8.3 Adding and Subtracting Rational Expressions with the Same Denominator

To add or subtract rational expressions with the same denominator, add or subtract the numerators and place the result over the common denominator. If possible, factor and simplify the resulting expression.

$$\frac{y^2 - 3y + 4}{y^2 + 8y + 15} - \frac{y^2 - 5y - 2}{y^2 + 8y + 15}$$

$$= \frac{y^2 - 3y + 4 - (y^2 - 5y - 2)}{y^2 + 8y + 15}$$

$$= \frac{y^2 - 3y + 4 - y^2 + 5y + 2}{y^2 + 8y + 15}$$

$$= \frac{2y + 6}{(y + 5)(y + 3)}$$

$$= \frac{2\overset{1}{\cancel{(y + 3)}}}{(y + 5)\underset{1}{\cancel{(y + 3)}}} = \frac{2}{y + 5}$$

To add or subtract rational expressions with opposite denominators, multiply either rational expression by $\frac{-1}{-1}$ to obtain a common denominator.

$$\frac{7}{x - 6} + \frac{x + 4}{6 - x}$$

$$= \frac{7}{x - 6} + \frac{(-1)}{(-1)} \cdot \frac{x + 4}{6 - x}$$

$$= \frac{7}{x + 6} + \frac{-x - 4}{x - 6}$$

$$= \frac{7 - x - 4}{x - 6} = \frac{3 - x}{x - 6}$$

Section 8.4 Adding and Subtracting Rational Expressions with Different Denominators

Finding the Least Common Denominator (LCD)

1. Factor denominators completely.
2. List factors of the first denominator.
3. Add to the list factors of the second denominator that are not already in the list.
4. The LCD is the product of factors in step 3.

Find the LCD of

$$\frac{x + 1}{2x - 2} \quad \text{and} \quad \frac{2x}{x^2 + 2x - 3}.$$

$$2x - 2 = 2(x - 1)$$

$$x^2 + 2x - 3 = (x - 1)(x + 3)$$

Factors of first denominator: $2, x - 1$
Factors of second denominator not in the list: $x + 3$
LCD: $2(x - 1)(x + 3)$

Adding and Subtracting Rational Expressions with Different Denominators

1. Find the LCD.
2. Rewrite each rational expression as an equivalent expression with the LCD.
3. Add or subtract numerators, placing the resulting expression over the LCD.
4. If possible, simplify.

$$\frac{x + 1}{2x - 2} - \frac{2x}{x^2 + 2x - 3}$$

$$= \frac{x + 1}{2(x - 1)} - \frac{2x}{(x - 1)(x + 3)}$$

LCD is $2(x - 1)(x + 3)$.

$$= \frac{(x + 1)(x + 3)}{2(x - 1)(x + 3)} - \frac{2x \cdot 2}{2(x - 1)(x + 3)}$$

$$= \frac{x^2 + 4x + 3 - 4x}{2(x - 1)(x + 3)}$$

$$= \frac{x^2 + 3}{2(x - 1)(x + 3)}$$

| **Definitions and Concepts** | **Examples** |

Section 8.5 Complex Rational Expressions

Complex rational expressions have numerators or denominators containing one or more rational expressions. Complex rational expressions can be simplified by obtaining single expressions in the numerator and denominator and then dividing. They can also be simplified by multiplying the numerator and denominator by the LCD of all rational expressions within the complex rational expression.

Simplify by dividing: $\dfrac{\dfrac{1}{x} + 5}{\dfrac{1}{x} - \dfrac{1}{3}}$.

$$= \frac{\dfrac{1}{x} + \dfrac{5x}{x}}{\dfrac{3}{3x} - \dfrac{x}{3x}} = \frac{\dfrac{1 + 5x}{x}}{\dfrac{3 - x}{3x}} = \frac{1 + 5x}{\overset{1}{\cancel{x}}} \cdot \frac{\overset{1}{\cancel{3}x}}{3 - x}$$

$$= \frac{3(1 + 5x)}{3 - x} \quad \text{or} \quad \frac{3 + 15x}{3 - x}$$

Simplify by the LCD method: $\dfrac{\dfrac{1}{x} + 5}{\dfrac{1}{x} - \dfrac{1}{3}}$.

LCD is $3x$.

$$\frac{3x}{3x} \cdot \frac{\left(\dfrac{1}{x} + 5\right)}{\left(\dfrac{1}{x} - \dfrac{1}{3}\right)} = \frac{3x \cdot \dfrac{1}{x} + 3x \cdot 5}{3x \cdot \dfrac{1}{x} - 3x \cdot \dfrac{1}{3}}$$

$$= \frac{3 + 15x}{3 - x}$$

Section 8.6 Solving Rational Equations

A rational equation is an equation containing one or more rational expressions.

Solving Rational Equations

1. List restrictions on the variable.
2. Clear fractions by multiplying both sides by the LCD.
3. Solve the resulting equation.
4. Reject any proposed solution in the list of restrictions. Check other proposed solutions in the original equation.

Solve: $\dfrac{7x}{x^2 - 4} + \dfrac{5}{x - 2} = \dfrac{2x}{x^2 - 4}$

$$\frac{7x}{(x + 2)(x - 2)} + \frac{5}{x - 2} = \frac{2x}{(x + 2)(x - 2)}$$

Denominators would equal 0 if $x = -2$ or $x = 2$.
Restrictions: $x \neq -2$ and $x \neq 2$.

LCD is $(x + 2)(x - 2)$.

$$(x + 2)(x - 2)\left[\frac{7x}{(x + 2)(x - 2)} + \frac{5}{x - 2}\right]$$

$$= (x + 2)(x - 2) \cdot \frac{2x}{(x + 2)(x - 2)}$$

$$7x + 5(x + 2) = 2x$$
$$7x + 5x + 10 = 2x$$
$$12x + 10 = 2x$$
$$10 = -10x$$
$$-1 = x$$

The proposed solution, -1, is not part of the restriction $x \neq -2$ and $x \neq 2$. It checks. The solution is -1 and the solution set is $\{-1\}$.

Definitions and Concepts	**Examples**

Section 8.7 Applications Using Rational Equations and Variation

Motion problems involving time are solved using

$$t = \frac{d}{r}.$$

$$\text{Time traveled} = \frac{\text{Distance traveled}}{\text{Rate of travel}}$$

It takes a cyclist who averages 16 miles per hour in still air the same time to travel 48 miles with the wind as 16 miles against the wind. What is the wind's rate?

$$x = \text{wind's rate}$$
$$16 + x = \text{cyclist's rate with wind}$$
$$16 - x = \text{cyclist's rate against wind}$$

	Distance	Rate	Time = $\dfrac{\text{Distance}}{\text{Rate}}$
With wind	48	$16 + x$	$\dfrac{48}{16 + x}$
Against wind	16	$16 - x$	$\dfrac{16}{16 - x}$

Two times are equal.

$$\frac{48}{16 + x} = \frac{16}{16 - x}$$

$$(16 + x)(16 - x) \cdot \frac{48}{16 + x} = \frac{16}{16 - x} \cdot (16 + x)(16 - x)$$

$$48(16 - x) = 16(16 + x)$$

Solving this equation, $x = 8$.
The wind's rate is 8 miles per hour.

Work problems are solved using the following condition:

Fraction of job done by the first	+	fraction of job done by the second	=	1.

One pipe fills a pool in 20 hours and a second pipe in 15 hours. How long will it take to fill the pool using both pipes?

$$x = \text{time using both pipes}$$

Fraction of pool filled by pipe 1 in x hours	+	fraction of pool filled by pipe 2 in x hours	=	1.
$\dfrac{x}{20}$	+	$\dfrac{x}{15}$	=	1

$$60\left(\frac{x}{20} + \frac{x}{15}\right) = 60 \cdot 1$$

$$3x + 4x = 60$$

$$7x = 60$$

$$x = \frac{60}{7} = 8\frac{4}{7} \text{ hours}$$

It will take $8\frac{4}{7}$ hours for both pipes to fill the pool.

Definitions and Concepts	**Examples**

Similar triangles have the same shape, but not necessarily the same size. Corresponding angles have the same measure, and corresponding sides are proportional. If the measures of two angles of one triangle are equal to those of two angles of a second triangle, then the two triangles are similar.

Find x for these similar triangles.

Corresponding sides are proportional:

$$\frac{7}{x} = \frac{10}{5}. \quad \left(\text{or } \frac{7}{x} = \frac{14}{7}\right)$$

$$5x \cdot \frac{7}{x} = \frac{10}{5} \cdot 5x$$

$$35 = 10x$$

$$x = \frac{35}{10} = 3.5$$

Variation

y varies directly as x: $y = kx$.

y varies inversely as x: $y = \dfrac{k}{x}$. ← constant of variation

Solving a Variation Problem

1. Write the variation equation.
2. Substitute the pair of values and find k.
3. Substitute the value of k into the variation equation.
4. Use the variation equation with k to answer the problem's question.

The time that it takes you to drive a certain distance varies inversely as your driving rate. Averaging 30 miles per hour, it takes you 10 hours to drive the distance. How long would the trip take averaging 50 miles per hour?

1. $t = \dfrac{k}{r}$ ← Time, t, varies inversely as rate, r.

2. It takes 10 hours at 30 miles per hour.

$$10 = \frac{k}{30}$$

$$k = 10(30) = 300$$

3. $t = \dfrac{300}{r}$

4. How long at 50 miles per hour? Substitute 50 for r.

$$t = \frac{300}{50} = 6$$

It takes 6 hours at 50 miles per hour.

CHAPTER 8 REVIEW EXERCISES

8.1 *In Exercises 1–4, find all numbers for which each rational expression is undefined. If the rational expression is defined for all real numbers, so state.*

1. $\dfrac{5x}{6x - 24}$

2. $\dfrac{x + 3}{(x - 2)(x + 5)}$

3. $\dfrac{x^2 + 3}{x^2 - 3x + 2}$

4. $\dfrac{7}{x^2 + 81}$

In Exercises 5–12, simplify each rational expression. If the rational expression cannot be simplified, so state.

5. $\dfrac{16x^2}{12x}$

6. $\dfrac{x^2 - 4}{x - 2}$

7. $\dfrac{x^3 + 2x^2}{x + 2}$

8. $\dfrac{x^2 + 3x - 18}{x^2 - 36}$

9. $\dfrac{x^2 - 4x - 5}{x^2 + 8x + 7}$

10. $\dfrac{y^2 + 2y}{y^2 + 4y + 4}$

11. $\dfrac{x^2}{x^2 + 4}$

12. $\dfrac{2x^2 - 18y^2}{3y - x}$

8.2 *In Exercises 13–17, multiply as indicated.*

13. $\dfrac{x^2 - 4}{12x} \cdot \dfrac{3x}{x + 2}$

14. $\dfrac{5x + 5}{6} \cdot \dfrac{3x}{x^2 + x}$

15. $\dfrac{x^2 + 6x + 9}{x^2 - 4} \cdot \dfrac{x - 2}{x + 3}$

16. $\dfrac{y^2 - 2y + 1}{y^2 - 1} \cdot \dfrac{2y^2 + y - 1}{5y - 5}$

17. $\dfrac{2y^2 + y - 3}{4y^2 - 9} \cdot \dfrac{3y + 3}{5y - 5y^2}$

In Exercises 18–22, divide as indicated.

18. $\dfrac{x^2 + x - 2}{10} \div \dfrac{2x + 4}{5}$

19. $\dfrac{6x + 2}{x^2 - 1} \div \dfrac{3x^2 + x}{x - 1}$

20. $\dfrac{1}{y^2 + 8y + 15} \div \dfrac{7}{y + 5}$

21. $\dfrac{y^2 + y - 42}{y - 3} \div \dfrac{y + 7}{(y - 3)^2}$

22. $\dfrac{8x + 8y}{x^2} \div \dfrac{x^2 - y^2}{x^2}$

8.3 *In Exercises 23–28, add or subtract as indicated. Simplify the result, if possible.*

23. $\dfrac{4x}{x + 5} + \dfrac{20}{x + 5}$

24. $\dfrac{8x - 5}{3x - 1} + \dfrac{4x + 1}{3x - 1}$

25. $\dfrac{3x^2 + 2x}{x - 1} - \dfrac{10x - 5}{x - 1}$

26. $\dfrac{6y^2 - 4y}{2y - 3} - \dfrac{12 - 3y}{2y - 3}$

27. $\dfrac{x}{x - 2} + \dfrac{x - 4}{2 - x}$

28. $\dfrac{x + 5}{x - 3} - \dfrac{x}{3 - x}$

8.4 *In Exercises 29–31, find the least common denominator of the rational expressions.*

29. $\dfrac{7}{9x^3}$ and $\dfrac{5}{12x}$

30. $\dfrac{3}{x^2(x - 1)}$ and $\dfrac{11}{x(x - 1)^2}$

31. $\dfrac{x}{x^2 + 4x + 3}$ and $\dfrac{17}{x^2 + 10x + 21}$

In Exercises 32–42, add or subtract as indicated. Simplify the result, if possible.

32. $\dfrac{7}{3x} + \dfrac{5}{2x^2}$

33. $\dfrac{5}{x + 1} + \dfrac{2}{x}$

34. $\dfrac{7}{x + 3} + \dfrac{4}{(x + 3)^2}$

35. $\dfrac{6y}{y^2 - 4} - \dfrac{3}{y + 2}$

36. $\dfrac{y - 1}{y^2 - 2y + 1} - \dfrac{y + 1}{y - 1}$

37. $\dfrac{x + y}{y} - \dfrac{x - y}{x}$

38. $\dfrac{2x}{x^2 + 2x + 1} + \dfrac{x}{x^2 - 1}$

39. $\dfrac{5x}{x + 1} - \dfrac{2x}{1 - x^2}$

40. $\dfrac{4}{x^2 - x - 6} - \dfrac{4}{x^2 - 4}$

41. $\dfrac{7}{x + 3} + 2$

42. $\dfrac{2y - 5}{6y + 9} - \dfrac{4}{2y^2 + 3y}$

8.5 *In Exercises 43–47, simplify each complex rational expression.*

43. $\dfrac{\dfrac{1}{2} + \dfrac{3}{8}}{\dfrac{3}{4} - \dfrac{1}{2}}$

44. $\dfrac{\dfrac{1}{x}}{1 - \dfrac{1}{x}}$

45. $\dfrac{\dfrac{1}{x} + \dfrac{1}{y}}{\dfrac{1}{xy}}$

46. $\dfrac{\dfrac{1}{x} - \dfrac{1}{2}}{\dfrac{1}{3} - \dfrac{x}{6}}$

47. $\dfrac{3 + \dfrac{12}{x}}{1 - \dfrac{16}{x^2}}$

8.6 *In Exercises 48–55, solve each rational equation.*

48. $\dfrac{3}{x} - \dfrac{1}{6} = \dfrac{1}{x}$

49. $\dfrac{3}{4x} = \dfrac{1}{x} + \dfrac{1}{4}$

50. $x + 5 = \dfrac{6}{x}$

51. $4 - \dfrac{x}{x+5} = \dfrac{5}{x+5}$

52. $\dfrac{2}{x-3} = \dfrac{4}{x+3} + \dfrac{8}{x^2-9}$

53. $\dfrac{2}{x} = \dfrac{2}{3} + \dfrac{x}{6}$

54. $\dfrac{13}{y-1} - 3 = \dfrac{1}{y-1}$

55. $\dfrac{1}{x+3} - \dfrac{1}{x-1} = \dfrac{x+1}{x^2+2x-3}$

56. Park rangers introduce 50 elk into a wildlife preserve. The formula

$$P = \frac{250(3t + 5)}{t + 25}$$

models the elk population, P, after t years. How many years will it take for the population to increase to 125 elk?

57. The formula

$$S = \frac{C}{1 - r}$$

describes the selling price, S, of a product in terms of its cost to the retailer, C, and its markup, r, usually expressed as a percent. A small television cost a retailer \$140 and was sold for \$200. Find the markup. Express the answer as a percent.

8.7

58. In still water, a paddle boat averages 20 miles per hour. It takes the boat the same amount of time to travel 72 miles downstream, with the current, as 48 miles upstream, against the current. What is the rate of the water's current?

59. A car travels 60 miles in the same time that a car traveling 10 miles per hour faster travels 90 miles. What is the rate of each car?

60. A painter can paint a fence around a house in 6 hours. Working alone, the painter's apprentice can paint the same fence in 12 hours. How many hours would it take them to do the job if they worked together?

61. The triangles shown in the figure are similar. Find the length of the side marked with an x.

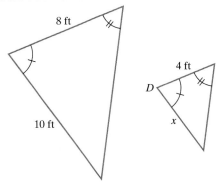

62. Find the height of the lamppost in the figure.

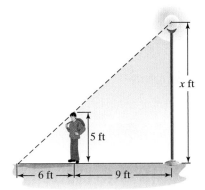

63. An electric bill varies directly as the amount of electricity used. The bill for 1400 kilowatts of electricity is \$98. What is the bill for 2200 kilowatts of electricity?

64. The time it takes to drive a certain distance varies inversely as the rate of travel. If it takes 4 hours at 50 miles per hour to drive the distance, how long will it take at 40 miles per hour?

CHAPTER 8 TEST

Remember to use your Chapter Test Prep Video CD to see the worked-out solutions to the test questions you want to review.

1. Find all numbers for which

$$\frac{x + 7}{x^2 + 5x - 36}$$

is undefined.

In Exercises 2–3, simplify each rational expression.

2. $\dfrac{x^2 + 2x - 3}{x^2 - 3x + 2}$

3. $\dfrac{4x^2 - 20x}{x^2 - 4x - 5}$

In Exercises 4–16, perform the indicated operations. Simplify the result, if possible.

4. $\dfrac{x^2 - 16}{10} \cdot \dfrac{5}{x + 4}$

5. $\dfrac{x^2 - 7x + 12}{x^2 - 4x} \cdot \dfrac{x^2}{x^2 - 9}$

6. $\dfrac{2x + 8}{x - 3} \div \dfrac{x^2 + 5x + 4}{x^2 - 9}$

7. $\dfrac{5y + 5}{(y - 3)^2} \div \dfrac{y^2 - 1}{y - 3}$

8. $\dfrac{2y^2 + 5}{y + 3} + \dfrac{6y - 5}{y + 3}$

9. $\dfrac{y^2 - 2y + 3}{y^2 + 7y + 12} - \dfrac{y^2 - 4y - 5}{y^2 + 7y + 12}$

10. $\dfrac{x}{x + 3} + \dfrac{5}{x - 3}$

11. $\dfrac{2}{x^2 - 4x + 3} + \dfrac{6}{x^2 + x - 2}$

12. $\dfrac{4}{x - 3} + \dfrac{x + 5}{3 - x}$

13. $1 + \dfrac{3}{x - 1}$

14. $\dfrac{2x + 3}{x^2 - 7x + 12} - \dfrac{2}{x - 3}$

15. $\dfrac{8y}{y^2 - 16} - \dfrac{4}{y - 4}$

16. $\dfrac{(x - y)^2}{x + y} \div \dfrac{x^2 - xy}{3x + 3y}$

In Exercises 17–18, simplify each complex rational expression.

17. $\dfrac{5 + \dfrac{5}{x}}{2 + \dfrac{1}{x}}$

18. $\dfrac{\dfrac{1}{x} - \dfrac{1}{y}}{\dfrac{1}{x}}$

In Exercises 19–21, solve each rational equation.

19. $\dfrac{5}{x} + \dfrac{2}{3} = 2 - \dfrac{2}{x} - \dfrac{1}{6}$

20. $\dfrac{3}{y + 5} - 1 = \dfrac{4 - y}{2y + 10}$

21. $\dfrac{2}{x - 1} = \dfrac{3}{x^2 - 1} + 1$

22. In still water, a boat averages 30 miles per hour. It takes the boat the same amount of time to travel 16 miles downstream, with the current, as 14 miles upstream, against the current. What is the rate of the water's current?

23. One pipe can fill a hot tub in 20 minutes and a second pipe can fill it in 30 minutes. If the hot tub is empty, how long will it take both pipes to fill it?

24. The triangles in the figure are similar. Find the length of the side marked with an x.

25. The amount of current flowing in an electrical circuit varies inversely as the resistance in the circuit. When the resistance in a particular circuit is 5 ohms, the current is 42 amperes. What is the current when the resistance is 4 ohms?

CUMULATIVE REVIEW EXERCISES (CHAPTERS 1–8)

In Exercises 1–6, solve each equation, inequality, or system of equations.

1. $2(x - 3) + 5x = 8(x - 1)$

2. $-3(2x - 4) > 2(6x - 12)$

3. $x^2 + 3x = 18$

4. $\dfrac{2x}{x^2 - 4} + \dfrac{1}{x - 2} = \dfrac{2}{x + 2}$

5. $y = 2x - 3$
    $x + 2y = 9$

6. $\ \ \ 3x + 2y = -2$
    $-4x + 5y = 18$

In Exercises 7–9, graph each equation or inequality in a rectangular coordinate system.

7. $3x - 2y = 6$ **8.** $y > -2x + 3$

9. $y = -3$

In Exercises 10–12, simplify each expression.

10. $-21 - 16 - 3(2 - 8)$

11. $\left(\dfrac{4x^5}{2x^2}\right)^3$

12. $\dfrac{\dfrac{1}{x} - 2}{4 - \dfrac{1}{x}}$

In Exercises 13–15, factor completely.

13. $4x^2 - 13x + 3$

14. $4x^2 - 20x + 25$

15. $3x^2 - 75$

In Exercises 16–18, perform the indicated operations.

16. $(4x^2 - 3x + 2) - (5x^2 - 7x - 6)$

17. $\dfrac{-8x^6 + 12x^4 - 4x^2}{4x^2}$

18. $\dfrac{x + 6}{x - 2} + \dfrac{2x + 1}{x + 3}$

19. You invested $4000, part at 5% and the remainder at 9% annual interest. At the end of the year, the total interest from these investments was $311. How much was invested at each rate?

20. A 68-inch board is to be cut into two pieces. If one piece must be three times as long as the other, find the length of each piece.

Roots and Radicals

W e are bombarded with advertising based on the philosophy of P.T. Barnum (1810–1881), cofounder of the circus called "The Greatest Show on Earth." In Barnum's words:

There is a sucker born every minute.
Each time the second hand sweeps to the top,
Like dandelions up they pop,
Their ears so big, their eyes so wide.
And though I feed 'em bonafide baloney,
With no truth in it,
Why you can bet I'll find some rube to buy my corn.
'Cause there's a sure-as-shootin' sucker born a minute,
And I'm referrin' to the minute you was born.

A few advertising factoids for us "corn-buying rubes:"

• Top ad-spending categories: cars, drugs, and movies (*Source: Advertising Age*) • Number of commercials watched by the average American child each year: 20,000 (*Source: Broadcasting and Cable*) • Number of brands recognized by the average American 3-year-old: 100 (*Source: Albany Times Union*) • Percentage of e-mail cluttered by junk advertising: 72% (*Source:* IronPort) • Portion of every prime-time cable TV hour devoted to advertising: 14.6 minutes (*Source:* Nielsen Monitor-Plus)

What does algebra have to say about advertising? In particular, what sorts of mathematical models can be used to describe spam as a percentage of e-mail or the number of minutes of commercial clutter on prime-time television? In this chapter, you will see how models with square roots describe these advertising phenomena that are continuing to grow, but whose growth is leveling off. By learning about roots and radicals, you will have new algebraic tools for describing your world.

You'll model spam in Exercise 91 of Exercise Set 9.1 and work with a model describing advertising clutter on cable TV in Review Exercises 9 and 10.

9.1

Objectives

1 Find square roots.

2 Evaluate models containing square roots.

3 Use a calculator to find decimal approximations for irrational square roots.

4 Find higher roots.

Finding Roots

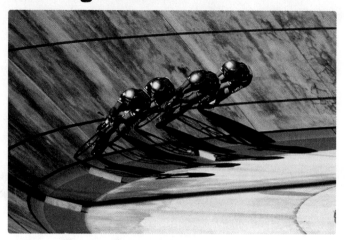

What is the maximum speed at which a racing cyclist can turn a corner without tipping over? The answer, in miles per hour, is given by the algebraic expression $4\sqrt{x}$, where x is the radius of the corner, in feet. Algebraic expressions containing roots describe phenomena as diverse as a wild animal's territorial area, evaporation on a lake's surface, and Albert Einstein's bizarre concept of how an astronaut moving close to the speed of light would barely age relative to friends watching from Earth. No description of your world can be complete without roots and radicals. In this section, we develop the basics of radical expressions through a notation that takes us from a number raised to a power back to the number itself.

1 Find square roots.

Square Roots

From our earlier work with exponents, we are aware that the square of both 5 and -5 is 25:

$$5^2 = 25 \quad \text{and} \quad (-5)^2 = 25.$$

The reverse operation of squaring a number is finding the *square root* of the number. For example,

- A square root of 25 is 5 because $5^2 = 25$.
- A square root of 25 is also -5 because $(-5)^2 = 25$.

In general, **if $b^2 = a$, then b is a square root of a.**

The symbol $\sqrt{}$ is used to denote the nonnegative or *principal square root* of a number. For example,

- $\sqrt{25} = 5$ because $5^2 = 25$ and 5 is positive.
- $\sqrt{100} = 10$ because $10^2 = 100$ and 10 is positive.

The symbol $\sqrt{}$ that we use to denote the principal square root is called a **radical sign**. The number under the radical sign is called the **radicand**. Together we refer to the radical sign and its radicand as a **radical**.

Radical sign $\sqrt{a}$ Radicand

Radical

Definition of the Principal Square Root

If a is a nonnegative real number, the nonnegative number b such that $b^2 = a$, denoted by $b = \sqrt{a}$, is the **principal square root** of a.

The symbol $-\sqrt{}$ is used to denote the negative square root of a number. For example,

- $-\sqrt{25} = -5$ because $(-5)^2 = 25$ and -5 is negative.
- $-\sqrt{100} = -10$ because $(-10)^2 = 100$ and -10 is negative.

EXAMPLE 1 Evaluating Expressions Containing Radicals

Evaluate:

a. $\sqrt{64}$ b. $-\sqrt{49}$ c. $\sqrt{\dfrac{1}{4}}$ d. $\sqrt{9 + 16}$ e. $\sqrt{9} + \sqrt{16}$.

Solution

Study Tip

In Example 1, parts (d) and (e), observe that $\sqrt{9 + 16}$ is not equal to $\sqrt{9} + \sqrt{16}$. In general,

$$\sqrt{a + b} \neq \sqrt{a} + \sqrt{b}$$

and

$$\sqrt{a - b} \neq \sqrt{a} - \sqrt{b}.$$

a. $\sqrt{64} = 8$
The principal square root of 64 is 8. Check: $8^2 = 64$.

b. $-\sqrt{49} = -7$
The negative square root of 49 is -7. Check: $(-7)^2 = 49$.

c. $\sqrt{\dfrac{1}{4}} = \dfrac{1}{2}$
The principal square root of $\frac{1}{4}$ is $\frac{1}{2}$. Check: $\left(\frac{1}{2}\right)^2 = \frac{1}{4}$.

d. $\sqrt{9 + 16} = \sqrt{25}$
$\qquad\quad\ = 5$
First simplify the expression under the radical sign. Then take the principal square root of 25, which is 5.

e. $\sqrt{9} + \sqrt{16} = 3 + 4$
$\qquad\qquad\quad\ = 7$
$\sqrt{9} = 3$ because $3^2 = 9$. $\sqrt{16} = 4$ because $4^2 = 16$.

✓ CHECK POINT 1 Evaluate:

a. $\sqrt{81}$ b. $-\sqrt{9}$ c. $\sqrt{\dfrac{1}{25}}$

d. $\sqrt{36 + 64}$ e. $\sqrt{36} + \sqrt{64}$.

---

2 Evaluate models containing square roots.

EXAMPLE 2 An Application: A Mathematical Model Containing a Radical

The amount of evaporation, in inches per day, from the surface of a large body of water can be modeled by the formula

$$E = \frac{w}{20\sqrt{a}}$$

where a = area of the water, in square miles
$\qquad\quad\ w$ = average wind speed of the air over the water, in miles per hour
$\qquad\quad\ E$ = surface evaporation, in inches per day.

Determine the evaporation from the surface of a lake whose area is 9 square miles on a day when the wind speed over the water is 10 miles per hour.

Study Tip

Some square roots occur so frequently that you may want to memorize them.

$\sqrt{1} = 1$	$\sqrt{4} = 2$
$\sqrt{9} = 3$	$\sqrt{16} = 4$
$\sqrt{25} = 5$	$\sqrt{36} = 6$
$\sqrt{49} = 7$	$\sqrt{64} = 8$
$\sqrt{81} = 9$	$\sqrt{100} = 10$
$\sqrt{121} = 11$	$\sqrt{144} = 12$

Solution

$$E = \frac{w}{20\sqrt{a}}$$ This is the given formula.

$$E = \frac{10}{20\sqrt{9}}$$ Substitute the given values: w (wind speed) = 10 miles per hour and a (area) = 9 square miles.

$$E = \frac{10}{20 \cdot 3}$$ $\sqrt{9} = 3$ because $3^2 = 9$.

$$= \frac{1}{6}$$ $\dfrac{\overset{1}{\cancel{10}}}{\underset{2}{\cancel{20}} \cdot 3} = \dfrac{1}{6}$

The surface evaporation is $\frac{1}{6}$ of an inch on that day.   ▬

☑ **CHECK POINT 2**   Use the model in Example 2 to solve this problem. Determine the evaporation from the surface of a lake whose area is 4 square miles on a day when the wind speed is 20 miles per hour.

Modeling Data with Square Roots

Many real-world phenomena start with rapid growth, and then the growth begins to level off. This type of behavior can be modeled by equations involving square roots. For example, the bar graph in **Figure 9.1(a)** displays the percentage of Americans 25 and older who had completed at least four years of high school in each of six selected years. The data are displayed as a set of six points in the scatter plot in **Figure 9.1(b)**.

Percentage of Americans 25 and Older Completing at Least Four Years of High School

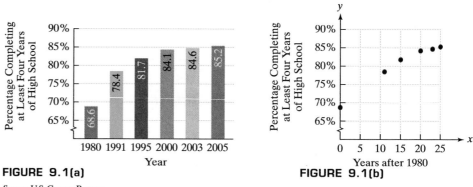

FIGURE 9.1(a)

FIGURE 9.1(b)

Source: U.S. Census Bureau

Can you see that the percentage completing at least four years of high school is slowing down for the period shown? Let's see why square roots can be used to model the data by looking at the shape of the graph of $y = \sqrt{x}$.

We graph $y = \sqrt{x}$ by first selecting integers for x. It is easiest to choose **perfect squares**, numbers that are squares of integers. **Table 9.1** shows five ordered pairs that are solutions of $y = \sqrt{x}$. We plot these ordered pairs as points in the rectangular coordinate system and connect the points with a smooth curve. The graph of $y = \sqrt{x}$ is shown in **Figure 9.2**.

Table 9.1

x	$y = \sqrt{x}$	(x, y)
0	$y = \sqrt{0} = 0$	$(0, 0)$
1	$y = \sqrt{1} = 1$	$(1, 1)$
4	$y = \sqrt{4} = 2$	$(4, 2)$
9	$y = \sqrt{9} = 3$	$(9, 3)$
16	$y = \sqrt{16} = 4$	$(16, 4)$

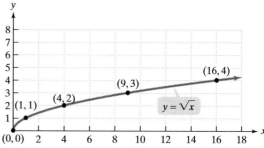

FIGURE 9.2   The graph of $y = \sqrt{x}$

The graph in **Figure 9.2** is increasing from left to right. However, the rate of increase is slowing down as the graph moves to the right. This is why equations with square roots are often used to model growing phenomena with growth that is leveling off.

3 Use a calculator to find decimal approximations for irrational square roots.

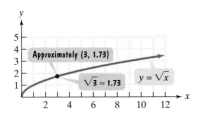

FIGURE 9.3 Visualizing $\sqrt{3} \approx 1.73$

Is it possible to choose values of x in **Table 9.1** that are not perfect squares? Yes. For example, we can let $x = 3$. Thus, $y = \sqrt{x} = \sqrt{3}$. Because 3 is not a perfect square, $\sqrt{3}$ is an irrational number, one that cannot be expressed as a quotient of integers. We can use a calculator to find a decimal approximation for $\sqrt{3}$.

Most Scientific Calculators	Most Graphing Calculators
3	3 ENTER

Rounding the displayed number to two decimal places, $\sqrt{3} \approx 1.73$. This information is shown visually as a point, approximately $(3, 1.73)$, on the graph of $y = \sqrt{x}$ in **Figure 9.3**.

EXAMPLE 3   Modeling Educational Attainment with a Square Root

The mathematical model

$$P = 3.32\sqrt{t} + 68.6$$

describes the percentage, P, of Americans ages 25 and over who had completed at least four years of high school t years after 1980.

a. Use the formula to find the percentage, to the nearest tenth of a percent, who had completed at least four years of high school in 2003.

b. Does the percentage obtained from the model underestimate or overestimate the actual percentage displayed in **Figure 9.1(a)**? By how much?

Solution

a. Because 2003 is 23 years after 1980, we substitute 23 for t in the given formula. Then we use a calculator to find P, the percentage who had completed at least four years of high school.

$P = 3.32\sqrt{t} + 68.6$	Use the given formula.
$P = 3.32\sqrt{23} + 68.6$	Substitute 23 for t.
≈ 84.5	Use a calculator. See the Using Technology box in the margin.

Rounding to the nearest tenth of a percent, approximately 84.5% of Americans ages 25 and over had completed at least four years of high school in 2003.

b. **Figure 9.1(a)** shows that 84.6% of Americans ages 25 and over had completed at least four years of high school in 2003. The percentage obtained from the model, 84.5%, underestimates this value by 84.6% − 84.5%, or by 0.1%. ◼

Using Technology

You can use your calculator to approximate

$$3.32\sqrt{23} + 68.6$$

as follows:

Scientific Calculator

3.32 ⊠ 23 √ + 68.6 =

Graphing Calculator

3.32 ⊠ √ 23 + 68.6 ENTER

Some calculators display an open parenthesis after √ is pressed. In this case, use a closed parenthesis after entering the radicand.

✓ **CHECK POINT 3**   Use the mathematical model in Example 3 to find the percentage, to the nearest tenth of a percent, who had completed at least four years of high school in 1995. Does this underestimate or overestimate the actual percentage displayed in **Figure 9.1(a)**? By how much?

Did you notice that in **Figure 9.2** we graphed $y = \sqrt{x}$ in quadrant I, where $x \geq 0$? What happens if x is negative? Is the square root of a negative number a real number? For example, consider $\sqrt{-25}$. Is there a real number whose square is -25? No. Thus, $\sqrt{-25}$ is not a real number. In general, **a square root of a negative number is not a real number.**

4 Find higher roots.

Higher Roots

Finding the square root of a number reverses the process of squaring a number. Similarly, the cube root of a number reverses the process of cubing a number. For example, $2^3 = 8$, and so the cube root of 8 is 2. The notation that we use is $\sqrt[3]{8} = 2$.

We define the **principal nth root** of a real number a, symbolized by $\sqrt[n]{a}$, as follows:

> ### Definition of the Principal nth Root of a Real Number
>
> $$\sqrt[n]{a} = b \text{ means that } b^n = a.$$
>
> The natural number n is called the **index**. The index for square roots, 2, is usually omitted.

For example,

$$\sqrt[3]{64} = 4 \text{ because } 4^3 = 64 \text{ and } \sqrt[5]{-32} = -2 \text{ because}(-2)^5 = -32.$$

The same vocabulary that we learned for square roots applies to nth roots. The symbol $\sqrt[n]{a}$ is called a **radical** and a is called the **radicand**.

Table 9.2 shows how various roots reverse raising numbers to powers.

Table 9.2	Reversing nth Powers with nth Roots		
	Powers	**Roots**	**Vocabulary**
Cube Roots	$4^3 = 64$	$\sqrt[3]{64} = 4$	3 is the index of the radical.
	$(-2)^3 = -8$	$\sqrt[3]{-8} = -2$	
	$5^3 = 125$	$\sqrt[3]{125} = 5$	
Fourth Roots	$1^4 = 1$	$\sqrt[4]{1} = 1$	Index = 4
	$3^4 = 81$	$\sqrt[4]{81} = 3$	
Fifth Roots	$2^5 = 32$	$\sqrt[5]{32} = 2$	Index = 5
nth Roots	$b^n = a$	$\sqrt[n]{a} = b$	Index = n

One of the entries in the table involves a negative radicand: $\sqrt[3]{-8} = -2$. By contrast to the square root of a negative number, the cube root of a negative number is a real number. In general, **if the index is even**, such as $\sqrt{}, \sqrt[4]{}, \sqrt[6]{}$, and so on, **the radicand must be nonnegative for the root to be a real number.** For example,

$$\sqrt[4]{81} = 3, \text{ but } \sqrt[4]{-81} \text{ is not a real number.}$$
$$\sqrt[6]{64} = 2, \text{ but } \sqrt[6]{-64} \text{ is not a real number.}$$

Furthermore, if the index is even, the principal nth root is nonnegative.

EXAMPLE 4   Finding Cube Roots

Find the cube roots:

a. $\sqrt[3]{27}$ b. $\sqrt[3]{-1}$ c. $\sqrt[3]{\dfrac{1}{8}}$.

Solution

a. $\sqrt[3]{27} = 3$ because $3^3 = 27$.

b. $\sqrt[3]{-1} = -1$ because $(-1)^3 = -1$.

c. $\sqrt[3]{\dfrac{1}{8}} = \dfrac{1}{2}$ because $\left(\dfrac{1}{2}\right)^3 = \dfrac{1}{8}$.

✓ **CHECK POINT 4**   Find the cube roots:

a. $\sqrt[3]{1}$ b. $\sqrt[3]{-27}$ c. $\sqrt[3]{\dfrac{1}{125}}$.

Study Tip

Some higher roots occur so frequently that you may want to memorize them.

Cube Roots	Fourth Roots
$\sqrt[3]{1} = 1$	$\sqrt[4]{1} = 1$
$\sqrt[3]{8} = 2$	$\sqrt[4]{16} = 2$
$\sqrt[3]{27} = 3$	$\sqrt[4]{81} = 3$
$\sqrt[3]{64} = 4$	$\sqrt[4]{256} = 4$
$\sqrt[3]{125} = 5$	$\sqrt[4]{625} = 5$
$\sqrt[3]{216} = 6$	

Fifth Roots
$\sqrt[5]{1} = 1$
$\sqrt[5]{32} = 2$
$\sqrt[5]{243} = 3$

EXAMPLE 5   Finding Higher Roots

Find the indicated root, or state that the expression is not a real number:

a. $\sqrt[4]{16}$ **b.** $\sqrt[4]{-16}$ **c.** $-\sqrt[4]{16}$ **d.** $\sqrt[5]{-32}$.

Solution

a. $\sqrt[4]{16} = 2$ because $2^4 = 16$.

b. $\sqrt[4]{-16}$ is not a real number because the index, 4, is even and the radicand, -16, is negative. No real number equals $\sqrt[4]{-16}$ because any real number raised to the fourth power gives a nonnegative result.

c. $-\sqrt[4]{16} = -2$

> Copy the negative sign and use the fact that $\sqrt[4]{16} = 2$.

d. $\sqrt[5]{-32} = -2$ because $(-2)^5 = -32$.

✓ **CHECK POINT 5**   Find the indicated root, or state that the expression is not a real number:

a. $\sqrt[4]{81}$ **b.** $\sqrt[4]{-81}$

c. $-\sqrt[4]{81}$ **d.** $\sqrt[5]{-\dfrac{1}{32}}$.

9.1 EXERCISE SET   *MyMathLab*   PRACTICE WATCH DOWNLOAD READ REVIEW

Practice Exercises

In Exercises 1–26, evaluate each expression, or state that the expression is not a real number.

1. $\sqrt{36}$

2. $\sqrt{16}$

3. $-\sqrt{36}$

4. $-\sqrt{16}$

5. $\sqrt{-36}$

6. $\sqrt{-16}$

7. $\sqrt{\dfrac{1}{9}}$

8. $\sqrt{\dfrac{1}{49}}$

9. $\sqrt{\dfrac{1}{100}}$

10. $\sqrt{\dfrac{1}{81}}$

11. $-\sqrt{\dfrac{1}{36}}$

12. $-\sqrt{\dfrac{1}{121}}$

13. $\sqrt{-\dfrac{1}{36}}$

14. $\sqrt{-\dfrac{1}{121}}$

15. $\sqrt{0.04}$

16. $\sqrt{0.64}$

17. $\sqrt{33 - 8}$

18. $\sqrt{51 + 13}$

19. $\sqrt{2 \cdot 32}$

20. $\sqrt{\dfrac{75}{3}}$

21. $\sqrt{144 + 25}$

22. $\sqrt{25 - 16}$

23. $\sqrt{144} + \sqrt{25}$

24. $\sqrt{25} - \sqrt{16}$

25. $\sqrt{25 - 144}$

26. $\sqrt{16 - 25}$

In Exercises 27–28, graph each equation. Begin by filling in the table and finding five solutions of the equation. Then plot these ordered pairs as points in the rectangular coordinate system and connect the points with a smooth curve.

27. $y = \sqrt{x - 1}$

> Because the radicand cannot be negative, the graph begins at the point with this x-coordinate.

x	$y = \sqrt{x - 1}$	(x, y)
1		
2		
5		
10		
17		

28. $y = \sqrt{x + 2}$

> Because the radicand cannot be negative, the graph begins at the point with this x-coordinate.

x	$y = \sqrt{x + 2}$	(x, y)
-2		
-1		
2		
7		
14		

29. Describe one similarity and one difference between your graph in Exercise 27 and the graph of $y = \sqrt{x}$, shown in **Figure 9.2** on page 552.

30. Describe one similarity and one difference between your graph in Exercise 28 and the graph of $y = \sqrt{x}$, shown in **Figure 9.2** on page 552.

In Exercises 31–36, use a calculator to approximate each square root. Round to three decimal places.

31. $\sqrt{7}$ **32.** $\sqrt{11}$

33. $\sqrt{23}$ **34.** $\sqrt{97}$

35. $-\sqrt{65}$ **36.** $-\sqrt{83}$

In Exercises 37–46, use a calculator to approximate each expression. Round to three decimal places. If the expression is not a real number and an approximation is not possible, so state.

37. $12 + \sqrt{11}$ **38.** $14 + \sqrt{13}$

39. $\dfrac{12 + \sqrt{11}}{2}$ **40.** $\dfrac{14 + \sqrt{13}}{2}$

41. $\dfrac{-5 + \sqrt{321}}{6}$ **42.** $\dfrac{-7 + \sqrt{839}}{5}$

43. $\sqrt{13 - 5}$ **44.** $\sqrt{21 - 4}$

45. $\sqrt{5 - 13}$

46. $\sqrt{4 - 21}$

In Exercises 47–56, find each cube root.

47. $\sqrt[3]{64}$ **48.** $\sqrt[3]{27}$

49. $\sqrt[3]{-27}$ **50.** $\sqrt[3]{-64}$

51. $-\sqrt[3]{8}$ **52.** $-\sqrt[3]{27}$

53. $\sqrt[3]{\dfrac{1}{125}}$ **54.** $\sqrt[3]{\dfrac{1}{1000}}$

55. $\sqrt[3]{-1000}$ **56.** $\sqrt[3]{-125}$

In Exercises 57–74, find the indicated root, or state that the expression is not a real number.

57. $\sqrt[4]{1}$ **58.** $\sqrt[5]{1}$

59. $\sqrt[4]{16}$ **60.** $\sqrt[4]{81}$

61. $-\sqrt[4]{16}$ **62.** $-\sqrt[4]{81}$

63. $\sqrt[4]{-16}$ **64.** $\sqrt[4]{-81}$

65. $\sqrt[5]{-1}$ **66.** $\sqrt[7]{-1}$

67. $\sqrt[6]{-1}$ **68.** $\sqrt[8]{-1}$

69. $-\sqrt[4]{256}$ **70.** $-\sqrt[4]{10,000}$

71. $\sqrt[6]{64}$ **72.** $\sqrt[5]{32}$

73. $-\sqrt[5]{32}$ **74.** $-\sqrt[6]{64}$

Practice PLUS

In Exercises 75–84, for which values of x is each radical expression a real number? Express your answer as an inequality or write "all real numbers."

75. $\sqrt{2x}$ **76.** $\sqrt{3x}$

77. $\sqrt{x - 2}$ **78.** $\sqrt{x - 3}$

79. $\sqrt{2 - x}$ **80.** $\sqrt{3 - x}$

81. $\sqrt{x^2 + 2}$ **82.** $\sqrt{x^2 + 3}$

83. $\sqrt{12 - 2x}$ **84.** $\sqrt{8 - 2x}$

Application Exercises

85. Racing cyclists use the formula $v = 4\sqrt{r}$ to determine the maximum velocity, v, in miles per hour, to turn a corner with radius r, in feet, without tipping over. What is the maximum velocity that a cyclist should travel around a corner of radius 9 feet without tipping over?

86. The formula $v = \sqrt{2.5r}$ models the maximum safe speed, v, in miles per hour, at which a car can travel on a curved road with radius of curvature r, in feet. A highway crew measures the radius of curvature at an exit ramp on a highway as 360 feet. What is the maximum safe speed?

Police use the formula $v = \sqrt{20L}$ to estimate the speed of a car, v, in miles per hour, based on the length, L, in feet, of its skid marks upon sudden braking on a dry asphalt road. Use the formula to solve Exercises 87–88.

87. A motorist is involved in an accident. A police officer measures the car's skid marks to be 245 feet long. Estimate the speed at which the motorist was traveling before braking. If the posted speed limit is 50 miles per hour and the motorist tells the officer he was not speeding, should the officer believe him? Explain.

88. A motorist is involved in an accident. A police officer measures the car's skid marks to be 45 feet long. Estimate the speed at which the motorist was traveling before braking. If the posted speed limit is 35 miles per hour and the motorist tells the officer she was not speeding, should the officer believe her? Explain.

Autism is a neurological disorder that impedes language and derails social and emotional development. New findings suggest that the condition is not a sudden calamity that strikes children at the age of 2 or 3, but a developmental problem linked to abnormally rapid brain growth during infancy. The graphs show that the heads of severely autistic children start out smaller than average and then go through a period of explosive growth. Exercises 89–90 involve mathematical models for the data shown by the graphs.

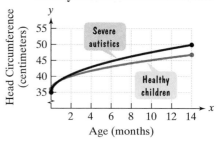

Developmental Differences between Healthy Children and Severe Autistics

Source: The Journal of the American Medical Association

89. The data for one of the two groups shown by the graphs can be modeled by

$$y = 2.9\sqrt{x} + 36,$$

where y is the head circumference, in centimeters, at age x months, $0 \le x \le 14$.

    a. According to the model, what is the head circumference at birth?

    b. According to the model, what is the head circumference at 9 months?

c. According to the model, what is the head circumference at 14 months? Use a calculator and round to the nearest tenth of a centimeter.

d. Use the values that you obtained in parts (a) through (c) and the graphs shown on the previous page to determine if the given model describes healthy children or severe autistics.

90. The data for one of the two groups shown by the graphs on the previous page can be modeled by

$$y = 4\sqrt{x} + 35,$$

where y is the head circumference, in centimeters, at age x months, $0 \le x \le 14$.

a. According to the model, what is the head circumference at birth?

b. According to the model, what is the head circumference at 9 months?

c. According to the model, what is the head circumference at 14 months? Use a calculator and round to the nearest centimeter.

d. Use the values that you obtained in parts (a) through (c) and the graphs shown on the previous page to determine if the given model describes healthy children or severe autistics.

91. The bar graph shows spam as a percentage of e-mail for four selected years. The data can be modeled by an equation in the form

$$y = a\sqrt{x} + b,$$

where y is the percentage of e-mail that is spam and x is the number of years after 1999. In this exercise, you will use the data from 1999 and 2005 to find values for a and b. This will give you an equation of a square root model for junk e-mail.

Spam as a Percentage of e-Mail

Source: IronPort

a. In 1999, 21% of e-mail was spam. Substitute 0 for x (1999 is 0 years after 1999) and 21 for y in $y = a\sqrt{x} + b$. Find the value for b and rewrite the model using this value.

b. In 2005, 72% of e-mail was spam. Substitute 6 for x (2005 is 6 years after 1999) and 72 for y in your model from part (a). Find the value for a, rounded to one decimal place, and rewrite the model using this value.

c. According to your model from part (b), what percentage of e-mail was spam in 2004? Use a calculator and round to the nearest tenth of a percent. Does this underestimate or overestimate the actual percentage given in the bar graph? By how much?

d. According to your model from part (b), what percentage of e-mail will be spam in 2011? Round to the nearest tenth of a percent.

Writing in Mathematics

92. What are the square roots of 36? Explain why each of these numbers is a square root.

93. What does the symbol $\sqrt{}$ denote? Which of your answers in Exercise 92 is given by this symbol? Write the symbol needed to obtain the other answer.

94. Explain why $\sqrt{-1}$ is not a real number.

95. Explain why $\sqrt[3]{8}$ is 2. Then describe what is meant by $\sqrt[n]{a} = b$.

96. Explain the meaning of the words *radical*, *radicand*, and *index*. Give an example with your explanation.

97. Suppose that a motorist first lightly applies the brakes and then forcefully applies the brakes after a few seconds. Does the model given in Exercises 87–88 underestimate or overestimate the speed at which the motorist was traveling before braking? Explain your answer.

98. The formula $d = \sqrt{1.5h}$ models the distance, d, in miles, that can be seen to the horizon from a height h, in feet. Write a word problem using the formula. Then solve the problem.

Critical Thinking Exercises

Make Sense? *In Exercises 99–102, determine whether each statement "makes sense" or "does not make sense" and explain your reasoning.*

99. Because of the negative sign, $\sqrt{-9}$ and $-\sqrt{9}$ are not real numbers.

100. A negative number has an nth root only if n is odd.

101. Using my calculator, I determined that $6^7 = 279{,}936$, so 6 must be a seventh root of 279,936.

102. Using my calculator to approximate $\dfrac{12 + \sqrt{11}}{2}$, I pressed $\boxed{=}$ after entering $12 + \sqrt{11}$ before I divided by 2, so in this case my calculator is not following the order of operations.

In Exercises 103–106, determine whether each statement is true or false. If the statement is false, make the necessary change(s) to produce a true statement.

103. $\sqrt{9} + \sqrt{16} = \sqrt{25}$

104. $\dfrac{\sqrt{64}}{2} = \sqrt{32}$

105. $\sqrt[3]{-27}$ is not a real number.

106. $\sqrt{\dfrac{1}{4}} + \sqrt{\dfrac{1}{9}} = \sqrt{\dfrac{25}{36}}$

In Exercises 107–109, simplify each expression.

107. $\sqrt{\sqrt[3]{64}}$ 108. $\sqrt[3]{-\sqrt{1}}$

109. $\sqrt{\sqrt{16}} - \sqrt[3]{\sqrt{64}}$

110. Between which two consecutive integers is $-\sqrt{47}$?

Technology Exercises

111. Use a graphing utility to graph $y_1 = \sqrt{x}$, $y_2 = \sqrt{x + 4}$, and $y_3 = \sqrt{x - 3}$ in the same $[-5, 10, 1]$ by $[0, 6, 1]$ viewing rectangle. Describe one similarity and one difference that you observe among the graphs. Use the word "shift" in your response.

112. Use a graphing utility to graph $y = \sqrt{x}$, $y = \sqrt{x} + 4$, and $y = \sqrt{x} - 3$ in the same $[-1, 10, 1]$ by $[-10, 10, 1]$ viewing rectangle. Describe one similarity and one difference that you observe among the graphs.

Review Exercises

113. Graph: $4x - 5y = 20$. (Section 4.2, Example 4)

114. Solve and graph the solution set on a number line: $2(x - 3) > 4x + 10$. (Section 2.6, Example 8)

115. Divide: $\dfrac{1}{x^2 - 17x + 30} \div \dfrac{1}{x^2 + 7x - 18}$.

(Section 8.2, Example 6)

Preview Exercises

Exercises 116–118 will help you prepare for the material covered in the next section.

116. a. Find $\sqrt{25} \cdot \sqrt{4}$.

b. Find $\sqrt{25 \cdot 4}$.

c. Based on your answers to parts (a) and (b), what can you conclude?

117. a. Use a calculator to approximate $\sqrt{500}$ to three decimal places.

b. Use a calculator to approximate $10\sqrt{5}$ to three decimal places.

c. Based on your answers to parts (a) and (b), what can you conclude?

118. a. Find $\dfrac{\sqrt{64}}{\sqrt{4}}$.

b. Find $\sqrt{\dfrac{64}{4}}$.

c. Based on your answers to parts (a) and (b), what can you conclude?

9.2

Multiplying and Dividing Radicals

Objectives

1 Multiply square roots.

2 Simplify square roots.

3 Use the quotient rule for square roots.

4 Use the product and quotient rules for other roots.

This photograph shows mathematical models used by Albert Einstein at a lecture on relativity. Notice the radicals that appear in many of the formulas. Among these models, there is one describing how an astronaut in a moving spaceship ages more slowly than friends who remain on Earth. In this section, in addition to learning how to multiply and divide radicals, you will see how radicals model time dilation for a futuristic high-speed trip to a nearby star.

1 Multiply square roots.

The Product Rule for Square Roots

A rule for multiplying square roots can be generalized by comparing $\sqrt{25} \cdot \sqrt{4}$ and $\sqrt{25 \cdot 4}$. Notice that

$$\sqrt{25} \cdot \sqrt{4} = 5 \cdot 2 = 10 \quad \text{and} \quad \sqrt{25 \cdot 4} = \sqrt{100} = 10.$$

Because we obtain 10 in both situations, the original radical expressions must be equal. That is,

$$\sqrt{25} \cdot \sqrt{4} = \sqrt{25 \cdot 4}.$$

This result is a special case of the **product rule for square roots** that can be generalized as follows:

The Product Rule for Square Roots

If a and b represent nonnegative real numbers, then

$$\sqrt{a} \cdot \sqrt{b} = \sqrt{ab} \quad \text{and} \quad \sqrt{ab} = \sqrt{a} \cdot \sqrt{b}.$$

> The product of two square roots is the square root of the product of the radicands.

> The square root of a product is the product of the square roots.

EXAMPLE 1 Using the Product Rule to Multiply Square Roots

Use the product rule for square roots to find each product:

  a. $\sqrt{2} \cdot \sqrt{5}$ **b.** $\sqrt{7x} \cdot \sqrt{11y}$ **c.** $\sqrt{7} \cdot \sqrt{7}$ **d.** $\sqrt{\dfrac{2}{5}} \cdot \sqrt{\dfrac{3}{7}}$.

Solution Because $\sqrt{a} \cdot \sqrt{b} = \sqrt{ab}$, we find each product by multiplying the radicands.

  a. $\sqrt{2} \cdot \sqrt{5} = \sqrt{2 \cdot 5} = \sqrt{10}$

  b. $\sqrt{7x} \cdot \sqrt{11y} = \sqrt{7x \cdot 11y} = \sqrt{77xy}$ Assume that $x \geq 0$ and $y \geq 0$.

  c. $\sqrt{7} \cdot \sqrt{7} = \sqrt{7 \cdot 7} = \sqrt{49} = 7$

  d. $\sqrt{\dfrac{2}{5}} \cdot \sqrt{\dfrac{3}{7}} = \sqrt{\dfrac{2}{5} \cdot \dfrac{3}{7}} = \sqrt{\dfrac{6}{35}}$

✓ **CHECK POINT 1** Use the product rule for square roots to find each product:

  a. $\sqrt{3} \cdot \sqrt{10}$ **b.** $\sqrt{2x} \cdot \sqrt{13y}$

  c. $\sqrt{5} \cdot \sqrt{5}$ **d.** $\sqrt{\dfrac{3}{2}} \cdot \sqrt{\dfrac{5}{11}}$.

---

2 Simplify square roots.

Using the Product Rule to Simplify Square Roots

We have seen that a number that is the square of an integer is called a **perfect square**. For example, 100 is a perfect square because $100 = 10^2$. Thus, $\sqrt{100} = 10$.

  A square root is **simplified** when its radicand has no factors other than 1 that are perfect squares. For example, $\sqrt{500}$ is not simplified because it can be expressed as $\sqrt{100 \cdot 5}$ and 100 is a perfect square. We can use the product rule in the form

$$\sqrt{ab} = \sqrt{a}\sqrt{b}$$

to simplify $\sqrt{500}$. We factor 500 so that one of its factors is the largest perfect square possible.

$$\sqrt{500} = \sqrt{100 \cdot 5} \qquad \text{Factor 500. 100 is the largest perfect square factor.}$$

$$= \sqrt{100}\sqrt{5} \qquad \text{Use the product rule: } \sqrt{ab} = \sqrt{a}\sqrt{b}.$$

$$= 10\sqrt{5} \qquad \text{Write } \sqrt{100} \text{ as 10. We read } 10\sqrt{5} \text{ as "ten times the square root of 5."}$$

Study Tip

When simplifying square roots, always look for the *greatest* perfect square factor possible. The following factorization will lead to further simplification:

$$\sqrt{500} = \sqrt{25 \cdot 20} = \sqrt{25}\sqrt{20} = 5\sqrt{20}.$$

> 25 is a perfect square factor of 500, but not the greatest perfect square factor.

Because 20 contains a perfect square factor, 4, the simplification is not complete.

$$5\sqrt{20} = 5\sqrt{4 \cdot 5} = 5\sqrt{4}\sqrt{5} = 5 \cdot 2\sqrt{5} = 10\sqrt{5}$$

Although the result checks with our simplification using $\sqrt{500} = \sqrt{100 \cdot 5}$, more work is required when the greatest perfect square factor is not used.

EXAMPLE 2 Using the Product Rule to Simplify Square Roots

Simplify:

   a. $\sqrt{75}$ **b.** $\sqrt{80}$ **c.** $\sqrt{38}$.

Solution In each case, try to factor the radicand as a product of the greatest perfect square factor and another factor.

  a. $\sqrt{75} = \sqrt{25 \cdot 3}$ *25 is the greatest perfect square that is a factor of 75.*

           $= \sqrt{25}\sqrt{3}$ *$\sqrt{ab} = \sqrt{a}\sqrt{b}$*

           $= 5\sqrt{3}$ *Write $\sqrt{25}$ as 5.*

  b. $\sqrt{80} = \sqrt{16 \cdot 5}$ *16 is the greatest perfect square that is a factor of 80.*

           $= \sqrt{16}\sqrt{5}$ *$\sqrt{ab} = \sqrt{a}\sqrt{b}$*

           $= 4\sqrt{5}$ *Write $\sqrt{16}$ as 4.*

  c. Although the radicand, 38, can be factored as $2 \cdot 19$, neither of these factors is a perfect square. Because 38 has no perfect square factors other than 1, $\sqrt{38}$ cannot be simplified. ▬

Using Technology

Numeric Connections

You can use a calculator to provide numerical support that your simplified answer is correct. For example, to support $\sqrt{75} = 5\sqrt{3}$, find decimal approximations for $\sqrt{75}$ and $5\sqrt{3}$. These approximations should be equal.

	$\sqrt{75}$:	$5\sqrt{3}$:
Scientific Calculator	75 $\boxed{\sqrt{}}$	55 $\boxed{\times}$ 3 $\boxed{\sqrt{}}$ $\boxed{=}$
Graphing Calculator	$\boxed{\sqrt{}}$ 75 $\boxed{\text{ENTER}}$	5 $\boxed{\sqrt{}}$ 3 $\boxed{\text{ENTER}}$

Correct to two decimal places, $\sqrt{75} \approx 8.66$ and $5\sqrt{3} \approx 8.66$. Use this technique to support the numerical results for the answers in this section. Caution: **A simplified square root does not mean a decimal approximation.**

☑ **CHECK POINT 2** Simplify, if possible:

   a. $\sqrt{12}$ **b.** $\sqrt{60}$ **c.** $\sqrt{55}$.

Simplifying Square Roots Containing Variables

Square roots can also contain variables. Here are three examples:

$$\sqrt{x^2} \qquad \sqrt{x^4} \qquad \sqrt{x^6}.$$

We know that the square root of a negative number is not a real number. Thus, we want to avoid values for variables in the radicand that would make the radicand negative. To avoid negative radicands, we assume throughout this chapter that **if a variable appears in the radicand of a radical expression, it represents nonnegative numbers only**.

Do you see that we can simplify each of the three square root expressions just shown?

$$\sqrt{x^2} = x \quad \text{Because } (x)^2 = x^2 \qquad \sqrt{x^4} = x^2 \quad \text{Because } (x^2)^2 = x^4 \qquad \sqrt{x^6} = x^3 \quad \text{Because } (x^3)^2 = x^6$$

Each simplified form is the variable in the radicand raised to one-half the power in the radicand.

To simplify square roots when the radicand contains x to an even power, we can use the following rule:

Simplifying Square Roots with Variables to Even Powers

$$\sqrt{x^{2n}} = x^n$$

The square root of a variable raised to an even power equals the variable raised to one-half that power.

For example,

$$\sqrt{x^{10}} = x^5 \qquad \text{and} \qquad \sqrt{y^{34}} = y^{17}.$$

$$\tfrac{1}{2} \cdot 10 = 5 \qquad\qquad \tfrac{1}{2} \cdot 34 = 17$$

EXAMPLE 3   Simplifying Square Roots with Variables to Even Powers

Simplify: $\sqrt{72x^{14}}$.

Solution   We write the radicand as the product of the greatest perfect square factor and another factor. Variables to even powers are part of the perfect square factor.

$$\sqrt{72x^{14}} = \sqrt{36x^{14} \cdot 2} \qquad 36x^{14} \text{ is the greatest perfect square factor of } 72x^{14}.$$

$$= \sqrt{36x^{14}}\sqrt{2} \qquad \text{Use the product rule: } \sqrt{ab} = \sqrt{a}\sqrt{b}.$$

$$= 6x^7\sqrt{2} \text{ or } 6\sqrt{2}x^7 \qquad \sqrt{36} = 6 \text{ and, using } \sqrt{x^{2n}} = x^n, \sqrt{x^{14}} = x^7. \quad \blacksquare$$

✓ **CHECK POINT 3**   Simplify: $\sqrt{40x^{16}}$.

How do we simplify the square root of a radicand containing a variable raised to an odd power, such as $\sqrt{x^5}$? Express the variable as the product of two factors, one of which has an even power. Then use the product rule to simplify. For example,

$$\sqrt{x^5} = \sqrt{x^4 \cdot x} = \sqrt{x^4}\sqrt{x} = x^2\sqrt{x}.$$

$$\sqrt{x^{2n}} = x^n$$

EXAMPLE 4 Simplifying Square Roots with Variables to Odd Powers

Simplify: $\sqrt{50x^7}$.

Solution

$$\sqrt{50x^7} = \sqrt{25x^6 \cdot 2x} \qquad \text{\small 25}x^6 \text{ \small is the greatest perfect square factor of 50}x^7.$$
$$= \sqrt{25x^6}\sqrt{2x} \qquad \text{\small Use the product rule: } \sqrt{ab} = \sqrt{a}\sqrt{b}.$$
$$= 5x^3\sqrt{2x} \qquad \text{\small} \sqrt{25} = 5 \text{ \small and, using } \sqrt{x^{2n}} = x^n, \sqrt{x^6} = x^3. \qquad \blacksquare$$

☑ **CHECK POINT 4** Simplify: $\sqrt{27x^9}$.

Now we look at an example where we use the product rule to multiply radicals and then simplify.

EXAMPLE 5 Multiplying and Simplifying Square Roots

Multiply and then simplify:

$$\sqrt{6x^8} \cdot \sqrt{2x^3}.$$

Solution

$$\sqrt{6x^8} \cdot \sqrt{2x^3} = \sqrt{6x^8 \cdot 2x^3} \qquad \text{\small Use the product rule: } \sqrt{a}\sqrt{b} = \sqrt{ab}.$$
$$= \sqrt{12x^{11}} \qquad \text{\small Multiply in the radicand: Multiply coefficients and add exponents.}$$
$$= \sqrt{4x^{10} \cdot 3x} \qquad \text{\small Simplify. 4}x^{10} \text{ \small is the greatest perfect square factor of 12}x^{11}.$$
$$= \sqrt{4x^{10}}\sqrt{3x} \qquad \text{\small Use the product rule: } \sqrt{ab} = \sqrt{a}\sqrt{b}.$$
$$= 2x^5\sqrt{3x} \qquad \text{\small} \sqrt{4} = 2 \text{ \small and, using } \sqrt{x^{2n}} = x^n, \sqrt{x^{10}} = x^5. \qquad \blacksquare$$

☑ **CHECK POINT 5** Multiply and then simplify:

$$\sqrt{15x^6} \cdot \sqrt{3x^7}.$$

3 Use the quotient rule for square roots.

The Quotient Rule for Square Roots

A rule for dividing square roots can be generalized by comparing

$$\sqrt{\frac{64}{4}} \quad \text{and} \quad \frac{\sqrt{64}}{\sqrt{4}}.$$

Note that

$$\sqrt{\frac{64}{4}} = \sqrt{16} = 4 \quad \text{and} \quad \frac{\sqrt{64}}{\sqrt{4}} = \frac{8}{2} = 4.$$

Because we obtain 4 in both situations, the original radical expressions must be equal:

$$\sqrt{\frac{64}{4}} = \frac{\sqrt{64}}{\sqrt{4}}.$$

This result is a special case of the **quotient rule for square roots** that is generalized in the box on the next page.

The Quotient Rule for Square Roots

If a and b represent nonnegative real numbers and $b \neq 0$, then

$$\frac{\sqrt{a}}{\sqrt{b}} = \sqrt{\frac{a}{b}} \quad \text{and} \quad \sqrt{\frac{a}{b}} = \frac{\sqrt{a}}{\sqrt{b}}.$$

> The quotient of two square roots is the square root of the quotient of the radicands.

> The square root of a quotient is the quotient of the square roots.

EXAMPLE 6 Using the Quotient Rule to Simplify Square Roots

Simplify:

a. $\sqrt{\dfrac{100}{9}}$ **b.** $\dfrac{\sqrt{48x^3}}{\sqrt{6x}}$.

Solution

a. $\sqrt{\dfrac{100}{9}} = \dfrac{\sqrt{100}}{\sqrt{9}} = \dfrac{10}{3}$

b. $\dfrac{\sqrt{48x^3}}{\sqrt{6x}} = \sqrt{\dfrac{48x^3}{6x}} = \sqrt{8x^2} = \sqrt{4x^2}\sqrt{2} = 2x\sqrt{2}$

■

☑ **CHECK POINT 6** Simplify:

a. $\sqrt{\dfrac{49}{25}}$ **b.** $\dfrac{\sqrt{48x^5}}{\sqrt{3x}}$.

---

4 Use the product and quotient rules for other roots.

The Product and Quotient Rules for Other Roots

The product and quotient rules apply to cube roots, fourth roots, and all higher roots.

The Product and Quotient Rules for nth Roots

For all real numbers, where the indicated roots represent real numbers,

$$\sqrt[n]{a} \cdot \sqrt[n]{b} = \sqrt[n]{ab} \quad \text{and} \quad \sqrt[n]{ab} = \sqrt[n]{a} \cdot \sqrt[n]{b}$$

> The product of two nth roots is the nth root of the product of the radicands.

> The nth root of a product is the product of the nth roots.

$$\frac{\sqrt[n]{a}}{\sqrt[n]{b}} = \sqrt[n]{\frac{a}{b}}, b \neq 0 \quad \text{and} \quad \sqrt[n]{\frac{a}{b}} = \frac{\sqrt[n]{a}}{\sqrt[n]{b}}, b \neq 0.$$

> The quotient of two nth roots is the nth root of the quotient of the radicands.

> The nth root of a quotient is the quotient of the nth roots.

EXAMPLE 7   Simplifying, Multiplying, and Dividing
Higher Roots

Simplify:

a. $\sqrt[3]{24}$ b. $\sqrt[4]{8} \cdot \sqrt[4]{4}$ c. $\sqrt[4]{\dfrac{81}{16}}$.

Solution

a. $\sqrt[3]{24} = \sqrt[3]{8 \cdot 3}$ Find the greatest perfect cube that is a factor of 24. $2^3 = 8$, so 8 is a perfect cube and is the greatest perfect cube factor of 24.

$\quad = \sqrt[3]{8} \cdot \sqrt[3]{3}$ $\sqrt[n]{ab} = \sqrt[n]{a} \cdot \sqrt[n]{b}$

$\quad = 2\sqrt[3]{3}$ $\sqrt[3]{8} = 2$

b. $\sqrt[4]{8} \cdot \sqrt[4]{4} = \sqrt[4]{8 \cdot 4}$ $\sqrt[n]{a} \cdot \sqrt[n]{b} = \sqrt[n]{ab}$

$\quad = \sqrt[4]{32}$ Find the greatest perfect fourth power that is a factor of 32.

$\quad = \sqrt[4]{16 \cdot 2}$ $2^4 = 16$, so 16 is a perfect fourth power and is the largest perfect fourth power that is a factor of 32.

$\quad = \sqrt[4]{16} \cdot \sqrt[4]{2}$ $\sqrt[n]{ab} = \sqrt[n]{a} \cdot \sqrt[n]{b}$

$\quad = 2\sqrt[4]{2}$ $\sqrt[4]{16} = 2$

c. $\sqrt[4]{\dfrac{81}{16}} = \dfrac{\sqrt[4]{81}}{\sqrt[4]{16}}$ $\sqrt[n]{\dfrac{a}{b}} = \dfrac{\sqrt[n]{a}}{\sqrt[n]{b}}$

$\quad = \dfrac{3}{2}$ $\sqrt[4]{81} = 2$ because $3^4 = 81$, and $\sqrt[4]{16} = 2$ because $2^4 = 16$. ∎

✓ **CHECK POINT 7**   Simplify:

a. $\sqrt[3]{40}$ b. $\sqrt[5]{8} \cdot \sqrt[5]{8}$ c. $\sqrt[3]{\dfrac{125}{27}}$.

Blitzer Bonus

A Radical Idea: Time Is Relative

Salvador Dali "The Persistence of Memory" 1931, oil on canvas, $9\frac{1}{2} \times 13$ in. (24.1 × 33 cm). The Museum of Modern Art/Licensed by Scala-Art Resource, NY. © 1999 Demart Pro Arte, Geneva/Artists Rights Society (ARS), New York.

What does travel in space have to do with radicals? Imagine that in the future we will be able to travel at velocities approaching the speed of light (approximately 186,000 miles per second). According to Einstein's theory of special relativity, time would pass more quickly on Earth than it would in the moving spaceship. The special-relativity equation

$$R_a = R_f \sqrt{1 - \left(\dfrac{v}{c}\right)^2}$$

gives the aging rate of an astronaut, R_a, relative to the aging rate of a friend, R_f, on Earth. In this formula, v is the astronaut's speed and c is the speed of light. As the astronaut's speed approaches the speed of light, we can substitute c for v.

$$R_a = R_f\sqrt{1 - \left(\frac{v}{c}\right)^2}$$ Einstein's equation gives the aging rate of an astronaut, R_a, relative to the aging rate of a friend, R_f, on Earth.

$$R_a = R_f\sqrt{1 - \left(\frac{c}{c}\right)^2}$$ The velocity, v, is approaching the speed of light, c, so let $v = c$.

$$= R_f\sqrt{1 - 1}$$ $\left(\dfrac{c}{c}\right)^2 = 1^2 = 1 \cdot 1 = 1$

$$= R_f\sqrt{0}$$ Simplify the radicand: $1 - 1 = 0$.

$$= R_f \cdot 0$$ $\sqrt{0} = 0$

$$= 0$$ Multiply: $R_f \cdot 0 = 0$.

Close to the speed of light, the astronaut's aging rate, R_a, relative to a friend, R_f, on Earth is nearly 0. What does this mean? As we age here on Earth, the space traveler would barely get older. The space traveler would return to an unknown futuristic world in which friends and loved ones would be long gone.

9.2 EXERCISE SET

PRACTICE   WATCH   DOWNLOAD   READ   REVIEW

Remember that throughout this chapter, variable expressions in radicands represent nonnegative real numbers.

Practice Exercises

In Exercises 1–14, use the product rule for square roots to find each product.

1. $\sqrt{2} \cdot \sqrt{7}$
2. $\sqrt{2} \cdot \sqrt{17}$
3. $\sqrt{3x} \cdot \sqrt{5y}$
4. $\sqrt{7x} \cdot \sqrt{11y}$
5. $\sqrt{5} \cdot \sqrt{5}$
6. $\sqrt{10} \cdot \sqrt{10}$
7. $\sqrt{\dfrac{2}{3}} \cdot \sqrt{\dfrac{5}{7}}$
8. $\sqrt{\dfrac{3}{5}} \cdot \sqrt{\dfrac{5}{7}}$
9. $\sqrt{0.1x} \cdot \sqrt{5y}$
10. $\sqrt{0.2x} \cdot \sqrt{3y}$
11. $\sqrt{\dfrac{1}{5}a} \cdot \sqrt{\dfrac{1}{5}b}$
12. $\sqrt{\dfrac{1}{7}a} \cdot \sqrt{\dfrac{1}{7}b}$
13. $\sqrt{\dfrac{2x}{9}} \cdot \sqrt{\dfrac{9}{2}}$
14. $\sqrt{\dfrac{5x}{11}} \cdot \sqrt{\dfrac{11}{5}}$

In Exercises 15–54, simplify each expression. If the expression cannot be simplified, so state.

15. $\sqrt{50}$
16. $\sqrt{27}$
17. $\sqrt{45}$
18. $\sqrt{28}$
19. $\sqrt{200}$
20. $\sqrt{300}$
21. $\sqrt{75x}$
22. $\sqrt{40x}$
23. $\sqrt{9x}$
24. $\sqrt{25x}$
25. $\sqrt{35}$
26. $\sqrt{22}$
27. $\sqrt{y^2}$
28. $\sqrt{z^2}$
29. $\sqrt{64x^2}$
30. $\sqrt{36x^2}$
31. $\sqrt{11x^2}$
32. $\sqrt{17x^2}$
33. $\sqrt{8x^2}$
34. $\sqrt{12x^2}$
35. $\sqrt{x^{20}}$
36. $\sqrt{x^{30}}$
37. $\sqrt{25y^{10}}$
38. $\sqrt{36y^{10}}$
39. $\sqrt{20x^6}$
40. $\sqrt{24x^8}$
41. $\sqrt{72y^{100}}$
42. $\sqrt{32y^{200}}$
43. $\sqrt{x^3}$
44. $\sqrt{y^3}$

45. $\sqrt{x^7}$
46. $\sqrt{x^9}$
47. $\sqrt{y^{17}}$
48. $\sqrt{y^{19}}$
49. $\sqrt{25x^5}$
50. $\sqrt{49x^5}$
51. $\sqrt{8x^{17}}$
52. $\sqrt{12x^7}$
53. $\sqrt{90y^{19}}$
54. $\sqrt{600y^{23}}$

In Exercises 55–68, multiply and, if possible, simplify.

55. $\sqrt{3} \cdot \sqrt{15}$
56. $\sqrt{3} \cdot \sqrt{6}$
57. $\sqrt{5x} \cdot \sqrt{10y}$
58. $\sqrt{8x} \cdot \sqrt{10y}$
59. $\sqrt{12x} \cdot \sqrt{3x}$
60. $\sqrt{20x} \cdot \sqrt{5x}$
61. $\sqrt{15x^2} \cdot \sqrt{3x}$
62. $\sqrt{2x^2} \cdot \sqrt{10x}$
63. $\sqrt{15x^4} \cdot \sqrt{5x^9}$
64. $\sqrt{50x^9} \cdot \sqrt{4x^4}$
65. $\sqrt{7x} \cdot \sqrt{3y}$
66. $\sqrt{5x} \cdot \sqrt{11y}$
67. $\sqrt{50xy} \cdot \sqrt{4xy^2}$
68. $\sqrt{5xy} \cdot \sqrt{10xy^2}$

In Exercises 69–92, simplify using the quotient rule for square roots.

69. $\sqrt{\dfrac{49}{16}}$
70. $\sqrt{\dfrac{100}{9}}$
71. $\sqrt{\dfrac{3}{4}}$
72. $\sqrt{\dfrac{7}{25}}$
73. $\sqrt{\dfrac{x^2}{36}}$
74. $\sqrt{\dfrac{x^2}{49}}$
75. $\sqrt{\dfrac{7}{x^4}}$
76. $\sqrt{\dfrac{13}{x^6}}$
77. $\sqrt{\dfrac{72}{y^{20}}}$
78. $\sqrt{\dfrac{300}{y^{30}}}$
79. $\dfrac{\sqrt{54}}{\sqrt{6}}$
80. $\dfrac{\sqrt{75}}{\sqrt{3}}$
81. $\dfrac{\sqrt{24}}{\sqrt{3}}$
82. $\dfrac{\sqrt{60}}{\sqrt{5}}$
83. $\dfrac{\sqrt{75}}{\sqrt{15}}$
84. $\dfrac{\sqrt{21}}{\sqrt{3}}$

85. $\dfrac{\sqrt{48x}}{\sqrt{3x}}$

86. $\dfrac{\sqrt{27x}}{\sqrt{3x}}$

87. $\dfrac{\sqrt{32x^3}}{\sqrt{8x}}$

88. $\dfrac{\sqrt{75x^3}}{\sqrt{3x}}$

89. $\dfrac{\sqrt{150x^4}}{\sqrt{3x}}$

90. $\dfrac{\sqrt{400x^4}}{\sqrt{2x}}$

91. $\dfrac{\sqrt{400x^{10}}}{\sqrt{10x^3}}$

92. $\dfrac{\sqrt{800x^{12}}}{\sqrt{10x^3}}$

In Exercises 93–108, simplify each radical expression.

93. $\sqrt[3]{16}$

94. $\sqrt[3]{32}$

95. $\sqrt[3]{54}$

96. $\sqrt[3]{250}$

97. $\sqrt[4]{32}$

98. $\sqrt[4]{48}$

99. $\sqrt[3]{4} \cdot \sqrt[3]{2}$

100. $\sqrt[3]{3} \cdot \sqrt[3]{9}$

101. $\sqrt[3]{9} \cdot \sqrt[3]{6}$

102. $\sqrt[3]{12} \cdot \sqrt[3]{4}$

103. $\sqrt[4]{4} \cdot \sqrt[4]{8}$

104. $\sqrt[5]{16} \cdot \sqrt[5]{4}$

105. $\sqrt[3]{\dfrac{27}{8}}$

106. $\sqrt[3]{\dfrac{125}{8}}$

107. $\sqrt[3]{\dfrac{3}{8}}$

108. $\sqrt[3]{\dfrac{5}{27}}$

Practice PLUS

In Exercises 109–114, simplify each expression.

109. $\sqrt{90(x + 4)^3}$

110. $\sqrt{150(x + 8)^3}$

111. $\sqrt{x^2 - 6x + 9}$

112. $\sqrt{x^2 - 10x + 25}$

113. $\sqrt{2^{43}x^{104}y^{13}}$

114. $\sqrt{3^{41}x^{102}y^{17}}$

In Exercises 115–116, use the fact that $\sqrt[3]{x^3} = x$ (why?) to simplify each expression.

115. $\sqrt[3]{24x^5}$

116. $\sqrt[3]{16x^4}$

Application Exercises

In Exercises 117–118, express the area of each rectangle as a square root in simplified form.

117.

$\sqrt{15}$ feet
$\sqrt{5}$ feet

118.

$\sqrt{6}$ feet
$\sqrt{3}$ feet

The mathematical model

$$A = \dfrac{\sqrt{h} \cdot \sqrt{w}}{56}$$

describes adult body surface area, A, in square meters, where h is the person's height, in inches, and w is the adult's weight, in pounds. Use this model to solve Exercises 119–120.

119. Consider an adult who is 68 inches tall and weighs 200 pounds.
    a. Determine this person's body surface area, A, in simplified radical form. Begin by simplifying each radical factor in the numerator of A using the given values for h and w.
    b. Use a calculator to approximate the surface area in part (a) correct to the nearest hundredth of a square meter.

120. Consider an adult who is 75 inches tall and weighs 220 pounds.
    a. Determine this person's body surface area, A, in simplified radical form. Begin by simplifying each radical factor in the numerator of A using the given values for h and w.
    b. Use a calculator to approximate the surface area in part (a) correct to the nearest hundredth of a square meter.

Writing in Mathematics

121. Use words to state the product rule for square roots, $\sqrt{a} \cdot \sqrt{b} = \sqrt{ab}$. Give an example with your description.

122. Explain why $\sqrt{50}$ is not simplified. What do we mean when we say that a square root is simplified?

123. Explain how to simplify square roots with variables to even powers.

124. Explain how to simplify square roots with variables to odd powers.

125. Explain how to simplify $\sqrt{10} \cdot \sqrt{5}$.

126. Use words to state the quotient rule for square roots, $\dfrac{\sqrt{a}}{\sqrt{b}} = \sqrt{\dfrac{a}{b}}$. Give an example with your description.

127. Read the Blitzer Bonus on pages 564–565. The future is now: You have the opportunity to explore the cosmos in a starship traveling near the speed of light. The experience will enable you to understand the mysteries of the universe in deeply personal ways, transporting you to unimagined levels of knowing and being. The down side: You return from your two-year journey to a futuristic world in which friends and loved ones are long gone. Do you explore space or stay here on Earth? What are the reasons for your choice?

Critical Thinking Exercises

Make Sense?   *In Exercises 128–131, determine whether each statement "makes sense" or "does not make sense" and explain your reasoning.*

128. I know that I've simplified a radical expression when it contains a single radical.

129. When multiplying $\sqrt{68}$ and $\sqrt{200}$, I find it easier to simplify each radical before multiplying rather than multiplying first and then simplifying the result.

130. Because the product rule for radicals applies when $\sqrt[n]{a}$ and $\sqrt[n]{b}$ are real numbers, I can use the rule to find $\sqrt[3]{3} \cdot \sqrt[3]{-2}$, but not to find $\sqrt{3} \cdot \sqrt{-2}$.

131. I multiply nth roots by taking the nth root of the product of the radicands.

In Exercises 132–135, determine whether each statement is true or false. If the statement is false, make the necessary change(s) to produce a true statement.

132. $\sqrt{20} = 4\sqrt{5}$

133. If $y \geq 0$, $\sqrt{y^9} = y^3$.

134. $\sqrt{2x}\sqrt{6y} = 2\sqrt{3xy}$ if x and y are nonnegative real numbers.

135. $\sqrt{2} \cdot \sqrt{8} = 16$

In Exercises 136–137, fill in the missing coefficients and exponents to make each statement true.

136. $\sqrt{x^{}} = 5x^7$

137. $\sqrt{15x^{}} \cdot \sqrt{x^5} = 3x^3\sqrt{10x}$

Technology Exercises

138. Use a calculator to provide numerical support for your simplifications in Exercises 15–20. In each case, find a decimal approximation for the given expression. Then find a decimal approximation for your simplified expression. The approximations should be the same.

In Exercises 139–141, determine if each simplification is correct by graphing each side of the equation with your graphing utility. Use the given viewing rectangle. The graphs should be the same. If they are not, correct the right side of the equation and then use your graphing utility to verify the simplification.

139. $\sqrt{x^4} = x^2$ $[0, 5, 1]$ by $[0, 20, 1]$

140. $\sqrt{8x^2} = 4x\sqrt{2}$ $[0, 5, 1]$ by $[0, 20, 1]$

141. $\sqrt{x^3} = x\sqrt{x}$ $[0, 5, 1]$ by $[0, 10, 1]$

Review Exercises

142. Solve the system:

$$4x + 3y = 18$$
$$5x - 9y = 48.$$

(Section 5.3, Example 3)

143. Subtract: $\dfrac{6x}{x^2 - 4} - \dfrac{3}{x + 2}$.

(Section 8.4, Example 7)

144. Factor completely: $2x^3 - 16x^2 + 30x$.
(Section 7.5, Example 2)

Preview Exercises

Exercises 145–147 will help you prepare for the material covered in the next section.

145. a. Simplify: $7x + 5x$.

b. Simplify: $7\sqrt{2} + 5\sqrt{2}$.

146. Simplify: $4\sqrt{50x}$.

147. a. Multiply: $3(x + 5)$.

b. Multiply: $\sqrt{3}(\sqrt{7} + \sqrt{5})$.

9.3

Operations with Radicals

Objectives

1 Add and subtract radicals.

2 Multiply radical expressions with more than one term.

3 Multiply conjugates.

If you are a full-time college student with a job, you're part of the data for a radical model. Before turning to this model and its graph in the exercise set, we focus our attention on adding, subtracting, and multiplying radicals.

1 Add and subtract radicals.

Adding and Subtracting Like Radicals

We know that like terms have exactly the same variable factors and can be combined. For example,

$$7x + 6x = (7 + 6)x = 13x.$$

Two or more square roots that have the same radicands are called **like radicals**. Like radicals are combined using the distributive property in exactly the same way that we combine like terms. For example,

$$7\sqrt{11} + 6\sqrt{11} = (7 + 6)\sqrt{11} = 13\sqrt{11}.$$

7 square roots of 11 plus 6 square roots of 11 result in 13 square roots of 11.

EXAMPLE 1 Adding and Subtracting Like Radicals

Add or subtract as indicated:

a. $7\sqrt{2} + 5\sqrt{2}$ b. $\sqrt{5x} - 7\sqrt{5x}$.

Solution

a. $7\sqrt{2} + 5\sqrt{2} = (7 + 5)\sqrt{2}$ Apply the distributive property.

$= 12\sqrt{2}$ Simplify.

b. $\sqrt{5x} - 7\sqrt{5x} = 1\sqrt{5x} - 7\sqrt{5x}$ Write $\sqrt{5x}$ as $1\sqrt{5x}$.

$= (1 - 7)\sqrt{5x}$ Apply the distributive property.

$= -6\sqrt{5x}$ Simplify.

✓ **CHECK POINT 1** Add or subtract as indicated:

a. $8\sqrt{13} + 9\sqrt{13}$ b. $\sqrt{17x} - 20\sqrt{17x}$.

In some cases, radicals can be combined once they have been simplified. For example, to add $\sqrt{2}$ and $\sqrt{8}$, we can write $\sqrt{8}$ as $\sqrt{4 \cdot 2}$ because 4 is a perfect square factor of 8.

$$\sqrt{2} + \sqrt{8} = \sqrt{2} + \sqrt{4 \cdot 2} = 1\sqrt{2} + 2\sqrt{2} = (1 + 2)\sqrt{2} = 3\sqrt{2}$$

Always begin by simplifying radical terms. This makes it possible to identify and combine any like radicals.

EXAMPLE 2 Combining Radicals That First Require Simplification

Add or subtract as indicated:

a. $7\sqrt{3} + \sqrt{12}$ b. $4\sqrt{50x} - 6\sqrt{32x}$.

Solution

a. $7\sqrt{3} + \sqrt{12}$

$= 7\sqrt{3} + \sqrt{4 \cdot 3}$ Split 12 into two factors such that one is a perfect square.

$= 7\sqrt{3} + 2\sqrt{3}$ $\sqrt{4 \cdot 3} = \sqrt{4}\sqrt{3} = 2\sqrt{3}$

$= (7 + 2)\sqrt{3}$ Apply the distributive property. You will find that this step is usually done mentally.

$= 9\sqrt{3}$ Simplify.

b. $4\sqrt{50x} - 6\sqrt{32x}$

$= 4\sqrt{25 \cdot 2x} - 6\sqrt{16 \cdot 2x}$ *25 is the greatest perfect square factor of 50x and 16 is the greatest perfect square factor of 32x.*

$= 4 \cdot 5\sqrt{2x} - 6 \cdot 4\sqrt{2x}$ $\sqrt{25 \cdot 2x} = \sqrt{25}\sqrt{2x} = 5\sqrt{2x}$ *and* $\sqrt{16 \cdot 2x} = \sqrt{16}\sqrt{2x} = 4\sqrt{2x}.$

$= 20\sqrt{2x} - 24\sqrt{2x}$ *Multiply:* $4 \cdot 5 = 20$ *and* $6 \cdot 4 = 24.$

$= (20 - 24)\sqrt{2x}$ *Apply the distributive property.*

$= -4\sqrt{2x}$ *Simplify.* ■

☑ **CHECK POINT 2**   Add or subtract as indicated:

a. $5\sqrt{27} + \sqrt{12}$ **b.** $6\sqrt{18x} - 4\sqrt{8x}.$

---

2 Multiply radical expressions with more than one term.

Multiplying Radical Expressions with More Than One Term

Radical expressions with more than one term are multiplied in much the same way that polynomials with more than one term are multiplied. Example 3 uses the distributive property and the FOIL method to perform multiplications.

EXAMPLE 3   **Multiplying Radicals**

Multiply:

a. $\sqrt{3}(\sqrt{7} + \sqrt{5})$ **b.** $(3 + \sqrt{2})(5 + \sqrt{2})$ **c.** $(5 + \sqrt{7})(6 - 3\sqrt{7}).$

Solution

a. $\sqrt{3}(\sqrt{7} + \sqrt{5}) = \sqrt{3} \cdot \sqrt{7} + \sqrt{3} \cdot \sqrt{5}$ *Use the distributive property.*

$= \sqrt{21} + \sqrt{15}$ *Use the product rule:* $\sqrt{a}\sqrt{b} = \sqrt{ab}.$

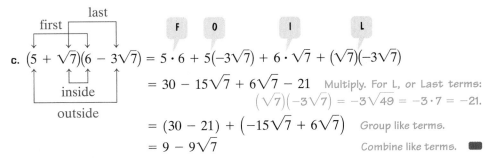

b. $(3 + \sqrt{2})(5 + \sqrt{2}) = 3 \cdot 5 + 3 \cdot \sqrt{2} + 5 \cdot \sqrt{2} + \sqrt{2} \cdot \sqrt{2}$

$= 15 + 3\sqrt{2} + 5\sqrt{2} + 2$ *Multiply.*

$= (15 + 2) + (3\sqrt{2} + 5\sqrt{2})$ *Group terms to be added. This step can be done mentally.*

$= 17 + 8\sqrt{2}$ *Add like terms.*

c. $(5 + \sqrt{7})(6 - 3\sqrt{7}) = 5 \cdot 6 + 5(-3\sqrt{7}) + 6 \cdot \sqrt{7} + (\sqrt{7})(-3\sqrt{7})$

$= 30 - 15\sqrt{7} + 6\sqrt{7} - 21$ *Multiply. For L, or Last terms:* $(\sqrt{7})(-3\sqrt{7}) = -3\sqrt{49} = -3 \cdot 7 = -21.$

$= (30 - 21) + (-15\sqrt{7} + 6\sqrt{7})$ *Group like terms.*

$= 9 - 9\sqrt{7}$ *Combine like terms.* ■

☑ **CHECK POINT 3**   Multiply:

a. $\sqrt{2}(\sqrt{5} + \sqrt{11})$ **b.** $(4 + \sqrt{3})(2 + \sqrt{3})$

c. $(3 + \sqrt{5})(8 - 4\sqrt{5}).$

3 Multiply conjugates.

Radical expressions that involve the sum and difference of the same two terms are called **conjugates**. For example,

$$5 + \sqrt{2} \quad \text{and} \quad 5 - \sqrt{2}$$

are conjugates of each other. Although you can multiply these conjugates using the FOIL method, there is an even faster way. Use the special-product formula for the sum and difference of the same two terms:

$$(A + B)(A - B) = A^2 - B^2.$$

Thus,

$$(A + B) \ (A - B) \ = \ A^2 - B^2$$

$$(5 + \sqrt{2})(5 - \sqrt{2}) = 5^2 - (\sqrt{2})^2 = 25 - 2 = 23.$$

EXAMPLE 4   **Multiplying Conjugates**

Multiply:

a. $\left(2 + \sqrt{7}\right)\left(2 - \sqrt{7}\right)$ b. $\left(\sqrt{5} - \sqrt{3}\right)\left(\sqrt{5} + \sqrt{3}\right).$

Solution   Use the special-product formula shown.

$$(A + B)(A - B) = A^2 - B^2$$

First term squared $-$ second term squared $=$ product

a. $\left(2 + \sqrt{7}\right)\left(2 - \sqrt{7}\right) = 2^2 - \left(\sqrt{7}\right)^2 = 4 - 7 = -3$

b. $\left(\sqrt{5} - \sqrt{3}\right)\left(\sqrt{5} + \sqrt{3}\right) = \left(\sqrt{5}\right)^2 - \left(\sqrt{3}\right)^2 = 5 - 3 = 2$

In the next section, we will use conjugates to simplify quotients.

☑ **CHECK POINT 4**   Multiply:

a. $\left(3 + \sqrt{11}\right)\left(3 - \sqrt{11}\right)$ b. $\left(\sqrt{7} - \sqrt{2}\right)\left(\sqrt{7} + \sqrt{2}\right).$

9.3 EXERCISE SET   **MyMathLab** Math XL PRACTICE   WATCH   DOWNLOAD   READ   REVIEW

Practice Exercises

In Exercises 1–22, add or subtract as indicated. If terms are not like radicals and cannot be combined, so state.

1. $8\sqrt{3} + 5\sqrt{3}$

2. $9\sqrt{5} + 6\sqrt{5}$

3. $17\sqrt{6} - 2\sqrt{6}$

4. $19\sqrt{7} - 2\sqrt{7}$

5. $3\sqrt{13} - 8\sqrt{13}$

6. $5\sqrt{17} - 11\sqrt{17}$

7. $12\sqrt{x} + 3\sqrt{x}$

8. $9\sqrt{x} + 2\sqrt{x}$

9. $70\sqrt{y} - 76\sqrt{y}$

10. $8\sqrt{y} - 28\sqrt{y}$

11. $7\sqrt{10x} + 2\sqrt{10x}$

12. $4\sqrt{6x} + 3\sqrt{6x}$

13. $7\sqrt{5y} - \sqrt{5y}$

14. $8\sqrt{3y} - \sqrt{3y}$

15. $\sqrt{5} + \sqrt{5}$

16. $\sqrt{3} + \sqrt{3}$

17. $4\sqrt{2} + 3\sqrt{2} + 5\sqrt{2}$

18. $6\sqrt{3} + 2\sqrt{3} + 5\sqrt{3}$

19. $4\sqrt{7} - 5\sqrt{7} + 8\sqrt{7}$

20. $5\sqrt{7} - 6\sqrt{7} + 10\sqrt{7}$

21. $4\sqrt{11} - 6\sqrt{11} + 2\sqrt{11}$

22. $7\sqrt{17} - 10\sqrt{17} + 3\sqrt{17}$

In Exercises 23–42, add or subtract as indicated. You will need to simplify terms before they can be combined. If terms cannot be simplified so that they can be combined, so state.

23. $\sqrt{5} + \sqrt{20}$

24. $\sqrt{3} + \sqrt{27}$

25. $\sqrt{8} - \sqrt{2}$

26. $\sqrt{50} - \sqrt{2}$

27. $\sqrt{50} + \sqrt{18}$

28. $\sqrt{28} + \sqrt{63}$

29. $7\sqrt{12} + \sqrt{75}$

30. $5\sqrt{12} + \sqrt{75}$

31. $3\sqrt{27} - 2\sqrt{18}$

32. $5\sqrt{27} - 3\sqrt{18}$

33. $2\sqrt{45x} - 2\sqrt{20x}$

34. $2\sqrt{50x} - 2\sqrt{18x}$

35. $\sqrt{8} + \sqrt{16} + \sqrt{18} + \sqrt{25}$

36. $\sqrt{6} + \sqrt{9} + \sqrt{24} + \sqrt{25}$

37. $\sqrt{2} + \sqrt{11}$

38. $\sqrt{3} + \sqrt{13}$

39. $2\sqrt{80} + 3\sqrt{75}$

40. $2\sqrt{75} + 3\sqrt{125}$

41. $3\sqrt{54} - 2\sqrt{20} + 4\sqrt{45} - \sqrt{24}$

42. $4\sqrt{8} - \sqrt{128} + 2\sqrt{48} + 3\sqrt{18}$

In Exercises 43–78, multiply as indicated. If possible, simplify any square roots that appear in the product.

43. $\sqrt{2}\left(\sqrt{3} + \sqrt{5}\right)$

44. $\sqrt{5}\left(\sqrt{3} + \sqrt{6}\right)$

45. $\sqrt{7}\left(\sqrt{6} - \sqrt{10}\right)$

46. $\sqrt{7}\left(\sqrt{5} - \sqrt{11}\right)$

47. $\sqrt{3}\left(5 + \sqrt{3}\right)$

48. $\sqrt{6}\left(7 + \sqrt{6}\right)$

49. $\sqrt{3}\left(\sqrt{6} - \sqrt{3}\right)$

50. $\sqrt{6}\left(\sqrt{6} - \sqrt{2}\right)$

51. $\left(5 + \sqrt{2}\right)\left(6 + \sqrt{2}\right)$

52. $\left(7 + \sqrt{2}\right)\left(8 + \sqrt{2}\right)$

53. $\left(4 + \sqrt{5}\right)\left(10 - 3\sqrt{5}\right)$

54. $\left(6 + \sqrt{5}\right)\left(9 - 4\sqrt{5}\right)$

55. $\left(6 - 3\sqrt{7}\right)\left(2 - 5\sqrt{7}\right)$

56. $\left(7 - 2\sqrt{7}\right)\left(5 - 3\sqrt{7}\right)$

57. $\left(\sqrt{10} - 3\right)\left(\sqrt{10} - 5\right)$

58. $\left(\sqrt{10} - 4\right)\left(\sqrt{10} - 6\right)$

59. $\left(\sqrt{3} + \sqrt{6}\right)\left(\sqrt{3} + 2\sqrt{6}\right)$

60. $\left(\sqrt{6} + \sqrt{3}\right)\left(\sqrt{6} + 5\sqrt{3}\right)$

61. $\left(\sqrt{2} + 1\right)\left(\sqrt{3} - 6\right)$

62. $\left(\sqrt{5} + 3\right)\left(\sqrt{2} - 8\right)$

63. $\left(3 + \sqrt{5}\right)\left(3 - \sqrt{5}\right)$

64. $\left(4 + \sqrt{7}\right)\left(4 - \sqrt{7}\right)$

65. $\left(1 - \sqrt{6}\right)\left(1 + \sqrt{6}\right)$

66. $\left(1 - \sqrt{5}\right)\left(1 + \sqrt{5}\right)$

67. $\left(\sqrt{11} + 5\right)\left(\sqrt{11} - 5\right)$

68. $\left(\sqrt{11} + 6\right)\left(\sqrt{11} - 6\right)$

69. $\left(\sqrt{7} - \sqrt{5}\right)\left(\sqrt{7} + \sqrt{5}\right)$

70. $\left(\sqrt{10} - \sqrt{7}\right)\left(\sqrt{10} + \sqrt{7}\right)$

71. $\left(2\sqrt{3} + 7\right)\left(2\sqrt{3} - 7\right)$

72. $\left(5\sqrt{3} + 6\right)\left(5\sqrt{3} - 6\right)$

73. $\left(2\sqrt{3} + \sqrt{5}\right)\left(2\sqrt{3} - \sqrt{5}\right)$

74. $\left(4\sqrt{5} + \sqrt{2}\right)\left(4\sqrt{5} - \sqrt{2}\right)$

75. $\left(\sqrt{2} + \sqrt{3}\right)^2$

76. $\left(\sqrt{3} + \sqrt{5}\right)^2$

77. $\left(\sqrt{x} - \sqrt{10}\right)^2$

78. $\left(\sqrt{x} - \sqrt{11}\right)^2$

Practice PLUS

In Exercises 79–86, add or subtract as indicated. You will need to simplify terms before they can be combined. Assume all variables represent nonnegative real numbers.

79. $5\sqrt{27x^3} - 3x\sqrt{12x}$

80. $7\sqrt{32x^3} - 3x\sqrt{50x}$

81. $6y^2\sqrt{x^5y} + 2x^2\sqrt{xy^5}$

82. $9y^2\sqrt{x^5y} + x^2\sqrt{xy^5}$

83. $3\sqrt[3]{54} - 4\sqrt[3]{16}$

84. $7\sqrt[3]{24} - 5\sqrt[3]{81}$

85. $x\sqrt[3]{32x} + 9\sqrt[3]{4x^4}$ Hint: $\sqrt[3]{x^3} = x$

86. $x\sqrt[3]{48x} + 11\sqrt[3]{6x^4}$ Hint: $\sqrt[3]{x^3} = x$

Application Exercises

In Exercises 87–92, write expressions for the perimeter and area of each figure. Then simplify these expressions. Assume that all measures are given in inches.

87.

$\sqrt{3} + \sqrt{5}$

$\sqrt{3} + \sqrt{5}$

88.

$\sqrt{2} + \sqrt{3}$

$\sqrt{2} + \sqrt{3}$

89.

$\sqrt{6} - 1$

$\sqrt{6} + 1$

90.

$\sqrt{7} - 1$

$\sqrt{7} + 1$

91.

92.

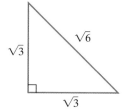

There is a formula for adding $\sqrt{a}$ and $\sqrt{b}$. The formula is $\sqrt{a} + \sqrt{b} = \sqrt{(a+b) + 2\sqrt{ab}}$. Use this formula to add the radicals in Exercises 93–94. Then work the problem again by the methods discussed in this section. Which method do you prefer? Why?

93. $\sqrt{2} + \sqrt{8}$

94. $\sqrt{5} + \sqrt{20}$

The bar graph shows the percentage of full-time college students in the United States who had jobs for four selected years. The data can be described by the mathematical model

$$J = 1.4\sqrt{x} + 55 - (20 - 1.2\sqrt{x}),$$

where J is the percentage of full-time college students with jobs x years after 1975. Use this information to solve Exercises 95–96.

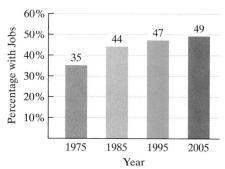

Percentage of Full-Time College Students with Jobs

Source: Department of Education

95. a. Simplify the mathematical model for *J*.

   b. Use the simplified form of the model to find the percentage of full-time college students who had jobs in 2005. Round to the nearest percent. How does this compare with the actual percentage displayed by the bar graph?

96. a. Simplify the mathematical model for *J*.

   b. Use the simplified form of the model to find the percentage of full-time college students who had jobs in 1995. Round to the nearest percent. How does this compare with the actual percentage displayed by the bar graph?

The graph of the model for the percentage of full-time college students with jobs is shown in a rectangular coordinate system. Use the graph to solve Exercises 97–98.

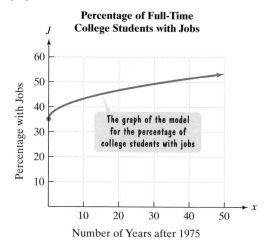

Percentage of Full-Time College Students with Jobs

97. a. Identify your solution from Exercise 95(b) as a point on the graph.

   b. Use the graph to estimate the percentage of full-time college students who will have jobs in 2015.

98. a. Identify your solution from Exercise 96(b) as a point on the graph.

   b. Use the graph to estimate the percentage of full-time college students who will have jobs in 2025.

Writing in Mathematics

99. What are like radicals? Give an example with your explanation.

100. Explain how to add like radicals. Give an example with your explanation.

101. If only like radicals can be added, why is it possible to add $\sqrt{2}$ and $\sqrt{8}$?

102. Explain how to perform this multiplication:
$$\sqrt{2}\left(\sqrt{7} + \sqrt{10}\right).$$

103. Explain how to perform this multiplication:
$$\left(2 + \sqrt{3}\right)\left(4 + \sqrt{3}\right).$$

104. What are conjugates? Give an example with your explanation.

105. Describe how to multiply conjugates.

Critical Thinking Exercises

Make Sense? *In Exercises 106–109, determine whether each statement "makes sense" or "does not make sense" and explain your reasoning.*

106. The unlike radicals $5\sqrt{2}$ and $4\sqrt{3}$ remind me of the unlike terms $5x$ and $4y$ that cannot be combined by addition or subtraction.

107. I simplified the terms of $2\sqrt{20} + 4\sqrt{75}$, and then I was able to identify and add the like radicals.

108. I use the same ideas to find $\left(3 + \sqrt{2}\right)\left(5 + \sqrt{2}\right)$ that I did to find the binomial product $(3 + x)(5 + x)$.

109. In some cases when I multiply a radical expression and its conjugate, the simplified product contains a radical.

In Exercises 110–113, determine whether each statement is true or false. If the statement is false, make the necessary change(s) to produce a true statement.

110. $\sqrt{2} + \sqrt{8} = \sqrt{10}$

111. $\left(\sqrt{5} + \sqrt{3}\right)^2 = 5 + 3$

112. $4\sqrt{3} + 5\sqrt{3} = 9\sqrt{6}$

113. $\left(\sqrt{7} + \sqrt{3}\right)\left(\sqrt{7} - \sqrt{3}\right) = 40$

114. Simplify: $\sqrt{5} \cdot \sqrt{15} + 6\sqrt{3}$.

115. Multiply: $\left(\sqrt[3]{4} + 1\right)\left(\sqrt[3]{2} - 3\right)$.

116. Fill in the boxes to make the statement true:

$\left(5 + \sqrt{}\right)\left(5 - \sqrt{}\right) = 22.$

117. Multiply: $\left(4\sqrt{3x} + \sqrt{2y}\right)\left(4\sqrt{3x} - \sqrt{2y}\right)$.

Technology Exercises

In Exercises 118–121, determine if each operation is performed correctly by graphing each side of the equation with your graphing utility. Use the given viewing rectangle. The graphs should be the same. If they are not, correct the right side of the equation and then use your graphing utility to verify the correction.

118. $\sqrt{4x} + \sqrt{9x} = 5\sqrt{x}$

$[0, 5, 1]$ by $[0, 10, 1]$

119. $\sqrt{16x} - \sqrt{9x} = \sqrt{7x}$

$[0, 5, 1]$ by $[0, 5, 1]$

120. $\left(\sqrt{x} - 1\right)\left(\sqrt{x} - 1\right) = x + 1$

$[0, 5, 1]$ by $[-1, 2, 1]$

121. $\left(\sqrt{x} + 2\right)\left(\sqrt{x} - 2\right) = x^2 - 4$

$[-10, 10, 1]$ by $[-10, 10, 1]$

Review Exercises

122. Multiply: $(5x + 3)(5x - 3)$. (Section 6.3, Example 4)

123. Factor completely: $64x^3 - x$. (Section 7.5, Example 3)

124. Graph: $y = -\dfrac{1}{4}x + 3$. (Section 4.4, Example 3)

Preview Exercises

Exercises 125–127 will help you prepare for the material covered in the next section.

125. Multiply and simplify: $\dfrac{\sqrt{3}}{\sqrt{5}} \cdot \dfrac{\sqrt{5}}{\sqrt{5}}$.

126. Multiply: $\left(5 + \sqrt{3}\right)\left(5 - \sqrt{3}\right)$.

127. Multiply: $\left(\sqrt{6} - \sqrt{2}\right)\left(\sqrt{6} + \sqrt{2}\right)$.

MID-CHAPTER CHECK POINT Section 9.1–Section 9.3

✓ **What You Know:** We learned to find roots of numbers. We saw that if an index is even, the radicand must be nonnegative for a root to be a real number. We learned to simplify radicals. We performed various operations with radicals including multiplication, division, addition, and subtraction.

In Exercises 1–20, simplify the given expression or perform the indicated operation and, if possible, simplify. Assume that all variables represent positive real numbers.

1. $\sqrt{50} \cdot \sqrt{6}$

2. $\sqrt{6} + 9\sqrt{6}$

3. $\sqrt{96x^3}$

4. $\sqrt[5]{\dfrac{4}{32}}$

5. $\sqrt{27} + 3\sqrt{12}$

6. $\left(\sqrt{10} + \sqrt{3}\right)\left(\sqrt{10} - \sqrt{3}\right)$

7. $\dfrac{\sqrt{5}}{2}\left(4\sqrt{3} - 6\sqrt{20}\right)$

8. $-\sqrt{32x^{21}}$

9. $\sqrt{6x^3} \cdot \sqrt{2x^4}$

10. $\dfrac{\sqrt[3]{32}}{\sqrt[3]{2}}$

11. $-3\sqrt{90} - 5\sqrt{40}$

12. $\left(2 - \sqrt{3}\right)\left(5 + 2\sqrt{3}\right)$

13. $\dfrac{\sqrt{56x^5}}{\sqrt{7x^3}}$

14. $-\sqrt[4]{32}$

15. $\left(\sqrt{2} + \sqrt{7}\right)^2$

16. $\sqrt[3]{\dfrac{1}{2}} \cdot \sqrt[3]{32}$

17. $\sqrt{5} + \sqrt{20} + \sqrt{45}$

18. $\dfrac{1}{3}\sqrt{\dfrac{90}{16}}$

19. $3\sqrt{2}\left(\sqrt{2} + \sqrt{5}\right)$

20. $\left(5 - \sqrt{2}\right)\left(5 + \sqrt{2}\right)$

9.4

Rationalizing the Denominator

Objectives

1 Rationalize denominators containing one term.

2 Rationalize denominators containing two terms.

"You don't pay taxes; they take taxes."
—*Chris Rock*

They say a fool and his money are soon parted and the rest of us just wait to be taxed. Of course, there's more complaining when you're not expecting a refund and more taxes must be paid. In most tax years, approximately 59% of all taxpayers receive a tax refund, whereas 41% must pay more taxes than were withheld.

In this section's exercise set, you will use a formula that models the percentage of taxpayers who must pay more taxes based on the taxpayer's age. In the section itself, you will learn how to rewrite the formula without a radical in its denominator using a technique called *rationalizing the denominator*.

1 Rationalize denominators containing one term.

Rationalizing Denominators Containing One Term

A calculator makes it fairly easy to find an approximation for divisions involving radicals, such as $\frac{1}{\sqrt{3}}$. However, it is sometimes easier to work with radical expressions if the denominators do not contain any radicals.

The calculator screen in **Figure 9.4** shows the approximate values for $\frac{1}{\sqrt{3}}$ and $\frac{\sqrt{3}}{3}$. The two approximations are the same. This is not a coincidence:

$$\frac{1}{\sqrt{3}} = \frac{1}{\sqrt{3}} \cdot \boxed{\frac{\sqrt{3}}{\sqrt{3}}} = \frac{\sqrt{3}}{\sqrt{9}} = \frac{\sqrt{3}}{3}.$$

Any number divided by itself is 1.
Multiplication by 1 does not change the value of $\frac{1}{\sqrt{3}}$.

```
1/√(3)
          .5773502692
√(3)/3
          .5773502692
```

FIGURE 9.4 The calculator screen shows approximate values for $\frac{1}{\sqrt{3}}$ and $\frac{\sqrt{3}}{3}$.

This process involves rewriting a radical expression as an equivalent expression in which the denominator no longer contains any radicals. The process is called **rationalizing the denominator**. If the denominator contains the square root of a natural number that is not a perfect square, **multiply the numerator and the denominator by the smallest number that produces the square root of a perfect square in the denominator**.

EXAMPLE 1   Rationalizing Denominators

Rationalize the denominator:

a. $\frac{15}{\sqrt{6}}$ **b.** $\sqrt{\frac{3}{5}}$.

Solution

a. If we multiply the numerator and denominator of $\dfrac{15}{\sqrt{6}}$ by $\sqrt{6}$, the denominator becomes $\sqrt{6} \cdot \sqrt{6} = \sqrt{36} = 6$. Therefore, we multiply by 1, choosing $\dfrac{\sqrt{6}}{\sqrt{6}}$ for 1.

$$\dfrac{15}{\sqrt{6}} = \dfrac{15}{\sqrt{6}} \cdot \dfrac{\sqrt{6}}{\sqrt{6}}$$ Multiply the numerator and denominator by $\sqrt{6}$ to remove the square root in the denominator.

$$= \dfrac{15\sqrt{6}}{6}$$ $\sqrt{6} \cdot \sqrt{6} = \sqrt{36} = 6$

$$= \dfrac{5\sqrt{6}}{2}$$ Simplify, dividing the numerator and denominator by 3.

b. $\sqrt{\dfrac{3}{5}} = \dfrac{\sqrt{3}}{\sqrt{5}}$ The square root of a quotient is the quotient of the square roots.

$$= \dfrac{\sqrt{3}}{\sqrt{5}} \cdot \dfrac{\sqrt{5}}{\sqrt{5}}$$ Because $\sqrt{5}$ is the smallest factor that will produce a perfect square in the denominator, multiply by 1, choosing $\dfrac{\sqrt{5}}{\sqrt{5}}$ for 1.

$$= \dfrac{\sqrt{15}}{5}$$ $\sqrt{5} \cdot \sqrt{5} = \sqrt{25} = 5$ ■

☑ **CHECK POINT 1**   Rationalize the denominator:

a. $\dfrac{25}{\sqrt{10}}$ b. $\sqrt{\dfrac{2}{7}}$.

It is a good idea to simplify a radical expression before attempting to rationalize the denominator.

EXAMPLE 2   **Simplifying and Then Rationalizing Denominators**

Rationalize the denominator:

a. $\dfrac{12}{\sqrt{8}}$ b. $\sqrt{\dfrac{7x}{75}}$.

Solution

a. We begin by simplifying $\sqrt{8}$.

$$\dfrac{12}{\sqrt{8}} = \dfrac{12}{\sqrt{4 \cdot 2}}$$ 4 is the greatest perfect square factor of 8.

$$= \dfrac{12}{2\sqrt{2}}$$ $\sqrt{4 \cdot 2} = \sqrt{4}\sqrt{2} = 2\sqrt{2}$

$$= \dfrac{6}{\sqrt{2}}$$ Simplify, dividing the numerator and denominator by 2.

$$= \dfrac{6}{\sqrt{2}} \cdot \dfrac{\sqrt{2}}{\sqrt{2}}$$ Rationalize the denominator.

$$= \dfrac{6\sqrt{2}}{2}$$ $\sqrt{2} \cdot \sqrt{2} = \sqrt{4} = 2$

$$= 3\sqrt{2}$$ Simplify.

b. $\sqrt{\dfrac{7x}{75}} = \dfrac{\sqrt{7x}}{\sqrt{75}}$ The square root of a quotient is the quotient of the square roots.

$= \dfrac{\sqrt{7x}}{\sqrt{25 \cdot 3}}$ Simplify the denominator. 25 is the greatest perfect square factor of 75.

$= \dfrac{\sqrt{7x}}{5\sqrt{3}}$ $\sqrt{25 \cdot 3} = \sqrt{25}\sqrt{3} = 5\sqrt{3}$

$= \dfrac{\sqrt{7x}}{5\sqrt{3}} \cdot \dfrac{\sqrt{3}}{\sqrt{3}}$ Rationalize the denominator, choosing $\dfrac{\sqrt{3}}{\sqrt{3}}$ for 1.

$= \dfrac{\sqrt{21x}}{5 \cdot 3}$ $\sqrt{3} \cdot \sqrt{3} = \sqrt{9} = 3$

$= \dfrac{\sqrt{21x}}{15}$

Study Tip

You cannot simplify

$$\dfrac{\sqrt{21x}}{15}$$

by dividing the numerator and the denominator by 3. Although 3 is a factor of 15, it is *not* a factor of $\sqrt{21x}$. (The factor is $\sqrt{3}$.) Thus, the answer in Example 2(b) *cannot* be written as

$$\dfrac{\sqrt{7x}}{5}.$$

☑ **CHECK POINT 2**   Rationalize the denominator:

a. $\dfrac{15}{\sqrt{18}}$ **b.** $\sqrt{\dfrac{7x}{20}}.$

---

2 Rationalize denominators containing two terms.

Rationalizing Denominators Containing Two Terms

How can we rationalize a denominator if the denominator contains two terms with one or more square roots? **Multiply the numerator and the denominator by the conjugate of the denominator.** Here are two examples of such expressions:

$$\dfrac{7}{5 + \sqrt{3}} \qquad\qquad \dfrac{2}{\sqrt{6} - \sqrt{2}}.$$

The conjugate of the denominator is $5 - \sqrt{3}$.

The conjugate of the denominator is $\sqrt{6} + \sqrt{2}$.

The product of the denominator and its conjugate is found using the formula

$$(A + B)(A - B) = A^2 - B^2.$$

The simplified product will not contain a radical.

EXAMPLE 3   Rationalizing a Denominator Containing Two Terms

Rationalize the denominator: $\dfrac{7}{5 + \sqrt{3}}.$

Solution The conjugate of the denominator is $5 - \sqrt{3}$. If we multiply the numerator and the denominator by $5 - \sqrt{3}$, the denominator will not contain a radical. Therefore, we multiply by 1, choosing $\dfrac{5 - \sqrt{3}}{5 - \sqrt{3}}$ for 1.

$\dfrac{7}{5 + \sqrt{3}} = \dfrac{7}{5 + \sqrt{3}} \cdot \dfrac{5 - \sqrt{3}}{5 - \sqrt{3}}$ Multiply by 1.

$= \dfrac{7(5 - \sqrt{3})}{5^2 - (\sqrt{3})^2}$ $(A + B)(A - B) = A^2 - B^2$

$= \dfrac{7(5 - \sqrt{3})}{25 - 3}$ $(\sqrt{3})^2 = \sqrt{3} \cdot \sqrt{3} = \sqrt{9} = 3$

$= \dfrac{7(5 - \sqrt{3})}{22}$ or $\dfrac{35 - 7\sqrt{3}}{22}$

☑ **CHECK POINT 3**   Rationalize the denominator: $\dfrac{8}{4 + \sqrt{5}}$.

EXAMPLE 4   Rationalizing a Denominator Containing Two Terms

Rationalize the denominator: $\dfrac{2}{\sqrt{6} - \sqrt{2}}$.

Solution   The conjugate of the denominator is $\sqrt{6} + \sqrt{2}$. If we multiply the numerator and the denominator by $\sqrt{6} + \sqrt{2}$, the denominator will not contain a radical. Therefore, we multiply by 1, choosing $\dfrac{\sqrt{6} + \sqrt{2}}{\sqrt{6} + \sqrt{2}}$ for 1.

$$\dfrac{2}{\sqrt{6} - \sqrt{2}} = \dfrac{2}{\sqrt{6} - \sqrt{2}} \cdot \dfrac{\sqrt{6} + \sqrt{2}}{\sqrt{6} + \sqrt{2}}$$ Multiply by 1.

$$= \dfrac{2(\sqrt{6} + \sqrt{2})}{(\sqrt{6})^2 - (\sqrt{2})^2}$$ $(A - B)(A + B) = A^2 - B^2$

$$= \dfrac{2(\sqrt{6} + \sqrt{2})}{6 - 2}$$ $(\sqrt{6})^2 = \sqrt{6} \cdot \sqrt{6} = \sqrt{36} = 6$
and $(\sqrt{2})^2 = \sqrt{2} \cdot \sqrt{2} = \sqrt{4} = 2$

$$= \dfrac{2(\sqrt{6} + \sqrt{2})}{4}$$

$$= \dfrac{\overset{1}{2}(\sqrt{6} + \sqrt{2})}{\underset{2}{4}}$$ Divide the numerator and denominator by the common factor, 2.

$$= \dfrac{\sqrt{6} + \sqrt{2}}{2}$$

☑ **CHECK POINT 4**   Rationalize the denominator: $\dfrac{8}{\sqrt{7} - \sqrt{3}}$.

9.4 EXERCISE SET   *MyMathLab*   Math XL
PRACTICE   WATCH   DOWNLOAD   READ   REVIEW

Practice Exercises

In Exercises 1–24, rationalize each denominator. If possible, simplify the rationalized expression by dividing the numerator and denominator by the greatest common factor.

1. $\dfrac{1}{\sqrt{10}}$

2. $\dfrac{1}{\sqrt{2}}$

3. $\dfrac{5}{\sqrt{5}}$

4. $\dfrac{7}{\sqrt{7}}$

5. $\dfrac{2}{\sqrt{6}}$

6. $\dfrac{4}{\sqrt{6}}$

7. $\dfrac{28}{\sqrt{7}}$

8. $\dfrac{40}{\sqrt{5}}$

9. $\sqrt{\dfrac{3}{5}}$

10. $\sqrt{\dfrac{3}{7}}$

11. $\sqrt{\dfrac{7}{3}}$

12. $\sqrt{\dfrac{5}{2}}$

13. $\sqrt{\dfrac{x^2}{3}}$

14. $\sqrt{\dfrac{x^2}{7}}$

15. $\sqrt{\dfrac{11}{x}}$

16. $\sqrt{\dfrac{6}{x}}$

17. $\sqrt{\dfrac{x}{y}}$

18. $\sqrt{\dfrac{a}{b}}$

19. $\sqrt{\dfrac{x^4}{2}}$

20. $\sqrt{\dfrac{x^4}{3}}$

21. $\dfrac{\sqrt{7}}{\sqrt{5}}$

22. $\dfrac{\sqrt{11}}{\sqrt{5}}$

23. $\dfrac{\sqrt{3x}}{\sqrt{14}}$

24. $\dfrac{\sqrt{2x}}{\sqrt{17}}$

In Exercises 25–52, begin by simplifying the expression. Then rationalize the denominator using the simplified expression.

25. $\dfrac{1}{\sqrt{20}}$

26. $\dfrac{1}{\sqrt{18}}$

27. $\dfrac{12}{\sqrt{32}}$

28. $\dfrac{15}{\sqrt{50}}$

29. $\dfrac{15}{\sqrt{12}}$

30. $\dfrac{13}{\sqrt{40}}$

31. $\sqrt{\dfrac{5}{18}}$

32. $\sqrt{\dfrac{7}{12}}$

33. $\sqrt{\dfrac{x}{32}}$

34. $\sqrt{\dfrac{x}{40}}$

35. $\sqrt{\dfrac{1}{45}}$

36. $\sqrt{\dfrac{1}{54}}$

37. $\dfrac{\sqrt{7}}{\sqrt{12}}$

38. $\dfrac{\sqrt{5}}{\sqrt{18}}$

39. $\dfrac{8x}{\sqrt{8}}$

40. $\dfrac{27x}{\sqrt{27}}$

41. $\dfrac{\sqrt{7y}}{\sqrt{8}}$

42. $\dfrac{\sqrt{3y}}{\sqrt{125}}$

43. $\sqrt{\dfrac{7x}{12}}$

44. $\sqrt{\dfrac{11x}{18}}$

45. $\sqrt{\dfrac{45}{x}}$

46. $\sqrt{\dfrac{27}{x}}$

47. $\dfrac{5}{\sqrt{x^3}}$

48. $\dfrac{11}{\sqrt{x^3}}$

49. $\sqrt{\dfrac{27}{y^3}}$

50. $\sqrt{\dfrac{45}{y^3}}$

51. $\dfrac{\sqrt{50x^2}}{\sqrt{12y^3}}$

52. $\dfrac{\sqrt{27x^2}}{\sqrt{12y^3}}$

In Exercises 53–74, rationalize each denominator. Simplify, if possible.

53. $\dfrac{1}{4+\sqrt{3}}$

54. $\dfrac{1}{5+\sqrt{2}}$

55. $\dfrac{9}{2-\sqrt{7}}$

56. $\dfrac{12}{2-\sqrt{7}}$

57. $\dfrac{16}{\sqrt{11}+3}$

58. $\dfrac{15}{\sqrt{7}+2}$

59. $\dfrac{18}{3-\sqrt{3}}$

60. $\dfrac{40}{5-\sqrt{5}}$

61. $\dfrac{\sqrt{2}}{\sqrt{2}+1}$

62. $\dfrac{\sqrt{3}}{\sqrt{3}+1}$

63. $\dfrac{\sqrt{10}}{\sqrt{10}-\sqrt{7}}$

64. $\dfrac{\sqrt{5}}{\sqrt{5}-\sqrt{3}}$

65. $\dfrac{6}{\sqrt{6}+\sqrt{3}}$

66. $\dfrac{8}{\sqrt{7}+\sqrt{3}}$

67. $\dfrac{2}{\sqrt{5}-\sqrt{3}}$

68. $\dfrac{5}{\sqrt{7}-\sqrt{2}}$

69. $\dfrac{2}{4+\sqrt{x}}$

70. $\dfrac{6}{4-\sqrt{x}}$

71. $\dfrac{2\sqrt{3}}{\sqrt{15}+2}$

72. $\dfrac{3\sqrt{2}}{\sqrt{10}+2}$

73. $\dfrac{\sqrt{5}+\sqrt{2}}{\sqrt{5}-\sqrt{2}}$

74. $\dfrac{\sqrt{5}+\sqrt{3}}{\sqrt{5}-\sqrt{3}}$

Practice PLUS

In Exercises 75–82, rationalize each denominator. Simplify, if possible.

75. $\dfrac{\sqrt{36x^2y^5}}{\sqrt{2x^3y}}$

76. $\dfrac{\sqrt{100x^5y^2}}{\sqrt{2xy^3}}$

77. $\dfrac{2}{\sqrt{x+2}-\sqrt{x}}$

78. $\dfrac{3}{\sqrt{x+3}-\sqrt{x}}$

79. $\dfrac{\sqrt{2}}{\sqrt{3}}+\dfrac{\sqrt{3}}{\sqrt{2}}$

80. $\dfrac{\sqrt{2}}{\sqrt{7}}+\dfrac{\sqrt{7}}{\sqrt{2}}$

81. $\dfrac{2x+4-2h}{\sqrt{x+2}-h}$

82. $\dfrac{4x+12-4h}{\sqrt{x+3}-h}$

Application Exercises

83. Do you expect to pay more taxes than were withheld? Would you be surprised to know that the percentage of taxpayers who receive a refund and the percentage of taxpayers who pay more taxes vary according to age? The formula

$$P = \frac{x(13+\sqrt{x})}{5\sqrt{x}}$$

models the percentage, P, of taxpayers who are x years old who must pay more taxes.

a. What percentage of 25-year-olds must pay more taxes?

b. Rewrite the formula by rationalizing the denominator.

c. Use the rationalized form of the formula from part (b) to find the percentage of 25-year-olds who must pay more taxes. Do you get the same answer as you did in part (a)? If so, does this prove that you correctly rationalized the denominator? Explain.

84. In the Peanuts cartoon, Woodstock appears to be working steps mentally. Fill in the missing steps that show how to go from $\dfrac{7\sqrt{2 \cdot 2 \cdot 3}}{6}$ to $\dfrac{7}{3}\sqrt{3}$.

PEANUTS © United Feature Syndicate, Inc.

85. The early Greeks believed that the most pleasing of all rectangles were *golden rectangles*, whose ratio of width to height is
$$\frac{w}{h} = \frac{2}{\sqrt{5} - 1}.$$
Rationalize the denominator for this ratio and then use a calculator to approximate the answer correct to the nearest hundredth.

Writing in Mathematics

86. Describe what it means to rationalize a denominator. Use both $\dfrac{1}{\sqrt{5}}$ and $\dfrac{1}{5 + \sqrt{5}}$ in your explanation.

87. When a radical expression has its denominator rationalized, we change the denominator so that it no longer contains a radical. Doesn't this change the value of the radical expression? Explain.

88. Square the real number $\dfrac{2}{\sqrt{3}}$. Observe that the radical is eliminated from the denominator. Explain whether this process is equivalent to rationalizing the denominator.

89. Use the model in Exercise 83 and a calculator to find the percentage of taxpayers ages 30, 40, and 50 who expect to pay more taxes. Describe the trend that you observe. What explanation can you offer for this trend?

Critical Thinking Exercises

Make Sense?   *In Exercises 90–93, determine whether each statement "makes sense" or "does not make sense" and explain your reasoning.*

90. I rationalized a numerical denominator and the simplified denominator still contained an irrational number.

91. Without using a calculator and knowing that $\sqrt{2} \approx 1.4142$, rationalizing the denominator of $\dfrac{1}{\sqrt{2}}$ makes division to to obtain a decimal approximation for $\dfrac{1}{\sqrt{2}}$ easier to perform.

92. Because 10 and 8 share a common factor of 2, I simplified $\dfrac{\sqrt{10}}{8}$ to $\dfrac{\sqrt{5}}{4}$.

93. I use the fact that 1 is the multiplicative identity to both rationalize denominators and rewrite rational expressions with a common denominator.

In Exercises 94–97, determine whether each statement is true or false. If the statement is false, make the necessary change(s) to produce a true statement.

94. $\dfrac{4 + 8\sqrt{3}}{4} = 1 + 8\sqrt{3}$

95. $\dfrac{3\sqrt{x}}{x\sqrt{6}} = \dfrac{\sqrt{6x}}{2x}$ for $x > 0$

96. Conjugates are used to rationalize the denominator of $\dfrac{2 - \sqrt{5}}{\sqrt{3}}$.

97. Radical expressions with rationalized denominators require less space to write than before they are rationalized.

98. Rationalize the denominator: $\dfrac{1}{\sqrt[3]{2}}$.

99. Simplify: $\sqrt{2} + \sqrt{\dfrac{1}{2}}$.

100. Simplify: $\sqrt{13 + \sqrt{2} + \dfrac{7}{3 + \sqrt{2}}}$.

101. Fill in the box to make the statement true:
$$\frac{4}{2 + \sqrt{\boxed{}}} = 8 - 4\sqrt{3}.$$

Review Exercises

102. Solve: $6x^2 - 11x + 5 = 0$. (Section 7.6, Example 2)

103. Simplify: $(2x^2)^{-3}$. (Section 6.7, Example 6)

104. Multiply: $\dfrac{x^2 - 6x + 9}{12} \cdot \dfrac{3}{x^2 - 9}$. (Section 8.2, Example 3)

Preview Exercises

Exercises 105–107 will help you prepare for the material covered in the next section.

105. Solve: $7 = -2.5x + 17$.

106. Solve: $2x - 1 = x^2 - 4x + 4$.

107. Which solution of the equation in Exercise 106 is a solution of $\sqrt{2x - 1} + 2 = x$?

<table>
<tr><td></td></tr>
</table>

9.5

Radical Equations

Objectives

1 Solve radical equations.

2 Solve problems involving square-root models.

Long before the passage of laws prohibiting marriage between people of the same sex, a substantial percentage of Americans were in favor of laws prohibiting interracial marriage. Since 1972, the General Social Survey has asked

> Do you think that there should be laws against marriages between negroes/ blacks/African Americans and whites?

In 1972, 39.3% of the adult U.S. population was in favor of such laws. The bar graph in **Figure 9.5** shows the percentage of Americans in favor of legislation prohibiting interracial marriage for five selected years from 1993 through 2002.

Percentage of Americans in Favor of Laws Prohibiting Interracial Marriage

FIGURE 9.5

Source: Ben Schott, *Schott's Almanac 2007*, Donnelley and Sons

The bar graph indicates that the percentage of Americans in favor of laws prohibiting interracial marriage is decreasing for the period shown. However, the rate of decrease is slowing down. For this reason, a model in the form $y = a\sqrt{x} + b$, where a is *negative*, is appropriate for describing the data. The formula

$$p = -2.5\sqrt{t} + 17$$

models the percentage of Americans, p, in favor of legislation prohibiting interracial marriage t years after 1993.

If trends shown from 1993 through 2002 continue, in what year will the percentage favoring such legislation decrease to 7%? Substitute 7 for p in the formula and solve for t:

$$7 = -2.5\sqrt{t} + 17.$$

The resulting equation contains a variable in the radicand and is called a *radical equation*. A **radical equation** is an equation in which the variable occurs in a square root, cube root, or any higher root.

$$7 = -2.5\sqrt{t} + 17$$

The variable occurs in a sqaure root.

In this section, you will learn how to solve radical equations with square roots. Solving such equations will enable you to solve new kinds of problems using square-root models, such as the model for the percentage of Americans in favor of laws prohibiting interracial marriage. We will return to this model at the end of the section.

1 Solve radical equations.

Radical Equations with Square Roots

Consider the following radical equation:

$$\sqrt{x} = 5.$$

We solve the equation by squaring both sides:

Squaring both sides eliminates the square root.

$$\left(\sqrt{x}\right)^2 = 5^2$$
$$x = 25.$$

The proposed solution, 25, can be checked in the original equation, $\sqrt{x} = 5$. Because $\sqrt{25} = 5$, the solution is 25.

In general, we solve radical equations with square roots by squaring both sides of the equation. Unfortunately, all the solutions of the squared equation may not be solutions of the original equation. Consider, for example, the equation

$$x = 4.$$

If we square both sides, we obtain

$$x^2 = 16.$$

This new equation has two solutions, -4 and 4. By contrast, only 4 is a solution of the original equation, $x = 4$. For this reason, we must **always check proposed solutions of radical equations in the original equation**.

Here is a general method for solving radical equations with square roots:

Solving Radical Equations Containing Square Roots

1. If necessary, arrange terms so that one radical is isolated on one side of the equation.
2. Square both sides of the equation to eliminate the square root.
3. Solve the resulting equation.
4. Check all proposed solutions in the original equation.

EXAMPLE 1 Solving a Radical Equation

Solve: $\sqrt{3x + 4} = 8$.

Solution

Step 1. Isolate the radical. The radical, $\sqrt{3x + 4}$, is already isolated on the left side of the equation, so we can skip this step.

Step 2. Square both sides.

$$\sqrt{3x + 4} = 8 \qquad \text{This is the given equation.}$$
$$\left(\sqrt{3x + 4}\right)^2 = 8^2 \qquad \text{Square both sides to eliminate the radical.}$$
$$3x + 4 = 64 \qquad \text{Simplify.}$$

Step 3. Solve the resulting equation.

$$3x + 4 = 64 \qquad \text{The resulting equation is a linear equation.}$$
$$3x = 60 \qquad \text{Subtract 4 from both sides.}$$
$$x = 20 \qquad \text{Divide both sides by 3.}$$

Using Technology

Graphic Connections

We can use a graphing utility to verify the solution to the equation in Example 1:

$$\sqrt{3x + 4} = 8.$$

Graph each side of the equation:

$$y_1 = \sqrt{3x + 4}$$
$$y_2 = 8.$$

$\boxed{\text{TRACE}}$ along the curves or use the utility's intersection feature. The solution, as shown below, is the first coordinate of the intersection point. Thus, the solution is 20.

[0, 22, 1] by [0, 10, 1]

Step 4. Check the proposed solution in the original equation.

Check 20:

$$\sqrt{3x + 4} = 8$$
$$\sqrt{3 \cdot 20 + 4} \stackrel{?}{=} 8$$
$$\sqrt{60 + 4} \stackrel{?}{=} 8$$
$$\sqrt{64} \stackrel{?}{=} 8$$
$$8 = 8, \quad \text{true}$$

The solution is 20 and the solution set is $\{20\}$.

☑ **CHECK POINT 1** Solve: $\sqrt{2x + 3} = 5$.

EXAMPLE 2 Solving a Radical Equation

Solve: $\sqrt{3x + 9} - 2\sqrt{x} = 0$.

Solution

Step 1. Isolate each radical. We can get each radical by itself on one side of the equation by adding $2\sqrt{x}$ to both sides.

$$\sqrt{3x + 9} - 2\sqrt{x} = 0 \qquad \text{This is the given equation.}$$
$$\sqrt{3x + 9} - 2\sqrt{x} + 2\sqrt{x} = 0 + 2\sqrt{x} \qquad \text{Add } 2\sqrt{x} \text{ to both sides.}$$
$$\sqrt{3x + 9} = 2\sqrt{x} \qquad \text{Simplify.}$$

Step 2. Square both sides.

$$\left(\sqrt{3x + 9}\right)^2 = \left(2\sqrt{x}\right)^2$$
$$3x + 9 = 4x \qquad \text{Simplify.}$$

Step 3. Solve the resulting equation.

$$3x + 9 = 4x \qquad \text{The resulting equation is a linear equation.}$$
$$9 = x \qquad \text{Subtract } 3x \text{ from both sides.}$$

Step 4. Check the proposed solution in the original equation.

Check 9:

$$\sqrt{3x + 9} - 2\sqrt{x} = 0$$
$$\sqrt{3 \cdot 9 + 9} - 2\sqrt{9} \stackrel{?}{=} 0$$
$$\sqrt{36} - 2\sqrt{9} \stackrel{?}{=} 0$$
$$6 - 2 \cdot 3 \stackrel{?}{=} 0$$
$$6 - 6 \stackrel{?}{=} 0$$
$$0 = 0, \quad \text{true}$$

The solution is 9 and the solution set is $\{9\}$.

☑ **CHECK POINT 2** Solve: $\sqrt{x + 32} - 3\sqrt{x} = 0$.

EXAMPLE 3 Solving a Radical Equation

Solve: $\sqrt{x} + 5 = 0$.

Solution

Step 1. Isolate the radical. We isolate the radical, $\sqrt{x}$, on the left side by subtracting 5 from both sides.

$$\sqrt{x} + 5 = 0 \qquad \text{This is the given equation.}$$
$$\sqrt{x} = -5 \qquad \text{Subtract 5 from both sides.}$$

> A principal square root cannot be negative. This equation has no solution. Let's continue the solution procedure to see what happens.

Step 2. Square both sides.

$$\left(\sqrt{x}\right)^2 = \left(-5\right)^2$$
$$x = 25 \qquad \text{Simplify.}$$

Step 3. Solve the resulting equation. We immediately see that 25 is the proposed solution.

Step 4. Check the proposed solution in the original equation.

Check 25:
$$\sqrt{x} + 5 = 0$$
$$\sqrt{25} + 5 \stackrel{?}{=} 0$$
$$5 + 5 \stackrel{?}{=} 0$$
$$10 = 0, \quad \text{false}$$

This false statement indicates that 25 is not a solution. Thus, the equation has no solution. The solution set is $\varnothing$, the empty set. ■

Example 3 illustrates that extra solutions may be introduced when you raise both sides of a radical equation to an even power. Such solutions, which are not solutions of the given equation, are called **extraneous solutions**. Thus, 25 is an extraneous solution of the equation $\sqrt{x} + 5 = 0$.

☑ **CHECK POINT 3** Solve: $\sqrt{x} + 1 = 0$.

Using Technology

Graphic Connections

The graphs of

$$y_1 = \sqrt{x}$$

and

$$y_2 = -5$$

do not intersect, as shown below. Thus, $\sqrt{x} = -5$ has no real number solution.

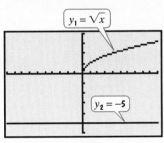

[−10, 10, 1] by [−6, 4, 1]

EXAMPLE 4 Solving a Radical Equation

Solve: $\sqrt{2x - 1} + 2 = x$.

Solution

Step 1. Isolate the radical. We isolate the radical, $\sqrt{2x - 1}$, by subtracting 2 from both sides.

$$\sqrt{2x - 1} + 2 = x \qquad \text{This is the given equation.}$$
$$\sqrt{2x - 1} = x - 2 \qquad \text{Subtract 2 from both sides.}$$

Step 2. **Square both sides.**

$$\left(\sqrt{2x-1}\right)^2 = (x-2)^2$$
$$2x - 1 = x^2 - 4x + 4$$

Simplify. Use the formula
$(A - B)^2 = A^2 - 2AB + B^2$
on the right side.

Step 3. **Solve the resulting equation.** Because of the x^2-term, the resulting equation is a quadratic equation. We can obtain 0 on the left side by subtracting $2x$ and adding 1 on both sides.

$$2x - 1 = x^2 - 4x + 4$$
$$0 = x^2 - 6x + 5$$
$$0 = (x - 1)(x - 5)$$
$$x - 1 = 0 \quad \text{or} \quad x - 5 = 0$$
$$x = 1 \qquad\qquad x = 5$$

The resulting equation is quadratic.
Write in standard form, subtracting $2x$ and adding 1 on both sides.
Factor.
Set each factor equal to 0.
Solve the resulting equations.

Step 4. **Check the proposed solutions in the original equation.**

Check 1:	Check 5:
$\sqrt{2x - 1} + 2 = x$	$\sqrt{2x - 1} + 2 = x$
$\sqrt{2 \cdot 1 - 1} + 2 \stackrel{?}{=} 1$	$\sqrt{2 \cdot 5 - 1} + 2 \stackrel{?}{=} 5$
$\sqrt{1} + 2 \stackrel{?}{=} 1$	$\sqrt{9} + 2 \stackrel{?}{=} 5$
$1 + 2 \stackrel{?}{=} 1$	$3 + 2 \stackrel{?}{=} 5$
$3 = 1, \quad$ false	$5 = 5, \quad$ true

Thus, 1 is an extraneous solution. The only solution is 5 and the solution set is $\{5\}$. ▄

Study Tip

If a radical equation leads to a quadratic equation with two solutions, there will not always be one solution that checks and one that does not. Both solutions of the quadratic equation can satisfy the original radical equation. It is also possible that neither solution of the quadratic equation satisfies the given radical equation.

Using Technology

Numeric Connections

A graphing utility's $\boxed{\text{TABLE}}$ feature provides a numeric check that 1 is not a solution and 5 is a solution of $\sqrt{2x - 1} + 2 = x$.

$y_1 = \sqrt{2x-1} + 2$ $y_2 = x$

When $x = 1$, y_1 and y_2 are not equal. 1 is an extraneous solution.

When $x = 5$, y_1 and y_2 are equal.

☑ **CHECK POINT 4** Solve: $\sqrt{x + 3} + 3 = x$.

---

2 Solve problems involving square-root models.

Applications of Radical Equations

Radical equations can be solved to answer questions about variables contained in square-root models.

EXAMPLE 5 Using a Square-Root Model

The formula

$$p = -2.5\sqrt{t} + 17$$

models the percentage of Americans, p, in favor of laws prohibiting interracial marriage t years after 1993. If trends indicated by this model continue, in what year will the percentage favoring such legislation decrease to 7%?

Solution Because we are interested in finding when 7% of Americans will favor laws prohibiting interracial marriage, we substitute 7 for p in the given formula. Then we solve for t, the number of years after 1993.

$p = -2.5\sqrt{t} + 17$	This is the given formula.
$7 = -2.5\sqrt{t} + 17$	Substitute 7 for p.
$-10 = -2.5\sqrt{t}$	Subtract 17 from both sides.
$\dfrac{-10}{-2.5} = \dfrac{-2.5\sqrt{t}}{-2.5}$	Divide both sides by -2.5.
$4 = \sqrt{t}$	Simplify: $\dfrac{-10}{-2.5} = 4$.
$4^2 = (\sqrt{t})^2$	Square both sides.
$16 = t$	Simplify.

The model indicates that the percentage of Americans favoring laws prohibiting interracial marriage will decrease to 7% by 16 years after 1993. Because $1993 + 16 = 2009$, this will occur in 2009.

✓ CHECK POINT 5 Use the formula in Example 5 to find by which year only 4.5% of Americans will favor laws prohibiting interracial marriage.

9.5 EXERCISE SET *MyMathLab*

Practice Exercises

In Exercises 1–44, solve each radical equation.

1. $\sqrt{x} = 5$
2. $\sqrt{x} = 6$
3. $\sqrt{x} - 4 = 0$
4. $\sqrt{x} - 9 = 0$
5. $\sqrt{x + 2} = 3$
6. $\sqrt{x - 6} = 5$
7. $\sqrt{x - 3} - 11 = 0$
8. $\sqrt{x + 5} - 8 = 0$
9. $\sqrt{3x - 5} = 4$
10. $\sqrt{5x - 6} = 8$
11. $\sqrt{x + 5} + 2 = 5$
12. $\sqrt{x + 6} + 1 = 3$
13. $\sqrt{x + 3} = \sqrt{4x - 3}$
14. $\sqrt{x + 8} = \sqrt{5x - 4}$
15. $\sqrt{6x - 2} = \sqrt{4x + 4}$
16. $\sqrt{5x - 4} = \sqrt{3x + 6}$
17. $11 = 6 + \sqrt{x + 1}$
18. $7 = 5 + \sqrt{x + 1}$
19. $\sqrt{x} + 10 = 0$
20. $\sqrt{x} + 8 = 0$
21. $\sqrt{x - 1} = -3$
22. $\sqrt{x - 2} = -5$
23. $3\sqrt{x} = \sqrt{8x + 16}$
24. $3\sqrt{x - 2} = \sqrt{7x + 4}$
25. $\sqrt{2x - 3} + 5 = 0$
26. $\sqrt{2x - 8} + 4 = 0$
27. $\sqrt{3x + 4} - 2 = 3$
28. $\sqrt{5x - 4} - 2 = 4$
29. $3\sqrt{x - 1} = \sqrt{3x + 3}$
30. $2\sqrt{3x + 4} = \sqrt{5x + 9}$

31. $\sqrt{x + 7} = x + 5$
32. $\sqrt{x + 10} = x - 2$
33. $\sqrt{2x + 13} = x + 7$
34. $\sqrt{6x + 1} = x - 1$
35. $\sqrt{9x^2 + 2x - 4} = 3x$
36. $\sqrt{9x^2 - 2x + 8} = 3x$
37. $x = \sqrt{2x - 2} + 1$
38. $x = \sqrt{3x + 7} - 3$
39. $x = \sqrt{8 - 7x} + 2$
40. $x = \sqrt{1 - 8x} + 2$
41. $\sqrt{3x + 10} = x + 4$
42. $\sqrt{x - 3} = x - 9$
43. $3\sqrt{x + 5} = 2$
44. $3\sqrt{x + 8} = 5$

Practice PLUS

45. Two added to the square root of the product of 4 and a number is equal to 10. Find the number.

46. Five added to the square root of the product of 6 and a number is equal to 8. Find the number.

47. A number is 4 more than the principal square root of twice the number. Find the number.

48. A number is 6 more than the principal square root of 3 times the number. Find the number.

49. Solve for h: $v = \sqrt{2gh}$.

50. Solve for l: $t = \dfrac{\pi}{2}\sqrt{\dfrac{l}{2}}$.

Solve the equations in Exercises 51–54. You will need to square both sides of each equation twice.

51. $\sqrt{x} + 2 = \sqrt{x + 8}$

52. $\sqrt{x} + 6 = \sqrt{x + 72}$

53. $\sqrt{x - 8} = \sqrt{x} - 2$

54. $\sqrt{x - 4} = \sqrt{x} - 2$

Application Exercises

When firefighters are working to put out a fire, the rate at which they spray water on the fire depends on the nozzle pressure. The formula

$$f = 120\sqrt{p}$$

models the water's flow rate, f, in gallons per minute, in terms of the nozzle pressure, p, in pounds per square inch. Use this formula to solve Exercises 55–56.

55. What nozzle pressure is needed to achieve a water flow rate of 840 gallons per minute?

56. What nozzle pressure is needed to achieve a water flow rate of 720 gallons per minute?

Use the graph of the firefighting formula to solve Exercises 57–58.

Firefighting: Nozzle Pressure and Flow Rate

$f = 120\sqrt{p}$

57. How is your answer to Exercise 55 shown on the graph?

58. How is your answer to Exercise 56 shown on the graph?

The graph shows the percentage of overweight or obese adults in the United States for five selected years from 1980 through 2004. The formula

$$p = 3.6\sqrt{t} + 46$$

models the percentage of Americans, p, who were overweight or obese t years after 1980. Use this mathematical model to solve Exercises 59–60.

Percentage of Overweight or Obese United States Adults

Source: Centers for Disease Control

59. a. According to the formula, what percentage of adults were overweight or obese in 2004? Round to the nearest tenth of a percent.

b. Does your answer in part (a) underestimate or overestimate the actual percentage given by the bar graph? By how much?

c. If trends indicated by the data continue, use the formula to determine when 66% of U.S. adults will be overweight or obese. Use a calculator and round to the nearest year.

60. a. According to the formula, what percentage of adults were overweight or obese in 2002? Round to the nearest tenth of a percent.

b. Does your answer in part (a) underestimate or overestimate the actual percentage given by the bar graph? By how much?

c. If trends indicated by the data continue, use the formula to determine when 69% of U.S. adults will be overweight or obese. Use a calculator and round to the nearest year.

Writing in Mathematics

61. What is a radical equation?

62. In solving $\sqrt{2x - 1} + 2 = x$, why is it a good idea to isolate the radical term? What if we don't do this and simply square each side? Describe what happens.

63. What is an extraneous solution of a radical equation?

64. Explain why $\sqrt{x} = -1$ has no solution.

Critical Thinking Exercises

Make Sense? *In Exercises 65–68, determine whether each statement "makes sense" or "does not make sense" and explain your reasoning.*

65. When checking a radical equation's proposed solution, I can substitute into the original equation or any equation that is part of the solution process.

66. After squaring both sides of a radical equation, the only solution that I obtained was extraneous, so $\varnothing$ must be the solution set of the original equation.

67. Whenever a radical equation leads to a quadratic equation with two solutions, one solution satisfies the original radical equation and the other does not.

68. I squared both sides of $\sqrt{x+7} = x + 5$ and obtained $x + 7 = x^2 + 25$.

In Exercises 69–72, determine whether each statement is true or false. If the statement is false, make the necessary change(s) to produce a true statement.

69. The equation $y^2 = 25$ has the same solutions as the equation $y = 5$.

70. The equation $\sqrt{x^2 + 2x} = -1$ has no real number solution.

71. The first step in solving $\sqrt{x} + 3 = 4$ is to square each side.

72. When an extraneous solution is substituted into an equation with radicals, a denominator of zero results.

73. The square root of the sum of two consecutive integers is one less than the smaller integer. Find the integers.

74. If $w = 2$, find x, y, and z if $y = \sqrt{x-2} + 2$, $z = \sqrt{y-2} + 2$, and $w = \sqrt{z-2} + 2$.

Technology Exercises

In Exercises 75–79, use a graphing utility to solve each radical equation. Graph each side of the equation in the given viewing rectangle. The equation's solution is given by the x-coordinate of the point(s) of intersection. Check by substitution.

75. $\sqrt{2x+2} = \sqrt{3x-5}$
$[-1, 10, 1]$ by $[-1, 5, 1]$

76. $\sqrt{x} + 3 = 5$
$[-1, 6, 1]$ by $[-1, 6, 1]$

77. $\sqrt{x^2 + 3} = x + 1$
$[-1, 6, 1]$ by $[-1, 6, 1]$

78. $4\sqrt{x} = x + 3$
$[-1, 10, 1]$ by $[-1, 13, 1]$

79. $\sqrt{x} + 4 = 2$
$[-2, 18, 1]$ by $[0, 10, 1]$

80. Use a graphing utility's ⎡TABLE⎤ feature to provide a numeric check for the solutions you obtained in any three radical equations from Exercises 29–38.

Review Exercises

81. A total of $9000 was invested for 1 year, part at 6% and the remainder at 4% simple interest. At the end of the year, the investments earned $500 in interest How much was invested at each rate? (Section 3.1, Example 2)

82. Producers of *The Sopranos*, the musical, are a bit worried about their basic concept and decide to sell tickets for previews at cut-rate prices. If four orchestra and two mezzanine seats sell for $22, while two orchestra and three mezzanine seats sell for $16, what is the price of an orchestra seat? (Section 5.4, Example 2)

83. Solve by graphing:
$$2x + y = -4$$
$$x + y = -3.$$
(Section 5.1, Example 2)

Preview Exercises

Exercises 84–86 will help you prepare for the material covered in the next section. In each exercise, evaluate the given expression.

84. $-\left(\sqrt[5]{32}\right)^4$

85. $\left(\sqrt[3]{27}\right)^2$

86. $\dfrac{1}{\left(\sqrt[4]{81}\right)^3}$

Rational Exponents

Objectives

1 Evaluate expressions with rational exponents.

2 Solve problems using models with rational exponents.

Animals in the wild have regions to which they confine their movement, called their territorial area. Territorial area, in square miles, is related to an animal's body weight. If an animal weighs W pounds, its territorial area is

$$W^{\frac{141}{100}}$$

square miles.

W to the *what* power?! How can we interpret the information given by this algebraic expression? In this section, we turn our attention to rational exponents such as $\frac{141}{100}$ and their relationship to roots of real numbers.

1 Evaluate expressions with rational exponents.

Defining Rational Exponents

We define rational exponents so that their properties are the same as the properties for integer exponents. For example, we know that exponents are multiplied when an exponential expression is raised to a power. For this to be true,

$$\left(7^{\frac{1}{2}}\right)^2 = 7^{\frac{1}{2} \cdot 2} = 7^1 = 7.$$

We also know that

$$\left(\sqrt{7}\right)^2 = \sqrt{7} \cdot \sqrt{7} = \sqrt{49} = 7.$$

Can you see that the square of both $7^{\frac{1}{2}}$ and $\sqrt{7}$ is 7? It is reasonable to conclude that

$$7^{\frac{1}{2}} \text{ means } \sqrt{7}.$$

We can generalize this idea with the following definition:

The Definition of $a^{\frac{1}{n}}$

If $\sqrt[n]{a}$ represents a real number and $n \geq 2$ is an integer, then

$$a^{\frac{1}{n}} = \sqrt[n]{a}.$$

> The denominator of the rational exponent is the radical's index.

EXAMPLE 1 Using the Definition of $a^{\frac{1}{n}}$

Simplify:

a. $64^{\frac{1}{2}}$ **b.** $125^{\frac{1}{3}}$ **c.** $-16^{\frac{1}{4}}$ **d.** $(-27)^{\frac{1}{3}}$.

Solution

a. $64^{\frac{1}{2}} = \sqrt{64} = 8$

b. $125^{\frac{1}{3}} = \sqrt[3]{125} = 5$

> The denominator is the index.

c. $-16^{\frac{1}{4}} = -\left(\sqrt[4]{16}\right) = -2$

> The base is 16 and the negative sign is not affected by the exponent.

d. $(-27)^{\frac{1}{3}} = \sqrt[3]{-27} = -3$

> Parentheses show that the base is −27 and that the negative sign is affected by the exponent.

☑ **CHECK POINT 1** Simplify:

a. $25^{\frac{1}{2}}$ **b.** $8^{\frac{1}{3}}$ **c.** $-81^{\frac{1}{4}}$ **d.** $(-8)^{\frac{1}{3}}$.

In Example 1 and Check Point 1, each rational exponent had a numerator of 1. If the numerator is some other integer, we still want to multiply exponents when raising a power to a power. For this reason,

$$a^{\frac{2}{3}} = \left(a^{\frac{1}{3}}\right)^2 \quad \text{and} \quad a^{\frac{2}{3}} = \left(a^2\right)^{\frac{1}{3}}.$$

> This means $(\sqrt[3]{a})^2$.

> This means $\sqrt[3]{a^2}$.

Thus,

$$a^{\frac{2}{3}} = \left(\sqrt[3]{a}\right)^2 = \sqrt[3]{a^2}.$$

Do you see that the denominator, 3, of the rational exponent is the same as the index of the radical? The numerator, 2, of the rational exponent serves as an exponent in each of the two radical forms. We generalize these ideas with the following definition:

The Definition of $a^{\frac{m}{n}}$

If $\sqrt[n]{a}$ represents a real number, and $\dfrac{m}{n}$ is a positive rational number, $n \geq 2$, then

> First take the nth root of a.

$$a^{\frac{m}{n}} = \left(\sqrt[n]{a}\right)^m.$$

Also,

> First raise a to the m power.

$$a^{\frac{m}{n}} = \sqrt[n]{a^m}.$$

The first form of the definition, shown again below, involves taking the root first. This form is often preferable because smaller numbers are involved. Notice that the rational exponent consists of two parts, indicated by the following voice balloons:

> The numerator is the exponent.

$$a^{\frac{m}{n}} = \left(\sqrt[n]{a}\right)^m.$$

> The denominator is the radical's index.

Using Technology

Here are the calculator key-stroke sequences for $27^{\frac{2}{3}}$:

Many Scientific Calculators

27 [y^x] [( ] [2] [÷] [3] [ )] [=]

Many Graphing Calculators

27 [∧] [( ] [2] [÷] [3] [ )] [ENTER]

EXAMPLE 2 Using the Definition of $a^{\frac{m}{n}}$

Simplify:

a. $27^{\frac{2}{3}}$ **b.** $9^{\frac{3}{2}}$ **c.** $-32^{\frac{4}{5}}$.

Solution

a. $27^{\frac{2}{3}} = \left(\sqrt[3]{27}\right)^2 = 3^2 = 9$

b. $9^{\frac{3}{2}} = \left(\sqrt{9}\right)^3 = 3^3 = 27$

c. $-32^{\frac{4}{5}} = -\left(\sqrt[5]{32}\right)^4 = -2^4 = -16$

> The base is 32 and the negative sign is not affected by the exponent.

✓ **CHECK POINT 2** Simplify:

a. $27^{\frac{4}{3}}$ **b.** $4^{\frac{3}{2}}$ **c.** $-16^{\frac{3}{4}}$.

Can a rational exponent be negative? Yes. The way that negative rational exponents are defined is similar to the way that negative integer exponents are defined:

> ### The Definition of $a^{-\frac{m}{n}}$
>
> If $a^{-\frac{m}{n}}$ is a nonzero real number, then
>
> $$a^{-\frac{m}{n}} = \frac{1}{a^{\frac{m}{n}}}.$$

EXAMPLE 3 Using the Definition of $a^{-\frac{m}{n}}$

Simplify:

    a. $100^{-\frac{1}{2}}$ **b.** $27^{-\frac{1}{3}}$ **c.** $81^{-\frac{3}{4}}$.

Solution

a. $100^{-\frac{1}{2}} = \dfrac{1}{100^{\frac{1}{2}}} = \dfrac{1}{\sqrt{100}} = \dfrac{1}{10}$

b. $27^{-\frac{1}{3}} = \dfrac{1}{27^{\frac{1}{3}}} = \dfrac{1}{\sqrt[3]{27}} = \dfrac{1}{3}$

c. $81^{-\frac{3}{4}} = \dfrac{1}{81^{\frac{3}{4}}} = \dfrac{1}{\left(\sqrt[4]{81}\right)^3} = \dfrac{1}{3^3} = \dfrac{1}{27}$

☑ **CHECK POINT 3**   Simplify:

    a. $25^{-\frac{1}{2}}$ **b.** $64^{-\frac{1}{3}}$ **c.** $32^{-\frac{4}{5}}$.

Using Technology

Here are the calculator key-stroke sequences for $81^{-\frac{3}{4}}$:

Many Scientific Calculators

81 $\boxed{y^x}$ $\boxed{(}$ 3 $\boxed{+/-}$ $\boxed{\div}$ 4 $\boxed{)}$ $\boxed{=}$

Many Graphing Calculators

81 $\boxed{\wedge}$ $\boxed{(}$ $\boxed{(-)}$ 3 $\boxed{\div}$ 4 $\boxed{)}$ $\boxed{\text{ENTER}}$

---

2 Solve problems using models with rational exponents.

Applications

Now that you know the meaning of rational exponents, you can work with mathematical models that contain these exponents.

EXAMPLE 4 Youth on the Web

The bar graph in **Figure 9.6** shows the percentage of people in the United States who kept a personal blog or maintained their own Web site in 2003 and in 2005.

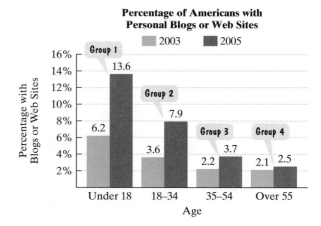

FIGURE 9.6

Source: Center for the Digital Future, USC Annenberg School

The formula

$$p = 15x^{-\frac{5}{4}}$$

models the percentage of Americans, p, with blogs or Web sites in 2005, where x is the group number designating the age ranges shown on the horizontal axis in **Figure 9.6**.

   a. According to the formula, what percentage of Americans between the ages of 35 and 54 kept blogs or maintained Web sites in 2005? Use a calculator and round to the nearest tenth of a percent.

   b. Does the answer in part (a) underestimate or overestimate the actual percentage given by the bar graph? By how much?

   c. Rewrite the mathematical model in radical notation.

Solution

   a. We are interested in Americans between the ages of 35 and 54, inclusive. The horizontal axis in **Figure 9.6** shows that this age range is designated by group 3. We substitute 3 for x in the given formula. Then we use a calculator to find p, the percentage with blogs or Web sites.

$$p = 15x^{-\frac{5}{4}} \qquad \text{This is the given formula.}$$
$$p = 15 \cdot 3^{-\frac{5}{4}} \qquad \text{Substitute 3 for } x.$$
$$p \approx 3.8 \qquad \text{Use a calculator. See the Using Technology box in the margin.}$$

The mathematical model indicates that approximately 3.8% of Americans between the ages of 35 and 54 kept blogs or maintained Web sites in 2005.

   b. The bar graph in **Figure 9.6** shows that 3.7% of Americans between the ages of 35 and 54 kept blogs or maintained Web sites in 2005. Because we obtained 3.8% in part (a), the model overestimates the percentage by 3.8% – 3.7%, or by 0.1%.

   c. Now we are ready to express the mathematical model in radical notation.

$$p = 15x^{-\frac{5}{4}} \qquad \text{This is the given formula.}$$
$$p = \frac{15}{x^{\frac{5}{4}}} \qquad a^{-\frac{m}{n}} = \frac{1}{a^{\frac{m}{n}}}$$
$$p = \frac{15}{\left(\sqrt[4]{x}\right)^5} \quad \text{or} \quad \frac{15}{\sqrt[4]{x^5}} \qquad a^{\frac{m}{n}} = \left(\sqrt[n]{a}\right)^m \quad \text{or} \quad \sqrt[n]{a^m}$$

Using Technology

Here are the calculator keystroke sequences for $15 \cdot 3^{-\frac{5}{4}}$:

Many Scientific Calculators

Many Graphing Calculators

15 × 3 ∧ ((−) 5 ÷
4) ENTER

☑ **CHECK POINT 4**

   a. According to the formula in Example 4, what percentage of Americans between the ages of 18 and 34 kept blogs or maintained Web sites in 2005? Use a calculator and round to the nearest tenth of a percent.

   b. Does the answer in part (a) underestimate or overestimate the actual percentage given by the bar graph in **Figure 9.6**? By how much?

9.6 EXERCISE SET *MyMathLab* Math XL PRACTICE WATCH DOWNLOAD READ REVIEW

In Exercises 1–48, simplify by first writing the expression in radical form. If applicable, use a calculator to verify your answer.

1. $49^{\frac{1}{2}}$

2. $100^{\frac{1}{2}}$

3. $121^{\frac{1}{2}}$

4. $1^{\frac{1}{2}}$

5. $27^{\frac{1}{3}}$

6. $64^{\frac{1}{3}}$

7. $-125^{\frac{1}{3}}$

8. $-27^{\frac{1}{3}}$

9. $16^{\frac{1}{4}}$

10. $81^{\frac{1}{4}}$

11. $-32^{\frac{1}{5}}$

12. $-243^{\frac{1}{5}}$

13. $\left(\frac{1}{9}\right)^{\frac{1}{2}}$

14. $\left(\frac{1}{25}\right)^{\frac{1}{2}}$

15. $\left(\frac{27}{64}\right)^{\frac{1}{3}}$

16. $\left(\frac{64}{125}\right)^{\frac{1}{3}}$

17. $81^{\frac{3}{2}}$

18. $25^{\frac{3}{2}}$

19. $125^{\frac{2}{3}}$

20. $1000^{\frac{2}{3}}$

21. $9^{\frac{3}{2}}$

22. $16^{\frac{3}{2}}$

23. $(-32)^{\frac{3}{5}}$

24. $(-27)^{\frac{2}{3}}$

25. $9^{-\frac{1}{2}}$

26. $49^{-\frac{1}{2}}$

27. $125^{-\frac{1}{3}}$

28. $27^{-\frac{1}{3}}$

29. $32^{-\frac{1}{5}}$

30. $243^{-\frac{1}{5}}$

31. $\left(\frac{1}{4}\right)^{-\frac{1}{2}}$

32. $\left(\frac{1}{9}\right)^{-\frac{1}{2}}$

33. $16^{-\frac{3}{4}}$

34. $625^{-\frac{3}{4}}$

35. $81^{-\frac{5}{4}}$

36. $32^{-\frac{4}{5}}$

37. $8^{-\frac{2}{3}}$

38. $625^{-\frac{5}{4}}$

39. $\left(\frac{4}{25}\right)^{-\frac{1}{2}}$

40. $\left(\frac{8}{27}\right)^{-\frac{1}{3}}$

41. $\left(\frac{8}{125}\right)^{-\frac{1}{3}}$

42. $\left(\frac{9}{100}\right)^{-\frac{1}{2}}$

43. $(-8)^{-\frac{2}{3}}$

44. $(-64)^{-\frac{2}{3}}$

45. $27^{\frac{2}{3}} + 16^{\frac{3}{4}}$

46. $4^{\frac{5}{2}} - 8^{\frac{2}{3}}$

47. $25^{\frac{3}{2}} \cdot 81^{\frac{1}{4}}$

48. $16^{-\frac{3}{4}} \cdot 16^{\frac{3}{2}}$

Practice PLUS

In Exercises 49–56, simplify each expression. Write answers in exponential form with positive exponents only. Assume that all variables represent positive real numbers.

49. $x^{\frac{1}{3}} \cdot x^{\frac{1}{4}}$

50. $x^{\frac{1}{4}} \cdot x^{\frac{1}{5}}$

51. $\dfrac{x^{\frac{1}{6}}}{x^{\frac{5}{6}}}$

52. $\dfrac{x^{\frac{1}{4}}}{x^{\frac{3}{4}}}$

53. $\left(x^{\frac{1}{4}}y^3\right)^{\frac{2}{3}}$

54. $\left(x^{\frac{1}{6}}y^{15}\right)^{\frac{3}{5}}$

55. $\left(\dfrac{x^{\frac{2}{5}}}{x^{\frac{6}{5}} \cdot x^{\frac{3}{5}}}\right)^5$

56. $\left(\dfrac{x^{\frac{4}{7}}}{x^{\frac{3}{7}} \cdot x^{\frac{2}{7}}}\right)^{49}$

Application Exercises

57. It is difficult to measure the height of a tall tree, particularly when it is growing in a dense forest. However, it is relatively easy to measure its base diameter. The formula

$$h = 0.84d^{\frac{2}{3}}$$

models a tree's height, h, in meters, in terms of its base diameter, d, in centimeters. (*Source:* Thomas McMahon, *Scientific American*, July, 1975)

  a. The largest known sequoia, the General Sherman in California, has a base diameter of 985 centimeters (about the size of a small house). Use a calculator to approximate the height of the General Sherman to the nearest tenth of a meter.

  b. Rewrite the formula in radical notation.

58. The formula

$$A = 1.83w^{\frac{2}{3}}$$

models the area of an egg shell, A, in square centimeters, in terms of its weight, w, in grams.

  a. An ostrich egg weighs approximately 1600 grams. Use a calculator to approximate the shell's area to the nearest tenth of a square centimeter.

  b. Rewrite the formula in radical notation.

Artificial gravity can be created in a space station to prevent bone and muscle loss for astronauts on lengthy space voyages. A rotating bedlike apparatus is used to create the artificial gravity. The required number of rotations is given by the following formula:

$$N = \frac{a^{\frac{1}{2}}}{2\pi r^{\frac{1}{2}}}.$$

Artificial gravity produced, in meters per second squared

Rate of rotation of the apparatus, in revolutions per second

Radius of the apparatus, in meters

Use this formula to solve Exercises 59–60.

59. a. How fast, in revolutions per second, would an apparatus with a radius of 1.7 meters have to rotate to simulate gravity on Earth, 9.8 meters per second squared? Express the answer correct to three decimal places.

  b. The required rotation that you obtained in part (a) is expressed in revolutions per second. Express the rotation needed for the apparatus in revolutions per minute.

  c. Rewrite the formula in radical notation without a radical in the denominator.

60. a. How fast, in revolutions per second, would an apparatus with a radius of 2.4 meters have to rotate to simulate gravity on Earth, 9.8 meters per second squared? Express the answer correct to three decimal places.

  b. The required rotation that you obtained in part (a) is expressed in revolutions per second. Express the rotation needed for the apparatus in revolutions per minute.

  c. Rewrite the formula in radical notation without a radical in the denominator.

Writing in Mathematics

61. What is the meaning of $a^{\frac{1}{n}}$? Give an example to support your explanation.

62. What is the meaning of $a^{\frac{m}{n}}$? Give an example.

63. What is the meaning of $a^{-\frac{m}{n}}$? Give an example.

64. Explain why $a^{\frac{1}{n}}$ is negative when n is odd and a is negative. What happens if n is even and a is negative? Why?

65. In simplifying $36^{\frac{3}{2}}$, is it better to use $a^{\frac{m}{n}} = \sqrt[n]{a^m}$ or $a^{\frac{m}{n}} = (\sqrt[n]{a})^m$? Explain.

Critical Thinking Exercises

Make Sense? *In Exercises 66–69, determine whether each statement "makes sense" or "does not make sense" and explain your reasoning.*

66. When I used my calculator to approximate $5^{\frac{2}{3}}$, I found it easier to first rewrite the expression in radical form, using the radical form for the keystroke sequence.

67. By adding the exponents, I simplified $5^{\frac{1}{2}} \cdot 5^{\frac{1}{2}}$ and obtained 25.

68. When I use the definition for $a^{\frac{m}{n}}$, I prefer to first raise a to the m power because smaller numbers are involved.

69. There's no question that $(-64)^{\frac{1}{3}} = -64^{\frac{1}{3}}$, so I can conclude that $(-64)^{\frac{1}{2}} = -64^{\frac{1}{2}}$.

In Exercises 70–77, determine whether each statement is true or false. If the statement is false, make the necessary change(s) to produce a true statement.

70. $2^{\frac{1}{2}} \cdot 2^{\frac{1}{2}} = 4^{\frac{1}{2}}$

71. $8^{-\frac{1}{2}} = \dfrac{1}{4}$

72. $25^{-\frac{1}{2}} = -5$

73. $-3^{-2} = \dfrac{1}{9}$

74. $2^{\frac{1}{2}} \cdot 2^{\frac{3}{2}} = \left(\dfrac{1}{4}\right)^{-1}$

75. $81^{\frac{1}{4}} \cdot 125^{\frac{1}{3}}$ is an integer.

76. $16^{-\frac{1}{4}} = -2$

77. $(-8)^{\frac{1}{3}} = -8^{\frac{1}{3}}$

Without using a calculator, simplify the expressions in Exercises 78–79 completely.

78. $25^{\frac{1}{4}} \cdot 25^{-\frac{3}{4}}$

79. $\dfrac{3^{-1} \cdot 3^{\frac{1}{2}}}{3^{-\frac{3}{2}}}$

Technology Exercises

80. The territorial area of an animal in the wild is the area of the region to which the animal confines its movements. The formula
$$T = W^{1.41} = W^{\frac{141}{100}} = \sqrt[100]{W^{141}}$$
models the territorial area, T, in square miles, in terms of an animal's body weight, W, in pounds.

a. Use a calculator to fill in the table of values, rounding T to the nearest whole square mile.

W	0	25	50	150	200	250	300
$T = W^{1.41}$							

b. Use the table of values to graph $T = W^{1.41}$. What does the shape of the graph indicate about the relationship between body weight and territorial area?

c. Verify your hand-drawn graph by using a graphing utility to graph the model.

81. If A is the surface area of a cube and V is its volume, then $A = 6V^{\frac{2}{3}}$.

a. Graph the equation $\left(y = 6x^{\frac{2}{3}}\right)$ relating a cube's surface area and volume using a graphing utility and a $[0, 30, 3]$ by $[0, 60, 3]$ viewing rectangle.

b. TRACE or ZOOM IN along the curve and verify the numbers shown in the figure below. In particular, show that a cube whose volume is 27 cubic units has a surface area of 54 square units.

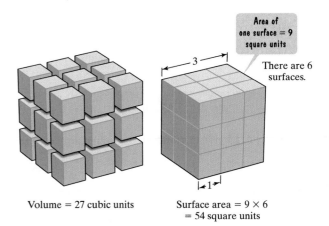

Volume = 27 cubic units   Surface area = 9 × 6 = 54 square units

Area of one surface = 9 square units

There are 6 surfaces.

c. TRACE or ZOOM IN along the curve and find the surface area of a cube whose volume is 8 cubic units.

Review Exercises

82. Solve the system:
$$7x - 3y = -14$$
$$y = 3x + 6.$$
(Section 5.2, Example 1)

83. Graph the solution set of the system:
$$-3x + 4y \le 12$$
$$x \ge 2.$$
(Section 5.5, Example 3)

84. Simplify: $\dfrac{(2x)^5}{x^3}$. (Section 6.7, Example 5)

Preview Exercises

Exercises 85–87 will help you prepare for the material covered in the first section of the next chapter.

85. Use substitution to determine if $-\sqrt{5}$ is a solution of the quadratic equation $4x^2 = 20$.

86. Use substitution to determine if $1 + \sqrt{3}$ is a solution of $(x - 1)^2 = 5$.

87. Simplify: $\sqrt{[6 - (-4)]^2 + [2 - (-3)]^2}$.

GROUP PROJECT

The following topics related to irrational roots are appropriate for a group research project. A group report should be given to the class on the researched topic.

a. A History of How Irrational Numbers Developed
b. Proving that $\sqrt{2}$ Is Irrational
c. Golden Rectangles in Art and Architecture (See Exercise Set 9.4, Exercise 85.)
d. Golden Ratios in Proportions of the Human Body (See Exercise Set 9.4, Exercise 85.)
e. Radicals in Nature

Chapter 9 Summary

Definitions and Concepts	**Examples**

Section 9.1 Finding Roots

If $b^2 = a$, then b is a square root of a. The principal square root of a, designated $\sqrt{a}$, is the nonnegative number satisfying $b^2 = a$. The negative square root of a is written $-\sqrt{a}$. A square root of a negative number is not a real number.

- $\sqrt{100} = 10$ because $10^2 = 100$.
- $-\sqrt{100} = -10$
- $\sqrt{-100}$ is not a real number.

The principal nth root of a real number a, symbolized by $\sqrt[n]{a}$, is defined as follows:

$$\sqrt[n]{a} = b \text{ means that } b^n = a.$$

The natural number n is called the index, the symbol $\sqrt[n]{}$ is called a radical sign, and the expression under the radical sign is called the radicand. If the index is even, the radicand must be nonnegative for the root to be a real number and the principal nth root is nonnegative.

- $\sqrt[3]{8} = 2$ because $2^3 = 8$.
- $\sqrt[3]{-125} = -5$ because $(-5)^3 = -125$.
- $\sqrt[4]{81} = 3$ because $3^4 = 81$.
- $\sqrt[4]{-81}$ is not a real number.

Section 9.2 Multiplying and Dividing Radicals

The Product Rule for Roots
$$\sqrt[n]{a} \cdot \sqrt[n]{b} = \sqrt[n]{ab} \quad \text{and} \quad \sqrt[n]{ab} = \sqrt[n]{a} \cdot \sqrt[n]{b}$$
(The roots represent real numbers.)

- $\sqrt{11} \cdot \sqrt{7} = \sqrt{11 \cdot 7} = \sqrt{77}$
- $\sqrt[3]{4} \cdot \sqrt[3]{2} = \sqrt[3]{4 \cdot 2} = \sqrt[3]{8} = 2$

A square root is simplified when its radicand has no factors other than 1 that are perfect squares. To simplify a square root, factor the radicand so that one of its factors is a perfect square. Simplify square roots with a variable to an even power using $\sqrt{x^{2n}} = x^n$. Simplify square roots with a variable to an odd power by expressing the variable as the product of two factors, one of which has an even power.

- $\sqrt{63} = \sqrt{9 \cdot 7} = \sqrt{9} \cdot \sqrt{7} = 3\sqrt{7}$
- $\sqrt{x^{26}} = x^{13}$
- $\sqrt{x^{27}} = \sqrt{x^{26} \cdot x}$
  $= \sqrt{x^{26}} \cdot \sqrt{x} = x^{13}\sqrt{x}$
- $\sqrt{45x^3} = \sqrt{9x^2 \cdot 5x}$
  $= \sqrt{9x^2} \cdot \sqrt{5x} = 3x\sqrt{5x}$
- $\sqrt{8} \cdot \sqrt{3} = \sqrt{24} = \sqrt{4 \cdot 6}$
  $= \sqrt{4} \cdot \sqrt{6} = 2\sqrt{6}$

Definitions and Concepts	Examples

Section 9.2 Multiplying and Dividing Radicals (continued)

The Quotient Rule for Roots

$$\frac{\sqrt[n]{a}}{\sqrt[n]{b}} = \sqrt[n]{\frac{a}{b}} \quad \text{and} \quad \sqrt[n]{\frac{a}{b}} = \frac{\sqrt[n]{a}}{\sqrt[n]{b}}$$

(The roots represent real numbers and no denominators are 0.)

- $\sqrt{\dfrac{36}{x^2}} = \dfrac{\sqrt{36}}{\sqrt{x^2}} = \dfrac{6}{x}$

- $\sqrt[3]{\dfrac{2}{125}} = \dfrac{\sqrt[3]{2}}{\sqrt[3]{125}} = \dfrac{\sqrt[3]{2}}{5}$

- $\dfrac{\sqrt{50x^4}}{\sqrt{2x}} = \sqrt{\dfrac{50x^4}{2x}} = \sqrt{25x^3}$
  $= \sqrt{25x^2 \cdot x} = 5x\sqrt{x}$

Section 9.3 Operations with Radicals

Square roots with the same radicand are like radicals. Like radicals can be added or subtracted using the distributive property. In some cases, radicals can be combined once they have been simplified.

- $7\sqrt{2} + 9\sqrt{2} = (7+9)\sqrt{2}$
  $= 16\sqrt{2}$
- $6\sqrt{8} - \sqrt{2} = 6\sqrt{4 \cdot 2} - \sqrt{2}$
  $= 6\sqrt{4}\sqrt{2} - \sqrt{2}$
  $= 6 \cdot 2\sqrt{2} - \sqrt{2}$
  $= 12\sqrt{2} - \sqrt{2}$
  $= (12-1)\sqrt{2}$
  $= 11\sqrt{2}$

Radical expressions with more than one term are multiplied in much the same way that polynomials with more than one term are multiplied.

- $\sqrt{5}(\sqrt{2} + \sqrt{5}) = \sqrt{10} + \sqrt{25}$
  $= \sqrt{10} + 5$
- $(2 + \sqrt{3})(4 + 5\sqrt{3})$
  F   O   I   L
  $= 2 \cdot 4 + 2(5\sqrt{3}) + 4\sqrt{3} + \sqrt{3}(5\sqrt{3})$
  $= 8 + 10\sqrt{3} + 4\sqrt{3} + 15$
  $= 23 + 14\sqrt{3}$

Radical expressions that involve the sum and difference of the same two terms are called conjugates. Use the special product
$$(A+B)(A-B) = A^2 - B^2$$
to multiply conjugates.

$(\sqrt{11} + 5)(\sqrt{11} - 5)$
$= (\sqrt{11})^2 - 5^2 = 11 - 25 = -14$

Section 9.4 Rationalizing the Denominator

The process of rewriting a radical expression as an equivalent expression in which the denominator no longer contains any radicals is called rationalizing the denominator. If the denominator contains the square root of a natural number that is not a perfect square, multiply the numerator and the denominator by the smallest number that produces the square root of a perfect square in the denominator.

- Rationalize the denominator:
$$\frac{4}{\sqrt{7}}.$$
$$\frac{4}{\sqrt{7}} = \frac{4}{\sqrt{7}} \cdot \frac{\sqrt{7}}{\sqrt{7}} = \frac{4\sqrt{7}}{7}$$

Definitions and Concepts	**Examples**

Section 9.4 Rationalizing the Denominator (continued)

If the denominator contains two terms, rationalize the denominator by multiplying the numerator and the denominator by the conjugate of the denominator.

$$\frac{9}{7 - \sqrt{5}} = \frac{9}{7 - \sqrt{5}} \cdot \frac{7 + \sqrt{5}}{7 + \sqrt{5}}$$

$$= \frac{9(7 + \sqrt{5})}{7^2 - (\sqrt{5})^2}$$

$$= \frac{9(7 + \sqrt{5})}{49 - 5} = \frac{9(7 + \sqrt{5})}{44}$$

Section 9.5 Radical Equations

A radical equation is an equation in which the variable occurs in a radicand.

Solving Radical Equations Containing Square Roots

1. Isolate the radical.
2. Square both sides.
3. Solve the resulting equation.
4. Check proposed solutions in the original equation. Solutions of the squared equation, but not the original equation, are called extraneous solutions.

Solve: $\sqrt{2x + 1} - x = -7$.

$\sqrt{2x + 1} = x - 7$	Isolate the radical.
$(\sqrt{2x + 1})^2 = (x - 7)^2$	Square both sides.
$2x + 1 = x^2 - 14x + 49$	Use $(A - B)^2 = A^2 - 2AB + B^2$.
$0 = x^2 - 16x + 48$	Write in standard form.
$0 = (x - 12)(x - 4)$	Factor.
$x - 12 = 0$ or $x - 4 = 0$	Set each factor equal to 0.
$x = 12 \qquad x = 4$	Solve the resulting equations.

Check both proposed solutions. 12 checks but 4 is extraneous. The only solution is 12 and the solution set is $\{12\}$.

Section 9.6 Rational Exponents

- $a^{\frac{1}{n}} = \sqrt[n]{a}$
- $a^{\frac{m}{n}} = (\sqrt[n]{a})^m$ or $\sqrt[n]{a^m}$
- $a^{-\frac{m}{n}} = \frac{1}{a^{\frac{m}{n}}}$

- $16^{\frac{1}{2}} = \sqrt{16} = 4$
- $27^{\frac{1}{3}} = \sqrt[3]{27} = 3$
- $8^{\frac{5}{3}} = (\sqrt[3]{8})^5 = 2^5 = 32$
- $81^{-\frac{3}{4}} = \frac{1}{81^{\frac{3}{4}}} = \frac{1}{(\sqrt[4]{81})^3} = \frac{1}{3^3} = \frac{1}{27}$

CHAPTER 9 REVIEW EXERCISES

9.1 *In Exercises 1–6, find the indicated root, or state that the expression is not a real number.*

1. $\sqrt{121}$

2. $-\sqrt{121}$

3. $\sqrt{-121}$

4. $\sqrt[3]{\frac{8}{125}}$

5. $\sqrt[5]{-32}$

6. $-\sqrt[4]{81}$

In Exercises 7–8, use a calculator to approximate each square root. Round to three decimal places.

7. $\sqrt{75}$

8. $\sqrt{398 - 5}$

9. The formula $M = 0.7\sqrt{t} + 12.5$ models the average number of non-program minutes in an hour of prime-time cable TV t years after 1996. (Nonprogram minutes consist of commercials and plugs for other shows.) According to the model, how many nonprogram minutes disrupted an hour of cable TV action in 2005?

10. Use the model in Exercise 9 to describe what is happening to the number of minutes disrupting an hour of cable TV over time.

11. The formula

$$d = \sqrt{\frac{3h}{2}}$$

models the distance, d, in miles, that you can see to the horizon at a height of h feet. The height of the Sears Tower in Chicago is 1450 feet. How far to the horizon can visitors see from the top of the building? Use a calculator and round to the nearest mile.

9.2 *In Exercises 12–19, simplify each expression.*

12. $\sqrt{54}$

13. $6\sqrt{20}$

14. $\sqrt{63x^2}$

15. $\sqrt{48x^3}$

16. $\sqrt{x^8}$ **17.** $\sqrt{75x^9}$

18. $\sqrt{45x^{23}}$ **19.** $\sqrt[3]{24}$

In Exercises 20–25, multiply and, if possible, simplify.

20. $\sqrt{7}\cdot\sqrt{11}$ **21.** $\sqrt{3}\cdot\sqrt{12}$

22. $\sqrt{5x}\cdot\sqrt{10x}$ **23.** $\sqrt{3x^2}\cdot\sqrt{4x^3}$

24. $\sqrt[3]{6}\cdot\sqrt[3]{9}$ **25.** $\sqrt{\dfrac{5}{2}}\cdot\sqrt{\dfrac{3}{8}}$

In Exercises 26–33, simplify using the quotient rule.

26. $\sqrt{\dfrac{121}{4}}$ **27.** $\sqrt{\dfrac{7x}{25}}$

28. $\sqrt{\dfrac{18}{x^2}}$ **29.** $\dfrac{\sqrt{200}}{\sqrt{2}}$

30. $\dfrac{\sqrt{96}}{\sqrt{3}}$ **31.** $\dfrac{\sqrt{72x^8}}{\sqrt{x^3}}$

32. $\sqrt[3]{\dfrac{5}{64}}$ **33.** $\sqrt[3]{\dfrac{40}{27}}$

9.3 *In Exercises 34–39, add or subtract as indicated.*

34. $7\sqrt{5}+13\sqrt{5}$ **35.** $\sqrt{8}+\sqrt{50}$

36. $\sqrt{75}-\sqrt{48}$ **37.** $2\sqrt{80}+3\sqrt{45}$

38. $4\sqrt{72}-2\sqrt{48}$

39. $2\sqrt{18}+3\sqrt{27}-\sqrt{12}$

In Exercises 40–47, multiply as indicated and, if possible, simplify.

40. $\sqrt{10}\left(\sqrt{5}+\sqrt{6}\right)$

41. $\sqrt{3}\left(\sqrt{6}-\sqrt{12}\right)$

42. $\left(9+\sqrt{2}\right)\left(10+\sqrt{2}\right)$

43. $\left(1+3\sqrt{7}\right)\left(4-\sqrt{7}\right)$

44. $\left(\sqrt{3}+2\right)\left(\sqrt{6}-4\right)$

45. $\left(2+\sqrt{7}\right)\left(2-\sqrt{7}\right)$

46. $\left(\sqrt{11}-\sqrt{5}\right)\left(\sqrt{11}+\sqrt{5}\right)$

47. $\left(1+\sqrt{2}\right)^2$

9.4 *In Exercises 48–56, rationalize each denominator and, if possible, simplify.*

48. $\dfrac{30}{\sqrt{5}}$ **49.** $\dfrac{13}{\sqrt{50}}$

50. $\sqrt{\dfrac{2}{3}}$ **51.** $\sqrt{\dfrac{3}{8}}$

52. $\sqrt{\dfrac{17}{x}}$ **53.** $\dfrac{11}{\sqrt{5}+2}$

54. $\dfrac{21}{4-\sqrt{3}}$ **55.** $\dfrac{12}{\sqrt{5}+\sqrt{3}}$

56. $\dfrac{7\sqrt{2}}{\sqrt{2}-4}$

9.5 *In Exercises 57–63, solve each radical equation.*

57. $\sqrt{x+3}=4$ **58.** $\sqrt{2x+3}=5$

59. $3\sqrt{x}=\sqrt{6x+15}$ **60.** $\sqrt{5x+1}=x+1$

61. $\sqrt{x+1}+5=x$ **62.** $\sqrt{x-2}+5=1$

63. $x=\sqrt{x^2+4x+4}$

64. Paleontologists use the formula

$$W=4\sqrt{2x}$$

to estimate the walking speed of a dinosaur, W, in feet per second, where x is the length, in feet, of the dinosaur's leg. What was the leg length of a dinosaur whose walking speed was 16 feet per second?

65. The formula

$$p=3.5\sqrt{t}+38$$

models the percentage of U.S. women in the labor force, p, t years after 1960. According to the formula, when will 66% of U.S. women participate in the labor force?

9.6 *In Exercises 66–71, simplify by first writing the expression in radical form.*

66. $16^{\frac{1}{2}}$ **67.** $125^{\frac{1}{3}}$

68. $64^{\frac{2}{3}}$ **69.** $25^{-\frac{1}{2}}$

70. $27^{-\frac{1}{3}}$ **71.** $(-8)^{-\frac{4}{3}}$

72. The formula $S=28.6A^{\frac{1}{3}}$ models the number of plant species, S, on the various islands of the Galápagos chain in terms of the area, A, in square miles, of a particular island. Approximately how many species of plants are there on a Galápagos island whose area is 8 square miles?

CHAPTER 9 TEST   Remember to use your Chapter Test Prep Video CD to help you study and view solutions to the test questions you need help with.

In Exercises 1–2, find the indicated root, or state that the expression is not a real number.

1. $-\sqrt{64}$ **2.** $\sqrt[3]{64}$

In Exercises 3–8, simplify each expression.

3. $\sqrt{48}$ **4.** $\sqrt{72x^3}$

5. $\sqrt{x^{29}}$ **6.** $\sqrt{\dfrac{25}{x^2}}$

7. $\sqrt{\dfrac{75}{27}}$ **8.** $\sqrt[3]{\dfrac{5}{8}}$

In Exercises 9–18, perform the indicated operation and, if possible, simplify.

9. $\dfrac{\sqrt{80x^4}}{\sqrt{2x^2}}$ **10.** $\sqrt{10}\cdot\sqrt{5}$

11. $\sqrt{6x} \cdot \sqrt{6y}$ **12.** $\sqrt{10x^2} \cdot \sqrt{2x^3}$

13. $\sqrt{24} + 3\sqrt{54}$ **14.** $7\sqrt{8} - 2\sqrt{32}$

15. $\sqrt{3}(\sqrt{10} + \sqrt{3})$

16. $(7 - \sqrt{5})(10 + 3\sqrt{5})$

17. $(\sqrt{6} + 2)(\sqrt{6} - 2)$ **18.** $(3 + \sqrt{7})^2$

In Exercises 19–20, rationalize each denominator and, if possible, simplify.

19. $\dfrac{4}{\sqrt{5}}$ **20.** $\dfrac{5}{4 + \sqrt{3}}$

In Exercises 21–22, solve each radical equation.

21. $\sqrt{3x} + 5 = 11$ **22.** $\sqrt{2x - 1} = x - 2$

23. The less income people have, the more likely they are to report that their health is fair or poor. The formula

$$p = -4.4\sqrt{x} + 38$$

models the percentage of Americans reporting fair or poor health, p, in terms of annual income, x, in thousands of dollars. According to the formula, what is the income, in thousands of dollars, for the group in which 16% report fair or poor health?

In Exercises 24–25, simplify by first writing the expression in radical form.

24. $8^{\frac{2}{3}}$ **25.** $9^{-\frac{1}{2}}$

CUMULATIVE REVIEW EXERCISES (CHAPTERS 1–9)

In Exercises 1–6, solve each equation or system of equations.

1. $2x + 3x - 5 + 7 = 10x + 3 - 6x - 4$

2. $2x^2 + 5x = 12$

3. $8x - 5y = -4$
$2x + 15y = -66$

4. $\dfrac{15}{x} - 4 = \dfrac{6}{x} + 3$

5. $-3x - 7 = 8$

6. $\sqrt{2x - 1} - x = -2$

In Exercises 7–11, simplify each expression.

7. $\dfrac{8x^3}{-4x^7}$ **8.** $6\sqrt{75} - 4\sqrt{12}$

9. $\dfrac{\dfrac{1}{x} - \dfrac{1}{2}}{\dfrac{1}{3} - \dfrac{x}{6}}$ **10.** $\dfrac{4 - x^2}{3x^2 - 5x - 2}$

11. $-5 - (-8) - (4 - 6)$

In Exercises 12–13, factor completely.

12. $x^2 - 18x + 77$

13. $x^3 - 25x$

In Exercises 14–18, perform the indicated operations. If possible, simplify the answer.

14. $\dfrac{6x^3 - 19x^2 + 16x - 4}{x - 2}$

15. $(2x - 3)(4x^2 + 6x + 9)$

16. $\dfrac{3x}{x^2 + x - 2} - \dfrac{2}{x + 2}$

17. $\dfrac{5x^2 - 6x + 1}{x^2 - 1} \div \dfrac{16x^2 - 9}{4x^2 + 7x + 3}$

18. $\sqrt{12} - 4\sqrt{75}$

In Exercises 19–21, graph each equation or inequality in a rectangular coordinate system.

19. $2x - y = 4$

20. $y = -\dfrac{2}{3}x$

21. $x \geq -1$

22. Find the slope of the line through $(-1, 5)$ and $(2, -3)$.

23. Write the point-slope form of the equation of the line with slope 5, passing through $(-2, -3)$. Then use the point-slope equation to write the slope-intercept form of the line's equation.

24. Seven subtracted from five times a number is 208. Find the number.

25. Park rangers catch, tag, and then release 318 deer back into a state park. Two weeks later, they select a sample of 168 deer, 56 of which are tagged. Assuming the ratio of tagged deer in the sample holds for all deer in the park, approximately how many deer are in the park?

Quadratic Equations and Introduction to Functions

America is a nation of immigrants. Since 1820, more than 40 million people have immigrated to the United States from all over the world. They chose to come for various reasons, such as to live in freedom, to practice religion without persecution, to escape poverty or oppression, and to make better lives for themselves and their children. As a result, in 2005, 11.7% of the United States population was foreign-born. How can we use mathematical models to project when this percentage might grow to one in every five Americans, or even more?

In this chapter, you will learn techniques for solving quadratic equations,

$$ax^2 + bx + c = 0,$$

that will give you new ways of exploring mathematical models. The United States foreign-born model appears as Example 3 in Section 10.3.

10.1

Solving Quadratic Equations by the Square Root Property

Objectives

1. Solve quadratic equations using the square root property.

2. Solve problems using the Pythagorean Theorem.

3. Find the distance between two points.

Pythagoras

Shown here is Renaissance artist Raphael Sanzio's (1483–1520) image of Pythagoras from The School of Athens *mural. Detail of left side.*

Raphael (1483–1520) "School of Athens" detail of left side. Stanza della Segnatura, Vatican Palace, Vatican State. Scala/Art Resource, NY

For the followers of the Greek mathematician Pythagoras in the sixth century B.C., numbers took on a life-and-death importance. The "Pythagorean Brotherhood" was a secret group whose members were convinced that properties of whole numbers were the key to understanding the universe. Members of the Brotherhood (which admitted women) thought that all numbers that were not whole numbers could be represented as the ratio of whole numbers. A crisis occurred for the Pythagoreans when they discovered the existence of a number that was not rational. Because the Pythagoreans viewed numbers with reverence and awe, the punishment for speaking about this irrational number was death. However, a member of the Brotherhood revealed the secret of the irrational number's existence. When he later died in a shipwreck, his death was viewed as punishment from the gods.

In this section, you will work with the triangle that led the Pythagoreans to the discovery of irrational numbers. You will find the lengths of one of the triangle's sides using a property other than factoring that can be used to solve quadratic equations.

1. Solve quadratic equations using the square root property.

The Square Root Property

Let's begin with a relatively simple quadratic equation:

$$x^2 = 9.$$

The value of x must be a number whose square is 9. There are two numbers whose square is 9:

$$x = \sqrt{9} = 3 \quad \text{or} \quad x = -\sqrt{9} = -3.$$

Thus, the solutions of $x^2 = 9$ are 3 and -3. This is an example of the **square root property**.

> ### The Square Root Property
> If u is an algebraic expression and d is a positive real number, then $u^2 = d$ has exactly two solutions:
>
> $$\text{If } u^2 = d, \quad \text{then } u = \sqrt{d} \text{ or } u = -\sqrt{d}.$$
>
> Equivalently,
>
> $$\text{If } u^2 = d, \quad \text{then } u = \pm\sqrt{d}.$$

Notice that $u = \pm\sqrt{d}$ is a shorthand notation to indicate that $u = \sqrt{d}$ or $u = -\sqrt{d}$. Although we usually read $u = \pm\sqrt{d}$ as "u equals plus or minus the square root of d," we actually mean that u is the positive square root of d or the negative square root of d.

EXAMPLE 1   Solving Quadratic Equations by the Square Root Property

Solve by the square root property:

a. $x^2 = 49$ **b.** $4x^2 = 20$ **c.** $2x^2 - 5 = 0$.

Solution

a.

$x^2 = 49$	This is the original equation.
$x = \sqrt{49}$ or $x = -\sqrt{49}$	Apply the square root property. You can also write $x = \pm\sqrt{49}$.
$x = 7$ or $x = -7$	In abbreviated notation, $x = \pm 7$.

Substitute both values into the original equation and confirm that the solutions are 7 and -7. The solution set is $\{\pm 7\}$.

b. To apply the square root property, we need a squared expression by itself on one side of the equation.

$$4x^2 = 20$$

We want x^2 by itself.

We can get x^2 by itself if we divide both sides by 4.

$4x^2 = 20$	This is the original equation.
$\dfrac{4x^2}{4} = \dfrac{20}{4}$	Divide both sides by 4.
$x^2 = 5$	Simplify.
$x = \sqrt{5}$ or $x = -\sqrt{5}$	Apply the square root property.

Now let's check these proposed solutions in the original equation. Because the equation has an x^2-term and no x-term, we can check both values, $\pm\sqrt{5}$, at once.

Check $\sqrt{5}$ and $-\sqrt{5}$:

$4x^2 = 20$	This is the original equation.
$4\left(\pm\sqrt{5}\right)^2 \stackrel{?}{=} 20$	Substitute the proposed solutions.
$4 \cdot 5 \stackrel{?}{=} 20$	$(\pm\sqrt{5})^2 = 5$
$20 = 20$,	true

The solutions are $-\sqrt{5}$ and $\sqrt{5}$. The solution set is $\{-\sqrt{5}, \sqrt{5}\}$ or $\{\pm\sqrt{5}\}$.

c. To solve $2x^2 - 5 = 0$ by the square root property, we must isolate the squared expression by itself on one side of the equation.

$$2x^2 - 5 = 0$$

We want x^2 by itself.

$$2x^2 - 5 = 0 \qquad \text{This is the original equation.}$$

$$2x^2 = 5 \qquad \text{Add 5 to both sides.}$$

$$x^2 = \frac{5}{2} \qquad \text{Divide both sides by 2.}$$

$$x = \sqrt{\frac{5}{2}} \quad \text{or} \quad x = -\sqrt{\frac{5}{2}} \qquad \text{Apply the square root property.}$$

In this section, we will express irrational solutions in simplified radical form, rationalizing denominators when possible. Because the proposed solutions are opposites, we can rationalize both denominators at once:

$$\pm\sqrt{\frac{5}{2}} = \pm\frac{\sqrt{5}}{\sqrt{2}} = \pm\frac{\sqrt{5}}{\sqrt{2}} \cdot \frac{\sqrt{2}}{\sqrt{2}} = \pm\frac{\sqrt{10}}{2}.$$

Substitute these values into the original equation and verify that the solutions are $-\dfrac{\sqrt{10}}{2}$ and $\dfrac{\sqrt{10}}{2}$. The solution set is $\left\{ -\dfrac{\sqrt{10}}{2}, \dfrac{\sqrt{10}}{2} \right\}$ or $\left\{ \pm\dfrac{\sqrt{10}}{2} \right\}$. ■

✓ **CHECK POINT 1** Solve by the square root property:

a. $x^2 = 36$

b. $5x^2 = 15$

c. $2x^2 - 7 = 0$.

Can we solve an equation such as $(x - 5)^2 = 16$ using the square root property? Yes. The equation is in the form $u^2 = d$, where u^2, the squared expression, is by itself on the left side.

$$(x - 5)^2 \qquad = \qquad 16$$

This is u^2 in $u^2 = d$ with $u = x - 5$. This is d in $u^2 = d$ with $d = 16$.

EXAMPLE 2 **Solving a Quadratic Equation by the Square Root Property**

Solve by the square root property: $(x - 5)^2 = 16$.

Solution

$$(x - 5)^2 = 16 \qquad \text{This is the original equation.}$$

$$x - 5 = \sqrt{16} \quad \text{or} \quad x - 5 = -\sqrt{16} \qquad \text{Apply the square root property.}$$

$$x - 5 = 4 \qquad \text{or} \quad x - 5 = -4 \qquad \text{Simplify.}$$

$$x = 9 \qquad\qquad x = 1 \qquad \text{Add 5 to both sides in each equation.}$$

Substitute both values into the original equation and confirm that the solutions are 9 and 1. The solutions are visually confirmed in the Using Technology box on the next page. The solution set is $\{1, 9\}$. ■

Using Technology

Graphic Connections

We can use graphs to confirm that the solution set of $(x - 5)^2 = 16$ is $\{1, 9\}$.

The graph of $y = (x - 5)^2 - 16$ has x-intercepts at 1 and 9. The solutions of $(x - 5)^2 - 16 = 0$, or $(x - 5)^2 = 16$, are 1 and 9.

Another option is to graph each side of the equation and find the x-coordinates of the intersection points. Graphing

$$y_1 = (x - 5)^2$$

and

$$y_2 = 16,$$

the x-coordinates of the intersection points are 1 and 9.

x-intercepts are 1 and 9.

$y = (x - 5)^2 - 16$

[-2, 10, 1] by [-20, 10, 1]

$y_2 = 16$

$y_1 = (x - 5)^2$

$x = 1$ $x = 9$

[-2, 10, 1] by [0, 20, 1]

✓ **CHECK POINT 2** Solve by the square root property: $(x - 3)^2 = 25$.

EXAMPLE 3 Solving a Quadratic Equation by the Square Root Property

Solve by the square root property: $(x - 1)^2 = 5$.

Solution

$$(x - 1)^2 = 5 \qquad \text{This is the original equation.}$$
$$x - 1 = \sqrt{5} \quad \text{or} \quad x - 1 = -\sqrt{5} \qquad \text{Apply the square root property.}$$
$$x = 1 + \sqrt{5} \qquad x = 1 - \sqrt{5} \qquad \text{Add 1 to both sides in each equation.}$$

Check $1 + \sqrt{5}$:

$$(x - 1)^2 = 5$$
$$\left(1 + \sqrt{5} - 1\right)^2 \overset{?}{=} 5$$
$$\left(\sqrt{5}\right)^2 \overset{?}{=} 5$$
$$5 = 5, \quad \text{true}$$

Check $1 - \sqrt{5}$:

$$(x - 1)^2 = 5$$
$$\left(1 - \sqrt{5} - 1\right)^2 \overset{?}{=} 5$$
$$\left(-\sqrt{5}\right)^2 \overset{?}{=} 5$$
$$5 = 5, \quad \text{true}$$

The solutions are $1 + \sqrt{5}$ and $1 - \sqrt{5}$, expressed in abbreviated notation as $1 \pm \sqrt{5}$. The solution set is $\{1 \pm \sqrt{5}\}$. ∎

✓ **CHECK POINT 3** Solve by the square root property: $(x - 2)^2 = 7$.

---

2 Solve problems using the Pythagorean Theorem.

Hypotenuse

Leg

Leg

The Pythagorean Theorem and the Square Root Property

The ancient Greek philosopher and mathematician Pythagoras (approximately 582–500 B.C.) founded a school whose motto was "All is number." Pythagoras is best remembered for his work with the **right triangle**, a triangle with one angle measuring 90°. The side opposite the 90° angle is called the **hypotenuse**. The other sides are called **legs**.

Pythagoras found that if he constructed squares on each of the legs, as well as a larger square on the hypotenuse, the sum of the areas of the smaller squares is equal to the area of the larger square. This is illustrated in **Figure 10.1**.

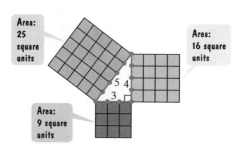

Area: 25 square units

Area: 16 square units

Area: 9 square units

FIGURE 10.1 The area of the large square equals the sum of the areas of the smaller squares.

This relationship is usually stated in terms of the lengths of the three sides of a right triangle and is called the **Pythagorean Theorem**.

The Pythagorean Theorem

The sum of the squares of the lengths of the legs of a right triangle equals the square of the length of the hypotenuse.

If the legs have lengths a and b, and the hypotenuse has length c, then

$$a^2 + b^2 = c^2.$$

FIGURE 10.2 A right triangle is formed by the television screen's height, width, and diagonal.

EXAMPLE 4   Using the Pythagorean Theorem

In a 25-inch television set, the length of the screen's diagonal is 25 inches. If the screen's height is 15 inches, what is its width?

Solution **Figure 10.2** shows a right triangle that is formed by the height, width, and diagonal. We can find w, the screen's width, using the Pythagorean Theorem.

(leg)² plus (leg)² equals (hypotenuse)².

$$w^2 \quad + \quad 15^2 \quad = \quad 25^2$$

This is the equation resulting from the Pythagorean Theorem.

The equation $w^2 + 15^2 = 25^2$ can be solved by the square root property.

$w^2 + 15^2 = 25^2$ — This is the equation that models the verbal conditions.

$w^2 + 225 = 625$ — Square 15 and 25.

$w^2 = 400$ — Isolate w^2 by subtracting 225 from both sides.

$w = \sqrt{400}$ or $w = -\sqrt{400}$ — Apply the square root property.

$w = 20$ $\qquad$ $w = -20$ — Simplify.

Because w represents the width of the television's screen, this dimension must be positive. We reject -20. Thus, the width of the television screen is 20 inches.

✓ **CHECK POINT 4** What is the width of a rectangle whose height is 20 inches and whose diagonal measures 50 inches? Express the answer in simplified radical form. Then find an approximation correct to the nearest tenth of an inch.

3 Find the distance between two points.

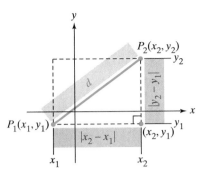

FIGURE 10.3

The Distance Formula

Using the Pythagorean Theorem, we can find the distance between the two points $P_1(x_1, y_1)$ and $P_2(x_2, y_2)$ in the rectangular coordinate system. The two points are illustrated in **Figure 10.3**.

The distance that we need to find is represented by d and shown in blue. Notice that the distance between two points on the dashed horizontal line is the absolute value of the difference between the x-coordinates of the two points. This distance, $|x_2 - x_1|$, is shown in pink. Similarly, the distance between two points on the dashed vertical line is the absolute value of the difference between the y-coordinates of the two points. This distance, $|y_2 - y_1|$, is also shown in pink.

Because the dashed lines are horizontal and vertical, a right triangle is formed. Thus, we can use the Pythagorean Theorem to find the distance d. Squaring the lengths of the triangle's sides results in positive numbers, so absolute value notation is not necessary.

$$d^2 = (x_2 - x_1)^2 + (y_2 - y_1)^2 \quad \text{Apply the Pythagorean Theorem to the right triangle in Figure 10.3.}$$

$$d = \pm\sqrt{(x_2 - x_1)^2 + (y_2 - y_1)^2} \quad \text{Apply the square root property.}$$

$$d = \sqrt{(x_2 - x_1)^2 + (y_2 - y_1)^2} \quad \text{Because distance is nonnegative, write only the principal square root.}$$

This result is called the **distance formula**.

The Distance Formula

The distance, d, between the points (x_1, y_1) and (x_2, y_2) in the rectangular coordinate system is

$$d = \sqrt{(x_2 - x_1)^2 + (y_2 - y_1)^2}.$$

To compute the distance between two points, find the square of the difference between the x-coordinates plus the square of the difference between the y-coordinates. The principal square root of this sum is the distance.

When using the distance formula, it does not matter which point you call (x_1, y_1) and which you call (x_2, y_2).

EXAMPLE 5 Using the Distance Formula

Find the distance between $(-4, -3)$ and $(6, 2)$.

Solution We will let $(x_1, y_1) = (-4, -3)$ and $(x_2, y_2) = (6, 2)$.

$$d = \sqrt{(x_2 - x_1)^2 + (y_2 - y_1)^2} \quad \text{Use the distance formula.}$$

$$= \sqrt{[6 - (-4)]^2 + [2 - (-3)]^2} \quad \text{Substitute the given values.}$$

$$= \sqrt{10^2 + 5^2} \quad \text{Perform operations within grouping symbols: } 6 - (-4) = 6 + 4 = 10 \text{ and } 2 - (-3) = 2 + 3 = 5.$$

$$= \sqrt{100 + 25} \quad \begin{array}{c}\text{Caution! This is \textbf{not} equal to } \sqrt{100} + \sqrt{25}.\end{array} \quad \text{Square 10 and 5.}$$

$$= \sqrt{125} \quad \text{Add.}$$

$$= 5\sqrt{5} \approx 11.18 \quad \sqrt{125} = \sqrt{25 \cdot 5} = \sqrt{25}\sqrt{5} = 5\sqrt{5}$$

The distance between the given points is $5\sqrt{5}$ units, or approximately 11.18 units. The situation is illustrated in **Figure 10.4**.

FIGURE 10.4 Finding the distance between two points

(6, 2)

(−4, −3)

Distance is $5\sqrt{5}$ units.

✓ **CHECK POINT 5** Find the distance between $(-4, 9)$ and $(1, -3)$.

Practice Exercises

In Exercises 1–30, solve each quadratic equation by the square root property. If possible, simplify radicals or rationalize denominators.

1. $x^2 = 16$ **2.** $x^2 = 100$

3. $y^2 = 81$ **4.** $y^2 = 144$

5. $x^2 = 7$ **6.** $x^2 = 13$

7. $x^2 = 50$ **8.** $x^2 = 27$

9. $5x^2 = 20$ **10.** $3x^2 = 75$

11. $4y^2 = 49$ **12.** $16y^2 = 25$

13. $2x^2 + 1 = 51$ **14.** $3x^2 - 1 = 47$

15. $3x^2 - 2 = 0$

16. $3x^2 - 5 = 0$

17. $5z^2 - 7 = 0$

18. $5z^2 - 2 = 0$

19. $(x - 3)^2 = 16$

20. $(x - 2)^2 = 25$

21. $(x + 5)^2 = 121$

22. $(x + 6)^2 = 144$

23. $(3x + 2)^2 = 9$

24. $(2x + 1)^2 = 49$

25. $(x - 5)^2 = 3$

26. $(x - 3)^2 = 15$

27. $(y + 8)^2 = 11$

28. $(y + 7)^2 = 5$

29. $(z - 4)^2 = 18$

30. $(z - 6)^2 = 12$

In Exercises 31–40, solve each quadratic equation by first factoring the perfect square trinomial on the left side. Then apply the square root property. Simplify radicals, if possible.

31. $x^2 + 4x + 4 = 16$

32. $x^2 + 4x + 4 = 25$

33. $x^2 - 6x + 9 = 36$

34. $x^2 - 6x + 9 = 49$

35. $x^2 - 10x + 25 = 2$

36. $x^2 - 10x + 25 = 3$

37. $x^2 + 2x + 1 = 5$

38. $x^2 + 2x + 1 = 7$

39. $y^2 - 14y + 49 = 12$

40. $y^2 - 14y + 49 = 18$

In Exercises 41–48, use the Pythagorean Theorem to find the missing length in each right triangle. Express the answer in radical form and simplify, if possible.

41.

8 m, c, 15 m

42.

7 m, c, 24 m

43.

15 m, c, 36 m

44. 5 in.

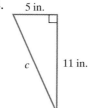

c, 11 in.

45. 16 cm

a, 20 cm

46. a

13 ft, 5 ft

47.

16 m, 9 m, b

48.

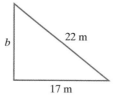

b, 22 m, 17 m

In Exercises 49–58, find the distance between each pair of points. Express answers in simplified radical form and, if necessary, round to two decimal places.

49. $(3, 5)$ and $(4, 1)$

50. $(1, 5)$ and $(6, 2)$

51. $(-4, 2)$ and $(4, 17)$

52. $(2, -2)$ and $(5, 2)$

53. $(6, -1)$ and $(9, 5)$

54. $(-4, -1)$ and $(2, -3)$

55. $(-7, -5)$ and $(-2, -1)$

56. $(-8, -4)$ and $(-3, -8)$

57. $(-2\sqrt{7}, 10)$ and $(4\sqrt{7}, 8)$

58. $(-\sqrt{3}, 4\sqrt{6})$ and $(2\sqrt{3}, \sqrt{6})$

Practice PLUS

59. The square of the difference between a number and 3 is 25. Find the number(s).

60. The square of the difference between a number and 7 is 16. Find the number(s).

61. If 3 times a number is increased by 2 and this sum is squared, the result is 49. Find the number(s).

62. If 4 times a number is decreased by 3 and this difference is squared, the result is 9. Find the number(s).

In Exercises 63–66, solve the formula for the specified variable. Because each variable is nonnegative, list only the principal square root. If possible, simplify radicals or rationalize denominators.

63. $A = \pi r^2$ for r

64. $ax^2 - b = 0$ for x

65. $I = \dfrac{k}{d^2}$ for d

66. $A = p(1 + r)^2$ for r

Application Exercises

Use the Pythagorean Theorem to solve Exercises 67–72. Express the answer in radical form and simplify, if possible.

67. Find the length of the ladder.

68. How high is the airplane above the ground?

69. A baseball diamond is actually a square with 90-foot sides. What is the distance from home plate to second base?

70. The base of a 20-foot ladder is 15 feet from the house. How far up the house does the ladder reach?

Use the formula for the area of a circle, $A = \pi r^2$, to solve Exercises 71–72.

71. If the area of a circle is 36π square inches, find its radius.

72. If the area of a circle is 49π square inches, find its radius.

The weight of a human fetus is modeled by the formula $W = 3t^2$, where W is the weight, in grams, and t is the time, in weeks, with $0 \le t \le 39$. Use this formula to solve Exercises 73–74.

73. After how many weeks does the fetus weigh 108 grams?

74. After how many weeks does the fetus weigh 192 grams?

The distance, d, in feet, that an object falls in t seconds is modeled by the formula $d = 16t^2$. Use this formula to solve Exercises 75–76.

75. If you drop a rock from a cliff 400 feet above the water how long will it take for the rock to hit the water?

76. If you drop a rock from a cliff 576 feet above the water, how long will it take for the rock to hit the water?

77. A square flower bed is to be enlarged by adding 2 meters on each side. If the larger square has an area of 144 square meters, what is the length of the original square?

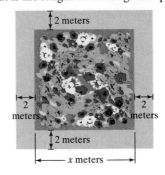

78. A square flower bed is to be enlarged by adding 3 feet on each side. If the larger square has an area of 169 square feet, what is the length of the original square?

79. A machine produces open boxes using square sheets of metal. The figure illustrates that the machine cuts equal-sized squares measuring 2 inches on a side from the corners and then shapes the metal into an open box by turning up the sides. If each box must have a volume of 200 cubic inches, find the size of the length and width of the open box.

80. A machine produces open boxes using square sheets of metal. The machine cuts equal-sized squares measuring 3 inches on a side from the corners and then shapes the metal into an open box by turning up the sides. If each box must have a volume of 75 cubic inches, find the size of the length and width of the open box.

Writing in Mathematics

81. What is the square root property?

82. Explain how to solve $(x - 1)^2 = 16$ using the square root property.

83. In your own words, state the Pythagorean Theorem.

84. In the 1939 movie *The Wizard of Oz*, upon being presented with a Th.D. (Doctor of Thinkology), the Scarecrow proudly exclaims, "The sum of the square roots of any two sides of an isosceles triangle is equal to the square root of the remaining side." Did the Scarecrow get the Pythagorean Theorem right? In particular, describe four errors in the Scarecrow's statement.

Critical Thinking Exercises

Make Sense? *In Exercises 85–88, determine whether each statement "makes sense" or "does not make sense" and explain your reasoning.*

85. I have a graphing calculator, so I used the x-coordinates of the intersection points of the graphs of $y_1 = (x - 2)^2$ and $y_2 = 25$ to verify the solution set for $(x - 2)^2 = 25$.

86. When I'm given the picture of a right triangle, the hypotenuse is always the side on top.

87. I've noticed that in mathematics there is a connection between topics, such as using the Pythagorean Theorem to derive the distance formula.

88. When I use the square root property to determine the length of a right triangle's side, I don't even bother to list the negative square root.

In Exercises 89–92, determine whether each statement is true or false. If the statement is false, make the necessary change(s) to produce a true statement.

89. The equation $(x + 5)^2 = 8$ is equivalent to $x + 5 = 2\sqrt{2}$.

90. The equation $x^2 = 0$ has no solution.

91. The equation $x^2 = -1$ has no solutions that are real numbers.

92. The solutions of $3x^2 - 5 = 0$ are $\dfrac{\sqrt{5}}{3}$ and $-\dfrac{\sqrt{5}}{3}$.

93. Find the value(s) of x if the distance between $(-3, -2)$ and $(x, -5)$ is 5 units.

Technology Exercises

94. Use a graphing utility to solve $4 - (x + 1)^2 = 0$. Graph $y = 4 - (x + 1)^2$ in a $[-5, 5, 1]$ by $[-5, 5, 1]$ viewing rectangle. The equation's solutions are the graph's x-intercepts. Check by substitution in the given equation.

95. Use a graphing utility to solve $(x - 1)^2 - 9 = 0$. Graph $y = (x - 1)^2 - 9$ in a $[-5, 5, 1]$ by $[-9, 3, 1]$ viewing rectangle. The equation's solutions are the graph's x-intercepts. Check by substitution in the given equation.

Review Exercises

96. Factor completely: $12x^2 + 14x - 6$. (Section 7.5, Example 2)

97. Divide: $\dfrac{x^2 - x - 6}{3x - 3} \div \dfrac{x^2 - 4}{x - 1}$. (Section 8.2, Example 6)

98. Solve: $4(x - 5) = 22 + 2(6x + 3)$. (Section 2.3, Example 3)

Preview Exercises

Exercises 99–101 will help you prepare for the material covered in the next section.

99. Factor: $x^2 + 8x + 16$.

100. Factor: $x^2 - 14x + 49$.

101. Factor: $x^2 + 5x + \dfrac{25}{4}$.

SECTION

10.2

Solving Quadratic Equations by Completing the Square

Objectives

1 Complete the square of a binomial.

2 Solve quadratic equations by completing the square.

There is a lack of completion in both the Escher image and the unfinished square on the left. Completion for the geometric figure can be obtained by adding a small square to its upper-right-hand corner. Understanding this process algebraically will give you a new method, appropriately called *completing the square*, for solving quadratic equations.

1 Complete the square of a binomial.

Completing the Square

How do we solve a quadratic equation, $ax^2 + bx + c = 0$, if the trinomial $ax^2 + bx + c$ cannot be factored? We can convert the equation into an equivalent equation that can be solved using the square root property. This is accomplished by **completing the square**.

> ### Completing the Square
>
> If $x^2 + bx$ is a binomial, then by adding $\left(\dfrac{b}{2}\right)^2$, which is the square of half the coefficient of x, a perfect square trinomial will result. That is,
>
> | The coefficient of x^2 must be 1 to complete the square. |
>
> $$x^2 + bx + \left(\frac{b}{2}\right)^2 = \left(x + \frac{b}{2}\right)^2.$$

> **EXAMPLE 1** Completing the Square

Complete the square for each binomial. Then factor the resulting perfect square trinomial:

a. $x^2 + 8x$ **b.** $x^2 - 14x$ **c.** $x^2 + 5x$.

Solution To complete the square, we must add a term to each binomial. The term that should be added is the square of half the coefficient of x.

$$x^2 + 8x \qquad x^2 - 14x \qquad x^2 + 5x$$

| Add $\left(\frac{8}{2}\right)^2 = 4^2$.
Add 16 to complete the square. | Add $\left(\frac{-14}{2}\right)^2 = (-7)^2$.
Add 49 to complete the square. | Add $\left(\frac{5}{2}\right)^2$, or $\frac{25}{4}$,
to complete the square. |

a. The coefficient of the x-term of $x^2 + 8x$ is 8. Half of 8 is 4, and $4^2 = 16$. Add 16.

$$x^2 + 8x + 16 = (x + 4)^2$$

b. The coefficient of the x-term of $x^2 - 14x$ is -14. Half of -14 is -7, and $(-7)^2 = 49$. Add 49.

$$x^2 - 14x + 49 = (x - 7)^2$$

c. The coefficient of the x-term of $x^2 + 5x$ is 5. Half of 5 is $\frac{5}{2}$, and $\left(\frac{5}{2}\right)^2 = \frac{25}{4}$. Add $\frac{25}{4}$.

$$x^2 + 5x + \frac{25}{4} = \left(x + \frac{5}{2}\right)^2$$

Geometric figures make it possible to visualize completing the square.

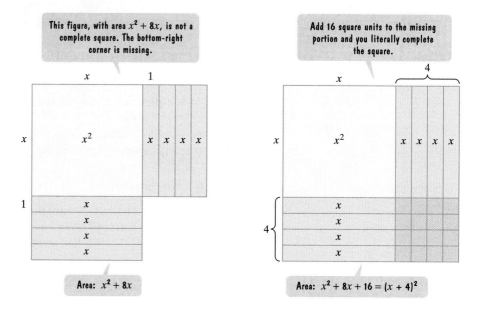

This figure, with area $x^2 + 8x$, is not a complete square. The bottom-right corner is missing.

Add 16 square units to the missing portion and you literally complete the square.

Area: $x^2 + 8x$

Area: $x^2 + 8x + 16 = (x + 4)^2$

✓ **CHECK POINT 1** Complete the square for each binomial. Then factor the resulting perfect square trinomial:

a. $x^2 + 10x$ **b.** $x^2 - 6x$

c. $x^2 + 3x$.

---

2 Solve quadratic equations by completing the square.

Solving Quadratic Equations by Completing the Square

We can solve *any* quadratic equation by completing the square. If the coefficient of the x^2-term is one, we add the square of half the coefficient of x to both sides of the equation. **When you add a constant term to one side of the equation to complete the square, be certain to add the same constant to the other side of the equation.** These ideas are illustrated in Example 2.

EXAMPLE 2 Solving Quadratic Equations by Completing the Square

Solve by completing the square:

a. $x^2 + 8x = -15$ **b.** $x^2 - 6x + 2 = 0$.

Study Tip

When you complete the square for the binomial expression $x^2 + bx$, you obtain a different polynomial. When you solve a quadratic equation by completing the square, you obtain an equation with the same solution set because the constant needed to complete the square is added to *both sides*.

Solution

a. To complete the square on the binomial $x^2 + 8x$, we take half of 8, which is 4, and square 4, giving 16. We add 16 to both sides of the equation. This makes the left side a perfect square trinomial.

$x^2 + 8x = -15$	This is the given equation.
$x^2 + 8x + 16 = -15 + 16$	Add 16 to both sides to complete the square.
$(x + 4)^2 = 1$	Factor and simplify.
$x + 4 = \sqrt{1}$ or $x + 4 = -\sqrt{1}$	Apply the square root property.
$x + 4 = 1$ $x + 4 = -1$	Simplify.
$x = -3$ $x = -5$	Subtract 4 from both sides in each equation.

The solutions are -5 and -3. The solution set is $\{-5, -3\}$.

b. To solve $x^2 - 6x + 2 = 0$ by completing the square, we first subtract 2 from both sides. This is done to isolate the binomial $x^2 - 6x$ so that we can complete the square.

$x^2 - 6x + 2 = 0$	This is the original equation.
$x^2 - 6x = -2$	Subtract 2 from both sides.

Next, we complete the square. Find half the coefficient of the x-term and square it. The coefficient of the x-term is -6. Half of -6 is -3, and $(-3)^2 = 9$. Thus, we add 9 to both sides of the equation.

$x^2 - 6x + 9 = -2 + 9$	Add 9 to both sides to complete the square.
$(x - 3)^2 = 7$	Factor and simplify.
$x - 3 = \sqrt{7}$ or $x - 3 = -\sqrt{7}$	Apply the square root property.
$x = 3 + \sqrt{7}$ $x = 3 - \sqrt{7}$	Add 3 to both sides in each equation.

The solutions are $3 + \sqrt{7}$ and $3 - \sqrt{7}$, expressed in abbreviated notation as $3 \pm \sqrt{7}$. The solution set is $\{3 \pm \sqrt{7}\}$. ▬

If you solve a quadratic equation by completing the square and the solutions are rational numbers, the equation can also be solved by factoring. By contrast, quadratic equations with irrational solutions cannot be solved by factoring. However, all quadratic equations can be solved by completing the square.

✓ **CHECK POINT 2**   Solve by completing the square:

a. $x^2 + 6x = 7$

b. $x^2 - 10x + 18 = 0$.

We have seen that the leading coefficient must be 1 in order to complete the square. If the coefficient of the x^2-term in a quadratic equation is not 1, you must divide each side of the equation by this coefficient before completing the square. For example, to solve $2x^2 + 5x - 4 = 0$ by completing the square, first divide every term by 2:

$$\frac{2x^2}{2} + \frac{5x}{2} - \frac{4}{2} = \frac{0}{2}$$

$$x^2 + \frac{5}{2}x - 2 = 0.$$

Now that the coefficient of the x^2-term is 1, we can solve by completing the square.

EXAMPLE 3   **Solving a Quadratic Equation by Completing the Square**

Solve by completing the square: $2x^2 + 5x - 4 = 0$.

Solution

$$2x^2 + 5x - 4 = 0$$

This is the original equation.

$$x^2 + \frac{5}{2}x - 2 = 0$$

Divide both sides by 2.

$$x^2 + \frac{5}{2}x = 2$$

Add 2 to both sides and isolate the binomial.

$$x^2 + \frac{5}{2}x + \frac{25}{16} = 2 + \frac{25}{16}$$

Complete the square: Half of $\frac{5}{2}$ is $\frac{5}{4}$ and $\left(\frac{5}{4}\right)^2 = \frac{25}{16}$.

$$\left(x + \frac{5}{4}\right)^2 = \frac{57}{16}$$

Factor and simplify. On the right: $2 + \frac{25}{16} = \frac{32}{16} + \frac{25}{16} = \frac{57}{16}$.

$$x + \frac{5}{4} = \sqrt{\frac{57}{16}} \quad \text{or} \quad x + \frac{5}{4} = -\sqrt{\frac{57}{16}}$$

Apply the square root property.

$$x + \frac{5}{4} = \frac{\sqrt{57}}{4} \qquad x + \frac{5}{4} = -\frac{\sqrt{57}}{4}$$

$\sqrt{\frac{57}{16}} = \frac{\sqrt{57}}{\sqrt{16}} = \frac{\sqrt{57}}{4}$

$$x = -\frac{5}{4} + \frac{\sqrt{57}}{4} \qquad x = -\frac{5}{4} - \frac{\sqrt{57}}{4}$$

Solve the equations, subtracting $\frac{5}{4}$ from both sides.

$$x = \frac{-5 + \sqrt{57}}{4} \qquad x = \frac{-5 - \sqrt{57}}{4}$$

Express solutions with a common denominator.

The solutions are $\dfrac{-5 \pm \sqrt{57}}{4}$ and the solution set is $\left\{\dfrac{-5 \pm \sqrt{57}}{4}\right\}$.   ∎

Using Technology

Graphic Connections

Obtain a decimal approximation for each solution of $2x^2 + 5x - 4 = 0$, the equation in Example 3:

$y = 2x^2 + 5x - 4$

$$\frac{-5 + \sqrt{57}}{4} \approx 0.6$$

$$\frac{-5 - \sqrt{57}}{4} \approx -3.1.$$

x-intercept: ≈ −3.1   x-intercept: ≈ 0.6

[−4, 2, 1] by [−10, 10, 1]

The x-intercepts of $y = 2x^2 + 5x - 4$ verify these solutions.

☑ **CHECK POINT 3**   Solve by completing the square: $2x^2 - 10x - 1 = 0$.

10.2 EXERCISE SET MyMathLab
PRACTICE WATCH DOWNLOAD READ REVIEW

Practice Exercises

In Exercises 1–12, complete the square for each binomial. Then factor the resulting perfect square trinomial.

1. $x^2 + 10x$

2. $x^2 + 12x$

3. $x^2 - 2x$

4. $x^2 - 4x$

5. $x^2 + 5x$

6. $x^2 + 3x$

7. $x^2 - 7x$

8. $x^2 - x$

9. $x^2 + \dfrac{1}{2}x$

10. $x^2 + \dfrac{1}{3}x$

11. $x^2 - \dfrac{4}{3}x$

12. $x^2 - \dfrac{4}{5}x$

In Exercises 13–34, solve each quadratic equation by completing the square.

13. $x^2 + 4x = 5$

14. $x^2 + 6x = -8$

15. $x^2 - 10x = -24$

16. $x^2 - 2x = 8$

17. $x^2 - 2x = 5$

18. $x^2 - 4x = -2$

19. $x^2 + 4x + 1 = 0$

20. $x^2 + 6x - 5 = 0$

21. $x^2 - 3x = 28$

22. $x^2 - 5x = -6$

23. $x^2 + 3x - 1 = 0$

24. $x^2 - 3x - 5 = 0$

25. $x^2 = 7x - 3$

26. $x^2 = 5x - 3$

27. $2x^2 - 2x - 6 = 0$

28. $2x^2 - 4x - 2 = 0$

29. $2x^2 - 3x + 1 = 0$

30. $2x^2 - x - 1 = 0$

31. $2x^2 + 10x + 11 = 0$

32. $2x^2 + 8x + 5 = 0$

33. $4x^2 - 2x - 3 = 0$

34. $3x^2 - 2x - 4 = 0$

Practice PLUS

In Exercises 35–40, solve each quadratic equation by completing the square.

35. $\dfrac{x^2}{6} - \dfrac{x}{3} - 1 = 0$

36. $\dfrac{x^2}{6} + x - \dfrac{3}{2} = 0$

37. $(x + 2)(x - 3) = 1$

38. $(x - 5)(x - 3) = 1$

39. $x^2 + 4bx = 5b^2$

40. $x^2 + 6bx = 7b^2$

Writing in Mathematics

41. Explain how to complete the square for a binomial. Use $x^2 + 6x$ to illustrate your explanation.

42. Explain how to solve $x^2 + 6x + 8 = 0$ by completing the square.

Critical Thinking Exercises

Make Sense? *In Exercises 43–46, determine whether each statement "makes sense" or "does not make sense" and explain your reasoning.*

43. When I complete the square, I convert a quadratic equation into an equivalent equation that can be solved by the square root property.

44. When the coefficient of the x-term in a quadratic equation is negative and I'm solving by completing the square, I add a negative constant to each side of the equation.

45. When I complete the square for the binomial $x^2 + bx$, I obtain a different polynomial, but when I solve a quadratic equation by completing the square, I obtain an equation with the same solution set.

46. I solved $4x^2 + 10x = 0$ by completing the square and added 25 to both sides of the equation.

In Exercises 47–50, determine whether each statement is true or false. If the statement is false, make the necessary change(s) to produce a true statement.

47. Completing the square is a method for finding the area and perimeter of a square.

48. The trinomial $x^2 - 3x + 9$ is a perfect square trinomial.

49. Although not every quadratic equation can be solved by completing the square, they can all be solved using factoring.

50. In completing the square for $x^2 - 7x = 5$, we should add $\frac{49}{4}$ to both sides.

51. Write a perfect square trinomial whose x-term is $-20x$.

52. Solve by completing the square: $x^2 + x + c = 0$.

53. Solve by completing the square: $x^2 + bx + c = 0$.

Technology Exercise

54. Use the technique shown in the Using Technology box on page 612 to verify the solutions of any two quadratic equations in Exercises 17–20.

Review Exercises

In Exercises 55–56, perform the indicated operations. If possible, simplify the answer.

55. $\dfrac{2x + 3}{x^2 - 7x + 12} - \dfrac{2}{x - 3}$ (Section 8.4, Example 7)

56. $\dfrac{x - \dfrac{1}{3}}{3 - \dfrac{1}{x}}$ (Section 8.5, Example 2 or Example 5)

57. Solve: $\sqrt{2x + 3} = 2x - 3$. (Section 9.5, Example 4)

Preview Exercises

Exercises 58–60 will help you prepare for the material covered in the next section. In each exercise, evaluate

$$\frac{-b \pm \sqrt{b^2 - 4ac}}{2a}$$

for the given values of a, b, and c. Where necessary, express answers in simplified radical form.

58. $a = 2, b = 9, c = -5$

59. $a = 9, b = -12, c = 4$

60. $a = 1, b = -2, c = -6$

The Quadratic Formula

Objectives

1 Solve quadratic equations using the quadratic formula.

2 Determine the most efficient method to use when solving a quadratic equation.

3 Solve problems using quadratic equations.

In the chapter opener, we observed that a substantial percentage of the United States population is foreign-born. In this section, we will explore the foreign-born model using a formula that will enable you to solve quadratic equations more quickly than using the method of completing the square. We begin by deriving this formula.

1 Solve quadratic equations using the quadratic formula.

Solving Quadratic Equations Using the Quadratic Formula

We can apply the method of completing the square to derive a formula that can be used to solve all quadratic equations. The derivation given here also shows a particular quadratic equation, $3x^2 - 2x - 4 = 0$, to specifically illustrate each of the steps.

Deriving the Quadratic Formula

Standard Form of a Quadratic Equation	Comment	A Specific Example
$ax^2 + bx + c = 0, a > 0$	This is the given equation.	$3x^2 - 2x - 4 = 0$
$x^2 + \dfrac{b}{a}x + \dfrac{c}{a} = 0$	Divide both sides by a so that the coefficient of x^2 is 1.	$x^2 - \dfrac{2}{3}x - \dfrac{4}{3} = 0$
$x^2 + \dfrac{b}{a}x = -\dfrac{c}{a}$	Isolate the binomial by adding $-\dfrac{c}{a}$ on both sides of the equation.	$x^2 - \dfrac{2}{3}x = \dfrac{4}{3}$
$x^2 + \dfrac{b}{a}x + \underbrace{\left(\dfrac{b}{2a}\right)^2}_{\text{(half)}^2} = -\dfrac{c}{a} + \left(\dfrac{b}{2a}\right)^2$ $x^2 + \dfrac{b}{a}x + \dfrac{b^2}{4a^2} = -\dfrac{c}{a} + \dfrac{b^2}{4a^2}$	Complete the square. Add the square of half the coefficient of x to both sides.	$x^2 - \dfrac{2}{3}x + \underbrace{\left(-\dfrac{1}{3}\right)^2}_{\text{(half)}^2} = \dfrac{4}{3} + \left(-\dfrac{1}{3}\right)^2$ $x^2 - \dfrac{2}{3}x + \dfrac{1}{9} = \dfrac{4}{3} + \dfrac{1}{9}$
$\left(x + \dfrac{b}{2a}\right)^2 = -\dfrac{c}{a}\cdot\dfrac{4a}{4a} + \dfrac{b^2}{4a^2}$	Factor on the left side and obtain a common denominator on the right side.	$\left(x - \dfrac{1}{3}\right)^2 = \dfrac{4}{3}\cdot\dfrac{3}{3} + \dfrac{1}{9}$
$\left(x + \dfrac{b}{2a}\right)^2 = \dfrac{-4ac + b^2}{4a^2}$ $\left(x + \dfrac{b}{2a}\right)^2 = \dfrac{b^2 - 4ac}{4a^2}$	Add fractions on the right side.	$\left(x - \dfrac{1}{3}\right)^2 = \dfrac{12 + 1}{9}$ $\left(x - \dfrac{1}{3}\right)^2 = \dfrac{13}{9}$
$x + \dfrac{b}{2a} = \pm\sqrt{\dfrac{b^2 - 4ac}{4a^2}}$	Apply the square root property.	$x - \dfrac{1}{3} = \pm\sqrt{\dfrac{13}{9}}$
$x + \dfrac{b}{2a} = \pm\dfrac{\sqrt{b^2 - 4ac}}{2a}$	Take the square root of the quotient, simplifying the denominator.	$x - \dfrac{1}{3} = \pm\dfrac{\sqrt{13}}{3}$
$x = \dfrac{-b}{2a} \pm \dfrac{\sqrt{b^2 - 4ac}}{2a}$	Solve for x by subtracting $\dfrac{b}{2a}$ from both sides.	$x = \dfrac{1}{3} \pm \dfrac{\sqrt{13}}{3}$
$x = \dfrac{-b \pm \sqrt{b^2 - 4ac}}{2a}$	Combine fractions on the right side.	$x = \dfrac{1 \pm \sqrt{13}}{3}$

The formula shown at the bottom of the left column is called the *quadratic formula*. A similar proof shows that the same formula can be used to solve quadratic equations if a, the coefficient of the x^2-term, is negative.

The Quadratic Formula

The solutions of a quadratic equation in standard form $ax^2 + bx + c = 0$, with $a \neq 0$, are given by the **quadratic formula**

$$x = \frac{-b \pm \sqrt{b^2 - 4ac}}{2a}.$$

x equals negative *b* plus or minus the square root of $b^2 - 4ac$, all divided by $2a$.

To use the quadratic formula, begin by writing the quadratic equation in standard form, if necessary. Then determine the numerical values for a (the coefficient of the x^2-term), b (the coefficient of the x-term), and c (the constant term). Substitute the values of a, b, and c into the quadratic formula and evaluate the expression. The $\pm$ sign indicates that if $b^2 - 4ac$ is not zero, there are two solutions of the equation.

EXAMPLE 1   Solving a Quadratic Equation Using the Quadratic Formula

Solve using the quadratic formula: $2x^2 + 9x - 5 = 0$.

Solution   The given equation is in standard form. Begin by identifying the values for a, b, and c.

$$2x^2 + 9x - 5 = 0$$

$$a = 2 \quad b = 9 \quad c = -5$$

Substituting these values into the quadratic formula and simplifying gives the equation's solutions.

$$x = \frac{-b \pm \sqrt{b^2 - 4ac}}{2a}$$ Use the quadratic formula.

$$x = \frac{-9 \pm \sqrt{9^2 - 4(2)(-5)}}{2(2)}$$ Substitute the values for a, b, and c: $a = 2$, $b = 9$, and $c = -5$.

$$= \frac{-9 \pm \sqrt{81 + 40}}{4}$$ $9^2 - 4(2)(-5) = 81 - (-40) = 81 + 40$

$$= \frac{-9 \pm \sqrt{121}}{4}$$ Add under the radical sign.

$$= \frac{-9 \pm 11}{4}$$ $\sqrt{121} = 11$

Now we will evaluate this expression in two different ways to obtain the two solutions. At the left, we will *add* 11 to -9. At the right, we will *subtract* 11 from -9.

$$x = \frac{-9 + 11}{4} \quad \text{or} \quad x = \frac{-9 - 11}{4}$$

$$= \frac{2}{4} = \frac{1}{2} \qquad\qquad = \frac{-20}{4} = -5$$

The solutions are -5 and $\frac{1}{2}$, and the solution set is $\left\{-5, \frac{1}{2}\right\}$.

In Example 1, the solutions of $2x^2 + 9x - 5 = 0$ are rational numbers. This means that the equation can also be solved using factoring. The reason that the solutions are rational numbers is that $b^2 - 4ac$, the radicand in the quadratic formula, is 121, which is a perfect square.

✓ **CHECK POINT 1**   Solve using the quadratic formula: $8x^2 + 2x - 1 = 0$.

Using Technology

Graphic Connections

The graph of $y = 2x^2 + 9x - 5$ has x-intercepts at -5 and $\frac{1}{2}$. This verifies that -5 and $\frac{1}{2}$ are the solutions of

$$2x^2 + 9x - 5 = 0.$$

x-intercept: -5

$y = 2x^2 + 9x - 5$

x-intercept: $\frac{1}{2}$

$[-6, 3, 1]$ by $[-16, 5, 1]$

EXAMPLE 2   Solving a Quadratic Equation Using the Quadratic Formula

Solve using the quadratic formula: $2x^2 = 6x - 1$.

Solution   The quadratic equation must be in standard form to identify the values for a, b, and c. We need to move all terms to one side and obtain zero on the other side. To obtain zero on the right, we subtract $6x$ and add 1 on both sides. Then we can identify the values for a, b, and c.

$$2x^2 = 6x - 1 \qquad \text{This is the given equation.}$$

$$2x^2 - 6x + 1 = 6x - 6x - 1 + 1 \qquad \text{This step is usually performed mentally.}$$

$$2x^2 - 6x + 1 = 0 \qquad \text{Identify } a, \text{ the } x^2\text{-coefficient, } b, \text{ the}$$
$$\underline{a = 2} \quad \underline{b = -6} \quad \underline{c = 1} \qquad x\text{-coefficient, and } c, \text{ the constant.}$$

Substituting these values into the quadratic formula and simplifying gives the equation's solutions.

$$x = \frac{-b \pm \sqrt{b^2 - 4ac}}{2a} \qquad \text{Use the quadratic formula.}$$

$$x = \frac{-(-6) \pm \sqrt{(-6)^2 - 4(2)(1)}}{2 \cdot 2} \qquad \text{Substitute the values for } a, b, \text{ and } c: \\ a = 2, b = -6, \text{ and } c = 1.$$

$$= \frac{6 \pm \sqrt{36 - 8}}{4} \qquad -(-6) = 6 \text{ and } (-6)^2 = (-6)(-6) = 36.$$

$$= \frac{6 \pm \sqrt{28}}{4} \qquad \text{Complete the subtraction under the radical.}$$

$$= \frac{6 \pm 2\sqrt{7}}{4} \qquad \sqrt{28} = \sqrt{4 \cdot 7} = \sqrt{4}\sqrt{7} = 2\sqrt{7}$$

$$= \frac{2(3 \pm \sqrt{7})}{4} \qquad \text{Factor out 2 from the numerator.}$$

$$= \frac{3 \pm \sqrt{7}}{2} \qquad \text{Divide the numerator and denominator by 2.}$$

The solutions are $\dfrac{3 + \sqrt{7}}{2}$ and $\dfrac{3 - \sqrt{7}}{2}$, abbreviated $\dfrac{3 \pm \sqrt{7}}{2}$. The solution set is is $\left\{\dfrac{3 \pm \sqrt{7}}{2}\right\}$. ■

In Example 2, the solutions of $2x^2 = 6x - 1$ are irrational numbers. This means that the equation cannot be solved using factoring. The reason that the solutions are irrational numbers is that $b^2 - 4ac$, the radicand in the quadratic formula, is 28, which is not a perfect square.

Study Tip

Many students use the quadratic formula correctly until the last step, where they make an error in simplifying the solutions. Be sure to factor the numerator before dividing the numerator and denominator by the greatest common factor:

$$\frac{6 \pm 2\sqrt{7}}{4} = \frac{2(3 \pm \sqrt{7})}{4} = \frac{\overset{1}{\cancel{2}}(3 \pm \sqrt{7})}{\underset{2}{\cancel{4}}} = \frac{3 \pm \sqrt{7}}{2}.$$

Factor first. Then divide by the GCF.

You cannot divide just one term in the numerator and the denominator by their greatest common factor.

Incorrect!

Can all irrational solutions of quadratic equations be simplified? No. The following solutions cannot be simplified.

$$\frac{5 \pm 2\sqrt{7}}{2} \qquad \frac{-4 \pm 3\sqrt{7}}{2}$$

> Other than 1, terms in each numerator have no common factor.

☑ **CHECK POINT 2** Solve using the quadratic formula: $x^2 = 6x - 4$.

---

2 Determine the most efficient method to use when solving a quadratic equation.

Determining Which Method to Use

All quadratic equations can be solved by the quadratic formula. However, if an equation is in the form $u^2 = d$, such as $x^2 = 5$ or $(2x + 3)^2 = 8$, it is faster to use the square root property, taking the square root of both sides. If the equation is not in the form $u^2 = d$, write the quadratic equation in standard form ($ax^2 + bx + c = 0$). Try to solve the equation by factoring. If $ax^2 + bx + c$ cannot be factored, then solve the quadratic equation by the quadratic formula.

Because we used the method of completing the square to derive the quadratic formula, we no longer need it for solving quadratic equations. However, you will use completing the square in more advanced algebra courses to help graph certain kinds of equations.

Table 10.1 summarizes our observations about which technique to use when solving a quadratic equation.

Table 10.1 Determining the Most Efficient Technique to Use When Solving a Quadratic Equation

Description and Form of the Quadratic Equation	Most Efficient Solution Method	Example
$ax^2 + bx + c = 0$ and $ax^2 + bx + c$ can be factored easily.	Factor and use the zero-product principle.	$3x^2 + 5x - 2 = 0$ $(3x - 1)(x + 2) = 0$ $3x - 1 = 0$ or $x + 2 = 0$ $x = \dfrac{1}{3} \qquad x = -2$
$ax^2 + c = 0$ The quadratic equation has no x-term. ($b = 0$)	Solve for x^2 and apply the square root property.	$4x^2 - 7 = 0$ $4x^2 = 7$ $x^2 = \dfrac{7}{4}$ $x = \pm\dfrac{\sqrt{7}}{2}$
$u^2 = d$; u is a first-degree polynomial.	Use the square root property.	$(x + 4)^2 = 5$ $x + 4 = \pm\sqrt{5}$ $x = -4 \pm \sqrt{5}$
$ax^2 + bx + c = 0$ and $ax^2 + bx + c$ cannot be factored or the factoring is too difficult.	Use the quadratic formula: $x = \dfrac{-b \pm \sqrt{b^2 - 4ac}}{2a}$.	$x^2 - 2x - 6 = 0$ $a = 1 \quad b = -2 \quad c = -6$ $x = \dfrac{-(-2) \pm \sqrt{(-2)^2 - 4(1)(-6)}}{2(1)}$ $= \dfrac{2 \pm \sqrt{4 + 24}}{2(1)}$ $= \dfrac{2 \pm \sqrt{28}}{2} = \dfrac{2 \pm \sqrt{4}\sqrt{7}}{2}$ $= \dfrac{2 \pm 2\sqrt{7}}{2} = \dfrac{2(1 \pm \sqrt{7})}{2}$ $= 1 \pm \sqrt{7}$

3 Solve problems using quadratic equations.

Application

Quadratic equations can be solved by any efficient method to answer questions about variables contained in mathematical models.

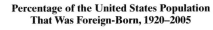

EXAMPLE 3 Making Predictions About the U.S. Foreign-Born Population

A substantial percentage of the United States population is foreign-born. The graph in **Figure 10.5** shows the percentage of foreign-born Americans for selected years from 1920 through 2005.

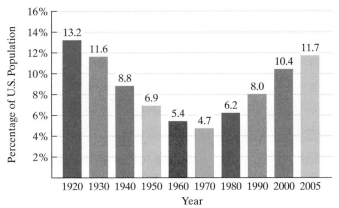

Percentage of the United States Population That Was Foreign-Born, 1920–2005

FIGURE 10.5

Source: U.S. Census Bureau

The percentage, p, of the United States population that was foreign-born x years after 1920 can be modeled by the formula

$$p = 0.004x^2 - 0.37x + 14.1.$$

According to this model, in which year will 15% of the United States population be foreign-born?

Solution   Because we are interested in when the foreign-born percentage will reach 15%, we substitute 15 for p in the given formula. Then we solve for x, the number of years after 1920.

$$p = 0.004x^2 - 0.37x + 14.1$$ This is the given formula.

$$15 = 0.004x^2 - 0.37x + 14.1$$ Substitute 15 for p.

$$0 = 0.004x^2 - 0.37x - 0.9$$ Subtract 15 from both sides and write the quadratic equation in standard form.

$a = 0.004$ $b = -0.37$ $c = -0.9$

$$0 = 0.004x^2 - 0.37x - 0.9$$

a = 0.004   b = -0.37   c = -0.9

The equation we need to solve
(repeated)

The standard form of the quatratic equation that we need to solve is repeated in the margin. Because the trinomial on the right side of the equation, $0.004x^2 - 0.37x - 0.9$, is not easily factored, if it can be factored at all, we solve using the quadratic formula.

$$x = \frac{-b \pm \sqrt{b^2 - 4ac}}{2a}$$

Use the quadratic formula.

$$x = \frac{-(-0.37) \pm \sqrt{(-0.37)^2 - 4(0.004)(-0.9)}}{2(0.004)}$$

Substitute the values for a, b, and c: a = 0.004, b = -0.37, and c = -0.9.

$$= \frac{0.37 \pm \sqrt{0.1513}}{0.008}$$

Perform the indicated computations. Use a calculator to simplify the radicand.

$$x = \frac{0.37 + \sqrt{0.1513}}{0.008} \quad \text{or} \quad x = \frac{0.37 - \sqrt{0.1513}}{0.008}$$

$$x \approx 95 \qquad\qquad\qquad x \approx -2$$

Use a calculator and round to the nearest whole number.

Reject this solution. The model applies to years *after*, not *before*, 1920.

According to the model, approximately 95 years after 1920, or in 2015, 15% of the United States population will be foreign-born. ■

Using Technology

Numeric Connections

A graphing utility's |TABLE| feature can be used to verify the solution to Example 3.

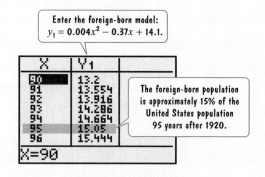

Enter the foreign-born model:
$y_1 = 0.004x^2 - 0.37x + 14.1$.

X	Y₁
90	13.2
91	13.554
92	13.916
93	14.286
94	14.664
95	15.05
96	15.444

X=90

The foreign-born population is approximately 15% of the United States population 95 years after 1920.

☑ **CHECK POINT 3** Use the model in Example 3 to determine in which year 25% of the United States population will be foreign-born.

Blitzer Bonus

Art, Nature, and Quadratic Equations

A **golden rectangle** can be a rectangle of any size, but its long side must be Φ times as long as its short side, where Φ ≈ 1.6. Artists often use golden rectangles in their work because they are considered to be more visually pleasing than other rectangles.

In *The Bathers at Asnieres*, by the French impressionist Georges Seurat (1859–1891), the artist positions parts of the painting as though they were inside golden rectangles.

Georges Seurat, "The Bathers at Asnieres," 1883–1884. Oil on canvas, $79\frac{1}{2} \times 118\frac{1}{2}$ in. Erich Lessing/ Art Resource, NY.

If a golden rectangle is divided into a square and a rectangle, as in **Figure 10.6(a)**, the smaller rectangle is a golden rectangle. If the smaller golden rectangle is divided again, the same is true of the yet smaller rectangle, and so on. The process of repeatedly dividing each golden rectangle in this manner is illustrated in **Figure 10.6(b)**. We've also created a spiral by connecting the opposite corners of all the squares with a smooth curve. This spiral matches the spiral shape of the chambered nautilus shell shown in **Figure 10.6(c)**. The shell spirals out at an ever-increasing rate that is governed by this geometry.

Golden Rectangle *A*

FIGURE 10.6(a)

FIGURE 10.6(b)

FIGURE 10.6(c)

We can use the golden rectangles in **Figure 10.7** to obtain an exact value for Φ, the ratio of the long side to the short side in a golden rectangle of any size.

Golden Rectangle *A*

FIGURE 10.7

Golden ratio in rectangle *A*   Golden ratio in rectangle *B*

$$\frac{\Phi}{1} = \frac{1}{\Phi - 1}$$

$$(\Phi - 1)\left(\frac{\Phi}{1}\right) = (\Phi - 1)\left(\frac{1}{\Phi - 1}\right) \qquad \text{Multiply both sides by } \Phi - 1, \text{ the LCD.}$$

$$\Phi^2 - \Phi = 1 \qquad \text{Perform the multiplications.}$$

$$\Phi^2 - \Phi - 1 = 0 \qquad \text{Write the quadratic equation in standard form.}$$

$a = 1$ $b = -1$ $c = -1$

(continued)

Golden Rectangle A

FIGURE 10.7 (repeated)

$$\Phi = \frac{-b \pm \sqrt{b^2 - 4ac}}{2a}$$

Use the quadratic formula to solve $\Phi^2 - \Phi - 1 = 0$.

$$\Phi = \frac{-(-1) \pm \sqrt{(-1)^2 - 4(1)(-1)}}{2(1)}$$

Substitute the values for a, b, and c: $a = 1$, $b = -1$, and $c = -1$.

$$\Phi = \frac{1 \pm \sqrt{5}}{2}$$

Simplify.

$$\Phi = \frac{1 + \sqrt{5}}{2} \approx 1.618$$

This is the exact value of the golden ratio.

$$\Phi = \frac{1 - \sqrt{5}}{2} \approx -0.618$$

Reject this value. The length of golden rectangle A cannot be negative.

The ratio of the long side to the short side in a golden rectangle of any size is $\frac{1 + \sqrt{5}}{2}$ to 1.

10.3 EXERCISE SET
MyMathLab

 PRACTICE   WATCH   DOWNLOAD   READ   REVIEW

Practice Exercises

In Exercises 1–22, solve each equation using the quadratic formula. Simplify irrational solutions, if possible.

1. $x^2 + 5x + 6 = 0$ **2.** $x^2 + 7x + 10 = 0$

3. $x^2 + 5x + 3 = 0$ **4.** $x^2 + 5x + 2 = 0$

5. $x^2 + 4x - 6 = 0$ **6.** $x^2 + 2x - 4 = 0$

7. $x^2 + 4x - 7 = 0$ **8.** $x^2 + 4x + 1 = 0$

9. $x^2 - 3x - 18 = 0$ **10.** $x^2 - 3x - 10 = 0$

11. $6x^2 - 5x - 6 = 0$ **12.** $9x^2 - 12x - 5 = 0$

13. $x^2 - 2x - 10 = 0$ **14.** $x^2 + 6x - 10 = 0$

15. $x^2 - x = 14$

16. $x^2 - 5x = 10$

17. $6x^2 + 6x + 1 = 0$

18. $3x^2 - 5x + 1 = 0$

19. $9x^2 - 12x + 4 = 0$

20. $4x^2 + 12x + 9 = 0$

21. $4x^2 = 2x + 7$

22. $3x^2 = 6x - 1$

In Exercises 23–44, solve each equation by the method of your choice. Simplify irrational solutions, if possible.

23. $2x^2 - x = 1$ **24.** $3x^2 - 4x = 4$

25. $5x^2 + 2 = 11x$

26. $5x^2 = 6 - 13x$

27. $3x^2 = 60$

28. $2x^2 = 250$

29. $x^2 - 2x = 1$

30. $2x^2 + 3x = 1$

31. $(2x + 3)(x + 4) = 1$

32. $(2x - 5)(x + 1) = 2$

33. $(3x - 4)^2 = 16$

34. $(2x + 7)^2 = 25$

35. $3x^2 - 12x + 12 = 0$

36. $9 - 6x + x^2 = 0$

37. $4x^2 - 16 = 0$

38. $3x^2 - 27 = 0$

39. $x^2 + 9x = 0$

40. $x^2 - 6x = 0$

41. $\frac{3}{4}x^2 - \frac{5}{2}x - 2 = 0$

42. $\frac{1}{3}x^2 - \frac{1}{2}x - \frac{3}{2} = 0$

43. $(3x - 2)^2 = 10$

44. $(4x - 1)^2 = 15$

Practice PLUS

In Exercises 45–52, solve each equation by the method of your choice. Simplify irrational solutions, if possible.

45. $\frac{x^2}{x + 7} - \frac{3}{x + 7} = 0$

46. $\frac{x^2}{x + 9} - \frac{11}{x + 9} = 0$

47. $(x + 2)^2 + x(x + 1) = 4$

48. $(x - 1)(3x + 2) = -7(x - 1)$

49. $2x^2 - 9x - 3 = 9 - 9x$

50. $3x^2 - 6x - 3 = 12 - 6x$

51. $\frac{1}{x} + \frac{1}{x + 3} = \frac{1}{4}$

52. $\frac{1}{x} + \frac{2}{x + 3} = \frac{1}{4}$

Application Exercises

53. A football is kicked straight up from a height of 4 feet with an initial speed of 60 feet per second. The formula

$$h = -16t^2 + 60t + 4$$

describes the ball's height above the ground, h, in feet, t seconds after it is kicked. How long will it take for the football to hit the ground? Use a calculator and round to the nearest tenth of a second.

54. Standing on a platform 50 feet high, a person accidentally fires a gun straight into the air. The formula

$$h = -16t^2 + 100t + 50$$

describes the bullet's height above the ground, h, in feet, t seconds after the gun is fired. How long will it take for the bullet to hit the ground? Use a calculator and round to the nearest tenth of a second.

The height of the arch supporting the bridge shown in the figure is modeled by

$$h = -0.05x^2 + 27,$$

where x is the distance, in feet, from the center of the arch. Use this formula to solve Exercises 55–56.

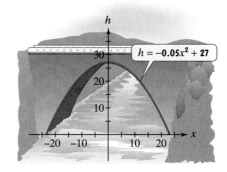

55. How far to the right of the center is the height 22 feet?

56. How far to the right of the center is the height 7 feet?

The bar graph shows the number of visitors, in millions, to America's national parks for seven selected years. The data can be modeled by the formula

$$N = -0.095x^2 + 1.74x + 60,$$

where N is the number of visitors to national parks, in millions, x years after 1988. Use this information to solve Exercises 57–58.

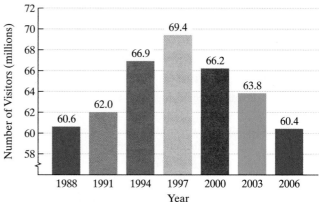

Number of Visitors to America's National Parks

Source: National Park Service

(In Exercises 57–58, be sure to refer to the mathematical model and the bar graph shown at the bottom of the previous page.)

57. a. According to the model, how many visitors to national parks were there in 2006? Round to the nearest tenth of a million. Does the model underestimate or overestimate the actual number displayed by the bar graph? By how much?

 b. If trends shown by the model continue, in which year will there be 41 million visitors to national parks?

58. a. According to the model, how many visitors to national parks were there in 2003? Round to the nearest tenth of a million. Does the model underestimate or overestimate the actual number displayed by the bar graph? By how much?

 b. According to the model, in which year will there be no visitors to America's national parks? Does this prediction seem reasonable? Explain your answer.

59. The length of a rectangle is 3 meters longer than the width. If the area is 36 square meters, find the rectangle's dimensions. Round to the nearest tenth of a meter.

60. The length of a rectangle is 2 meters longer than the width. If the area is 10 square meters, find the rectangle's dimensions. Round to the nearest tenth of a meter.

61. The hypotenuse of a right triangle is 4 feet long. One leg is 1 foot longer than the other. Find the lengths of the legs. Round to the nearest tenth of a foot.

62. The hypotenuse of a right triangle is 6 feet long. One leg is 1 foot shorter than the other. Find the lengths of the legs. Round to the nearest tenth of a foot.

63. In Exercise Set 8.4, we considered two formulas that approximate the dosage of a drug prescribed for children:

$$\text{Young's rule:} \quad C = \frac{DA}{A + 12}$$

$$\text{Cowling's rule:} \quad C = \frac{D(A + 1)}{24}.$$

In each formula, A = the child's age, in years, D = an adult dosage, and C = the proper child's dosage. The formulas apply for ages 2 through 13, inclusive. At which age, to the nearest tenth of a year, do the two formulas give the same dosage?

Writing in Mathematics

64. What is the quadratic formula and why is it useful?

65. Without going into specific details for each step, describe how the quadratic formula is derived.

66. Explain how to solve $x^2 + 6x + 8 = 0$ using the quadratic formula.

67. If you are given a quadratic equation, how do you determine which method to use to solve it?

Critical Thinking Exercises

Make Sense? *In Exercises 68–71, determine whether each statement "makes sense" or "does not make sense" and explain your reasoning.*

68. Because I want to solve $25x^2 - 49 = 0$ fairly quickly, I'll use the quadratic formula.

69. The fastest way for me to solve $x^2 - x - 2 = 0$ is to use the quadratic formula.

70. The data showing the percentage of the United States population that was foreign-born (**Figure 10.5** on page 619) could be described more accurately by a linear model than by the quadratic model that was given.

71. I simplified $\dfrac{3 + 2\sqrt{3}}{2}$ to $3 + \sqrt{3}$ because 2 is a factor of $2\sqrt{3}$.

In Exercises 72–75, determine whether each statement is true or false. If the statement is false, make the necessary change(s) to produce a true statement.

72. The quadratic formula can be expressed as

$$x = -b \pm \frac{\sqrt{b^2 - 4ac}}{2a}.$$

73. The solutions $\dfrac{4 \pm \sqrt{3}}{2}$ can be simplified to $2 \pm \sqrt{3}$.

74. For the quadratic equation $-2x^2 + 3x = 0$, we have $a = -2, b = 3$, and $c = 0$.

75. Any quadratic equation that can be solved by completing the square can be solved by the quadratic formula.

76. The radicand of the quadratic formula, $b^2 - 4ac$, can be used to determine whether $ax^2 + bx + c = 0$ has solutions that are rational, irrational, or not real numbers. Explain how this works. Is it possible to determine the kinds of answers that one will obtain to a quadratic equation without actually solving the equation? Explain.

77. Solve: $x^2 + 2\sqrt{3}x - 9 = 0$.

78. A rectangular vegetable garden is 5 feet wide and 9 feet long. The garden is to be surrounded by a tile border of uniform width. If there are 40 square feet of tile for the border, how wide, to the nearest tenth of a foot, should it be?

Technology Exercises

79. Graph the formula in Exercise 53,

$$y = -16x^2 + 60x + 4,$$

in a $[0, 4, 1]$ by $[0, 65, 5]$ viewing rectangle. Use the graph to verify your solution to the exercise.

80. Graph the formula in Exercise 57,

$$y = -0.095x^2 + 1.74x + 60,$$

in a $[0, 30, 1]$ by $[0, 70, 5]$ viewing rectangle. Use the graph to verify your solution to part (b) of the exercise.

Review Exercises

81. Evaluate: $125^{-\frac{2}{3}}$. (Section 9.6, Example 3)

82. Rationalize the denominator: $\dfrac{12}{3 + \sqrt{5}}$.
(Section 9.4, Example 3)

83. Multiply: $(x - y)(x^2 + xy + y^2)$. (Section 6.2, Example 7)

Preview Exercises

Exercises 84–86 will help you prepare for the material covered in the next section.

84. Can squaring a real number result in a negative number? Based on your answer, are $\sqrt{-1}$ and $\sqrt{-4}$ real numbers?

In Exercises 85–86, list the numbers from each set that are:
a. *rational numbers;* **b.** *irrational numbers;* **c.** *real numbers;*
d. *not real numbers. (Hint: Your answer to each question in Exercise 84 should be "no.")*

85. $\{-\sqrt{5}, -\sqrt{4}, \sqrt{-1}, \sqrt{-4}, \sqrt{1}, \sqrt{5}\}$

86. $\{-\sqrt{9}, -\sqrt{7}, \sqrt{-9}, \sqrt{-7}, \sqrt{0}, \sqrt{7}, \sqrt{9}\}$

MID-CHAPTER CHECK POINT Section 10.1–Section 10.3

 What You Know: We saw that not all quadratic equations can be solved by factoring. We learned three new methods for solving these equations—the square root property, completing the square, and the quadratic formula. We also learned to determine the most efficient technique to use when solving a quadratic equation.

In Exercises 1–12, solve each equation by the method of your choice. Simplify irrational solutions, if possible.

1. $(3x - 2)^2 = 100$

2. $15x^2 = 5x$

3. $x^2 - 2x - 10 = 0$

4. $x^2 - 8x + 16 = 7$

5. $3x^2 - x - 2 = 0$

6. $6x^2 = 10x - 3$

7. $x^2 + (x + 1)^2 = 25$

8. $(x + 5)^2 = 40$

9. $2(x^2 - 8) = 11 - x^2$

10. $2x^2 + 5x + 1 = 0$

11. $(x - 8)(2x - 3) = 34$

12. $x + \dfrac{16}{x} = 8$

13. Solve by completing the square: $x^2 + 14x - 32 = 0$.

14. Find the missing length in the right triangle. Express the answer in simplified radical form.

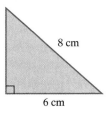

15. Find the distance between $(-3, 2)$ and $(9, -3)$.

16. The figure shows a right triangle whose hypotenuse measures 20 inches and whose leg measurements are in the ratio 3:4. Find the length of each leg.

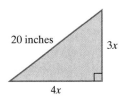

SECTION 10.4

Imaginary Numbers as Solutions of Quadratic Equations

Objectives

1 Express square roots of negative numbers in terms of *i*.

2 Solve quadratic equations with imaginary solutions.

THE KID WHO LEARNED ABOUT MATH ON THE STREET

If you divide 6,973 by 0, you die.

Once, this guy tried to find the square root of -9, and his eyeballs turned black.

This girl my brother knows found out exactly what π equals, but she went nuts.

Who is this kid warning us about our eyeballs turning black if we attempt to find the square root of −9? Don't believe what you hear on the street. Although square roots of negative numbers are not real numbers, they do play a significant role in algebra. In this section, we move beyond the real numbers and discuss square roots with negative radicands.

1 Express square roots of negative numbers in terms of *i*.

The Imaginary Unit *i*

Throughout this chapter, we have avoided quadratic equations that have no real numbers as solutions. A fairly simple example is the equation

$$x^2 = -1.$$

Because the square of a real number is never negative, there is no real number x such that $x^2 = -1$. To provide a setting in which such equations have solutions, mathematicians invented an expanded system of numbers, the *complex numbers*. The *imaginary number i*, defined to be a solution to the equation $x^2 = -1$, is the basis of this new set.

The Imaginary Unit *i*

The **imaginary unit** *i* is defined as

$$i = \sqrt{-1}, \quad \text{where} \quad i^2 = -1.$$

Using the imaginary unit *i*, we can express the square root of any negative number as a real multiple of *i*. For example,

$$\sqrt{-25} = \sqrt{25(-1)} = \sqrt{25}\sqrt{-1} = 5i.$$

We can check this result by squaring $5i$ and obtaining -25.

$$(5i)^2 = 5^2 i^2 = 25(-1) = -25$$

The Square Root of a Negative Number

If b is a positive real number, then

$$\sqrt{-b} = \sqrt{b(-1)} = \sqrt{b}\sqrt{-1} = \sqrt{b}\,i \quad \text{or} \quad i\sqrt{b}.$$

EXAMPLE 1 Expressing Square Roots of Negative Numbers as Multiples of i

Write as a multiple of i:

a. $\sqrt{-9}$ b. $\sqrt{-7}$ c. $\sqrt{-8}$.

Solution

a. $\sqrt{-9} = \sqrt{9(-1)} = \sqrt{9}\sqrt{-1} = 3i$

b. $\sqrt{-7} = \sqrt{7(-1)} = \sqrt{7}\sqrt{-1} = \sqrt{7}i$ Be sure not to write i under the radical.

c. $\sqrt{-8} = \sqrt{8(-1)} = \sqrt{8}\sqrt{-1} = \sqrt{4\cdot2}\sqrt{-1} = 2\sqrt{2}i$

In order to avoid writing i under a radical, let's agree to write i *before* any radical. Consequently, we express the multiple of i in part (b) as $i\sqrt{7}$ and the multiple of i in part (c) as $2i\sqrt{2}$. ▬

✓ **CHECK POINT 1** Write as a multiple of i:

a. $\sqrt{-16}$ b. $\sqrt{-5}$ c. $\sqrt{-50}$.

A new system of numbers, called *complex numbers*, is based on adding multiples of i, such as $5i$, to the real numbers.

Complex Numbers and Imaginary Numbers

The set of all numbers in the form

$$a + bi$$

with real numbers a and b, and i, the imaginary unit, is called the set of **complex numbers**. The real number a is called the **real part** and the real number b is called the **imaginary part** of the complex number $a + bi$. If $b \neq 0$, then the complex number is called an **imaginary number** (see **Figure 10.8**).

Here are some examples of complex numbers. Each number can be written in the form $a + bi$.

$-4 + 6i$	$2i = 0 + 2i$	$3 = 3 + 0i$
a, the real part, is −4. b, the imaginary part, is 6.	a, the real part, is 0. b, the imaginary part, is 2.	a, the real part, is 3. b, the imaginary part, is 0.

Can you see that b, the imaginary part, is not zero in the first two complex numbers? Because $b \neq 0$, these complex numbers are imaginary numbers. By contrast, the imaginary part of the complex number on the right, $3 + 0i$, is zero. This complex number is not an imaginary number. The number 3, or $3 + 0i$, is a real number.

Solving Quadratic Equations with Imaginary Solutions

The equation $x^2 = -25$ has no real solutions, but it does have imaginary solutions.

$$x^2 = -25 \qquad \text{No real number squared results in a negative number.}$$

$$x = \pm\sqrt{-25} \qquad \text{Apply the square root property.}$$

$$x = \pm 5i \qquad \sqrt{-25} = \sqrt{25(-1)} = \sqrt{25}\sqrt{-1} = 5i$$

The solutions are $5i$ and $-5i$. The next examples involve quadratic equations that have no real solutions, but do have imaginary solutions.

Study Tip

We allow the use of the product rule

$$\sqrt{ab} = \sqrt{a}\sqrt{b}$$

when a is positive and b is −1. However, you cannot use $\sqrt{ab} = \sqrt{a}\sqrt{b}$ when both a and b are negative.

FIGURE 10.8
The complex number system

2 Solve quadratic equations with imaginary solutions.

EXAMPLE 2 Solving a Quadratic Equation Using the Square Root Property

Solve: $(x + 4)^2 = -36$.

Solution

$$(x + 4)^2 = -36$$ This is the given equation.

$$x + 4 = \sqrt{-36} \quad \text{or} \quad x + 4 = -\sqrt{-36}$$ Apply the square root property. Equivalently, $x + 4 = \pm\sqrt{-36}$.

$$x + 4 = 6i \qquad\qquad x + 4 = -6i$$ $\sqrt{-36} = \sqrt{36(-1)} = \sqrt{36}\sqrt{-1} = 6i$

$$x = -4 + 6i \qquad\qquad x = -4 - 6i$$ Solve the equations by subtracting 4 from both sides.

Check $-4 + 6i$:	**Check** $-4 - 6i$:
$(x + 4)^2 = -36$	$(x + 4)^2 = -36$
$(-4 + 6i + 4)^2 \overset{?}{=} -36$	$(-4 - 6i + 4)^2 \overset{?}{=} -36$
$(6i)^2 \overset{?}{=} -36$	$(-6i)^2 \overset{?}{=} -36$
$36i^2 \overset{?}{=} -36$	$36i^2 \overset{?}{=} -36$
$36(-1) \overset{?}{=} -36$	$36(-1) \overset{?}{=} -36$
$-36 = -36,$ true	$-36 = -36,$ true

The imaginary solutions are $-4 + 6i$ and $-4 - 6i$. The solution set is $\{-4 + 6i, -4 - 6i\}$ or $\{-4 \pm 6i\}$. ▬

✓ **CHECK POINT 2** Solve: $(x + 2)^2 = -25$.

EXAMPLE 3 Solving a Quadratic Equation Using the Quadratic Formula

Solve: $x^2 - 2x + 2 = 0$.

Solution Because the trinomial on the left side is prime, we solve using the quadratic formula.

$$x^2 - 2x + 2 = 0$$

$a = 1$ $b = -2$ $c = 2$

$$x = \frac{-b \pm \sqrt{b^2 - 4ac}}{2a}$$ Use the quadratic formula.

$$x = \frac{-(-2) \pm \sqrt{(-2)^2 - 4(1)(2)}}{2(1)}$$ Substitute the values for a, b, and c: $a = 1$, $b = -2$, and $c = 2$.

$$= \frac{2 \pm \sqrt{-4}}{2}$$ $(-2)^2 - 4(1)(2) = 4 - 8 = -4$

$$= \frac{2 \pm \sqrt{4(-1)}}{2}$$

$$= \frac{2 \pm 2i}{2}$$ $\sqrt{4(-1)} = \sqrt{4}\sqrt{-1} = 2i$

$$= \frac{2(1 \pm i)}{2}$$ Factor out 2 from the numerator.

$$= 1 \pm i$$ Divide the numerator and denominator by 2.

The imaginary solutions are $1 + i$ and $1 - i$. The solution set is $\{1 \pm i\}$. ▬

✓ **CHECK POINT 3** Solve: $x^2 + 6x + 13 = 0$.

Using Technology

Graphic Connections

The graph of

$$y = x^2 - 2x + 2$$

has no x-intercepts, so

$$x^2 - 2x + 2 = 0$$

has no real solutions. However, the equation does have imaginary solutions.

$y = x^2 - 2x + 2$

$[-10, 10, 1]$ by $[-10, 10, 1]$

10.4 EXERCISE SET *MyMathLab*

Practice Exercises

In Exercises 1–16, express each number in terms of i.

1. $\sqrt{-36}$

2. $\sqrt{-49}$

3. $\sqrt{-13}$

4. $\sqrt{-19}$

5. $\sqrt{-50}$

6. $\sqrt{-12}$

7. $\sqrt{-20}$

8. $\sqrt{-300}$

9. $-\sqrt{-28}$

10. $-\sqrt{-150}$

11. $7 + \sqrt{-16}$

12. $9 + \sqrt{-4}$

13. $10 + \sqrt{-3}$

14. $5 + \sqrt{-5}$

15. $6 - \sqrt{-98}$

16. $6 - \sqrt{-18}$

In Exercises 17–24, solve each quadratic equation using the square root property. Express imaginary solutions in $a + bi$ form.

17. $(x - 3)^2 = -9$

18. $(x - 5)^2 = -36$

19. $(x + 7)^2 = -64$

20. $(x + 12)^2 = -100$

21. $(x - 2)^2 = -7$

22. $(x - 1)^2 = -13$

23. $(y + 3)^2 = -18$

24. $(y + 4)^2 = -48$

In Exercises 25–36, solve each quadratic equation using the quadratic formula.

25. $x^2 + 4x + 5 = 0$

26. $x^2 + 2x + 2 = 0$

27. $x^2 - 6x + 13 = 0$

28. $x^2 - 6x + 10 = 0$

29. $x^2 - 12x + 40 = 0$

30. $x^2 - 4x + 29 = 0$

31. $x^2 = 10x - 27$

32. $x^2 = 4x - 7$

33. $5x^2 = 2x - 3$

34. $6x^2 = -2x - 1$

35. $2y^2 = 4y - 5$

36. $5y^2 = 6y - 7$

Practice PLUS

In Exercises 37–42, solve each equation by the method of your choice.

37. $12x^2 + 35 = 8x^2 + 15$

38. $8x^2 - 9 = 5x^2 - 30$

39. $\dfrac{x + 3}{5} = \dfrac{x - 2}{x}$

40. $\dfrac{x + 4}{4} = \dfrac{x - 5}{x - 2}$

41. $\dfrac{1}{x + 1} - \dfrac{1}{2} = \dfrac{1}{x}$

42. $\dfrac{4}{x} = \dfrac{8}{x^2} + 1$

Application Exercises

43. The personnel manager of a roller skate company knows that the company's weekly revenue, R, in thousands of dollars, can be modeled by the formula

$$R = -2x^2 + 36x,$$

where x is the price of a pair of skates, in dollars. A job applicant promises the personnel manager an advertising campaign guaranteed to generate \$200,000 in weekly revenue. Substitute 200 for R in the given formula and solve the equation. Are the solutions real numbers? Explain why the applicant will or will not be hired in the advertising department.

44. A football is kicked straight up from a height of 4 feet with an initial speed of 60 feet per second. The formula

$$h = -16t^2 + 60t + 4$$

describes the ball's height above the ground, h, in feet, t seconds after it is kicked. Will the ball reach a height of 80 feet? Substitute 80 for h in the given formula and solve the equation. Are the solutions real numbers? Explain why the ball will or will not reach 80 feet.

Writing in Mathematics

45. What is the imaginary unit i?

46. Explain how to write $\sqrt{-64}$ as a multiple of i.

47. What is a complex number?

48. What is an imaginary number?

49. Why is every real number also a complex number?

50. Explain each of the three jokes in the cartoon on page 626.

Critical Thinking Exercises

Make Sense? *In Exercises 51–54, determine whether each statement "makes sense" or "does not make sense" and explain your reasoning.*

51. The word *imaginary* in imaginary numbers tells me that these numbers are undefined.

52. Writing i before any radical helps me to avoid placing i in the radicand.

53. By writing the imaginary number $6i$, I can immediately see that 6 is the constant and i is the variable.

54. When I use the quadratic formula to solve a quadratic equation and $b^2 - 4ac$ is negative, I can be certain that the equation has two imaginary solutions.

In Exercises 55–58, determine whether each statement is true or false. If the statement is false, make the necessary change(s) to produce a true statement.

55. $-\sqrt{-9} = -(-3) = 3$

56. The complex number $a + 0i$ is the real number a.

57. $2 + \sqrt{-4} = 2 - 2i$

58. $\dfrac{2 \pm 4i}{2} = 1 \pm 4i$

59. Show that $1 + i$ is a solution of $x^2 - 2x + 2 = 0$ by substituting $1 + i$ for x. You should obtain

$$(1 + i)^2 - 2(1 + i) + 2.$$

Square $1 + i$ as you would a binomial. Distribute -2 as indicated. Then simplify the resulting expression by combining like terms and replacing i^2 with -1. You should obtain 0. Use this procedure to show that $1 - i$ is the equation's other solution.

60. Prove that there is no real number such that when twice the number is subtracted from its square, the difference is -5.

Technology Exercises

61. Reread Exercise 43. Use your graphing utility to illustrate the problem's solution by graphing $y = -2x^2 + 36x$ and $y = 200$ in a $[0, 20, 1]$ by $[0, 210, 30]$ viewing rectangle. Explain how the two graphs show that weekly revenue will not reach $200,000 and, therefore, the applicant will not be hired.

62. Reread Exercise 44. Use your graphing utility to illustrate the problem's solution by graphing $y = -16x^2 + 60x + 4$ and $y = 80$ in a $[0, 4, 1]$ by $[0, 100, 10]$ viewing rectangle. Explain how the graphs show that the ball will not reach a height of 80 feet.

Review Exercises

In Exercises 63–65, graph each equation in a rectangular coordinate system.

63. $y = \dfrac{1}{3}x - 2$ (Section 4.4, Example 3)

64. $2x - 3y = 6$ (Section 4.2, Example 4)

65. $x = -2$ (Section 4.2, Example 8)

Preview Exercises

Exercises 66–68 will help you prepare for the material covered in the next section.

66. Use point plotting to graph $y = x^2 + 4x + 3$. Select integers from -5 to 1, inclusive, for x.

67. Replace y with 0 and find the x-intercepts for the graph of $y = x^2 - 2x - 3$.

68. Replace x with 0 and find the y-intercept for the graph of $y = x^2 - 2x - 3$.

SECTION 10.5

Graphs of Quadratic Equations

Objectives

1 Understand the characteristics of graphs of quadratic equations.

2 Find a parabola's intercepts.

3 Find a parabola's vertex.

4 Graph quadratic equations.

5 Solve problems using a parabola's vertex.

Many sports involve objects that are thrown, kicked, or hit, and then proceed with no additional force of their own. Such objects are called **projectiles**. Paths of projectiles, as well as their heights over time, can be modeled by quadratic equations. In this section, you will learn to use graphs of quadratic equations to gain a visual understanding of the algebra that describes football, baseball, basketball, the shot put, and other projectile sports.

1 Understand the characteristics of graphs of quadratic equations.

Graphs of Quadratic Equations

The graph of any quadratic equation

$$y = ax^2 + bx + c, \quad a \neq 0$$

is called a **parabola**. Parabolas are shaped like bowls or inverted bowls, as shown in **Figure 10.9**. If the coefficient of x^2 (the value of a in $ax^2 + bx + c$) is positive, the parabola opens upward. If the coefficient of x^2 is negative, the parabola opens downward. The **vertex** (or turning point) of the parabola is the lowest point, or minimum point, on the graph when it opens upward and the highest point, or maximum point, on the graph when it opens downward.

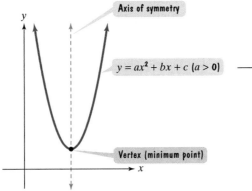

FIGURE 10.9 Characteristics of graphs of quadratic equations

$a > 0$: Parabola opens upward.

$a < 0$: Parabola opens downward.

Look at the unusual image of the word *mirror* shown on the right. The artist, Scott Kim, has created the image so that the two halves of the whole are mirror images of each other. A parabola shares this kind of symmetry. A "mirror line" through the vertex, called the **axis of symmetry**, divides the figure in half. If a parabola is folded along its axis of symmetry, the two halves match exactly.

EXAMPLE 1 Using Point Plotting to Graph a Parabola

Consider the equation $y = x^2 + 4x + 3$.

  a. Is the graph a parabola that opens upward or downward?

  b. Use point plotting to graph the parabola. Select integers from -5 to 1, inclusive, for x.

Solution

  a. To determine whether a parabola opens upward or downward, we begin by identifying a, the coefficient of x^2. The following voice balloons show the values for a, b, and c in $y = x^2 + 4x + 3$. Notice that we wrote x^2 as $1x^2$.

$$y = 1x^2 + 4x + 3$$

a, the coefficient of x^2, is 1. b, the coefficient of x, is 4. c, the constant term, is 3.

When a is greater than 0, a parabola opens upward. When a is less than 0, a parabola opens downward. Because $a = 1$, which is greater than 0, the parabola opens upward.

b. To use point plotting to graph the parabola, we first make a table of x- and y-coordinates.

x	$y = x^2 + 4x + 3$	(x, y)
-5	$y = (-5)^2 + 4(-5) + 3 = 8$	$(-5, 8)$
-4	$y = (-4)^2 + 4(-4) + 3 = 3$	$(-4, 3)$
-3	$y = (-3)^2 + 4(-3) + 3 = 0$	$(-3, 0)$
-2	$y = (-2)^2 + 4(-2) + 3 = -1$	$(-2, -1)$
-1	$y = (-1)^2 + 4(-1) + 3 = 0$	$(-1, 0)$
0	$y = 0^2 + 4(0) + 3 = 3$	$(0, 3)$
1	$y = 1^2 + 4(1) + 3 = 8$	$(1, 8)$

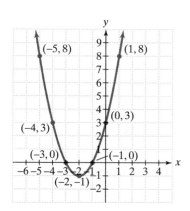

FIGURE 10.10
The graph of $y = x^2 + 4x + 3$

Then we plot the points and connect them with a smooth curve. The graph of $y = x^2 + 4x + 3$ is shown in **Figure 10.10**.

✓ **CHECK POINT 1**   Consider the equation $y = x^2 - 6x + 8$.

    a. Is the graph a parabola that opens upward or downward?

    b. Use point plotting to graph the parabola. Select integers from 0 to 6, inclusive, for x.

Several points are important when graphing a quadratic equation. These points, labeled in **Figure 10.11**, are the x-intercepts (although not every parabola has two x-intercepts), the y-intercept, and the vertex. Let's see how we can locate these points.

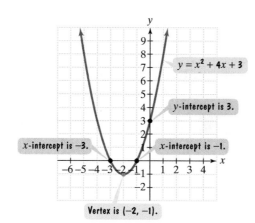

FIGURE 10.11   Useful points in graphing a parabola

2 Find a parabola's intercepts.

Finding a Parabola's x-Intercepts

At each point where a parabola crosses the x-axis, the value of y equals 0. Thus, the x-intercepts can be found by replacing y with 0 in $y = ax^2 + bx + c$. Use factoring or the quadratic formula to solve the resulting quadratic equation for x.

EXAMPLE 2   Finding a Parabola's x-Intercepts

Find the x-intercepts for the parabola whose equation is $y = x^2 + 4x + 3$.

Solution   Replace y with 0 in $y = x^2 + 4x + 3$. We obtain $0 = x^2 + 4x + 3$, or $x^2 + 4x + 3 = 0$. We can solve this equation by factoring.

$$x^2 + 4x + 3 = 0$$
$$(x + 3)(x + 1) = 0$$
$$x + 3 = 0 \quad \text{or} \quad x + 1 = 0$$
$$x = -3 \qquad x = -1$$

Thus, the x-intercepts are -3 and -1. The parabola passes through $(-3, 0)$ and $(-1, 0)$, as shown in **Figure 10.11**. ∎

✓ **CHECK POINT 2** Find the x-intercepts for the parabola whose equation is $y = x^2 - 6x + 8$.

Finding a Parabola's y-Intercept

At the point where a parabola crosses the y-axis, the value of x equals 0. Thus, the y-intercept can be found by replacing x with 0 in $y = ax^2 + bx + c$. Simple arithmetic will produce a value for y, which is the y-intercept.

EXAMPLE 3 Finding a Parabola's y-Intercept

Find the y-intercept for the parabola whose equation is $y = x^2 + 4x + 3$.

Solution Replace x with 0 in $y = x^2 + 4x + 3$.

$$y = 0^2 + 4 \cdot 0 + 3 = 0 + 0 + 3 = 3$$

The y-intercept is 3. The parabola passes through $(0, 3)$, as shown in **Figure 10.11**. ∎

✓ **CHECK POINT 3** Find the y-intercept for the parabola whose equation is $y = x^2 - 6x + 8$.

3 Find a parabola's vertex.

Finding a Parabola's Vertex

Keep in mind that a parabola's vertex is its turning point. The x-coordinate of the vertex for the parabola in **Figure 10.11**, -2, is midway between the x-intercepts, -3 and -1. If a parabola has two x-intercepts, they are found by solving $ax^2 + bx + c = 0$. The solutions of this equation,

$$x = \frac{-b - \sqrt{b^2 - 4ac}}{2a} \quad \text{and} \quad x = \frac{-b + \sqrt{b^2 - 4ac}}{2a},$$

are the x-intercepts. The value of x midway between these intercepts is $x = \dfrac{-b}{2a}$. This equation can be used to find the x-coordinate of the vertex even when no x-intercepts exist.

The Vertex of a Parabola

For a parabola whose equation is $y = ax^2 + bx + c$,

1. The x-coordinate of the vertex is $\dfrac{-b}{2a}$.

2. The y-coordinate of the vertex is found by substituting the x-coordinate into the parabola's equation and evaluating.

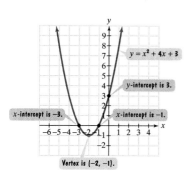

FIGURE 10.11 (repeated)

EXAMPLE 4 Finding a Parabola's Vertex

Find the vertex for the parabola whose equation is $y = x^2 + 4x + 3$.

Solution In the equation $y = x^2 + 4x + 3$, $a = 1$ and $b = 4$.

$$x\text{-coordinate of vertex} = \frac{-b}{2a} = \frac{-4}{2 \cdot 1} = \frac{-4}{2} = -2$$

To find the y-coordinate of the vertex, we substitute -2 for x in $y = x^2 + 4x + 3$ and then evaluate.

$$y\text{-coordinate of vertex} = (-2)^2 + 4(-2) + 3 = 4 + (-8) + 3 = -1$$

The vertex is $(-2, -1)$, as shown in **Figure 10.11**. ■

☑ **CHECK POINT 4** Find the vertex for the parabola whose equation is $y = x^2 - 6x + 8$.

---

4 Graph quadratic equations.

A Strategy for Graphing Quadratic Equations

Here is a procedure to sketch the graph of the quadratic equation, $y = ax^2 + bx + c$:

Graphing Quadratic Equations

The graph of $y = ax^2 + bx + c$, called a parabola, can be graphed using the following steps:

1. Determine whether the parabola opens upward or downward. If $a > 0$, it opens upward. If $a < 0$, it opens downward.

2. Determine the vertex of the parabola. The x-coordinate is $\dfrac{-b}{2a}$.

   The y-coordinate is found by substituting the x-coordinate into the parabola's equation and evaluating.

3. Find any x-intercepts by replacing y with 0. Solve the resulting quadratic equation for x. The real solutions are the x-intercepts.

4. Find the y-intercept by replacing x with 0. Because $y = a \cdot 0^2 + b \cdot 0 + c$ simplifies to $y = c$, the y-intercept is c, the constant term, and the parabola passes through $(0, c)$.

5. Plot the intercepts and the vertex.

6. Connect these points with a smooth curve.

EXAMPLE 5 Graphing a Parabola

Graph the quadratic equation: $y = x^2 - 2x - 3$.

Solution We can graph this equation by following the steps in the box.

Step 1. Determine how the parabola opens. Note that a, the coefficient of x^2, is 1. Thus, $a > 0$; this positive value tells us that the parabola opens upward.

Step 2. Find the vertex. We know that the x-coordinate of the vertex is $\dfrac{-b}{2a}$. Let's identify the numbers a, b, and c in the given equation, which is in the form $y = ax^2 + bx + c$.

$$y = x^2 - 2x - 3$$

$$a = 1 \qquad b = -2 \qquad c = -3$$

Now we substitute the values of a and b, $a = 1$ and $b = -2$, into the expression for the x-coordinate:

$$x\text{-coordinate of vertex} = \frac{-b}{2a} = \frac{-(-2)}{2(1)} = \frac{2}{2} = 1.$$

The x-coordinate of the vertex is 1. We can substitute 1 for x in the equation $y = x^2 - 2x - 3$ to find the y-coordinate:

$$y\text{-coordinate of vertex} = 1^2 - 2 \cdot 1 - 3 = 1 - 2 - 3 = -4.$$

The vertex is $(1, -4)$.

Step 3. Find the x-intercepts. Replace y with 0 in $y = x^2 - 2x - 3$. We obtain $0 = x^2 - 2x - 3$ or $x^2 - 2x - 3 = 0$. We can solve this equation by factoring.

$$x^2 - 2x - 3 = 0$$
$$(x - 3)(x + 1) = 0$$
$$x - 3 = 0 \quad \text{or} \quad x + 1 = 0$$
$$x = 3 \qquad\qquad x = -1$$

The x-intercepts are 3 and -1. The parabola passes through $(3, 0)$ and $(-1, 0)$.

Step 4. Find the y-intercept. Replace x with 0 in $y = x^2 - 2x - 3$:

$$y = 0^2 - 2 \cdot 0 - 3 = 0 - 0 - 3 = -3.$$

The y-intercept is -3. The parabola passes through $(0, -3)$.

Steps 5 and 6. Plot the intercepts and the vertex. Connect these points with a smooth curve. The intercepts and the vertex are shown as the four labeled points in **Figure 10.12**. Also shown is the graph of the quadratic equation, obtained by connecting the points with a smooth curve. ■

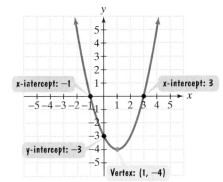

FIGURE 10.12
The graph of $y = x^2 - 2x - 3$

✓ **CHECK POINT 5** Graph the quadratic equation: $y = x^2 + 6x + 5$.

EXAMPLE 6 Graphing a Parabola

Graph the quadratic equation: $y = -x^2 + 4x - 1$.

Solution

Step 1. Determine how the parabola opens. Note that a, the coefficient of x^2, is -1. Thus, $a < 0$; this negative value tells us that the parabola opens downward.

Step 2. Find the vertex. The x-coordinate of the vertex is $\dfrac{-b}{2a}$.

$$y = -x^2 + 4x - 1$$

$a = -1 \qquad b = 4 \qquad c = -1$

$$x\text{-coordinate of vertex} = \frac{-b}{2a} = \frac{-4}{2(-1)} = \frac{-4}{-2} = 2$$

Substitute 2 for x in $y = -x^2 + 4x - 1$ to find the y-coordinate:

$$y\text{-coordinate of vertex} = -2^2 + 4 \cdot 2 - 1 = -4 + 8 - 1 = 3.$$

The vertex is $(2, 3)$.

Step 3. Find the x-intercepts. Replace y with 0 in $y = -x^2 + 4x - 1$. We obtain $0 = -x^2 + 4x - 1$ or $-x^2 + 4x - 1 = 0$. This equation cannot be solved by factoring. We will use the quadratic formula to solve it.

$$a = -1, \quad b = 4, \quad c = -1$$

$$x = \frac{-b \pm \sqrt{b^2 - 4ac}}{2a} = \frac{-4 \pm \sqrt{4^2 - 4(-1)(-1)}}{2(-1)} = \frac{-4 \pm \sqrt{16 - 4}}{-2}$$

$$x = \frac{-4 + \sqrt{12}}{-2} \approx 0.3 \quad \text{or} \quad x = \frac{-4 - \sqrt{12}}{-2} \approx 3.7$$

The x-intercepts are approximately 0.3 and 3.7. The parabola passes through $(0.3, 0)$ and $(3.7, 0)$.

Step 4. Find the y-intercept. Replace x with 0 in $y = -x^2 + 4x - 1$:

$$y = -0^2 + 4 \cdot 0 - 1 = -1.$$

The y-intercept is -1. The parabola passes through $(0, -1)$.

Steps 5 and 6. Plot the intercepts and the vertex. Connect these points with a smooth curve. The intercepts and the vertex are shown as the four labeled points in **Figure 10.13**. Also shown is the graph of the quadratic equation, obtained by connecting the points with a smooth curve.

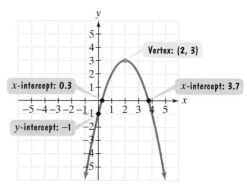

FIGURE 10.13
The graph of $y = -x^2 + 4x - 1$

☑ **CHECK POINT 6** Graph the quadratic equation: $y = -x^2 - 2x + 5$.

---

5 Solve problems using a parabola's vertex.

Applications

Many applied problems involve finding the maximum or minimum value of equations in the form $y = ax^2 + bx + c$. The vertex of the graph is the point of interest. If $a < 0$, the parabola opens downward and the vertex is its highest point. If $a > 0$, the parabola opens upward and the vertex is its lowest point.

EXAMPLE 7 **The Parabolic Path of a Punted Football**

Figure 10.14 at the top of the next page shows that when a football is kicked, the nearest defensive player is 6 feet from the point of impact with the kicker's foot. The height of the punted football, y, in feet, can be modeled by

$$y = -0.01x^2 + 1.18x + 2,$$

where x is the ball's horizontal distance, in feet, from the point of impact with the kicker's foot.

FIGURE 10.14

a. What is the maximum height of the punt and how far from the point of impact does this occur?

b. How far must the nearest defensive player, who is 6 feet from the kicker's point of impact, reach to block the punt?

c. If the ball is not blocked by the defensive player, how far down the field will it go before hitting the ground?

d. Graph the equation that models the football's parabolic path.

Solution

a. We begin by identifying the numbers a, b, and c in the given model.

$$y = -0.01x^2 + 1.18x + 2$$

$a = -0.01$ $b = 1.18$ $c = 2$

Because the coefficient of x^2, -0.01, is negative, the parabola opens downward and the vertex is the highest point on the graph. The y-coordinate of the vertex gives the maximum height of the punt and the x-coordinate reveals how far from the point of impact this occurs.

$$x\text{-coordinate of vertex} = \frac{-b}{2a} = \frac{-1.18}{2(-0.01)} = \frac{-1.18}{-0.02} = 59$$

Substitute 59 for x in $y = -0.01x^2 + 1.18x + 2$ to find the y-coordinate.

$$y\text{-coordinate of vertex} = -0.01(59)^2 + 1.18(59) + 2$$
$$= -34.81 + 69.62 + 2 = 36.81$$

The vertex is $(59, 36.81)$. The maximum height of the punt is 36.81 feet and this occurs 59 feet from the kicker's point of impact.

b. **Figure 10.14** shows that the defensive player is 6 feet from the kicker's point of impact. To block the punt, he must touch the football along its parabolic path. This means that we must find the height of the ball 6 feet from the kicker. Replace x with 6 in the given model, $y = -0.01x^2 + 1.18x + 2$.

$$y = -0.01(6)^2 + 1.18(6) + 2 = -0.36 + 7.08 + 2 = 8.72$$

The defensive player must reach 8.72 feet above the ground to block the punt.

c. Assuming that the ball is not blocked by the defensive player, we are interested in how far down the field it will go before hitting the ground. We are looking for the ball's horizontal distance, x, when its height above the ground, y, is 0 feet. To find this x-intercept, replace y with 0 in $y = -0.01x^2 + 1.18x + 2$. We obtain $0 = -0.01x^2 + 1.18x + 2$, or $-0.01x^2 + 1.18x + 2 = 0$. The equation cannot be solved by factoring. We will use the quadratic formula to solve it.

$$-0.01x^2 + 1.18x + 2 = 0$$

$a = -0.01$	$b = 1.18$	$c = 2$

The equation for determining the ball's maximum horizontal distance (repeated)

> Use a calculator to evaluate the radicand.

$$x = \frac{-b \pm \sqrt{b^2 - 4ac}}{2a} = \frac{-1.18 \pm \sqrt{(1.18)^2 - 4(-0.01)(2)}}{2(-0.01)} = \frac{-1.18 \pm \sqrt{1.4724}}{-0.02}$$

$$x = \frac{-1.18 + \sqrt{1.4724}}{-0.02} \quad \text{or} \quad x = \frac{-1.18 - \sqrt{1.4724}}{-0.02}$$

$$x \approx -1.7 \qquad\qquad x \approx 119.7$$

> Use a calculator and round to the nearest tenth.

> Reject this value. We are interested in the football's height corresponding to horizontal distances from its point of impact onward, or $x \geq 0$.

If the football is not blocked by the defensive player, it will go approximately 119.7 feet down the field before hitting the ground.

d. In terms of graphing the model for the football's parabolic path, $y = -0.01x^2 + 1.18x + 2$, we have already determined the vertex and the appropriate x-intercept.

vertex: $(59, 36.81)$

> The ball's maximum height, 36.81 feet, occurs at a horizontal distance of 59 feet.

x-intercept: 119.7

> The ball's maximum horizontal distance is approximately 119.7 feet.

Figure 10.14 indicates that the y-intercept is 2, meaning that the ball is kicked from a height of 2 feet. Let's verify this value by replacing x with 0 in $y = -0.01x^2 + 1.18x + 2$.

$$y = -0.01 \cdot 0^2 + 1.18 \cdot 0 + 2 = 0 + 0 + 2 = 2$$

Using the vertex, $(59, 36.81)$, the x-intercept, 119.7, and the y-intercept, 2, the graph of the equation that models the football's parabolic path is shown in **Figure 10.15**. The graph is shown only for $x \geq 0$, indicating horizontal distances that begin at the football's impact with the kicker's foot and end with the ball hitting the ground.

Height of the Punted Football (feet)

Distance from the Point of Impact (feet)

FIGURE 10.14 (repeated)

Football's Height (feet)

Vertex: (59, 36.81)
The ball's maximum height is 36.81 feet.

x-intercept: 119.7
The ball hits the ground after traveling a maximum horizontal distance of 119.7 feet.

$y = -0.01x^2 + 1.18x + 2$

y-intercept: 2
The ball is kicked from a height of 2 feet.

Football's Horizontal Distance (feet)

FIGURE 10.15
The parabolic path of a punted football

☑ **CHECK POINT 7** An archer's arrow follows a parabolic path. The height of the arrow, y, in feet, can be modeled by

$$y = -0.005x^2 + 2x + 5,$$

where x is the arrow's horizontal distance, in feet.

a. What is the maximum height of the arrow and how far from its release does this occur?

b. Find the horizontal distance the arrow travels before it hits the ground. Round to the nearest foot.

c. Graph the equation that models the arrow's parabolic path.

10.5 EXERCISE SET *MyMathLab*

PRACTICE   WATCH   DOWNLOAD   READ   REVIEW

Practice Exercises

In Exercises 1–4 determine if the parabola whose equation is given opens upward or downward.

1. $y = x^2 - 4x + 3$

2. $y = x^2 - 6x + 5$

3. $y = -2x^2 + x + 6$

4. $y = -2x^2 - 4x + 6$

In Exercises 5–10, find the x-intercepts for the parabola whose equation is given. If the x-intercepts are irrational numbers, round your answers to the nearest tenth.

5. $y = x^2 - 4x + 3$

6. $y = x^2 - 6x + 5$

7. $y = -x^2 + 8x - 12$

8. $y = -x^2 - 2x + 3$

9. $y = x^2 + 2x - 4$

10. $y = x^2 + 8x + 14$

In Exercises 11–18, find the y-intercept for the parabola whose equation is given.

11. $y = x^2 - 4x + 3$ **12.** $y = x^2 - 6x + 5$

13. $y = -x^2 + 8x - 12$ **14.** $y = -x^2 - 2x + 3$

15. $y = x^2 + 2x - 4$ **16.** $y = x^2 + 8x + 14$

17. $y = x^2 + 6x$ **18.** $y = x^2 + 8x$

In Exercises 19–24, find the vertex for the parabola whose equation is given.

19. $y = x^2 - 4x + 3$ **20.** $y = x^2 - 6x + 5$

21. $y = 2x^2 + 4x - 6$

22. $y = -2x^2 - 4x - 2$

23. $y = x^2 + 6x$ **24.** $y = x^2 + 8x$

In Exercises 25–36, graph the parabola whose equation is given.

25. $y = x^2 + 8x + 7$ **26.** $y = x^2 + 10x + 9$

27. $y = x^2 - 2x - 8$ **28.** $y = x^2 + 4x - 5$

29. $y = -x^2 + 4x - 3$ **30.** $y = -x^2 + 2x + 3$

31. $y = x^2 - 1$ **32.** $y = x^2 - 4$

33. $y = x^2 + 2x + 1$ **34.** $y = x^2 - 2x + 1$

35. $y = -2x^2 + 4x + 5$ **36.** $y = -3x^2 + 6x - 2$

Practice PLUS

In Exercises 37–44, find the vertex for the parabola whose equation is given by first writing the equation in the form $y = ax^2 + bx + c$.

37. $y = (x - 3)^2 + 2$ **38.** $y = (x - 4)^2 + 3$

39. $y = (x + 5)^2 - 4$

40. $y = (x + 6)^2 - 5$

41. $y = 2(x - 1)^2 - 3$

42. $y = 2(x - 1)^2 - 4$

43. $y = -3(x + 2)^2 + 5$

44. $y = -3(x + 4)^2 + 6$

45. Generalize your work in Exercises 37–44 and complete the following statement: For a parabola whose equation is $y = a(x - h)^2 + k$, the vertex is the point _____.

Application Exercises

An athlete whose event is the shot put releases the shot with the same initial velocity, but at different angles. The figure shows the parabolic paths for shots released at angles of 35° and 65°. Exercises 46–47 are based on the equations that model the parabolic paths.

46. When the shot is released at an angle of 65°, its height, y, in feet, can be modeled by

$$y = -0.04x^2 + 2.1x + 6.1,$$

where x is the shot's horizontal distance, in feet, from its point of release. Use this model to solve parts (a) through (c) and verify your answers using the red graph.

a. What is the maximum height, to the nearest tenth of a foot, of the shot and how far from its point of release does this occur?

b. What is the shot's maximum horizontal distance, to the nearest tenth of a foot, or the distance of the throw?

c. From what height was the shot released?

47. When the shot whose path is shown by the blue graph on the previous page is released at an angle of 35°, its height, y, in feet, can be modeled by

$$y = -0.01x^2 + 0.7x + 6.1,$$

where x is the shot's horizontal distance, in feet, from its point of release. Use this model to solve parts (a) through (c) and verify your answers using the blue graph.

a. What is the maximum height of the shot and how far from its point of release does this occur?

b. What is the shot's maximum horizontal distance, to the nearest tenth of a foot, or the distance of the throw?

c. From what height was the shot released?

The Food Stamp Program is the first line of defense against hunger for millions of American families. The program provides benefits for eligible participants to purchase approved food items at food stores. Over half of all participants are children; one out of six is a low-income older adult. The bar graph shows the number of households, in millions, participating in the Food Stamp Program from 1999 through 2004.

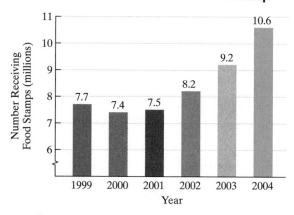

United States Households on Food Stamps

Source: Food Stamp Program

The formula

$$y = 0.22x^2 - 0.50x + 7.68$$

models the number of households, y, in millions, participating in the Food Stamp Program x years after 1999. Use this model to solve Exercises 48–49.

48. a. According to the model, how many households received food stamps in 2004? Does this underestimate or overestimate the actual number shown by the graph? By how much?

b. According to the graph, in which year from 1999 through 2004 was the number of households receiving food stamps at a minimum and how many households received food stamps for that year?

c. According to the model, in which year, to the nearest whole year, was the number of households receiving food stamps at a minimum and how many households received food stamps for that year? How does this compare with the actual data in part (b)?

49. a. According to the model, how many households received food stamps in 1999? Does this underestimate or overestimate the actual number shown by the graph? By how much?

b. According to the graph, in which year from 1999 through 2004 was the number of households receiving food stamps at a minimum and how many households received food stamps for that year?

c. According to the model, in which year, to the nearest whole year, was the number of households receiving food stamps at a minimum and how many households received food stamps for that year? How does this compare with the actual data in part (b)?

50. You have 120 feet of fencing to enclose a rectangular plot that borders on a river. If you do not fence the side along the river, find the length and width of the plot that will maximize the area. What is the largest area that can be enclosed?

51. The figure shown indicates that you have 100 yards of fencing to enclose a rectangular area. Find the dimensions of the rectangle that maximize the enclosed area. What is the maximum area?

Writing in Mathematics

52. What is a parabola? Describe its shape.

53. Explain how to decide whether a parabola opens upward or downward.

54. If a parabola has two x-intercepts, explain how to find them.

55. Explain how to find a parabola's y-intercept.

56. Describe how to find a parabola's vertex.

57. A parabola that opens upward has its vertex at $(1, 2)$. Describe as much as you can about the parabola based on this information. Include in your discussion the number of x-intercepts, if any, for the parabola.

Critical Thinking Exercises

Make Sense? *In Exercises 58–61, determine whether each statement "makes sense" or "does not make sense" and explain your reasoning.*

58. I must have made an error when graphing this parabola because it is symmetric with respect to the y-axis.

59. Parabolas that open up appear to form smiles ($a > 0$), while parabolas that open down frown ($a < 0$).

60. I threw a baseball vertically upward and its path was a parabola.

61. **Figure 10.14** on page 637 shows that a linear model provides a better description of the football's path than a quadratic model.

In Exercises 62–65, determine whether each statement is true or false. If the statement is false, make the necessary change(s) to produce a true statement.

62. The x-coordinate of the vertex of the parabola whose equation is $y = ax^2 + bx + c$ is $\dfrac{b}{2a}$.

63. If a parabola has only one x-intercept, then the x-intercept is also the vertex.

64. There is no relationship between the graph of $y = ax^2 + bx + c$ and the number of real solutions of the equation $ax^2 + bx + c = 0$.

65. If $y = 4x^2 - 40x + 4$, then the vertex is the highest point on the graph.

66. Find two numbers whose sum is 200 and whose product is a maximum.

67. Graph $y = 2x^2 - 8$ and $y = -2x^2 + 8$ in the same rectangular coordinate system. What are the coordinates of the points of intersection?

68. A parabola has x-intercepts at 3 and 7, a y-intercept at -21, and $(5, 4)$ for its vertex. Write the parabola's equation.

Technology Exercises

69. Use a graphing utility to verify any five of your hand-drawn graphs in Exercises 25–36.

70. a. Use a graphing utility to graph $y = 2x^2 - 82x + 720$ in a standard viewing rectangle. What do you observe?

b. Find the coordinates of the vertex for the given quadratic equation.

c. The answer to part (b) is $(20.5, -120.5)$. Because the leading coefficient, 2 of $y = 2x^2 - 82x + 720$ is positive, the vertex is a minimum point on the graph. Use this fact to help find a viewing rectangle that will give a relatively complete picture of the parabola. With an axis of symmetry at $x = 20.5$, the setting for x should extend past this, so try Xmin = 0 and Xmax = 30. The setting for y should include (and probably go below) the y-coordinate of the graph's minimum point, so try Ymin = -130. Experiment with Ymax until your utility shows the parabola's major features.

d. In general, explain how knowing the coordinates of a parabola's vertex can help determine a reasonable viewing rectangle on a graphing utility for obtaining a complete picture of the parabola.

In Exercises 71–74, find the vertex for each parabola. Then determine a reasonable viewing rectangle on your graphing utility and use it to graph the parabola.

71. $y = -0.25x^2 + 40x$

72. $y = -4x^2 + 20x + 160$

73. $y = 5x^2 + 40x + 600$

74. $y = 0.01x^2 + 0.6x + 100$

Review Exercises

In Exercises 75–77, solve each equation or system of equations.

75. $7(x - 2) = 10 - 2(x + 3)$ (Section 2.3, Example 3)

76. $\dfrac{7}{x + 2} + \dfrac{2}{x + 3} = \dfrac{1}{x^2 + 5x + 6}$ (Section 8.6, Example 4)

77. $5x - 3y = -13$
$x = 2 - 4y$ (Section 5.2, Example 1)

Preview Exercises

Exercises 78–80 will help you prepare for the material covered in the next section.

78. Here are two sets of ordered pairs:

$$\text{set 1: } \{(1, 5), (2, 5)\}$$
$$\text{set 2: } \{(5, 1), (5, 2)\}.$$

In which set is each x-coordinate paired with one and only one y-coordinate?

79. Here are two graphs:

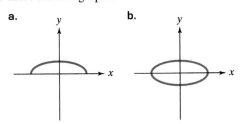

a. **b.**

In which graph is each x-coordinate paired with one and only one y-coordinate?

80. Evaluate $x^2 + 3x + 5$ for $x = -3$.

SECTION

10.6

Objectives

1 Find the domain and range of a relation.

2 Determine whether a relation is a function.

3 Evaluate a function.

4 Use the vertical line test to identify functions.

5 Find function values for functions that model data.

Introduction to Functions

The number of calories that you burn per hour depends on the activity in which you are engaged. The average number of takeout meals per person depends on the year in which this average is computed. In both of these situations, the relationship between variables can be illustrated with the notion of a *function*. Understanding this concept will give you a new perspective on many ordinary situations. Much of your work in subsequent algebra courses will be devoted to the important topic of functions and how they model your world.

1 Find the domain and range of a relation.

Relations

Studies show that exercise can promote good long-term health no matter how much you weigh. A brisk half-hour walk each day is enough to get the benefits. Combined with a healthy diet, it also helps to stave off obesity. How many calories does your workout burn? The graph in **Figure 10.16** shows the calories burned per hour in six activities.

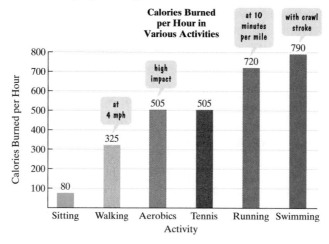

FIGURE 10.16 Counting calories
Source: FITRESOURCE.COM

The information shown in the bar graph indicates a correspondence between the activities and calories burned per hour. We can write this correspondence using a set of ordered pairs:

$$\{(\text{sitting, } 80), (\text{walking, } 325), (\text{aerobics, } 505), (\text{tennis, } 505),$$
$$(\text{running, } 720), (\text{swimming, } 790)\}.$$

These braces indicate that we are representing a set.

The mathematical term for a set of ordered pairs is a **relation**.

> **Definition of a Relation**
>
> A **relation** is any set of ordered pairs. The set of all first components of the ordered pairs is called the **domain** of the relation and the set of all second components is called the **range** of the relation.

EXAMPLE 1 Finding the Domain and Range of a Relation

Find the domain and range of the relation:

$$\{(\text{sitting}, 80), (\text{walking}, 325), (\text{aerobics}, 505), (\text{tennis}, 505),$$
$$(\text{running}, 720), (\text{swimming}, 790)\}.$$

Solution The domain is the set of all first components. Thus, the domain is

$$\{\text{sitting, walking, aerobics, tennis, running, swimming}\}.$$

> Parentheses and square brackets are not used to represent sets.

The range is the set of all second components. Thus, the range is

$$\{80, 325, 505, 720, 790\}.$$

> Although both aerobics and tennis burn 505 calories per hour, it is not necessary to list 505 twice.

☑ **CHECK POINT 1** The following set shows calories burned per hour in activities not included in **Figure 10.16**. Find the domain and range of the relation:

$$\{(\text{golf}, 250), (\text{lawn mowing}, 325), (\text{water skiing}, 430),$$
$$(\text{hiking}, 430), (\text{bicycling}, 720)\}.$$

> with cart

> at 15 mph

---

2 Determine whether a relation is a function.

Functions

Shown in the margin are the calories burned per hour for the activities in the bar graph in **Figure 10.16**. We've used this information to define two relations. **Figure 10.17(a)** shows a correspondence between activities and calories burned. **Figure 10.17(b)** shows a correspondence between calories burned and activities.

Activity	Calories Burned per hour
Sitting	80
Walking	325
Aerobics	505
Tennis	505
Running	720
Swimming	790

FIGURE 10.17(a) Activities correspond to calories burned.

FIGURE 10.17(b) Calories burned correspond to activities.

FIGURE 10.17(a) (repeated)

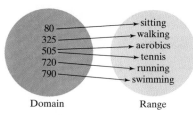

FIGURE 10.17(b) (repeated)

A relation in which each member of the domain corresponds to exactly one member of the range is a **function**. Can you see that the relation in **Figure 10.17(a)** is a function? Each activity in the domain corresponds to exactly one number representing calories burned per hour in the range. If we know the activity, we know the calories burned per hour. Notice that more than one element in the domain can correspond to the same element in the range: Aerobics and tennis both burn 505 calories per hour.

Is the relation in **Figure 10.17(b)** a function? Does each member of the domain correspond to precisely one member of the range? This relation is not a function because there is a member of the domain that corresponds to two members of the range:

$$(505, \text{aerobics}) \quad (505, \text{tennis}).$$

The member of the domain, 505, corresponds to both aerobics and tennis. If we know the calories burned per hour, 505, we cannot be sure of the activity. Because **a function is a relation in which no two ordered pairs have the same first component and different second components**, the ordered pairs (505, aerobics) and (505, tennis) are not ordered pairs of a function.

Same first component

(505, aerobics)   (505, tennis)

Different second components

Definition of a Function

A **function** is a relation in which each member of the domain corresponds to exactly one member of the range. No two ordered pairs of a function can have the same first component and different second components.

EXAMPLE 2   Determining Whether a Relation Is a Function

Determine whether each relation is a function:

a. $\{(1, 6), (2, 6), (3, 8), (4, 9)\}$ **b.** $\{(6, 1), (6, 2), (8, 3), (9, 4)\}.$

Solution   We begin by making a figure for each relation that shows the domain and the range (**Figure 10.18**).

FIGURE 10.18(a)

FIGURE 10.18(b)

a. Figure 10.18(a) shows that every element in the domain corresponds to exactly one element in the range. The element 1 in the domain corresponds to the element 6 in the range. Furthermore, 2 corresponds to 6, 3 corresponds to 8, and 4 corresponds to 9. No two ordered pairs in the given relation have the same first component and different second components. Thus, the relation is a function.

b. Figure 10.18(b) shows that 6 corresponds to both 1 and 2. If any element in the domain corresponds to more than one element in the range, the relation is not a function. This relation is not a function; two ordered pairs have the same first component and different second components.

Same first component

(6, 1)   (6, 2)

Different second components

Look at **Figure 10.18(a)** on the previous page again. The fact that 1 and 2 in the domain correspond to the same number, 6, in the range does not violate the definition of a function. **A function can have two different first components with the same second component.** By contrast, a relation is not a function when two different ordered pairs have the same first component and different second components. Thus, the relation in **Figure 10.18(b)** is not a function.

✓ **CHECK POINT 2** Determine whether each relation is a function:

    a. $\{(1, 2), (3, 4), (5, 6), (5, 8)\}$

    b. $\{(1, 2), (3, 4), (6, 5), (8, 5)\}$.

3 Evaluate a function.

Functions as Equations and Function Notation

Functions are usually given in terms of equations rather than as sets of ordered pairs. For example, here is an equation that shows a woman's cholesterol level as a function of her age:

$$y = 1.17x + 147.15.$$

The variable x represents a woman's age, in years. The variable y represents her cholesterol level. For each age, x, the equation gives one and only one cholesterol level, y. The variable y is a function of the variable x.

When an equation represents a function, the function is often named by a letter such as f, g, h, F, G, or H. Any letter can be used to name a function. Suppose that f names a function. Think of the domain as the set of the function's inputs and the range as the set of the function's outputs. As shown in **Figure 10.19**, the input is represented by x and the output by $f(x)$. The special notation $f(x)$, read "f of x" or "f at x," represents the **value of the function at the number x**.

Let's make this clearer by considering a specific example. We know the equation

$$y = 1.17x + 147.15$$

defines y as a function of x. We'll name the function f. Now, we can apply our new function notation.

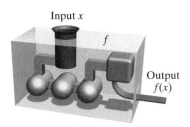

Input x

f

Output $f(x)$

FIGURE 10.19 A function as a machine with inputs and outputs

Input	Output	Equation
x	$f(x)$	$f(x) = 1.17x + 147.15$

We read this equation as "f of x equals $1.17x + 147.15$."

Suppose we are interested in finding $f(30)$, the function's output when the input is 30. To find the value of the function at 30, we substitute 30 for x. We are **evaluating the function** at 30.

$f(x) = 1.17x + 147.15$ This is the given function.

$f(30) = 1.17(30) + 147.15$ The input is 30.

       $= 35.1 + 147.15$ Multiply.

       $= 182.25$ Add.

Study Tip

The notation $f(x)$ does *not* mean "f times x." The notation describes the value of the function at x.

The statement $f(30) = 182.25$, read "f of 30 equals 182.25," tells us that the value of the function at 30 is 182.25. When the function's input is 30, its output is 182.25. The model gives us exactly one cholesterol level for a 30-year-old woman, namely 182.25. **Figure 10.20** illustrates the input and output in terms of a function machine.

Input $x = 30$

$f(x) = 1.17x + 147.15$

$1.17(30) + 147.15$

Output $f(30) = 182.25$

FIGURE 10.20 A function machine at work

EXAMPLE 3 Evaluating a Function

If $f(x) = 3x + 1$, find each of the following:

   a. $f(5)$ **b.** $f(-4)$ **c.** $f(0)$.

Solution We substitute $5, -4,$ and 0, respectively, for x in the function's equation, $f(x) = 3x + 1$.

   a. $f(5) = 3 \cdot 5 + 1 = 15 + 1 = 16$ *f of 5 equals 16.*

   b. $f(-4) = 3(-4) + 1 = -12 + 1 = -11$ *f of −4 equals −11.*

   c. $f(0) = 3 \cdot 0 + 1 = 0 + 1 = 1$ *f of 0 equals 1.* ■

The function $f(x) = 3x + 1$ is an example of a *linear function*. **Linear functions** have equations of the form $f(x) = mx + b$.

✓ **CHECK POINT 3** If $f(x) = 4x + 3$, find each of the following:

   a. $f(5)$ **b.** $f(-2)$ **c.** $f(0)$.

EXAMPLE 4 Evaluating a Function

If $g(x) = x^2 + 3x + 5$, find each of the following:

   a. $g(2)$ **b.** $g(-3)$ **c.** $g(0)$.

Solution We substitute $2, -3,$ and 0, respectively, for x in the function's equation, $g(x) = x^2 + 3x + 5$.

   a. $g(2) = 2^2 + 3 \cdot 2 + 5 = 4 + 6 + 5 = 15$ *g of 2 is 15.*

   b. $g(-3) = (-3)^2 + 3(-3) + 5 = 9 + (-9) + 5 = 5$ *g of −3 is 5.*

   c. $g(0) = 0^2 + 3 \cdot 0 + 5 = 0 + 0 + 5 = 5$ *g of 0 is 5.* ■

The function in Example 4 is a *quadratic function*. **Quadratic functions** have equations of the form $f(x) = ax^2 + bx + c, a \neq 0$.

✓ **CHECK POINT 4** If $g(x) = x^2 + 4x + 3$, find each of the following:

   a. $g(5)$ **b.** $g(-4)$ **c.** $g(0)$.

---

4 Use the vertical line test to identify functions.

Graphs of Functions and the Vertical Line Test

The **graph of a function** is the graph of its ordered pairs. For example, the graph of $f(x) = 3x + 1$ is the set of points (x, y) in the rectangular coordinate system satisfying the equation $y = 3x + 1$. Thus, the graph of f is a line with slope 3 and y-intercept 1. Similarly, the graph of $f(x) = x^2 + 3x + 5$ is the set of points (x, y) in the rectangular coordinate system satisfying the equation $y = x^2 + 3x + 5$. Thus, the graph of g is a parabola.

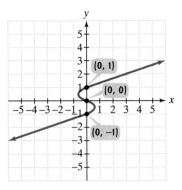

FIGURE 10.21
y is not a function of *x* because 0 is paired with three values of *y*, namely, 1, 0, and −1.

Not every graph in the rectangular coordinate system is the graph of a function. The definition of a function specifies that no value of *x* can be paired with two or more different values of *y*. Consequently, if a graph contains two or more different points with the same first coordinate, the graph cannot represent a function. This is illustrated in **Figure 10.21**. Observe that points sharing a common first coordinate are vertically above or below each other.

This observation is the basis of a useful test for determining whether a graph defines *y* as a function of *x*. The test is called the **vertical line test**.

> ### The Vertical Line Test for Functions
>
> If any vertical line intersects a graph in more than one point, the graph does not define *y* as a function of *x*.

EXAMPLE 5   Using the Vertical Line Test

Use the vertical line test to identify graphs in which *y* is a function of *x*.

a.   **b.**   **c.**   **d.**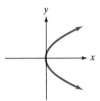

Solution   *y* is a function of *x* for the graphs in (b) and (c).

a.   **b.**   **c.**   **d.**

y **is not a function** of *x*.
Two values of *y*
correspond to one *x*-value.

y **is a function** of *x*.

y **is a function** of *x*.

y **is not a function** of *x*.
Two values of *y*
correspond to one *x*-value.

☑ **CHECK POINT 5**   Use the vertical line test to identify graphs in which *y* is a function of *x*.

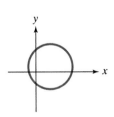

5 Find function values for functions that model data.

Applications

Like formulas, functions can be obtained from verbal conditions or from actual data. Throughout your next algebra course, you'll have lots of practice doing this. For now, let's make sure that we can find and interpret function values for functions that were obtained from modeling data.

EXAMPLE 6 Sixty and Beyond

Figure 10.22 shows your chances of surviving to various ages once you reach 60.

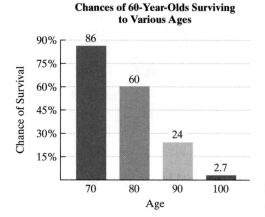

FIGURE 10.22

Source: National Center for Health Statistics, 2003 data

The functions

$$f(x) = -2.9x + 286$$

$$\text{and} \quad g(x) = 0.01x^2 - 4.9x + 370$$

model the chance, as a percent, that a 60-year-old will survive to age x. Which model serves as a better description for the chance of surviving to age 70?

Solution   We evaluate each function at 70 by substituting 70 for x in the function's equation. The graph in **Figure 10.22** indicates that 60-year-olds have an 86% chance of surviving to 70. The evaluation that comes closer to 86 will serve as the better description.

$f(x) = -2.9x + 286$	This is the given linear function.
$f(70) = -2.9(70) + 286$	Replace x with 70.
$= 83$	Use a calculator.

We see that $f(70) = 83$. The linear model indicates an 83% chance of surviving to 70.

$g(x) = 0.01x^2 - 4.9x + 370$	This is the given quadratic function.
$g(70) = 0.01(70)^2 - 4.9(70) + 370$	Replace each occurrence of x with 70.
$= 76$	Use a calculator.

We see that $g(70) = 76$. The quadratic model indicates a 76% chance of surviving to 70.

• $f(70) = 83$	• $g(70) = 76$	• Chance of surviving to 70 is 86 percent.
Linear function	Quadratic function	Actual data

We see that the linear function f serves as a better description of the data.

Using Technology

Graphing utilities can be used to evaluate functions. The screens below show the evaluation of

$$f(x) = -2.9x + 286 \quad \text{and} \quad g(x) = 0.01x^2 - 4.9x + 370$$

at 70 on a TI-84 Plus graphing calculator. The function f is named Y_1 and the function g is named Y_2.

✓ **CHECK POINT 6**   Use the functions in Example 6 to solve this exercise.
   a. According to the linear function, what is the chance that a 60-year-old will survive to age 90?
   b. According to the quadratic function, what is the chance that a 60-year-old will survive to age 90?
   c. Which model serves as a better description for the chance of survival displayed by the bar graph in **Figure 10.22**?

10.6 EXERCISE SET   *MyMathLab*   Math XL PRACTICE   WATCH   DOWNLOAD   READ   REVIEW

Practice Exercises

In Exercises 1–8, determine whether each relation is a function. Give the domain and range for each relation.

1. $\{(1, 2), (3, 4), (5, 5)\}$

2. $\{(4, 5), (6, 7), (8, 8)\}$

3. $\{(3, 4), (3, 5), (4, 4), (4, 5)\}$

4. $\{(5, 6), (5, 7), (6, 6), (6, 7)\}$

5. $\{(-3, -3), (-2, -2), (-1, -1), (0, 0)\}$

6. $\{(-7, -7), (-5, -5), (-3, -3), (0, 0)\}$

7. $\{(1, 4), (1, 5), (1, 6)\}$

8. $\{(4, 1), (5, 1), (6, 1)\}$

In Exercises 9–24, evaluate each function at the given values.

9. $f(x) = x + 5$
   a. $f(7)$ **b.** $f(-6)$ **c.** $f(0)$

10. $f(x) = x + 6$
   a. $f(4)$ **b.** $f(-8)$ **c.** $f(0)$

11. $f(x) = 7x$
   a. $f(10)$ **b.** $f(-4)$ **c.** $f(0)$

12. $f(x) = 9x$
   a. $f(10)$ **b.** $f(-5)$ **c.** $f(0)$

13. $f(x) = 8x - 3$
   a. $f(12)$ **b.** $f\left(-\frac{1}{2}\right)$ **c.** $f(0)$

14. $f(x) = 6x - 5$
   a. $f(12)$ **b.** $f\left(-\frac{1}{2}\right)$ **c.** $f(0)$

15. $g(x) = x^2 + 3x$
   a. $g(2)$ **b.** $g(-2)$ **c.** $g(0)$

16. $g(x) = x^2 + 7x$
   a. $g(2)$ **b.** $g(-2)$ **c.** $g(0)$

17. $h(x) = x^2 - 2x + 3$
   a. $h(4)$ **b.** $h(-4)$ **c.** $h(0)$

18. $h(x) = x^2 - 4x + 5$
   a. $h(4)$ **b.** $h(-4)$ **c.** $h(0)$

19. $f(x) = 5$
   a. $f(9)$ **b.** $f(-9)$ **c.** $f(0)$

20. $f(x) = 7$
   a. $f(10)$ **b.** $f(-10)$ **c.** $f(0)$

21. $f(r) = \sqrt{r + 6} + 3$
   a. $f(-6)$ **b.** $f(10)$

22. $f(r) = \sqrt{25 - r} - 6$
   a. $f(16)$ **b.** $f(-24)$

23. $f(x) = \dfrac{x}{|x|}$
   a. $f(6)$ **b.** $f(-6)$

24. $f(x) = \dfrac{x}{|x|}$
   a. $f(5)$ **b.** $f(-5)$

In Exercises 25–32, use the vertical line test to identify graphs in which y is a function of x.

25.

26.

27.

28.

29.

30.

31.

32.

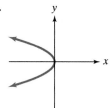

Practice PLUS

In Exercises 33–36, express each function as a set of ordered pairs.

33. $f(x) = 2x + 3$; domain: $\{-1, 0, 1\}$

34. $f(x) = 3x + 5$; domain: $\{-1, 0, 1\}$

35. $g(x) = x - x^2$; domain: the set of integers from -2 to 2, inclusive

36. $g(x) = x - |x|$; domain: the set of integers from -2 to 2, inclusive

In Exercises 37–40, find

$$\frac{f(x) - f(h)}{x - h}$$

and simplify.

37. $f(x) = 6x + 7$

38. $f(x) = 8x + 9$

39. $f(x) = x^2 - 1$

40. $f(x) = x^3 - 1$

Application Exercises

41. The bar graph shows the percentage of children in the world's leading industrial countries who daydream about being rich.

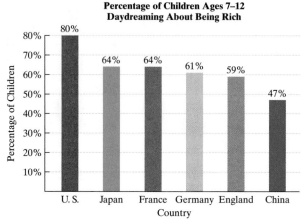

Source: Roper Starch Worldwide for A.B.C. Research

a. Write a set of six ordered pairs in which countries correspond to the percentage of children daydreaming about being rich. Each ordered pair should be in the form (country, percent).

b. Is the relation in part (a) a function? Explain your answer.

c. Write a set of six ordered pairs in which the percentage of children daydreaming about being rich corresponds to a country. Each ordered pair should be in the form (percent, country).

d. Is the relation in part (c) a function? Explain your answer.

42. The bar graph shows the number of days in a school year for the six countries with the longest school years. (To put these numbers in perspective, a school year in the United States is 180 days.)

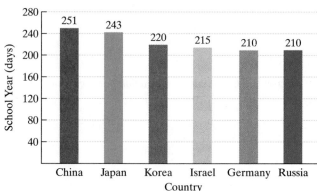

Source: UNESCO

a. Write a set of six ordered pairs in which countries correspond to the number of days in the school year. Each ordered pair should be in the form

(country, days in school year).

b. Is the relation in part (a) a function? Explain your answer.

c. Write a set of six ordered pairs in which the number of days in the school year corresponds to a country. Each ordered pair should be in the form

(days in school year, country).

d. Is the relation in part (c) a function? Explain your answer.

The function $f(x) = 0.76x + 171.4$ models the cholesterol level of an American man as a function of his age, x, in years. Use the function to solve Exercises 43–44.

43. Find and interpret $f(20)$.

44. Find and interpret $f(50)$.

To go, please. The graphs show that more Americans are ordering their food to go, instead of dining inside restaurants. A quadratic function models the average number of meals per person that Americans ordered for takeout and a linear function models the average number of meals ordered for eating in restaurants. In each model, x represents the number of years after 1984. Use the graphs and the displayed equations to solve Exercises 45–47.

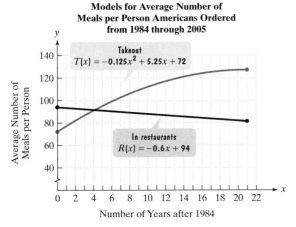

Models for Average Number of Meals per Person Americans Ordered from 1984 through 2005

Takeout
$T(x) = -0.125x^2 + 5.25x + 72$

In restaurants
$R(x) = -0.6x + 94$

Source: NPD Group

45. a. Use the equation for function T to find and interpret $T(20)$. How is this shown on the graph of T?

b. Use the equation for function R to find and interpret $R(0)$. How is this shown on the graph of R?

46. a. Use the equation for function T to find and interpret $T(18)$. How is this shown on the graph of T?

b. Use the equation for function R to find and interpret $R(20)$. How is this shown on the graph of R?

47. According to the graphs, in which year did the average number of takeout orders approximately equal the average number of in-restaurant orders? Use the equations for T and R to find the average number of meals per person for each kind of order in that year.

Writing in Mathematics

48. If a relation is represented by a set of ordered pairs, explain how to determine whether the relation is a function.

49. Your friend heard that functions are studied in intermediate and college algebra courses. He asks you what a function is. Provide him with a clear, relatively concise response.

50. Does $f(x)$ mean f times x when referring to a function f? If not, what does $f(x)$ mean? Provide an example with your explanation.

51. Explain how the vertical line test is used to determine whether a graph is a function.

52. For people filing a single return, federal income tax is a function of adjusted gross income. For each value of adjusted gross income, there is a specific tax to be paid. By contrast, the price of a house is not a function of the lot size on which the house sits. Houses on same-sized lots can sell for many different prices.

a. Describe an everyday situation between variables that is a function.

b. Describe an everyday situation between variables that is not a function.

Critical Thinking Exercises

Make Sense? *In Exercises 53–56, determine whether each statement "makes sense" or "does not make sense" and explain your reasoning.*

53. My body temperature is a function of the time of day.

54. My height is a function of my age.

55. Using $f(x) = 3x + 2$, I found $f(50)$ by applying the distributive property to $(3x + 2)50$.

56. A function models how a woman's cholesterol level depends on her age, so the domain is the set of various cholesterol levels.

In Exercises 57–60, determine whether each statement is true or false. If the statement is false, make the necessary change(s) to produce a true statement.

57. All relations are functions.

58. No two ordered pairs of a function can have the same second component and different first components.

59. The graph of every line is a function.

60. A horizontal line can intersect the graph of a function at more than one point.

61. Write a linear function, $f(x) = mx + b$, satisfying the following conditions:

$$f(0) = 7 \quad \text{and} \quad f(1) = 10.$$

62. If $f(x) = ax^2 + bx + c$ and $r = \dfrac{-b + \sqrt{b^2 - 4ac}}{2a}$, find $f(r)$ without doing any algebra and explain how you arrived at your result.

63. A car was purchased for $22,500. The value of the car decreases by $3200 per year for the first seven years. Write a function V that describes the value of the car after x years, where $0 \le x \le 7$. Then find and interpret $V(3)$.

Review Exercises

64. Write 0.00397 in scientific notation. (Section 6.7, Example 8)

65. Divide: $\dfrac{x^3 + 7x^2 - 2x + 3}{x - 2}$. (Section 6.6, Example 2)

66. Solve:

$$3x + 2y = 6$$
$$8x - 3y = 1.$$

(Section 5.3, Example 4)

GROUP PROJECT

CHAPTER 10

The bar graphs in Exercises 41 and 42 in Exercise Set 10.6 (see page 650) illustrate that if a relation is a function, reversing the components in each of its ordered pairs may result in a relation that is no longer a function. Group members should find examples of bar graphs, like the ones in Exercises 41 and 42, that illustrate this idea. Consult almanacs, newspapers, magazines, or the Internet. The group should select the graph with the most intriguing data. For the graph selected, write and solve a problem with four parts similar to Exercise 41 or 42.

Chapter 10 Summary

Definitions and Concepts	Examples

Section 10.1 Solving Quadratic Equations by the Square Root Property

The Square Root Property

If u is an algebraic expression and d is a positive real number, then

If $u^2 = d$, then $u = \sqrt{d}$ or $u = -\sqrt{d}$.

Equivalently,

If $u^2 = d$, then $u = \pm\sqrt{d}$.

Solve:

$$(x - 1)^2 = 7.$$
$$x - 1 = \pm\sqrt{7}$$
$$x = 1 \pm \sqrt{7}$$

The solution set is $\{1 \pm \sqrt{7}\}$.

Definitions and Concepts	**Examples**

The Pythagorean Theorem

The sum of the squares of the lengths of the legs of a right triangle equals the square of the length of the hypotenuse.

Hypotenuse c Leg a $a^2 + b^2 = c^2$

Leg b

Find a.

$c = 6$ a

$b = 2$

$$a^2 + b^2 = c^2$$
$$a^2 + 2^2 = 6^2$$
$$a^2 + 4 = 36$$
$$a^2 = 32$$

a must be positive, so do not use $a = -\sqrt{32}$. $a = \sqrt{32} = \sqrt{16 \cdot 2} = 4\sqrt{2}$

The Distance Formula

The distance, d, between the points (x_1, y_1) and (x_2, y_2) is given by

$$d = \sqrt{(x_2 - x_1)^2 + (y_2 - y_1)^2}.$$

Find the distance between $(-3, -5)$ and $(6, -2)$.

$$d = \sqrt{[6 - (-3)]^2 + [-2 - (-5)]^2}$$
$$= \sqrt{9^2 + 3^2} = \sqrt{81 + 9} = \sqrt{90} = \sqrt{9 \cdot 10} = 3\sqrt{10} \approx 9.49$$

Completing the Square

If $x^2 + bx$ is a binomial, then by adding $\left(\dfrac{b}{2}\right)^2$, the square of half the coefficient of x, you will obtain a perfect square trinomial. That is,

$$x^2 + bx + \left(\frac{b}{2}\right)^2 = \left(x + \frac{b}{2}\right)^2.$$

Complete the square: $x^2 + 10x$.

Add $\left(\frac{10}{2}\right)^2 = 5^2$, or 25.

$$x^2 + 10x + 25 = (x + 5)^2$$

Solving Quadratic Equations by Completing the Square

1. If the coefficient of x^2 is not 1, divide both sides by this coefficient.
2. Isolate variable terms on one side.
3. Complete the square by adding the square of half the coefficient of x to both sides.
4. Factor the perfect square trinomial.
5. Solve by applying the square root property.

Solve by completing the square:

$$2x^2 + 12x - 4 = 0.$$

$$\frac{2x^2}{2} + \frac{12x}{2} - \frac{4}{2} = \frac{0}{2} \qquad \text{Divide by 2.}$$

$$x^2 + 6x - 2 = 0 \qquad \text{Simplify.}$$

$$x^2 + 6x = 2 \qquad \text{Add 2.}$$

The coefficient of x is 6. Half of 6 is 3, and $3^2 = 9$. Add 9 to both sides.

$$x^2 + 6x + 9 = 2 + 9$$
$$(x + 3)^2 = 11$$
$$x + 3 = \pm\sqrt{11}$$
$$x = -3 \pm \sqrt{11}$$

The solution set is $\{-3 \pm \sqrt{11}\}$.

Definitions and Concepts	**Examples**

Section 10.3 The Quadratic Formula

The solutions of a quadratic equation in standard form

$$ax^2 + bx + c = 0, \quad a \neq 0,$$

are given by the quadratic formula

$$x = \frac{-b \pm \sqrt{b^2 - 4ac}}{2a}.$$

Solve by the quadratic formula:

$$2x^2 + 4x = 5.$$

First write in standard form by subtracting 5 from both sides.

$$2x^2 + 4x - 5 = 0$$

$$\boxed{a = 2} \quad \boxed{b = 4} \quad \boxed{c = -5}$$

$$x = \frac{-4 \pm \sqrt{4^2 - 4 \cdot 2(-5)}}{2 \cdot 2}$$

$$= \frac{-4 \pm \sqrt{16 - (-40)}}{4}$$

$$= \frac{-4 \pm \sqrt{56}}{4} = \frac{-4 \pm \sqrt{4 \cdot 14}}{4}$$

$$= \frac{-4 \pm 2\sqrt{14}}{4}$$

$$= \frac{2(-2 \pm \sqrt{14})}{2 \cdot 2} = \frac{-2 \pm \sqrt{14}}{2}$$

Section 10.4 Imaginary Numbers as Solutions of Quadratic Equations

The imaginary unit i is defined as

$$i = \sqrt{-1}, \quad \text{where} \quad i^2 = -1.$$

The set of numbers in the form $a + bi$ is called the set of complex numbers. If $b = 0$, the complex number is a real number. If $b \neq 0$, the complex number is called an imaginary number.

- $\sqrt{-36} = \sqrt{36(-1)} = \sqrt{36}\sqrt{-1} = 6i$
- $\sqrt{-50} = \sqrt{50(-1)} = \sqrt{25 \cdot 2}\sqrt{-1} = 5i\sqrt{2}$

Some quadratic equations have complex solutions that are imaginary numbers.

Solve: $x^2 - 2x + 2 = 0$.

$$\boxed{a = 1} \quad \boxed{b = -2} \quad \boxed{c = 2}$$

$$x = \frac{-b \pm \sqrt{b^2 - 4ac}}{2a} = \frac{-(-2) \pm \sqrt{(-2)^2 - 4 \cdot 1 \cdot 2}}{2 \cdot 1}$$

$$= \frac{2 \pm \sqrt{4 - 8}}{2} = \frac{2 \pm \sqrt{-4}}{2} = \frac{2 \pm \sqrt{4(-1)}}{2}$$

$$= \frac{2 \pm 2i}{2} = \frac{2(1 \pm i)}{2} = 1 \pm i$$

The solution set is $\{1 \pm i\}$.

Definitions and Concepts

Examples

Section 10.5 Graphs of Quadratic Equations

The graph of $y = ax^2 + bx + c$, called a parabola, can be graphed using the following steps:

1. If $a > 0$, the parabola opens upward. If $a < 0$, it opens downward.
2. Find the vertex, the lowest point if the parabola opens upward and the highest point if it opens downward. The x-coordinate of the vertex is $\dfrac{-b}{2a}$. Substitute this value into the parabola's equation to find the y-coordinate.
3. Find any x-intercepts by letting $y = 0$ and solving the resulting equation.
4. Find the y-intercept by letting $x = 0$.
5. Plot the intercepts and the vertex.
6. Connect these points with a smooth curve.

Graph: $y = x^2 - 2x - 8$.

$a = 1$ $b = -2$ $c = -8$

- $a > 0$, so the parabola opens upward.

- Vertex: x-coordinate $= \dfrac{-b}{2a} = \dfrac{-(-2)}{2 \cdot 1} = 1$

  y-coordinate $= 1^2 - 2 \cdot 1 - 8 = -9$

  Vertex is $(1, -9)$.

- x-intercepts: Let $y = 0$.
$$x^2 - 2x - 8 = 0$$
$$(x - 4)(x + 2) = 0$$
$$x - 4 = 0 \quad \text{or} \quad x + 2 = 0$$
$$x = 4 \qquad\qquad x = -2$$

  The parabola passes through $(4, 0)$ and $(-2, 0)$.

- y-intercept: Let $x = 0$.
$$y = 0^2 - 2 \cdot 0 - 8 = 0 - 0 - 8 = -8$$

  The parabola passes through $(0, -8)$.

Section 10.6 Introduction to Functions

A relation is any set of ordered pairs. The set of first components is the domain and the set of second components is the range. A function is a relation in which each member of the domain corresponds to exactly one member of the range. No two ordered pairs of a function can have the same first component and different second components.

The domain of the relation $\{(1, 2), (3, 4), (3, 7)\}$ is $\{1, 3\}$. The range is $\{2, 4, 7\}$. The relation is not a function: 3, in the domain, corresponds to both 4 and 7 in the range.

If a function is defined as an equation, the notation $f(x)$, read "f of x" or "f at x," describes the value of the function at the number x.

If $f(x) = x^2 - 5x + 4$, find $f(3)$.

$$f(3) = 3^2 - 5 \cdot 3 + 4 = 9 - 15 + 4 = -2$$

Thus, f of 3 is -2.

Definitions and Concepts	**Examples**

Section 10.6 Introduction to Functions (continued)

The Vertical Line Test for Functions

If any vertical line intersects a graph in more than one point, the graph does not define y as a function of x.

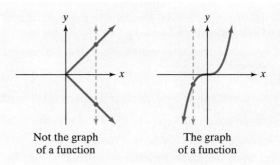

Not the graph of a function   The graph of a function

CHAPTER 10 REVIEW EXERCISES

10.1 *In Exercises 1–8, solve each quadratic equation by the square root property. If possible, simplify radicals or rationalize denominators.*

1. $x^2 = 64$

2. $x^2 = 17$

3. $2x^2 = 150$

4. $(x - 3)^2 = 9$

5. $(y + 4)^2 = 5$

6. $3y^2 - 5 = 0$

7. $(2x - 7)^2 = 25$

8. $(x + 5)^2 = 12$

In Exercises 9–11, use the Pythagorean Theorem to find the missing length in each right triangle. Express the answer in radical form and simplify, if possible.

9.

6 ft

8 ft

10.

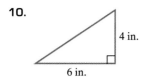

4 in.

6 in.

11.

15 cm

11 cm

Use the Pythagorean Theorem to solve Exercises 12–13.

12. How far away from the building shown in the figure is the bottom of the ladder?

20 ft   25 ft

?

13. A vertical pole is to be supported by three wires. Each wire is 13 yards long and is anchored 5 yards from the base of the pole. How far up the pole will the wires be attached?

14. The weight of a human fetus is modeled by the formula $W = 3t^2$, where W is the weight, in grams, and t is the time, in weeks, $0 \le t \le 39$. After how many weeks does the fetus weigh 1200 grams?

15. The distance, d, in feet, that an object falls in t seconds is modeled by the formula $d = 16t^2$. If you dive from a height of 100 feet, how long will it take to hit the water?

In Exercises 16–17, find the distance between each pair of points. Express answers in simplified radical form and, if necessary, round to two decimal places.

16. $(-3, -2)$ and $(1, -5)$

17. $(3, 8)$ and $(5, 4)$

10.2 *In Exercises 18–21, complete the square for each binomial. Then factor the resulting perfect square trinomial.*

18. $x^2 + 16x$

19. $x^2 - 6x$

20. $x^2 + 3x$

21. $x^2 - 5x$

In Exercises 22–24, solve each quadratic equation by completing the square.

22. $x^2 - 12x + 27 = 0$

23. $x^2 - 6x + 4 = 0$

24. $3x^2 - 12x + 11 = 0$

10.3 *In Exercises 25–27, solve each equation using the quadratic formula. Simplify irrational solutions, if possible.*

25. $2x^2 + 5x - 3 = 0$

26. $x^2 = 2x + 4$

27. $3x^2 + 5 = 9x$

In Exercises 28–32, solve each equation by the method of your choice. Simplify irrational solutions, if possible.

28. $2x^2 - 11x + 5 = 0$

29. $(3x + 5)(x - 3) = 5$

30. $3x^2 - 7x + 1 = 0$

31. $x^2 - 9 = 0$

32. $(2x - 3)^2 = 5$

33. The amount spent by governments and businesses worldwide on goods and services to thwart terrorists is increasing. The formula
$$T = 0.6x^2 + 5.4x + 28$$
models the amount spent to fight terrorism, T, in billions of dollars, x years after 2003. (*Source:* Homeland Security Research)
a. How much was spent to fight terrorism in 2007?

b. By which year is the amount spent to fight terrorism expected to reach \$160 billion?

10.4 *In Exercises 34–37, express each number in terms of i.*

34. $\sqrt{-81}$ **35.** $\sqrt{-23}$

36. $\sqrt{-48}$ **37.** $3 + \sqrt{-49}$

Exercises 38–42 involve quadratic equations with imaginary solutions. Solve each equation.

38. $x^2 = -100$

39. $5x^2 = -125$

40. $(2x + 1)^2 = -8$

41. $x^2 - 4x + 13 = 0$

42. $3x^2 - x + 2 = 0$

10.5 *In Exercises 43–46,*
  a. *Determine if the parabola whose equation is given opens upward or downward.*
  b. *Find the parabola's x-intercepts. If they are irrational, round to the nearest tenth.*
  c. *Find the parabola's y-intercept.*
  d. *Find the parabola's vertex.*
  e. *Graph the parabola.*

43. $y = x^2 - 6x - 7$

44. $y = -x^2 - 2x + 3$

45. $y = -3x^2 + 6x + 1$

46. $y = x^2 - 4x$

47. Fireworks are launched into the air. The formula
$$y = -16x^2 + 200x + 4$$
models the fireworks' height, y, in feet, x seconds after they are launched. When should the fireworks explode so that they go off at the greatest height? What is that height?

48. A quarterback tosses a football to a receiver 40 yards downfield. The formula
$$y = -0.025x^2 + x + 6$$
models the football's height, y, in feet, when it is x yards from the quarterback.
a. How many yards from the quarterback does the football reach its greatest height? What is that height?

b. If a defender is 38 yards from the quarterback, how far must he reach to deflect or catch the ball?

c. If the football is neither deflected by the defender nor caught by the receiver, how many yards will it go, to the nearest tenth of a yard, before hitting the ground?

d. Graph the equation that models the football's parabolic path.

10.6 *In Exercises 49–51, determine whether each relation is a function. Give the domain and range for each relation.*

49. $\{(2, 7), (3, 7), (5, 7)\}$

50. $\{(1, 10), (2, 500), (3, \pi)\}$

51. $\{(12, 13), (14, 15), (12, 19)\}$

In Exercises 52–53, evaluate each function at the given values.

52. $f(x) = 3x - 4$
  a. $f(-5)$ **b.** $f(6)$ **c.** $f(0)$

53. $g(x) = x^2 - 5x + 2$
  a. $g(-4)$ **b.** $g(3)$ **c.** $g(0)$

In Exercises 54–57, use the vertical line test to identify graphs in which y is a function of x.

54.

55.

56.

57.
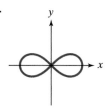

A Model for Sales of Organic Foods in the United States, 1997–2005

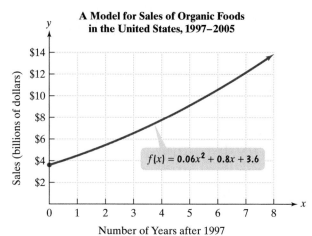

Source: Nutrition Business Journal

a. Find and interpret $f(0)$. How is this shown on the graph of f?

b. Find and interpret $f(8)$. How is this shown on the graph of f?

58. The graph shows the rising market for food grown without pesticides, herbicides, antibiotics, or hormones. The graph represents the quadratic function

$$f(x) = 0.06x^2 + 0.8x + 3.6,$$

which models the sales of organic foods in the United States, $f(x)$, in billions of dollars, x years after 1997.

CHAPTER 10 TEST

Remember to use your Chapter Test Prep Video CD to see the worked-out solutions to the test questions you want to review.

In Exercises 1–2, solve by the square root property.

1. $3x^2 = 48$

2. $(x - 3)^2 = 5$

3. To find the distance across a lake, a surveyor inserts poles at P and Q, measuring the respective distances to point R, as shown in the figure. Use the surveyor's measurements given in the figure to find the distance PQ across the lake in simplified radical form.

4. Find the distance between $(3, -2)$ and $(-4, 1)$. Express the answer in radical form and then round to two decimal places.

5. Solve by completing the square: $x^2 + 4x - 3 = 0$.

In Exercises 6–10, solve each equation by the method of your choice. Simplify irrational solutions, if possible.

6. $3x^2 + 5x + 1 = 0$

7. $(3x - 5)(x + 2) = -6$

8. $(2x + 1)^2 = 36$

9. $2x^2 = 6x - 1$

10. $2x^2 + 9x = 5$

In Exercises 11–12, express each number in terms of i.

11. $\sqrt{-121}$ **12.** $\sqrt{-75}$

In Exercises 13–15, solve each quadratic equation. Express imaginary solutions in a + bi form.

13. $x^2 + 36 = 0$ **14.** $(x - 5)^2 = -25$

15. $x^2 - 2x + 5 = 0$

In Exercises 16–17, graph each parabola whose equation is given. Label the x-intercepts, the y-intercept, and the vertex.

16. $y = x^2 + 2x - 8$

17. $y = -2x^2 + 16x - 24$

A batter hits a baseball into the air. The formula

$$y = -16x^2 + 64x + 5$$

models the baseball's height above the ground, y, in feet, x seconds after it is hit. Use the formula to solve Exercises 18–19.

18. When does the baseball reach its maximum height? What is that height?

19. After how many seconds does the baseball hit the ground? Round to the nearest tenth of a second.

In Exercises 20–21, determine whether each relation is a function. Give the domain and range for each relation.

20. $\{(1, 2), (3, 4), (5, 6), (6, 6)\}$

21. $\{(2, 1), (4, 3), (6, 5), (6, 6)\}$

22. If $f(x) = 7x - 3$, find $f(10)$.

23. If $g(x) = x^2 - 3x + 7$, find $g(-2)$.

In Exercises 24–25, identify the graph or graphs in which y is a function of x.

24.

25.

26. In a round-robin chess tournament, each player is paired with every other player once. The function

$$f(x) = \frac{x^2 - x}{2}$$

models the number of chess games, $f(x)$, that must be played in a round-robin tournament with x chess players. Find and interpret $f(9)$.

CUMULATIVE REVIEW EXERCISES (CHAPTERS 1–10)

In Exercises 1–10, solve each equation, inequality, or system of equations.

1. $2 - 4(x + 2) = 5 - 3(2x + 1)$

2. $\dfrac{x}{2} - 3 = \dfrac{x}{5}$

3. $3x + 9 \geq 5(x - 1)$

4. $2x + 3y = 6$
$x + 2y = 5$

5. $3x - 2y = 1$
$y = 10 - 2x$

6. $\dfrac{3}{x + 5} - 1 = \dfrac{4 - x}{2x + 10}$

7. $x + \dfrac{6}{x} = -5$

8. $x - 5 = \sqrt{x + 7}$

9. $(x - 2)^2 = 20$

10. $3x^2 - 6x + 2 = 0$

11. Solve for t: $A = \dfrac{5r + 2}{t}$.

In Exercises 12–24, perform the indicated operations. If possible, simplify the answer.

12. $\dfrac{12x^3}{3x^{12}}$

13. $4 \cdot 6 \div 2 \cdot 3 + (-5)$

14. $(6x^2 - 8x + 3) - (-4x^2 + x - 1)$

15. $(7x + 4)(3x - 5)$

16. $(5x - 2)^2$

17. $(x + y)(x^2 - xy + y^2)$

18. $\dfrac{x^2 + 6x + 8}{x^2} \div (3x^2 + 6x)$

19. $\dfrac{x}{x^2 + 2x - 3} - \dfrac{x}{x^2 - 5x + 4}$

20. $\dfrac{x - \dfrac{1}{5}}{5 - \dfrac{1}{x}}$

21. $3\sqrt{20} + 2\sqrt{45}$

22. $\sqrt{3x} \cdot \sqrt{6x}$

23. $\dfrac{2}{\sqrt{3}}$

24. $\dfrac{8}{3 - \sqrt{5}}$

In Exercises 25–30, factor completely.

25. $4x^2 - 49$

26. $x^3 + 3x^2 - x - 3$

27. $2x^2 + 8x - 42$

28. $x^5 - 16x$

29. $x^3 - 10x^2 + 25x$

30. $x^3 - 8$

31. Evaluate: $8^{-\frac{2}{3}}$.

In Exercises 32–37, graph each equation, inequality, or system of inequalities in a rectangular coordinate system.

32. $y = \dfrac{1}{3}x - 1$

33. $3x + 2y = -6$

34. $y = -2$

35. $3x - 4y \le 12$

36. $y = x^2 - 2x - 3$

37. $2x + y < 4$
$\quad\quad x > 2$

38. Find the slope of the line passing through the points $(-1, 3)$ and $(2, -3)$.

39. Write the point-slope form of the equation of the line passing through the points $(1, 2)$ and $(3, 6)$. Then use the point-slope equation to write the slope-intercept form of the line's equation.

In Exercises 40–50, use an equation or a system of equations to solve each problem.

40. Seven subtracted from five times a number is 208. Find the number.

41. After a 20% reduction, a digital camera sold for $256. What was the price before the reduction?

42. A rectangular field is three times as long as it is wide. If the perimeter of the field is 400 yards, what are the field's dimensions?

43. You invested $20,000 in two accounts paying 7% and 9% annual interest. If the total interest earned for the year is $1550, how much was invested at each rate?

44. A chemist needs to mix a 40% acid solution with a 70% acid solution to obtain 12 liters of a 50% acid solution. How many liters of each solution should be used?

45. Two boats started at the same time from the same port. One traveled due east at 13 miles per hour and the other traveled due west at 19 miles per hour. After how many hours will the boats be 232 miles apart?

46. A university with 176 people on the faculty wants to maintain a student-to-faculty ratio of 23:2. How many students should they enroll to maintain that ratio?

47. A sailboat has a triangular sail with an area of 120 square feet and a base that is 15 feet long. Find the height of the sail.

48. In a triangle, the measure of the first angle is 10° more than the measure of the second angle. The measure of the third angle is 20° more than four times that of the second angle. What is the measure of each angle?

49. A salesperson works in the TV and stereo department of an electronics store. One day she sold 3 TVs and 4 stereos for $2530. The next day, she sold 4 of the same TVs and 3 of the same stereos for $2510. Find the price of a TV and a stereo.

50. The length of a rectangle is 6 meters more than the width. The area is 55 square meters. Find the rectangle's dimensions.

Mean, Median, and Mode

Objectives

1 Determine the mean of a set of data items.

2 Determine the median of a set of data items.

3 Determine the mode of a set of data items.

1 Determine the mean of a set of data items.

One way to analyze a list of data items is to determine a single value that represents what is "average" or "typical" of the data. Such values are known as **measures of central tendency** because they are located toward the center of the data. Three such measures are discussed in this appendix: the *mean* (or *average*), the *median,* and the *mode*.

The Mean

By far the most commonly used measure of central tendency is the *mean*. The **mean** is obtained by adding all the data items and then dividing the sum by the number of items. The Greek letter sigma, Σ, called **a symbol of summation**, is used to indicate the sum of data items. The notation Σx, read "the sum of x," means to add all the data items in a given data set. We can use this symbol to give a formula for calculating the mean.

The Mean

The **mean** is the sum of the data items divided by the number of items.

$$\text{Mean} = \frac{\Sigma x}{n},$$

where Σx represents the sum of all the data items and n represents the number of items.

EXAMPLE 1   Calculating the Mean

Table A.1 shows the ten youngest male singers in the United States to have a number 1 single. Find the mean age of these male singers at the time of their number 1 single.

Table A.1	Youngest U.S. Male Singers to have a Number 1 Single	
Artist/Year	**Title**	**Age**
Stevie Wonder, 1963	"Fingertips"	13
Donny Osmond, 1971	"Go Away Little Girl"	13
Michael Jackson, 1972	"Ben"	14
Laurie London, 1958	"He's Got the Whole World in His Hands"	14
Paul Anka, 1957	"Diana"	16
Brian Hyland, 1960	"Itsy Bitsy Teenie Weenie Yellow Polkadot Bikini"	16
Shaun Cassidy, 1977	"Da Doo Ron Ron"	17
Bobby Vee, 1961	"Take Good Care of My Baby"	18
Usher, 1998	"Nice & Slow"	19
Andy Gibb, 1977	"I Just Want to Be Your Everything"	19

Source: Russell Ash, *The Top 10 of Everything*

Ages of Youngest U.S. Male Singers (repeated)

Age	Age
13	16
13	17
14	18
14	19
16	19

2 Determine the median of a set of data items.

Solution We find the mean by adding the ages and dividing this sum by 10, the number of data items.

$$\text{Mean} = \frac{\Sigma x}{n} = \frac{13 + 13 + 14 + 14 + 16 + 16 + 17 + 18 + 19 + 19}{10} = \frac{159}{10} = 15.9$$

The mean age of the ten youngest singers to have a number 1 single is 15.9. ∎

One and only one mean can be calculated for any set of data items. The mean may or may not be one of the actual data items. In Example 1, the mean was 15.9, although no data item is 15.9.

✓ **CHECK POINT 1** Find the mean for each group of data items:

a. 10, 20, 30, 40, 50   **b.** 3, 10, 10, 10, 117.

The Median

The *median* age in the United States is 35.3. The oldest state by median age is Florida (38.7) and the youngest state is Utah (27.1). To find these values, researchers begin with appropriate random samples. The data items—that is, the ages—are arranged from youngest to oldest. The median age is the data item in the middle of each set of ranked, or ordered, data.

The Median

To find the **median** of a group of data items,

1. Arrange the data items in order, from smallest to largest.
2. If the number of data items is odd, the median is the data item in the middle of the list.
3. If the number of data items is even, the median is the mean of the two middle data items.

EXAMPLE 2 Finding the Median

Find the median for each of the following groups of data:

a. 84, 90, 98, 95, 88
b. 68, 74, 7, 13, 15, 25, 28, 59, 34, 47.

Solution

a. Arrange the data items in order, from smallest to largest. The number of data items in the list, five, is odd. Thus, the median is the middle number.

84, 88, 90, 95, 98

Middle data item

The median is 90. Notice that two data items lie above 90 and two data items lie below 90.

b. Arrange the data items in order, from smallest to largest. The number of data items in the list, ten, is even. Thus, the median is the mean of the two middle data items.

7, 13, 15, 25, 28, 34, 47, 59, 68, 74

Middle data items are 28 and 34.

$$\text{Median} = \frac{28 + 34}{2} = \frac{62}{2} = 31$$

The median is 31. Five data items lie above 31 and five data items lie below 31.

7 13 15 25 28 31 34 47 59 68 74

Five data items lie below 31.   Median is 31.   Five data items lie above 31.

☑ **CHECK POINT 2**   Find the median for each of the following groups of data:

a. 28, 42, 40, 25, 35

b. 72, 61, 85, 93, 79, 87.

Statisticians generally use the median, rather than the mean, when reporting income. Why? Our next example will help to answer this question.

EXAMPLE 3   **Comparing the Median and the Mean**

Five employees in the assembly section of a television manufacturing company earn salaries of $19,700, $20,400, $21,500, $22,600, and $23,000 annually. The section manager has an annual salary of $95,000.

a. Find the median annual salary for the six people.

b. Find the mean annual salary for the six people.

Solution

a. To compute the median, first arrange the salaries in order:

$19,700, $20,400, $21,500, $22,600, $23,000, $95,000.

Because the list contains an even number of data items, six, the median is the mean of the two middle items.

$$\text{Median} = \frac{\$21,500 + \$22,600}{2} = \frac{\$44,100}{2} = \$22,050$$

The median annual salary is $22,050.

b. We find the mean annual salary by adding the six annual salaries and dividing by 6.

$$\text{Mean} = \frac{\$19,700 + \$20,400 + \$21,500 + \$22,600 + \$23,000 + \$95,000}{6}$$

$$= \frac{\$202,200}{6} = \$33,700$$

The mean annual salary is $33,700.

In Example 3, the median annual salary is $22,050 and the mean annual salary is $33,700. Why such a big difference between these two measures of central tendency? The relatively high annual salary of the section manager, $95,000, pulls the mean salary to a value considerably higher than the median salary. When one or more data items are much greater than the other items, these extreme values can greatly influence the mean. In cases like this, the median is often more representative of the data.

☑ **CHECK POINT 3** The ten countries in **Table A.2** accounted for 75% of the nearly $900 billion spent in 2003 on defense worldwide.

Table A.2	Top Ten Military Spenders, 2003			
Country	National Expenditure (billions of dollars)	Country	National Expenditure (billions of dollars)
United States	$417.4	Germany	$27.2
Japan	$46.9	Italy	$20.8
UK	$37.1	Iran	$19.2
France	$35.0	Saudi Arabia	$19.1
China	$32.8	South Korea	$13.9

Source: SIPRI, "The Major Spenders in 2003"

a. Find the mean national expenditure on defense, in billions of dollars, for the ten countries.

b. Find the median national expenditure on defense, in billions of dollars, for the ten countries.

c. Describe why one of the measures of central tendency is so much greater than the other.

3 Determine the mode of a set of data items.

The Mode

A value that occurs most often in a set of data items is called the *mode*.

The Mode

The **mode** in a set of data items is the value that occurs most often or the values that occur most often. If each data item occurs the same number of times, then the data set has no mode.

EXAMPLE 4 Finding the Mode

Find the mode for each of the following groups of data:

a. 7, 2, 4, 7, 8, 10   b. 2, 1, 4, 5, 3   c. 3, 3, 4, 5, 6, 6.

Solution

a. 7, 2, 4, 7, 8, 10

7 occurs most often.

The mode is 7.

b. 2, 1, 4, 5, 3

Each data item occurs the same number of times.

There is no mode.

c. 3, 3, 4, 5, 6, 6

Both 3 and 6 occur most often.

The modes are 3 and 6.

☑ **CHECK POINT 4** Find the mode for each of the following groups of data:

a. 3, 8, 5, 8, 9, 10   b. 3, 8, 5, 8, 9, 3   c. 3, 8, 5, 6, 9, 10.

EXAMPLE 5   Finding Three Measures of Central Tendency

Suppose your six exam grades in a course are

$$52, \quad 69, \quad 75, \quad 86, \quad 86, \quad \text{and} \quad 92.$$

Compute your final course grade ($90\text{--}100 = A, 80\text{--}89 = B, 70\text{--}79 = C, 60\text{--}69 = D$, below $60 = F$) using the

a. mean.   **b.** median.   **c.** mode.

Solution

a. The mean is the sum of the data items divided by the number of items, 6.

$$\text{Mean} = \frac{52 + 69 + 75 + 86 + 86 + 92}{6} = \frac{460}{6} \approx 76.67$$

Using the mean, your final course grade is C.

b. The six data items, 52, 69, 75, 86, 86, and 92, are arranged in order. Because the number of data items is even, the median is the mean of the two middle items.

$$\text{Median} = \frac{75 + 86}{2} = \frac{161}{2} = 80.5$$

Using the median, your final course grade is B.

c. The mode is the data value that occurs most frequently. Because 86 occurs most often, the mode is 86. Using the mode, your final course grade is B. ■

☑ **CHECK POINT 5**   *Consumer Reports* magazine gave the following data for the number of calories in a meat hot dog for each of eight brands:

$$107, \quad 136, \quad 138, \quad 138, \quad 172, \quad 173, \quad 190, \quad 191.$$

Find the mean, median, and mode for the number of calories in a meat hot dog for the eight brands. If necessary, round answers to the nearest tenth of a calorie.

APPENDIX EXERCISE SET   *MyMathLab*   Math XL PRACTICE   WATCH   DOWNLOAD   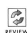 READ   REVIEW

Practice Exercises

In Exercises 1–8, find the mean for each group of data items.

1. 7, 4, 3, 2, 8, 5, 1, 3

2. 11, 6, 4, 0, 2, 1, 12, 0, 0

3. 91, 95, 99, 97, 93, 95

4. 100, 100, 90, 30, 70, 100

5. 100, 40, 70, 40, 60

6. 1, 3, 5, 10, 8, 5, 6, 8

7. 1.6, 3.8, 5.0, 2.7, 4.2, 4.2, 3.2, 4.7, 3.6, 2.5, 2.5

8. 1.4, 2.1, 1.6, 3.0, 1.4, 2.2, 1.4, 9.0, 9.0, 1.8

In Exercises 9–16, find the median for each group of data items.

9. 7, 4, 3, 2, 8, 5, 1, 3

10. 11, 6, 4, 0, 2, 1, 12, 0, 0

11. 91, 95, 99, 97, 93, 95

12. 100, 100, 90, 30, 70, 100

13. 100, 40, 70, 40, 60

14. 1, 3, 5, 10, 8, 5, 6, 8

15. 1.6, 3.8, 5.0, 2.7, 4.2, 4.2, 3.2, 4.7, 3.6, 2.5, 2.5

16. 1.4, 2.1, 1.6, 3.0, 1.4, 2.2, 1.4, 9.0, 9.0, 1.8

In Exercises 17–24, find the mode for each group of data items. If there is no mode, so state.

17. 7, 4, 3, 2, 8, 5, 1, 3

18. 11, 6, 4, 0, 2, 1, 12, 0, 0

19. 91, 95, 99, 97, 93, 95

20. 100, 100, 90, 30, 70, 100

21. 100, 40, 70, 40, 60

22. 1, 3, 5, 10, 8, 5, 6, 8

23. 1.6, 3.8, 5.0, 2.7, 4.2, 4.2, 3.2, 4.7, 3.6, 2.5, 2.5

24. 1.4, 2.1, 1.6, 3.0, 1.4, 2.2, 1.4, 9.0, 9.0, 1.8

Application Exercises

Exercises 25–26 present data related to age. For each data set described in boldface, find the mean, median, and mode (or state that there is no mode). If necessary, round answers to the nearest tenth.

25. Ages of the Justices of the United States Supreme Court in 2007

Roberts (52), Stevens (87), Scalia (71), Kennedy (71), Souter (68), Thomas (59), Ginsburg (74), Breyer (69), Alito (56)

26. Ages of the First Six U.S. Presidents at Inauguration

Washington (57), J. Adams (61), Jefferson (57), Madison (57), Monroe (58), J. Q. Adams (57)

27. The annual salaries of four salespeople and the owner of a bookstore are

$17,500, $19,000, $22,000, $27,500, $98,500.

Find the mean and the median. Is the mean or the median more representative of the five annual salaries? Briefly explain your answer.

28. In one common system for finding a grade-point average, or GPA,

$$A = 4, B = 3, C = 2, D = 1, F = 0.$$

The GPA is calculated by multiplying the number of credit hours for a course and the number assigned to each grade, and then adding these products. Then divide this sum by the total number of credit hours. Because each course grade is weighted according to the number of credits of the course, GPA is called a *weighted mean*. Calculate the GPA, rounded to the nearest tenth, for this transcript:

Sociology: 3 cr. A;   Biology: 3.5 cr. C;   Music: 1 cr. B;

Math: 4 cr. B;   English: 3 cr. C.

ANSWERS TO SELECTED EXERCISES
CHAPTER 1

Section 1.1

Check Point Exercises

1. a. 26   **b.** 32   **2. a.** 37   **b.** 2   **3. a.** $6x$ **b.** $4 + x$ **c.** $3x + 5$ **d.** $12 - 2x$ **e.** $\dfrac{15}{x}$ **4. a.** not a solution   **b.** solution

5. a. $\dfrac{x}{6} = 5$ **b.** $7 - 2x = 1$ **6. a.** 65%   **b.** 60%   **c.** overestimates by 5%

Exercise Set 1.1

1. 12   **3.** 8   **5.** 20   **7.** 7   **9.** 17   **11.** 18   **13.** 5   **15.** 19   **17.** 24   **19.** 13   **21.** 10   **23.** 5   **25.** $x + 4$ **27.** $x - 4$

29. $x + 4$ **31.** $x - 9$ **33.** $9 - x$ **35.** $3x - 5$ **37.** $12x - 1$ **39.** $\dfrac{10}{x} + \dfrac{x}{10}$ **41.** $\dfrac{x}{30} + 6$ **43.** solution   **45.** solution

47. not a solution   **49.** solution   **51.** not a solution   **53.** solution   **55.** solution   **57.** not a solution   **59.** $4x = 28$ **61.** $\dfrac{14}{x} = \dfrac{1}{2}$

63. $20 - x = 5$ **65.** $2x + 6 = 16$ **67.** $3x - 5 = 7$ **69.** $4x + 5 = 33$ **71.** $4(x + 5) = 33$ **73.** $5x = 24 - x$ **75.** 8   **77.** 59

79. a. 14   **b.** yes   **81. a.** 6   **b.** no   **83. a.** 11%; overestimates by 1%   **b.** 9%; It gives the percent displayed for 2004.

85. a. 35.6%; overestimates by 0.6%   **b.** 29.8%; overestimates by 3.8%   **87. a.** 44   **b.** 164   **99.** makes sense

101. makes sense   **103.** true   **105.** true   **107.** Choices of variables may vary. Samples are given: $h = $ hours worked,

$s = $ salary; $s = 20h$.   **109. a.** O'Donnell: \$0.65; Vieira: \$1.79; Couric: \$2.05   **b.** O'Donnell   **110.** $\dfrac{6}{35}$ **111.** $\dfrac{10}{21}$ **112.** $\dfrac{4}{17}$

Section 1.2

Check Point Exercises

1. $\dfrac{21}{8}$ **2.** $1\dfrac{2}{3}$ **3.** $2 \cdot 2 \cdot 3 \cdot 3$ **4. a.** $\dfrac{2}{3}$ **b.** $\dfrac{7}{4}$ **c.** $\dfrac{13}{15}$ **d.** $\dfrac{1}{5}$ **5. a.** $\dfrac{8}{33}$ **b.** $\dfrac{18}{5}$ or $3\dfrac{3}{5}$ **c.** $\dfrac{2}{7}$ **d.** $\dfrac{51}{10}$ or $5\dfrac{1}{10}$ **6. a.** $\dfrac{10}{3}$ or $3\dfrac{1}{3}$

b. $\dfrac{2}{9}$ **c.** $\dfrac{3}{2}$ or $1\dfrac{1}{2}$ **7. a.** $\dfrac{5}{11}$ **b.** $\dfrac{2}{3}$ **c.** $\dfrac{9}{4}$ or $2\dfrac{1}{4}$ **8.** $\dfrac{14}{21}$ **9. a.** $\dfrac{11}{10}$ or $1\dfrac{1}{10}$ **b.** $\dfrac{7}{12}$ **c.** $\dfrac{5}{4}$ or $1\dfrac{1}{4}$ **10.** $\dfrac{53}{60}$

11. a. solution   **b.** solution   **12. a.** $\dfrac{2}{3}(x - 6)$ **b.** $\dfrac{3}{4}x - 2 = \dfrac{1}{5}x$ **13.** 25°C

Exercise Set 1.2

1. $\dfrac{19}{8}$ **3.** $\dfrac{38}{5}$ **5.** $\dfrac{135}{16}$ **7.** $4\dfrac{3}{5}$ **9.** $8\dfrac{4}{9}$ **11.** $35\dfrac{11}{20}$ **13.** $2 \cdot 11$ **15.** $2 \cdot 2 \cdot 5$ **17.** prime   **19.** $2 \cdot 2 \cdot 3 \cdot 3$ **21.** $2 \cdot 2 \cdot 5 \cdot 7$

23. prime   **25.** $3 \cdot 3 \cdot 3 \cdot 3$ **27.** $2 \cdot 2 \cdot 2 \cdot 2 \cdot 3 \cdot 5$ **29.** $\dfrac{5}{8}$ **31.** $\dfrac{5}{6}$ **33.** $\dfrac{7}{10}$ **35.** $\dfrac{2}{5}$ **37.** $\dfrac{22}{25}$ **39.** $\dfrac{60}{43}$ **41.** $\dfrac{2}{15}$ **43.** $\dfrac{21}{88}$

45. $\dfrac{36}{7}$ **47.** $\dfrac{1}{12}$ **49.** $\dfrac{15}{14}$ **51.** 6   **53.** $\dfrac{15}{16}$ **55.** $\dfrac{9}{5}$ **57.** $\dfrac{5}{9}$ **59.** 3   **61.** $\dfrac{7}{10}$ **63.** $\dfrac{1}{2}$ **65.** 6   **67.** $\dfrac{6}{11}$ **69.** $\dfrac{2}{3}$ **71.** $\dfrac{5}{4}$

73. $\dfrac{1}{6}$ **75.** 2   **77.** $\dfrac{7}{10}$ **79.** $\dfrac{9}{10}$ **81.** $\dfrac{19}{24}$ **83.** $\dfrac{7}{18}$ **85.** $\dfrac{7}{12}$ **87.** $\dfrac{41}{80}$ **89.** $1\dfrac{5}{12}$ or $\dfrac{17}{12}$ **91.** solution   **93.** solution

95. not a solution   **97.** solution   **99.** not a solution   **101.** solution   **103.** $\dfrac{1}{5}x$ **105.** $x - \dfrac{1}{4}x$ **107.** $x - \dfrac{1}{4} = \dfrac{1}{2}x$ **109.** $\dfrac{1}{7}x + \dfrac{1}{8}x = 12$

111. $\dfrac{2}{3}(x + 6)$ **113.** $\dfrac{2}{3}x + 6 = x - 3$ **115.** $\dfrac{3a}{20}$ **117.** $\dfrac{20}{x}$ **119.** $\dfrac{4}{15}$ **121.** not a solution   **123.** 20°C   **125. a.** 140 beats per minute

b. 160 beats per minute   **127. a.** $H = \dfrac{9}{10}(220 - a)$ **b.** 162 beats per minute   **129.** 474 billion; underestimates by 2 billion

141. makes sense   **143.** makes sense   **145.** false   **147.** true

149.

say   does   that   Star-span-gled   Ban-ner   yet   wave   O'er   the

150. 5   **151.** $\dfrac{5}{2}$ **152.** −4

Section 1.3

Check Point Exercises

1. a. -500 **b.** -282 **2.**

(a) (b) (c)

3.

(b) (a)

4. a. 0.375 **b.** $0.\overline{45}$ **5. a.** $\sqrt{9}$ **b.** $0, \sqrt{9}$ **c.** $-9, 0, \sqrt{9}$ **d.** $-9, -1.3, 0, 0.\overline{3}, \sqrt{9}$ **e.** $\dfrac{\pi}{2}, \sqrt{10}$ **f.** $-9, -1.3, 0, 0.\overline{3}, \dfrac{\pi}{2}, \sqrt{9}, \sqrt{10}$

6. a. $>$ **b.** $<$ **c.** $<$ **d.** $<$ **7. a.** true **b.** true **c.** false **8. a.** 4 **b.** 6 **c.** $\sqrt{2}$

Exercise Set 1.3

1. -20 **3.** 8 **5.** -3000 **7.** -4 billion **9–19.**

21. 0.75 **23.** 0.35 **25.** 0.875 **27.** $0.\overline{81}$

29. -0.5 **31.** $-0.8\overline{3}$ **33. a.** $\sqrt{100}$ **b.** $0, \sqrt{100}$ **c.** $-9, 0, \sqrt{100}$ **d.** $-9, -\dfrac{4}{5}, 0, 0.25, 9.2, \sqrt{100}$ **e.** $\sqrt{3}$

f. $-9, -\dfrac{4}{5}, 0, 0.25, \sqrt{3}, 9.2, \sqrt{100}$ **35. a.** $\sqrt{64}$ **b.** $0, \sqrt{64}$ **c.** $-11, 0, \sqrt{64}$ **d.** $-11, -\dfrac{5}{6}, 0, 0.75, \sqrt{64}$ **e.** $\sqrt{5}, \pi$

f. $-11, -\dfrac{5}{6}, 0, 0.75, \sqrt{5}, \pi, \sqrt{64}$ **37.** 0 **39.** Answers will vary; $\dfrac{1}{2}$ is an example. **41.** Answers will vary; 6 is an example.

43. Answers will vary; π is an example. **45.** $<$ **47.** $>$ **49.** $>$ **51.** $<$ **53.** $>$ **55.** $<$ **57.** $<$ **59.** $>$

61. $>$ **63.** true **65.** true **67.** true **69.** false **71.** 6 **73.** 7 **75.** $\dfrac{5}{6}$ **77.** $\sqrt{11}$ **79.** $>$ **81.** $=$ **83.** $<$ **85.** $=$

87. rational numbers **89.** integers **91.** all real numbers **93.** whole numbers

95. a.

b. Rhode Island, Georgia, Louisiana, Florida, Hawaii

109. does not make sense **111.** makes sense **113.** false **115.** false **117.** false **119.** $-\dfrac{1}{2}d$ **121.** $-3.464; -4$ and -3

123. $-0.236; -1$ and 0 **124.** 27; 27; Both expressions have the same value. **125.** 32; 32; Both expressions have the same value.

126. 28; 28; Both expressions have the same value.

Section 1.4

Check Point Exercises

1. a. 3 terms **b.** 6 **c.** 11 **d.** $6x$ and $2x$ **2. a.** $14 + x$ **b.** $y7$ **3. a.** $17 + 5x$ **b.** $x5 + 17$ **4. a.** $20 + x$ or $x + 20$ **b.** $30x$
5. $12 + x$ or $x + 12$ **6.** $5x + 15$ **7.** $24y + 42$ **8. a.** $10x$ **b.** $5a$ **9. a.** $18x + 10$ **b.** $14x + 8y$ **10.** $25x + 21$ **11.** $38x + 23y$

Exercise Set 1.4

1. a. 2 **b.** 3 **c.** 5 **d.** no **3. a.** 3 **b.** 1 **c.** 2 **d.** yes; x and $5x$ **5. a.** 3 **b.** 4 **c.** 1 **d.** no **7.** $4 + y$ **9.** $3x + 5$
11. $5y + 4x$ **13.** $5(3 + x)$ **15.** $x9$ **17.** $x + 6y$ **19.** $x7 + 23$ **21.** $(x + 3)5$ **23.** $(7 + 5) + x = 12 + x$ **25.** $(7 \cdot 4)x = 28x$
27. $3x + 15$ **29.** $16x + 24$ **31.** $4 + 2r$ **33.** $5x + 5y$ **35.** $3x - 6$ **37.** $8x - 10$ **39.** $\dfrac{5}{2}x - 6$ **41.** $8x + 28$ **43.** $6x + 18 + 12y$
45. $15x - 10 + 20y$ **47.** $17x$ **49.** $8a$ **51.** $14 + x$ **53.** $11y + 3$ **55.** $9x + 1$ **57.** $14a + 14$ **59.** $15x + 6$ **61.** $15x + 2$
63. $41a + 24b$ **65.** Distributive property; Commutative property of addition; Associative property of addition; Commutative property of addition
67. $7x + 2x; 9x$ **69.** $12x - 3x; 9x$ **71.** $6(4x); 24x$ **73.** $6(4 + x); 24 + 6x$ **75.** $8 + 5(x - 1); 5x + 3$ **77. a.** $M = 4.5n + 51$
b. 73.5%; underestimates by 1.5% **89.** does not make sense **91.** makes sense **93.** false **95.** false **97.** 60 **98.** -40 **99.** -5

Mid-Chapter Check Point Exercises

1. 62 **2.** 2 **3.** 43 **4.** $\dfrac{1}{4}x - 2$ **5.** $\dfrac{x}{6} + 5 = 19$ **6.** not a solution **7.** not a solution **8. a.** 537; overestimates by 30 per

100,000 inhabitants **b.** 327 per 100,000 inhabitants **9.** $\dfrac{1}{6}$ **10.** $\dfrac{1}{2}$ **11.** $\dfrac{25}{66}$ **12.** $\dfrac{2}{3}$ **13.** $\dfrac{1}{5}$ **14.** $\dfrac{25}{27}$ **15.** $\dfrac{37}{18}$ or $2\dfrac{1}{18}$ **16.** $10°C$

17. $<$ **18.** $0.\overline{09}$ **19.** 19.3 **20.** $-11, -\dfrac{3}{7}, 0, 0.45, \sqrt{25}$ **21.** $(x + 3)5$ **22.** $5(3 + x)$ **23.** $5x + 15$ **24.** $65x + 21$

25. $10x + 34y$

Section 1.5

Check Point Exercises

1. $4 + (-7) = -3$ **2. a.** $-1 + (-3) = -4$

b. $-5 + 3 = -2$ **3. a.** -35 **b.** -1.5 **c.** $-\dfrac{5}{6}$ **4. a.** -13 **b.** 1.2 **c.** $-\dfrac{1}{2}$

5. a. $-17x$ **b.** $-7y + 6z$ **c.** $-20x + 15$ **6.** down 3 ft

Exercise Set 1.5

1. 4   **3.** -7 **5.** -4

7. 0   **9.** -7 **11.** 0   **13.** -60 **15.** -18 **17.** -1.3 **19.** -1 **21.** -5 **23.** 4   **25.** -3

27. -1.5 **29.** -5.7 **31.** $\dfrac{3}{10}$ **33.** $\dfrac{1}{8}$ **35.** $-\dfrac{43}{35}$ **37.** -8 **39.** 62   **41.** 8   **43.** -21

45. 22.1   **47.** $-8x$ **49.** $13y$ **51.** $-23a$ **53.** $-6y + 4z$ **55.** $-8b + 4$ **57.** $-2x + 14y$ **59.** $-3y + 24$ **61.** 12   **63.** -30

65. $>$ **67.** $-6x + (-13x);\ -19x$ **69.** $\dfrac{-20}{x} + \dfrac{3}{x};\ \dfrac{-17}{x}$ **71.** 44°F   **73.** 600 ft below sea level   **75.** 3°F   **77.** the 25-yard line

79. a. $-\$413$ billion   **b.** $-\$318$ billion   **c.** $-\$731$ billion   **89.** makes sense   **91.** makes sense   **93.** true   **95.** false   **97.** negative

99. positive   **102. a.** $\sqrt{4}$ **b.** $0, \sqrt{4}$ **c.** $-6, 0, \sqrt{4}$ **d.** $-6, 0, 0.\overline{7}, \sqrt{4}$ **e.** $-\pi, \sqrt{3}$ **f.** $-6, -\pi, 0, 0.\overline{7}, \sqrt{3}, \sqrt{4}$ **103.** true

104. solution   **105.** -3 **106.** -21 **107.** 5

Section 1.6

Check Point Exercises

1. a. -8 **b.** 9   **c.** -5 **2. a.** 9.2   **b.** $-\dfrac{14}{15}$ **c.** 7π **3.** 15   **4.** $-6, 4a, -7ab$ **5. a.** $4 - 7x$ **b.** $-9x + 4y$ **6.** 19,763 m

Exercise Set 1.6

1. a. -12 **b.** $5 + (-12)$ **3. a.** 7   **b.** $5 + 7$ **5.** 6   **7.** -6 **9.** 23   **11.** 11   **13.** -11 **15.** -38 **17.** 0   **19.** 0   **21.** 26

23. -13 **25.** 13   **27.** $-\dfrac{2}{7}$ **29.** $\dfrac{4}{5}$ **31.** -1 **33.** $-\dfrac{3}{5}$ **35.** $\dfrac{3}{4}$ **37.** $\dfrac{1}{4}$ **39.** 7.6   **41.** -2 **43.** 2.6   **45.** 0   **47.** 3π

49. 13π **51.** 19   **53.** -3 **55.** -15 **57.** 0   **59.** -52 **61.** -187 **63.** $\dfrac{7}{6}$ **65.** -4.49 **67.** $-\dfrac{3}{8}$ **69.** $-3x, -8y$

71. $12x, -5xy, -4$ **73.** $-6x$ **75.** $4 - 10y$ **77.** $5 - 7a$ **79.** $-4 - 9b$ **81.** $24 + 11x$ **83.** $3y - 8x$ **85.** 9   **87.** $\dfrac{7}{8}$

89. -2 **91.** $6x - (-5x);\ 11x$ **93.** $\dfrac{-5}{x} - \left(\dfrac{-2}{x}\right);\ \dfrac{-3}{x}$ **95.** 19,757 ft   **97.** 21°F   **99.** 3°F   **101. a.** $\$1.0$ billion

b. $\$1.4$ billion   **c.** $-\$0.4$ billion   **103. a.** $\$1.02$ billion   **b.** $\$1.52$ billion   **c.** $-\$0.5$ billion   **d.** $\$0.1$ billion   **111.** makes sense

113. makes sense   **115.** false   **117.** true   **119.** negative   **121.** positive   **125.** not a solution   **126.** $15x + 40y$

127. Answers will vary; -1 is an example.   **128.** -12 **129.** -9 **130.** 12

Section 1.7

Check Point Exercises

1. a. -40 **b.** $-\dfrac{4}{21}$ **c.** 36   **d.** 5.5   **e.** 0   **2. a.** 24   **b.** -30 **3. a.** $\dfrac{1}{7}$ **b.** 8   **c.** $-\dfrac{1}{6}$ **d.** $-\dfrac{13}{7}$ **4. a.** -4 **b.** 8

5. a. 8   **b.** $-\dfrac{8}{15}$ **c.** -7.3 **d.** 0   **6. a.** $-20x$ **b.** $10x$ **c.** $-b$ **d.** $-21x + 28$ **e.** $-7y + 6$ **7.** $-y - 26$

8. not a solution   **9.** 49.4%; overestimates by 0.4%

Exercise Set 1.7

1. -45 **3.** 24   **5.** -21 **7.** 19   **9.** 0   **11.** -12 **13.** 9   **15.** $\dfrac{12}{35}$ **17.** $-\dfrac{14}{27}$ **19.** -3.6 **21.** 0.12   **23.** 30   **25.** -72

27. 24   **29.** -27 **31.** 90   **33.** 0   **35.** $\dfrac{1}{4}$ **37.** 5   **39.** $-\dfrac{1}{10}$ **41.** $-\dfrac{5}{2}$ **43. a.** $-32 \cdot \left(\dfrac{1}{4}\right)$ **b.** -8 **45. a.** $-60 \cdot \left(-\dfrac{1}{5}\right)$

b. 12   **47.** -3 **49.** -7 **51.** 30   **53.** 0   **55.** undefined   **57.** -5 **59.** -12 **61.** 6   **63.** 0   **65.** undefined   **67.** -4.3

69. $\frac{5}{6}$ **71.** $-\frac{16}{9}$ **73.** -1 **75.** -15 **77.** $-10x$ **79.** $3y$ **81.** $9x$ **83.** $-4x$ **85.** $-b$ **87.** $3y$ **89.** $-8x + 12$

91. $6x - 12$ **93.** $-2y + 5$ **95.** $y - 14$ **97.** solution **99.** solution **101.** not a solution **103.** solution **105.** not a solution

107. solution **109.** $4(-10) + 8 = -32$ **111.** $(-9)(-3) - (-2) = 29$ **113.** $\frac{-18}{-15 + 12} = 6$ **115.** $-6 - \left(\frac{12}{-4}\right) = -3$ **117.** $-30°C$

119. a. 50% **b.** 50%; The model provides an exact description of the data. **c.** Answers will vary; approximately 22%.

d. 18%; It's less than the estimate. **e.** $R = \frac{-0.5n + 28}{-n + 70}$ **121. a.** $\frac{1}{4}$; $2500 **b.** 0.25 **c.** $\frac{10.85}{44}$ **d.** 0.25; perfectly well; The rounded

decimal gives the value obtained from the data. **131.** makes sense **133.** does not make sense **135.** false **137.** true **139.** $50x$

141. $\frac{40}{x + 40}$ **146.** -9 **147.** -3 **148.** 2 **149.** 36 **150.** -125 **151.** 16

Section 1.8

Check Point Exercises

1. a. 36 **b.** -64 **c.** 1 **d.** -1 **2. a.** $21x^2$ **b.** $8x^3$ **c.** cannot simplify **3.** 15 **4.** 32 **5. a.** 36 **b.** 12 **6.** $\frac{3}{4}$ **7.** -40

8. -31 **9.** $\frac{5}{7}$ **10.** -5 **11.** $7x^2 + 15$ **12.** 37% **13. a.** $330 **b.** $60 **c.** $33

Exercise Set 1.8

1. 81 **3.** 64 **5.** 16 **7.** -64 **9.** 625 **11.** -625 **13.** -100 **15.** $19x^2$ **17.** $15x^3$ **19.** $9x^4$ **21.** $-x^2$ **23.** x^3

25. cannot be simplified **27.** 0 **29.** 25 **31.** 27 **33.** 12 **35.** 5 **37.** 45 **39.** -24 **41.** 300 **43.** 0 **45.** -32 **47.** 64

49. 30 **51.** $\frac{4}{3}$ **53.** 2 **55.** 2 **57.** 3 **59.** 88 **61.** -60 **63.** -36 **65.** 14 **67.** $-\frac{3}{4}$ **69.** $-\frac{9}{40}$ **71.** $-\frac{37}{36}$ or $-1\frac{1}{36}$

73. 24 **75.** 28 **77.** 9 **79.** -7 **81.** $15x - 27$ **83.** $15 - 3y$ **85.** $16y - 25$ **87.** $-2x^2 - 9$ **89.** $-10 - (-2)^3 = -2$

91. $[2(7 - 10)]^2 = 36$ **93.** $x - (5x + 8) = -4x - 8$ **95.** $5(x^3 - 4) = 5x^3 - 20$ **97.** 836 thousand; overestimates by 62 thousand

99. a. 92%, although answers may vary by ± 1% **b.** 91.65%; very close to the estimate **101. a.** 2000: 34%; 2002: 45%; 2004: 52%; 2005: 62%;

Answers may vary by ± 1%. **b.** model 2 **103. a.** $3,000,000 (or $300 tens of thousands) **b.** $18,000,000 (or $1800 tens of thousands)

c. Cost increases. **107.** makes sense **109.** makes sense **111.** false **113.** true **115.** $-\frac{79}{4}$ **117.** $\left(2 \cdot 5 - \frac{1}{2} \cdot 10\right) \cdot 9 = 45$

118. 6 **119.** -24 **120.** Answers will vary; -3 is an example. **121.** solution **122.** solution **123.** not a solution

Review Exercises

1. 40 **2.** 50 **3.** 3 **4.** 84 **5.** $7x - 6$ **6.** $\frac{x}{5} - 2 = 18$ **7.** $9 - 2x = 14$ **8.** $3(x + 7)$ **9.** not a solution **10.** solution

11. solution **12.** 11, although answers may vary by ± 1 **13.** 11 Latin words; very well; It is the same. **14. a.** 59%; underestimates by 1%

b. 93% **15.** $\frac{23}{7}$ **16.** $\frac{64}{11}$ **17.** $1\frac{8}{9}$ **18.** $5\frac{2}{5}$ **19.** $2 \cdot 2 \cdot 3 \cdot 5$ **20.** $3 \cdot 3 \cdot 7$ **21.** prime **22.** $\frac{5}{11}$ **23.** $\frac{8}{15}$ **24.** $\frac{21}{50}$ **25.** $\frac{8}{3}$

26. $\frac{1}{4}$ **27.** $\frac{2}{3}$ **28.** $\frac{29}{18}$ **29.** $\frac{37}{60}$ **30.** solution **31.** not a solution **32.** $2 - \frac{1}{2}x = \frac{1}{4}x$ **33.** $\frac{3}{5}(x + 6)$ **34.** 152 beats per minute

35.

36.

37. 0.625 **38.** $0.\overline{27}$ **39. a.** $\sqrt{81}$ **b.** $0, \sqrt{81}$

c. $-17, 0, \sqrt{81}$ **d.** $-17, -\frac{9}{13}, 0, 0.75, \sqrt{81}$

e. $\sqrt{2}, \pi$ **f.** $-17, -\frac{9}{13}, 0, 0.75, \sqrt{2}, \pi, \sqrt{81}$

40. Answers will vary; -2 is an example. **41.** Answers will vary; $\frac{1}{2}$ is an example. **42.** Answers will vary; π is an example. **43.** $<$

44. $>$ **45.** $>$ **46.** $<$ **47.** false **48.** true **49.** 58 **50.** 2.75 **51.** $13y + 7$ **52.** $(x + 7)9$ **53.** $(6 + 4) + y = 10 + y$

54. $(7 \cdot 10)x = 70x$ **55.** $24x - 12 + 30y$ **56.** $7a + 2$ **57.** $28x + 19$

58. 2

59. -3 **60.** $-\frac{11}{20}$ **61.** -7 **62.** $-4x + 5y$ or $5y - 4x$ **63.** $-10y + 40$ or $40 - 10y$

64. 800 ft below sea level **65.** 23 ft **66.** $9 + (-13)$ **67.** 4

68. $-\frac{6}{5}$ **69.** -1.5 **70.** -7 **71.** -3 **72.** $-5 - 8a$ **73.** 27,150 ft **74.** 84 **75.** $-\frac{3}{11}$ **76.** -120 **77.** -9 **78.** undefined

79. 2 **80.** $3x$ **81.** $-x - 1$ **82.** not a solution **83.** solution **84. a.** 2734; underestimates by 7 **b.** 1110 **85.** 36

86. -36 **87.** -32 **88.** $6x^3$ **89.** cannot be simplified **90.** -16 **91.** -16 **92.** 10 **93.** -2 **94.** 17 **95.** -88 **96.** 14

97. $-\frac{20}{3}$ **98.** $-\frac{2}{5}$ **99.** 6 **100.** 10 **101.** $28a - 20$ **102.** $6y - 12$ **103.** 65%; underestimates by 0.4%

Chapter Test

1. 4   **2.** -11 **3.** -51 **4.** $\frac{1}{5}$ **5.** $-\frac{35}{6}$ or $-5\frac{5}{6}$ **6.** -5 **7.** 1   **8.** -4 **9.** -32 **10.** 1   **11.** $4x + 4$ **12.** $13x - 19y$

13. $10 - 6x$ **14.** $-7, -\frac{4}{5}, 0, 0.25, \sqrt{4}, \frac{22}{7}$ **15.** $>$ **16.** 12.8   **17.** -15 **18.** 150   **19.** $2(3 + x)$ **20.** $(-6 \cdot 4)x = -24x$

21. $35x - 7 + 14y$ **22.** 17,030 ft   **23.** not a solution   **24.** solution   **25.** $\frac{1}{4}x - 5 = 32$ **26.** $5(x + 4) - 7$

27. 11 million; underestimates by 0.3 million   **28.** 144 beats per minute, although answers may vary by ± 1 **29.** 144 beats per minute; very well; It is the same.

CHAPTER 2

Section 2.1

Check Point Exercises

1. 17 or {17}   **2.** 2.9 or {2.29}   **3.** $\frac{1}{4}$ or $\left\{\frac{1}{4}\right\}$ **4.** 13 or {13}   **5.** 12 or {12}   **6.** 11 or {11}   **7.** 2100 words

Exercise Set 2.1

1. linear   **3.** not linear   **5.** not linear   **7.** linear   **9.** not linear   **11.** 23 or {23}   **13.** -20 or $\{-20\}$ **15.** -16 or $\{-16\}$

17. -12 or $\{-12\}$ **19.** 4 or {4}   **21.** -11 or $\{-11\}$ **23.** 2 or {2}   **25.** $-\frac{17}{12}$ or $\left\{-\frac{17}{12}\right\}$ **27.** $\frac{21}{4}$ or $\left\{\frac{21}{4}\right\}$ **29.** $-\frac{11}{20}$ or $\left\{-\frac{11}{20}\right\}$

31. 4.3 or {4.3}   **33.** $-\frac{21}{4}$ or $\left\{-\frac{21}{4}\right\}$ **35.** 18 or {18}   **37.** $\frac{9}{10}$ or $\left\{\frac{9}{10}\right\}$ **39.** -310 or $\{-310\}$ **41.** 4.3 or {4.3}   **43.** 0 or {0}

45. 11 or {11}   **47.** 5 or {5}   **49.** -13 or $\{-13\}$ **51.** 6 or {6}   **53.** -12 or $\{-12\}$ **55.** $x = \triangle + \square$ **57.** $\triangle - \square = x$

59. $x - 12 = -2; 10$ **61.** $\frac{2}{5}x - 8 = \frac{7}{5}x; -8$ **63.** \$1700   **65.** 94; underestimates by 2   **67. a.** 47% $\pm 1\%$ **b.** 46.8%; very well

73. does not make sense   **75.** makes sense   **77.** false   **79.** true   **81.** Answers will vary; example: $x - 100 = -101$.

83. 2.7529 or {2.7529}   **84.** $\frac{9}{x} - 4x$ **85.** -12 **86.** $6 - 9x$ **87.** x **88.** y **89.** yes

Section 2.2

Check Point Exercises

1. 36 or {36}   **2. a.** 21 or {21}   **b.** -4 or $\{-4\}$ **c.** -3.1 or $\{-3.1\}$ **3. a.** 24 or {24}   **b.** -16 or $\{-16\}$ **4. a.** -5 or $\{-5\}$ **b.** 3 or {3}   **5.** 6 or {6}   **6.** -10 or $\{-10\}$ **7.** 6 or {6}   **8. a.** overestimates by 0.08 million   **b.** 2010

Exercise Set 2.2

1. 30 or {30}   **3.** -33 or $\{-33\}$ **5.** 7 or {7}   **7.** -9 or $\{-9\}$ **9.** $-\frac{7}{2}$ or $\left\{-\frac{7}{2}\right\}$ **11.** 6 or {6}   **13.** $-\frac{3}{4}$ or $\left\{-\frac{3}{4}\right\}$ **15.** 0 or {0}

17. 18 or {18}   **19.** -8 or $\{-8\}$ **21.** -17 or $\{-17\}$ **23.** 47 or {47}   **25.** 45 or {45}   **27.** -5 or $\{-5\}$ **29.** 5 or {5}   **31.** 6 or {6}

33. -1 or $\{-1\}$ **35.** -2 or $\{-2\}$ **37.** $\frac{9}{4}$ or $\left\{\frac{9}{4}\right\}$ **39.** -6 or $\{-6\}$ **41.** -3 or $\{-3\}$ **43.** -3 or $\{-3\}$ **45.** 4 or {4}   **47.** $-\frac{3}{2}$ or $\left\{-\frac{3}{2}\right\}$

49. 2 or {2}   **51.** -4 or $\{-4\}$ **53.** -6 or $\{-6\}$ **55.** $x = \square \cdot \triangle$ **57.** $-\triangle = x$ **59.** $6x = 10; \frac{5}{3}$ **61.** $\frac{x}{-9} = 5; -45$

63. $4x - 8 = 56; 16$ **65.** $-3x + 15 = -6; 7$ **67.** 10 sec   **69.** 1502.2 mph   **71. a.** overestimates by \$655   **b.** 2011

77. does not make sense   **79.** does not make sense   **81.** false   **83.** true   **85.** Answers will vary; example: $\frac{5}{4}x = -20$.

87. 6.5 or {6.5}   **88.** 100   **89.** -100 **90.** 3   **91.** $-3x + 7$ **92.** yes   **93.** $2x - 78$

Section 2.3

Check Point Exercises

1. 6 or {6}   **2.** 2 or {2}   **3.** 5 or {5}   **4.** -2 or $\{-2\}$ **5.** no solution or $\varnothing$ **6.** all real numbers or $\{x \mid x \text{ is a real number}\}$
7. 3.7; shown as the point whose corresponding value on the vertical axis is 10 and whose value on the horizontal axis is 3.7

Exercise Set 2.3

1. 3 or {3}   **3.** -1 or $\{-1\}$ **5.** 4 or {4}   **7.** 4 or {4}   **9.** $\frac{7}{2}$ or $\left\{\frac{7}{2}\right\}$ **11.** -3 or $\{-3\}$ **13.** 6 or {6}   **15.** 8 or {8}   **17.** 4 or {4}

19. 1 or {1}   **21.** -4 or $\{-4\}$ **23.** 5 or {5}   **25.** 6 or {6}   **27.** 1 or {1}   **29.** -57 or $\{-57\}$ **31.** -10 or $\{-10\}$ **33.** 18 or {18}

35. $\frac{7}{4}$ or $\left\{\frac{7}{4}\right\}$ **37.** 1 or {1}   **39.** 24 or {24}   **41.** -6 or $\{-6\}$ **43.** 20 or {20}   **45.** -7 or $\{-7\}$ **47.** no solution or $\varnothing$

49. all real numbers or $\{x \mid x \text{ is a real number}\}$ **51.** $\frac{2}{3}$ or $\left\{\frac{2}{3}\right\}$ **53.** all real numbers or $\{x \mid x \text{ is a real number}\}$ **55.** no solution or $\varnothing$

57. 0 or {0}   **59.** no solution or ∅   **61.** 0 or {0}   **63.** $\frac{4}{3}$ or $\left\{\frac{4}{3}\right\}$ **65.** $x = \square\$ - \square\triangle$ **67.** 240   **69.** $\frac{x}{5} + \frac{x}{3} = 16; 30$

71. $\frac{3x}{4} - 3 = \frac{x}{2}; 12$ **73.** 85 mph   **75.** 142 lb; 13 lb   **77.** 409.2 ft   **83.** makes sense   **85.** does not make sense

87. false   **89.** true   **91.** Yes, because the skeleton's height is about 5 ft 1 in.   **93.** $\frac{10}{3}$ or $\left\{\frac{10}{3}\right\}$ **95.** 6.5 or {6.5}   **96.** <

97. <   **98.** −10   **99. a.** $T - D = pm$ **b.** $\frac{T - D}{p} = m$ **100.** 16 or {16}   **101.** 0.05 or {0.05}

Section 2.4

Check Point Exercises

1. $l = \frac{A}{w}$ **2.** $l = \frac{P - 2w}{2}$ **3.** $m = \frac{T - D}{p}$ **4.** $x = 15 + 12y$ **5. a.** 0.67   **b.** 2.5   **6.** 2.3%   **7.** 4.5   **8.** 15   **9.** 36%

10. 35%   **11. a.** $1152   **b.** 4% decrease

Exercise Set 2.4

1. $r = \frac{d}{t}$ **3.** $P = \frac{I}{rt}$ **5.** $r = \frac{C}{2\pi}$ **7.** $m = \frac{E}{c^2}$ **9.** $m = \frac{y - b}{x}$ **11.** $D = T - pm$ **13.** $b = \frac{2A}{h}$ **15.** $n = 5M$ **17.** $c = 4F - 160$

19. $a = 2A - b$ **21.** $r = \frac{S - P}{Pt}$ **23.** $b = \frac{2A}{h} - a$ **25.** $x = \frac{C - By}{A}$ **27.** 0.27   **29.** 0.634   **31.** 1.7   **33.** 0.03   **35.** 0.005

37. 89%   **39.** 0.2%   **41.** 478%   **43.** 10,000%   **45.** 6   **47.** 7.2   **49.** 5   **51.** 170   **53.** 20%   **55.** 12%   **57.** 60%   **59.** 75%

61. $x = \frac{y}{a + b}$ **63.** $x = \frac{y - 5}{a - b}$ **65.** $x = \frac{y}{c + d}$ **67.** $x = \frac{y + C}{A + B}$ **69. a.** $z = 3A - x - y$ **b.** 96%   **71. a.** $t = \frac{d}{r}$ **b.** 2.5 hr

73. 408   **75.** 500 billion pounds   **77. a.** 27%   **b.** 46%   **c.** 117%   **79.** 12.5%   **81.** $9   **83. a.** $1008   **b.** $17,808

85. a. $103.20   **b.** $756.80   **87.** 15%   **89.** no; 2% loss   **97.** does not make sense   **99.** does not make sense   **101.** false

103. true   **105.** 12 or {12}   **106.** 20 or {20}   **107.** 0.7x   **108.** $\frac{13}{x} - 7x$ **109.** $8(x + 14)$ **110.** $9(x - 5)$

Mid-Chapter Check Point Exercises

1. 16 or {16}   **2.** −3 or {−3}   **3.** $C = \frac{825H}{E}$ **4.** 8.4   **5.** 30 or {30}   **6.** $-\frac{8}{9}$ or $\left\{-\frac{8}{9}\right\}$ **7.** $r = \frac{S}{2\pi h}$ **8.** 40

9. 20 or {20}   **10.** −1 or {−1}   **11.** $x = \frac{By + C}{A}$ **12.** no solution or ∅   **13.** 40 or {40}   **14.** 12.5%   **15.** $-\frac{6}{5}$ or $\left\{-\frac{6}{5}\right\}$

16. 25%   **17.** all real numbers or {$x \mid x$ is a real number}   **18. a.** underestimates by 0.7%   **b.** 2014

Section 2.5

Check Point Exercises

1. 12   **2.** lawyers: $98,930; teachers: $44,040   **3.** pages 72 and 73   **4.** 32; 4 mi   **5.** 40 ft wide and 120 ft long   **6.** $940

Exercise Set 2.5

1. $x + 60 = 410; 350$ **3.** $x - 23 = 214; 237$ **5.** $7x = 126; 18$ **7.** $\frac{x}{19} = 5; 95$ **9.** $4 + 2x = 56; 26$ **11.** $5x - 7 = 178; 37$

13. $x + 5 = 2x; 5$ **15.** $2(x + 4) = 36; 14$ **17.** $9x = 3x + 30; 5$ **19.** $\frac{3x}{5} + 4 = 34; 50$ **21.** radio: 974 hr; TV: 1555 hr

23. carpenters: $35,580; computer programmers: $63,420   **25.** pages 314 and 315   **27.** Ruth: 59; Maris: 61   **29.** 800 mi

31. 10 years after 2004; 2014   **33.** 50 yd wide and 200 yd long   **35.** 160 ft wide and 360 ft long   **37.** 12 ft long and 4 ft high   **39.** $400

41. $46,500   **43.** $22,500   **45.** 11 hr   **51.** does not make sense   **53.** makes sense   **55.** false   **57.** true   **59.** 5 ft 7 in.

61. The uncle is 60 years old, and the woman is 20 years old.   **63.** −20 or {20}   **64.** 0 or {0}   **65.** $w = \frac{3V}{lh}$ **66.** yes   **67.** yes

68. 2 or {2}

Section 2.6

Check Point Exercises

1. a.   **b.**   **c.**

2. a. $[0, \infty)$;   **b.** $(-\infty, 5)$;

3. $(-\infty, 3)$ or $\{x | x < 3\}$

4. $[-2, \infty)$ or $\{x | x \geq -2\}$

5. a. $(-\infty, 8)$ or $\{x | x < 8\}$

b. $(-3, \infty)$ or $\{x | x > -3\}$

6. $[4, \infty)$ or $\{y | y \geq 4\}$

7. $[1, \infty)$ or $\{x | x \geq 1\}$

8. $[1, \infty)$ or $\{x | x \geq 1\}$ **9.** no solution or $\varnothing$ **10.** $(-\infty, \infty)$ or $\{x | x \text{ is a real number}\}$ **11.** at least 83%

Exercise Set 2.6

1.

3.

5.

7.

9.

11.

13. $(-\infty, 3]$

15. $\left(\dfrac{5}{2}, \infty\right)$

17. $(-\infty, 0]$

19. $(-\infty, 4)$

21. $(7, \infty)$ or $\{x | x > 7\}$

23. $(-\infty, 6]$ or $\{x | x \leq 6\}$

25. $(-\infty, 2)$ or $\{y | y < 2\}$

27. $(-\infty, 3]$ or $\{x | x \leq 3\}$

29. $(-\infty, 16)$ or $\{x | x < 16\}$

31. $(4, \infty)$ or $\{x | x > 4\}$

33. $\left(\dfrac{7}{6}, \infty\right)$ or $\left\{x \,\middle|\, x > \dfrac{7}{6}\right\}$

35. $\left(-\infty, -\dfrac{3}{8}\right]$ or $\left\{y \,\middle|\, y \leq -\dfrac{3}{8}\right\}$

37. $(0, \infty)$ or $\{y | y > 0\}$

39. $(-\infty, 8)$ or $\{x | x < 8\}$

41. $(-6, \infty)$ or $\{x | x > -6\}$

43. $(-\infty, 5)$ or $\{x | x < 5\}$

45. $[-7, \infty)$ or $\{x | x \geq -7\}$

47. $(-5, \infty)$ or $\{x | x > -5\}$

49. $(-\infty, -5]$ or $\{x | x \leq -5\}$

51. $(-\infty, 3)$ or $\{x | x < 3\}$

53. $\left[-\dfrac{1}{8}, \infty\right)$ or $\left\{y \,\middle|\, y \geq -\dfrac{1}{8}\right\}$

55. $(-4, \infty)$ or $\{x | x > -4\}$

57. $(5, \infty)$ or $\{x | x > 5\}$

59. $(-\infty, 5)$ or $\{x | x < 5\}$

61. $[-2, \infty)$ or $\{x | x \geq -2\}$

63. $(-3, \infty)$ or $\{x | x > -3\}$

65. $[4, \infty)$ or $\{x | x \geq 4\}$

67. $\left(\dfrac{11}{3}, \infty\right)$ or $\left\{x \,\middle|\, x > \dfrac{11}{3}\right\}$

69. $(2, \infty)$ or $\{y | y > 2\}$

71. $(-\infty, 2)$ or $\{y | y < 2\}$

73. $(-\infty, 3)$ or $\{x | x < 3\}$

75. $\left(\dfrac{5}{3}, \infty\right)$ or $\left\{x \,\middle|\, x > \dfrac{5}{3}\right\}$

77. $[9, \infty)$ or $\{x | x \geq 9\}$

79. $(-\infty, -6)$ or $\{x | x < -6\}$

81. no solution or $\varnothing$ **83.** $(-\infty, \infty)$ or $\{x|x$ is a real number$\}$ **85.** no solution or $\varnothing$ **87.** $(-\infty, \infty)$ or $\{x|x$ is a real number$\}$

89. $(-\infty, 0]$ or $\{x|x \le 0\}$ **91.** $x > \dfrac{b-a}{3}$ **93.** $\dfrac{y-b}{m} \ge x$ **95.** x is between -2 and 2; $|x| < 2$ **97.** x is greater than 2 or less than -2; $|x| > 2$

99. Denmark, Netherlands, Norway **101.** Japan, Mexico **103.** Netherlands, Norway, Canada, U.S. **105.** 20 yr; from 2009 onward

107. a. at least 96 **b.** if you get less than 66 on the final **109.** up to 1280 mi **111.** up to 29 bags of cement **117.** makes sense

119. makes sense **121.** false **123.** false **125.** more than 720 mi **127.** $(-\infty, 0.4)$ or $\{x|x < 0.4\}$ **129.** 20 **130.** length: 11 in.; width: 6 in.

131. 4 or $\{4\}$ **132.** 240 **133.** 5000 or $\{5000\}$ **134.** 4 or $\{4\}$

Review Exercises

1. 32 or $\{32\}$ **2.** -22 or $\{-22\}$ **3.** 12 or $\{12\}$ **4.** -22 or $\{-22\}$ **5.** 5 or $\{5\}$ **6.** 80 or $\{80\}$ **7.** -56 or $\{-56\}$ **8.** 11 or $\{11\}$

9. 4 or $\{4\}$ **10.** -15 or $\{-15\}$ **11.** -12 or $\{-12\}$ **12.** -25 or $\{-25\}$ **13.** 10 or $\{10\}$ **14.** 6 or $\{6\}$ **15.** -5 or $\{-5\}$

16. -10 or $\{-10\}$ **17.** 2 or $\{2\}$ **18.** 1 or $\{1\}$ **19. a.** underestimates by 8 million barrels **b.** 2015 **20.** -1 or $\{-1\}$ **21.** 12 or $\{12\}$

22. -13 or $\{-13\}$ **23.** -3 or $\{-3\}$ **24.** -10 or $\{-10\}$ **25.** 2 or $\{2\}$ **26.** 2 or $\{2\}$ **27.** no solution or $\varnothing$ **28.** all real numbers or

$\{x|x$ is a real number$\}$ **29.** 30 years old **30.** $r = \dfrac{I}{P}$ **31.** $h = \dfrac{3V}{B}$ **32.** $w = \dfrac{P-2l}{2}$ **33.** $B = 2A - C$ **34.** $m = \dfrac{T-D}{p}$

35. 0.65 **36.** 1.5 **37.** 0.03 **38.** 72% **39.** 0.35% **40.** 9.6 **41.** 200 **42.** 48% **43.** 100% **44.** 40% **45.** 12.5%

46. no; 1% **47. a.** $h = 7r$ **b.** 5 ft 3 in. **48.** 350 gallons **49.** 10 **50.** cats: \$188; dogs: \$217 **51.** pages 46 and 47

52. females: 49%; males: 51% **53.** 9 yr; 2012 **54.** 18 checks **55.** length: 150 yd; width: 50 yd **56.** \$240

57.

58.

59. $\left[\dfrac{3}{2}, \infty\right)$

60. $(-\infty, 0)$

61. $(-\infty, 4)$ or $\{x|x < 4\}$

62. $(-8, \infty)$ or $\{x|x > -8\}$

63. $[-3, \infty)$ or $\{x|x \ge -3\}$

64. $(6, \infty)$ or $\{x|x > 6\}$

65. $[4, \infty)$ or $\{x|x \ge 4\}$

66. $(-\infty, 2]$ or $\{x|x \le 2\}$

67. $(-\infty, \infty)$ or $\{x|x$ is a real number$\}$ **68.** no solution or $\varnothing$ **69.** at least 64 **70.** no more than 99 min

Chapter Test

1. $\dfrac{9}{2}$ or $\left\{\dfrac{9}{2}\right\}$ **2.** -5 or $\{-5\}$ **3.** $-\dfrac{4}{3}$ or $\left\{-\dfrac{4}{3}\right\}$ **4.** 2 or $\{2\}$ **5.** -20 or $\{-20\}$ **6.** $-\dfrac{5}{3}$ or $\left\{-\dfrac{5}{3}\right\}$ **7.** 60 yr; 2020 **8.** $h = \dfrac{V}{\pi r^2}$

9. $w = \dfrac{P-2l}{2}$ **10.** 8.4 **11.** 150 **12.** 5% **13.** 63 **14.** \$50,000–\$74,999: 5.8 million; \$75,000 or more: 8.4 million

15. 600 min **16.** length: 150 yd; width: 75 yd **17.** \$35

18. $(-2, \infty)$

19. $(-\infty, 3]$

20. $(-\infty, -6)$ or $\{x|x < -6\}$

21. $(-\infty, -3]$ or $\{x|x \le -3\}$

22. $(7, \infty)$ or $\{x|x > 7\}$ **23.** at least 92 **24.** widths greater than 8 in.

Cumulative Review Exercises

1. -4 **2.** -2 **3.** -128 **4.** $-103 - 20x$ **5.** $-4, -\dfrac{1}{3}, 0, \sqrt{4}, 1063$ **6.** $\dfrac{5}{x} - (x+2)$ **7.** $<$ **8.** $24x - 6 - 30y$

9. underestimates by 0.3% **10.** 2018 **11.** 1 or $\{1\}$ **12.** -15 or $\{-15\}$ **13.** $A = \dfrac{3V}{h}$ **14.** 160 **15.** length: 130 yd; width: 70 yd

16. 75,000 gal

17. $\left(-\infty, \dfrac{1}{2}\right]$

18. $(-\infty, -3)$ or $\{x|x < -3\}$

19. $[-6, \infty)$ or $\{x|x \ge -6\}$ **20.** more than \$47,500

CHAPTER 3

Section 3.1

Check Point Exercises

1. $150   **2.** $15,000 at 9% and $10,000 at 12%   **3.** 27 ml of acid   **4.** 30 ml of 10% and 20 ml of 60% acid solution   **5.** 9 mph   **6.** 4 hr

Exercise Set 3.1

1. $160   **3.** $0.08(20,000 - x)$; $8000 at 7% and $12,000 at 8%   **5.** $120,000 - x$; $0.08x$; $0.18(120,000 - x)$; $116,000 at 8% and $4000 at 18%
7. $3600 at 6% and $2400 at 9%   **9.** $31,250 at 15% and $18,750 at 7%   **11.** 6 ml   **13.** $0.10(50 - x)$; $0.08(50)$; 20 l of 5% and 30 l of 10%
15. 6 oz   **17.** The north had 600 students, and the south had 400 students.   **19.** $10x$; $12x$; 3 hr   **21.** $5(x + 5)$; $5x$; faster truck: 62.5 mph; slower
truck: 57.5 mph   **23.** 2.4 hr (or 2 hr and 24 min)   **25.** 2 hr at 50 mph and 3 hr at 40 mph   **27.** 12 lb of grade A and 8 lb of grade B   **29.** 12 lb
31. 20 dimes; 12 quarters   **33.** $10 bills: 5; $5 bills: 30   **35.** 225 adult tickets; 80 children tickets   **45.** makes sense   **47.** does not make sense
49. 25 ml   **51.** -71 **52.** $-100, 0, \sqrt{16}$ **53.** -3 or $\{-3\}$ **54.** $\frac{4}{25}$ **55.** yes   **56.** $\frac{63}{x} = \frac{7}{5}$

Section 3.2

Check Point Exercises

1. a. 30:7, or 30 to 7   **b.** 7:50, or 7 to 50   **2. a.** 15 or $\{15\}$ **b.** -27 or $\{-27\}$ **3.** $5880   **4.** 720 deer

Exercise Set 3.2

1. $\frac{1}{2}$ **3.** $\frac{12}{5}$ **5.** $\frac{3}{4}$ **7.** $\frac{2}{1}$; 2:1   **9.** $\frac{1}{3}$; 1:3   **11.** 12 or $\{12\}$ **13.** 7.5 or $\{7.5\}$ **15.** -9 or $\{-9\}$ **17.** 34 or $\{34\}$

19. -49 or $\{-49\}$ **21.** $-\frac{149}{6}$ or $\left\{-\frac{149}{6}\right\}$ **23.** $x = \frac{ab}{c}$ **25.** $x = \frac{ad + bd}{c}$ **27.** $x = \frac{ab}{c}$ **29.** $\frac{5}{2}$; 5:2   **31.** $\frac{3}{4}$; 3:4

33. $1800   **35.** 20,489 fur seal pups   **37.** 0.9 billion   **39.** 154.1 in.   **41.** $\frac{62}{123}$ (or 62:123); $\frac{10}{9}$ (or 10:9)   **49.** makes sense
51. does not make sense   **53.** true   **55.** false   **57.** 5 consecutive times at bat   **59.** approximately 66.34 mi
60. $[-2, \infty)$ or $\{x|x \geq -2\}$ ⊢┼┼┼┼[┼┼┼┼→ $-5\ -4\ -3\ -2\ -1\ \ 0\ \ 1\ \ 2\ \ 3\ \ 4\ \ 5$ **61.** 280   **62.** $19 - 8x$ **63.** 5   **64.** 91   **65.** 64 or $\{64\}$

Mid-Chapter Check Point Exercises

1. 8.05 days   **2.** 3 l   **3.** $-\frac{3}{4}$ or $\left\{-\frac{3}{4}\right\}$ **4.** $11,500 at 8% and $13,500 at 9%   **5.** $\frac{5}{12}$; 5:12, or 5 to 12   **6.** 375 mph and 475 mph
7. 800 elk   **8.** 8 qt   **9.** -3 or $\{-3\}$ **10.** $2500 at 4% and $1500 at 3%   **11.** 4 hours by plane and 2 hours by car   **12.** $5\frac{1}{3}$ in.

Section 3.3

Check Point Exercises

1. 12 ft   **2.** 400π ft$^2 \approx 1256$ ft^2 or 1257 ft^2; 40π ft ≈ 126 ft   **3.** large pizza   **4.** 2 times   **5.** no; About 32 more cubic inches are needed.
6. 120°, 40°, 20°   **7.** 60°

Exercise Set 3.3

1. 18 m; 18 m^2 **3.** 56 in.^{2} **5.** 91 m^2 **7.** 50 ft   **9.** 8 ft   **11.** 50 cm   **13.** 16π cm$^2 \approx 50$ cm^2; 8π cm ≈ 25 cm
15. 36π yd$^2 \approx 113$ yd^2; 12π yd ≈ 38 yd   **17.** 7 in.; 14 in.   **19.** 36 in.^{3} **21.** 150π cm$^3 \approx 471$ cm^3 **23.** 972π cm$^3 \approx 3054$ cm^3
25. 48π m$^3 \approx 151$ m^3 **27.** $h = \frac{V}{\pi r^2}$ **29.** 9 times   **31.** $x = 50$; $x + 30 = 80$; 50°, 50°, 80°   **33.** $4x = 76$; $3x + 4 = 61$; $2x + 5 = 43$; 76°, 61°, 43°
35. 40°, 80°, 60°   **37.** 32°   **39.** 2°   **41.** 48°   **43.** 90°   **45.** 75°   **47.** 135°   **49.** 50°   **51.** 72 m^2 **53.** 70.5 cm^2 **55.** 448 in.^{3}
57. $698.18   **59.** large pizza   **61.** $2262   **63.** approx 19.7 ft   **65.** 21,000 yd^3 **67.** the can with diameter of 6 in. and height of 5 in.
69. Yes, the water tank is a little over one cubic foot too small.   **79.** does not make sense   **81.** does not make sense   **83.** true   **85.** false
87. 2.25 times   **89.** Volume increases 8 times.   **91.** 35°   **92.** $s = \frac{P - b}{2}$ **93.** 8 or $\{8\}$ **94.** 0   **95.** yes   **96.** no   **97.** -3

Review Exercises

1. $3000 at 8% and $7000 at 10%   **2.** $8000 at 10% and $14,000 at 12%   **3.** 4 gal of 75% and 6 gal of 50%   **4.** No, it is not possible.
5. North had 60 students, and south had 90 students.   **6.** 3 hr   **7.** 30 mph and 40 mph   **8.** $5 bills: 6; $10 bills: 9   **9.** exactly enough information
10. not enough information   **11.** exactly enough information   **12.** $\frac{2}{5}$; 2:5, or 2 to 5   **13.** $\frac{21}{29}$; 21:29, or 21 to 29   **14.** $\frac{17}{8}$; or 17:8, or 17 to 8
15. 5 or $\{5\}$ **16.** -65 or $\{-65\}$ **17.** $\frac{2}{5}$ or $\left\{\frac{2}{5}\right\}$ **18.** 3 or $\{3\}$ **19.** 324 teachers   **20.** 287 trout   **21.** 19,200 skates   **22.** 32.5 ft^2

23. 50 cm²   **24.** 135 yd²   **25.** 7608 m²   **26.** 20π m $\approx$ 63 m; 100π m² $\approx$ 314 m²   **27.** 6 ft   **28.** 156 ft²   **29.** \$1890   **30.** medium pizza
31. 60 cm³   **32.** 128π yd³ $\approx$ 402 yd³   **33.** 288π m³ $\approx$ 905 m³   **34.** 4800 m³   **35.** 16 fish   **36.** $x = 30, 3x = 90, 2x = 60$; 30°, 60°, 90°
37. 85°, 35°, 60°   **38.** 33°   **39.** 105°   **40.** 57.5°   **41.** 45° and 135°

Chapter Test

1. \$4000 at 9% and \$2000 at 6%   **2.** 40 ml of 50% and 60 ml of 80%   **3.** 5 hr   **4.** $\frac{4}{7}$; 4:7, or 4 to 7   **5.** $-\frac{15}{2}$ or $\left\{-\frac{15}{2}\right\}$
6. 6000 tule elk   **7.** 137.5 lb/in.²   **8.** 517 m²   **9.** 525 in.²   **10.** 66 ft²   **11.** 18 in.³   **12.** 175π cm³ $\approx$ 550 cm³   **13.** \$650   **14.** 14 ft
15. 126°, 42°, 12°   **16.** 53°

Cumulative Review Exercises

1. $-\frac{27}{68}$ **2.** $-1 - 6x$ **3.** $-3, 0, \sqrt{9}$ **4.** 39   **5.** $<$ **6.** 6°F   **7.** 8 or {8}   **8.** $-\frac{1}{2}$ or $\left\{-\frac{1}{2}\right\}$ **9.** $m = 2A - n$
10. 633   **11.** 50   **12.** 240   **13.** width: 53 m; length: 120 m   **14.** 200 lb

15. $(-\infty, 0]$ or $\{x \mid x \le 0\}$ **16.** $(-\infty, 1)$ or $\{x \mid x < 1\}$

17. $1\frac{1}{2}$ hr   **18.** $-\frac{21}{4}$ or $\left\{-\frac{21}{4}\right\}$ **19.** Reagan: 69 years old; Buchanan: 65 years old
20. No, the percent increase is 225%, which is not more than 300%.

CHAPTER 4

Section 4.1

Check Point Exercises

1.   **2.** $E(-4, -2), F(-2, 0), G(6, 0)$ **3. a.** solution   **b.** not a solution   **4.** $(-2, -4), (-1, -1), (0, 2), (1, 5)$, and $(2, 8)$

5.   **6.**   **7.**

8. a. $(0, 1), (5, 8), (10, 15)$, and $(15, 22)$ **b.**   **c.** approximately 29%, although answers may vary by $\pm 1\%$ **d.** 29%

Exercise Set 4.1

1. I   **3.** II   **5.** III   **7.** IV

9–23.   **25.** $(5, 2)$ **27.** $(-6, 5)$ **29.** $(-2, -3)$ **31.** $(5, -3)$ **33.** I and II   **35.** I and III

37. $(2, 3)$ and $(3, 2)$ are not solutions; $(-4, -12)$ is a solution.   **39.** $(-5, -20)$ is not a solution; $(0, 0)$ and $(9, -36)$ are solutions.
41. $(2, -2)$ is not a solution; $(0, 6)$ and $(-3, 0)$ are solutions.   **43.** $(0, 5)$ is not a solution; $(-5, 6)$ and $(10, -3)$ are solutions.

45. $\left(1, \dfrac{1}{3}\right)$ is not a solution; $(0, 0)$ and $\left(2, -\dfrac{2}{3}\right)$ are solutions. **47.** $(3, 4)$ and $(0, -4)$ are not solutions; $(4, 7)$ is a solution.

49.

x	(x, y)
-2	$(-2, -24)$
-1	$(-1, -12)$
0	$(0, 0)$
1	$(1, 12)$
2	$(2, 24)$

51.

x	(x, y)
-2	$(-2, 20)$
-1	$(-1, 10)$
0	$(0, 0)$
1	$(1, -10)$
2	$(2, -20)$

53.

x	(x, y)
-2	$(-2, -21)$
-1	$(-1, -13)$
0	$(0, -5)$
1	$(1, 3)$
2	$(2, 11)$

55.

x	(x, y)
-2	$(-2, 13)$
-1	$(-1, 10)$
0	$(0, 7)$
1	$(1, 4)$
2	$(2, 1)$

57.

59.

61.

63.

65.

67.

69.

71.

73.

75.

77.

79.

81.

83.

85. a. $8x + 6y = 14.50$ **b.** \$1.25 **87.** $(2, 7)$; The football is 7 ft above ground when it is 2 yd from the quarterback. **89.** $(6, 9.25)$

91. 12 ft; 15 yd

93. a. $(0, 28), (10, 41), (20, 54), (30, 67),$ and $(40, 80)$

b.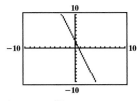

c. approximately 99 pounds per person, although answers may vary by ± 1 **d.** 99.5 pounds per person

103. makes sense **105.** does not make sense **107.** false **109.** false

111. a. 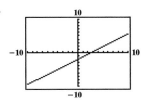 **b.** Change the sign of each x-coordinate. **c.** Change the sign of each y-coordinate.

d. Change the sign of both coordinates.

113.

Answers will vary.

115.

Answers will vary.

117. 2 or $\{2\}$ **118.** 1 **119.** $h = \dfrac{3V}{A}$ **120.** $(8, 0)$ **121.** $(0, -6)$ **122.** $(0, 0)$

Section 4.2

Check Point Exercises

1. a. x-intercept: -3; y-intercept: 5 **b.** y-intercept: 4; no x-intercept **c.** x-intercept: 0; y-intercept: 0 **2.** 3 **3.** -4

4.

5.

6.

7.

8.

Exercise Set 4.2

1. a. 3   **b.** 4   **3. a.** −4   **b.** −2   **5. a.** 0   **b.** 0   **7. a.** no *x*-intercept   **b.** −2   **9.** *x*-intercept: 10; *y*-intercept: 4

11. *x*-intercept: $\frac{15}{2}$, or $7\frac{1}{2}$; *y*-intercept: −5   **13.** *x*-intercept: 8; *y*-intercept: $-\frac{8}{3}$, or $-2\frac{2}{3}$ **15.** *x*-intercept: 0; *y*-intercept: 0

17. *x*-intercept: $-\frac{11}{2}$, or $-5\frac{1}{2}$; *y*-intercept: $\frac{11}{3}$, or $3\frac{2}{3}$

19.

21.

23.

25.

27.

29.

31.

33.   **35.**

37.

39.   **41.** $y = 3$ **43.** $x = -3$ **45.** $y = 0$ **47.**   **49.**

51.   **53.**   **55.**   **57.**   **59.**

61.   **63.** Exercise 4   **65.** Exercise 7   **67.** Exercise 1   **69.**

71. from 3 to 12 sec   **73.** 45; The eagle was 45 m above the ground when the observation started.   **75.** 12, 13, 14, 15, 16; The eagle is on the ground at this time.   **77. a.** 9; After 9 years, the car is worth nothing.   **b.** 45,000; The new car is worth $45,000.

c. 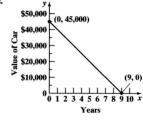   **d.** $20,000; Estimates will vary.   **87.** makes sense   **89.** makes sense   **91.** 2; 5

95.

97.

x-intercept: 3; *y*-intercept: −9   *x*-intercept: −10; *y*-intercept: 20

98. 13.4   **99.** $4x + 5$ **100.** $[-5, \infty)$ or $\{x \mid x \geq -5\}$ **101.** 2   **102.** −1   **103.** 0

Section 4.3

Check Point Exercises

1. a. 6　**b.** $-\dfrac{7}{5}$　**2. a.** 0　**b.** undefined　**3.** Both slopes equal 2, so the lines are parallel.　**4.** Product of slopes is $-\dfrac{1}{2}(2) = -1$,

so the lines are perpendicular.　**5.** $m \approx 0.25$; The number of men living alone increased by 0.25 million per year. The rate of change is

0.25 million men per year.

Exercise Set 4.3

1. $\dfrac{3}{4}$; rises　**3.** $\dfrac{1}{4}$; rises　**5.** 0; horizontal　**7.** -5; falls　**9.** undefined; vertical　**11.** $\dfrac{1}{2}$　**13.** $-\dfrac{1}{3}$　**15.** $-\dfrac{1}{2}$　**17.** $-\dfrac{2}{3}$

19. 0　**21.** undefined　**23.** parallel　**25.** not parallel　**27.** perpendicular　**29.** not perpendicular　**31.** parallel

33. neither　**35.** perpendicular

37.

39. Slopes of corresponding opposite sides are equal: $-\dfrac{2}{5}$ and $\dfrac{4}{3}$.　**41.** 2

43. 4　**45. a.** 0.16　**b.** 0.16; 0.16; minute of brisk walking　**47.** $\dfrac{1}{3}$

49. 8.3%　**57.** does not make sense　**59.** makes sense　**61.** false　**63.** false

65. m_1, m_3, m_2, m_4　**67.** $m = 2$　**69.** $m = -\dfrac{1}{2}$　**71.** The line's slope is the coefficient of x.

72. 12 in. and 24 in.　**73.** -42　**74.** $(-\infty, 4]$ or $\{x \mid x \le 4\}$　

75. $(1, 1)$　**76.** $(3, -1)$　**77.** $y = -\dfrac{2}{5}x$

Section 4.4

Check Point Exercises

1. a. 5; -3　**b.** $\dfrac{2}{3}$; 4　**c.** -7; 6　**2.** 　**3.** 　**4.** 　**5. a.** $y = 0.13x + 3.9$　**b.** 9.1%

Exercise Set 4.4

1. 3; 2　**3.** 3; -5　**5.** $-\dfrac{1}{2}$; 5　**7.** 7; 0　**9.** 0; 10　**11.** -1; 4　**13.** $y = 5x + 7$; 5; 7　**15.** $y = -x + 6$; -1; 6　**17.** $y = -6x$; -6; 0

19. $y = 2x$; 2; 0　**21.** $y = -\dfrac{2}{7}x$; $-\dfrac{2}{7}$; 0　**23.** $y = -\dfrac{3}{2}x + \dfrac{3}{2}$; $-\dfrac{3}{2}$; $\dfrac{3}{2}$　**25.** $y = \dfrac{3}{4}x - 3$; $\dfrac{3}{4}$; -3

27. 　**29.** 　**31.** 　**33.** 　**35.**

37.

39. a. $y = -3x$　**b.** -3; 0　**c.**

41. a. $y = \dfrac{4}{3}x$　**b.** $\dfrac{4}{3}$; 0　**c.**

43. a. $y = -2x + 3$　**b.** -2; 3　**c.**

45. a. $y = -\dfrac{7}{2}x + 7$　**b.** $-\dfrac{7}{2}$; 7　**c.**

47.

parallel; $m = 3$

49.

not parallel or perpendicular; $m = -3$ and $m = 3$

51. 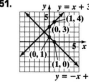 ; perpendicular, $m = 1$ and $m = -1$, and $1(-1) = -1$

53.

parallel; $m = \dfrac{1}{2}$

55. ; perpendicular, $m = 2$ and $m = -\dfrac{1}{2}$, and $2\left(-\dfrac{1}{2}\right) = -1$ **57.** $y = -3x + 5$

59. $y = -\dfrac{1}{5}x + 6$ **61.** $y = -x + 2$ **63.** $y = x$ **65. a.** $y = -0.36x + 68$ **b.** 32%   **71.** does not make sense

73. does not make sense   **75.** false   **77.** false   **79.** 8 or {8}   **80.** 0   **81.** 56   **82.** $y = 4x + 7$ **83.** $y = -\dfrac{3}{2}x + 3$

84. $y = 0.265x + 27.35$

Mid-Chapter Check Point Exercises

1. a. 4   **b.** 2   **c.** $-\dfrac{1}{2}$ **2. a.** -5 **b.** no y-intercept   **c.** undefined slope   **3. a.** 0   **b.** 0   **c.** $\dfrac{3}{5}$

4. $y = -2x$ **5.** $y = -2$ **6.** $x + y = -2$ **7.** $y = \dfrac{1}{3}x - 2$ **8.** $x = 3.5$

9. $4x - 2y = 8$ **10.** $y = 3x + 2$ **11.** $3x + y = 0$ **12.** $y = -x + 4$ **13.** $y = x - 4$

14. $5y = -3x$ **15.** $5y = 20$ **16.** $\dfrac{5}{2}; -5$ **17.** perpendicular

18. neither   **19.** parallel

20. a. $y = 131x + 1098$ **b.** The average annual household gasoline bill increased at an average of \$131 per year.   **c.** \$3194

Section 4.5

Check Point Exercises

1. $y + 5 = 6(x - 2); y = 6x - 17$ **2. a.** $y + 1 = -5(x + 2)$ or $y + 6 = -5(x + 1)$ **b.** $y = -5x - 11$ **3.** $y = 0.28x + 27.2; 41.2$

Exercise Set 4.5

1. $y - 5 = 3(x - 2); y = 3x - 1$ **3.** $y - 6 = 5(x + 2); y = 5x + 16$ **5.** $y + 2 = -8(x + 3); y = -8x - 26$

7. $y - 0 = -12(x + 8); y = -12x - 96$ **9.** $y + 2 = -1\left(x + \dfrac{1}{2}\right); y = -x - \dfrac{5}{2}$ **11.** $y - 0 = \dfrac{1}{2}(x - 0); y = \dfrac{1}{2}x$

13. $y + 2 = -\dfrac{2}{3}(x - 6); y = -\dfrac{2}{3}x + 2$ **15.** $y - 2 = 2(x - 1)$ or $y - 10 = 2(x - 5); y = 2x$

17. $y - 0 = 1(x + 3)$ or $y - 3 = 1(x - 0); y = x + 3$ **19.** $y + 1 = 1(x + 3)$ or $y - 4 = 1(x - 2); y = x + 2$

21. $y + 1 = \dfrac{5}{7}(x + 4)$ or $y - 4 = \dfrac{5}{7}(x - 3); y = \dfrac{5}{7}x + \dfrac{13}{7}$ **23.** $y + 1 = 0(x + 3)$ or $y + 1 = 0(x - 4); y = -1$

25. $y - 4 = 1(x - 2)$ or $y - 0 = 1(x + 2); y = x + 2$ **27.** $y - 0 = 8\left(x + \dfrac{1}{2}\right)$ or $y - 4 = 8(x - 0); y = 8x + 4$ **29.** $y = 4x + 14$

31. $y = -3x - 8$ **33.** $y = -2x + 1$ **35.** $y = 3x - 2$ **37.** $y = -5x - 20$ **39. a.** $y = 0.68x + 61.2$ **b.** 85%

41. a & b.

b. $y = -0.2x + 17.8$ **c.** overestimates by 0.7%

47. does not make sense   **49.** makes sense   **51.** false

53. false   **57.** at most 12 sheets of paper   **58.** 1, $\sqrt{4}$

59.

60. yes   **61.** no   **62.** no

Section 4.6

Check Point Exercises

1. a. solution   **b.** not a solution   **2.**   **3.**   **4. a.**   **b.**

5. a. $B(60, 20)$; A region that has an average annual temperature of 60°F and an average annual precipitation of 20 inches is a grassland.
b. $5(60) - 7(20) \geq 70, 160 \geq 70$ true; $3(60) - 35(20) \leq -140, -520 \leq -140$ true

Exercise Set 4.6

1. no; yes; yes   **3.** yes; yes; no   **5.** yes; yes; no   **7.** yes; no; yes   **9.**   **11.**

13.   **15.**   **17.**   **19.**   **21.**

23.   **25.**   **27.**   **29.**   **31.**

33.   **35.**   **37.**   **39.**   **41.**

43.   **45. a.**   **b.** Answers will vary; (1000, 2000) is an example. The plane can carry 1000 bottles of water and 2000 medical kits.

47. a. $50x + 150y > 2000$ **b.**   **c.** Answers will vary; (20, 15) is an example. The elevator cannot carry 20 children and 15 adults.

49. a. 27.1   **b.** overweight   **59.** does not make sense   **61.** makes sense   **63.** false   **65.** true   **67.** $2x + y \leq 4$

69.

71.

72. $h = \dfrac{V}{lw}$ **73.** $-\dfrac{8}{15}$ **74.** 5

75. yes   **76.** no   **77.** $(-3, 4)$

Review Exercises

1. IV

2. IV

3. I

4. II

5. $A(5, 6)$; $B(-3, 0)$; $C(-5, 2)$; $D(-4, -2)$; $E(0, -5)$; $F(3, -1)$ **6.** $(-3, 3)$ is not a solution; $(0, 6)$ and $(1, 9)$ are solutions.

7. $(0, 4)$ and $(-1, 15)$ are not solutions; $(4, 0)$ is a solution.

8. a.

x	(x, y)
-2	$(-2, -7)$
-1	$(-1, -5)$
0	$(0, -3)$
1	$(1, -1)$
2	$(2, 1)$

b.

9. a.

x	(x, y)
-2	$(-2, 0)$
-1	$\left(-1, \frac{1}{2}\right)$
0	$(0, 1)$
1	$\left(1, \frac{3}{2}\right)$
2	$(2, 2)$

b.

10. a. -2 **b.** -4 **11. a.** no x-intercept   **b.** 2   **12. a.** 0   **b.** 0

13.

14.

15.

16.

17.

18.

19.

20.

21. a. 5:00 P.M.; $-4°$F   **b.** 8:00 P.M.; $16°$F   **c.** 4 and 6; At 4:00 P.M. and 6:00 P.M., the temperature was $0°$F.   **d.** 12; At noon, the temperature was $12°$F.   **e.** The temperature stayed the same, $12°$F.   **22.** $-\dfrac{1}{2}$; falls   **23.** 3; rises   **24.** 0; horizontal   **25.** undefined; vertical

26. $\dfrac{3}{5}$ **27.** undefined   **28.** $-\dfrac{1}{3}$ **29.** 0   **30.** neither   **31.** perpendicular   **32.** parallel   **33. a.** 1   **b.** 1%; 1%; year

34. 5; -7 **35.** -4; 6   **36.** 0; 3   **37.** $-\dfrac{2}{3}$; 2

38.

39.

40.

41.

42.

43.

Yes, they are parallel since both have slopes of $-\dfrac{1}{2}$ and different y-intercepts.

44. a. $y = 3.85x + 12.6$ **b.** 89.6%   **45.** $y - 7 = 6(x + 4); y = 6x + 31$ **46.** $y - 4 = 3(x - 3)$ or $y - 1 = 3(x - 2); y = 3x - 5$

47. a. $y = 0.08x + 2.1$ **b.** 8.1 billion   **48.** $(0, 0)$ and $(-3, 4)$ are not solutions; $(3, -6)$ and $(-2, -5)$ are solutions.

49. $x - 2y > 6$

50. $4x - 6y \le 12$

51. $y > 3x + 2$

52. $y \le \frac{1}{3}x - 1$

53. $y < -\frac{1}{2}x$

54. $x < 4$

55. $y \ge -2$

56. $x + 2y \le 0$

Chapter Test

1. $(-2, 1)$ is not a solution; $(0, -5)$ and $(4, 3)$ are solutions.

2.

x	(x, y)
-2	$(-2, -5)$
-1	$(-1, -2)$
0	$(0, 1)$
1	$(1, 4)$
2	$(2, 7)$

 $y = 3x + 1$

3. a. 2   **b.** -3 **4.** $4x - 2y = -8$ **5.** $y = 4$

6. 3; rises   **7.** undefined; vertical   **8.** $\dfrac{3}{2}$ **9.** perpendicular   **10.** parallel   **11.** $-1; 10$ **12.** $-2; 6$

13. $y = \frac{2}{3}x - 1$

14. $y = -2x + 3$

15. $y - 4 = -2(x + 1); y = -2x + 2$

16. $y - 1 = 3(x - 2)$ or $y + 8 = 3(x + 1); y = 3x - 5$

17. $3x - 2y < 6$

18. $y \ge 2x - 2$

19. $x > -1$

20. a. 182.5   **b.** $182.5 billion; $182.5 billion; year

Cumulative Review Exercises

1. 2   **2.** $4 - 6x$ **3.** $\sqrt{5}$ **4.** 19 or $\{19\}$ **5.** $\dfrac{5}{4}$ or $\left\{\dfrac{5}{4}\right\}$ **6.** $x = \dfrac{y - b}{m}$ **7.** 800   **8.** 40 mph

9. $(-\infty, -2]$ or $\{x | x \le -2\}$

10. $(-\infty, 0)$ or $\{x | x < 0\}$

11. 6 hr   **12.** 8 l of 40% and 4 l of 70%   **13.** 3.5 hr   **14.** $40°, 60°, 80°$ **15.** 16 or $\{16\}$

16. $2x - y = 4$

17. $x = -5$

18. $y = -4x + 3$

19. $3x - 2y < -6$

20. $y \ge -1$

CHAPTER 5

Section 5.1

Check Point Exercises

1. a. solution   **b.** not a solution   **2.** $(1, 4)$ or $\{(1, 4)\}$ **3.** $(3, 3)$ or $\{(3, 3)\}$ **4.** no solution or $\varnothing$

5. infinitely many solutions; $\{(x, y) | x + y = 3\}$ or $\{(x, y) | 2x + 2y = 6\}$ **6. a.** $(10, 20)$ or $\{(10, 20)\}$ **b.** If the bridge is used 10 times in a month, the total monthly cost without the coupon book is the same as the monthly cost with the coupon book, namely $20.

Exercise Set 5.1

1. solution   **3.** solution   **5.** not a solution   **7.** solution   **9.** not a solution   **11.** $(4, 2)$ or $\{(4, 2)\}$ **13.** $(-1, 2)$ or $\{(-1, 2)\}$
15. $(3, 0)$ or $\{(3, 0)\}$ **17.** $(1, 0)$ or $\{(1, 0)\}$ **19.** $(-1, 4)$ or $\{(-1, 4)\}$ **21.** $(2, 4)$ or $\{(2, 4)\}$ **23.** $(2, -1)$ or $\{(2, -1)\}$ **25.** no solution or $\varnothing$
27. $(-2, 6)$ or $\{(-2, 6)\}$ **29.** infinitely many solutions; $\{(x, y) \mid x - 2y = 4\}$ or $\{(x, y) \mid 2x - 4y = 8\}$ **31.** $(2, 3)$ or $\{(2, 3)\}$ **33.** no solution or $\varnothing$
35. infinitely many solutions; $\{(x, y) \mid x - y = 0\}$ or $\{(x, y) \mid y = x\}$ **37.** $(2, 4)$ or $\{(2, 4)\}$ **39.** no solution or $\varnothing$ **41.** no solution or $\varnothing$
43. $m = \frac{1}{2}$; $b = -3$ and -5; no solution   **45.** $m = -\frac{1}{2}$ and 3; $b = 4$; one solution   **47.** $m = 3$; $b = -6$; infinite number of solutions
49. $m = -3$; $b = 0$ and 1; no solution   **51. a.** 40; Both companies charge the same for 40 miles driven.   **b.** Answers will vary: 55.
c. 54; Both companies charge $54 for 40 miles driven.   **53. a.** $(5, 20)$ or $\{(5, 20)\}$ **b.** Nonmembers and members pay the same amount
per month for taking 5 classes, namely $20.   **63.** does not make sense   **65.** makes sense   **67.** false   **69.** false   **73.** $(1, 4)$ or $\{(1, 4)\}$
75. $(2, 0)$ or $\{(2, 0)\}$ **77.** $(0, -5)$ or $\{(0, -5)\}$ **79.** $(7, 3)$ or $\{(7, 3)\}$ **81.** -12 **82.** 6   **83.** 27   **84.** 3 or $\{3\}$ **85.** 4 or $\{4\}$
86. no solution or $\varnothing$

Section 5.2

Check Point Exercises

1. $(3, 2)$ or $\{(3, 2)\}$ **2.** $(1, -2)$ or $\{(1, -2)\}$ **3.** no solution or $\varnothing$ **4.** infinitely many solutions; $\{(x, y) \mid y = 3x - 4\}$ or $\{(x, y) \mid 9x - 3y = 12\}$
5. a. $(30, 900)$ or $\{(30, 900)\}$; equilibrium quantity: 30,000; equilibrium price: $900   **b.** $900; 30,000; 30,000

Exercise Set 5.2

1. $(1, 3)$ or $\{(1, 3)\}$ **3.** $(5, 1)$ or $\{(5, 1)\}$ **5.** $(2, 1)$ or $\{(2, 1)\}$ **7.** $(-1, 3)$ or $\{(-1, 3)\}$ **9.** $(4, 5)$ or $\{(4, 5)\}$
11. $\left(-\frac{2}{5}, -\frac{11}{5}\right)$ or $\left\{\left(-\frac{2}{5}, -\frac{11}{5}\right)\right\}$ **13.** no solution or $\varnothing$ **15.** infinitely many solutions; $\{(x, y) \mid y = 3x - 5\}$ or $\{(x, y) \mid 21x - 35 = 7y\}$
17. $(0, 0)$ or $\{(0, 0)\}$ **19.** $\left(\frac{17}{7}, -\frac{8}{7}\right)$ or $\left\{\left(\frac{17}{7}, -\frac{8}{7}\right)\right\}$ **21.** no solution or $\varnothing$ **23.** $\left(\frac{43}{5}, -\frac{1}{5}\right)$ or $\left\{\left(\frac{43}{5}, -\frac{1}{5}\right)\right\}$ **25.** $(200, 700)$ or $\{(200, 700)\}$
27. $(7, 3)$ or $\{(7, 3)\}$ **29.** $(-1, -1)$ or $\{(-1, -1)\}$ **31.** $(5, 4)$ or $\{(5, 4)\}$ **33.** $x + y = 81, y = x + 41$; 20 and 61
35. $x - y = 5, 4x = 6y$; 10 and 15   **37.** $x - y = 1, x + 2y = 7$; 2 and 3   **39.** $(2, 8)$ or $\{(2, 8)\}$ **41. a.** $(4, 4.5)$ or $\{(4, 4.5)\}$; equilibrium number
of workers: 4 million; equilibrium hourly wage: $4.50   **b.** $4.50; 4; 4   **c.** 2 million   **d.** 5.7 million   **e.** 3.7 million   **49.** does not make sense
51. does not make sense   **53.** true   **55.** false   **57.** $m = \frac{5}{2}$

58.

$4x + 6y = 12$; $(0, 2)$; $(3, 0)$

59. 12 or $\{12\}$ **60.** $-73, 0, \frac{3}{1}$
61. 8; The same value of x is obtained in both cases, so $(8, 12)$ is the solution.
62. 12 or $\{12\}$ **63.** $-4x + 20y = -12$

Section 5.3

Check Point Exercises

1. $(7, -2)$ or $\{(7, -2)\}$ **2.** $(6, 2)$ or $\{(6, 2)\}$ **3.** $(2, -1)$ or $\{(2, -1)\}$ **4.** $\left(\frac{60}{17}, -\frac{11}{17}\right)$ or $\left\{\left(\frac{60}{17}, -\frac{11}{17}\right)\right\}$ **5.** no solution or $\varnothing$
6. infinitely many solutions; $\{(x, y) \mid x - 5y = 7\}$ or $\{(x, y) \mid 3x - 15y = 21\}$

Exercise Set 5.3

1. $(4, -7)$ or $\{(4, -7)\}$ **3.** $(3, 0)$ or $\{(3, 0)\}$ **5.** $(-3, 5)$ or $\{(-3, 5)\}$ **7.** $(3, 1)$ or $\{(3, 1)\}$ **9.** $(2, 1)$ or $\{(2, 1)\}$ **11.** $(-2, 2)$ or $\{(-2, 2)\}$
13. $(-7, -1)$ or $\{(-7, -1)\}$ **15.** $(2, 0)$ or $\{(2, 0)\}$ **17.** $(1, -2)$ or $\{(1, -2)\}$ **19.** $(-1, 1)$ or $\{(-1, 1)\}$ **21.** $(3, 1)$ or $\{(3, 1)\}$
23. $(-5, -2)$ or $\{(-5, -2)\}$ **25.** $\left(\frac{11}{12}, -\frac{7}{6}\right)$ or $\left\{\left(\frac{11}{12}, -\frac{7}{6}\right)\right\}$ **27.** $\left(\frac{23}{16}, \frac{3}{8}\right)$ or $\left\{\left(\frac{23}{16}, \frac{3}{8}\right)\right\}$ **29.** no solution or $\varnothing$
31. infinitely many solutions; $\{(x, y) \mid x + 3y = 2\}$ or $\{(x, y) \mid 3x + 9y = 6\}$ **33.** no solution or $\varnothing$ **35.** $\left(\frac{1}{2}, -\frac{1}{2}\right)$ or $\left\{\left(\frac{1}{2}, -\frac{1}{2}\right)\right\}$
37. infinitely many solutions; $\{(x, y) \mid x = 5 - 3y\}$ or $\{(x, y) \mid 2x + 6y = 10\}$ **39.** $\left(\frac{1}{3}, 1\right)$ or $\left\{\left(\frac{1}{3}, 1\right)\right\}$ **41.** $(-10, 21)$ or $\{(-10, 21)\}$
43. $(0, 1)$ or $\{(0, 1)\}$ **45.** $(2, -1)$ or $\{(2, -1)\}$ **47.** $(1, -3)$ or $\{(1, -3)\}$ **49.** $(4, 3)$ or $\{(4, 3)\}$ **51.** no solution or $\varnothing$
53. infinitely many solutions; $\{(x, y) \mid 2(x + 2y) = 6\}$ or $\{(x, y) \mid 3(x + 2y - 3) = 0\}$ **55.** $(3, 2)$ or $\{(3, 2)\}$ **57.** $(-1, 2)$ or $\{(-1, 2)\}$
59. $(-1, 0)$ or $\{(-1, 0)\}$ **61.** $(3, 1)$ or $\{(3, 1)\}$ **63.** dependent   **65.** 2009; 18.5% pro-choice and 18.5% pro-life
73. does not make sense   **75.** does not make sense   **77.** false   **79.** false   **81.** $A = 2, B = 4$ **82.** 10   **83.** 11   **84.** 26 or $\{26\}$
85. a. $x + y = 28; x - y = 6$ **b.** 17 and 11   **86.** $3x + 2y$ **87. a.** $30   **b.** $y = 20 + 0.05x$

Mid-Chapter Check Point Exercises

1. $(2, 0)$ or $\{(2, 0)\}$ **2.** $(1, 1)$ or $\{(1, 1)\}$ **3.** no solution or $\varnothing$ **4.** $(5, 8)$ or $\{(5, 8)\}$ **5.** $(2, -1)$ or $\{(2, -1)\}$ **6.** $(2, 9)$ or $\{(2, 9)\}$
7. $(-2, -3)$ or $\{(-2, -3)\}$ **8.** $(10, -1)$ or $\{(10, -1)\}$ **9.** no solution or $\varnothing$ **10.** $(-12, -1)$ or $\{(-12, -1)\}$ **11.** $(-7, -23)$ or $\{(-7, -23)\}$
12. $\left(\dfrac{16}{17}, -\dfrac{12}{17}\right)$ or $\left\{\left(\dfrac{16}{17}, -\dfrac{12}{17}\right)\right\}$ **13.** infinitely many solutions; $\{(x, y)\,|\,y - 2x = 7\}$ or $\{(x, y)\,|\,4x = 2y - 14\}$ **14.** $(7, 11)$ or $\{(7, 11)\}$
15. $(6, 10)$ or $\{(6, 10)\}$

Section 5.4

Check Point Exercises

1. men: 65 minutes per day; women: 73 minutes per day   **2.** Quarter Pounder: 420 calories; Whopper with cheese: 589 calories
3. length: 100 ft; width: 80 ft   **4.** 17.5 yr; $24,250

Exercise Set 5.4

1. $x + y = 17, x - y = -3$; 7 and 10   **3.** $3x - y = -1, x + 2y = 23$; 3 and 10   **5.** women: 49 minutes; men: 37 minutes
7. Mr. Goodbar: 264 calories; Mounds bar: 258 calories   **9.** 3 Mr. Goodbars and 2 Mounds bars   **11.** sweater: $12; shirt: $10
13. 44 ft by 20 ft   **15.** 90 ft by 70 ft   **17. a.** 300 min; $35   **b.** plan B; Answers will vary.   **19.** $600 of merchandise; $580
21. adult: 93 tickets; student: 208 tickets   **23.** each item in column A: $3.99; each item in column B: $1.50
25. 2 servings of macaroni and 4 servings of broccoli   **27.** A: $100°$; B: $40°$; C: $40°$ **29.** still water rate: 6 mph; current rate: 2 mph
35. does not make sense   **37.** does not make sense   **39.** 10 birds and 20 lions   **41.** There are 5 people downstairs and 7 people upstairs.
44. $\dfrac{5}{2}$ or $\left\{\dfrac{5}{2}\right\}$ **45.** $6x + 11$ **46.** $y = -2x - 4$

47. $2x - y < 4$

48. $y \geq x + 1$

49. $x \geq 2$

Section 5.5

Check Point Exercises

1. Point $B = (66, 130)$; $4.9(66) - 130 \geq 165$, or $193.4 \geq 165$, is true; $3.7(66) - 130 \leq 125$, or $114.2 \leq 125$, is true.

2. $x + 2y > 4$
$2x - 3y \leq -6$

3. $y \geq x + 2$
$x \geq 1$

Exercise Set 5.5

1. $x + y \leq 4$
$x - y \leq 2$

3. $2x - 4y \leq 8$
$x + y \geq -1$

5. $x + 3y \leq 6$
$x - 2y \leq 4$

7. $x - 2y > 4$
$2x + y \geq 6$

9. $x + y > 1$
$x + y < 4$

11. $y \geq 2x + 1$
$y \leq 4$

13. $y > x - 1$
$x > 5$

15. $y \geq 2x - 3$
$y \leq 2x + 1$

17. $y > 2x - 3$
$y \leq -x + 6$

19. $x + 2y \leq 4$
$y \geq x - 3$

21. $x \leq 3$
$y \geq -2$

23. $x \geq 3$
$y < 2$

25. $x \geq 0$
$y \leq 0$

27. $x \geq 0$
$y > 0$

29. $x + y \leq 5$
$x \geq 0$
$y \geq 0$

31.

$4x - 3y > 12$
$x \geq 0$
$y \leq 0$

33.

$0 \leq x \leq 3$
$0 \leq y \leq 3$

35.

$x - y \leq 4$
$x + 2y \leq 4$

37.

$x + y \geq 1$
$x - y \geq 1$
$x \geq 4$

39.

$x + 2y < 6$
$y > 2x - 2$
$y \geq 2$

41.

$y \leq -3x + 3$
$y \geq -x - 1$
$y < x + 7$

43. no solution

$y \geq 2x + 2$
$y < 2x - 3$
$x \geq 2$

45. Point $A = (66, 160)$; $5.3(66) - 160 \geq 180$, or $189.8 \geq 180$, is true; $4.1(66) - 160 \leq 140$, or $110.6 \leq 140$, is true.   **47.** no

49. Answers will vary.; $165x + 110y \leq 330$

   53. makes sense   **55.** makes sense   **57.** $y \geq -x + 3, x > 5$ **59.** $x < 0, y < 0$

61.

$x + y \geq 25,000$
$35x + 50y \geq 1,025,000$

62. $-\dfrac{1}{4}$ **63.** $-\dfrac{11}{20}$ **64.** 4 or $\{4\}$ **65.** $17x^3$ **66.** $-2x^2$ **67.** $-7y^4$

Review Exercises

1. solution   **2.** not a solution   **3.** no; $(-1, 3)$ does not satisfy $2x + y = -5$.   **4.** $(4, -2)$ or $\{(4, -2)\}$ **5.** $(3, -2)$ or $\{(3, -2)\}$
6. $(2, 0)$ or $\{(2, 0)\}$ **7.** $(2, 1)$ or $\{(2, 1)\}$ **8.** $(4, -1)$ or $\{(4, -1)\}$ **9.** no solution or $\varnothing$ **10.** infinitely many solutions;
$\{(x, y) | 2x - 4y = 8\}$ or $\{(x, y) | x - 2y = 4\}$ **11.** no solution or $\varnothing$ **12.** $(-2, -6)$ or $\{(-2, -6)\}$ **13.** $(2, 5)$ or $\{(2, 5)\}$
14. no solution or $\varnothing$ **15.** $(2, -1)$ or $\{(2, -1)\}$ **16.** $(5, 4)$ or $\{(5, 4)\}$ **17.** $(-2, -1)$ or $\{(-2, -1)\}$ **18.** $(1, -4)$ or $\{(1, -4)\}$
19. $(20, -21)$ or $\{(20, -21)\}$ **20.** infinitely many solutions; $\{(x, y) | 4x + y = 5\}$ or $\{(x, y) | 12x + 3y = 15\}$ **21.** no solution or $\varnothing$
22. $(4, 18)$ or $\{(4, 18)\}$ **23.** $\left(-1, -\dfrac{1}{2}\right)$ or $\left\{\left(-1, -\dfrac{1}{2}\right)\right\}$ **24. a.** $(20, 1000)$ or $\{(20, 1000)\}$; equilibrium quantity: 20,000; equilibrium price: $1000
b. $1000; 20,000; 20,000   **25.** $(2, 4)$ or $\{(2, 4)\}$ **26.** $(-1, -1)$ or $\{(-1, -1)\}$ **27.** $(2, -1)$ or $\{(2, -1)\}$ **28.** $(3, 2)$ or $\{(3, 2)\}$
29. $(2, 1)$ or $\{(2, 1)\}$ **30.** $(0, 0)$ or $\{(0, 0)\}$ **31.** $\left(\dfrac{17}{7}, -\dfrac{15}{7}\right)$ or $\left\{\left(\dfrac{17}{7}, -\dfrac{15}{7}\right)\right\}$ **32.** no solution or $\varnothing$
33. infinitely many solutions; $\{(x, y) | 3x - 4y = -1\}$ or $\{(x, y) | -6x + 8y = 2\}$ **34.** $(4, -2)$ or $\{(4, -2)\}$ **35.** $(-8, -6)$ or $\{(-8, -6)\}$
36. $(-4, 1)$ or $\{(-4, 1)\}$ **37.** $\left(\dfrac{5}{2}, 3\right)$ or $\left\{\left(\dfrac{5}{2}, 3\right)\right\}$ **38.** $(3, 2)$ or $\{(3, 2)\}$ **39.** $\left(\dfrac{1}{2}, -2\right)$ or $\left\{\left(\dfrac{1}{2}, -2\right)\right\}$ **40.** no solution or $\varnothing$
41. $(3, 7)$ or $\{(3, 7)\}$ **42.** Klimt: $135 million; Picasso: $104 million   **43.** shrimp: 42 mg; scallops: 15 mg   **44.** 9 ft by 5 ft   **45.** 7 yd by 5 yd
46. room: $80; car: $60   **47.** 200 min; $25   **48.** orchestra: 6 tickets; balcony: 3 tickets

49.

$3x - y \leq 6$
$x + y \geq 2$

50.

$x + y < 4$
$x - y < 4$

51.

$y < 2x - 2$
$x \geq 3$

52.

$4x + 6y \leq 24$
$y > 2$

53.

$x \leq 3$
$y \geq -2$

54.

$y \geq \dfrac{1}{2}x - 2$
$y \leq \dfrac{1}{2}x + 1$

55.

$x \leq 0$
$y \geq 0$

Chapter Test

1. solution   **2.** not a solution   **3.** $(2, 4)$ or $\{(2, 4)\}$ **4.** $(2, 4)$ or $\{(2, 4)\}$ **5.** $(1, -3)$ or $\{(1, -3)\}$ **6.** $(2, -3)$ or $\{(2, -3)\}$

7. no solution or $\varnothing$ **8.** $(-1, 4)$ or $\{(-1, 4)\}$ **9.** $(-4, 3)$ or $\{(-4, 3)\}$ **40.** infinitely many solutions;

$\{(x, y) \mid 3x - 2y = 2\}$ or $\{(x, y) \mid -9x + 6y = -6\}$ **11.** Mary: 2.6%; Patricia: 1.1%   **12.** 12 yd by 5 yd   **13.** 500 min; $40

14.   **15.**

Cumulative Review Exercises

1. -36 **2.** $17x - 11$ **3.** -6 or $\{-6\}$ **4.** 20 or $\{20\}$ **5.** $t = \dfrac{A - P}{Pr}$ **6.** $(2, \infty)$ or $\{x \mid x > 2\}$;

7.   **8.**   **9.**

10. $(0, -2)$ or $\{(0, -2)\}$ **11.** $\left(\dfrac{3}{2}, -2\right)$ or $\left\{\left(\dfrac{3}{2}, -2\right)\right\}$ **12.** 1 **13.** $y - 6 = -4(x + 1)$; $y = -4x + 2$ **14.** 10 ft

15. pen: $0.80; pad: $1.20   **16.** $-93, 0, \dfrac{7}{1}, \sqrt{100}$ **17.** skim milk; Consumption, in gallons per person, increased from 1945 through 2003.

18. whole milk; Consumption, in gallons per person, decreased from 1945 through 2003.   **19.** 13 years; 2016   **20.** 2013

CHAPTER 6

Section 6.1

Check Point Exercises

1. $5x^3 + 4x^2 - 8x - 20$ **2.** $5x^3 + 4x^2 - 8x - 20$ **3.** $7x^2 + 11x + 4$ **4.** $7x^3 + 3x^2 + 12x - 8$ **5.** $3y^3 - 10y^2 - 11y - 8$

6.

Exercise Set 6.1

1. binomial, 1   **3.** binomial, 3   **5.** monomial, 2   **7.** monomial, 0   **9.** trinomial, 2   **11.** trinomial, 4   **13.** binomial, 3

15. monomial, 23   **17.** $-8x + 13$ **19.** $12x^2 + 15x - 9$ **21.** $10x^2 - 12x$ **23.** $5x^2 - 3x + 13$ **25.** $4y^3 + 10y^2 + y - 2$

27. $3x^3 + 2x^2 - 9x + 7$ **29.** $-2y^3 + 4y^2 + 13y + 13$ **31.** $-3y^6 + 8y^4 + y^2$ **33.** $10x^3 + 1$ **35.** $-\dfrac{2}{5}x^4 + x^3 - \dfrac{1}{8}x^2$

37. $0.01x^5 + x^4 - 0.1x^3 + 0.3x + 0.33$ **39.** $11y^3 - 3y^2$ **41.** $-2x^2 - x + 1$ **43.** $-\dfrac{1}{4}x^4 - \dfrac{7}{15}x^3 - 0.3$ **45.** $-y^3 + 8y^2 - 3y - 14$

47. $-5x^3 - 6x^2 + x - 4$ **49.** $7x^4 - 2x^3 + 4x - 2$ **51.** $8x^2 + 7x - 5$ **53.** $9x^3 - 4.9x^2 + 11.1$ **55.** $-2x - 10$ **57.** $-5x^2 - 9x - 12$

59. $-5x^2 - x$ **61.** $-4x^2 - 4x - 6$ **63.** $-2y - 6$ **65.** $6y^3 + y^2 + 7y - 20$ **67.** $n^3 + 2$ **69.** $y^6 - y^3 - y^2 + y$

71. $26x^4 + 9x^2 + 6x$ **73.** $\dfrac{5}{7}x^3 - \dfrac{9}{20}x$ **75.** $4x + 6$ **77.** $10x^2 - 7$ **79.** $-4y^2 - 7y + 5$ **81.** $9x^3 + 11x^2 - 8$

83. $-y^3 + 8y^2 + y + 14$ **85.** $7x^4 - 2x^3 + 3x^2 - x + 2$ **87.** $0.05x^3 + 0.02x^2 + 1.02x$

89.   **91.**   **93.**

95. $x^2 + 12x$ **97.** $y^2 - 19y + 16$ **99.** $2x^3 + 3x^2 + 7x - 5$ **101.** $-10y^3 + 2y^2 + y + 3$

103. a. $M - W = -35x^3 + 1373x^2 - 15{,}995x + 63{,}210$ **b.** \$12,348   **c.** \$10,923; overestimates by \$1425   **105. a.** \$56,995; underestimates by \$225

b. $(16, 56{,}995)$ on the graph for men   **c.** \$42,000, although estimates may vary by $\pm$\$1000   **115.** does not make sense   **117.** does not make sense

119. false   **121.** false   **123.** $\frac{2}{3}t^3 - 2t^2 + 4t$ **125.** -10 **126.** 5.6   **127.** -4 or $\{-4\}$ **128.** 7   **129.** $3x^2 + 15x$ **130.** $x^2 + 5x + 6$

Section 6.2

Check Point Exercises

1. a. 2^6 or 64 **b.** x^{10} **c.** y^8 **d.** y^9 **2. a.** 3^{20} **b.** x^{90} **c.** $(-5)^{21}$ **3. a.** $16x^4$ **b.** $-64y^6$ **4. a.** $70x^3$ **b.** $-20x^9$

5. a. $3x^2 + 15x$ **b.** $30x^5 - 12x^3 + 18x^2$ **6. a.** $x^2 + 9x + 20$ **b.** $10x^2 - 29x - 21$ **7.** $5x^3 - 18x^2 + 7x + 6$

8. $6x^5 - 19x^4 + 22x^3 - 8x^2$

Exercise Set 6.2

1. x^{18} **3.** y^{12} **5.** x^{11} **7.** 7^{19} **9.** 6^{90} **11.** x^{45} **13.** $(-20)^9$ **15.** $8x^3$ **17.** $25x^2$ **19.** $16x^6$ **21.** $16y^{24}$ **23.** $-32x^{35}$

25. $14x^2$ **27.** $24x^3$ **29.** $-15y^7$ **31.** $\frac{1}{8}a^5$ **33.** $-48x^7$ **35.** $4x^2 + 12x$ **37.** $x^2 - 3x$ **39.** $2x^2 - 12x$ **41.** $-12y^2 - 20y$

43. $4x^3 + 8x^2$ **45.** $2y^4 + 6y^3$ **47.** $6y^4 - 8y^3 + 14y^2$ **49.** $6x^4 + 8x^3$ **51.** $-2x^3 - 10x^2 + 6x$ **53.** $12x^4 - 3x^3 + 15x^2$

55. $x^2 + 8x + 15$ **57.** $2x^2 + 9x + 4$ **59.** $x^2 - 2x - 15$ **61.** $x^2 - 2x - 99$ **63.** $2x^2 + 3x - 20$ **65.** $\frac{3}{16}x^2 + \frac{11}{4}x - 4$

67. $x^3 + 3x^2 + 5x + 3$ **69.** $y^3 - 6y^2 + 13y - 12$ **71.** $2a^3 - 9a^2 + 19a - 15$ **73.** $x^4 + 3x^3 + 5x^2 + 7x + 4$

75. $4x^4 - 4x^3 + 6x^2 - \frac{17}{2}x + 3$ **77.** $x^4 + x^3 + x^2 + 3x + 2$ **79.** $x^3 + 3x^2 - 37x + 24$ **81.** $2x^3 - 9x^2 + 27x - 27$

83. $2x^4 + 9x^3 + 6x^2 + 11x + 12$ **85.** $12z^4 - 14z^3 + 19z^2 - 22z + 8$ **87.** $21x^5 - 43x^4 + 38x^3 - 24x^2$

89. $4y^6 - 2y^5 - 6y^4 + 5y^3 - 5y^2 + 8y - 3$ **91.** $x^4 + 6x^3 - 11x^2 - 4x + 3$ **93.** $2x - 2$ **95.** $15x^5 + 42x^3 - 8x^2$

97. $2y^3$ **99.** $16y + 32$ **101.** $2x^2 + 7x - 15$ ft^2 **103. a.** $(2x + 1)(x + 2)$ **b.** $2x^2 + 5x + 2$ **c.** $(2x + 1)(x + 2) = 2x^2 + 5x + 2$

113. makes sense   **115.** makes sense   **117.** false   **119.** true   **121.** $8x + 16$ **123.** $-8x^4$ **124.** $(-\infty, -1)$ or $\{x | x < -1\}$

125.

126. $-\frac{2}{3}$ **127. a.** $x^2 + 7x + 12$ **b.** $x^2 + 25x + 100$ **128. a.** $x^2 - 9$ **b.** $x^2 - 25$

129. a. $x^2 + 6x + 9$ **b.** $x^2 + 10x + 25$

Section 6.3

Check Point Exercises

1. $x^2 + 11x + 30$ **2.** $28x^2 - x - 15$ **3.** $6x^2 - 22x + 20$ **4. a.** $49y^2 - 64$ **b.** $16x^2 - 25$ **c.** $4a^6 - 9$ **5. a.** $x^2 + 20x + 100$

b. $25x^2 + 40x + 16$ **6. a.** $x^2 - 18x + 81$ **b.** $49x^2 - 42x + 9$

Exercise Set 6.3

1. $x^2 + 10x + 24$ **3.** $y^2 - 4y - 21$ **5.** $2x^2 + 7x - 15$ **7.** $4y^2 - y - 3$ **9.** $10x^2 - 9x - 9$ **11.** $12y^2 - 43y + 35$

13. $-15x^2 - 32x + 7$ **15.** $6y^2 - 28y + 30$ **17.** $15x^4 - 47x^2 + 28$ **19.** $-6x^2 + 17x - 10$ **21.** $x^3 + 5x^2 + 3x + 15$

23. $8x^5 + 40x^3 + 3x^2 + 15$ **25.** $x^2 - 9$ **27.** $9x^2 - 4$ **29.** $9r^2 - 16$ **31.** $9 - r^2$ **33.** $25 - 49x^2$ **35.** $4x^2 - \frac{1}{4}$ **37.** $y^4 - 1$

39. $r^6 - 4$ **41.** $1 - y^8$ **43.** $x^{20} - 25$ **45.** $x^2 + 4x + 4$ **47.** $4x^2 + 20x + 25$ **49.** $x^2 - 6x + 9$ **51.** $9y^2 - 24y + 16$

53. $16x^4 - 8x^2 + 1$ **55.** $49 - 28x + 4x^2$ **57.** $4x^2 + 2x + \frac{1}{4}$ **59.** $16y^2 - 2y + \frac{1}{16}$ **61.** $x^{16} + 6x^8 + 9$ **63.** $x^3 - 1$

65. $x^2 - 2x + 1$ **67.** $9y^2 - 49$ **69.** $12x^4 + 3x^3 + 27x^2$ **71.** $70y^2 + 2y - 12$ **73.** $x^4 + 2x^2 + 1$ **75.** $x^4 + 3x^2 + 2$

77. $x^4 - 16$ **79.** $4 - 12x^5 + 9x^{10}$ **81.** $\frac{3}{16}x^4 + 7x^2 - 96$ **83.** $x^2 + 2x + 1$ **85.** $4x^2 - 9$ **87.** $6x + 22$

89. $16x^4 - 72x^2 + 81$ **91.** $16x^4 - 1$ **93.** $x^3 + 6x^2 + 12x + 8$ **95.** $x^2 + 6x + 9 - y^2$ **97.** $(x + 1)(x + 2)$ yd^2 **99.** 56 yd^2; $(6, 56)$

101. $(x^2 + 4x + 4)$ in^2 **109.** makes sense   **111.** makes sense, although answers may vary   **113.** true   **115.** false   **117.** $4x^3 - 36x^2 + 80x$

119. Change $x^2 + 1$ to $x^2 + 2x + 1$.   **121.** Graphs coincide.   **123.** $(2, -1)$ or $\{(2, -1)\}$ **124.** $(1, 1)$ or $\{(1, 1)\}$

125.

126. -72 **127.** $11xy$ **128.** $3x^2 + 11xy + 10y^2$

Section 6.4

Check Point Exercises

1. -9 **2.** polynomial degree: 9;

Term	Coefficient	Degree
$8x^4y^5$	8	9
$-7x^3y^2$	-7	5
$-x^2y$	-1	3
$-5x$	-5	1
11	11	0

3. $2x^2y + 2xy - 4$ **4.** $3x^3 + 2x^2y + 5xy^2 - 10$ **5.** $60x^5y^5$ **6.** $60x^5y^7 - 12x^3y^3 + 18xy^2$
7. a. $21x^2 - 25xy + 6y^2$ **b.** $4x^2 + 16xy + 16y^2$ **8. a.** $36x^2y^4 - 25x^2$ **b.** $x^3 - y^3$

Exercise Set 6.4

1. 1   **3.** -47 **5.** -6 **7.** polynomial degree: 9;

Term	Coefficient	Degree
x^3y^2	1	5
$-5x^2y^7$	-5	9
$6y^2$	6	2
-3	-3	0

9. $7x^2y - 4xy$ **11.** $2x^2y + 13xy + 13$ **13.** $-11x^4y^2 - 11x^2y^2 + 2xy$ **15.** $-5x^3 + 8xy - 9y^2$ **17.** $x^4y^2 + 8x^3y + y - 6x$
19. $5x^3 + x^2y - xy^2 - 4y^3$ **21.** $-3x^2y^2 + xy^2 + 5y^2$ **23.** $8a^2b^4 + 3ab^2 + 8ab$ **25.** $-30x + 37y$ **27.** $40x^3y^2$ **29.** $-24x^5y^9$
31. $45x^2y + 18xy^2$ **33.** $50x^3y^2 - 15xy^3$ **35.** $28a^3b^5 + 8a^2b^3$ **37.** $-a^2b + ab^2 - b^3$ **39.** $7x^2 + 38xy + 15y^2$ **41.** $2x^2 + xy - 21y^2$
43. $15x^2y^2 + xy - 2$ **45.** $4x^2 + 12xy + 9y^2$ **47.** $x^2y^2 - 6xy + 9$ **49.** $x^4 + 2x^2y^2 + y^4$ **51.** $x^4 - 4x^2y^2 + 4y^4$ **53.** $9x^2 - y^2$
55. $a^2b^2 - 1$ **57.** $x^2 - y^4$ **59.** $9a^4b^2 - a^2$ **61.** $9x^2y^4 - 16y^2$ **63.** $a^3 - ab^2 + a^2b - b^3$ **65.** $x^3 + 4x^2y + 4xy^2 + y^3$
67. $x^3 - 4x^2y + 4xy^2 - y^3$ **69.** $x^2y^2 - a^2b^2$ **71.** $x^6y + x^4y + x^4 + 2x^2 + 1$ **73.** $x^4y^4 - 6x^2y^2 + 9$ **75.** $x^2 + 2xy + y^2 - 1$
77. $3x^2 + 8xy + 5y^2$ **79.** $2xy + y^2$ **81.** $x^{12}y^{12} - 2x^6y^6 + 1$ **83.** $x^4y^4 - 18x^2y^2 + 81$ **85.** $x^2 - y^2 - 2yz - z^2$
87. no; need 120 more board feet   **89.** 192 ft   **91.** 0 ft; The ball hits the ground.   **93.** 2.5 to 6 sec   **95.** $(2, 192)$
97. 2.5 sec; 196 ft   **101.** makes sense   **103.** does not make sense   **105.** false   **107.** true   **109.** $-x^2 + 18xy + 80y^2$
112. $W = \dfrac{2R - L}{3}$ **113.** 3.8   **114.** 105 or $\{105\}$ **115.** 4   **116.** $\dfrac{x^6}{125}$ **117.** $\dfrac{32a^{15}}{b^{20}}$

Mid-Chapter Check Point Exercises

1. $-55x^4y^6$ **2.** $6x^2y^3$ **3.** $12x^2 - x - 35$ **4.** $-x + 12$ **5.** $2x^3 - 11x^2 + 17x - 5$ **6.** $x^2 - x - 4$ **7.** $64x^2 - 48x + 9$ **8.** $70x^9$
9. $x^4 - 4$ **10.** $x^4 + 4x^2 + 4$ **11.** $18a^2 - 11ab - 10b^2$ **12.** $70x^5 - 14x^3 + 21x^2$ **13.** $5a^2b^3 + 2ab - b^2$ **14.** $18y^2 - 50$
15. $2x^3 - x^2 + x$ **16.** $10x^2 - 5xy - 3y^2$ **17.** $-4x^5 + 7x^4 - 10x + 23$ **18.** $x^3 + 27y^3$ **19.** $10x^7 - 5x^4 + 8x^3 - 4$
20. $y^2 - 12yz + 36z^2$ **21.** $-21x^2 + 7$

22.

$y = 1 - x^2$

Section 6.5

Check Point Exercises

1. a. 5^8 **b.** x^7 **c.** y^{19} **2. a.** 1   **b.** 1   **c.** -1 **d.** 20   **e.** 1   **3. a.** $\dfrac{x^2}{25}$ **b.** $\dfrac{x^{12}}{8}$ **c.** $\dfrac{16a^{40}}{b^{12}}$

4. a. $-2x^8$ **b.** $\dfrac{1}{5}$ **c.** $3x^5y^3$ **5.** $-5x^7 + 2x^3 - 3x$ **6.** $5x^6 - \dfrac{7}{5}x + 2$ **7.** $3x^6y^4 - xy + 10$

Exercise Set 6.5

1. 3^{15} **3.** x^4 **5.** y^8 **7.** $5^3 \cdot 2^4$ **9.** $x^{75}y^{40}$ **11.** 1   **13.** 1   **15.** -1 **17.** 100   **19.** 1   **21.** 0   **23.** -2 **25.** $\dfrac{x^2}{9}$ **27.** $\dfrac{x^6}{64}$
29. $\dfrac{4x^6}{25}$ **31.** $-\dfrac{64}{27a^9}$ **33.** $-\dfrac{32a^{35}}{b^{20}}$ **35.** $\dfrac{x^8y^{12}}{16z^4}$ **37.** $3x^5$ **39.** $-2x^{20}$ **41.** $-\dfrac{1}{2}y^3$ **43.** $\dfrac{7}{5}y^{12}$ **45.** $6x^5y^4$ **47.** $-\dfrac{1}{2}x^{12}$ **49.** $\dfrac{9}{7}$
51. $-\dfrac{1}{10}x^8y^9z^4$ **53.** $5x^4 + x^3$ **55.** $2x^3 - x^2$ **57.** $y^6 - 9y + 1$ **59.** $-8x^2 + 5x$ **61.** $6x^3 + 2x^2 + 3x$ **63.** $3x^3 - 2x^2 + 10x$
65. $4x - 6$ **67.** $-6z^2 - 2z$ **69.** $4x^2 + 3x - 1$ **71.** $5x^4 - 3x^2 - x$ **73.** $-9x^3 + \dfrac{9}{2}x^2 - 10x + 5$ **75.** $4xy + 2x - 5y$

77. $-4x^5y^3 + 3xy + 2$ **79.** $4x^2 - x + 6$ **81.** $-xy^2$ **83.** $y + 5$ **85.** $3x^{12n} - 6x^{9n} + 2$

87. a. \$5.39   **b.** $\dfrac{3.6x^2 + 158x + 2790}{-0.2x^2 + 21x + 1015}$ **c.** \$5.45; overestimates by \$0.06   **d.** no; The divisor is not a monomial.

97. does not make sense   **99.** does not make sense   **101.** false   **103.** true   **105.** $\dfrac{9x^8 - 12x^6}{3x^3}$ **107.** 20.3   **108.** 0.875

109.

$y = \frac{1}{3}x + 2$

110. quotient: 26; remainder: 0   **111.** quotient: 123; remainder: 6   **112.** quotient: 257; remainder: 1

Section 6.6

Check Point Exercises

1. $x + 5$ **2.** $4x - 3 - \dfrac{3}{2x + 3}$ **3.** $x^2 + x + 1$

Exercise Set 6.6

1. $x + 4$ **3.** $2x + 5$ **5.** $x - 2$ **7.** $2y + 1$ **9.** $x - 5 + \dfrac{14}{x + 2}$ **11.** $y + 3 + \dfrac{4}{y + 2}$ **13.** $x^2 - 5x + 2$ **15.** $6y - 1$

17. $2a + 3$ **19.** $y^2 - y + 2$ **21.** $3x + 5 - \dfrac{5}{2x - 5}$ **23.** $x^2 + 2x + 8 + \dfrac{13}{x - 2}$ **25.** $2y^2 + y + 1 + \dfrac{6}{2y + 3}$

27. $2y^2 - 3y + 2 + \dfrac{1}{3y + 2}$ **29.** $9x^2 + 3x + 1$ **31.** $y^3 - 9y^2 + 27y - 27$ **33.** $2y + 4 + \dfrac{4}{2y - 1}$ **35.** $y^3 + y^2 - y - 1 + \dfrac{4}{y - 1}$

37. $4x - 3 + \dfrac{-7x + 7}{x^2 + 2}$ **39.** $x^2 + ax + a^2$ **41.** $2x^2 - 3x + 2$ **43.** $x^2 + 2x + 3$ **45.** $x^2 + 2x + 3$ units   **47. a.** $\dfrac{30{,}000x^3 - 30{,}000}{x - 1}$

b. $30{,}000x^2 + 30{,}000x + 30{,}000$ **c.** \$94,575   **53.** makes sense   **55.** does not make sense   **57.** false   **59.** true   **61.** $2x^2 - 2x + 5$

63. Answers will vary; $x^6 - x^5 + x^4 - x^3 + x^2 - x + 1 - \dfrac{2}{x + 1}$ **65.** $x - 5$ should be $x + 5$.   **67.** $2x - 4$ should be $2x + 4$.

69.
$2x - y \geq 4$
$x + y \leq -1$

70. 1.2   **71.** -6 or $\{-6\}$ **72. a.** $2; -2$ **b.** $\dfrac{1}{7^2} = 7^{-2}$ **73.** $16x^2$ **74.** x^9

Section 6.7

Check Point Exercises

1. a. $\dfrac{1}{6^2} = \dfrac{1}{36}$ **b.** $\dfrac{1}{5^3} = \dfrac{1}{125}$ **c.** $\dfrac{1}{(-3)^4} = \dfrac{1}{81}$ **d.** $-\dfrac{1}{3^4} = -\dfrac{1}{81}$ **e.** $\dfrac{1}{8^1} = \dfrac{1}{8}$ **2. a.** $\dfrac{7^2}{2^3} = \dfrac{49}{8}$ **b.** $\dfrac{5^2}{4^2} = \dfrac{25}{16}$ **c.** $\dfrac{y^2}{7}$ **d.** $\dfrac{y^8}{x^1} = \dfrac{y^8}{x}$

3. $\dfrac{1}{x^{10}}$ **4. a.** $\dfrac{1}{x^8}$ **b.** $\dfrac{15}{x^6}$ **c.** $-\dfrac{2}{y^6}$ **5.** $\dfrac{36}{x^3}$ **6.** $\dfrac{1}{x^{20}}$ **7. a.** 7,400,000,000   **b.** 0.000003017   **8. a.** 7.41×10^9 **b.** 9.2×10^{-8}

9. a. 6×10^{10} **b.** 2.1×10^{11} **c.** 6.4×10^{-5} **10.** \$2500

Exercise Set 6.7

1. $\dfrac{1}{8^2} = \dfrac{1}{64}$ **3.** $\dfrac{1}{5^3} = \dfrac{1}{125}$ **5.** $\dfrac{1}{(-6)^2} = \dfrac{1}{36}$ **7.** $-\dfrac{1}{6^2} = -\dfrac{1}{36}$ **9.** $\dfrac{1}{4^1} = \dfrac{1}{4}$ **11.** $\dfrac{1}{2^1} + \dfrac{1}{3^1} = \dfrac{1}{2} + \dfrac{1}{3} = \dfrac{5}{6}$ **13.** $3^2 = 9$ **15.** $(-3)^2 = 9$

17. $\dfrac{8^2}{2^3} = 8$ **19.** $\dfrac{4^2}{1^2} = 16$ **21.** $\dfrac{5^3}{3^3} = \dfrac{125}{27}$ **23.** $\dfrac{x^5}{6}$ **25.** $\dfrac{y^1}{x^8} = \dfrac{y}{x^8}$ **27.** $3 \cdot (-5)^3 = -375$ **29.** $\dfrac{1}{x^5}$ **31.** $\dfrac{8}{x^3}$ **33.** $\dfrac{1}{x^6}$ **35.** $\dfrac{1}{y^{99}}$

37. $\dfrac{3}{z^5}$ **39.** $-\dfrac{4}{x^4}$ **41.** $-\dfrac{1}{3a^3}$ **43.** $\dfrac{7}{5w^8}$ **45.** $\dfrac{1}{x^5}$ **47.** $\dfrac{1}{y^{11}}$ **49.** $\dfrac{16}{x^2}$ **51.** $216y^{17}$ **53.** $\dfrac{1}{x^6}$ **55.** $\dfrac{1}{16x^{12}}$ **57.** $\dfrac{x^2}{9}$ **59.** $-\dfrac{y^3}{8}$

61. $\dfrac{2x^6}{5}$ **63.** x^8 **65.** $16y^6$ **67.** $\dfrac{1}{y^2}$ **69.** $\dfrac{1}{y^{50}}$ **71.** $\dfrac{1}{a^{12}b^{15}}$ **73.** $\dfrac{a^8}{b^{24}}$ **75.** $\dfrac{4}{x^4}$ **77.** $\dfrac{y^9}{x^6}$ **79.** 870   **81.** 923,000   **83.** 3.4

85. 0.79   **87.** 0.0215   **89.** 0.000786   **91.** 3.24×10^4 **93.** 2.2×10^8 **95.** 7.13×10^2 **97.** 6.751×10^3 **99.** 2.7×10^{-3}

101. 2.02×10^{-5} **103.** 5×10^{-3} **105.** 3.14159×10^0 **107.** 6×10^5 **109.** 1.6×10^9 **111.** 3×10^4 **113.** 3×10^6 **115.** 3×10^{-6}

117. 9×10^4 **119.** 2.5×10^6 **121.** 1.25×10^8 **123.** 8.1×10^{-7} **125.** 2.5×10^{-7} **127.** 1   **129.** $\dfrac{y}{16x^8z^6}$ **131.** $\dfrac{1}{x^{12}y^{16}z^{20}}$

133. $\dfrac{x^{18}y^6}{4}$ **135.** 2.5×10^{-3} **137.** 8×10^{-5} **139.** 9.2×10^3 **141.** 2.5×10^{-16} **143.** 9.05×10^7 **145.** 4.65×10^{11}

147. a. 2.27×10^{12} **b.** 2.98×10^8 **c.** \$7617   **149.** $\$3.72 \times 10^{10}$ **151.** 1.25 sec   **159.** does not make sense

161. does not make sense   **163.** false   **165.** false   **167.** false   **169.** false   **171.** $\dfrac{1}{4}$ **176.** 225 deer   **177.** 5   **178.** $0, \sqrt{16}$

179. $16x^5 - 12x^4 + 4x^3$ **180.** $27x^2y^3 - 9xy^2 + 81xy$ **181.** $x^3 + 3x^2 + 5x + 15$

Review Exercises

1. binomial, 4 **2.** trinomial, 2 **3.** monomial, 1 **4.** $8x^3 + 10x^2 - 20x - 4$ **5.** $13y^3 - 8y^2 + 7y - 5$ **6.** $11y^2 - 4y - 4$
7. $8x^4 - 5x^3 + 6$ **8.** $-14x^4 - 13x^2 + 16x$ **9.** $7y^4 - 5y^3 + 3y^2 - y - 4$ **10.** $3x^2 - 7x + 9$ **11.** $10x^3 - 9x^2 + 2x + 11$

12.

13.

14. x^{23} **15.** y^{14} **16.** x^{100} **17.** $100y^2$ **18.** $-64x^{30}$ **19.** $50x^4$ **20.** $-36y^{11}$ **21.** $30x^{12}$ **22.** $21x^3 + 63x$ **23.** $20x^5 - 55x^4$
24. $-21y^4 + 9y^3 - 18y^2$ **25.** $16y^8 - 20y^7 + 2y^5$ **26.** $x^3 - 2x^2 - 13x + 6$ **27.** $12y^3 + y^2 - 21y + 10$ **28.** $3y^3 - 17y^2 + 41y - 35$
29. $8x^4 + 8x^3 - 18x^2 - 20x - 3$ **30.** $x^2 + 8x + 12$ **31.** $6y^2 - 7y - 5$ **32.** $4x^4 - 14x^2 + 6$ **33.** $25x^2 - 16$ **34.** $49 - 4y^2$
35. $y^4 - 1$ **36.** $x^2 + 6x + 9$ **37.** $9y^2 + 24y + 16$ **38.** $y^2 - 2y + 1$ **39.** $25y^2 - 20y + 4$ **40.** $x^4 + 8x^2 + 16$ **41.** $x^4 - 16$
42. $x^4 - x^2 - 20$ **43.** $x^2 + 7x + 12$ **44.** $x^2 + 50x + 600 \text{ yd}^2$ **45.** 28
46. polynomial degree: 5;

Term	Coefficient	Degree
$4x^2y$	4	3
$9x^3y^2$	9	5
$-17x^4$	-17	4
-12	-12	0

47. $-x^2 - 17xy + 5y^2$ **48.** $2x^3y^2 + x^2y - 6x^2 - 4$ **49.** $-35x^6y^9$ **50.** $15a^3b^5 - 20a^2b^3$ **51.** $3x^2 + 16xy - 35y^2$
52. $36x^2y^2 - 31xy + 3$ **53.** $9x^2 + 30xy + 25y^2$ **54.** $x^2y^2 - 14xy + 49$ **55.** $49x^2 - 16y^2$ **56.** $a^3 - b^3$ **57.** 6^{30} **58.** x^{15}
59. 1 **60.** -1 **61.** 400 **62.** $\dfrac{x^{12}}{8}$ **63.** $\dfrac{81}{16y^{24}}$ **64.** $-5y^6$ **65.** $8x^7y^3$ **66.** $3x^3 - 2x + 6$ **67.** $-6x^5 + 5x^4 + 8x^2$

68. $9x^2y - 3x - 6y$ **69.** $2x + 7$ **70.** $x^2 - 3x + 5$ **71.** $x^2 + 5x + 2 + \dfrac{7}{x - 7}$ **72.** $y^2 + 3y + 9$ **73.** $\dfrac{1}{7^2} = \dfrac{1}{49}$ **74.** $\dfrac{1}{(-4)^3} = -\dfrac{1}{64}$

75. $\dfrac{1}{2^1} + \dfrac{1}{4^1} = \dfrac{1}{2} + \dfrac{1}{4} = \dfrac{3}{4}$ **76.** $5^2 = 25$ **77.** $\dfrac{5^3}{2^3} = \dfrac{125}{8}$ **78.** $\dfrac{1}{x^6}$ **79.** $\dfrac{6}{y^2}$ **80.** $\dfrac{30}{x^5}$ **81.** x^8 **82.** $81y^2$ **83.** $\dfrac{1}{y^{19}}$

84. $\dfrac{x^3}{8}$ **85.** $\dfrac{1}{x^6}$ **86.** y^{20} **87.** 23,000 **88.** 0.00176 **89.** 0.9 **90.** 7.39×10^7 **91.** 6.2×10^{-4} **92.** 3.8×10^{-1} **93.** 3.8×10^0
94. 9×10^3 **95.** 5×10^4 **96.** 1.6×10^{-3} **97.** 2.57×10^{11} **98.** 1.75×10^8 **99.** $1469

Chapter Test

1. trinomial, 2 **2.** $13x^3 + x^2 - x - 24$ **3.** $5x^3 + 2x^2 + 2x - 9$

4.

5. $-35x^{11}$ **6.** $48x^5 - 30x^3 - 12x^2$ **7.** $3x^3 - 10x^2 - 17x - 6$ **8.** $6y^2 - 13y - 63$ **9.** $49x^2 - 25$ **10.** $x^4 + 6x^2 + 9$
11. $25x^2 - 30x + 9$ **12.** 30 **13.** $2x^2y^3 + 3xy + 12y^2$ **14.** $12a^2 - 13ab - 35b^2$ **15.** $4x^2 + 12xy + 9y^2$ **16.** $-5x^{12}$
17. $3x^3 - 2x^2 + 5x$ **18.** $x^2 - 2x + 3 + \dfrac{1}{2x + 1}$ **19.** $\dfrac{1}{10^2} = \dfrac{1}{100}$ **20.** $4^3 = 64$ **21.** $-27x^6$ **22.** $\dfrac{4}{x^5}$ **23.** $-\dfrac{21}{x^6}$

24. $16y^4$ **25.** $\dfrac{x^8}{25}$ **26.** $\dfrac{1}{x^{15}}$ **27.** 0.00037 **28.** 7.6×10^6 **29.** 1.23×10^{-2} **30.** 2.1×10^8 **31.** $x^2 + 10x + 16$

Cumulative Review Exercises

1. $\dfrac{35}{9}$ **2.** -128 **3.** 15,050 ft **4.** $-\dfrac{2}{3}$ or $\left\{-\dfrac{2}{3}\right\}$ **5.** $-\dfrac{5}{3}$ or $\left\{-\dfrac{5}{3}\right\}$ **6.** 4 ft by 10 ft

7. $[6, \infty)$ or $\{x | x \geq 6\}$;

8. 3 hr

9. 104 lb; 6 bags **10.**

11.

12. $-\dfrac{6}{5}$; falling **13.** $y + 1 = -2(x - 3); y = -2x + 5$

14. $(0, 5)$ or $\{(0, 5)\}$ **15.** $(3, -4)$ or $\{(3, -4)\}$

16. 500 min; $40   **17.**

18. $3x^5 - 6x^3 + 9x + 2$ **19.** $x^2 + 2x + 3$ **20.** $\dfrac{81}{x^2}$

CHAPTER 7

Section 7.1

Check Point Exercises

1. a. $3x^2$ **b.** $4x^2$ **c.** x^2y **2.** $6(x^2 + 3)$ **3.** $5x^2(5 + 7x)$ **4.** $3x^3(5x^2 + 4x - 9)$ **5.** $2xy(4x^2y - 7x + 1)$
6. a. $(x + 1)(x^2 + 7)$ **b.** $(y + 4)(x - 7)$ **7.** $(x + 5)(x^2 + 2)$ **8.** $(y + 3)(x - 5)$

Exercise Set 7.1

1. 4   **3.** $4x$ **5.** $2x^3$ **7.** $3y$ **9.** xy **11.** $4x^4y^3$ **13.** $8(x + 1)$ **15.** $4(y - 1)$ **17.** $5(x + 6)$ **19.** $6(5x - 2)$ **21.** $x(x + 5)$
23. $6(3y^2 + 2)$ **25.** $7x^2(2x + 3)$ **27.** $y(13y - 25)$ **29.** $9y^4(1 + 3y^2)$ **31.** $4x^2(2 - x^2)$ **33.** $4(3y^2 + 4y - 2)$ **35.** $3x^2(3x^2 + 6x + 2)$
37. $50y^2(2y^3 - y + 2)$ **39.** $5x(2 - 4x + x^2)$ **41.** cannot be factored   **43.** $3xy(2x^2y + 3)$ **45.** $10xy(3xy^2 - y + 2)$
47. $8x^2y(4xy - 3x - 2)$ **49.** $(x + 5)(x + 3)$ **51.** $(x + 2)(x - 4)$ **53.** $(y + 6)(x - 7)$ **55.** $(x + y)(3x - 1)$ **57.** $(3x + 1)(4x + 1)$
59. $(5x + 4)(7x^2 + 1)$ **61.** $(x + 2)(x + 4)$ **63.** $(x - 5)(x + 3)$ **65.** $(x^2 + 5)(x - 2)$ **67.** $(x^2 + 2)(x - 1)$ **69.** $(y + 5)(x + 9)$
71. $(y - 1)(x + 5)$ **73.** $(x - 2y)(3x + 5y)$ **75.** $(3x - 2)(x^2 - 2)$ **77.** $(x - a)(x - b)$ **79.** $6x^2yz(4xy^2z^2 + 5y + 3z)$
81. $(x^3 - 4)(1 + 3y)$ **83.** $2x^2(x + 1)(2x^3 - 3x - 4)$ **85.** $(x - 1)(3x^4 + x^2 + 5)$ **87.** $36x^2 - 4\pi x^2; 4x^2(9 - \pi)$
89. a. 48 ft   **b.** $16x(4 - x)$ **c.** 48 ft; yes; no; Answers will vary.   **91.** $x^3 - 2$ **99.** makes sense   **101.** does not make sense
103. false   **105.** false   **107.** $4(x + 150)$ **111.** $(x - 2)(x - 5)$ should be $(x - 2)(x + 5)$ **113.** $x^2 + 17x + 70$
114. $(-3, -2)$ or $\{(-3, -2)\}$ **115.** $y - 2 = 1(x + 7)$ or $y - 5 = 1(x + 4); y = x + 9$ **116.** 2 and 4   **117.** -3 and -2 **118.** -5 and 7

Section 7.2

Check Point Exercises

1. $(x + 2)(x + 3)$ **2.** $(x - 2)(x - 4)$ **3.** $(x + 5)(x - 2)$ **4.** $(y - 9)(y + 3)$ **5.** cannot factor over the integers; prime
6. $(x - 3y)(x - y)$ **7.** $2x(x - 4)(x + 7)$

Exercise Set 7.2

1. $(x + 6)(x + 1)$ **3.** $(x + 2)(x + 5)$ **5.** $(x + 1)(x + 10)$ **7.** $(x - 4)(x - 3)$ **9.** $(x - 6)(x - 6)$ **11.** $(y - 3)(y - 5)$
13. $(x + 5)(x - 2)$ **15.** $(y + 13)(y - 3)$ **17.** $(x - 5)(x + 3)$ **19.** $(x - 4)(x + 2)$ **21.** prime   **23.** $(y - 4)(y - 12)$ **25.** prime
27. $(w - 32)(w + 2)$ **29.** $(y - 5)(y - 13)$ **31.** $(r + 3)(r + 9)$ **33.** prime   **35.** $(x + 6y)(x + y)$ **37.** $(x - 3y)(x - 5y)$
39. $(x - 6y)(x + 3y)$ **41.** $(a - 15b)(a - 3b)$ **43.** $3(x + 2)(x + 3)$ **45.** $4(y - 2)(y + 1)$ **47.** $10(x - 10)(x + 6)$
49. $3(x - 2)(x - 9)$ **51.** $2r(r + 2)(r + 1)$ **53.** $4x(x + 6)(x - 3)$ **55.** $2r(r + 8)(r - 4)$ **57.** $y^2(y + 10)(y - 8)$
59. $x^2(x - 5)(x + 2)$ **61.** $2w^2(w - 16)(w + 3)$ **63.** $15x(y - 1)(y + 4)$ **65.** $x^3(x - y)(x + 4y)$ **67.** $2x^2(y - 15z)(y - z)$
69. $(a + b)(x + 5)(x - 4)$ **71.** $(x + 0.3)(x + 0.2)$ **73.** $\left(x - \dfrac{1}{5}\right)\left(x - \dfrac{1}{5}\right)$ **75.** $-(x + 8)(x - 5)$ **77. a.** $-16(t - 2)(t + 1)$
b. 0; yes; After 2 seconds, you hit the water.   **83.** does not make sense   **85.** does not make sense   **87.** false   **89.** true   **91.** 8, 16
95. $x(x + 1)(x + 2)$; the product of three consecutive integers   **97.** correctly factored   **99.** $(x + 1)(x - 1)$ should be $(x - 1)(x - 1)$.

101. 13 or $\{13\}$ **102.**

103.   **104.** $2x^2 - x - 6$ **105.** $9x^2 + 15x + 4$ **106.** $(4x - 1)(2x - 5)$

Section 7.3

Check Point Exercises

1. $(5x - 4)(x - 2)$ **2.** $(3x - 1)(2x + 7)$ **3.** $(3x - y)(x - 4y)$ **4.** $(3x + 5)(x - 2)$ **5.** $(2x - 1)(4x - 3)$ **6.** $y^2(5y + 3)(y + 2)$

Exercise Set 7.3

1. $(2x + 3)(x + 1)$ **3.** $(3x + 1)(x + 4)$ **5.** $(2x + 3)(x + 4)$ **7.** $(5y - 1)(y - 3)$ **9.** $(3y + 4)(y - 1)$ **11.** $(3x - 2)(x + 5)$
13. $(3x - 1)(x - 7)$ **15.** $(5y - 1)(y - 3)$ **17.** $(3x - 2)(x - 5)$ **19.** $(3w - 4)(2w - 1)$ **21.** $(8x + 1)(x + 4)$
23. $(5x - 2)(x + 7)$ **25.** $(7y - 3)(2y + 3)$ **27.** prime   **29.** $(5z - 3)(5z - 3)$ **31.** $(3y + 1)(5y - 2)$ **33.** prime

35. $(5y - 1)(2y + 9)$ **37.** $(4x + 1)(2x - 1)$ **39.** $(3y - 1)(3y - 2)$ **41.** $(5x + 8)(4x - 1)$ **43.** $(2x + y)(x + y)$
45. $(3x + 2y)(x + y)$ **47.** $(2x - 3y)(x - 3y)$ **49.** $(2x - 3y)(3x + 2y)$ **51.** $(3x - 2y)(5x + 7y)$ **53.** $(2a + 5b)(a + b)$
55. $(3a - 2b)(5a + 3b)$ **57.** $(3x - 4y)(4x - 3y)$ **59.** $2(2x + 3)(x + 5)$ **61.** $3(3x + 4)(x - 2)$ **63.** $2(2y - 5)(y + 3)$
65. $3(3y - 4)(y + 5)$ **67.** $x(3x + 1)(x + 1)$ **69.** $x(2x - 5)(x + 1)$ **71.** $3y(3y - 1)(y - 4)$ **73.** $5z(6z + 1)(2z + 1)$
75. $3x^2(5x - 3)(x - 2)$ **77.** $x^3(2x - 3)(5x - 1)$ **79.** $3(2x + 3y)(x - 2y)$ **81.** $2(2x - y)(3x + 4y)$ **83.** $2y(4x - 7)(x + 6)$
85. $2b(2a - 7b)(3a - b)$ **87.** $10(y + 1)(x + 1)(3x - 2)$ **89.** $-4y^4(8x + 3)(x - 1)$ **91. a.** $(2x + 1)(x - 3)$ **b.** $(2y + 3)(y - 2)$
93. $(x - 2)(3x - 2)(x - 1)$ **95. a.** $x^2 + 3x + 2$ **b.** $(x + 2)(x + 1)$ **c.** $x^2 + 3x + 2 = (x + 2)(x + 1)$ **101.** makes sense
103. does not make sense   **105.** true   **107.** false   **109.** $5, 7, -5, -7$ **111.** $(3x^5 + 5)(x^5 - 3)$ **113.** $(25, -5)$ or $\{(25, -5)\}$
114. 8.6×10^{-4} **115.** -1 or $\{-1\}$ **116.** $81x^2 - 100$ **117.** $16x^2 + 40xy + 25y^2$ **118.** $x^3 + 8$

Mid-Chapter Check Point Exercises

1. $x^4(x + 1)$ **2.** $(x + 9)(x - 2)$ **3.** $x^2y(y^2 - y + 1)$ **4.** prime   **5.** $(7x - 1)(x - 3)$ **6.** $(x^2 + 3)(x + 5)$ **7.** $x(2x - 1)(x - 5)$
8. $(x - 4)(y - 7)$ **9.** $(x - 15y)(x - 2y)$ **10.** $(5x + 2)(5x - 7)$ **11.** $2(8x - 3)(x - 4)$ **12.** $(3x + 7y)(x + y)$

Section 7.4

Check Point Exercises

1. a. $(x + 9)(x - 9)$ **b.** $(6x + 5)(6x - 5)$ **2. a.** $(5 + 2x^5)(5 - 2x^5)$ **b.** $(10x + 3y)(10x - 3y)$ **3. a.** $2x(3x + 1)(3x - 1)$
b. $18(2 + x)(2 - x)$ **4.** $(9x^2 + 4)(3x + 2)(3x - 2)$ **5. a.** $(x + 7)^2$ **b.** $(x - 3)^2$ **c.** $(4x - 7)^2$ **6.** $(2x + 3y)^2$
7. $(x + 3)(x^2 - 3x + 9)$ **8.** $(1 - y)(1 + y + y^2)$ **9.** $(5x + 2)(25x^2 - 10x + 4)$

Exercise Set 7.4

1. $(x + 5)(x - 5)$ **3.** $(y + 1)(y - 1)$ **5.** $(2x + 3)(2x - 3)$ **7.** $(5 + x)(5 - x)$ **9.** $(1 + 7x)(1 - 7x)$ **11.** $(3 + 5y)(3 - 5y)$
13. $(x^2 + 3)(x^2 - 3)$ **15.** $(7y^2 + 4)(7y^2 - 4)$ **17.** $(x^5 + 3)(x^5 - 3)$ **19.** $(5x + 4y)(5x - 4y)$ **21.** $(x^2 + y^5)(x^2 - y^5)$
23. $(x^2 + 4)(x + 2)(x - 2)$ **25.** $(4x^2 + 9)(2x + 3)(2x - 3)$ **27.** $2(x + 3)(x - 3)$ **29.** $2x(x + 6)(x - 6)$ **31.** prime
33. $3x(x^2 + 9)$ **35.** $2(3 + y)(3 - y)$ **37.** $3y(y + 4)(y - 4)$ **39.** $2x(3x + 1)(3x - 1)$ **41.** $(x + 1)^2$ **43.** $(x - 7)^2$ **45.** $(x - 1)^2$
47. $(x + 11)^2$ **49.** $(2x + 1)^2$ **51.** $(5y - 1)^2$ **53.** prime   **55.** $(x + 7y)^2$ **57.** $(x - 6y)^2$ **59.** prime   **61.** $(4x - 5y)^2$
63. $3(2x - 1)^2$ **65.** $x(3x + 1)^2$ **67.** $2(y - 1)^2$ **69.** $2y(y + 7)^2$ **71.** $(x + 1)(x^2 - x + 1)$ **73.** $(x - 3)(x^2 + 3x + 9)$
75. $(2y - 1)(4y^2 + 2y + 1)$ **77.** $(3x + 2)(9x^2 - 6x + 4)$ **79.** $(xy - 4)(x^2y^2 + 4xy + 16)$ **81.** $y(3y + 2)(9y^2 - 6y + 4)$
83. $2(3 - 2y)(9 + 6y + 4y^2)$ **85.** $(4x + 3y)(16x^2 - 12xy + 9y^2)$ **87.** $(5x - 4y)(25x^2 + 20xy + 16y^2)$ **89.** $\left(5x + \dfrac{2}{7}\right)\left(5x - \dfrac{2}{7}\right)$

91. $y\left(y - \dfrac{1}{10}\right)\left(y^2 + \dfrac{y}{10} + \dfrac{1}{100}\right)$ **93.** $x(0.5 + x)(0.5 - x)$ **95.** $(x + 6)(x - 4)$ **97.** $(x - 3)(x + 1)^2$ **99.** $x^2 - 25 = (x + 5)(x - 5)$
101. $x^2 - 16 = (x + 4)(x - 4)$ **107.** does not make sense   **109.** does not make sense   **111.** false   **113.** false
115. $a - b = 0$ and division by 0 is not permitted.   **117.** $(x^n + 5y^n)(x^n - 5y^n)$ **119.** $[(x + 3) - 1]^2$ or $(x + 2)^2$ **121.** 1
123. correctly factored   **125.** $(x - 1)(x^2 - x + 1)$ should be $(x - 1)(x^2 + x + 1)$.   **126.** $80x^9y^{14}$ **127.** $-4x^2 + 3$ **128.** $2x + 5$
129. $3x(x + 5)(x - 5)$ **130.** $2(x - 5)^2$ **131.** $(x - 2)(x + 1)(x - 1)$

Section 7.5

Check Point Exercises

1. $5x^2(x + 3)(x - 3)$ **2.** $4(x - 6)(x + 2)$ **3.** $4x(x^2 + 4)(x + 2)(x - 2)$ **4.** $(x - 4)(x + 3)(x - 3)$ **5.** $3x(x - 5)^2$
6. $2x^2(x + 3)(x^2 - 3x + 9)$ **7.** $3y(x^2 + 4y^2)(x + 2y)(x - 2y)$ **8.** $3x(2x + 3y)^2$

Exercise Set 7.5

1. $5x(x + 2)(x - 2)$ **3.** $7x(x^2 + 1)$ **5.** $5(x - 3)(x + 2)$ **7.** $2(x^2 + 9)(x + 3)(x - 3)$ **9.** $(x + 2)(x + 3)(x - 3)$ **11.** $3x(x - 4)^2$
13. $2x^2(x + 1)(x^2 - x + 1)$ **15.** $2x(3x + 4)$ **17.** $2(y - 8)(y + 7)$ **19.** $7y^2(y + 1)^2$ **21.** prime   **23.** $2(4y + 1)(2y - 1)$
25. $r(r - 25)$ **27.** $(2w + 5)(2w - 1)$ **29.** $x(x + 2)(x - 2)$ **31.** prime   **33.** $(9y + 4)(y + 1)$ **35.** $(y + 2)(y + 2)(y - 2)$
37. $(4y + 3)^2$ **39.** $4y(y - 5)(y - 2)$ **41.** $y(y^2 + 9)(y + 3)(y - 3)$ **43.** $5a^2(2a + 3)(2a - 3)$ **45.** $3x^2(3x^2 + 6x + 2)$
47. $(4y - 1)(3y - 2)$ **49.** $(3y + 8)(3y - 8)$ **51.** prime   **53.** $(2y + 3)(y + 5)(y - 5)$ **55.** $2r(r + 17)(r - 2)$ **57.** $2x^3(2x + 1)(2x - 1)$
59. $3(x^2 + 81)$ **61.** $x(x + 2)(x^2 - 2x + 4)$ **63.** $2y^2(y - 1)(y^2 + y + 1)$ **65.** $2x(3x + 4y)$ **67.** $(y - 7)(x + 3)$ **69.** $(x - 4y)(x + y)$
71. $12a^2(6ab^2 + 1 - 2a^2b^2)$ **73.** $3(a + 6b)(a + 3b)$ **75.** $3x^2y(4x + 1)(4x - 1)$ **77.** $b(3a + 2)(2a - 1)$ **79.** $7xy(x^2 + y^2)(x + y)(x - y)$
81. $2xy(5x - 2y)(x - y)$ **83.** $2b(x + 11)^2$ **85.** $(5a + 7b)(3a - 2b)$ **87.** $2xy(9x - 2y)(2x - 3y)$ **89.** $(y - x)(a + b)(a - b)$
91. $ax(3x + 7)(3x - 2)$ **93.** $2x^2(x^2 + 3xy + y^2)$ **95.** $y(9x^2 + y^2)(3x + y)(3x - y)$ **97.** $(x + 1)(5x - 6)(2x + 1)$ **99.** $(x^2 + 6)(6x^2 - 1)$
101. $(x - 7 + 2a)(x - 7 - 2a)$ **103.** $(x + 4 + 5a)(x + 4 - 5a)$ **105.** $y(y^2 + 1)(y^4 - y^2 + 1)$ **107.** $16(4 + t)(4 - t)$
109. $\pi b^2 - \pi a^2$; $\pi(b + a)(b - a)$ **113.** makes sense   **115.** makes sense   **117.** false   **119.** false   **121.** $3x(x^2 + 2)(x + 3)(x - 3)$
123. $(4x^2 - 5)(x + 1)(x - 1)$ **125.** $3(x^n + 3y^n)(x^n - 3y^n)$ **127.** $3x(x + 5)(x - 1)$ should be $3x(x - 5)(x + 1)$.   **129.** correctly factored

131. $(3x + 4)(3x - 4)$ **132.**   **133.** $20°, 60°, 100°$ **134.** 0   **135.** 0   **136.** $(x + 4)(x - 3)$

Section 7.6

Check Point Exercises

1. $-\frac{1}{2}$ and 4, or $\left\{-\frac{1}{2}, 4\right\}$ **2.** 1 and 5, or $\{1, 5\}$ **3.** 0 and $\frac{1}{2}$, or $\left\{0, \frac{1}{2}\right\}$ **4.** 5 or $\{5\}$ **5.** $-\frac{5}{4}$ and $\frac{5}{4}$, or $\left\{-\frac{5}{4}, \frac{5}{4}\right\}$ **6.** -2 and 9, or $\{-2, 9\}$
7. 1 sec and 2 sec; $(1, 192)$ and $(2, 192)$ **8.** length: 9 ft; width: 6 ft

Exercise Set 7.6

1. -7 and 0, or $\{-7, 0\}$ **3.** -4 and 6, or $\{-4, 6\}$ **5.** $-\frac{4}{5}$ and 9, or $\left\{-\frac{4}{5}, 9\right\}$ **7.** $-\frac{9}{2}$ and 4, or $\left\{-\frac{9}{2}, 4\right\}$ **9.** -5 and -3, or $\{-5, -3\}$
11. -3 and 5, or $\{-3, 5\}$ **13.** -3 and 7, or $\{-3, 7\}$ **15.** -8 and -1, or $\{-8, -1\}$ **17.** -4 and 0, or $\{-4, 0\}$ **19.** 0 and 5, or $\{0, 5\}$
21. 0 and 4, or $\{0, 4\}$ **23.** 0 and $\frac{5}{2}$, or $\left\{0, \frac{5}{2}\right\}$ **25.** $-\frac{5}{3}$ and 0, or $\left\{-\frac{5}{3}, 0\right\}$ **27.** -2 or $\{-2\}$ **29.** 6 or $\{6\}$ **31.** $\frac{3}{2}$ or $\left\{\frac{3}{2}\right\}$
33. $-\frac{1}{2}$ and 4, or $\left\{-\frac{1}{2}, 4\right\}$ **35.** -2 and $\frac{9}{5}$, or $\left\{-2, \frac{9}{5}\right\}$ **37.** -7 and 7, or $\{-7, 7\}$ **39.** $-\frac{5}{2}$ and $\frac{5}{2}$, or $\left\{-\frac{5}{2}, \frac{5}{2}\right\}$ **41.** $-\frac{5}{9}$ and $\frac{5}{9}$, or $\left\{-\frac{5}{9}, \frac{5}{9}\right\}$
43. -3 and 7, or $\{-3, 7\}$ **45.** $-\frac{5}{2}$ and $\frac{3}{2}$, or $\left\{-\frac{5}{2}, \frac{3}{2}\right\}$ **47.** -6 and 3, or $\{-6, 3\}$ **49.** -2 and $-\frac{3}{2}$, or $\left\{-2, -\frac{3}{2}\right\}$ **51.** 4 or $\{4\}$
53. $-\frac{5}{2}$ or $\left\{-\frac{5}{2}\right\}$ **55.** $\frac{3}{8}$ or $\left\{\frac{3}{8}\right\}$ **57.** $-3, -2$, and 4, or $\{-3, -2, 4\}$ **59.** $-6, 0$, and 6, or $\{-6, 0, 6\}$ **61.** $-2, -1$, and 0, or $\{-2, -1, 0\}$
63. 2 and 8, or $\{2, 8\}$ **65.** 4 and 5, or $\{4, 5\}$ **67.** 5 sec; Each tick represents one second.   **69.** 2 sec; $(2, 276)$ **71.** $\frac{1}{2}$ sec and 4 sec
73. 2002 and 2004   **75.** $(2, 66)$ and $(4, 66)$ **77.** 10 yr   **79.** $(10, 7250)$ **81.** 10 teams   **83.** length: 15 yd; width: 12 yd
85. base: 5 cm; height: 6 cm   **87. a.** $4x^2 + 44x$ **b.** 3 ft   **91.** does not make sense   **93.** makes sense   **95.** false   **97.** false
99. $x^2 - 2x - 15 = 0$ **101.** 4 and 5, or $\{4, 5\}$ **103.** c   **105.** d   **107.** -4 and 1, or $\{-4, 1\}$ **109.** -4 and 3, or $\{-4, 3\}$
113.

114. $\dfrac{4}{x^6}$ **115.** -2 or $\{-2\}$ **116.** 375   **117.** When x is replaced with 4, the denominator is 0.
118. $\dfrac{(x + 5)(x + 1)}{(x + 5)(x - 5)} \cdot \dfrac{x + 1}{x - 5}$

Review Exercises

1. $15(2x - 3)$ **2.** $4x(3x^2 + 4x - 100)$ **3.** $5x^2y(6x^2 + 3x + 1)$ **4.** $5(x + 3)$ **5.** $(7x^2 - 1)(x + y)$ **6.** $(x^2 + 2)(x + 3)$
7. $(x + 1)(y + 4)$ **8.** $(x^2 + 5)(x + 1)$ **9.** $(x - 2)(y + 4)$ **10.** $(x - 2)(x - 1)$ **11.** $(x - 5)(x + 4)$ **12.** $(x + 3)(x + 16)$
13. $(x - 4y)(x - 2y)$ **14.** prime   **15.** $(x + 17y)(x - y)$ **16.** $3(x + 4)(x - 2)$ **17.** $3x(x - 11)(x - 1)$ **18.** $(x + 5)(3x + 2)$
19. $(y - 3)(5y - 2)$ **20.** $(2x + 5)(2x - 3)$ **21.** prime   **22.** $2(2x + 3)(2x - 1)$ **23.** $x(2x - 9)(x + 8)$ **24.** $4y(3y + 1)(y + 2)$
25. $(2x - y)(x - 3y)$ **26.** $(5x + 4y)(x - 2y)$ **27.** $(2x + 1)(2x - 1)$ **28.** $(9 + 10y)(9 - 10y)$ **29.** $(5a + 7b)(5a - 7b)$
30. $(z^2 + 4)(z + 2)(z - 2)$ **31.** $2(x + 3)(x - 3)$ **32.** prime   **33.** $x(3x + 1)(3x - 1)$ **34.** $2x(3y + 2)(3y - 2)$ **35.** $(x + 11)^2$
36. $(x - 8)^2$ **37.** $(3y + 8)^2$ **38.** $(4x - 5)^2$ **39.** prime   **40.** $(6x + 5y)^2$ **41.** $(5x - 4y)^2$ **42.** $(x - 3)(x^2 + 3x + 9)$
43. $(4x + 1)(16x^2 - 4x + 1)$ **44.** $2(3x - 2y)(9x^2 + 6xy + 4y^2)$ **45.** $y(3x + 2)(9x^2 - 6x + 4)$ **46.** $(a + 3)(a - 3)$
47. $(a + 2b)(a - 2b)$ **48.** $A^2 + 2A + 1 = (A + 1)^2$ **49.** $x(x - 7)(x - 1)$ **50.** $(5y + 2)(2y + 1)$ **51.** $2(8 + y)(8 - y)$
52. $(3x + 1)^2$ **53.** $4x^3(5x^4 - 9)$ **54.** $(x - 3)^2(x + 3)$ **55.** prime   **56.** $x(2x + 5)(x + 7)$ **57.** $3x(x - 5)^2$
58. $3x^2(x - 2)(x^2 + 2x + 4)$ **59.** $4y^2(y + 3)(y - 3)$ **60.** $5(x + 7)(x - 3)$ **61.** prime   **62.** $2x^3(5x - 2)(x - 4)$
63. $(10y + 7)(10y - 7)$ **64.** $9x^4(x - 2)$ **65.** $(x^2 + 1)(x + 1)(x - 1)$ **66.** $2(y - 2)(y^2 + 2y + 4)$ **67.** $(x + 4)(x^2 - 4x + 16)$
68. $(3x - 2)(2x + 5)$ **69.** $3x^2(x + 2)(x - 2)$ **70.** $(x - 10)(x + 9)$ **71.** $(5x + 2y)(5x + 3y)$ **72.** $x(x + 5)(x^2 - 5x + 25)$
73. $2y(4y + 3)(4y + 1)$ **74.** $2(y - 4)^2$ **75.** $(x + 5y)(x - 7y)$ **76.** $(x + y)(x + 7)$ **77.** $(3x + 4y)^2$ **78.** $2x^2y(x + 1)(x - 1)$
79. $(10y + 7z)(10y - 7z)$ **80.** prime   **81.** $3x^2y^2(x + 2y)(x - 2y)$ **82.** 0 and 12, or $\{0, 12\}$ **83.** $-\frac{9}{4}$ and 7, or $\left\{-\frac{9}{4}, 7\right\}$
84. -7 and 2, or $\{-7, 2\}$ **85.** -4 and 0, or $\{-4, 0\}$ **86.** -8 and $\frac{1}{2}$, or $\left\{-8, \frac{1}{2}\right\}$ **87.** -4 and 8, or $\{-4, 8\}$ **88.** -8 and 7, or $\{-8, 7\}$
89. 7 or $\{7\}$ **90.** $-\frac{10}{3}$ and $\frac{10}{3}$, or $\left\{-\frac{10}{3}, \frac{10}{3}\right\}$ **91.** -5 and -2, or $\{-5, -2\}$ **92.** $\frac{1}{3}$ and 7, or $\left\{\frac{1}{3}, 7\right\}$
93. 2 sec   **94.** width: 5 ft; length: 8 ft   **95.** 11 m by 11 m

Chapter Test

1. $(x - 3)(x - 6)$ **2.** $(x - 7)^2$ **3.** $5y^2(3y - 1)(y - 2)$ **4.** $(x^2 + 3)(x + 2)$ **5.** $x(x - 9)$ **6.** $x(x + 7)(x - 1)$
7. $2(7x - 3)(x + 5)$ **8.** $(5x + 3)(5x - 3)$ **9.** $(x + 2)(x^2 - 2x + 4)$ **10.** $(x + 3)(x - 7)$ **11.** prime   **12.** $3y(2y + 1)(y + 1)$

13. $4(y + 3)(y - 3)$ **14.** $4(2x + 3)^2$ **15.** $2(x^2 + 4)(x + 2)(x - 2)$ **16.** $(6x - 7)^2$ **17.** $(7x - 1)(x - 7)$ **18.** $(x^2 - 5)(x + 2)$
19. $3y(2y + 3)(2y - 5)$ **20.** $(y - 5)(y^2 + 5y + 25)$ **21.** $5(x - 3y)(x + 2y)$ **22.** -6 and 4, or $\{-6, 4\}$
23. $-\dfrac{1}{3}$ and 2, or $\left\{-\dfrac{1}{3}, 2\right\}$ **24.** -2 and 8, or $\{-2, 8\}$ **25.** 0 and $\dfrac{7}{2}$, or $\left\{0, \dfrac{7}{2}\right\}$ **26.** $-\dfrac{9}{4}$ and $\dfrac{9}{4}$, or $\left\{-\dfrac{9}{4}, \dfrac{9}{4}\right\}$
27. -1 and $\dfrac{6}{5}$, or $\left\{-1, \dfrac{6}{5}\right\}$ **28.** $x^2 - 4 = (x + 2)(x - 2)$ **29.** 6 sec   **30.** width: 5 ft; length: 11 ft

Cumulative Review Exercises

1. -48 **2.** 0 or $\{0\}$ **3.** 12 or $\{12\}$ **4.** $(-\infty, -2)$ or $\{x | x < -2\}$;   **5.** $38°, 38°, 104°$ **6.** 150

7.   **8.** $y + 4 = 5(x - 2)$ or $y - 1 = 5(x - 3)$; $y = 5x - 14$ **9.**

10. $(3, -1)$ or $\{(3, -1)\}$ **11.** $\left(-\dfrac{2}{13}, \dfrac{23}{13}\right)$ or $\left\{\left(-\dfrac{2}{13}, \dfrac{23}{13}\right)\right\}$ **12.** $-\dfrac{13}{40}$ **13.** $2x^4 - x^3 + 3x + 9$ **14.** $6x^2 + 17xy - 45y^2$
15. $2x^2 + 5x - 3 - \dfrac{2}{3x - 5}$ **16.** 7.1×10^{-3} **17.** $(3x + 2)(x + 3)$ **18.** $y(y^2 + 4)(y + 2)(y - 2)$ **19.** $(2x + 3)^2$
20. width: 4 ft; length: 6 ft

CHAPTER 8

Section 8.1

Check Point Exercises

1. a. $x = 5$ **b.** $x = -7$ and $x = 4$ **2.** $\dfrac{x + 4}{3x}$ **3.** $\dfrac{x^2}{7}$ **4.** $\dfrac{x - 1}{x + 1}$ **5.** $-\dfrac{3x + 7}{4}$

Exercise Set 8.1

1. $x = 0$ **3.** $x = 8$ **5.** $x = 4$ **7.** $x = -9$ and $x = 2$ **9.** $x = \dfrac{17}{3}$ and $x = -3$ **11.** $x = -4$ and $x = 3$
13. defined for all real numbers   **15.** $y = -1$ and $y = \dfrac{3}{4}$ **17.** $y = -5$ and $y = 5$ **19.** defined for all real numbers   **21.** $2x$
23. $\dfrac{x - 3}{5}$ **25.** $\dfrac{x - 4}{2x}$ **27.** $\dfrac{1}{x - 3}$ **29.** $-\dfrac{5}{x - 3}$ **31.** 3 **33.** $\dfrac{1}{x - 5}$ **35.** $\dfrac{2}{3}$ **37.** $\dfrac{1}{x - 3}$ **39.** $\dfrac{4}{x - 2}$ **41.** $\dfrac{y - 1}{y + 9}$
43. $\dfrac{y - 3}{y - 2}$ **45.** cannot be simplified   **47.** $\dfrac{x + 6}{x - 6}$ **49.** $x^2 + 1$ **51.** $x^2 + 2x + 4$ **53.** $\dfrac{x - 4}{x + 4}$ **55.** cannot be simplified
57. cannot be simplified   **59.** -1 **61.** -1 **63.** cannot be simplified   **65.** -2 **67.** $-\dfrac{2}{x}$ **69.** $-x - 1$ **71.** $-y - 3$ **73.** $-\dfrac{1}{x}$
75. $\dfrac{x - y}{2x - y}$ **77.** $\dfrac{x - 6}{x^2 + 3x + 9}$ **79.** $\dfrac{3 + y}{3 - y}$ **81.** $\dfrac{y + 3}{x + 3}$ **83.** $\dfrac{2}{1 - 2x}$ **85. a.** It costs $86.67 million to inoculate 40% of the population.
It costs $520 million to inoculate 80% of the population. It costs $1170 million to inoculate 90% of the population.   **b.** $x = 100$
c. The cost keeps rising.; No amount of money will be enough to inoculate 100% of the population.   **87.** 400 mg   **89. a.** $300   **b.** $125
c. decrease   **91.** 1.5 mg per liter; (3, 1.5)   **93. a.** 1924; underestimates by 76 calories   **b.** 2332; underestimates by 68 calories
c. $\dfrac{-41x^2 + 327x + 310}{-48x^2 + 401x + 330}$ **101.** makes sense   **103.** does not make sense   **105.** false   **107.** false   **111.** correctly simplified
113. $x^2 - 1$ should be $x - 1$.   **115.** $\dfrac{3}{10}$ **116.** $\dfrac{1}{6}$ **117.** $(4, 2)$ or $\{(4, 2)\}$ **118.** $\dfrac{6}{35}$ **119.** $\dfrac{3}{2}$ **120.** $\dfrac{2}{3}$

Section 8.2

Check Point Exercises

1. $\dfrac{9x - 45}{2x + 8}$ **2.** $\dfrac{3}{8}$ **3.** $\dfrac{x + 2}{9}$ **4.** $-\dfrac{5(x + 1)}{7x(2x - 3)}$ **5.** $\dfrac{(x + 3)(x + 7)}{x - 4}$ **6.** $\dfrac{x + 3}{x - 5}$ **7.** $\dfrac{y + 1}{5y(y^2 + 1)}$

Exercise Set 8.2

1. $\dfrac{4x - 20}{9x + 27}$ **3.** $\dfrac{4x}{x + 5}$ **5.** $\dfrac{4}{5}$ **7.** $\dfrac{4}{9}$ **9.** 1 **11.** $\dfrac{x + 5}{x}$ **13.** $\dfrac{2}{y}$ **15.** $\dfrac{y + 2}{y + 4}$ **17.** $4(y + 3)$ **19.** $\dfrac{x - 1}{x + 2}$ **21.** $\dfrac{x^2 + 2x + 4}{3x}$
23. $\dfrac{x - 2}{x - 1}$ **25.** $-\dfrac{2}{x(x + 1)}$ **27.** $-\dfrac{y - 10}{y - 7}$ **29.** $(x - y)(x + y)$ **31.** $\dfrac{4(x + y)}{3(x - y)}$ **33.** $\dfrac{3x}{35}$ **35.** $\dfrac{1}{4}$ **37.** 10 **39.** $\dfrac{7}{9}$ **41.** $\dfrac{3}{4}$

43. $\dfrac{(x-2)^2}{x}$　**45.** $\dfrac{(y-4)(y^2+4)}{y-1}$　**47.** $\dfrac{y}{3}$　**49.** $\dfrac{2(x+3)}{3}$　**51.** $\dfrac{x-5}{2}$　**53.** $\dfrac{y^2+1}{y^2}$　**55.** $\dfrac{y+1}{y-7}$　**57.** 6　**59.** $\dfrac{2}{3}$

61. $\dfrac{(x+y)^2}{32(x-y)^2}$　**63.** $\dfrac{y(x+y)}{(x+1)^2}$　**65.** $\dfrac{y+3}{y-3}$　**67.** $\dfrac{5(x-5)}{4}$　**69.** $\dfrac{(x+y)^2}{x-y}$　**71.** $\dfrac{2}{y^2-by+b^2}$　**73.** $\dfrac{x-2}{x(x-1)}$ sq in.　**75.** $\dfrac{3x}{x+2}$ sq in.

81. makes sense　**83.** makes sense　**85.** true　**87.** false　**89.** numerator: -3; denominator: $2x-3$　**91.** correct answer

93. $x-3$ should be $x+3$.　**95.** $(18,\infty)$ or $\{x\,|\,x>18\}$　**96.** $3(x-7)(x+2)$　**97.** -5 and $\dfrac{1}{2}$, or $\left\{-5,\dfrac{1}{2}\right\}$　**98.** $\dfrac{2}{3}$

99. x　**100.** $\dfrac{x-3}{x+3}$

Section 8.3

Check Point Exercises

1. $x+2$　**2.** $\dfrac{x-5}{x+5}$　**3. a.** $\dfrac{3x+5}{x+7}$　**b.** $3x-4$　**4.** $\dfrac{3y+2}{y-4}$　**5.** $x+3$　**6.** $\dfrac{-4x^2+12x}{x^2-2x-9}$

Exercise Set 8.3

1. $\dfrac{9x}{13}$　**3.** $\dfrac{3x}{5}$　**5.** $\dfrac{x+3}{2}$　**7.** $\dfrac{6}{x}$　**9.** $\dfrac{7}{3x}$　**11.** $\dfrac{9}{x+3}$　**13.** $\dfrac{5x+5}{x-3}$　**15.** 2　**17.** $\dfrac{2y+3}{y-5}$　**19.** $\dfrac{7}{5y}$　**21.** $\dfrac{2}{x+2}$　**23.** $\dfrac{x-2}{x-3}$

25. $\dfrac{3x-4}{5x-4}$　**27.** 1　**29.** 1　**31.** $\dfrac{x}{2x-1}$　**33.** $-\dfrac{1}{y}$　**35.** $\dfrac{y+3}{3y+8}$　**37.** $\dfrac{3y+2}{y-3}$　**39.** $\dfrac{2}{x-3}$　**41.** $\dfrac{3x+7}{x-6}$　**43.** $\dfrac{3x+1}{3x-4}$

45. $x+2$　**47.** 0　**49.** $\dfrac{11}{x-1}$　**51.** $\dfrac{12}{x+3}$　**53.** $\dfrac{y+1}{y-1}$　**55.** $\dfrac{x-2}{x-7}$　**57.** $\dfrac{2x-4}{x^2-25}$　**59.** 1　**61.** $\dfrac{2}{x+y}$　**63.** $\dfrac{x-3}{x-1}$

65. $\dfrac{4b}{4b-3}$　**67.** $\dfrac{y}{y-5}$　**69.** $-\dfrac{1}{c+d}$　**71.** $\dfrac{3y^2+8y-5}{(y+1)(y-4)}$　**73. a.** $\dfrac{100W}{L}$　**b.** round　**75.** 10 m　**81.** does not make sense

83. makes sense　**85.** false　**87.** false　**89.** $\dfrac{x+1}{x+2}$　**91.** $2x+1$　**93.** -7　**95.** $-4x-1$　**97.** $x-2$ should be $x-3$.

99. $\dfrac{31}{45}$　**100.** $(9x^2+1)(3x+1)(3x-1)$　**101.** $3x^2-7x-5$　**102.** $\dfrac{7}{6}$　**103.** $-\dfrac{17}{24}$　**104.** $\dfrac{y-2}{4y}$

Section 8.4

Check Point Exercises

1. $30x^2$　**2.** $(x+3)(x-3)$　**3.** $7x(x+4)(x+4)$ or $7x(x+4)^2$　**4.** $\dfrac{9+14x}{30x^2}$　**5.** $\dfrac{6x+6}{(x+3)(x-3)}$　**6.** $-\dfrac{5}{x+5}$

7. $-\dfrac{5+y}{5y}$　**8.** $\dfrac{x-15}{(x+5)(x-5)}$

Exercise Set 8.4

1. $120x^2$　**3.** $30x^5$　**5.** $(x-3)(x+1)$　**7.** $7y(y+2)$　**9.** $(x+4)(x-4)$　**11.** $y(y+3)(y-3)$　**13.** $(y+1)(y-1)(y-1)$

15. $(x-5)(x+4)(2x-1)$　**17.** $\dfrac{3x+5}{x^2}$　**19.** $\dfrac{37}{18x}$　**21.** $\dfrac{8x+7}{2x^2}$　**23.** $\dfrac{6x+1}{x}$　**25.** $\dfrac{2+9x}{x}$　**27.** $\dfrac{x+1}{2}$　**29.** $\dfrac{7x-20}{x(x-5)}$

31. $\dfrac{5x+1}{(x-1)(x+2)}$　**33.** $\dfrac{11y+15}{4y(y+5)}$　**35.** $-\dfrac{7}{x+7}$　**37.** $\dfrac{3x-55}{(x+5)(x-5)}$　**39.** $\dfrac{x^2+6x}{(x-4)(x+4)}$　**41.** $\dfrac{y+12}{(y+3)(y-3)}$

43. $\dfrac{7x-10}{(x-1)(x-1)}$　**45.** $\dfrac{9y}{4(y-5)}$　**47.** $\dfrac{8y+16}{y(y+4)}$　**49.** $\dfrac{17}{(x-3)(x-4)}$　**51.** $\dfrac{7x-1}{(x+1)(x+1)(x-1)}$　**53.** $\dfrac{x^2-x}{(x+3)(x-2)(x+5)}$

55. $\dfrac{y^2+8y+4}{(y+4)(y+1)(y+1)}$　**57.** $\dfrac{2x^2-4x+34}{(x+3)(x-5)}$　**59.** $\dfrac{5-3y}{2y(y-1)}$　**61.** $-\dfrac{x^2}{(x+3)(x-3)}$　**63.** $\dfrac{2}{y+5}$　**65.** $\dfrac{4x-11}{x-3}$　**67.** $\dfrac{3}{y+1}$

69. $\dfrac{-x^2+7x+3}{(x-3)(x+2)}$　**71.** $\dfrac{x^2+2x-1}{(x+1)(x-1)(x+2)}$　**73.** $\dfrac{-y^2+8y+9}{15y^2}$　**75.** $\dfrac{x^2-2x-6}{3(x+2)(x-2)}$　**77.** $\dfrac{-y-2}{(y+1)(y-1)}$

79. $\dfrac{x+2xy-y}{xy}$　**81.** $\dfrac{5x+2y}{(x+y)(x-y)}$　**83.** $\dfrac{-x+6}{(x+2)(x-2)}$　**85.** $\dfrac{6x+5}{(x-5)(x+5)(x-6)}$　**87.** $\dfrac{9}{(x-3)(x^2+3x+9)}$

89. $\dfrac{3}{y-3}$　**91.** $-\dfrac{22y}{(x+3y)(x+y)(x-3y)}$　**93.** $\dfrac{D}{5}$　**95.** $\dfrac{D}{24}$　**97.** no　**99.** 5 yr　**101.** $\dfrac{4x^2+14x}{(x+3)(x+4)}$

109. makes sense　**111.** does not make sense　**113.** false　**115.** true　**117.** $-\dfrac{1}{x(x+h)}$　**119.** $\dfrac{2}{x+1}$　**120.** $6x^2-11x-35$

121. $3x-y<3$ 　**122.** $y=x-1$　**123. a.** $\dfrac{11}{15}$　**b.** $\dfrac{1}{15}$　**c.** 11　**124. a.** $\dfrac{y+x}{xy}$　**b.** $\dfrac{1}{y+x}$　**125.** $y+x$

Mid-Chapter Check Point Exercises

1. $x = -2$ and $x = 4$ **2.** $\dfrac{x-2}{2x+1}$ **3.** $\dfrac{-3}{y-2}$ or $\dfrac{3}{2-y}$ **4.** $\dfrac{2}{w}$ **5.** $\dfrac{4}{x+4}$ **6.** $\dfrac{4}{(x-2)^2}$ **7.** $\dfrac{x+5}{x-2}$ **8.** $\dfrac{x+2}{x-4}$

9. $\dfrac{2x+1}{(x+2)(x+3)(x-1)}$ **10.** $\dfrac{9-x}{x-5}$ **11.** $-\dfrac{2y-1}{3y}$ or $\dfrac{1-2y}{3y}$ **12.** $\dfrac{-y^2}{(y+1)(y+2)}$ **13.** $\dfrac{w+1}{(w-3)(w+5)}$

14. $\dfrac{2z^2-3z+15}{(z+3)(z-3)(z+1)}$ **15.** $\dfrac{3z^2+5z+3}{(3z-1)^2}$ **16.** $\dfrac{3x-1}{(x+7)(x-3)}$ **17.** x **18.** $\dfrac{2x+1}{(x+1)(x-2)(x+3)}$ **19.** $\dfrac{5x-5y}{x}$ **20.** $\dfrac{x+3}{x+5}$

Section 8.5

Check Point Exercises

1. $\dfrac{11}{5}$ **2.** $\dfrac{2x-1}{2x+1}$ **3.** $y-x$ **4.** $\dfrac{11}{5}$ **5.** $\dfrac{2x-1}{2x+1}$ **6.** $y-x$

Exercise Set 8.5

1. $\dfrac{9}{10}$ **3.** $\dfrac{18}{23}$ **5.** $-\dfrac{4}{5}$ **7.** $\dfrac{3-4x}{3+4x}$ **9.** $\dfrac{7x-2}{5x+1}$ **11.** $\dfrac{2y+3}{y-7}$ **13.** $\dfrac{4-6y}{4+3y}$ **15.** $x-5$ **17.** $\dfrac{x}{x-1}$ **19.** $-\dfrac{1}{y}$

21. $\dfrac{xy+2}{x}$ **23.** $\dfrac{y+x}{x^2y^2}$ **25.** $\dfrac{x^2+y}{y(y+1)}$ **27.** $\dfrac{1}{y}$ **29.** $\dfrac{4-x}{5x-3}$ **31.** $\dfrac{2y}{y-3}$ **33.** $\dfrac{1}{x+3}$ **35.** $\dfrac{x^2-5x+3}{x^2-7x+2}$ **37.** $\dfrac{2y^2+3xy}{y^2+2x}$

39. $-\dfrac{6}{5}$ **41.** $\dfrac{1-x}{(x-3)(x+6)}$ **43.** $\dfrac{1}{y(y+5)}$ **45.** $\dfrac{1}{x-1}$ **47.** $\dfrac{x+1}{2x+1}$ **49.** $\dfrac{2r_1r_2}{r_1+r_2}$; $34\dfrac{2}{7}$ mph   **51. a.** $\dfrac{11x^2+40x+1040}{12x^2+230x+2190}$

b. 63%   **c.** 64%; overestimates by 1%   **57.** makes sense   **59.** does not make sense   **61.** true   **63.** false   **65.** $\dfrac{2y^2}{y+1}$

67. correct answer   **69.** $x+\dfrac{1}{3}$ should be $3+x$.   **70.** $2x(x-5)^2$ **71.** -2 or $\{-2\}$ **72.** x^3+y^3 **73.** 1 or $\{1\}$ **74.** 4 or $\{4\}$

75. $\dfrac{1}{2}$ and 2, or $\left\{\dfrac{1}{2}, 2\right\}$

Section 8.6

Check Point Exercises

1. 4 or $\{4\}$ **2.** 3 or $\{3\}$ **3.** -3 and -2, or $\{-3, -2\}$ **4.** 24 or $\{24\}$ **5.** no solution or $\varnothing$ **6.** 75%

Exercise Set 8.6

1. 12 or $\{12\}$ **3.** 0 or $\{0\}$ **5.** -2 or $\{-2\}$ **7.** 6 or $\{6\}$ **9.** -2 or $\{-2\}$ **11.** 2 or $\{2\}$ **13.** 15 or $\{15\}$ **15.** 4 or $\{4\}$

17. -6 and -1, or $\{-6, -1\}$ **19.** -5 and 5, or $\{-5, 5\}$ **21.** -3 and 3, or $\{-3, 3\}$ **23.** -8 and 1, or $\{-8, 1\}$ **25.** -1 and 16, or $\{-1, 16\}$

27. $\dfrac{3}{2}$ or $\left\{\dfrac{3}{2}\right\}$ **29.** 3 or $\{3\}$ **31.** no solution or $\varnothing$ **33.** $\dfrac{2}{3}$ and 4, or $\left\{\dfrac{2}{3}, 4\right\}$ **35.** no solution or $\varnothing$ **37.** 1 or $\{1\}$ **39.** $\dfrac{1}{5}$ or $\left\{\dfrac{1}{5}\right\}$

41. -3 or $\{-3\}$ **43.** no solution or $\varnothing$ **45.** -6 and $-\dfrac{1}{2}$, or $\left\{-6, -\dfrac{1}{2}\right\}$ **47.** -3 or $\{-3\}$ **49.** $\dfrac{-6x-18}{(x-5)(x+4)}$

51. 1 and 5, or $\{1, 5\}$ **53.** 3 or $\{3\}$ **55.** 10,000 wheelchairs   **57.** 50%   **59.** 5 yr old

61. either 50 or approx 67 cases; $(50, 350)$ or $\left(66\dfrac{2}{3}, 350\right)$ **69.** makes sense   **71.** does not make sense   **73.** true   **75.** false

77. $f_2 = \dfrac{-ff_1}{f-f_1}$ or $f_2 = \dfrac{ff_1}{f_1-f}$ **79.** no solution or $\varnothing$ **81.** 8 or $\{8\}$ **83.** -3 and -2, or $\{-3, -2\}$ **84.** $(x^3-3)(x+2)$

85. $-\dfrac{12}{x^8}$ **86.** $-20x+55$ **87.** 2 or $\{2\}$ **88.** $\dfrac{1}{5}, \dfrac{3}{5}, \dfrac{x}{5}$ **89. a.** 96   **b.** $y = \dfrac{96}{x}$ **c.** 32

Section 8.7

Check Point Exercises

1. 2 mph   **2.** $2\dfrac{2}{3}$ hr or 2 hr 40 min   **3.** 32 in.   **4.** 32 yd   **5.** 48 gal   **6.** 4.8 pounds per square inch

Exercise Set 8.7

1. walking rate: 6 mph; car rate: 9 mph   **3.** downhill rate: 10 mph; uphill rate: 6 mph   **5.** 5 mph

7. walking rate: 3 mph; jogging rate: 6 mph   **9.** 10 mph   **11.** It will take about 8.6 min, which is enough time.

13. It will take about 171.4 hr, which is not enough time.   **15.** 2.4 hr, or 2 hr 24 min   **17.** 5 in.   **19.** 6 m   **21.** 16 in.   **23.** 16 ft

25. $g = kh$ **27.** $w = \dfrac{k}{v}$ **29.** 20   **31.** 6000   **33.** 84   **35.** 25   **37. a.** $G = kW$ **b.** $G = 0.02W$ **c.** 1.04 in.   **39.** $60

41. 2442 mph   **43.** 0.5 hr or 30 min   **45.** 6.4 lb per cm^2 **47.** 5000 pens   **63.** does not make sense   **65.** makes sense

67. 30 mph   **69.** 7 hr   **71.** 2.5 hr   **73.** 7% and 8%   **74.** $(5x+9)(5x-9)$ **75.** 6 or $\{6\}$

76.

77. 4   **78.** 50   **79.** −31

Review Exercises

1. $x = 4$ **2.** $x = 2$ and $x = -5$ **3.** $x = 1$ and $x = 2$ **4.** defined for all real numbers   **5.** $\dfrac{4x}{3}$ **6.** $x + 2$ **7.** x^2 **8.** $\dfrac{x - 3}{x - 6}$

9. $\dfrac{x - 5}{x + 7}$ **10.** $\dfrac{y}{y + 2}$ **11.** cannot be simplified   **12.** $-2(x + 3y)$ **13.** $\dfrac{x - 2}{4}$ **14.** $\dfrac{5}{2}$ **15.** $\dfrac{x + 3}{x + 2}$ **16.** $\dfrac{2y - 1}{5}$ **17.** $\dfrac{-3(y + 1)}{5y(2y - 3)}$

18. $\dfrac{x - 1}{4}$ **19.** $\dfrac{2}{x(x + 1)}$ **20.** $\dfrac{1}{7(y + 3)}$ **21.** $(y - 3)(y - 6)$ **22.** $\dfrac{8}{x - y}$ **23.** 4   **24.** 4   **25.** $3x - 5$ **26.** $3y + 4$

27. $\dfrac{4}{x - 2}$ **28.** $\dfrac{2x + 5}{x - 3}$ **29.** $36x^3$ **30.** $x^2(x - 1)^2$ **31.** $(x + 3)(x + 1)(x + 7)$ **32.** $\dfrac{14x + 15}{6x^2}$ **33.** $\dfrac{7x + 2}{x(x + 1)}$ **34.** $\dfrac{7x + 25}{(x + 3)^2}$

35. $\dfrac{3}{y - 2}$ **36.** $-\dfrac{y}{y - 1}$ **37.** $\dfrac{x^2 + y^2}{xy}$ **38.** $\dfrac{3x^2 - x}{(x + 1)^2(x - 1)}$ **39.** $\dfrac{5x^2 - 3x}{(x + 1)(x - 1)}$ **40.** $\dfrac{4}{(x + 2)(x - 2)(x - 3)}$ **41.** $\dfrac{2x + 13}{x + 3}$

42. $\dfrac{y - 4}{3y}$ **43.** $\dfrac{7}{2}$ **44.** $\dfrac{1}{x - 1}$ **45.** $x + y$ **46.** $\dfrac{3}{x}$ **47.** $\dfrac{3x}{x - 4}$ **48.** 12 or {12}   **49.** −1 or {−1}   **50.** −6 and 1, or {−6, 1}

51. no solution or $\varnothing$ **52.** 5 or {5}   **53.** −6 and 2, or {−6, 2}   **54.** 5 or {5}   **55.** −5 or {−5}   **56.** 3 yr   **57.** 30%   **58.** 4 mph

59. Slower car's rate is 20 mph; faster car's rate is 30 mph.   **60.** 4 hr   **61.** 5 ft   **62.** $12\frac{1}{2}$ ft   **63.** \$154   **64.** 5 hr

Chapter Test

1. $x = -9$ and $x = 4$ **2.** $\dfrac{x + 3}{x - 2}$ **3.** $\dfrac{4x}{x + 1}$ **4.** $\dfrac{x - 4}{2}$ **5.** $\dfrac{x}{x + 3}$ **6.** $\dfrac{2x + 6}{x + 1}$ **7.** $\dfrac{5}{(y - 1)(y - 3)}$ **8.** $2y$

9. $\dfrac{2}{y + 3}$ **10.** $\dfrac{x^2 + 2x + 15}{(x + 3)(x - 3)}$ **11.** $\dfrac{8x - 14}{(x + 2)(x - 1)(x - 3)}$ **12.** $\dfrac{-x - 1}{x - 3}$ **13.** $\dfrac{x + 2}{x - 1}$ **14.** $\dfrac{11}{(x - 3)(x - 4)}$ **15.** $\dfrac{4}{y + 4}$

16. $\dfrac{3x - 3y}{x}$ **17.** $\dfrac{5x + 5}{2x + 1}$ **18.** $\dfrac{y - x}{y}$ **19.** 6 or {6}   **20.** −8 or {−8}   **21.** 0 and 2, or {0, 2}   **22.** 2 mph

23. 12 min   **24.** 3.2 in.   **25.** 52.5 amp

Cumulative Review Exercises

1. 2 or {2}   **2.** $(-\infty, 2)$ or $\{x \mid x < 2\}$ **3.** −6 and 3, or {−6, 3}   **4.** −6 or {−6}   **5.** (3, 3) or {(3, 3)}   **6.** $(-2, 2)$ or $\{(-2, 2)\}$

7. $3x - 2y = 6$

8. $y > -2x + 3$

9. $y = -3$

10. −19   **11.** $8x^9$ **12.** $\dfrac{1 - 2x}{4x - 1}$ **13.** $(4x - 1)(x - 3)$ **14.** $(2x - 5)^2$ **15.** $3(x + 5)(x - 5)$ **16.** $-x^2 + 4x + 8$

17. $-2x^4 + 3x^2 - 1$ **18.** $\dfrac{3x^2 + 6x + 16}{(x - 2)(x + 3)}$ **19.** \$1225 at 5% and \$2775 at 9%   **20.** 17 in. and 51 in.

CHAPTER 9

Section 9.1

Check Point Exercises

1. a. 9   **b.** −3   **c.** $\dfrac{1}{5}$ **d.** 10   **e.** 14   **2.** $\dfrac{1}{2}$ in.   **3.** 81.5%; underestimates by 0.2%

4. a. 1   **b.** −3   **c.** $\dfrac{1}{5}$ **5. a.** 3   **b.** not a real number   **c.** −3   **d.** $-\dfrac{1}{2}$

Exercise Set 9.1

1. 6   **3.** −6   **5.** not a real number   **7.** $\frac{1}{3}$ **9.** $\frac{1}{10}$ **11.** $-\frac{1}{6}$ **13.** not a real number   **15.** 0.2   **17.** 5   **19.** 8   **21.** 13

23. 17   **25.** not a real number   **27.**   **29.** Answers will vary; example: Exercise 27 graph is shaped like $y = \sqrt{x}$, but shifted 1 unit to the right.   **31.** 2.646   **33.** 4.796   **35.** −8.062

37. 15.317   **39.** 7.658   **41.** 2.153   **43.** 2.828

45. not a real number   **47.** 4   **49.** −3   **51.** −2   **53.** $\frac{1}{5}$ **55.** −10

57. 1   **59.** 2   **61.** −2   **63.** not a real number   **65.** −1   **67.** not a real number   **69.** −4   **71.** 2   **73.** −2   **75.** $x \geq 0$

77. $x \geq 2$ **79.** $x \leq 2$ **81.** all real numbers   **83.** $x \leq 6$ **85.** 12 mph   **87.** 70 mph; He was speeding.   **89. a.** 36 cm   **b.** 44.7 cm

c. 46.9 cm   **d.** describes healthy children   **91. a.** $y = a\sqrt{x} + 21$ **b.** $y = 20.8\sqrt{x} + 21$ **c.** 67.5%; underestimates by 0.5%   **d.** 93.1%

99. does not make sense   **101.** makes sense   **103.** false   **105.** false   **107.** 2   **109.** 0

111.   **113.**   **114.** $(-\infty, -8)$ or $\{x | x < -8\}$

115. $\frac{x+9}{x-15}$ **116. a.** 10   **b.** 10   **c.** $\sqrt{25} \cdot \sqrt{4} = \sqrt{25 \cdot 4}$ **117. a.** 22.361   **b.** 22.361   **c.** $\sqrt{500} = 10\sqrt{5}$

118. a. 4   **b.** 4   **c.** $\frac{\sqrt{64}}{\sqrt{4}} = \sqrt{\frac{64}{4}}$

Section 9.2

Check Point Exercises

1. a. $\sqrt{30}$ **b.** $\sqrt{26xy}$ **c.** 5   **d.** $\sqrt{\frac{15}{22}}$ **2. a.** $2\sqrt{3}$ **b.** $2\sqrt{15}$ **c.** cannot be simplified   **3.** $2x^8\sqrt{10}$ **4.** $3x^4\sqrt{3x}$

5. $3x^6\sqrt{5x}$ **6. a.** $\frac{7}{5}$ **b.** $4x^2$ **7. a.** $2\sqrt[3]{5}$ **b.** $2\sqrt[5]{2}$ **c.** $\frac{5}{3}$

Exercise Set 9.2

1. $\sqrt{14}$ **3.** $\sqrt{15xy}$ **5.** 5   **7.** $\sqrt{\frac{10}{21}}$ **9.** $\sqrt{0.5xy}$ **11.** $\frac{1}{5}\sqrt{ab}$ **13.** $\sqrt{x}$ **15.** $5\sqrt{2}$ **17.** $3\sqrt{5}$ **19.** $10\sqrt{2}$ **21.** $5\sqrt{3x}$

23. $3\sqrt{x}$ **25.** cannot be simplified   **27.** y **29.** $8x$ **31.** $x\sqrt{11}$ **33.** $2x\sqrt{2}$ **35.** x^{10} **37.** $5y^5$ **39.** $2x^3\sqrt{5}$ **41.** $6y^{50}\sqrt{2}$

43. $x\sqrt{x}$ **45.** $x^3\sqrt{x}$ **47.** $y^8\sqrt{y}$ **49.** $5x^2\sqrt{x}$ **51.** $2x^8\sqrt{2x}$ **53.** $3y^9\sqrt{10y}$ **55.** $3\sqrt{5}$ **57.** $5\sqrt{2xy}$ **59.** $6x$ **61.** $3x\sqrt{5x}$

63. $5x^6\sqrt{3x}$ **65.** $\sqrt{21xy}$ **67.** $10xy\sqrt{2y}$ **69.** $\frac{7}{4}$ **71.** $\frac{\sqrt{3}}{2}$ **73.** $\frac{x}{6}$ **75.** $\frac{\sqrt{7}}{x^2}$ **77.** $\frac{6\sqrt{2}}{y^{10}}$ **79.** 3   **81.** $2\sqrt{2}$ **83.** $\sqrt{5}$

85. 4   **87.** $2x$ **89.** $5x\sqrt{2x}$ **91.** $2x^3\sqrt{10x}$ **93.** $2\sqrt[3]{2}$ **95.** $3\sqrt[3]{2}$ **97.** $2\sqrt[4]{2}$ **99.** 2   **101.** $3\sqrt[3]{2}$ **103.** $2\sqrt[4]{2}$ **105.** $\frac{3}{2}$

107. $\frac{\sqrt[3]{3}}{2}$ **109.** $3(x+4)\sqrt{10(x+4)}$ **111.** $x - 3$ **113.** $2^{21}x^{52}y^6\sqrt{2y}$ **115.** $2x\sqrt[3]{3x^2}$ **117.** $5\sqrt{3}$ sq ft

119. a. $\frac{5\sqrt{34}}{14}$ m²   **b.** 2.08 m²   **129.** makes sense   **131.** makes sense   **133.** false   **135.** false   **137.** $\sqrt{15x^2} \cdot \sqrt{6x^5}$

139. correctly simplified   **141.** correctly simplified   **142.** $(6, -2)$ or $\{(6, -2)\}$ **143.** $\frac{3}{x-2}$ **144.** $2x(x-3)(x-5)$

145. a. $12x$ **b.** $12\sqrt{2}$ **146.** $20\sqrt{2x}$ **147. a.** $3x + 15$ **b.** $\sqrt{21} + \sqrt{15}$

Section 9.3

Check Point Exercises

1. a. $17\sqrt{13}$ **b.** $-19\sqrt{17x}$ **2. a.** $17\sqrt{3}$ **b.** $10\sqrt{2x}$ **3. a.** $\sqrt{10} + \sqrt{22}$ **b.** $11 + 6\sqrt{3}$ **c.** $4 - 4\sqrt{5}$ **4. a.** −2   **b.** 5

Exercise Set 9.3

1. $13\sqrt{3}$ **3.** $15\sqrt{6}$ **5.** $-5\sqrt{13}$ **7.** $15\sqrt{x}$ **9.** $-6\sqrt{y}$ **11.** $9\sqrt{10x}$ **13.** $6\sqrt{5y}$ **15.** $2\sqrt{5}$ **17.** $12\sqrt{2}$ **19.** $7\sqrt{7}$

21. 0   **23.** $3\sqrt{5}$ **25.** $\sqrt{2}$ **27.** $8\sqrt{2}$ **29.** $19\sqrt{3}$ **31.** $9\sqrt{3} - 6\sqrt{2}$ **33.** $2\sqrt{5x}$ **35.** $9 + 5\sqrt{2}$ **37.** cannot be combined

39. $8\sqrt{5} + 15\sqrt{3}$ **41.** $7\sqrt{6} + 8\sqrt{5}$ **43.** $\sqrt{6} + \sqrt{10}$ **45.** $\sqrt{42} - \sqrt{70}$ **47.** $5\sqrt{3} + 3$ **49.** $3\sqrt{2} - 3$ **51.** $32 + 11\sqrt{2}$

53. $25 - 2\sqrt{5}$ **55.** $117 - 36\sqrt{7}$ **57.** $25 - 8\sqrt{10}$ **59.** $15 + 9\sqrt{2}$ **61.** $\sqrt{6} - 6\sqrt{2} + \sqrt{3} - 6$ **63.** 4 **65.** -5 **67.** -14

69. 2 **71.** -37 **73.** 7 **75.** $5 + 2\sqrt{6}$ **77.** $x - 2\sqrt{10x} + 10$ **79.** $9x\sqrt{3x}$ **81.** $8x^2y^2\sqrt{xy}$ **83.** $\sqrt[3]{2}$ **85.** $11x\sqrt[3]{4x}$

87. perimeter $= 4(\sqrt{3} + \sqrt{5}) = (4\sqrt{3} + 4\sqrt{5})$ in.; area $= (\sqrt{3} + \sqrt{5})^2 = (8 + 2\sqrt{15})$ sq in.

89. perimeter $= 2(\sqrt{6} - 1) + 2(\sqrt{6} + 1) = 4\sqrt{6}$ in.; area $= (\sqrt{6} + 1)(\sqrt{6} - 1) = 5$ sq in.

91. perimeter $= \sqrt{2} + \sqrt{2} + 2 = (2 + 2\sqrt{2})$ in.; area $= \frac{1}{2}(\sqrt{2})(\sqrt{2}) = 1$ sq in. **93.** $3\sqrt{2}$; Answers will vary.

95. a. $J = 2.6\sqrt{x} + 35$ **b.** 49%; It's the same. **97. a.** $(30, 49)$ **b.** 51% and 52% are reasonable estimates.

107. does not make sense **109.** does not make sense **111.** false **113.** false **115.** $-1 - 3\sqrt[3]{4} + \sqrt[3]{2}$ **117.** $48x - 2y$

119. $\sqrt{7x}$ should be $\sqrt{x}$. **121.** $x^2 - 4$ should be $x - 4$.

122. $25x^2 - 9$ **123.** $x(8x + 1)(8x - 1)$ **124.** $y = -\frac{1}{4}x + 3$ **125.** $\frac{\sqrt{15}}{5}$ **126.** 22 **127.** 4

Mid-Chapter Check Point Exercises

1. $10\sqrt{3}$ **2.** $10\sqrt{6}$ **3.** $4x\sqrt{6x}$ **4.** $\frac{\sqrt[5]{4}}{2}$ **5.** $9\sqrt{3}$ **6.** 7 **7.** $2\sqrt{15} - 30$ **8.** $-4x^{10}\sqrt{2x}$ **9.** $2x^3\sqrt{3x}$ **10.** $2\sqrt[3]{2}$

11. $-19\sqrt{10}$ **12.** $4 - \sqrt{3}$ **13.** $2x\sqrt{2}$ **14.** $-2\sqrt[4]{2}$ **15.** $9 + 2\sqrt{14}$ **16.** $2\sqrt[3]{2}$ **17.** $6\sqrt{5}$ **18.** $\frac{\sqrt{10}}{4}$ **19.** $6 + 3\sqrt{10}$ **20.** 23

Section 9.4

Check Point Exercises

1. a. $\frac{5\sqrt{10}}{2}$ **b.** $\frac{\sqrt{14}}{7}$ **2. a.** $\frac{5\sqrt{2}}{2}$ **b.** $\frac{\sqrt{35x}}{10}$ **3.** $\frac{32 - 8\sqrt{5}}{11}$ **4.** $2\sqrt{7} + 2\sqrt{3}$

Exercise Set 9.4

1. $\frac{\sqrt{10}}{10}$ **3.** $\sqrt{5}$ **5.** $\frac{\sqrt{6}}{3}$ **7.** $4\sqrt{7}$ **9.** $\frac{\sqrt{15}}{5}$ **11.** $\frac{\sqrt{21}}{3}$ **13.** $\frac{x\sqrt{3}}{3}$ **15.** $\frac{\sqrt{11x}}{x}$ **17.** $\frac{\sqrt{xy}}{y}$ **19.** $\frac{x^2\sqrt{2}}{2}$ **21.** $\frac{\sqrt{35}}{5}$

23. $\frac{\sqrt{42x}}{14}$ **25.** $\frac{\sqrt{5}}{10}$ **27.** $\frac{3\sqrt{2}}{2}$ **29.** $\frac{5\sqrt{3}}{2}$ **31.** $\frac{\sqrt{10}}{6}$ **33.** $\frac{\sqrt{2x}}{8}$ **35.** $\frac{\sqrt{5}}{15}$ **37.** $\frac{\sqrt{21}}{6}$ **39.** $2x\sqrt{2}$ **41.** $\frac{\sqrt{14y}}{4}$ **43.** $\frac{\sqrt{21x}}{6}$

45. $\frac{3\sqrt{5x}}{x}$ **47.** $\frac{5\sqrt{x}}{x^2}$ **49.** $\frac{3\sqrt{3y}}{y^2}$ **51.** $\frac{5x\sqrt{6y}}{6y^2}$ **53.** $\frac{4 - \sqrt{3}}{13}$ **55.** $-6 - 3\sqrt{7}$ **57.** $8\sqrt{11} - 24$ **59.** $9 + 3\sqrt{3}$

61. $2 - \sqrt{2}$ **63.** $\frac{10 + \sqrt{70}}{3}$ **65.** $2\sqrt{6} - 2\sqrt{3}$ **67.** $\sqrt{5} + \sqrt{3}$ **69.** $\frac{8 - 2\sqrt{x}}{16 - x}$ **71.** $\frac{6\sqrt{5} - 4\sqrt{3}}{11}$ **73.** $\frac{7 + 2\sqrt{10}}{3}$

75. $\frac{3y^2\sqrt{2x}}{x}$ **77.** $\sqrt{x + 2} + \sqrt{x}$ **79.** $\frac{5\sqrt{6}}{6}$ **81.** $2\sqrt{x + 2} - h$ **83. a.** 18% **b.** $\frac{13\sqrt{x} + x}{5}$ **c.** 18%; yes; no; Answers will vary.

85. $\frac{\sqrt{5} + 1}{2} \approx 1.62$ **91.** makes sense **93.** makes sense **95.** true **97.** false **99.** $\frac{3\sqrt{2}}{2}$ **101.** $\frac{4}{2 + \sqrt{3}}$ **102.** $\frac{5}{6}$ and 1, or $\left\{\frac{5}{6}, 1\right\}$

103. $\frac{1}{8x^6}$ **104.** $\frac{x - 3}{4(x + 3)}$ **105.** 4 or $\{4\}$ **106.** 1 and 5, or $\{1, 5\}$ **107.** 5

Section 9.5

Check Point Exercises

1. 11 or $\{11\}$ **2.** 4 or $\{4\}$ **3.** no solution or $\varnothing$ **4.** 6 or $\{6\}$ **5.** 2018

Exercise Set 9.5

1. 25 or $\{25\}$ **3.** 16 or $\{16\}$ **5.** 7 or $\{7\}$ **7.** 124 or $\{124\}$ **9.** 7 or $\{7\}$ **11.** 4 or $\{4\}$ **13.** 2 or $\{2\}$ **15.** 3 or $\{3\}$ **17.** 24 or $\{24\}$

19. no solution or $\varnothing$ **21.** no solution or $\varnothing$ **23.** 16 or $\{16\}$ **25.** no solution or $\varnothing$ **27.** 7 or $\{7\}$ **29.** 2 or $\{2\}$ **31.** -3 or $\{-3\}$

33. -6 or $\{-6\}$ **35.** 2 or $\{2\}$ **37.** 1 and 3, or $\{1, 3\}$ **39.** no solution or $\varnothing$ **41.** 12 or $\{12\}$ **43.** no solution or $\varnothing$ **45.** 16

47. 8 **49.** $h = \frac{v^2}{2g}$ **51.** 1 or $\{1\}$ **53.** 9 or $\{9\}$ **55.** 49 pounds per square inch **57.** by the point $(49, 840)$ **59. a.** 63.6%

b. underestimates by 0.3% **c.** 2011 **65.** does not make sense **67.** does not make sense **69.** false **71.** false **73.** 4 and 5

75. 7 or $\{7\}$ **77.** 1 or $\{1\}$ **79.** no solution or $\varnothing$ **81.** \$7000 at 6% and \$2000 at 4% **82.** \$4.25 **83.** $(-1, -2)$ or $\{(-1, -2)\}$

84. -16 **85.** 9 **86.** $\frac{1}{27}$

Section 9.6

Check Point Exercises

1. a. 5   **b.** 2   **c.** -3 **d.** -2 **2. a.** 81   **b.** 8   **c.** -8 **3. a.** $\dfrac{1}{5}$ **b.** $\dfrac{1}{4}$ **c.** $\dfrac{1}{16}$ **4. a.** 6.3%   **b.** underestimates by 1.6%

Exercise Set 9.6

1. 7   **3.** 11   **5.** 3   **7.** -5 **9.** 2   **11.** -2 **13.** $\dfrac{1}{3}$ **15.** $\dfrac{3}{4}$ **17.** 729   **19.** 25   **21.** 27   **23.** -8 **25.** $\dfrac{1}{3}$ **27.** $\dfrac{1}{5}$

29. $\dfrac{1}{2}$ **31.** 2   **33.** $\dfrac{1}{8}$ **35.** $\dfrac{1}{243}$ **37.** $\dfrac{1}{4}$ **39.** $\dfrac{5}{2}$ **41.** $\dfrac{5}{2}$ **43.** $\dfrac{1}{4}$ **45.** 17   **47.** 375   **49.** $x^{\frac{7}{12}}$ **51.** $\dfrac{1}{x^{\frac{3}{3}}}$ **53.** $x^{\frac{1}{2}}y^2$ **55.** $\dfrac{1}{x^7}$

57. a. 83.2 m   **b.** $h = 0.84(\sqrt[3]{d})^2$ or $h = 0.84\sqrt[3]{d^2}$ **59. a.** approximately 0.382 revolution per second

b. approximately 22.92 revolutions per minute   **c.** $N = \dfrac{\sqrt{ar}}{2\pi r}$ **67.** does not make sense   **69.** does not make sense   **71.** false

73. false   **75.** true   **77.** true   **79.** 3

81. a–b.   **c.** 24 square units   **82.** $(-2, 0)$ or $\{(-2, 0)\}$ **83.**   **84.** $32x^2$

85. solution   **86.** not a solution   **87.** $5\sqrt{5}$

Review Exercises

1. 11   **2.** -11 **3.** not a real number   **4.** $\dfrac{2}{5}$ **5.** -2 **6.** -3 **7.** 8.660   **8.** 19.824   **9.** 14.6 min

10. Answers will vary. Example: The number of disruptive minutes is increasing, but the rate of increase is slowing down.   **11.** 47 mi

12. $3\sqrt{6}$ **13.** $12\sqrt{5}$ **14.** $3x\sqrt{7}$ **15.** $4x\sqrt{3x}$ **16.** x^4 **17.** $5x^4\sqrt{3x}$ **18.** $3x^{11}\sqrt{5x}$ **19.** $2\sqrt[3]{3}$ **20.** $\sqrt{77}$ **21.** 6

22. $5x\sqrt{2}$ **23.** $2x^2\sqrt[3]{3x}$ **24.** $3\sqrt[3]{2}$ **25.** $\dfrac{\sqrt{15}}{4}$ **26.** $\dfrac{11}{2}$ **27.** $\dfrac{\sqrt{7x}}{5}$ **28.** $\dfrac{3\sqrt{2}}{x}$ **29.** 10   **30.** $4\sqrt{2}$ **31.** $6x^2\sqrt{2x}$ **32.** $\dfrac{\sqrt[3]{5}}{4}$

33. $\dfrac{2\sqrt[3]{5}}{3}$ **34.** $20\sqrt{5}$ **35.** $7\sqrt{2}$ **36.** $\sqrt{3}$ **37.** $17\sqrt{5}$ **38.** $24\sqrt{2} - 8\sqrt{3}$ **39.** $6\sqrt{2} + 7\sqrt{3}$ **40.** $5\sqrt{2} + 2\sqrt{15}$ **41.** $3\sqrt{2} - 6$

42. $92 + 19\sqrt{2}$ **43.** $-17 + 11\sqrt{7}$ **44.** $3\sqrt{2} - 4\sqrt{3} + 2\sqrt{6} - 8$ **45.** -3 **46.** 6   **47.** $3 + 2\sqrt{2}$ **48.** $6\sqrt{5}$ **49.** $\dfrac{13\sqrt{2}}{10}$

50. $\dfrac{\sqrt{6}}{3}$ **51.** $\dfrac{\sqrt{6}}{4}$ **52.** $\dfrac{\sqrt{17x}}{x}$ **53.** $11\sqrt{5} - 22$ **54.** $\dfrac{84 + 21\sqrt{3}}{13}$ **55.** $6\sqrt{5} - 6\sqrt{3}$ **56.** $-1 - 2\sqrt{2}$ **57.** 13 or $\{13\}$

58. 11 or $\{11\}$ **59.** 5 or $\{5\}$ **60.** 0 and 3, or $\{0, 3\}$ **61.** 8 or $\{8\}$ **62.** no solution or $\varnothing$ **63.** no solution or $\varnothing$ **64.** 8 ft

65. 2024   **66.** $\sqrt{16} = 4$ **67.** $\sqrt[3]{125} = 5$ **68.** $(\sqrt[3]{64})^2 = 16$ **69.** $\dfrac{1}{\sqrt{25}} = \dfrac{1}{5}$ **70.** $\dfrac{1}{\sqrt[3]{27}} = \dfrac{1}{3}$ **71.** $\dfrac{1}{(\sqrt[4]{-8})^4} = \dfrac{1}{16}$

72. approx 57 species

Chapter Test

1. -8 **2.** 4   **3.** $4\sqrt{3}$ **4.** $6x\sqrt{2x}$ **5.** $x^{14}\sqrt{x}$ **6.** $\dfrac{5}{x}$ **7.** $\dfrac{5}{3}$ **8.** $\dfrac{\sqrt[3]{5}}{2}$ **9.** $2x\sqrt{10}$ **10.** $5\sqrt{2}$ **11.** $6\sqrt{xy}$ **12.** $2x^2\sqrt{5x}$

13. $11\sqrt{6}$ **14.** $6\sqrt{2}$ **15.** $\sqrt{30} + 3$ **16.** $55 + 11\sqrt{5}$ **17.** 2   **18.** $16 + 6\sqrt{7}$ **19.** $\dfrac{4\sqrt{5}}{5}$ **20.** $\dfrac{20 - 5\sqrt{3}}{13}$ **21.** 12 or $\{12\}$

22. 5 or $\{5\}$ **23.** $25 thousand   **24.** $(\sqrt[3]{8})^2 = 4$ **25.** $\dfrac{1}{\sqrt{9}} = \dfrac{1}{3}$

Cumulative Review Exercises

1. -3 or $\{-3\}$ **2.** -4 and $\dfrac{3}{2}$, or $\left\{-4, \dfrac{3}{2}\right\}$ **3.** $(-3, -4)$ or $\{(-3, -4)\}$ **4.** $\dfrac{9}{7}$ or $\left\{\dfrac{9}{7}\right\}$ **5.** -5 or $\{-5\}$ **6.** 5 or $\{5\}$

7. $-\dfrac{2}{x^4}$ **8.** $22\sqrt{3}$ **9.** $\dfrac{3}{x}$ **10.** $-\dfrac{2 + x}{3x + 1}$ **11.** 5   **12.** $(x - 7)(x - 11)$ **13.** $x(x + 5)(x - 5)$ **14.** $6x^2 - 7x + 2$ **15.** $8x^3 - 27$

16. $\dfrac{1}{x - 1}$ **17.** $\dfrac{5x - 1}{4x - 3}$ **18.** $-18\sqrt{3}$

19.

20.

21.

22. $-\dfrac{8}{3}$ **23.** $y + 3 = 5(x + 2)$; $y = 5x + 7$ **24.** 43 **25.** 954 deer

CHAPTER 10

Section 10.1

Check Point Exercises

1. a. -6 and 6, or $\{\pm 6\}$ **b.** $-\sqrt{3}$ and $\sqrt{3}$, or $\{\pm \sqrt{3}\}$ **c.** $\pm \dfrac{\sqrt{14}}{2}$ or $\left\{\pm \dfrac{\sqrt{14}}{2}\right\}$ **2.** -2 and 8, or $\{-2, 8\}$ **3.** $2 \pm \sqrt{7}$ or $\{2 \pm \sqrt{7}\}$

4. $10\sqrt{21}$ in. ≈ 45.8 in. **5.** 13 units

Exercise Set 10.1

1. ± 4 or $\{\pm 4\}$ **3.** ± 9 or $\{\pm 9\}$ **5.** $\pm \sqrt{7}$ or $\{\pm \sqrt{7}\}$ **7.** $\pm 5\sqrt{2}$ or $\{\pm 5\sqrt{2}\}$ **9.** ± 2 or $\{\pm 2\}$ **11.** $\pm \dfrac{7}{2}$ or $\left\{\pm \dfrac{7}{2}\right\}$ **13.** ± 5 or $\{\pm 5\}$

15. $\pm \dfrac{\sqrt{6}}{3}$ or $\left\{\pm \dfrac{\sqrt{6}}{3}\right\}$ **17.** $\pm \dfrac{\sqrt{35}}{5}$ or $\left\{\pm \dfrac{\sqrt{35}}{5}\right\}$ **19.** -1 and 7, or $\{-1, 7\}$ **21.** -16 and 6, or $\{-16, 6\}$ **23.** $-\dfrac{5}{3}$ and $\dfrac{1}{3}$, or $\left\{-\dfrac{5}{3}, \dfrac{1}{3}\right\}$

25. $5 \pm \sqrt{3}$ or $\{5 \pm \sqrt{3}\}$ **27.** $-8 \pm \sqrt{11}$ or $\{-8 \pm \sqrt{11}\}$ **29.** $4 \pm 3\sqrt{2}$ or $\{4 \pm 3\sqrt{2}\}$ **31.** -6 and 2, or $\{-6, 2\}$ **33.** -3 and 9, or $\{-3, 9\}$

35. $5 \pm \sqrt{2}$ or $\{5 \pm \sqrt{2}\}$ **37.** $-1 \pm \sqrt{5}$ or $\{-1 \pm \sqrt{5}\}$ **39.** $7 \pm 2\sqrt{3}$ or $\{7 \pm 2\sqrt{3}\}$ **41.** 17 m **43.** 39 m **45.** 12 cm **47.** $5\sqrt{7}$ m

49. $\sqrt{17}$ or 4.12 units **51.** 17 units **53.** $3\sqrt{5}$ or 6.71 units **55.** $\sqrt{41}$ or 6.40 units **57.** 16 units **59.** -2 and 8 **61.** -3 and $\dfrac{5}{3}$

63. $r = \dfrac{\sqrt{A\pi}}{\pi}$ **65.** $d = \dfrac{\sqrt{kI}}{I}$ **67.** $2\sqrt{41}$ ft **69.** $90\sqrt{2}$ ft **71.** 6 in. **73.** 6 weeks **75.** 5 sec **77.** 8 m **79.** 10 in.

85. makes sense **87.** makes sense **89.** false **91.** true **93.** -7 and 1 **95.** -2 and 4, or $\{-2, 4\}$ **96.** $2(2x + 3)(3x - 1)$

97. $\dfrac{x - 3}{3(x - 2)}$ **98.** -6 or $\{-6\}$ **99.** $(x + 4)^2$ **100.** $(x - 7)^2$ **101.** $\left(x + \dfrac{5}{2}\right)^2$

Section 10.2

Check Point Exercises

1. a. $x^2 + 10x + 25 = (x + 5)^2$ **b.** $x^2 - 6x + 9 = (x - 3)^2$ **c.** $x^2 + 3x + \dfrac{9}{4} = \left(x + \dfrac{3}{2}\right)^2$

2. a. -7 and 1, or $\{-7, 1\}$ **b.** $5 \pm \sqrt{7}$ or $\{5 \pm \sqrt{7}\}$ **3.** $\dfrac{5 \pm 3\sqrt{3}}{2}$ or $\left\{\dfrac{5 \pm 3\sqrt{3}}{2}\right\}$

Exercise Set 10.2

1. $x^2 + 10x + 25 = (x + 5)^2$ **3.** $x^2 - 2x + 1 = (x - 1)^2$ **5.** $x^2 + 5x + \dfrac{25}{4} = \left(x + \dfrac{5}{2}\right)^2$ **7.** $x^2 - 7x + \dfrac{49}{4} = \left(x - \dfrac{7}{2}\right)^2$

9. $x^2 + \dfrac{1}{2}x + \dfrac{1}{16} = \left(x + \dfrac{1}{4}\right)^2$ **11.** $x^2 - \dfrac{4}{3}x + \dfrac{4}{9} = \left(x - \dfrac{2}{3}\right)^2$ **13.** -5 and 1, or $\{-5, 1\}$ **15.** 4 and 6, or $\{4, 6\}$

17. $1 \pm \sqrt{6}$ or $\{1 \pm \sqrt{6}\}$ **19.** $-2 \pm \sqrt{3}$ or $\{-2 \pm \sqrt{3}\}$ **21.** -4 and 7, or $\{-4, 7\}$ **23.** $\dfrac{-3 \pm \sqrt{13}}{2}$ or $\left\{\dfrac{-3 \pm \sqrt{13}}{2}\right\}$

25. $\dfrac{7 \pm \sqrt{37}}{2}$ or $\left\{\dfrac{7 \pm \sqrt{37}}{2}\right\}$ **27.** $\dfrac{1 \pm \sqrt{13}}{2}$ or $\left\{\dfrac{1 \pm \sqrt{13}}{2}\right\}$ **29.** $\dfrac{1}{2}$ and 1, or $\left\{\dfrac{1}{2}, 1\right\}$ **31.** $\dfrac{-5 \pm \sqrt{3}}{2}$ or $\left\{\dfrac{-5 \pm \sqrt{3}}{2}\right\}$

33. $\dfrac{1 \pm \sqrt{13}}{4}$ or $\left\{\dfrac{1 \pm \sqrt{13}}{4}\right\}$ **35.** $1 \pm \sqrt{7}$ or $\{1 \pm \sqrt{7}\}$ **37.** $\dfrac{1 \pm \sqrt{29}}{2}$ or $\left\{\dfrac{1 \pm \sqrt{29}}{2}\right\}$ **39.** $-5b$ and b, or $\{-5b, b\}$

43. makes sense **45.** makes sense **47.** false **49.** false **51.** $x^2 - 20x + 100$

53. $-\dfrac{b}{2} \pm \sqrt{\dfrac{b^2}{4} - c}, \dfrac{-b \pm \sqrt{b^2 - 4ac}}{2}, \left\{-\dfrac{b}{2} \pm \sqrt{\dfrac{b^2}{4} - c}\right\}$, or $\left\{\dfrac{-b \pm \sqrt{b^2 - 4ac}}{2}\right\}$ **55.** $\dfrac{11}{(x - 3)(x - 4)}$ **56.** $\dfrac{x}{3}$

57. 3 or $\{3\}$ **58.** $\dfrac{1}{2}; -5$ **59.** $\dfrac{2}{3}$ **60.** $1 + \sqrt{7}; 1 - \sqrt{7}$

Section 10.3

Check Point Exercises

1. $-\dfrac{1}{2}$ and $\dfrac{1}{4}$, or $\left\{-\dfrac{1}{2}, \dfrac{1}{4}\right\}$ **2.** $3 \pm \sqrt{5}$ or $\{3 \pm \sqrt{5}\}$ **3.** 2036

Exercise Set 10.3

1. -3 and -2, or $\{-3, -2\}$ **3.** $\dfrac{-5 \pm \sqrt{13}}{2}$ or $\left\{\dfrac{-5 \pm \sqrt{13}}{2}\right\}$ **5.** $-2 \pm \sqrt{10}$ or $\{-2 \pm \sqrt{10}\}$ **7.** $-2 \pm \sqrt{11}$ or $\{-2 \pm \sqrt{11}\}$

9. -3 and 6, or $\{-3, 6\}$ **11.** $-\dfrac{2}{3}$ and $\dfrac{3}{2}$, or $\left\{-\dfrac{2}{3}, \dfrac{3}{2}\right\}$ **13.** $1 \pm \sqrt{11}$ or $\{1 \pm \sqrt{11}\}$ **15.** $\dfrac{1 \pm \sqrt{57}}{2}$ or $\left\{\dfrac{1 \pm \sqrt{57}}{2}\right\}$

17. $\dfrac{-3 \pm \sqrt{3}}{6}$ or $\left\{\dfrac{-3 \pm \sqrt{3}}{6}\right\}$ **19.** $\dfrac{2}{3}$ or $\left\{\dfrac{2}{3}\right\}$ **21.** $\dfrac{1 \pm \sqrt{29}}{4}$ or $\left\{\dfrac{1 \pm \sqrt{29}}{4}\right\}$ **23.** $-\dfrac{1}{2}$ and 1, or $\left\{-\dfrac{1}{2}, 1\right\}$ **25.** $\dfrac{1}{5}$ and 2, or $\left\{\dfrac{1}{5}, 2\right\}$

27. $\pm 2\sqrt{5}$ or $\{\pm 2\sqrt{5}\}$ **29.** $1 \pm \sqrt{2}$ or $\{1 \pm \sqrt{2}\}$ **31.** $\dfrac{-11 \pm \sqrt{33}}{4}$ or $\left\{\dfrac{-11 \pm \sqrt{33}}{4}\right\}$ **33.** 0 and $\dfrac{8}{3}$, or $\left\{0, \dfrac{8}{3}\right\}$ **35.** 2 or $\{2\}$

37. -2 and 2, or $\{-2, 2\}$ **39.** -9 and 0, or $\{-9, 0\}$ **41.** $-\dfrac{2}{3}$ and 4, or $\left\{-\dfrac{2}{3}, 4\right\}$ **43.** $\dfrac{2 \pm \sqrt{10}}{3}$ or $\left\{\dfrac{2 \pm \sqrt{10}}{3}\right\}$ **45.** $\pm \sqrt{3}$ or $\{\pm \sqrt{3}\}$

47. $-\dfrac{5}{2}$ and 0, or $\left\{-\dfrac{5}{2}, 0\right\}$ **49.** $\pm \sqrt{6}$ or $\{\pm \sqrt{6}\}$ **51.** $\dfrac{5 \pm \sqrt{73}}{2}$ or $\left\{\dfrac{5 \pm \sqrt{73}}{2}\right\}$ **53.** about 3.8 sec   **55.** 10 ft

57. a. 60.5 million; overestimates by 0.1 million   **b.** 2014   **59.** width: 4.7 m; length: 7.7 m   **61.** 2.3 and 3.3 ft   **63.** 9.8 yr

69. does not make sense   **71.** does not make sense   **73.** false   **75.** true   **77.** $-3\sqrt{3}$ and $\sqrt{3}$, or $\{-\sqrt{3}, \sqrt{3}\}$

79.

81. $\dfrac{1}{25}$ **82.** $9 - 3\sqrt{5}$ **83.** $x^3 - y^3$ **84.** no; no   **85. a.** $-\sqrt{4}, \sqrt{1}$ **b.** $-\sqrt{5}, \sqrt{5}$ **c.** $-\sqrt{5}, -\sqrt{4}, \sqrt{1}, \sqrt{5}$ **d.** $\sqrt{-4}, \sqrt{-1}$

86. a. $-\sqrt{9}, \sqrt{0}, \sqrt{9}$ **b.** $-\sqrt{7}, \sqrt{7}$ **c.** $-\sqrt{9}, -\sqrt{7}, \sqrt{0}, \sqrt{9}, \sqrt{7}$ **d.** $\sqrt{-9}, \sqrt{-7}$

Mid-Chapter Check Point Exercises

1. $-\dfrac{8}{3}$ and 4, or $\left\{-\dfrac{8}{3}, 4\right\}$ **2.** 0 and $\dfrac{1}{3}$, or $\left\{0, \dfrac{1}{3}\right\}$ **3.** $1 \pm \sqrt{11}$ or $\{1 \pm \sqrt{11}\}$ **4.** $4 \pm \sqrt{7}$ or $\{4 \pm \sqrt{7}\}$ **5.** $-\dfrac{2}{3}$ and 1, or $\left\{-\dfrac{2}{3}, 1\right\}$

6. $\dfrac{5 \pm \sqrt{7}}{6}$ or $\left\{\dfrac{5 \pm \sqrt{7}}{6}\right\}$ **7.** -4 and 3, or $\{-4, 3\}$ **8.** $-5 \pm 2\sqrt{10}$ or $\{-5 \pm 2\sqrt{10}\}$ **9.** -3 and 3, or $\{-3, 3\}$

10. $\dfrac{-5 \pm \sqrt{17}}{4}$ or $\left\{\dfrac{-5 \pm \sqrt{17}}{4}\right\}$ **11.** $-\dfrac{1}{2}$ and 10, or $\left\{-\dfrac{1}{2}, 10\right\}$ **12.** 4 or $\{4\}$ **13.** -16 and 2, or $\{-16, 2\}$ **14.** $2\sqrt{7}$ cm

15. 13 units   **16.** 12 in. and 16 in.

Section 10.4

Check Point Exercises

1. a. $4i$ **b.** $i\sqrt{5}$ **c.** $5i\sqrt{2}$ **2.** $-2 + 5i$ and $-2 - 5i$, or $\{-2 \pm 5i\}$ **3.** $-3 \pm 2i$ or $\{-3 \pm 2i\}$

Exercise Set 10.4

1. $6i$ **3.** $i\sqrt{13}$ **5.** $5i\sqrt{2}$ **7.** $2i\sqrt{5}$ **9.** $-2i\sqrt{7}$ **11.** $7 + 4i$ **13.** $10 + i\sqrt{3}$ **15.** $6 - 7i\sqrt{2}$ **17.** $3 \pm 3i$ or $\{3 \pm 3i\}$

19. $-7 \pm 8i$ or $\{-7 \pm 8i\}$ **21.** $2 \pm i\sqrt{7}$ or $\{2 \pm i\sqrt{7}\}$ **23.** $-3 \pm 3i\sqrt{2}$ or $\{-3 \pm 3i\sqrt{2}\}$ **25.** $-2 \pm i$ or $\{-2 \pm i\}$ **27.** $3 \pm 2i$ or $\{3 \pm 2i\}$

29. $6 \pm 2i$ or $\{6 \pm 2i\}$ **31.** $5 \pm i\sqrt{2}$ or $\{5 \pm i\sqrt{2}\}$ **33.** $\dfrac{1 \pm i\sqrt{14}}{5}$ or $\left\{\dfrac{1 \pm i\sqrt{14}}{5}\right\}$ **35.** $\dfrac{2 \pm i\sqrt{6}}{2}$ or $\left\{\dfrac{2 \pm i\sqrt{6}}{2}\right\}$ **37.** $\pm i\sqrt{5}$ or $\{\pm i\sqrt{5}\}$

39. $1 \pm 3i$ or $\{1 \pm 3i\}$ **41.** $\dfrac{-1 \pm i\sqrt{7}}{2}$ or $\left\{\dfrac{-1 \pm i\sqrt{7}}{2}\right\}$ **43.** $9 \pm i\sqrt{19}$ are not real numbers. Applicant will not be hired. Answers will vary.

51. does not make sense   **53.** does not make sense   **55.** false   **57.** false

63.

$y = \dfrac{1}{3}x - 2$

64.

$2x - 3y = 6$

65.

$x = -2$

66.

$y = x^2 + 4x + 3$

67. -1 and 3 **68.** -3

Section 10.5

Check Point Exercises

1. a. upward   **b.**   **2.** 2 and 4   **3.** 8   **4.** $(3, -1)$

5.   **6.**   **7. a.** 205 ft; 200 ft   **b.** 402 ft   **c.**

Exercise Set 10.5

1. upward   **3.** downward   **5.** 1 and 3   **7.** 2 and 6   **9.** 1.2 and -3.2 **11.** 3   **13.** -12 **15.** -4 **17.** 0   **19.** $(2, -1)$
21. $(-1, -8)$ **23.** $(-3, -9)$

25.   **27.**   **29.**

31.   **33.**   **35.**

37. $(3, 2)$ **39.** $(-5, -4)$ **41.** $(1, -3)$ **43.** $(-2, 5)$ **45.** (h, k) **47. a.** 18.35 ft; 35 ft   **b.** 77.8 ft   **c.** 6.1 ft
49. a. 7.68 million; underestimates by 0.02 million   **b.** 2000; 7.4 million   **c.** 2000; 7.4 million; It's the same.   **51.** 25 yd by 25 yd; 625 sq yd
59. makes sense   **61.** does not make sense   **63.** true   **65.** false

67. $(-2, 0)$ and $(2, 0)$

71. $(80, 1600)$; Answers will vary.   **73.** $(-4, 520)$; Answers will vary.   **75.** 2 or $\{2\}$ **76.** $-\dfrac{8}{3}$ or $\left\{-\dfrac{8}{3}\right\}$ **77.** $(-2, 1)$ or $\{(-2, 1)\}$
78. set 1   **79.** graph a   **80.** 5

Section 10.6

Check Point Exercises

1. domain: {golf, lawn mowing, water skiing, hiking, bicycling}; range: {250, 325, 430, 720}   **2. a.** not a function   **b.** function
3. a. 23   **b.** -5 **c.** 3   **4. a.** 48   **b.** 3   **c.** 3   **5. a.** function   **b.** function   **c.** not a function
6. a. 25%   **b.** 10%   **c.** function f, the linear function

Exercise Set 10.6

1. function; domain: {1, 3, 5}; range: {2, 4, 5}   **3.** not a function; domain: {3, 4}; range: {4, 5}   **5.** function; domain: {$-3, -2, -1, 0$};
range: {$-3, -2, -1, 0$}   **7.** not a function; domain: {1}; range: {4, 5, 6}   **9. a.** 12   **b.** -1 **c.** 5   **11. a.** 70   **b.** -28 **c.** 0
13. a. 93   **b.** -7 **c.** -3 **15. a.** 10   **b.** -2 **c.** 0   **17. a.** 11   **b.** 27   **c.** 3   **19. a.** 5   **b.** 5   **c.** 5   **21. a.** 3   **b.** 7

23. a. 1   **b.** −1   **25.** function   **27.** function   **29.** not a function   **31.** function   **33.** $\{(-1, 1), (0, 3), (1, 5)\}$

35. $\{(-2, -6), (-1, -2), (0, 0), (1, 0), (2, -2)\}$ **37.** 6   **39.** $x + h$ **41. a.** $\{(\text{U.S.}, 80\%), (\text{Japan}, 64\%), (\text{France}, 64\%), (\text{Germany}, 61\%),$

$(\text{England}, 59\%), (\text{China}, 47\%)\}$ **b.** Yes; Each country corresponds to exactly one percent.   **c.** $\{(80\%, \text{U.S.}), (64\%, \text{Japan}), (64\%, \text{France}),$

$(61\%, \text{Germany}), (59\%, \text{England}), (47\%, \text{China})\}$ **d.** No; 64% in the domain corresponds to two members of the range, Japan and France.

43. 186.6; At age 20, an American man's cholesterol level is 186.6, or approx 187.   **45. a.** 127; In 2004, Americans ordered an average of 127

takeout meals per person.; by the point $(20, 127)$ **b.** 94; In 1984, Americans ordered an average of 94 meals in restaurants per person.;

by the point $(0, 94)$ **47.** 1988; 91 takeout meals and 91.6 meals in restaurants   **53.** makes sense   **55.** does not make sense   **57.** false

59. false   **61.** $f(x) = 3x + 7$ **63.** $V(x) = 22{,}500 - 3200x; V(3) = 12{,}900;$ After 3 years, the value of the car is $12,900.

64. 3.97×10^{-3} **65.** $x^2 + 9x + 16 + \dfrac{35}{x - 2}$ **66.** $\left(\dfrac{4}{5}, \dfrac{9}{5}\right)$ or $\left\{\left(\dfrac{4}{5}, \dfrac{9}{5}\right)\right\}$

Review Exercises

1. −8 and 8, or $\{-8, 8\}$ **2.** $\pm\sqrt{17}$ or $\{\pm\sqrt{17}\}$ **3.** $\pm5\sqrt{3}$ or $\{\pm5\sqrt{3}\}$ **4.** 0 and 6, or $\{0, 6\}$ **5.** $-4 \pm \sqrt{5}$ or $\{-4 \pm \sqrt{5}\}$

6. $\pm\dfrac{\sqrt{15}}{3}$ or $\left\{\pm\dfrac{\sqrt{15}}{3}\right\}$ **7.** 1 and 6, or $\{1, 6\}$ **8.** $-5 \pm 2\sqrt{3}$ or $\{-5 \pm 2\sqrt{3}\}$ **9.** 10 ft   **10.** $2\sqrt{13}$ in.   **11.** $2\sqrt{26}$ cm

12. 15 ft   **13.** 12 yd   **14.** 20 weeks   **15.** 2.5 sec   **16.** 5 units   **17.** $2\sqrt{5}$ or 4.47 units   **18.** $x^2 + 16x + 64 = (x + 8)^2$

19. $x^2 - 6x + 9 = (x - 3)^2$ **20.** $x^2 + 3x + \dfrac{9}{4} = \left(x + \dfrac{3}{2}\right)^2$ **21.** $x^2 - 5x + \dfrac{25}{4} = \left(x - \dfrac{5}{2}\right)^2$ **22.** 3 and 9, or $\{3, 9\}$

23. $3 \pm \sqrt{5}$ or $\{3 \pm \sqrt{5}\}$ **24.** $\dfrac{6 \pm \sqrt{3}}{3}$ or $\left\{\dfrac{6 \pm \sqrt{3}}{3}\right\}$ **25.** −3 and $\dfrac{1}{2}$, or $\left\{-3, \dfrac{1}{2}\right\}$ **26.** $1 \pm \sqrt{5}$ or $\{1 \pm \sqrt{5}\}$

27. $\dfrac{9 \pm \sqrt{21}}{6}$ or $\left\{\dfrac{9 \pm \sqrt{21}}{6}\right\}$ **28.** $\dfrac{1}{2}$ and 5, or $\left\{\dfrac{1}{2}, 5\right\}$ **29.** −2 and $\dfrac{10}{3}$, or $\left\{-2, \dfrac{10}{3}\right\}$ **30.** $\dfrac{7 \pm \sqrt{37}}{6}$ or $\left\{\dfrac{7 \pm \sqrt{37}}{6}\right\}$ **31.** −3 and 3, or $\{-3, 3\}$

32. $\dfrac{3 \pm \sqrt{5}}{2}$ or $\left\{\dfrac{3 \pm \sqrt{5}}{2}\right\}$ **33. a.** $59.2 billion   **b.** 2014   **34.** $9i$ **35.** $i\sqrt{23}$ **36.** $4i\sqrt{3}$ **37.** $3 + 7i$ **38.** $\pm10i$ or $\{\pm10i\}$

39. $\pm5i$ or $\{\pm5i\}$ **40.** $\dfrac{-1 \pm 2i\sqrt{2}}{2}$ or $\left\{\dfrac{-1 \pm 2i\sqrt{2}}{2}\right\}$ **41.** $2 \pm 3i$ or $\{2 \pm 3i\}$ **42.** $\dfrac{1 \pm i\sqrt{23}}{6}$ or $\left\{\dfrac{1 \pm i\sqrt{23}}{6}\right\}$

43. a. upward
b. −1 and 7
c. −7
d. $(3, -16)$
e.

44. a. downward
b. −3 and 1
c. 3
d. $(-1, 4)$
e.

$y = -x^2 - 2x + 3$

45. a. downward
b. −0.2 and 2.2
c. 1
d. $(1, 4)$
e.

$y = -3x^2 + 6x + 1$

46. a. upward
b. 0 and 4
c. 0
d. $(2, -4)$
e.

$y = x^2 - 4x$

47. 6.25 sec; 629 ft   **48. a.** 20 yd; 16 ft   **b.** 7.9 ft   **c.** 45.3 yd   **d.**

$y = -0.025x^2 + x + 6$
$(20, 16)$
$(38, 7.9)$
$(0, 6)$
Football's Height (feet)
Horizontal Distance from Quarterback (yards)

49. function; domain: $\{2, 3, 5\}$; range: $\{7\}$ **50.** function; domain: $\{1, 2, 3\}$; range: $\{10, 500, \pi\}$ **51.** not a function; domain: $\{12, 14\}$;

range: $\{13, 15, 19\}$ **52. a.** −19   **b.** 14   **c.** −4   **53. a.** 38   **b.** −4   **c.** 2   **54.** not a function   **55.** function   **56.** function

57. not a function   **58. a.** 3.6; In 1997, sales of organic foods were $3.6 billion.; by the point $(0, 3.6)$ **b.** 13.84; In 2005, sales of organic foods

were $13.84 billion.; by the point $(8, 13.84)$

Chapter Test

1. −4 and 4, or $\{-4, 4\}$ **2.** $3 \pm \sqrt{5}$ or $\{3 \pm \sqrt{5}\}$ **3.** $4\sqrt{5}$ yd   **4.** $\sqrt{58}$ or 7.62 units   **5.** $-2 \pm \sqrt{7}$ or $\{-2 \pm \sqrt{7}\}$

6. $\dfrac{-5 \pm \sqrt{13}}{6}$ or $\left\{\dfrac{-5 \pm \sqrt{13}}{6}\right\}$ **7.** $-\dfrac{4}{3}$ and 1, or $\left\{-\dfrac{4}{3}, 1\right\}$ **8.** $-\dfrac{7}{2}$ and $\dfrac{5}{2}$, or $\left\{-\dfrac{7}{2}, \dfrac{5}{2}\right\}$ **9.** $\dfrac{3 \pm \sqrt{7}}{2}$ or $\left\{\dfrac{3 \pm \sqrt{7}}{2}\right\}$

10. −5 and $\dfrac{1}{2}$, or $\left\{-5, \dfrac{1}{2}\right\}$ **11.** $11i$ **12.** $5i\sqrt{3}$ **13.** $\pm6i$ or $\{\pm6i\}$ **14.** $5 \pm 5i$ or $\{5 \pm 5i\}$ **15.** $1 \pm 2i$ or $\{1 \pm 2i\}$

16.

17.

18. 2 sec; 69 ft   **19.** 4.1 sec   **20.** function; domain: $\{1, 3, 5, 6\}$; range: $\{2, 4, 6\}$ **21.** not a function; domain: $\{2, 4, 6\}$; range: $\{1, 3, 5, 6\}$

22. 67   **23.** 17   **24.** function   **25.** not a function   **26.** 36; In a tournament with 9 chess players, 36 games must be played.

Cumulative Review Exercises

1. 4 or $\{4\}$ **2.** 10 or $\{10\}$ **3.** $(-\infty, 7]$ or $\{x \mid x \le 7\}$ **4.** $(-3, 4)$ or $\{(-3, 4)\}$ **5.** $(3, 4)$ or $\{(3, 4)\}$ **6.** -8 or $\{-8\}$

7. -3 and -2, or $\{-3, -2\}$ **8.** 9 or $\{9\}$ **9.** $2 \pm 2\sqrt{5}$ or $\{2 \pm 2\sqrt{5}\}$ **10.** $\dfrac{3 \pm \sqrt{3}}{3}$ or $\left\{\dfrac{3 \pm \sqrt{3}}{3}\right\}$ **11.** $t = \dfrac{5r + 2}{A}$ **12.** $\dfrac{4}{x^9}$

13. 31   **14.** $10x^2 - 9x + 4$ **15.** $21x^2 - 23x - 20$ **16.** $25x^2 - 20x + 4$ **17.** $x^3 + y^3$ **18.** $\dfrac{x + 4}{3x^3}$ **19.** $\dfrac{-7x}{(x + 3)(x - 1)(x - 4)}$

20. $\dfrac{x}{5}$ **21.** $12\sqrt{5}$ **22.** $3x\sqrt{2}$ **23.** $\dfrac{2\sqrt{3}}{3}$ **24.** $6 + 2\sqrt{5}$ **25.** $(2x + 7)(2x - 7)$ **26.** $(x + 3)(x + 1)(x - 1)$ **27.** $2(x - 3)(x + 7)$

28. $x(x^2 + 4)(x + 2)(x - 2)$ **29.** $x(x - 5)^2$ **30.** $(x - 2)(x^2 + 2x + 4)$ **31.** $\dfrac{1}{4}$

32.

33.

34.

35.

36.

37.

38. -2 **39.** $y - 2 = 2(x - 1)$ or $y - 6 = 2(x - 3)$; $y = 2x$ **40.** 43   **41.** \$320   **42.** length: 150 yd; width: 50 yd

43. \$12,500 at 7% and \$7500 at 9%   **44.** 8 l of 40% and 4 l of 70%   **45.** 7.25 hr or 7 hr, 15 min   **46.** 2024 students   **47.** 16 ft

48. $35°, 25°, 120°$ **49.** TV: \$350; stereo: \$370   **50.** length: 11 m; width: 5 m

APPENDIX

Check Point Exercises

1. a. 30   **b.** 30   **2. a.** 35   **b.** 82   **3. a.** \$66.94 billion   **b.** \$30 billion   **c.** One data item is so much greater than the other items.

4. a. 8   **b.** 3 and 8   **c.** no mode   **5.** mean: 155.6; median: 155; mode: 138

Appendix Exercise Set

1. 4.125   **3.** 95   **5.** 62   **7.** ≈ 3.45 **9.** 3.5   **11.** 95   **13.** 60   **15.** 3.6   **17.** 3   **19.** 95   **21.** 40   **23.** 2.5 and 4.2

25. mean: 67.4; median: 69; mode: 71   **27.** mean: \$36,900; median: \$22,000; The median is more representative.; Answers will vary.

APPLICATIONS INDEX

SUBJECT INDEX

PHOTO CREDITS

Definitions, Rules, and Formulas

The Real Numbers

Natural Numbers: $\{1, 2, 3, \ldots\}$

Whole Numbers: $\{0, 1, 2, 3, \ldots\}$

Integers: $\{\ldots, -3, -2, -1, 0, 1, 2, 3, \ldots\}$

Rational Numbers: $\{\frac{a}{b} \mid a \text{ and } b \text{ are integers}, b \neq 0\}$

Irrational Numbers: $\{x \mid x \text{ is real and not rational}\}$

Basic Rules of Algebra

Commutative: $a + b = b + a$; $ab = ba$

Associative: $(a + b) + c = a + (b + c)$;
$(ab)c = a(bc)$

Distributive: $a(b + c) = ab + ac$;
$a(b - c) = ab - ac$

Identity: $a + 0 = a$; $a \cdot 1 = a$

Inverse: $a + (-a) = 0$; $a \cdot \frac{1}{a} = 1 \ (a \neq 0)$

Multiplication Properties: $(-1)a = -a$;
$(-1)(-a) = a$; $a \cdot 0 = 0$; $(-a)(b) = (a)(-b) = -ab$;
$(-a)(-b) = ab$

Order of Operations

1. Perform operations above and below any fraction bar, following steps (2) through (5).

2. Perform operations inside grouping symbols, innermost grouping symbols first, following steps (3) through (5).

3. Simplify exponential expressions..

4. Do multiplication and division as they occur, working from left to right.

5. Do addition and subtraction as they occur, working from left to right.

Set-Builder Notation and Graphs

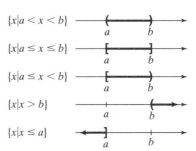

Slope Formula

$$\text{slope } (m) = \frac{\text{Change in } y}{\text{Change in } x} = \frac{y_2 - y_1}{x_2 - x_1}, \ x_2 - x_1 \neq 0$$

1. If $m > 0$, the line rises from left to right. If $m < 0$, the line falls from left to right.

2. The slope of a horizontal line is 0. The slope of a vertical line is undefined.

3. Parallel lines have equal slopes.

Equations of Lines

1. *Standard form:* $Ax + By = C$

2. *Slope-intercept form:* $y = mx + b$

   m is the line's slope and b is its y-intercept.

3. *Point-slope form:* $y - y_1 = m(x - x_1)$

   m is the line's slope and (x_1, y_1) is a fixed point on the line.

4. *Horizontal line parallel to the x-axis:* $y = b$

5. *Vertical line parallel to the y-axis:* $x = a$

Definitions, Rules, and Formulas (continued)

Properties of Exponents

1. $b^m \cdot b^n = b^{m+n}$
2. $(b^m)^n = b^{mn}$
3. $(ab)^m = a^m b^m$
4. $\dfrac{b^m}{b^n} = b^{m-n}$
5. $\left(\dfrac{a}{b}\right)^m = \dfrac{a^m}{b^m}$
6. $b^0 = 1$, where $b \neq 0$
7. $b^{-n} = \dfrac{1}{b^n}$ and $\dfrac{1}{b^{-n}} = b^n$, where $b \neq 0$

Special Factorizations

1. *Difference of two squares:*
   $$A^2 - B^2 = (A + B)(A - B)$$
2. *Perfect square trinomials:*
   $$A^2 + 2AB + B^2 = (A + B)^2$$
   $$A^2 - 2AB + B^2 = (A - B)^2$$
3. *Sum of two cubes:*
   $$A^3 + B^3 = (A + B)(A^2 - AB + B^2)$$
4. *Difference of two cubes:*
   $$A^3 - B^3 = (A - B)(A^2 + AB + B^2)$$

Variation

English Statement	Equation
y varies directly as x.	$y = kx$
y varies inversely as x.	$y = \dfrac{k}{x}$

Properties of Radicals

All roots represent real numbers.

1. The product rule:
   $$\sqrt[n]{a} \cdot \sqrt[n]{b} = \sqrt[n]{ab}$$
2. The quotient rule:
   $$\dfrac{\sqrt[n]{a}}{\sqrt[n]{b}} = \sqrt[n]{\dfrac{a}{b}}, b \neq 0$$

Triangles

1. The sum of the measures of the interior angles of a triangle is $180°$.
2. Similar triangles have corresponding angles with the same measure and corresponding sides that are proportional. Two triangles are similar if two angles of one are equal in measure to two corresponding angles of the other.

Rational Exponents

1. $a^{\frac{1}{n}} = \sqrt[n]{a}$
2. $a^{\frac{m}{n}} = \left(\sqrt[n]{a}\right)^m = \sqrt[n]{a^m}$
3. $a^{-\frac{m}{n}} = \dfrac{1}{a^{\frac{m}{n}}} = \dfrac{1}{\left(\sqrt[n]{a}\right)^m} = \dfrac{1}{\sqrt[n]{a^m}}$

The Quadratic Formula

The solutions of $ax^2 + bx + c = 0$ with $a \neq 0$ are
$$x = \dfrac{-b \pm \sqrt{b^2 - 4ac}}{2a}.$$

Imaginary and Complex Numbers

1. $i = \sqrt{-1}$ and $i^2 = -1$
2. The set of numbers in the form $a + bi$ is the set of complex numbers. If $b = 0$, the complex number $a + bi$ is a real number. If $b \neq 0$, the complex number $a + bi$ is an imaginary number.

The Graph of $y = ax^2 + bx + c$

1. The graph of $y = ax^2 + bx + c$ is called a parabola, shaped like a cup. If $a > 0$, the parabola opens upward, and if $a < 0$, the parabola opens downward. The turning point of the parabola is the vertex.
2. Graph $y = ax^2 + bx + c$ by finding any x-intercepts (replace y with 0), the y-intercept (replace x with 0), the vertex, and additional points near the vertex and intercepts. The x-coordinate of the vertex is $-\dfrac{b}{2a}$. The y-coordinate is found by substituting $-\dfrac{b}{2a}$ for x in $y = ax^2 + bx + c$ and solving for y.
3. The vertex of $y = ax^2 + bx + c$ is a minimum point when $a > 0$ and a maximum point when $a < 0$.

3. The Pythagorean Theorem

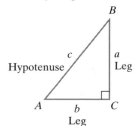

In any right triangle with hypotenuse of length c and legs of length a and b, $c^2 = a^2 + b^2$.